Rational (Reciprocal) Function

$f(x) = \dfrac{1}{x}$

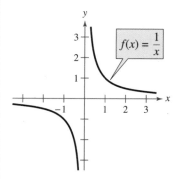

Domain: $(-\infty, 0) \cup (0, \infty)$
Range: $(-\infty, 0) \cup (0, \infty)$
No intercepts
Decreasing on $(-\infty, 0)$ and $(0, \infty)$
Odd function
Origin symmetry
Vertical asymptote: y-axis
Horizontal asymptote: x-axis

Exponential Function

$f(x) = a^x, \ a > 1$

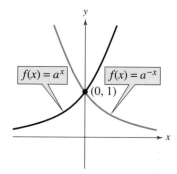

Domain: $(-\infty, \infty)$
Range: $(0, \infty)$
Intercept: $(0, 1)$
Increasing on $(-\infty, \infty)$
 for $f(x) = a^x$
Decreasing on $(-\infty, \infty)$
 for $f(x) = a^{-x}$
Horizontal asymptote: x-axis
Continuous

Logarithmic Function

$f(x) = \log_a x, \ a > 1$

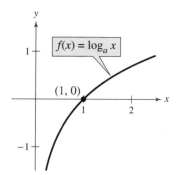

Domain: $(0, \infty)$
Range: $(-\infty, \infty)$
Intercept: $(1, 0)$
Increasing on $(0, \infty)$
Vertical asymptote: y-axis
Continuous
Reflection of graph of $f(x) = a^x$
 in the line $y = x$

SYMMETRY

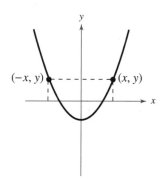

y-Axis Symmetry

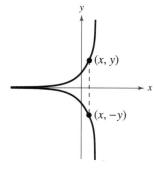

x-Axis Symmetry

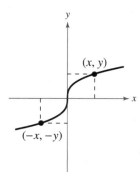

Origin Symmetry

College Algebra Alternate

with Enhanced WebAssign

Eighth Edition

Ron Larson

The Pennsylvania State University
The Behrend College

With the assistance of

David C. Falvo

The Pennsylvania State University
The Behrend College

BROOKS/COLE
CENGAGE Learning™

Australia • Brazil • Japan • Korea • Mexico • Singapore • Spain • United Kingdom • United States

with Enhanced WebAssign,

ary Whalen
...cy Green
Assistant Editor: Cynthia Ashton
Editorial Assistant: Guanglei Zhang
Associate Media Editor: Lynh Pham
Marketing Manager: Myriah FitzGibbon
Marketing Coordinator: Angela Kim
Marketing Communications Manager: Katy Malatesta
Content Project Manager: Susan Miscio
Senior Art Director: Jill Ort
Senior Print Buyer: Diane Gibbons
Production Editor: Carol Merrigan
Text Designer: Walter Kopek
Rights Acquiring Account Manager, Photos: Don Schlotman
Photo Researcher: Prepress PMG
Cover Designer: Harold Burch
Cover Image: Richard Edelman/Woodstock Graphics Studio
Compositor: Larson Texts, Inc.

For product information and technology assistance, contact us at
Cengage Learning Customer & Sales Support, 1-800-354-9706

For permission to use material from this text or product, submit all requests online at **www.cengage.com/permissions**. Further permissions questions can be emailed to **permissionrequest@cengage.com**

Library of Congress Control Number: 2009930252
Student Edition:
ISBN-13: 978-0-495-97065-1
ISBN-10: 0-495-97065-4

Enhanced WebAssign:
ISBN-13: 978-0-538-73810-1
ISBN-10: 0-538-73810-3

Brooks/Cole
10 Davis Drive
Belmont, CA 94002-3098
USA

Cengage Learning is a leading provider of customized learning solutions with office locations around the globe, including Singapore, the United Kingdom, Australia, Mexico, Brazil, and Japan. Locate your local office at: **international.cengage.com/region**

Cengage Learning products are represented in Canada by Nelson Education, Ltd.

For your course and learning solutions, visit **www.cengage.com**

Purchase any of our products at your local college store or at our preferred online store **www.ichapters.com**

Printed in the United States of America
1 2 3 4 5 6 7 13 12 11 10 09

Contents

chapter 3

Polynomial Functions   259

chapter 4

Rational Functions and Conics   331

chapter 5

Exponential and Logarithmic Functions   379

A Word from the Author

Welcome to the Eighth Edition of *College Algebra*! We are proud to offer you a new and revised version of our textbook. With each edition, we have listened to you, our users, and have incorporated many of your suggestions for improvement.

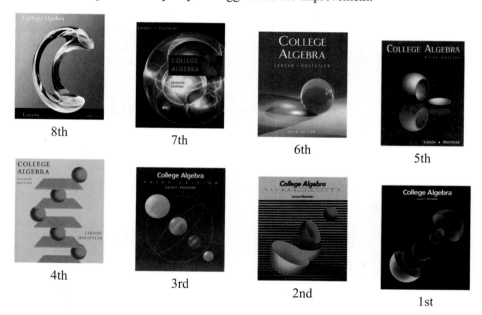

8th   7th   6th   5th

4th   3rd   2nd   1st

In the Eighth Edition, we continue to offer instructors and students a text that is pedagogically sound, mathematically precise, and still comprehensible. There are many changes in the mathematics, art, and design; the more significant changes are noted here.

- **New Chapter Openers**   Each *Chapter Opener* has three parts, *In Mathematics, In Real Life*, and *In Careers*. *In Mathematics* describes an important mathematical topic taught in the chapter. *In Real Life* tells students where they will encounter this topic in real-life situations. *In Careers* relates application exercises to a variety of careers.

- **New Study Tips and Warning/Cautions**   Insightful information is given to students in two new features. The *Study Tip* provides students with useful information or suggestions for learning the topic. The *Warning/Caution* points out common mathematical errors made by students.

- **New Algebra Helps**   *Algebra Help* directs students to sections of the textbook where they can review algebra skills needed to master the current topic.

- **New Side-by-Side Examples**   Throughout the text, we present solutions to many examples from multiple perspectives—algebraically, graphically, and numerically. The side-by-side format of this pedagogical feature helps students to see that a problem can be solved in more than one way and to see that different methods yield the same result. The side-by-side format also addresses many different learning styles.

- *New Capstone Exercises*   *Capstones* are conceptual problems that synthesize key topics and provide students with a better understanding of each section's concepts. Capstone exercises are excellent for classroom discussion or test prep, and teachers may find value in integrating these problems into their reviews of the section.

- *New Chapter Summaries*   The *Chapter Summary* now includes an explanation and/or example of each objective taught in the chapter.

- *Revised Exercise Sets*   The exercise sets have been carefully and extensively examined to ensure they are rigorous and cover all topics suggested by our users. Many new skill-building and challenging exercises have been added.

For the past several years, we've maintained an independent website—**CalcChat.com**—that provides free solutions to all odd-numbered exercises in the text. Thousands of students using our textbooks have visited the site for practice and help with their homework. For the Eighth Edition, we were able to use information from CalcChat.com, including which solutions students accessed most often, to help guide the revision of the exercises.

I hope you enjoy the Eighth Edition of *College Algebra*. As always, I welcome comments and suggestions for continued improvements.

Ron Larson

Acknowledgments

I would like to thank the many people who have helped me prepare the text and the supplements package. Their encouragement, criticisms, and suggestions have been invaluable.

Thank you to all of the instructors who took the time to review the changes in this edition and to provide suggestions for improving it. Without your help, this book would not be possible.

Reviewers

Chad Pierson, *University of Minnesota-Duluth*; Sally Shao, *Cleveland State University*; Ed Stumpf, *Central Carolina Community College*; Fuzhen Zhang, *Nova Southeastern University*; Dennis Shepherd, *University of Colorado, Denver*; Rhonda Kilgo, *Jacksonville State University*; C. Altay Özgener, *Manatee Community College Bradenton*; William Forrest, *Baton Rouge Community College*; Tracy Cook, *University of Tennessee Knoxville*; Charles Hale, *California State Poly University Pomona*; Samuel Evers, *University of Alabama*; Seongchun Kwon, *University of Toledo*; Dr. Arun K. Agarwal, *Grambling State University*; Hyounkyun Oh, *Savannah State University*; Michael J. McConnell, *Clarion University*; Martha Chalhoub, *Collin County Community College*; Angela Lee Everett, *Chattanooga State Tech Community College*; Heather Van Dyke, *Walla Walla Community College*; Gregory Buthusiem, *Burlington County Community College*; Ward Shaffer, *College of Coastal Georgia*; Carmen Thomas, *Chatham University*; Emily J. Keaton

My thanks to David Falvo, The Behrend College, The Pennsylvania State University, for his contributions to this project. My thanks also to Robert Hostetler, The Behrend College, The Pennsylvania State University, and Bruce Edwards, University of Florida, for their significant contributions to previous editions of this text.

I would also like to thank the staff at Larson Texts, Inc. who assisted with proofreading the manuscript, preparing and proofreading the art package, and checking and typesetting the supplements.

On a personal level, I am grateful to my spouse, Deanna Gilbert Larson, for her love, patience, and support. Also, a special thanks goes to R. Scott O'Neil. If you have suggestions for improving this text, please feel free to write to me. Over the past two decades I have received many useful comments from both instructors and students, and I value these comments very highly.

Ron Larson

Supplements

Supplements for the Instructor

Annotated Instructor's Edition   This AIE is the complete student text plus point-of-use annotations for the instructor, including extra projects, classroom activities, teaching strategies, and additional examples. Answers to even-numbered text exercises, Vocabulary Checks, and Explorations are also provided.

Complete Solutions Manual   This manual contains solutions to all exercises from the text, including Chapter Review Exercises and Chapter Tests.

Instructor's Companion Website   This free companion website contains an abundance of instructor resources.

PowerLecture™ with ExamView®   The CD-ROM provides the instructor with dynamic media tools for teaching college algebra. PowerPoint® lecture slides and art slides of the figures from the text, together with electronic files for the test bank and a link to the Solution Builder, are available. The algorithmic ExamView allows you to create, deliver, and customize tests (both print and online) in minutes with this easy-to-use assessment system. Enhance how your students interact with you, your lecture, and each other.

Solutions Builder   This is an electronic version of the complete solutions manual available via the PowerLecture and Instructor's Companion Website. It provides instructors with an efficient method for creating solution sets to homework or exams that can then be printed or posted.

Supplements for the Student

Student Companion Website   This free companion website contains an abundance of student resources.

Instructional DVDs   Keyed to the text by section, these DVDs provide comprehensive coverage of the course—along with additional explanations of concepts, sample problems, and applications—to help students review essential topics.

Student Study and Solutions Manual   This guide offers step-by-step solutions for all odd-numbered text exercises, Chapter and Cumulative Tests, and Practice Tests with solutions.

Premium eBook   The Premium eBook offers an interactive version of the textbook with search features, highlighting and note-making tools, and direct links to videos or tutorials that elaborate on the text discussions.

Enhanced WebAssign   Enhanced WebAssign is designed for you to do your homework online. This proven and reliable system uses pedagogy and content found in Larson's text, and then enhances it to help you learn College Algebra more effectively. Automatically graded homework allows you to focus on your learning and get interactive study assistance outside of class.

P Prerequisites

In Mathematics

Real numbers, exponents, radicals, and polynomials are used in many different branches of mathematics.

In Real Life

The concepts in this chapter are used to model compound interest, volumes, rates of change, and other real-life applications. For instance, polynomials can be used to model the stopping distance of an automobile. (See Exercise 116, page 36.)

Darren McCollester/ Getty Images News /Getty Images

IN CAREERS

There are many careers that use prealgebra concepts. Several are listed below.

What you should learn

- Represent and classify real numbers.
- Order real numbers and use inequalities.
- Find the absolute values of real numbers and find the distance between two real numbers.
- Evaluate algebraic expressions.
- Use the basic rules and properties of algebra.

Why you should learn it

Real numbers are used to represent many real-life quantities. For example, in Exercises 83–88 on page 13, you will use real numbers to represent the federal deficit.

P.1 REVIEW OF REAL NUMBERS AND THEIR PROPERTIES

Real Numbers

Real numbers are used in everyday life to describe quantities such as age, miles per gallon, and population. Real numbers are represented by symbols such as

$$-5, 9, 0, \frac{4}{3}, 0.666\ldots, 28.21, \sqrt{2}, \pi, \text{ and } \sqrt[3]{-32}.$$

Here are some important **subsets** (each member of subset B is also a member of set A) of the real numbers. The three dots, called *ellipsis points*, indicate that the pattern continues indefinitely.

$$\{1, 2, 3, 4, \ldots\}$$
$$\{0, 1, 2, 3, 4, \ldots\}$$
$$\{\ldots, -3, -2, -1, 0, 1, 2, 3, \ldots\}$$

A real number is **rational** if it can be written as the ratio p/q of two integers, where $q \neq 0$. For instance, the numbers

$$\frac{1}{3} = 0.3333\ldots = 0.\overline{3}, \frac{1}{8} = 0.125, \text{ and } \frac{125}{111} = 1.126126\ldots = 1.\overline{126}$$

are rational. The decimal representation of a rational number either repeats $\left(\text{as in } \frac{173}{55} = 3.1\overline{45}\right)$ or terminates $\left(\text{as in } \frac{1}{2} = 0.5\right)$. A real number that cannot be written as the ratio of two integers is called **irrational.** Irrational numbers have infinite nonrepeating decimal representations. For instance, the numbers

$$\sqrt{2} = 1.4142135\ldots \approx 1.41 \quad \text{and} \quad \pi = 3.1415926\ldots \approx 3.14$$

are irrational. (The symbol $\approx$ means "is approximately equal to.") Figure P.1 shows subsets of real numbers and their relationships to each other.

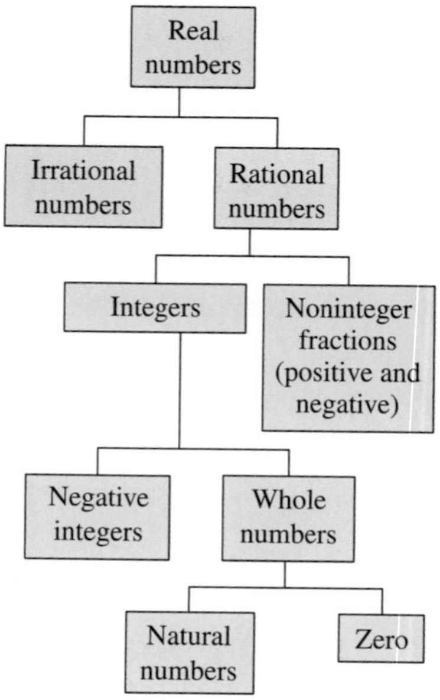

Example 1 Classifying Real Numbers

Determine which numbers in the set

$$\left\{-13, -\sqrt{5}, -1, -\frac{1}{3}, 0, \frac{5}{8}, \sqrt{2}, \pi, 7\right\}$$

are (a) natural numbers, (b) whole numbers, (c) integers, (d) rational numbers, and (e) irrational numbers.

Solution

a. Natural numbers: $\{7\}$

b. Whole numbers: $\{0, 7\}$

c. Integers: $\{-13, -1, 0, 7\}$

d. Rational numbers: $\left\{-13, -1, -\frac{1}{3}, 0, \frac{5}{8}, 7\right\}$

e. Irrational numbers: $\left\{-\sqrt{5}, \sqrt{2}, \pi\right\}$

CHECK**Point** Now try Exercise 11.

FIGURE P.1 Subsets of real numbers

Real numbers are represented graphically on the **real number line.** When you draw a point on the real number line that corresponds to a real number, you are **plotting** the real number. The point 0 on the real number line is the **origin.** Numbers to the right of 0 are positive, and numbers to the left of 0 are negative, as shown in Figure P.2. The term **nonnegative** describes a number that is either positive or zero.

FIGURE P.2   The real number line

As illustrated in Figure P.3, there is a *one-to-one correspondence* between real numbers and points on the real number line.

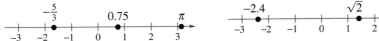

Every real number corresponds to exactly one point on the real number line.

Every point on the real number line corresponds to exactly one real number.

FIGURE P.3   One-to-one correspondence

Example 2   Plotting Points on the Real Number Line

Plot the real numbers on the real number line.

a. $-\dfrac{7}{4}$

b. 2.3

c. $\dfrac{2}{3}$

d. -1.8

Solution

All four points are shown in Figure P.4.

FIGURE P.4

a. The point representing the real number $-\frac{7}{4} = -1.75$ lies between -2 and -1, but closer to -2, on the real number line.

b. The point representing the real number 2.3 lies between 2 and 3, but closer to 2, on the real number line.

c. The point representing the real number $\frac{2}{3} = 0.666\ldots$ lies between 0 and 1, but closer to 1, on the real number line.

d. The point representing the real number -1.8 lies between -2 and -1, but closer to -2, on the real number line. Note that the point representing -1.8 lies slightly to the left of the point representing $-\frac{7}{4}$.

CHECK*Point*   Now try Exercise 17.

Ordering Real Numbers

One important property of real numbers is that they are *ordered*.

> **Definition of Order on the Real Number Line**
>
> If a and b are real numbers, a is less than b if $b - a$ is positive. The **order** of a and b is denoted by the **inequality** $a < b$. This relationship can also be described by saying that b is *greater than* a and writing $b > a$. The inequality $a \leq b$ means that a is *less than or equal to b*, and the inequality $b \geq a$ means that b is *greater than or equal to a*. The symbols $<$, $>$, $\leq$, and $\geq$ are *inequality symbols.*

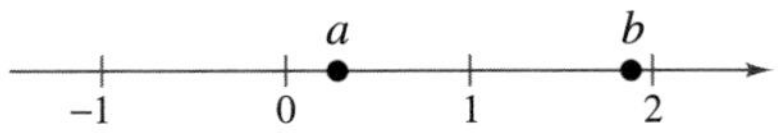

FIGURE P.5 $a < b$ if and only if a lies to the left of b.

Geometrically, this definition implies that $a < b$ if and only if a lies to the *left* of b on the real number line, as shown in Figure P.5.

Example 3 Ordering Real Numbers

Place the appropriate inequality symbol ($<$ or $>$) between the pair of real numbers.

a. $-3, 0$ **b.** $-2, -4$ **c.** $\dfrac{1}{4}, \dfrac{1}{3}$ **d.** $-\dfrac{1}{5}, -\dfrac{1}{2}$

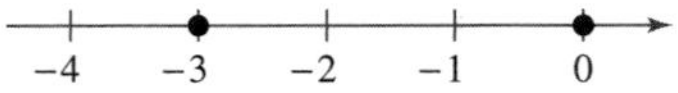

Solution

a. Because -3 lies to the left of 0 on the real number line, as shown in Figure P.6, you can say that -3 is *less than* 0, and write $-3 < 0$.

b. Because -2 lies to the right of -4 on the real number line, as shown in Figure P.7, you can say that -2 is *greater than* -4, and write $-2 > -4$.

c. Because $\frac{1}{4}$ lies to the left of $\frac{1}{3}$ on the real number line, as shown in Figure P.8, you can say that $\frac{1}{4}$ is *less than* $\frac{1}{3}$, and write $\frac{1}{4} < \frac{1}{3}$.

d. Because $-\frac{1}{5}$ lies to the right of $-\frac{1}{2}$ on the real number line, as shown in Figure P.9, you can say that $-\frac{1}{5}$ is *greater than* $-\frac{1}{2}$, and write $-\frac{1}{5} > -\frac{1}{2}$.

CHECK*Point* Now try Exercise 25.

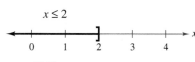

FIGURE P.6

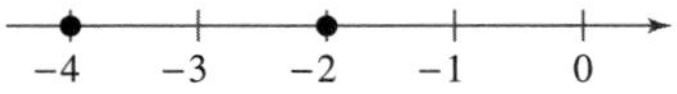

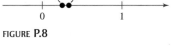

FIGURE P.7

FIGURE P.8

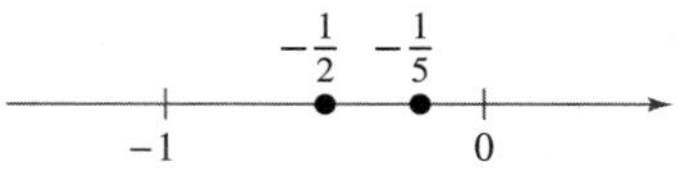

FIGURE P.9

Example 4 Interpreting Inequalities

Describe the subset of real numbers represented by each inequality.

a. $x \leq 2$ **b.** $-2 \leq x < 3$

Solution

a. The inequality $x \leq 2$ denotes all real numbers less than or equal to 2, as shown in Figure P.10.

b. The inequality $-2 \leq x < 3$ means that $x \geq -2$ *and* $x < 3$. This "double inequality" denotes all real numbers between -2 and 3, including -2 but not including 3, as shown in Figure P.11.

CHECK*Point* Now try Exercise 31.

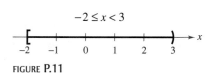

$x \leq 2$

FIGURE P.10

$-2 \leq x < 3$

FIGURE P.11

Inequalities can be used to describe subsets of real numbers called **intervals.** In the bounded intervals below, the real numbers a and b are the **endpoints** of each interval. The endpoints of a closed interval are included in the interval, whereas the endpoints of an open interval are not included in the interval.

Bounded Intervals on the Real Number Line

Notation	Interval Type	Inequality	Graph
$[a, b]$	Closed	$a \leq x \leq b$	
(a, b)	Open	$a < x < b$	
$[a, b)$		$a \leq x < b$	
$(a, b]$		$a < x \leq b$	

Study Tip

The reason that the four types of intervals at the right are called *bounded* is that each has a finite length. An interval that does not have a finite length is *unbounded* (see below).

The symbols ∞, **positive infinity,** and $-\infty$, **negative infinity,** do not represent real numbers. They are simply convenient symbols used to describe the unboundedness of an interval such as $(1, \infty)$ or $(-\infty, 3]$.

⚠ WARNING / CAUTION

Whenever you write an interval containing ∞ or $-\infty$, always use a parenthesis and never a bracket. This is because ∞ and $-\infty$ are never an endpoint of an interval and therefore are not included in the interval.

Unbounded Intervals on the Real Number Line

Notation	Interval Type	Inequality	Graph
$[a, \infty)$		$x \geq a$	
(a, ∞)	Open	$x > a$	
$(-\infty, b]$		$x \leq b$	
$(-\infty, b)$	Open	$x < b$	
$(-\infty, \infty)$	Entire real line	$-\infty < x < \infty$	

Example 5   Using Inequalities to Represent Intervals

Use inequality notation to describe each of the following.

a. c is at most 2.   **b.** m is at least -3.   **c.** All x in the interval $(-3, 5]$

Solution

a. The statement "c is at most 2" can be represented by $c \leq 2$.

b. The statement "m is at least -3" can be represented by $m \geq -3$.

c. "All x in the interval $(-3, 5]$" can be represented by $-3 < x \leq 5$.

CHECK *Point*   Now try Exercise 45.

Example 6   **Interpreting Intervals**

Give a verbal description of each interval.

a. $(-1, 0)$ **b.** $[2, \infty)$ **c.** $(-\infty, 0)$

Solution

a. This interval consists of all real numbers that are greater than -1 and less than 0.

b. This interval consists of all real numbers that are greater than or equal to 2.

c. This interval consists of all negative real numbers.

CHECK*Point*   Now try Exercise 41.

Absolute Value and Distance

The **absolute value** of a real number is its *magnitude*, or the distance between the origin and the point representing the real number on the real number line.

> **Definition of Absolute Value**
>
> If a is a real number, then the absolute value of a is
>
> $$|a| = \begin{cases} a, & \text{if } a \geq 0 \\ -a, & \text{if } a < 0 \end{cases}.$$

Notice in this definition that the absolute value of a real number is never negative. For instance, if $a = -5$, then $|-5| = -(-5) = 5$. The absolute value of a real number is either positive or zero. Moreover, 0 is the only real number whose absolute value is 0. So, $|0| = 0$.

Example 7   **Finding Absolute Values**

a. $|-15| = 15$ **b.** $\left|\dfrac{2}{3}\right| = \dfrac{2}{3}$

c. $|-4.3| = 4.3$ **d.** $-|-6| = -(6) = -6$

CHECK*Point*   Now try Exercise 51.

Example 8   **Evaluating the Absolute Value of a Number**

Evaluate $\dfrac{|x|}{x}$ for (a) $x > 0$ and (b) $x < 0$.

Solution

a. If $x > 0$, then $|x| = x$ and $\dfrac{|x|}{x} = \dfrac{x}{x} = 1$.

b. If $x < 0$, then $|x| = -x$ and $\dfrac{|x|}{x} = \dfrac{-x}{x} = -1$.

CHECK*Point*   Now try Exercise 59.

The **Law of Trichotomy** states that for any two real numbers a and b, *precisely* one of three relationships is possible:

$$a = b, \quad a < b, \quad \text{or} \quad a > b.$$

Example 9   Comparing Real Numbers

Place the appropriate symbol ($<$, $>$, or $=$) between the pair of real numbers.

a. $|-4|$ ___ $|3|$ **b.** $|-10|$ ___ $|10|$ **c.** $-|-7|$ ___ $|-7|$

Solution

a. $|-4| > |3|$ because $|-4| = 4$ and $|3| = 3$, and 4 is greater than 3.
b. $|-10| = |10|$ because $|-10| = 10$ and $|10| = 10$.
c. $-|-7| < |-7|$ because $-|-7| = -7$ and $|-7| = 7$, and -7 is less than 7.

CHECK*Point* ▶ Now try Exercise 61. ▌

Properties of Absolute Values

1. $|a| \geq 0$ **2.** $|-a| = |a|$

3. $|ab| = |a||b|$ **4.** $\left|\dfrac{a}{b}\right| = \dfrac{|a|}{|b|}, \quad b \neq 0$

Absolute value can be used to define the distance between two points on the real number line. For instance, the distance between -3 and 4 is

$$|-3 - 4| = |-7|$$
$$= 7$$

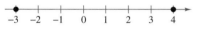

FIGURE P.12   The distance between -3 and 4 is 7.

as shown in Figure P.12.

Distance Between Two Points on the Real Number Line

Let a and b be real numbers. The **distance between a and b** is

$$d(a, b) = |b - a| = |a - b|.$$

Example 10   Finding a Distance

Find the distance between -25 and 13.

Solution

The distance between -25 and 13 is given by

$$|-25 - 13| = |-38| = 38.$$

The distance can also be found as follows.

$$|13 - (-25)| = |38| = 38$$

CHECK*Point* ▶ Now try Exercise 67. ▌

Algebraic Expressions

One characteristic of algebra is the use of letters to represent numbers. The letters are **variables,** and combinations of letters and numbers are **algebraic expressions.** Here are a few examples of algebraic expressions.

$$5x, \qquad 2x - 3, \qquad \frac{4}{x^2 + 2}, \qquad 7x + y$$

Definition of an Algebraic Expression

An **algebraic expression** is a collection of letters (**variables**) and real numbers (**constants**) combined using the operations of addition, subtraction, multiplication, division, and exponentiation.

The **terms** of an algebraic expression are those parts that are separated by *addition.* For example,

$$x^2 - 5x + 8 = x^2 + (-5x) + 8$$

has three terms: x^2 and $-5x$ are the **variable terms** and 8 is the **constant term.** The numerical factor of a term is called the **coefficient.** For instance, the coefficient of $-5x$ is -5, and the coefficient of x^2 is 1.

Example 11 Identifying Terms and Coefficients

Algebraic Expression	Terms	Coefficients
a. $5x - \dfrac{1}{7}$	$5x, -\dfrac{1}{7}$	$5, -\dfrac{1}{7}$
b. $2x^2 - 6x + 9$	$2x^2, -6x, 9$	$2, -6, 9$
c. $\dfrac{3}{x} + \dfrac{1}{2}x^4 - y$	$\dfrac{3}{x}, \dfrac{1}{2}x^4, -y$	$3, \dfrac{1}{2}, -1$

CHECK Point Now try Exercise 89.

To **evaluate** an algebraic expression, substitute numerical values for each of the variables in the expression, as shown in the next example.

Example 12 Evaluating Algebraic Expressions

	Expression	Value of Variable	Substitute	Value of Expression
a.	$-3x + 5$	$x = 3$	$-3(\ ) + 5$	$-9 + 5 = -4$
b.	$3x^2 + 2x - 1$	$x = -1$	$3(\ )^2 + 2(\ ) - 1$	$3 - 2 - 1 = 0$
c.	$\dfrac{2x}{x + 1}$	$x = -3$	$\dfrac{2(\ )}{(\ ) + 1}$	$\dfrac{-6}{-2} = 3$

Note that you must substitute the value for *each* occurrence of the variable.

CHECK Point Now try Exercise 95.

When an algebraic expression is evaluated, the **Substitution Principle** is used. It states that "If $a = b$, then a can be replaced by b in any expression involving a." In Example 12(a), for instance, 3 is *substituted* for x in the expression $-3x + 5$.

Basic Rules of Algebra

There are four arithmetic operations with real numbers: *addition, multiplication, subtraction,* and *division,* denoted by the symbols $+$, $\times$ or $\cdot$, $-$, and $\div$ or $/$. Of these, addition and multiplication are the two primary operations. Subtraction and division are the inverse operations of addition and multiplication, respectively.

Definitions of Subtraction and Division

Subtraction: Add the opposite.   **Division:** Multiply by the reciprocal.

$$a - b = a + (-b)$$ $$\text{If } b \neq 0, \text{ then } a/b = a\left(\frac{1}{b}\right) = \frac{a}{b}.$$

In these definitions, $-b$ is the **additive inverse** (or opposite) of b, and $1/b$ is the **multiplicative inverse** (or reciprocal) of b. In the fractional form a/b, a is the **numerator** of the fraction and b is the **denominator.**

Because the properties of real numbers below are true for variables and algebraic expressions as well as for real numbers, they are often called the **Basic Rules of Algebra.** Try to formulate a verbal description of each property. For instance, the first property states that *the order in which two real numbers are added does not affect their sum.*

Basic Rules of Algebra

Let a, b, and c be real numbers, variables, or algebraic expressions.

Property		*Example*
Commutative Property of Addition:	$a + b = b + a$	$4x + x^2 = x^2 + 4x$
Commutative Property of Multiplication:	$ab = ba$	$(4 - x)x^2 = x^2(4 - x)$
Associative Property of Addition:	$(a + b) + c = a + (b + c)$	$(x + 5) + x^2 = x + (5 + x^2)$
Associative Property of Multiplication:	$(ab)c = a(bc)$	$(2x \cdot 3y)(8) = (2x)(3y \cdot 8)$
Distributive Properties:	$a(b + c) = ab + ac$	$3x(5 + 2x) = 3x \cdot 5 + 3x \cdot 2x$
	$(a + b)c = ac + bc$	$(y + 8)y = y \cdot y + 8 \cdot y$
Additive Identity Property:	$a + 0 = a$	$5y^2 + 0 = 5y^2$
Multiplicative Identity Property:	$a \cdot 1 = a$	$(4x^2)(1) = 4x^2$
Additive Inverse Property:	$a + (-a) = 0$	$5x^3 + (-5x^3) = 0$
Multiplicative Inverse Property:	$a \cdot \dfrac{1}{a} = 1, \quad a \neq 0$	$(x^2 + 4)\left(\dfrac{1}{x^2 + 4}\right) = 1$

Because subtraction is defined as "adding the opposite," the Distributive Properties are also true for subtraction. For instance, the "subtraction form" of $a(b + c) = ab + ac$ is $a(b - c) = ab - ac$. Note that the operations of subtraction and division are neither commutative nor associative. The examples

$$7 - 3 \neq 3 - 7 \quad \text{and} \quad 20 \div 4 \neq 4 \div 20$$

show that subtraction and division are not commutative. Similarly

$$5 - (3 - 2) \neq (5 - 3) - 2 \quad \text{and} \quad 16 \div (4 \div 2) \neq (16 \div 4) \div 2$$

demonstrate that subtraction and division are not associative.

Example 13 Identifying Rules of Algebra

Identify the rule of algebra illustrated by the statement.

a. $(5x^3)2 = 2(5x^3)$

b. $\left(4x + \dfrac{1}{3}\right) - \left(4x + \dfrac{1}{3}\right) = 0$

c. $7x \cdot \dfrac{1}{7x} = 1, \quad x \neq 0$

d. $(2 + 5x^2) + x^2 = 2 + (5x^2 + x^2)$

Solution

a. This statement illustrates the Commutative Property of Multiplication. In other words, you obtain the same result whether you multiply $5x^3$ by 2, or 2 by $5x^3$.

b. This statement illustrates the Additive Inverse Property. In terms of subtraction, this property simply states that when any expression is subtracted from itself the result is 0.

c. This statement illustrates the Multiplicative Inverse Property. Note that it is important that x be a nonzero number. If x were 0, the reciprocal of x would be undefined.

d. This statement illustrates the Associative Property of Addition. In other words, to form the sum

$$2 + 5x^2 + x^2$$

it does not matter whether 2 and $5x^2$, or $5x^2$ and x^2 are added first.

CHECK*Point* Now try Exercise 101.

Properties of Negation and Equality

Let a, b, and c be real numbers, variables, or algebraic expressions.

Property	Example
1. $(-1)a = -a$	$(-1)7 = -7$
2. $-(-a) = a$	$-(-6) = 6$
3. $(-a)b = -(ab) = a(-b)$	$(-5)3 = -(5 \cdot 3) = 5(-3)$
4. $(-a)(-b) = ab$	$(-2)(-x) = 2x$
5. $-(a + b) = (-a) + (-b)$	$-(x + 8) = (-x) + (-8)$
	$= -x - 8$
6. If $a = b$, then $a \pm c = b \pm c$.	$\frac{1}{2} \quad = 0.5$
7. If $a = b$, then $ac = bc$.	$4^2 \quad = 16$
8. If $a \pm c = b \pm c$, then $a = b$.	$1.4 \quad = \frac{7}{5} \quad \Rightarrow 1.4 = \frac{7}{5}$
9. If $ac = bc$ and $c \neq 0$, then $a = b$.	$x = \quad \cdot 4 \Rightarrow x = 4$

Study Tip

The "or" in the Zero-Factor Property includes the possibility that either or both factors may be zero. This is an **inclusive or,** and it is the way the word "or" is generally used in mathematics.

Properties of Zero

Let a and b be real numbers, variables, or algebraic expressions.

1. $a + 0 = a$ and $a - 0 = a$ 2. $a \cdot 0 = 0$

3. $\dfrac{0}{a} = 0, \quad a \neq 0$ 4. $\dfrac{a}{0}$ is undefined.

5. **Zero-Factor Property:** If $ab = 0$, then $a = 0$ or $b = 0$.

Properties and Operations of Fractions

Let a, b, c, and d be real numbers, variables, or algebraic expressions such that $b \neq 0$ and $d \neq 0$.

1. **Equivalent Fractions:** $\dfrac{a}{b} = \dfrac{c}{d}$ if and only if $ad = bc$.

2. **Rules of Signs:** $-\dfrac{a}{b} = \dfrac{-a}{b} = \dfrac{a}{-b}$ and $\dfrac{-a}{-b} = \dfrac{a}{b}$

3. **Generate Equivalent Fractions:** $\dfrac{a}{b} = \dfrac{ac}{bc}, \quad c \neq 0$

4. **Add or Subtract with Like Denominators:** $\dfrac{a}{b} \pm \dfrac{c}{b} = \dfrac{a \pm c}{b}$

5. **Add or Subtract with Unlike Denominators:** $\dfrac{a}{b} \pm \dfrac{c}{d} = \dfrac{ad \pm bc}{bd}$

6. **Multiply Fractions:** $\dfrac{a}{b} \cdot \dfrac{c}{d} = \dfrac{ac}{bd}$

7. **Divide Fractions:** $\dfrac{a}{b} \div \dfrac{c}{d} = \dfrac{a}{b} \cdot \dfrac{d}{c} = \dfrac{ad}{bc}, \quad c \neq 0$

Study Tip

In Property 1 of fractions, the phrase "if and only if" implies two statements. One statement is: If $a/b = c/d$, then $ad = bc$. The other statement is: If $ad = bc$, where $b \neq 0$ and $d \neq 0$, then $a/b = c/d$.

Example 14 **Properties and Operations of Fractions**

a. Equivalent fractions: $\dfrac{x}{5} = \dfrac{3 \cdot x}{3 \cdot 5} = \dfrac{3x}{15}$ **b.** Divide fractions: $\dfrac{7}{x} \div \dfrac{3}{2} = \dfrac{7}{x} \cdot \dfrac{2}{3} = \dfrac{14}{3x}$

c. Add fractions with unlike denominators: $\dfrac{x}{3} + \dfrac{2x}{5} = \dfrac{5 \cdot x + 3 \cdot 2x}{3 \cdot 5} = \dfrac{11x}{15}$

CHECK*Point*   Now try Exercise 119.

If a, b, and c are integers such that $ab = c$, then a and b are **factors** or **divisors** of c. A **prime number** is an integer that has exactly two positive factors—itself and 1—such as 2, 3, 5, 7, and 11. The numbers 4, 6, 8, 9, and 10 are **composite** because each can be written as the product of two or more prime numbers. The number 1 is neither prime nor composite. The **Fundamental Theorem of Arithmetic** states that every positive integer greater than 1 can be written as the product of prime numbers in precisely one way (disregarding order). For instance, the *prime factorization* of 24 is $24 = 2 \cdot 2 \cdot 2 \cdot 3$.

P.1 EXERCISES

See www.CalcChat.com for worked-out solutions to odd-numbered exercises.

VOCABULARY: Fill in the blanks.

1. A real number is _____ if it can be written as the ratio $\frac{p}{q}$ of two integers, where $q \neq 0$.

2. _____ numbers have infinite nonrepeating decimal representations.

3. The point 0 on the real number line is called the _____.

4. The distance between the origin and a point representing a real number on the real number line is the _____ _____ of the real number.

5. A number that can be written as the product of two or more prime numbers is called a _____ number.

6. An integer that has exactly two positive factors, the integer itself and 1, is called a _____ number.

7. An algebraic expression is a collection of letters called _____ and real numbers called _____.

8. The _____ of an algebraic expression are those parts separated by addition.

9. The numerical factor of a variable term is the _____ of the variable term.

10. The _____ _____ states that if $ab = 0$, then $a = 0$ or $b = 0$.

SKILLS AND APPLICATIONS

In Exercises 11–16, determine which numbers in the set are (a) natural numbers, (b) whole numbers, (c) integers, (d) rational numbers, and (e) irrational numbers.

11. $\left\{ -9, -\frac{7}{2}, 5, \frac{2}{3}, \sqrt{2}, 0, 1, -4, 2, -11 \right\}$

12. $\left\{ \sqrt{5}, -7, -\frac{7}{3}, 0, 3.12, \frac{5}{4}, -3, 12, 5 \right\}$

13. $\{2.01, 0.666 \ldots, -13, 0.010110111 \ldots, 1, -6\}$

14. $\{2.3030030003 \ldots, 0.7575, -4.63, \sqrt{10}, -75, 4\}$

15. $\left\{ -\pi, -\frac{1}{3}, \frac{6}{3}, \frac{1}{2}\sqrt{2}, -7.5, -1, 8, -22 \right\}$

16. $\left\{ 25, -17, -\frac{12}{5}, \sqrt{9}, 3.12, \frac{1}{2}\pi, 7, -11.1, 13 \right\}$

In Exercises 17 and 18, plot the real numbers on the real number line.

17. (a) 3 (b) $\frac{7}{2}$ (c) $-\frac{5}{2}$ (d) -5.2

18. (a) 8.5 (b) $\frac{4}{3}$ (c) -4.75 (d) $-\frac{8}{3}$

In Exercises 19–22, use a calculator to find the decimal form of the rational number. If it is a nonterminating decimal, write the repeating pattern.

19. $\frac{5}{8}$

20. $\frac{1}{3}$

21. $\frac{41}{333}$

22. $\frac{6}{11}$

In Exercises 23 and 24, approximate the numbers and place the correct symbol (< or >) between them.

23.

24.

In Exercises 25–30, plot the two real numbers on the real number line. Then place the appropriate inequality symbol (< or >) between them.

25. $-4, -8$

26. $-3.5, 1$

27. $\frac{3}{2}, 7$

28. $1, \frac{16}{3}$

29. $\frac{5}{6}, \frac{2}{3}$

30. $-\frac{8}{7}, -\frac{3}{7}$

In Exercises 31–42, (a) give a verbal description of the subset of real numbers represented by the inequality or the interval, (b) sketch the subset on the real number line, and (c) state whether the interval is bounded or unbounded.

31. $x \leq 5$

32. $x \geq -2$

33. $x < 0$

34. $x > 3$

35. $[4, \infty)$

36. $(-\infty, 2)$

37. $-2 < x < 2$

38. $0 \leq x \leq 5$

39. $-1 \leq x < 0$

40. $0 < x \leq 6$

41. $[-2, 5)$

42. $(-1, 2]$

In Exercises 43–50, use inequality notation and interval notation to describe the set.

43. y is nonnegative.

44. y is no more than 25.

45. x is greater than -2 and at most 4.

46. y is at least -6 and less than 0.

47. t is at least 10 and at most 22.

48. k is less than 5 but no less than -3.

49. The dog's weight W is more than 65 pounds.

50. The annual rate of inflation r is expected to be at least 2.5% but no more than 5%.

In Exercises 51–60, evaluate the expression.

51. $|-10|$

52. $|0|$

53. $|3 - 8|$

54. $|4 - 1|$

55. $|-1| - |-2|$

56. $-3 - |-3|$

57. $\dfrac{-5}{|-5|}$

58. $-3|-3|$

59. $\dfrac{|x + 2|}{x + 2}, \quad x < -2$

60. $\dfrac{|x - 1|}{x - 1}, \quad x > 1$

In Exercises 61–66, place the correct symbol ($<$, $>$, or $=$) between the two real numbers.

61. $|-3| \quad \quad -|-3|$

62. $|-4| \quad \quad |4|$

63. $-5 \quad \quad -|5|$

64. $-|-6| \quad \quad |-6|$

65. $-|-2| \quad \quad -|2|$

66. $-(-2) \quad \quad -2$

In Exercises 67–72, find the distance between a and b.

67. $a = 126, b = 75$

68. $a = -126, b = -75$

69. $a = -\frac{5}{2}, b = 0$

70. $a = \frac{1}{4}, b = \frac{11}{4}$

71. $a = \frac{16}{5}, b = \frac{112}{75}$

72. $a = 9.34, b = -5.65$

In Exercises 73–78, use absolute value notation to describe the situation.

73. The distance between x and 5 is no more than 3.

74. The distance between x and -10 is at least 6.

75. y is at least six units from 0.

76. y is at most two units from a.

77. While traveling on the Pennsylvania Turnpike, you pass milepost 57 near Pittsburgh, then milepost 236 near Gettysburg. How many miles do you travel during that time period?

78. The temperature in Bismarck, North Dakota was 60°F at noon, then 23°F at midnight. What was the change in temperature over the 12-hour period?

BUDGET VARIANCE   In Exercises 79–82, the accounting department of a sports drink bottling company is checking to see whether the actual expenses of a department differ from the budgeted expenses by more than \$500 or by more than 5%. Fill in the missing parts of the table, and determine whether each actual expense passes the "budget variance test."

		Budgeted Expense, b	Actual Expense, a	$|a - b|$	$0.05b$
79.	Wages	\$112,700	\$113,356		
80.	Utilities	\$9,400	\$9,772		
81.	Taxes	\$37,640	\$37,335		
82.	Insurance	\$2,575	\$2,613		

FEDERAL DEFICIT   In Exercises 83–88, use the bar graph, which shows the receipts of the federal government (in billions of dollars) for selected years from 1996 through 2006. In each exercise you are given the expenditures of the federal government. Find the magnitude of the surplus or deficit for the year.   (Source: U.S. Office of Management and Budget)

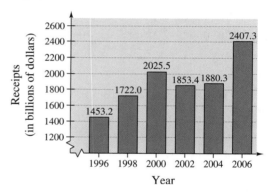

	Year	Receipts	Expenditures	$|$Receipts $-$ Expenditures$|$
83.	1996		\$1560.6 billion	
84.	1998		\$1652.7 billion	
85.	2000		\$1789.2 billion	
86.	2002		\$2011.2 billion	
87.	2004		\$2293.0 billion	
88.	2006		\$2655.4 billion	

In Exercises 89–94, identify the terms. Then identify the coefficients of the variable terms of the expression.

89. $7x + 4$

90. $6x^3 - 5x$

91. $\sqrt{3}x^2 - 8x - 11$

92. $3\sqrt{3}x^2 + 1$

93. $4x^3 + \dfrac{x}{2} - 5$

94. $3x^4 - \dfrac{x^2}{4}$

In Exercises 95–100, evaluate the expression for each value of x. (If not possible, state the reason.)

Expression	Values
95. $4x - 6$	(a) $x = -1$ (b) $x = 0$
96. $9 - 7x$	(a) $x = -3$ (b) $x = 3$
97. $x^2 - 3x + 4$	(a) $x = -2$ (b) $x = 2$
98. $-x^2 + 5x - 4$	(a) $x = -1$ (b) $x = 1$
99. $\dfrac{x + 1}{x - 1}$	(a) $x = 1$ (b) $x = -1$
100. $\dfrac{x}{x + 2}$	(a) $x = 2$ (b) $x = -2$

In Exercises 101–112, identify the rule(s) of algebra illustrated by the statement.

101. $x + 9 = 9 + x$

102. $2\left(\frac{1}{2}\right) = 1$

103. $\dfrac{1}{h + 6}(h + 6) = 1, \quad h \neq -6$

104. $(x + 3) - (x + 3) = 0$

105. $2(x + 3) = 2 \cdot x + 2 \cdot 3$

106. $(z - 2) + 0 = z - 2$

107. $1 \cdot (1 + x) = 1 + x$

108. $(z + 5)x = z \cdot x + 5 \cdot x$

109. $x + (y + 10) = (x + y) + 10$

110. $x(3y) = (x \cdot 3)y = (3x)y$

111. $3(t - 4) = 3 \cdot t - 3 \cdot 4$

112. $\frac{1}{7}(7 \cdot 12) = \left(\frac{1}{7} \cdot 7\right)12 = 1 \cdot 12 = 12$

In Exercises 113–120, perform the operation(s). (Write fractional answers in simplest form.)

113. $\frac{3}{16} + \frac{5}{16}$

114. $\frac{6}{7} - \frac{4}{7}$

115. $\frac{5}{8} - \frac{5}{12} + \frac{1}{6}$

116. $\frac{10}{11} + \frac{6}{33} - \frac{13}{66}$

117. $12 \div \frac{1}{4}$

118. $-\left(6 \cdot \frac{4}{8}\right)$

119. $\frac{2x}{3} - \frac{x}{4}$

120. $\frac{5x}{6} \cdot \frac{2}{9}$

EXPLORATION

In Exercises 121 and 122, use the real numbers A, B, and C shown on the number line. Determine the sign of each expression.

121. (a) $-A$

    (b) $B - A$

122. (a) $-C$

    (b) $A - C$

123. CONJECTURE

(a) Use a calculator to complete the table.

n	1	0.5	0.01	0.0001	0.000001
$5/n$					

(b) Use the result from part (a) to make a conjecture about the value of $5/n$ as n approaches 0.

124. CONJECTURE

(a) Use a calculator to complete the table.

n	1	10	100	10,000	100,000
$5/n$					

(b) Use the result from part (a) to make a conjecture about the value of $5/n$ as n increases without bound.

TRUE OR FALSE? In Exercises 125–128, determine whether the statement is true or false. Justify your answer.

125. If $a > 0$ and $b < 0$, then $a - b > 0$.

126. If $a > 0$ and $b < 0$, then $ab > 0$.

127. If $a < b$, then $\dfrac{1}{a} < \dfrac{1}{b}$, where $a \neq 0$ and $b \neq 0$.

128. Because $\dfrac{a + b}{c} = \dfrac{a}{c} + \dfrac{b}{c}$, then $\dfrac{c}{a + b} = \dfrac{c}{a} + \dfrac{c}{b}$.

129. THINK ABOUT IT Consider $|u + v|$ and $|u| + |v|$, where $u \neq 0$ and $v \neq 0$.

(a) Are the values of the expressions always equal? If not, under what conditions are they unequal?

(b) If the two expressions are not equal for certain values of u and v, is one of the expressions always greater than the other? Explain.

130. THINK ABOUT IT Is there a difference between saying that a real number is positive and saying that a real number is nonnegative? Explain.

131. THINK ABOUT IT Because every even number is divisible by 2, is it possible that there exist any even prime numbers? Explain.

132. THINK ABOUT IT Is it possible for a real number to be both rational and irrational? Explain.

133. WRITING Can it ever be true that $|a| = -a$ for a real number a? Explain.

134. CAPSTONE Describe the differences among the sets of natural numbers, whole numbers, integers, rational numbers, and irrational numbers.

P.2 EXPONENTS AND RADICALS

What you should learn

- Use properties of exponents.
- Use scientific notation to represent real numbers.
- Use properties of radicals.
- Simplify and combine radicals.
- Rationalize denominators and numerators.
- Use properties of rational exponents.

Why you should learn it

Real numbers and algebraic expressions are often written with exponents and radicals. For instance, in Exercise 121 on page 27, you will use an expression involving rational exponents to find the times required for a funnel to empty for different water heights.

Integer Exponents

Repeated *multiplication* can be written in **exponential form.**

Repeated Multiplication	Exponential Form
$a \cdot a \cdot a \cdot a \cdot a$	a^5
$(-4)(-4)(-4)$	$(-4)^3$
$(2x)(2x)(2x)(2x)$	$(2x)^4$

Exponential Notation

If a is a real number and n is a positive integer, then

$$a^n = a \cdot a \cdot a \cdots a$$

where n is the **exponent** and a is the **base.** The expression a^n is read "a to the nth **power.**"

An exponent can also be negative. In Property 3 below, be sure you see how to use a negative exponent.

Properties of Exponents

Let a and b be real numbers, variables, or algebraic expressions, and let m and n be integers. (All denominators and bases are nonzero.)

Property	*Example*
1. $a^m a^n = a^{m+n}$	$3^2 \cdot 3^4 = 3^{2+4} = 3^6 = 729$
2. $\dfrac{a^m}{a^n} = a^{m-n}$	$\dfrac{x^7}{x^4} = x^{7-4} = x^3$
3. $a^{-n} = \dfrac{1}{a^n} = \left(\dfrac{1}{a}\right)^n$	$y^{-4} = \dfrac{1}{y^4} = \left(\dfrac{1}{y}\right)^4$
4. $a^0 = 1, \quad a \neq 0$	$(x^2 + 1)^0 = 1$
5. $(ab)^m = a^m b^m$	$(5x)^3 = 5^3 x^3 = 125x^3$
6. $(a^m)^n = a^{mn}$	$(y^3)^{-4} = y^{3(-4)} = y^{-12} = \dfrac{1}{y^{12}}$
7. $\left(\dfrac{a}{b}\right)^m = \dfrac{a^m}{b^m}$	$\left(\dfrac{2}{x}\right)^3 = \dfrac{2^3}{x^3} = \dfrac{8}{x^3}$
8. $|a^2| = |a|^2 = a^2$	$|(-2)^2| = |-2|^2 = (2)^2 = 4$

TECHNOLOGY

You can use a calculator to evaluate exponential expressions. When doing so, it is important to know when to use parentheses because the calculator follows the order of operations. For instance, evaluate $(-2)^4$ as follows.

Scientific:

[(] 2 [+/−] [)] [y^x] 4 [=]

Graphing:

[(] [(−)] 2 [)] [^] 4 [ENTER]

The display will be 16. If you omit the parentheses, the display will be −16.

It is important to recognize the difference between expressions such as $(-2)^4$ and -2^4. In $(-2)^4$, the parentheses indicate that the exponent applies to the negative sign as well as to the 2, but in $-2^4 = -(2^4)$, the exponent applies only to the 2. So, $(-2)^4 = 16$ and $-2^4 = -16$.

The properties of exponents listed on the preceding page apply to *all* integers m and n, not just to positive integers, as shown in the examples in this section.

Example 1   Evaluating Exponential Expressions

a. $(-5)^2 = (-5)(-5) = 25$

b. $-5^2 = -(5)(5) = -25$

c. $2 \cdot 2^4 = 2^{1+4} = 2^5 = 32$

d. $\dfrac{4^4}{4^6} = 4^{4-6} = 4^{-2} = \dfrac{1}{4^2} = \dfrac{1}{16}$

CHECK Point   Now try Exercise 11.

Example 2   Evaluating Algebraic Expressions

Evaluate each algebraic expression when $x = 3$.

a. $5x^{-2}$ **b.** $\dfrac{1}{3}(-x)^3$

Solution

a. When $x = 3$, the expression $5x^{-2}$ has a value of

$$5x^{-2} = 5(\ \ )^{-2} = \frac{5}{3^2} = \frac{5}{9}.$$

b. When $x = 3$, the expression $\dfrac{1}{3}(-x)^3$ has a value of

$$\frac{1}{3}(-x)^3 = \frac{1}{3}(-\ \ )^3 = \frac{1}{3}(-27) = -9.$$

CHECK Point   Now try Exercise 23.

Example 3   Using Properties of Exponents

Use the properties of exponents to simplify each expression.

a. $(-3ab^4)(4ab^{-3})$ **b.** $(2xy^2)^3$ **c.** $3a(-4a^2)^0$ **d.** $\left(\dfrac{5x^3}{y}\right)^2$

Solution

a. $(-3ab^4)(4ab^{-3}) = (-3)(4)(a)(a)(b^4)(b^{-3}) = -12a^2b$

b. $(2xy^2)^3 = 2^3(x)^3(y^2)^3 = 8x^3y^6$

c. $3a(-4a^2)^0 = 3a(1) = 3a, \quad a \neq 0$

d. $\left(\dfrac{5x^3}{y}\right)^2 = \dfrac{5^2(x^3)^2}{y^2} = \dfrac{25x^6}{y^2}$

CHECK Point   Now try Exercise 31.

Study Tip

Rarely in algebra is there only one way to solve a problem. Don't be concerned if the steps you use to solve a problem are not exactly the same as the steps presented in this text. The important thing is to use steps that you understand and, of course, steps that are justified by the rules of algebra. For instance, you might prefer the following steps for Example 4(d).

$$\left(\frac{3x^2}{y}\right)^{-2} = \left(\frac{y}{3x^2}\right)^2 = \frac{y^2}{9x^4}$$

Note how Property 3 is used in the first step of this solution. The fractional form of this property is

$$\left(\frac{a}{b}\right)^{-m} = \left(\frac{b}{a}\right)^m.$$

| **Example 4** | **Rewriting with Positive Exponents** |

Rewrite each expression with positive exponents.

a. x^{-1} **b.** $\dfrac{1}{3x^{-2}}$ **c.** $\dfrac{12a^3b^{-4}}{4a^{-2}b}$ **d.** $\left(\dfrac{3x^2}{y}\right)^{-2}$

Solution

a. $x^{-1} = \dfrac{1}{x}$

b. $\dfrac{1}{3x^{-2}} = \dfrac{1(x^2)}{3} = \dfrac{x^2}{3}$

c. $\dfrac{12a^3b^{-4}}{4a^{-2}b} = \dfrac{12a^3 \cdot a^2}{4b \cdot b^4}$

$\qquad\qquad = \dfrac{3a^5}{b^5}$

d. $\left(\dfrac{3x^2}{y}\right)^{-2} = \dfrac{3^{-2}(x^2)^{-2}}{y^{-2}}$

$\qquad\qquad = \dfrac{3^{-2}x^{-4}}{y^{-2}}$

$\qquad\qquad = \dfrac{y^2}{3^2x^4}$

$\qquad\qquad = \dfrac{y^2}{9x^4}$

CHECK*Point*   Now try Exercise 41.

Scientific Notation

Exponents provide an efficient way of writing and computing with very large (or very small) numbers. For instance, there are about 359 billion billion gallons of water on Earth—that is, 359 followed by 18 zeros.

    359,000,000,000,000,000,000,000

It is convenient to write such numbers in **scientific notation.** This notation has the form $\pm c \times 10^n$, where $1 \le c < 10$ and n is an integer. So, the number of gallons of water on Earth can be written in scientific notation as

    $3.59 \times 100,000,000,000,000,000,000 = 3.59 \times 10^{20}$.

The *positive* exponent 20 indicates that the number is *large* (10 or more) and that the decimal point has been moved 20 places. A *negative* exponent indicates that the number is *small* (less than 1). For instance, the mass (in grams) of one electron is approximately

    $9.0 \times 10^{-28} = 0.0000000000000000000000000009$.

HISTORICAL NOTE

The French mathematician Nicolas Chuquet (ca. 1500) wrote *Triparty en la science des nombres*, in which a form of exponent notation was used. Our expressions $6x^3$ and $10x^2$ were written as .6.³ and .10.². Zero and negative exponents were also represented, so x^0 would be written as .1.⁰ and $3x^{-2}$ as .3.²ᵐ. Chuquet wrote that .72.¹ divided by .8.³ is .9.²ᵐ. That is, $72x \div 8x^3 = 9x^{-2}$.

Example 5 **Scientific Notation**

Write each number in scientific notation.

a. 0.0000782   **b.** 836,100,000

Solution

a. $0.0000782 = 7.82 \times 10^{-5}$

b. $836,100,000 = 8.361 \times 10^{8}$

CHECK*Point* Now try Exercise 45.

Example 6 **Decimal Notation**

Write each number in decimal notation.

a. -9.36×10^{-6} **b.** 1.345×10^{2}

Solution

a. $-9.36 \times 10^{-6} = -0.00000936$ **b.** $1.345 \times 10^{2} = 134.5$

CHECK*Point* Now try Exercise 55.

TECHNOLOGY

Most calculators automatically switch to scientific notation when they are showing large (or small) numbers that exceed the display range.

To *enter* numbers in scientific notation, your calculator should have an exponential entry key labeled

EE or EXP.

Consult the user's guide for your calculator for instructions on keystrokes and how numbers in scientific notation are displayed.

Example 7 **Using Scientific Notation**

Evaluate $\dfrac{(2,400,000,000)(0.0000045)}{(0.00003)(1500)}$.

Solution

Begin by rewriting each number in scientific notation and simplifying.

$$\frac{(2,400,000,000)(0.0000045)}{(0.00003)(1500)} = \frac{(2.4 \times 10^{9})(4.5 \times 10^{-6})}{(3.0 \times 10^{-5})(1.5 \times 10^{3})}$$

$$= \frac{(2.4)(4.5)(10^{3})}{(4.5)(10^{-2})}$$

$$= (2.4)(10^{5})$$

$$= 240,000$$

CHECK*Point* Now try Exercise 63(b).

Radicals and Their Properties

A **square root** of a number is one of its two equal factors. For example, 5 is a square root of 25 because 5 is one of the two equal factors of 25. In a similar way, a **cube root** of a number is one of its three equal factors, as in $125 = 5^3$.

Definition of *n*th Root of a Number

Let a and b be real numbers and let $n \geq 2$ be a positive integer. If

$$a = b^n$$

then b is an ***n*th root of *a*.** If $n = 2$, the root is a **square root.** If $n = 3$, the root is a **cube root.**

Some numbers have more than one *n*th root. For example, both 5 and -5 are square roots of 25. The *principal square root* of 25, written as $\sqrt{25}$, is the positive root, 5. The **principal *n*th root** of a number is defined as follows.

Principal *n*th Root of a Number

Let a be a real number that has at least one *n*th root. The **principal *n*th root of *a*** is the *n*th root that has the same sign as a. It is denoted by a **radical symbol**

$$\sqrt[n]{a}.$$

The positive integer n is the **index** of the radical, and the number a is the **radicand.** If $n = 2$, omit the index and write $\sqrt{a}$ rather than $\sqrt[2]{a}$. (The plural of index is *indices*.)

A common misunderstanding is that the square root sign implies both negative and positive roots. This is not correct. The square root sign implies only a positive root. When a negative root is needed, you must use the negative sign with the square root sign.

Incorrect: $\sqrt{4} = \pm 2$ *Correct:* $-\sqrt{4} = -2$ *and* $\sqrt{4} = 2$

Example 8   Evaluating Expressions Involving Radicals

a. $\sqrt{36} = 6$ because $6^2 = 36$.

b. $-\sqrt{36} = -6$ because $-\left(\sqrt{36}\right) = -\left(\sqrt{6^2}\right) = -(6) = -6$.

c. $\sqrt[3]{\dfrac{125}{64}} = \dfrac{5}{4}$ because $\left(\dfrac{5}{4}\right)^3 = \dfrac{5^3}{4^3} = \dfrac{125}{64}$.

d. $\sqrt[5]{-32} = -2$ because $(-2)^5 = -32$.

e. $\sqrt[4]{-81}$ is not a real number because there is no real number that can be raised to the fourth power to produce -81.

CHECK*Point*   Now try Exercise 65.

Here are some generalizations about the *n*th roots of real numbers.

Generalizations About *n*th Roots of Real Numbers

Real Number *a*	Integer *n*	Root(s) of *a*	Example
$a > 0$	$n > 0$, *n* is even.	$\sqrt[n]{a}, -\sqrt[n]{a}$	$\sqrt[4]{81} = 3, -\sqrt[4]{81} = -3$
$a > 0$ or $a < 0$	*n* is odd.	$\sqrt[n]{a}$	$\sqrt[3]{-8} = -2$
$a < 0$	*n* is even.	No real roots	$\sqrt{-4}$ is not a real number.
$a = 0$	*n* is even or odd.	$\sqrt[n]{0} = 0$	$\sqrt[5]{0} = 0$

Integers such as 1, 4, 9, 16, 25, and 36 are called **perfect squares** because they have integer square roots. Similarly, integers such as 1, 8, 27, 64, and 125 are called **perfect cubes** because they have integer cube roots.

Properties of Radicals

Let *a* and *b* be real numbers, variables, or algebraic expressions such that the indicated roots are real numbers, and let *m* and *n* be positive integers.

Property	*Example*
1. $\sqrt[n]{a^m} = \left(\sqrt[n]{a}\right)^m$	$\sqrt[3]{8^2} = \left(\sqrt[3]{8}\right)^2 = (2)^2 = 4$
2. $\sqrt[n]{a} \cdot \sqrt[n]{b} = \sqrt[n]{ab}$	$\sqrt{5} \cdot \sqrt{7} = \sqrt{5 \cdot 7} = \sqrt{35}$
3. $\dfrac{\sqrt[n]{a}}{\sqrt[n]{b}} = \sqrt[n]{\dfrac{a}{b}}, \quad b \neq 0$	$\dfrac{\sqrt[4]{27}}{\sqrt[4]{9}} = \sqrt[4]{\dfrac{27}{9}} = \sqrt[4]{3}$
4. $\sqrt[m]{\sqrt[n]{a}} = \sqrt[mn]{a}$	$\sqrt[3]{\sqrt{10}} = \sqrt[6]{10}$
5. $\left(\sqrt[n]{a}\right)^n = a$	$\left(\sqrt{3}\right)^2 = 3$
6. For *n* even, $\sqrt[n]{a^n} = |a|$.	$\sqrt{(-12)^2} = |-12| = 12$
For *n* odd, $\sqrt[n]{a^n} = a$.	$\sqrt[3]{(-12)^3} = -12$

A common special case of Property 6 is $\sqrt{a^2} = |a|$.

Example 9   Using Properties of Radicals

Use the properties of radicals to simplify each expression.

a. $\sqrt{8} \cdot \sqrt{2}$ b. $\left(\sqrt[3]{5}\right)^3$ c. $\sqrt[3]{x^3}$ d. $\sqrt[6]{y^6}$

Solution

a. $\sqrt{8} \cdot \sqrt{2} = \sqrt{8 \cdot 2} = \sqrt{16} = 4$

b. $\left(\sqrt[3]{5}\right)^3 = 5$

c. $\sqrt[3]{x^3} = x$

d. $\sqrt[6]{y^6} = |y|$

CHECK*Point* Now try Exercise 77.

Simplifying Radicals

An expression involving radicals is in **simplest form** when the following conditions are satisfied.

1. All possible factors have been removed from the radical.

2. All fractions have radical-free denominators (accomplished by a process called *rationalizing the denominator*).

3. The index of the radical is reduced.

To simplify a radical, factor the radicand into factors whose exponents are multiples of the index. The roots of these factors are written outside the radical, and the "leftover" factors make up the new radicand.

> ⚠ **WARNING / CAUTION**
>
> When you simplify a radical, it is important that both expressions are defined for the same values of the variable. For instance, in Example 10(b), $\sqrt{75x^3}$ and $5x\sqrt{3x}$ are both defined only for nonnegative values of x. Similarly, in Example 10(c), $\sqrt[4]{(5x)^4}$ and $5|x|$ are both defined for all real values of x.

Example 10   Simplifying Even Roots

a. $\sqrt[4]{48} = \sqrt[4]{16 \cdot 3} = \sqrt[4]{2^4 \cdot 3} = 2\sqrt[4]{3}$

b. $\sqrt{75x^3} = \sqrt{25x^2 \cdot 3x}$

$\qquad\quad = \sqrt{(5x)^2 \cdot 3x}$

$\qquad\quad = 5x\sqrt{3x}$

c. $\sqrt[4]{(5x)^4} = |5x| = 5|x|$

CHECK*Point* ➤ Now try Exercise 79(a).

Example 11   Simplifying Odd Roots

a. $\sqrt[3]{24} = \sqrt[3]{8 \cdot 3} = \sqrt[3]{2^3 \cdot 3} = 2\sqrt[3]{3}$

b. $\sqrt[3]{24a^4} = \sqrt[3]{8a^3 \cdot 3a}$

$\qquad\quad = \sqrt[3]{(2a)^3 \cdot 3a}$

$\qquad\quad = 2a\sqrt[3]{3a}$

c. $\sqrt[3]{-40x^6} = \sqrt[3]{(-8x^6) \cdot 5}$

$\qquad\quad = \sqrt[3]{(-2x^2)^3 \cdot 5}$

$\qquad\quad = -2x^2\sqrt[3]{5}$

CHECK*Point* ➤ Now try Exercise 79(b).

Radical expressions can be combined (added or subtracted) if they are **like radicals**—that is, if they have the same index and radicand. For instance, $\sqrt{2}$, $3\sqrt{2}$, and $\frac{1}{2}\sqrt{2}$ are like radicals, but $\sqrt{3}$ and $\sqrt{2}$ are unlike radicals. To determine whether two radicals can be combined, you should first simplify each radical.

Example 12   Combining Radicals

a. $2\sqrt{48} - 3\sqrt{27} = 2\sqrt{16 \cdot 3} - 3\sqrt{9 \cdot 3}$

$$= 8\sqrt{3} - 9\sqrt{3}$$

$$= (8 - 9)\sqrt{3}$$

$$= -\sqrt{3}$$

b. $\sqrt[3]{16x} - \sqrt[3]{54x^4} = \sqrt[3]{8 \cdot 2x} - \sqrt[3]{27 \cdot x^3 \cdot 2x}$

$$= 2\sqrt[3]{2x} - 3x\sqrt[3]{2x}$$

$$= (2 - 3x)\sqrt[3]{2x}$$

CHECK*Point* Now try Exercise 87.

Rationalizing Denominators and Numerators

To rationalize a denominator or numerator of the form $a - b\sqrt{m}$ or $a + b\sqrt{m}$, multiply both numerator and denominator by a **conjugate**: $a + b\sqrt{m}$ and $a - b\sqrt{m}$ are conjugates of each other. If $a = 0$, then the rationalizing factor for $\sqrt{m}$ is itself, $\sqrt{m}$. For cube roots, choose a rationalizing factor that generates a perfect cube.

Example 13   Rationalizing Single-Term Denominators

Rationalize the denominator of each expression.

a. $\dfrac{5}{2\sqrt{3}}$ b. $\dfrac{2}{\sqrt[3]{5}}$

Solution

a. $\dfrac{5}{2\sqrt{3}} = \dfrac{5}{2\sqrt{3}} \cdot$

$$= \dfrac{5\sqrt{3}}{2(3)}$$

$$= \dfrac{5\sqrt{3}}{6}$$

b. $\dfrac{2}{\sqrt[3]{5}} = \dfrac{2}{\sqrt[3]{5}} \cdot$

$$= \dfrac{2\sqrt[3]{5^2}}{\sqrt[3]{5^3}}$$

$$= \dfrac{2\sqrt[3]{25}}{5}$$

CHECK*Point* Now try Exercise 95.

Example 14   Rationalizing a Denominator with Two Terms

$$\frac{2}{3 + \sqrt{7}} = \frac{2}{3 + \sqrt{7}} \cdot$$

$$= \frac{2\left(3 - \sqrt{7}\right)}{3(3) + 3\left(-\sqrt{7}\right) + \sqrt{7}(3) - \left(\sqrt{7}\right)\left(\sqrt{7}\right)}$$

$$= \frac{2\left(3 - \sqrt{7}\right)}{(3)^2 - \left(\sqrt{7}\right)^2}$$

$$= \frac{2\left(3 - \sqrt{7}\right)}{9 - 7}$$

$$= \frac{2\left(3 - \sqrt{7}\right)}{2} = 3 - \sqrt{7}$$

CHECK **Point** ▶   Now try Exercise 97.

Sometimes it is necessary to rationalize the numerator of an expression. For instance, in Section P.5 you will use the technique shown in the next example to rationalize the numerator of an expression from calculus.

Example 15   Rationalizing a Numerator

$$\frac{\sqrt{5} - \sqrt{7}}{2} = \frac{\sqrt{5} - \sqrt{7}}{2} \cdot$$

$$= \frac{\left(\sqrt{5}\right)^2 - \left(\sqrt{7}\right)^2}{2\left(\sqrt{5} + \sqrt{7}\right)}$$

$$= \frac{5 - 7}{2\left(\sqrt{5} + \sqrt{7}\right)}$$

$$= \frac{-2}{2\left(\sqrt{5} + \sqrt{7}\right)} = \frac{-1}{\sqrt{5} + \sqrt{7}}$$

CHECK **Point** ▶   Now try Exercise 101.

Rational Exponents

Definition of Rational Exponents

If a is a real number and n is a positive integer such that the principal nth root of a exists, then $a^{1/n}$ is defined as

$$a^{1/n} = \sqrt[n]{a}, \text{ where } 1/n \text{ is the } \textbf{rational exponent} \text{ of } a.$$

Moreover, if m is a positive integer that has no common factor with n, then

$$a^{m/n} = (a^{1/n})^m = \left(\sqrt[n]{a}\right)^m \quad \text{and} \quad a^{m/n} = (a^m)^{1/n} = \sqrt[n]{a^m}.$$

The symbol ∫ indicates an example or exercise that highlights algebraic techniques specifically used in calculus.

The numerator of a rational exponent denotes the *power* to which the base is raised, and the denominator denotes the *index* or the *root* to be taken.

$$b^{m/n} = \left(\sqrt[n]{b}\right)^m = \sqrt[n]{b^m}$$

When you are working with rational exponents, the properties of integer exponents still apply. For instance, $2^{1/2}2^{1/3} = 2^{(1/2)+(1/3)} = 2^{5/6}$.

Example 16   Changing From Radical to Exponential Form

a. $\sqrt{3} = 3^{1/2}$

b. $\sqrt{(3xy)^5} = \sqrt[2]{(3xy)^5} = (3xy)^{5/2}$

c. $2x\sqrt[4]{x^3} = (2x)(x^{3/4}) = 2x^{1+(3/4)} = 2x^{7/4}$

CHECK *Point* ▶ Now try Exercise 103.

Example 17   Changing From Exponential to Radical Form

a. $(x^2 + y^2)^{3/2} = \left(\sqrt{x^2 + y^2}\right)^3 = \sqrt{(x^2 + y^2)^3}$

b. $2y^{3/4}z^{1/4} = 2(y^3z)^{1/4} = 2\sqrt[4]{y^3z}$

c. $a^{-3/2} = \dfrac{1}{a^{3/2}} = \dfrac{1}{\sqrt{a^3}}$

d. $x^{0.2} = x^{1/5} = \sqrt[5]{x}$

CHECK *Point* ▶ Now try Exercise 105.

Rational exponents are useful for evaluating roots of numbers on a calculator, for reducing the index of a radical, and for simplifying expressions in calculus.

Example 18   Simplifying with Rational Exponents

a. $(-32)^{-4/5} = \left(\sqrt[5]{-32}\right)^{-4} = (-2)^{-4} = \dfrac{1}{(-2)^4} = \dfrac{1}{16}$

b. $(-5x^{5/3})(3x^{-3/4}) = -15x^{(5/3)-(3/4)} = -15x^{11/12}, \quad x \neq 0$

c. $\sqrt[9]{a^3} = a^{3/9} = a^{1/3} = \sqrt[3]{a}$

d. $\sqrt[3]{\sqrt{125}} = \sqrt[6]{125} = \sqrt[6]{(5)^3} = 5^{3/6} = 5^{1/2} = \sqrt{5}$

e. $(2x - 1)^{4/3}(2x - 1)^{-1/3} = (2x - 1)^{(4/3)-(1/3)}$

$$= 2x - 1, \qquad x \neq \frac{1}{2}$$

CHECK *Point* ▶ Now try Exercise 115.

The expression in Example 18(e) is not defined when $x = \dfrac{1}{2}$ because

$$\left(2 \cdot \frac{1}{2} - 1\right)^{-1/3} = (0)^{-1/3}$$

is not a real number.

TECHNOLOGY

There are four methods of evaluating radicals on most graphing calculators. For square roots, you can use the *square root key* ⟦√⟧. For cube roots, you can use the *cube root key* ⟦∛⟧. For other roots, you can first convert the radical to exponential form and then use the *exponential key* ⟦^⟧, or you can use the *xth root key* ⟦ˣ√⟧ (or menu choice). Consult the user's guide for your calculator for specific keystrokes.

P.2   EXERCISES

See www.CalcChat.com for worked-out solutions to odd-numbered exercises.

VOCABULARY: Fill in the blanks.

1. In the exponential form a^n, n is the _____ and a is the _____.

2. A convenient way of writing very large or very small numbers is called _____ _____.

3. One of the two equal factors of a number is called a _____ _____ of the number.

4. The _____ _____ _____ of a number a is the nth root that has the same sign as a, and is denoted by $\sqrt[n]{a}$.

5. In the radical form $\sqrt[n]{a}$, the positive integer n is called the _____ of the radical and the number a is called the _____.

6. When an expression involving radicals has all possible factors removed, radical-free denominators, and a reduced index, it is in _____ _____.

7. Radical expressions can be combined (added or subtracted) if they are _____ _____.

8. The expressions $a + b\sqrt{m}$ and $a - b\sqrt{m}$ are _____ of each other.

9. The process used to create a radical-free denominator is known as _____ the denominator.

10. In the expression $b^{m/n}$, m denotes the _____ to which the base is raised and n denotes the _____ or root to be taken.

SKILLS AND APPLICATIONS

In Exercises 11–18, evaluate each expression.

11. (a) $3^2 \cdot 3$ (b) $3 \cdot 3^3$

12. (a) $\dfrac{5^5}{5^2}$ (b) $\dfrac{3^2}{3^4}$

13. (a) $(3^3)^0$ (b) -3^2

14. (a) $(2^3 \cdot 3^2)^2$ (b) $\left(-\frac{3}{5}\right)^3\left(\frac{5}{3}\right)^2$

15. (a) $\dfrac{3}{3^{-4}}$ (b) $48(-4)^{-3}$

16. (a) $\dfrac{4 \cdot 3^{-2}}{2^{-2} \cdot 3^{-1}}$ (b) $(-2)^0$

17. (a) $2^{-1} + 3^{-1}$ (b) $(2^{-1})^{-2}$

18. (a) $3^{-1} + 2^{-2}$ (b) $(3^{-2})^2$

In Exercises 19–22, use a calculator to evaluate the expression. (If necessary, round your answer to three decimal places.)

19. $(-4)^3(5^2)$ 20. $(8^{-4})(10^3)$

21. $\dfrac{3^6}{7^3}$ 22. $\dfrac{4^3}{3^{-4}}$

In Exercises 23–30, evaluate the expression for the given value of x.

23. $-3x^3$, $x = 2$ 24. $7x^{-2}$, $x = 4$

25. $6x^0$, $x = 10$ 26. $5(-x)^3$, $x = 3$

27. $2x^3$, $x = -3$ 28. $-3x^4$, $x = -2$

29. $-20x^2$, $x = -\frac{1}{2}$ 30. $12(-x)^3$, $x = -\frac{1}{3}$

In Exercises 31–38, simplify each expression.

31. (a) $(-5z)^3$ (b) $5x^4(x^2)$

32. (a) $(3x)^2$ (b) $(4x^3)^0$, $x \ne 0$

33. (a) $6y^2(2y^0)^2$ (b) $\dfrac{3x^5}{x^3}$

34. (a) $(-z)^3(3z^4)$ (b) $\dfrac{25y^8}{10y^4}$

35. (a) $\dfrac{7x^2}{x^3}$ (b) $\dfrac{12(x + y)^3}{9(x + y)}$

36. (a) $\dfrac{r^4}{r^6}$ (b) $\left(\dfrac{4}{y}\right)^3\left(\dfrac{3}{y}\right)^4$

37. (a) $\left[(x^2y^{-2})^{-1}\right]^{-1}$ (b) $\left(\dfrac{a^{-2}}{b^{-2}}\right)\left(\dfrac{b}{a}\right)^3$

38. (a) $(6x^7)^0$, $x \ne 0$ (b) $(5x^2z^6)^3(5x^2z^6)^{-3}$

In Exercises 39–44, rewrite each expression with positive exponents and simplify.

39. (a) $(x + 5)^0$, $x \ne -5$ (b) $(2x^2)^{-2}$

40. (a) $(2x^5)^0$, $x \ne 0$ (b) $(z + 2)^{-3}(z + 2)^{-1}$

41. (a) $(-2x^2)^3(4x^3)^{-1}$ (b) $\left(\dfrac{x}{10}\right)^{-1}$

42. (a) $(4y^{-2})(8y^4)$ (b) $\left(\dfrac{x^{-3}y^4}{5}\right)^{-3}$

43. (a) $3^n \cdot 3^{2n}$ (b) $\left(\dfrac{a^{-2}}{b^{-2}}\right)\left(\dfrac{b}{a}\right)^3$

44. (a) $\dfrac{x^2 \cdot x^n}{x^3 \cdot x^n}$ (b) $\left(\dfrac{a^{-3}}{b^{-3}}\right)\left(\dfrac{a}{b}\right)^3$

In Exercises 45–52, write the number in scientific notation.

45. 10,250.4

46. $-7,280,000$

47. -0.000125

48. 0.00052

49. Land area of Earth: 57,300,000 square miles

50. Light year: 9,460,000,000,000 kilometers

51. Relative density of hydrogen: 0.0000899 gram per cubic centimeter

52. One micron (millionth of a meter): 0.00003937 inch

In Exercises 53–60, write the number in decimal notation.

53. 1.25×10^5

54. -1.801×10^5

55. -2.718×10^{-3}

56. 3.14×10^{-4}

57. Interior temperature of the sun: 1.5×10^7 degrees Celsius

58. Charge of an electron: 1.6022×10^{-19} coulomb

59. Width of a human hair: 9.0×10^{-5} meter

60. Gross domestic product of the United States in 2007: 1.3743021×10^{13} dollars   (Source: U.S. Department of Commerce)

In Exercises 61 and 62, evaluate each expression without using a calculator.

61. (a) $(2.0 \times 10^9)(3.4 \times 10^{-4})$
(b) $(1.2 \times 10^7)(5.0 \times 10^{-3})$

62. (a) $\dfrac{6.0 \times 10^8}{3.0 \times 10^{-3}}$
(b) $\dfrac{2.5 \times 10^{-3}}{5.0 \times 10^2}$

In Exercises 63 and 64, use a calculator to evaluate each expression. (Round your answer to three decimal places.)

63. (a) $750\left(1 + \dfrac{0.11}{365}\right)^{800}$

(b) $\dfrac{67,000,000 + 93,000,000}{0.0052}$

64. (a) $(9.3 \times 10^6)^3(6.1 \times 10^{-4})$ (b) $\dfrac{(2.414 \times 10^4)^6}{(1.68 \times 10^5)^5}$

In Exercises 65–70, evaluate each expression without using a calculator.

65. (a) $\sqrt{9}$

(b) $\sqrt[3]{\dfrac{27}{8}}$

66. (a) $27^{1/3}$

(b) $36^{3/2}$

67. (a) $32^{-3/5}$

(b) $\left(\dfrac{16}{81}\right)^{-3/4}$

68. (a) $100^{-3/2}$

(b) $\left(\dfrac{9}{4}\right)^{-1/2}$

69. (a) $\left(-\dfrac{1}{64}\right)^{-1/3}$

(b) $\left(\dfrac{1}{\sqrt{32}}\right)^{-2/5}$

70. (a) $\left(-\dfrac{125}{27}\right)^{-1/3}$

(b) $-\left(\dfrac{1}{125}\right)^{-4/3}$

In Exercises 71–76, use a calculator to approximate the number. (Round your answer to three decimal places.)

71. (a) $\sqrt{57}$

(b) $\sqrt[5]{-27^3}$

72. (a) $\sqrt[3]{45^2}$

(b) $\sqrt[6]{125}$

73. (a) $(-12.4)^{-1.8}$

(b) $(5\sqrt{3})^{-2.5}$

74. (a) $\dfrac{7 - (4.1)^{-3.2}}{2}$

(b) $\left(\dfrac{13}{3}\right)^{-3/2} - \left(-\dfrac{3}{2}\right)^{13/3}$

75. (a) $\sqrt{4.5 \times 10^9}$

(b) $\sqrt[3]{6.3 \times 10^4}$

76. (a) $(2.65 \times 10^{-4})^{1/3}$

(b) $\sqrt{9 \times 10^{-4}}$

In Exercises 77 and 78, use the properties of radicals to simplify each expression.

77. (a) $(\sqrt[5]{2})^5$

(b) $\sqrt[5]{96x^5}$

78. (a) $\sqrt{12} \cdot \sqrt{3}$

(b) $\sqrt[4]{(3x^2)^4}$

In Exercises 79–90, simplify each radical expression.

79. (a) $\sqrt{20}$

(b) $\sqrt[3]{128}$

80. (a) $\sqrt[3]{\dfrac{16}{27}}$

(b) $\sqrt{\dfrac{75}{4}}$

81. (a) $\sqrt{72x^3}$

(b) $\sqrt{\dfrac{18^2}{z^3}}$

82. (a) $\sqrt{54xy^4}$

(b) $\sqrt{\dfrac{32a^4}{b^2}}$

83. (a) $\sqrt[3]{16x^5}$

(b) $\sqrt{75x^2y^{-4}}$

84. (a) $\sqrt[4]{3x^4y^2}$

(b) $\sqrt[5]{160x^8z^4}$

85. (a) $2\sqrt{50} + 12\sqrt{8}$

(b) $10\sqrt{32} - 6\sqrt{18}$

86. (a) $4\sqrt{27} - \sqrt{75}$

(b) $\sqrt[3]{16} + 3\sqrt[3]{54}$

87. (a) $5\sqrt{x} - 3\sqrt{x}$

(b) $-2\sqrt{9y} + 10\sqrt{y}$

88. (a) $8\sqrt{49x} - 14\sqrt{100x}$
(b) $-3\sqrt{48x^2} + 7\sqrt{75x^2}$

89. (a) $3\sqrt{x + 1} + 10\sqrt{x + 1}$
(b) $7\sqrt{80x} - 2\sqrt{125x}$

90. (a) $-\sqrt{x^3 - 7} + 5\sqrt{x^3 - 7}$
(b) $11\sqrt{245x^3} - 9\sqrt{45x^3}$

In Exercises 91–94, complete the statement with <, =, or >.

91. $\sqrt{5} + \sqrt{3}$ ▢ $\sqrt{5 + 3}$ **92.** $\sqrt{\dfrac{3}{11}}$ ▢ $\dfrac{\sqrt{3}}{\sqrt{11}}$

93. 5 ▢ $\sqrt{3^2 + 2^2}$ **94.** 5 ▢ $\sqrt{3^2 + 4^2}$

In Exercises 95–98, rationalize the denominator of the expression. Then simplify your answer.

95. $\dfrac{1}{\sqrt{3}}$

96. $\dfrac{8}{\sqrt[3]{2}}$

97. $\dfrac{5}{\sqrt{14} - 2}$

98. $\dfrac{3}{\sqrt{5} + \sqrt{6}}$

‖ In Exercises 99–102, rationalize the numerator of the expression. Then simplify your answer.

99. $\dfrac{\sqrt{8}}{2}$

100. $\dfrac{\sqrt{2}}{3}$

101. $\dfrac{\sqrt{5}+\sqrt{3}}{3}$

102. $\dfrac{\sqrt{7}-3}{4}$

In Exercises 103–110, fill in the missing form of the expression.

Radical Form	Rational Exponent Form
103. $\sqrt{2.5}$	
104. $\sqrt[3]{64}$	
105.	$81^{1/4}$
106.	$-(144^{1/2})$
107. $\sqrt[3]{-216}$	
108.	$(-243)^{1/5}$
109. $\left(\sqrt[4]{81}\right)^{3}$	
110.	$16^{5/4}$

In Exercises 111–114, perform the operations and simplify.

111. $\dfrac{(2x^2)^{3/2}}{2^{1/2}x^4}$

112. $\dfrac{x^{4/3}y^{2/3}}{(xy)^{1/3}}$

113. $\dfrac{x^{-3}\cdot x^{1/2}}{x^{3/2}\cdot x^{-1}}$

114. $\dfrac{5^{-1/2}\cdot 5x^{5/2}}{(5x)^{3/2}}$

In Exercises 115 and 116, reduce the index of each radical.

115. (a) $\sqrt[4]{3^2}$ (b) $\sqrt[6]{(x+1)^4}$

116. (a) $\sqrt[6]{x^3}$ (b) $\sqrt[4]{(3x^2)^4}$

In Exercises 117 and 118, write each expression as a single radical. Then simplify your answer.

117. (a) $\sqrt{\sqrt{32}}$ (b) $\sqrt{\sqrt[4]{2x}}$

118. (a) $\sqrt{\sqrt{243(x+1)}}$ (b) $\sqrt{\sqrt[3]{10a^7b}}$

119. **PERIOD OF A PENDULUM** The period T (in seconds) of a pendulum is $T = 2\pi\sqrt{L/32}$, where L is the length of the pendulum (in feet). Find the period of a pendulum whose length is 2 feet.

120. **EROSION** A stream of water moving at the rate of v feet per second can carry particles of size $0.03\sqrt{v}$ inches. Find the size of the largest particle that can be carried by a stream flowing at the rate of $\frac{3}{4}$ foot per second.

The symbol **‖** indicates an example or exercise that highlights algebraic techniques specifically used in calculus.

The symbol ∿ indicates an exercise or a part of an exercise in which you are instructed to use a graphing utility.

121. **MATHEMATICAL MODELING** A funnel is filled with water to a height of h centimeters. The formula

$$t = 0.03[12^{5/2} - (12 - h)^{5/2}], \quad 0 \le h \le 12$$

represents the amount of time t (in seconds) that it will take for the funnel to empty.

∿ (a) Use the *table* feature of a graphing utility to find the times required for the funnel to empty for water heights of $h = 0$, $h = 1$, $h = 2$, . . . , $h = 12$ centimeters.

(b) What value does t appear to be approaching as the height of the water becomes closer and closer to 12 centimeters?

122. **SPEED OF LIGHT** The speed of light is approximately 11,180,000 miles per minute. The distance from the sun to Earth is approximately 93,000,000 miles. Find the time for light to travel from the sun to Earth.

EXPLORATION

TRUE OR FALSE? In Exercises 123 and 124, determine whether the statement is true or false. Justify your answer.

123. $\dfrac{x^{k+1}}{x} = x^k$

124. $(a^n)^k = a^{n^k}$

125. Verify that $a^0 = 1$, $a \ne 0$. (*Hint:* Use the property of exponents $a^m/a^n = a^{m-n}$.)

126. Explain why each of the following pairs is not equal.

(a) $(3x)^{-1} \ne \dfrac{3}{x}$ (b) $y^3\cdot y^2 \ne y^6$

(c) $(a^2b^3)^4 \ne a^6b^7$ (d) $(a+b)^2 \ne a^2 + b^2$

(e) $\sqrt{4x^2} \ne 2x$ (f) $\sqrt{2} + \sqrt{3} \ne \sqrt{5}$

127. **THINK ABOUT IT** Is 52.7×10^5 written in scientific notation? Why or why not?

128. List all possible digits that occur in the units place of the square of a positive integer. Use that list to determine whether $\sqrt{5233}$ is an integer.

129. **THINK ABOUT IT** Square the real number $5/\sqrt{3}$ and note that the radical is eliminated from the denominator. Is this equivalent to rationalizing the denominator? Why or why not?

130. **CAPSTONE**

(a) Explain how to simplify the expression $(3x^3 y^{-2})^{-2}$.

(b) Is the expression $\sqrt{\dfrac{4}{x^3}}$ in simplest form? Why or why not?

What you should learn

- Write polynomials in standard form.
- Add, subtract, and multiply polynomials.
- Use special products to multiply polynomials.
- Use polynomials to solve real-life problems.

Why you should learn it

Polynomials can be used to model and solve real-life problems. For instance, in Exercise 106 on page 34, polynomials are used to model the cost, revenue, and profit for producing and selling hats.

David Noton/Masterfile

P.3 POLYNOMIALS AND SPECIAL PRODUCTS

Polynomials

The most common type of algebraic expression is the **polynomial.** Some examples are $2x + 5$, $3x^4 - 7x^2 + 2x + 4$, and $5x^2y^2 - xy + 3$. The first two are *polynomials in x* and the third is a *polynomial in x and y.* The terms of a polynomial in x have the form ax^k, where a is the **coefficient** and k is the **degree** of the term. For instance, the polynomial

$$2x^3 - 5x^2 + 1 = 2x^3 + (-5)x^2 + (0)x + 1$$

has coefficients 2, -5, 0, and 1.

> ### Definition of a Polynomial in x
>
> Let $a_0, a_1, a_2, \ldots, a_n$ be real numbers and let n be a nonnegative integer. A polynomial in x is an expression of the form
>
> $$a_nx^n + a_{n-1}x^{n-1} + \cdots + a_1x + a_0$$
>
> where $a_n \neq 0$. The polynomial is of **degree** n, a_n is the **leading coefficient,** and a_0 is the **constant term.**

Polynomials with one, two, and three terms are called **monomials, binomials,** and **trinomials,** respectively. In **standard form,** a polynomial is written with descending powers of x.

Example 1   Writing Polynomials in Standard Form

Polynomial	Standard Form	Degree	Leading Coefficient
a. $4x^2 - 5x^7 - 2 + 3x$	$-5x^7 + 4x^2 + 3x - 2$	7	-5
b. $4 - 9x^2$	$-9x^2 + 4$	2	-9
c. 8	$8 \ (8 = 8x^0)$	0	8

CHECK*Point* Now try Exercise 19.

A polynomial that has all zero coefficients is called the **zero polynomial,** denoted by 0. No degree is assigned to this particular polynomial. For polynomials in more than one variable, the degree of a *term* is the sum of the exponents of the variables in the term. The degree of the *polynomial* is the highest degree of its terms. For instance, the degree of the polynomial $-2x^3y^6 + 4xy - x^7y^4$ is 11 because the sum of the exponents in the last term is the greatest. The leading coefficient of the polynomial is the coefficient of the highest-degree term. Expressions are not polynomials if a variable is underneath a radical or if a polynomial expression (with degree greater than 0) is in the denominator of a term. The following expressions are not polynomials.

$$x^3 - \sqrt{3x} = x^3 - (3x)^{1/2}$$

$$x^2 + \frac{5}{x} = x^2 + 5x^{-1}$$

Operations with Polynomials

You can add and subtract polynomials in much the same way you add and subtract real numbers. Simply add or subtract the *like terms* (terms having the same variables to the same powers) by adding their coefficients. For instance, $-3xy^2$ and $5xy^2$ are like terms and their sum is

$$-3xy^2 + 5xy^2 = (-3 + 5)xy^2$$
$$= 2xy^2.$$

Example 2   Sums and Differences of Polynomials

a. $(5x^3 - 7x^2 - 3) + (x^3 + 2x^2 - x + 8)$

$= (5x^3 + x^3) + (-7x^2 + 2x^2) - x + (-3 + 8)$

$= 6x^3 - 5x^2 - x + 5$

b. $(7x^4 - x^2 - 4x + 2) - (3x^4 - 4x^2 + 3x)$

$= 7x^4 - x^2 - 4x + 2 - 3x^4 + 4x^2 - 3x$

$= (7x^4 - 3x^4) + (-x^2 + 4x^2) + (-4x - 3x) + 2$

$= 4x^4 + 3x^2 - 7x + 2$

CHECK Point Now try Exercise 41.

To find the *product* of two polynomials, use the left and right Distributive Properties. For example, if you treat $5x + 7$ as a single quantity, you can multiply $3x - 2$ by $5x + 7$ as follows.

$(3x - 2)(5x + 7) = 3x(5x + 7) - 2(5x + 7)$

$= (3x)(5x) + (3x)(7) - (2)(5x) - (2)(7)$

$= 15x^2 + 21x - 10x - 14$

|Product of First terms|Product of Outer terms|Product of Inner terms|Product of Last terms|

$= 15x^2 + 11x - 14$

Note in this **FOIL Method** (which can only be used to multiply two binomials) that the outer (O) and inner (I) terms are like terms and can be combined.

Example 3   Finding a Product by the FOIL Method

Use the FOIL Method to find the product of $2x - 4$ and $x + 5$.

Solution

$(2x - 4)(x + 5) = 2x^2 + 10x - 4x - 20$

$= 2x^2 + 6x - 20$

CHECK Point Now try Exercise 59.

When multiplying two polynomials, be sure to multiply each term of one polynomial by *each* term of the other. A vertical arrangement is helpful.

Example 4 A Vertical Arrangement for Multiplication

Multiply $x^2 - 2x + 2$ by $x^2 + 2x + 2$ using a vertical arrangement.

Solution

$$
\begin{array}{r}
x^2 - 2x + 2 \\
\times\ x^2 + 2x + 2 \\
\hline
x^4 - 2x^3 + 2x^2 \\
2x^3 - 4x^2 + 4x \\
2x^2 - 4x + 4 \\
\hline
x^4 + 0x^3 + 0x^2 + 0x + 4 = x^4 + 4
\end{array}
$$

So, $(x^2 - 2x + 2)(x^2 + 2x + 2) = x^4 + 4$.

CHECK Point Now try Exercise 61.

Special Products

Some binomial products have special forms that occur frequently in algebra. You do not need to memorize these formulas because you can use the Distributive Property to multiply. However, becoming familiar with these formulas will enable you to manipulate the algebra more quickly.

Special Products

Let u and v be real numbers, variables, or algebraic expressions.

Special Product	*Example*

Sum and Difference of Same Terms

$(u + v)(u - v) = u^2 - v^2$

$(x + 4)(x - 4) = x^2 - 4^2$
$ = x^2 - 16$

Square of a Binomial

$(u + v)^2 = u^2 + 2uv + v^2$

$(x + 3)^2 = x^2 + 2(x)(3) + 3^2$
$ = x^2 + 6x + 9$

$(u - v)^2 = u^2 - 2uv + v^2$

$(3x - 2)^2 = (3x)^2 - 2(3x)(2) + 2^2$
$ = 9x^2 - 12x + 4$

Cube of a Binomial

$(u + v)^3 = u^3 + 3u^2v + 3uv^2 + v^3$

$(x + 2)^3 = x^3 + 3x^2(2) + 3x(2^2) + 2^3$
$ = x^3 + 6x^2 + 12x + 8$

$(u - v)^3 = u^3 - 3u^2v + 3uv^2 - v^3$

$(x - 1)^3 = x^3 - 3x^2(1) + 3x(1^2) - 1^3$
$ = x^3 - 3x^2 + 3x - 1$

Example 5   Sum and Difference of Same Terms

Find the product of $5x + 9$ and $5x - 9$.

Solution

The product of a sum and a difference of the *same* two terms has no middle term and takes the form $(u + v)(u - v) = u^2 - v^2$.

$$(5x + 9)(5x - 9) = (5x)^2 - 9^2 = 25x^2 - 81$$

CHECK *Point*   Now try Exercise 67.

Study Tip

When squaring a binomial, note that the resulting middle term is always *twice* the product of the two terms.

Example 6   Square of a Binomial

Find $(6x - 5)^2$.

Solution

The square of a binomial has the form $(u - v)^2 = u^2 - 2uv + v^2$.

$$(6x - 5)^2 = (6x)^2 - 2(6x)(5) + 5^2 = 36x^2 - 60x + 25$$

CHECK *Point*   Now try Exercise 71.

Example 7   Cube of a Binomial

Find $(3x + 2)^3$.

Solution

The cube of a binomial has the form

$$(u + v)^3 = u^3 + 3u^2v + 3uv^2 + v^3.$$

Note the *decreasing* powers of $u = 3x$ and the *increasing* powers of $v = 2$.

$$(3x + 2)^3 = (3x)^3 + 3(3x)^2(2) + 3(3x)(2^2) + 2^3$$

$$= 27x^3 + 54x^2 + 36x + 8$$

CHECK *Point*   Now try Exercise 73.

Example 8   The Product of Two Trinomials

Find the product of $x + y - 2$ and $x + y + 2$.

Solution

By grouping $x + y$ in parentheses, you can write the product of the trinomials as a special product.

$$(x + y - 2)(x + y + 2) = [(x + y) - 2][(x + y) + 2]$$

$$= (x + y)^2 - 2^2$$

$$= x^2 + 2xy + y^2 - 4$$

CHECK *Point*   Now try Exercise 81.

Application

Example 9 Volume of a Box

An open box is made by cutting squares from the corners of a piece of metal that is 16 inches by 20 inches, as shown in Figure P.13. The edge of each cut-out square is x inches. Find the volume of the box when $x = 1$, $x = 2$, and $x = 3$.

Solution

The volume of a rectangular box is equal to the product of its length, width, and height. From the figure, the length is $20 - 2x$, the width is $16 - 2x$, and the height is x. So, the volume of the box is

$$\text{Volume} = (20 - 2x)(16 - 2x)(x)$$
$$= (320 - 72x + 4x^2)(x)$$
$$= 320x - 72x^2 + 4x^3.$$

When $x = 1$ inch, the volume of the box is

$$\text{Volume} = 320(\ ) - 72(\ )^2 + 4(\ )^3$$
$$= 252 \text{ cubic inches.}$$

When $x = 2$ inches, the volume of the box is

$$\text{Volume} = 320(\ ) - 72(\ )^2 + 4(\ )^3$$
$$= 384 \text{ cubic inches.}$$

When $x = 3$ inches, the volume of the box is

$$\text{Volume} = 320(\ ) - 72(\ )^2 + 4(\ )^3$$
$$= 420 \text{ cubic inches.}$$

CHECK*Point* Now try Exercise 109.

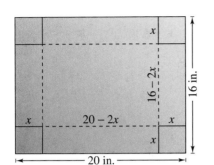

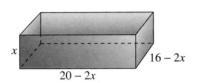

FIGURE P.13

CLASSROOM DISCUSSION

Mathematical Experiment In Example 9, the volume of the open box is given by

$$\text{Volume} = 320x - 72x^2 + 4x^3.$$

You want to create a box that has as much volume as possible. From Example 9, you know that by cutting one-, two-, and three-inch squares from the corners, you can create boxes whose volumes are 252, 384, and 420 cubic inches, respectively. What are the possible values of x that make sense in this problem? Write your answer as an interval. Try several other values of x to find the size of the squares that should be cut from the corners to produce a box that has maximum volume. Write a summary of your findings.

P.3   EXERCISES

See www.CalcChat.com for worked-out solutions to odd-numbered exercises.

VOCABULARY

In Exercises 1–5, fill in the blanks.

1. For the polynomial $a_n x^n + a_{n-1} x^{n-1} + \cdots + a_1 x + a_0$, $a_n \neq 0$, the degree is _____, the leading coefficient is _____, and the constant term is _____.

2. A polynomial in x in standard form is written with _____ powers of x.

3. A polynomial with one term is called a _____, while a polynomial with two terms is called a _____, and a polynomial with three terms is called a _____.

4. To add or subtract polynomials, add or subtract the _____ _____ by adding their coefficients.

5. The letters in "FOIL" stand for the following. F _____   O _____   I _____   L _____

In Exercises 6–8, match the special product form with its name.

6. $(u + v)(u - v) = u^2 - v^2$ (a) A binomial sum squared

7. $(u + v)^2 = u^2 + 2uv + v^2$ (b) A binomial difference squared

8. $(u - v)^2 = u^2 - 2uv + v^2$ (c) The sum and difference of same terms

SKILLS AND APPLICATIONS

In Exercises 9–14, match the polynomial with its description. [The polynomials are labeled (a), (b), (c), (d), (e), and (f).]

(a) $3x^2$ (b) $1 - 2x^3$

(c) $x^3 + 3x^2 + 3x + 1$ (d) 12

(e) $-3x^5 + 2x^3 + x$ (f) $\frac{2}{3}x^4 + x^2 + 10$

9. A polynomial of degree 0

10. A trinomial of degree 5

11. A binomial with leading coefficient -2

12. A monomial of positive degree

13. A trinomial with leading coefficient $\frac{2}{3}$

14. A third-degree polynomial with leading coefficient 1

In Exercises 15–18, write a polynomial that fits the description. (There are many correct answers.)

15. A third-degree polynomial with leading coefficient -2

16. A fifth-degree polynomial with leading coefficient 6

17. A fourth-degree binomial with a negative leading coefficient

18. A third-degree binomial with an even leading coefficient

In Exercises 19–30, (a) write the polynomial in standard form, (b) identify the degree and leading coefficient of the polynomial, and (c) state whether the polynomial is a monomial, a binomial, or a trinomial.

19. $14x - \frac{1}{2}x^5$ **20.** $2x^2 - x + 1$

21. $x^2 - 4 - 3x^4$ **22.** $7x$

23. $3 - x^6$ **24.** $-y + 25y^2 + 1$

25. 3 **26.** $-8 + t^2$

27. $1 + 6x^4 - 4x^5$ **28.** $3 + 2x$

29. $4x^3 y$ **30.** $-x^5 y + 2x^2 y^2 + xy^4$

In Exercises 31–36, determine whether the expression is a polynomial. If so, write the polynomial in standard form.

31. $2x - 3x^3 + 8$ **32.** $5x^4 - 2x^2 + x^{-2}$

33. $\dfrac{3x + 4}{x}$ **34.** $\dfrac{x^2 + 2x - 3}{2}$

35. $y^2 - y^4 + y^3$ **36.** $y^4 - \sqrt{y}$

In Exercises 37–54, perform the operation and write the result in standard form.

37. $(6x + 5) - (8x + 15)$

38. $(2x^2 + 1) - (x^2 - 2x + 1)$

39. $-(t^3 - 1) + (6t^3 - 5t)$

40. $-(5x^2 - 1) - (-3x^2 + 5)$

41. $(15x^2 - 6) - (-8.3x^3 - 14.7x^2 - 17)$

42. $(15.6w^4 - 14w - 17.4) - (16.9w^4 - 9.2w + 13)$

43. $5z - [3z - (10z + 8)]$

44. $(y^3 + 1) - [(y^2 + 1) + (3y - 7)]$

45. $3x(x^2 - 2x + 1)$ **46.** $y^2(4y^2 + 2y - 3)$

47. $-5z(3z - 1)$ **48.** $(-3x)(5x + 2)$

49. $(1 - x^3)(4x)$ **50.** $-4x(3 - x^3)$

51. $(1.5t^2 + 5)(-3t)$ **52.** $(2 - 3.5y)(2y^3)$

53. $-2x(0.1x + 17)$ **54.** $6y(5 - \frac{3}{8}y)$

In Exercises 55–62, perform the operation.

55. Add $7x^3 - 2x^2 + 8$ and $-3x^3 - 4$.

56. Add $2x^5 - 3x^3 + 2x + 3$ and $4x^3 + x - 6$.

57. Subtract $x - 3$ from $5x^2 - 3x + 8$.

58. Subtract $-t^4 + 0.5t^2 - 5.6$ from $0.6t^4 - 2t^2$.

59. Multiply $(x + 7)$ and $(2x + 3)$.

60. Multiply $(3x + 1)$ and $(x - 5)$.

61. Multiply $(x^2 + 2x + 3)$ and $(x^2 - 2x + 3)$.

62. Multiply $(x^2 + x - 4)$ and $(x^2 - 2x + 1)$.

In Exercises 63–100, multiply or find the special product.

63. $(x + 3)(x + 4)$ **64.** $(x - 5)(x + 10)$

65. $(3x - 5)(2x + 1)$ **66.** $(7x - 2)(4x - 3)$

67. $(x + 10)(x - 10)$ **68.** $(2x + 3)(2x - 3)$

69. $(x + 2y)(x - 2y)$ **70.** $(4a + 5b)(4a - 5b)$

71. $(2x + 3)^2$ **72.** $(5 - 8x)^2$

73. $(x + 1)^3$ **74.** $(x - 2)^3$

75. $(2x - y)^3$ **76.** $(3x + 2y)^3$

77. $(4x^3 - 3)^2$ **78.** $(8x + 3)^2$

79. $(x^2 - x + 1)(x^2 + x + 1)$

80. $(x^2 + 3x - 2)(x^2 - 3x - 2)$

81. $(-x^2 + x - 5)(3x^2 + 4x + 1)$

82. $(2x^2 - x + 4)(x^2 + 3x + 2)$

83. $[(m - 3) + n][(m - 3) - n]$

84. $[(x - 3y) + z][(x - 3y) - z]$

85. $[(x - 3) + y]^2$ **86.** $[(x + 1) - y]^2$

87. $(2r^2 - 5)(2r^2 + 5)$ **88.** $(3a^3 - 4b^2)(3a^3 + 4b^2)$

89. $\left(\frac{1}{4}x - 5\right)^2$ **90.** $\left(\frac{3}{5}t + 4\right)^2$

91. $\left(\frac{1}{5}x - 3\right)\left(\frac{1}{5}x + 3\right)$ **92.** $\left(3x + \frac{1}{6}\right)\left(3x - \frac{1}{6}\right)$

93. $(2.4x + 3)^2$

94. $(1.8y - 5)^2$

95. $(1.5x - 4)(1.5x + 4)$

96. $(2.5y + 3)(2.5y - 3)$

97. $5x(x + 1) - 3x(x + 1)$

98. $(2x - 1)(x + 3) + 3(x + 3)$

99. $(u + 2)(u - 2)(u^2 + 4)$

100. $(x + y)(x - y)(x^2 + y^2)$

In Exercises 101–104, find the product. (The expressions are not polynomials, but the formulas can still be used.)

101. $\left(\sqrt{x} + \sqrt{y}\right)\left(\sqrt{x} - \sqrt{y}\right)$

102. $\left(5 + \sqrt{x}\right)\left(5 - \sqrt{x}\right)$

103. $\left(x - \sqrt{5}\right)^2$

104. $\left(x + \sqrt{3}\right)^2$

105. COST, REVENUE, AND PROFIT An electronics manufacturer can produce and sell x MP3 players per week. The total cost C (in dollars) of producing x MP3 players is $C = 73x + 25,000$, and the total revenue R (in dollars) is $R = 95x$.

(a) Find the profit P in terms of x.

(b) Find the profit obtained by selling 5000 MP3 players per week.

106. COST, REVENUE, AND PROFIT An artisan can produce and sell x hats per month. The total cost C (in dollars) for producing x hats is $C = 460 + 12x$, and the total revenue R (in dollars) is $R = 36x$.

(a) Find the profit P in terms of x.

(b) Find the profit obtained by selling 42 hats per month.

107. COMPOUND INTEREST After 2 years, an investment of $500 compounded annually at an interest rate r will yield an amount of $500(1 + r)^2$.

(a) Write this polynomial in standard form.

(b) Use a calculator to evaluate the polynomial for the values of r shown in the table.

r	$2\frac{1}{2}\%$	3%	4%	$4\frac{1}{2}\%$	5%
$500(1 + r)^2$					

(c) What conclusion can you make from the table?

108. COMPOUND INTEREST After 3 years, an investment of $1200 compounded annually at an interest rate r will yield an amount of $1200(1 + r)^3$.

(a) Write this polynomial in standard form.

(b) Use a calculator to evaluate the polynomial for the values of r shown in the table.

r	2%	3%	$3\frac{1}{2}\%$	4%	$4\frac{1}{2}\%$
$1200(1 + r)^3$					

(c) What conclusion can you make from the table?

109. VOLUME OF A BOX A take-out fast-food restaurant is constructing an open box by cutting squares from the corners of a piece of cardboard that is 18 centimeters by 26 centimeters (see figure). The edge of each cut-out square is x centimeters.

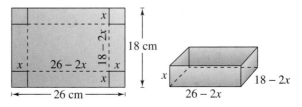

(a) Find the volume of the box in terms of x.

(b) Find the volume when $x = 1$, $x = 2$, and $x = 3$.

110. VOLUME OF A BOX   An overnight shipping company is designing a closed box by cutting along the solid lines and folding along the broken lines on the rectangular piece of corrugated cardboard shown in the figure. The length and width of the rectangle are 45 centimeters and 15 centimeters, respectively.

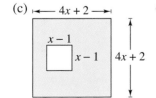

(a) Find the volume of the shipping box in terms of x.

(b) Find the volume when $x = 3$, $x = 5$, and $x = 7$.

111. GEOMETRY   Find the area of the shaded region in each figure. Write your result as a polynomial in standard form.

(a)

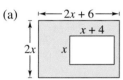

(b)

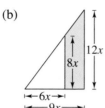

(c)

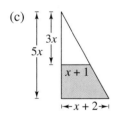

(d)

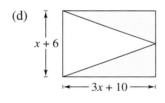

112. GEOMETRY   Find the area of the shaded region in each figure. Write your result as a polynomial in standard form.

(a)

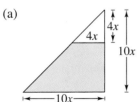

(b)

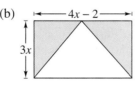

(c)

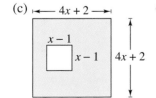

(d)

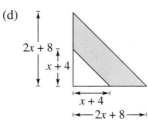

GEOMETRY   In Exercises 113 and 114, find a polynomial that represents the total number of square feet for the floor plan shown in the figure.

113.

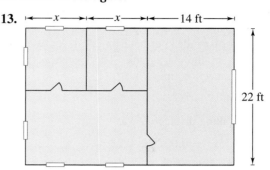

114.

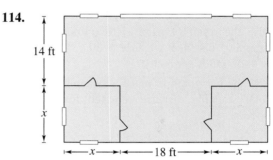

115. ENGINEERING   A uniformly distributed load is placed on a one-inch-wide steel beam. When the span of the beam is x feet and its depth is 6 inches, the safe load S (in pounds) is approximated by

$$S_6 = (0.06x^2 - 2.42x + 38.71)^2.$$

When the depth is 8 inches, the safe load is approximated by

$$S_8 = (0.08x^2 - 3.30x + 51.93)^2.$$

(a) Use the bar graph to estimate the difference in the safe loads for these two beams when the span is 12 feet.

(b) How does the difference in safe load change as the span increases?

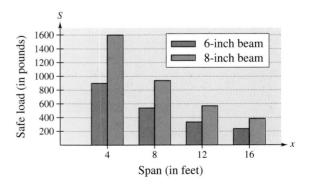

116. **STOPPING DISTANCE** The stopping distance of an automobile is the distance traveled during the driver's reaction time plus the distance traveled after the brakes are applied. In an experiment, these distances were measured (in feet) when the automobile was traveling at a speed of x miles per hour on dry, level pavement, as shown in the bar graph. The distance traveled during the reaction time R was

$$R = 1.1x$$

and the braking distance B was

$$B = 0.0475x^2 - 0.001x + 0.23.$$

(a) Determine the polynomial that represents the total stopping distance T.

(b) Use the result of part (a) to estimate the total stopping distance when $x = 30$, $x = 40$, and $x = 55$ miles per hour.

(c) Use the bar graph to make a statement about the total stopping distance required for increasing speeds.

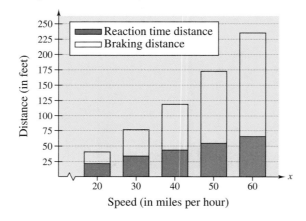

GEOMETRY In Exercises 117 and 118, use the area model to write two different expressions for the area. Then equate the two expressions and name the algebraic property that is illustrated.

117.

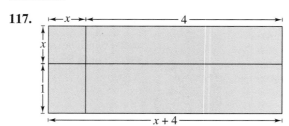

118.

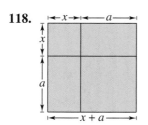

EXPLORATION

TRUE OR FALSE? In Exercises 119 and 120, determine whether the statement is true or false. Justify your answer.

119. The product of two binomials is always a second-degree polynomial.

120. The sum of two binomials is always a binomial.

121. Find the degree of the product of two polynomials of degrees m and n.

122. Find the degree of the sum of two polynomials of degrees m and n if $m < n$.

123. **WRITING** A student's homework paper included the following.

$$(x - 3)^2 = x^2 + 9$$

Write a paragraph fully explaining the error and give the correct method for squaring a binomial.

124. **CAPSTONE** A third-degree polynomial and a fourth-degree polynomial are added.

(a) Can the sum be a fourth-degree polynomial? Explain or give an example.

(b) Can the sum be a second-degree polynomial? Explain or give an example.

(c) Can the sum be a seventh-degree polynomial? Explain or give an example.

125. **THINK ABOUT IT** Must the sum of two second-degree polynomials be a second-degree polynomial? If not, give an example.

126. **THINK ABOUT IT** When the polynomial

$$-x^3 + 3x^2 + 2x - 1$$

is subtracted from an unknown polynomial, the difference is $5x^2 + 8$. If it is possible, find the unknown polynomial.

127. **LOGICAL REASONING** Verify that $(x + y)^2$ is not equal to $x^2 + y^2$ by letting $x = 3$ and $y = 4$ and evaluating both expressions. Are there any values of x and y for which $(x + y)^2 = x^2 + y^2$? Explain.

P.4   FACTORING POLYNOMIALS

What you should learn

- Remove common factors from polynomials.
- Factor special polynomial forms.
- Factor trinomials as the product of two binomials.
- Factor polynomials by grouping.

Why you should learn it

Polynomial factoring can be used to solve real-life problems. For instance, in Exercise 148 on page 44, factoring is used to develop an alternative model for the rate of change of an autocatalytic chemical reaction.

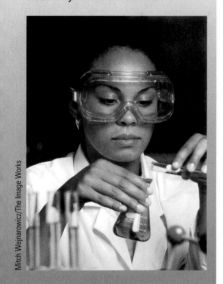

Polynomials with Common Factors

The process of writing a polynomial as a product is called **factoring.** It is an important tool for solving equations and for simplifying rational expressions.

Unless noted otherwise, when you are asked to factor a polynomial, you can assume that you are looking for factors with integer coefficients. If a polynomial cannot be factored using integer coefficients, then it is **prime** or **irreducible over the integers.** For instance, the polynomial $x^2 - 3$ is irreducible over the integers. Over the *real numbers*, this polynomial can be factored as

$$x^2 - 3 = \left(x + \sqrt{3}\right)\left(x - \sqrt{3}\right).$$

A polynomial is **completely factored** when each of its factors is prime. For instance

$$x^3 - x^2 + 4x - 4 = (x - 1)(x^2 + 4)$$

is completely factored, but

$$x^3 - x^2 - 4x + 4 = (x - 1)(x^2 - 4)$$

is not completely factored. Its complete factorization is

$$x^3 - x^2 - 4x + 4 = (x - 1)(x + 2)(x - 2).$$

The simplest type of factoring involves a polynomial that can be written as the product of a monomial and another polynomial. The technique used here is the Distributive Property, $a(b + c) = ab + ac$, in the *reverse* direction.

$$b + c = (b + c)$$

Removing (factoring out) any common factors is the first step in completely factoring a polynomial.

Example 1   Removing Common Factors

Factor each expression.

a. $6x^3 - 4x$

b. $-4x^2 + 12x - 16$

c. $(x - 2)(2x) + (x - 2)(3)$

Solution

a. $6x^3 - 4x = \quad (3x^2) - \quad (2)$

$= 2x(3x^2 - 2)$

b. $-4x^2 + 12x - 16 = \quad (x^2) + (\quad)(-3x) + (\quad)4$

$= -4(x^2 - 3x + 4)$

c. $(x - 2)(2x) + (x - 2)(3) = \qquad (2x + 3)$

CHECK**Point**   Now try Exercise 11.

Factoring Special Polynomial Forms

Some polynomials have special forms that arise from the special product forms on page 30. You should learn to recognize these forms so that you can factor such polynomials easily.

Factoring Special Polynomial Forms

Factored Form	*Example*
Difference of Two Squares	
$u^2 - v^2 = (u + v)(u - v)$	$9x^2 - 4 = (3x)^2 - 2^2 = (3x + 2)(3x - 2)$
Perfect Square Trinomial	
$u^2 + 2uv + v^2 = (u + v)^2$	$x^2 + 6x + 9 = x^2 + 2(x)(3) + 3^2 = (x + 3)^2$
$u^2 - 2uv + v^2 = (u - v)^2$	$x^2 - 6x + 9 = x^2 - 2(x)(3) + 3^2 = (x - 3)^2$
Sum or Difference of Two Cubes	
$u^3 + v^3 = (u + v)(u^2 - uv + v^2)$	$x^3 + 8 = x^3 + 2^3 = (x + 2)(x^2 - 2x + 4)$
$u^3 - v^3 = (u - v)(u^2 + uv + v^2)$	$27x^3 - 1 = (3x)^3 - 1^3 = (3x - 1)(9x^2 + 3x + 1)$

One of the easiest special polynomial forms to factor is the difference of two squares. The factored form is always a set of *conjugate pairs*.

$$u^2 - v^2 = (u + v)(u - v)$$

To recognize perfect square terms, look for coefficients that are squares of integers and variables raised to *even powers*.

Study Tip

In Example 2, note that the first step in factoring a polynomial is to check for any common factors. Once the common factors are removed, it is often possible to recognize patterns that were not immediately obvious.

Example 2   Removing a Common Factor First

$$3 - 12x^2 = 3(1 - 4x^2)$$
$$= 3[1^2 - (2x)^2]$$
$$= 3(1 + 2x)(1 - 2x)$$

CHECK*Point* Now try Exercise 25.

Example 3   Factoring the Difference of Two Squares

a. $(x + 2)^2 - y^2 = [(x + 2) + y][(x + 2) - y]$
$$= (x + 2 + y)(x + 2 - y)$$

b. $16x^4 - 81 = (4x^2)^2 - 9^2$
$$= (4x^2 + 9)(4x^2 - 9)$$
$$= (4x^2 + 9)[(2x)^2 - 3^2]$$
$$= (4x^2 + 9)(2x + 3)(2x - 3)$$

CHECK*Point* Now try Exercise 29.

A **perfect square trinomial** is the square of a binomial, and it has the following form.

$$u^2 + 2uv + v^2 = (u + v)^2 \qquad \text{or} \qquad u^2 - 2uv + v^2 = (u - v)^2$$

Note that the first and last terms are squares and the middle term is twice the product of u and v.

Example 4   Factoring Perfect Square Trinomials

Factor each trinomial.

a. $x^2 - 10x + 25$

b. $16x^2 + 24x + 9$

Solution

a. $x^2 - 10x + 25 = x^2 - 2(x)(5) + 5^2 = (x - 5)^2$

b. $16x^2 + 24x + 9 = (4x)^2 + 2(4x)(3) + 3^2 = (4x + 3)^2$

CHECK*Point*   Now try Exercise 35.

The next two formulas show the sums and differences of cubes. Pay special attention to the signs of the terms.

$$u^3 + v^3 = (u + v)(u^2 - uv + v^2) \qquad u^3 - v^3 = (u - v)(u^2 + uv + v^2)$$

Example 5   Factoring the Difference of Cubes

Factor $x^3 - 27$.

Solution

$$x^3 - 27 = x^3 - 3^3$$
$$= (x - 3)(x^2 + 3x + 9)$$

CHECK*Point*   Now try Exercise 45.

Example 6   Factoring the Sum of Cubes

a. $y^3 + 8 = y^3 + 2^3$
$$= (y + 2)(y^2 - 2y + 4)$$

b. $3(x^3 + 64) = 3(x^3 + 4^3)$
$$= 3(x + 4)(x^2 - 4x + 16)$$

CHECK*Point*   Now try Exercise 47.

Trinomials with Binomial Factors

To factor a trinomial of the form $ax^2 + bx + c$, use the following pattern.

$$ax^2 + bx + c = (\boxed{}\, x + \boxed{})(\boxed{}\, x + \boxed{})$$

The goal is to find a combination of factors of a and c such that the outer and inner products add up to the middle term bx. For instance, in the trinomial $6x^2 + 17x + 5$, you can write all possible factorizations and determine which one has outer and inner products that add up to $17x$.

$$(6x + 5)(x + 1), \ (6x + 1)(x + 5), \ (2x + 1)(3x + 5), \ (2x + 5)(3x + 1)$$

You can see that $(2x + 5)(3x + 1)$ is the correct factorization because the outer (O) and inner (I) products add up to $17x$.

$$\overset{\text{F}}{} \quad \overset{\text{O}}{} \quad \overset{\text{I}}{} \quad \overset{\text{L}}{} \qquad \overset{\text{O + I}}{}$$

$$(2x + 5)(3x + 1) = 6x^2 + \quad + \quad + 5 = 6x^2 + \quad + 5$$

Example 7 Factoring a Trinomial: Leading Coefficient Is 1

Factor $x^2 - 7x + 12$.

Solution

The possible factorizations are

$$(x - 2)(x - 6), \quad (x - 1)(x - 12), \quad \text{and} \quad (x - 3)(x - 4).$$

Testing the middle term, you will find the correct factorization to be

$$x^2 - 7x + 12 = (x - 3)(x - 4).$$

CHECK *Point*   Now try Exercise 57.

Example 8 Factoring a Trinomial: Leading Coefficient Is Not 1

Factor $2x^2 + x - 15$.

Solution

The eight possible factorizations are as follows.

$$(2x - 1)(x + 15) \qquad (2x + 1)(x - 15)$$
$$(2x - 3)(x + 5) \qquad (2x + 3)(x - 5)$$
$$(2x - 5)(x + 3) \qquad (2x + 5)(x - 3)$$
$$(2x - 15)(x + 1) \qquad (2x + 15)(x - 1)$$

Testing the middle term, you will find the correct factorization to be

$$2x^2 + x - 15 = (2x - 5)(x + 3).$$

CHECK *Point*   Now try Exercise 65.

Study Tip

Factoring a trinomial can involve trial and error. However, once you have produced the factored form, it is an easy matter to check your answer. For instance, you can verify the factorization in Example 7 by multiplying out the expression $(x - 3)(x - 4)$ to see that you obtain the original trinomial, $x^2 - 7x + 12$.

Factoring by Grouping

Sometimes polynomials with more than three terms can be factored by a method called **factoring by grouping.** It is not always obvious which terms to group, and sometimes several different groupings will work.

Example 9   Factoring by Grouping

Use factoring by grouping to factor $x^3 - 2x^2 - 3x + 6$.

Solution

$$x^3 - 2x^2 - 3x + 6 = (x^3 - 2x^2) - (3x - 6)$$
$$= x^2 \qquad - 3$$
$$= (x - 2)(x^2 - 3)$$

CHECK**Point**   Now try Exercise 73.

> ### Study Tip
>
> Another way to factor the polynomial in Example 9 is to group the terms as follows.
>
> $x^3 - 2x^2 - 3x + 6$
>
> $= (x^3 - 3x) - (2x^2 - 6)$
>
> $= x(x^2 - 3) - 2(x^2 - 3)$
>
> $= (x^2 - 3)(x - 2)$
>
> As you can see, you obtain the same result as in Example 9.

Factoring a trinomial can involve quite a bit of trial and error. Some of this trial and error can be lessened by using factoring by grouping. The key to this method of factoring is knowing how to rewrite the middle term. In general, to factor a trinomial $ax^2 + bx + c$ by grouping, choose factors of the product ac that add up to b and use these factors to rewrite the middle term. This technique is illustrated in Example 10.

Example 10   Factoring a Trinomial by Grouping

Use factoring by grouping to factor $2x^2 + 5x - 3$.

Solution

In the trinomial $2x^2 + 5x - 3$, $a = 2$ and $c = -3$, which implies that the product ac is -6. Now, -6 factors as $(6)(-1)$ and $6 - 1 = 5 = b$. So, you can rewrite the middle term as $5x = 6x - x$. This produces the following.

$$2x^2 + \qquad - 3 = 2x^2 + \qquad - 3$$
$$= (2x^2 + 6x) - (x + 3)$$
$$= 2x \qquad -$$
$$= (x + 3)(2x - 1)$$

So, the trinomial factors as $2x^2 + 5x - 3 = (x + 3)(2x - 1)$.

CHECK**Point**   Now try Exercise 79.

Guidelines for Factoring Polynomials

1. Factor out any common factors using the Distributive Property.

2. Factor according to one of the special polynomial forms.

3. Factor as $ax^2 + bx + c = (mx + r)(nx + s)$.

4. Factor by grouping.

P.4 EXERCISES

See www.CalcChat.com for worked-out solutions to odd-numbered exercises.

VOCABULARY

In Exercises 1–3, fill in the blanks.

1. The process of writing a polynomial as a product is called _____.
2. A polynomial is _____ _____ when each of its factors is prime.
3. If a polynomial has more than three terms, a method of factoring called _____ _____ _____ may be used.

4. Match the factored form of the polynomial with its name.
   (a) $u^2 - v^2 = (u + v)(u - v)$ (i) Perfect square trinomial
   (b) $u^3 - v^3 = (u - v)(u^2 + uv + v^2)$ (ii) Difference of two squares
   (c) $u^2 - 2uv + v^2 = (u - v)^2$ (iii) Difference of two cubes

SKILLS AND APPLICATIONS

In Exercises 5–8, find the greatest common factor of the expressions.

5. $80, 280$
6. $24, 96, 256$
7. $12x^2y^3, 18x^2y, 24x^3y^2$
8. $15(x + 2)^3, 42x(x + 2)^2$

In Exercises 9–16, factor out the common factor.

9. $4x + 16$
10. $5y - 30$
11. $2x^3 - 6x$
12. $3z^3 - 6z^2 + 9z$
13. $3x(x - 5) + 8(x - 5)$
14. $3x(x + 2) - 4(x + 2)$
15. $(x + 3)^2 - 4(x + 3)$
16. $(5x - 4)^2 + (5x - 4)$

In Exercises 17–22, find the greatest common factor such that the remaining factors have only integer coefficients.

17. $\frac{1}{2}x + 4$
18. $\frac{1}{3}y + 5$
19. $\frac{1}{2}x^3 + 2x^2 - 5x$
20. $\frac{1}{3}y^4 - 5y^2 + 2y$
21. $\frac{2}{3}x(x - 3) - 4(x - 3)$
22. $\frac{4}{5}y(y + 1) - 2(y + 1)$

In Exercises 23–32, completely factor the difference of two squares.

23. $x^2 - 81$
24. $x^2 - 64$
25. $48y^2 - 27$
26. $50 - 98z^2$
27. $16x^2 - \frac{1}{9}$
28. $\frac{4}{25}y^2 - 64$
29. $(x - 1)^2 - 4$
30. $25 - (z + 5)^2$
31. $9u^2 - 4v^2$
32. $25x^2 - 16y^2$

In Exercises 33–44, factor the perfect square trinomial.

33. $x^2 - 4x + 4$
34. $x^2 + 10x + 25$
35. $4t^2 + 4t + 1$
36. $9x^2 - 12x + 4$
37. $25y^2 - 10y + 1$
38. $36y^2 - 108y + 81$
39. $9u^2 + 24uv + 16v^2$
40. $4x^2 - 4xy + y^2$
41. $x^2 - \frac{4}{3}x + \frac{4}{9}$
42. $z^2 + z + \frac{1}{4}$
43. $4x^2 - \frac{4}{3}x + \frac{1}{9}$
44. $9y^2 - \frac{3}{2}y + \frac{1}{16}$

In Exercises 45–56, factor the sum or difference of cubes.

45. $x^3 - 8$
46. $27 - x^3$
47. $y^3 + 64$
48. $z^3 + 216$
49. $x^3 - \frac{8}{27}$
50. $y^3 + \frac{8}{125}$
51. $8t^3 - 1$
52. $27x^3 + 8$
53. $u^3 + 27v^3$
54. $64x^3 - y^3$
55. $(x + 2)^3 - y^3$
56. $(x - 3y)^3 - 8z^3$

In Exercises 57–70, factor the trinomial.

57. $x^2 + x - 2$
58. $x^2 + 5x + 6$
59. $s^2 - 5s + 6$
60. $t^2 - t - 6$
61. $20 - y - y^2$
62. $24 + 5z - z^2$
63. $x^2 - 30x + 200$
64. $x^2 - 13x + 42$
65. $3x^2 - 5x + 2$
66. $2x^2 - x - 1$
67. $5x^2 + 26x + 5$
68. $12x^2 + 7x + 1$
69. $-9z^2 + 3z + 2$
70. $-5u^2 - 13u + 6$

In Exercises 71–78, factor by grouping.

71. $x^3 - x^2 + 2x - 2$
72. $x^3 + 5x^2 - 5x - 25$
73. $2x^3 - x^2 - 6x + 3$
74. $5x^3 - 10x^2 + 3x - 6$
75. $6 + 2x - 3x^3 - x^4$
76. $x^5 + 2x^3 + x^2 + 2$
77. $6x^3 - 2x + 3x^2 - 1$
78. $8x^5 - 6x^2 + 12x^3 - 9$

In Exercises 79–84, factor the trinomial by grouping.

79. $3x^2 + 10x + 8$
80. $2x^2 + 9x + 9$
81. $6x^2 + x - 2$
82. $6x^2 - x - 15$
83. $15x^2 - 11x + 2$
84. $12x^2 - 13x + 1$

In Exercises 85–120, completely factor the expression.

85. $6x^2 - 54$
86. $12x^2 - 48$
87. $x^3 - x^2$
88. $x^3 - 4x^2$

89. $x^3 - 16x$

90. $x^3 - 9x$

91. $x^2 - 2x + 1$

92. $16 + 6x - x^2$

93. $1 - 4x + 4x^2$

94. $-9x^2 + 6x - 1$

95. $2x^2 + 4x - 2x^3$

96. $13x + 6 + 5x^2$

97. $\frac{1}{81}x^2 + \frac{2}{9}x - 8$

98. $\frac{1}{8}x^2 - \frac{1}{96}x - \frac{1}{16}$

99. $3x^3 + x^2 + 15x + 5$

100. $5 - x + 5x^2 - x^3$

101. $x^4 - 4x^3 + x^2 - 4x$

102. $3u - 2u^2 + 6 - u^3$

103. $2x^3 + x^2 - 8x - 4$

104. $3x^3 + x^2 - 27x - 9$

105. $\frac{1}{4}x^3 + 3x^2 + \frac{3}{4}x + 9$

106. $\frac{1}{5}x^3 + x^2 - x - 5$

107. $(t - 1)^2 - 49$

108. $(x^2 + 1)^2 - 4x^2$

109. $(x^2 + 8)^2 - 36x^2$

110. $2t^3 - 16$

111. $5x^3 + 40$

112. $4x(2x - 1) + (2x - 1)^2$

113. $5(3 - 4x)^2 - 8(3 - 4x)(5x - 1)$

114. $2(x + 1)(x - 3)^2 - 3(x + 1)^2(x - 3)$

115. $7(3x + 2)^2(1 - x)^2 + (3x + 2)(1 - x)^3$

116. $7x(2)(x^2 + 1)(2x) - (x^2 + 1)^2(7)$

117. $3(x - 2)^2(x + 1)^4 + (x - 2)^3(4)(x + 1)^3$

118. $2x(x - 5)^4 - x^2(4)(x - 5)^3$

119. $5(x^6 + 1)^4(6x^5)(3x + 2)^3 + 3(3x + 2)^2(3)(x^6 + 1)^5$

120. $\frac{x^2}{2}(x^2 + 1)^4 - (x^2 + 1)^5$

GEOMETRIC MODELING In Exercises 121–124, match the factoring formula with the correct "geometric factoring model." [The models are labeled (a), (b), (c), and (d).] For instance, a factoring model for

$$2x^2 + 3x + 1 = (2x + 1)(x + 1)$$

is shown in the following figure.

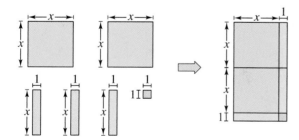

(a)

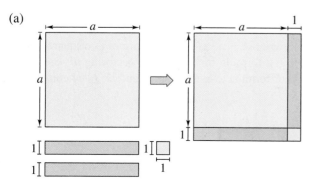

(b)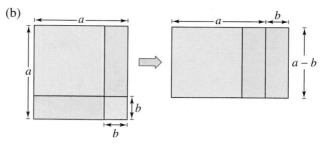

(c)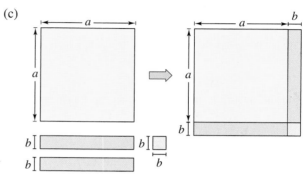

(d)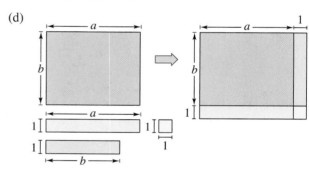

121. $a^2 - b^2 = (a + b)(a - b)$

122. $a^2 + 2ab + b^2 = (a + b)^2$

123. $a^2 + 2a + 1 = (a + 1)^2$

124. $ab + a + b + 1 = (a + 1)(b + 1)$

GEOMETRIC MODELING In Exercises 125–128, draw a "geometric factoring model" to represent the factorization.

125. $3x^2 + 7x + 2 = (3x + 1)(x + 2)$

126. $x^2 + 4x + 3 = (x + 3)(x + 1)$

127. $2x^2 + 7x + 3 = (2x + 1)(x + 3)$

128. $x^2 + 3x + 2 = (x + 2)(x + 1)$

GEOMETRY In Exercises 129–132, write an expression in factored form for the area of the shaded portion of the figure.

129.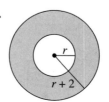

130.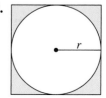

131.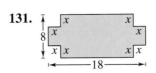

132.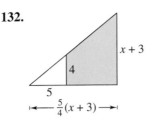

In Exercises 133–138, completely factor the expression.

133. $x^4(4)(2x + 1)^3(2x) + (2x + 1)^4(4x^3)$

134. $x^3(3)(x^2 + 1)^2(2x) + (x^2 + 1)^3(3x^2)$

135. $(2x - 5)^4(3)(5x - 4)^2(5) + (5x - 4)^3(4)(2x - 5)^3(2)$

136. $(x^2 - 5)^3(2)(4x + 3)(4) + (4x + 3)^2(3)(x^2 - 5)^2(x^2)$

137. $\dfrac{(5x - 1)(3) - (3x + 1)(5)}{(5x - 1)^2}$

138. $\dfrac{(2x + 3)(4) - (4x - 1)(2)}{(2x + 3)^2}$

In Exercises 139–142, find all values of b for which the trinomial can be factored.

139. $x^2 + bx - 15$ **140.** $x^2 + bx - 12$

141. $x^2 + bx + 50$ **142.** $x^2 + bx + 24$

In Exercises 143–146, find two integer values of c such that the trinomial can be factored. (There are many correct answers.)

143. $2x^2 + 5x + c$ **144.** $3x^2 - 10x + c$

145. $3x^2 - x + c$ **146.** $2x^2 + 9x + c$

147. GEOMETRY The volume V of concrete used to make the cylindrical concrete storage tank shown in the figure is $V = \pi R^2 h - \pi r^2 h$, where R is the outside radius, r is the inside radius, and h is the height of the storage tank.

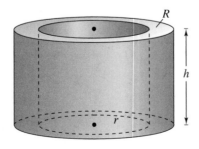

(a) Factor the expression for the volume.

(b) From the result of part (a), show that the volume of concrete is

$$2\pi(\text{average radius})(\text{thickness of the tank})h.$$

(c) An 80-pound bag of concrete mix yields $\frac{3}{5}$ cubic foot of concrete. Find the number of bags required to construct a concrete storage tank having the following dimensions.

Outside radius, $R = 4$ feet

Inside radius, $r = 3\frac{2}{3}$ feet

Height, h feet

 (d) Use the *table* feature of a graphing utility to create a table showing the number of bags of concrete required to construct the storage tank in part (c) with heights of $h = \frac{1}{2}$, $h = 1$, $h = \frac{3}{2}$, $h = 2$, . . . , $h = 6$ feet.

148. CHEMISTRY The rate of change of an autocatalytic chemical reaction is $kQx - kx^2$, where Q is the amount of the original substance, x is the amount of substance formed, and k is a constant of proportionality. Factor the expression.

EXPLORATION

TRUE OR FALSE? In Exercises 149 and 150, determine whether the statement is true or false. Justify your answer.

149. The difference of two perfect squares can be factored as the product of conjugate pairs.

150. The sum of two perfect squares can be factored as the binomial sum squared.

151. ERROR ANALYSIS Describe the error.

$$9x^2 - 9x - 54 = (3x + 6)(3x - 9)$$
$$= 3(x + 2)(x - 3)$$

152. THINK ABOUT IT Is $(3x - 6)(x + 1)$ completely factored? Explain.

153. Factor $x^{2n} - y^{2n}$ as completely as possible.

154. Factor $x^{3n} + y^{3n}$ as completely as possible.

155. Give an example of a polynomial that is prime with respect to the integers.

156. CAPSTONE Explain what is meant when it is said that a polynomial is in factored form.

157. Rewrite $u^6 - v^6$ as the difference of two squares. Then find a formula for completely factoring $u^6 - v^6$. Use your formula to factor $x^6 - 1$ and $x^6 - 64$ completely.

P.5   RATIONAL EXPRESSIONS

Domain of an Algebraic Expression

The set of real numbers for which an algebraic expression is defined is the **domain** of the expression. Two algebraic expressions are **equivalent** if they have the same domain and yield the same values for all numbers in their domain. For instance, $(x + 1) + (x + 2)$ and $2x + 3$ are equivalent because

$$(x + 1) + (x + 2) = x + 1 + x + 2$$
$$= x + x + 1 + 2$$
$$= 2x + 3.$$

Example 1   Finding the Domain of an Algebraic Expression

a. The domain of the polynomial

$$2x^3 + 3x + 4$$

is the set of all real numbers. In fact, the domain of any polynomial is the set of all real numbers, unless the domain is specifically restricted.

b. The domain of the radical expression

$$\sqrt{x - 2}$$

is the set of real numbers greater than or equal to 2, because the square root of a negative number is not a real number.

c. The domain of the expression

$$\frac{x + 2}{x - 3}$$

is the set of all real numbers except $x = 3$, which would result in division by zero, which is undefined.

CHECK*Point*   Now try Exercise 7.

The quotient of two algebraic expressions is a *fractional expression*. Moreover, the quotient of two *polynomials* such as

$$\frac{1}{x}, \qquad \frac{2x - 1}{x + 1}, \qquad \text{or} \qquad \frac{x^2 - 1}{x^2 + 1}$$

is a **rational expression.**

Simplifying Rational Expressions

Recall that a fraction is in simplest form if its numerator and denominator have no factors in common aside from ± 1. To write a fraction in simplest form, divide out common factors.

$$\frac{a \cdot c}{b \cdot c} = \frac{a}{b}, \quad c \neq 0$$

The key to success in simplifying rational expressions lies in your ability to *factor* polynomials. When simplifying rational expressions, be sure to factor each polynomial completely before concluding that the numerator and denominator have no factors in common.

Example 2 Simplifying a Rational Expression

Write $\dfrac{x^2 + 4x - 12}{3x - 6}$ in simplest form.

Solution

$$\frac{x^2 + 4x - 12}{3x - 6} = \frac{(x + 6)(x - 2)}{3(x - 2)}$$

$$= \frac{x + 6}{3}, \qquad x \neq 2$$

Note that the original expression is undefined when $x = 2$ (because division by zero is undefined). To make sure that the simplified expression is *equivalent* to the original expression, you must restrict the domain of the simplified expression by excluding the value $x = 2$.

CHECK*Point* Now try Exercise 33.

Sometimes it may be necessary to change the sign of a factor by factoring out (-1) to simplify a rational expression, as shown in Example 3.

Example 3 Simplifying Rational Expressions

Write $\dfrac{12 + x - x^2}{2x^2 - 9x + 4}$ in simplest form.

Solution

$$\frac{12 + x - x^2}{2x^2 - 9x + 4} = \frac{(4 - x)(3 + x)}{(2x - 1)(x - 4)}$$

$$= \frac{-(x - 4)(3 + x)}{(2x - 1)(x - 4)}$$

$$= -\frac{3 + x}{2x - 1}, \quad x \neq 4$$

CHECK*Point* Now try Exercise 39.

In this text, when a rational expression is written, the domain is usually not listed with the expression. It is *implied* that the real numbers that make the denominator zero are excluded from the expression. Also, when performing operations with rational expressions, this text follows the convention of listing *by the simplified expression* all values of x that must be specifically excluded from the domain in order to make the domains of the simplified and original expressions agree. In Example 3, for instance, the restriction $x \neq 4$ is listed with the simplified expression to make the two domains agree. Note that the value $x = \frac{1}{2}$ is excluded from *both* domains, so it is not necessary to list this value.

> **⚠ WARNING / CAUTION**
>
> In Example 2, do not make the mistake of trying to simplify further by dividing out terms.
>
> $$\frac{x + 6}{3} = \frac{x + \cancel{6}}{\cancel{3}} = x + 2$$
>
> Remember that to simplify fractions, divide out common *factors*, not terms. To learn about other common errors, see Appendix A.

Operations with Rational Expressions

To multiply or divide rational expressions, use the properties of fractions discussed in Section P.1. Recall that to divide fractions, you invert the divisor and multiply.

Example 4   Multiplying Rational Expressions

$$\frac{2x^2 + x - 6}{x^2 + 4x - 5} \cdot \frac{x^3 - 3x^2 + 2x}{4x^2 - 6x} = \frac{(2x - 3)(x + 2)}{(x + 5)(x - 1)} \cdot \frac{x(x - 2)(x - 1)}{2x(2x - 3)}$$

$$= \frac{(x + 2)(x - 2)}{2(x + 5)}, \quad x \neq 0, x \neq 1, x \neq \tfrac{3}{2}$$

CHECK *Point*   Now try Exercise 53.

In Example 4, the restrictions $x \neq 0$, $x \neq 1$, and $x \neq \tfrac{3}{2}$ are listed with the simplified expression in order to make the two domains agree. Note that the value $x = -5$ is excluded from both domains, so it is not necessary to list this value.

Example 5   Dividing Rational Expressions

$$\frac{x^3 - 8}{x^2 - 4} \div \frac{x^2 + 2x + 4}{x^3 + 8} = \frac{x^3 - 8}{x^2 - 4} \cdot \frac{x^3 + 8}{x^2 + 2x + 4}$$

$$= \frac{(x - 2)(x^2 + 2x + 4)}{(x + 2)(x - 2)} \cdot \frac{(x + 2)(x^2 - 2x + 4)}{(x^2 + 2x + 4)}$$

$$= x^2 - 2x + 4, \quad x \neq \pm 2$$

CHECK *Point*   Now try Exercise 55.

To add or subtract rational expressions, you can use the LCD (least common denominator) method or the *basic definition*

$$\frac{a}{b} \pm \frac{c}{d} = \frac{ad \pm bc}{bd}, \qquad b \neq 0, d \neq 0.$$

This definition provides an efficient way of adding or subtracting *two* fractions that have no common factors in their denominators.

Example 6   Subtracting Rational Expressions

$$\frac{x}{x - 3} - \frac{2}{3x + 4} = \frac{x(3x + 4) - 2(x - 3)}{(x - 3)(3x + 4)}$$

$$= \frac{3x^2 + 4x - 2x + 6}{(x - 3)(3x + 4)}$$

$$= \frac{3x^2 + 2x + 6}{(x - 3)(3x + 4)}$$

CHECK *Point*   Now try Exercise 65.

⚠ **WARNING / CAUTION**

When subtracting rational expressions, remember to distribute the negative sign to all the terms in the quantity that is being subtracted.

For three or more fractions, or for fractions with a repeated factor in the denominators, the LCD method works well. Recall that the least common denominator of several fractions consists of the product of all prime factors in the denominators, with each factor given the highest power of its occurrence in any denominator. Here is a numerical example.

$$\frac{1}{6} + \frac{3}{4} - \frac{2}{3} = \frac{1 \cdot}{6 \cdot} + \frac{3 \cdot}{4 \cdot} - \frac{2 \cdot}{3 \cdot}$$

$$= \frac{2}{12} + \frac{9}{12} - \frac{8}{12}$$

$$= \frac{3}{12}$$

$$= \frac{1}{4}$$

Sometimes the numerator of the answer has a factor in common with the denominator. In such cases the answer should be simplified. For instance, in the example above, $\frac{3}{12}$ was simplified to $\frac{1}{4}$.

Example 7   Combining Rational Expressions: The LCD Method

Perform the operations and simplify.

$$\frac{3}{x - 1} - \frac{2}{x} + \frac{x + 3}{x^2 - 1}$$

Solution

Using the factored denominators $(x - 1)$, x, and $(x + 1)(x - 1)$, you can see that the LCD is $x(x + 1)(x - 1)$.

$$\frac{3}{x - 1} - \frac{2}{x} + \frac{x + 3}{(x + 1)(x - 1)}$$

$$= \frac{3}{(x - 1)} - \frac{2}{x} + \frac{(x + 3)}{(x + 1)(x - 1)}$$

$$= \frac{3(x)(x + 1) - 2(x + 1)(x - 1) + (x + 3)(x)}{x(x + 1)(x - 1)}$$

$$= \frac{3x^2 + 3x - 2x^2 + 2 + x^2 + 3x}{x(x + 1)(x - 1)}$$

$$= \frac{3x^2 - 2x^2 + x^2 + 3x + 3x + 2}{x(x + 1)(x - 1)}$$

$$= \frac{2x^2 + 6x + 2}{x(x + 1)(x - 1)}$$

$$= \frac{2(x^2 + 3x + 1)}{x(x + 1)(x - 1)}$$

CHECK*Point*   Now try Exercise 67.

Complex Fractions and the Difference Quotient

Fractional expressions with separate fractions in the numerator, denominator, or both are called **complex fractions.** Here are two examples.

$$\frac{\left(\dfrac{1}{x}\right)}{x^2 + 1} \quad \text{and} \quad \frac{\left(\dfrac{1}{x}\right)}{\left(\dfrac{1}{x^2 + 1}\right)}$$

To simplify a complex fraction, combine the fractions in the numerator into a single fraction and then combine the fractions in the denominator into a single fraction. Then invert the denominator and multiply.

| Example 8   Simplifying a Complex Fraction

$$\frac{\left(\dfrac{2}{x} - 3\right)}{\left(1 - \dfrac{1}{x - 1}\right)} = \frac{\left[\dfrac{2 - 3(x)}{x}\right]}{\left[\dfrac{1(x - 1) - 1}{x - 1}\right]}$$

$$= \frac{\left(\dfrac{2 - 3x}{x}\right)}{\left(\dfrac{x - 2}{x - 1}\right)}$$

$$= \frac{2 - 3x}{x} \cdot \frac{x - 1}{x - 2}$$

$$= \frac{(2 - 3x)(x - 1)}{x(x - 2)}, \quad x \neq 1$$

CHECK *Point*   Now try Exercise 73.

Another way to simplify a complex fraction is to multiply its numerator and denominator by the LCD of all fractions in its numerator and denominator. This method is applied to the fraction in Example 8 as follows.

$$\frac{\left(\dfrac{2}{x} - 3\right)}{\left(1 - \dfrac{1}{x - 1}\right)} = \frac{\left(\dfrac{2}{x} - 3\right)}{\left(1 - \dfrac{1}{x - 1}\right)} \cdot$$

$$= \frac{\left(\dfrac{2 - 3x}{x}\right) \cdot x(x - 1)}{\left(\dfrac{x - 2}{x - 1}\right) \cdot x(x - 1)}$$

$$= \frac{(2 - 3x)(x - 1)}{x(x - 2)}, \quad x \neq 1$$

The next three examples illustrate some methods for simplifying rational expressions involving negative exponents and radicals. These types of expressions occur frequently in calculus.

To simplify an expression with negative exponents, one method is to begin by factoring out the common factor with the *smaller* exponent. Remember that when factoring, you *subtract* exponents. For instance, in $3x^{-5/2} + 2x^{-3/2}$ the smaller exponent is $-\frac{5}{2}$ and the common factor is $x^{-5/2}$.

$$3x^{-5/2} + 2x^{-3/2} = x^{-5/2}[3(1) + 2x^{-3/2-(-5/2)}]$$

$$= x^{-5/2}(3 + 2x^1)$$

$$= \frac{3 + 2x}{x^{5/2}}$$

Example 9 Simplifying an Expression

Simplify the following expression containing negative exponents.

$$x(1 - 2x)^{-3/2} + (1 - 2x)^{-1/2}$$

Solution

Begin by factoring out the common factor with the *smaller exponent*.

$$x(1 - 2x)^{-3/2} + (1 - 2x)^{-1/2} = (1 - 2x)^{-3/2}[x + (1 - 2x)^{(-1/2)-(-3/2)}]$$

$$= (1 - 2x)^{-3/2}[x + (1 - 2x)^1]$$

$$= \frac{1 - x}{(1 - 2x)^{3/2}}$$

CHECK Point Now try Exercise 81.

A second method for simplifying an expression with negative exponents is shown in the next example.

Example 10 Simplifying an Expression with Negative Exponents

$$\frac{(4 - x^2)^{1/2} + x^2(4 - x^2)^{-1/2}}{4 - x^2}$$

$$= \frac{(4 - x^2)^{1/2} + x^2(4 - x^2)^{-1/2}}{4 - x^2} \cdot$$

$$= \frac{(4 - x^2)^1 + x^2(4 - x^2)^0}{(4 - x^2)^{3/2}}$$

$$= \frac{4 - x^2 + x^2}{(4 - x^2)^{3/2}}$$

$$= \frac{4}{(4 - x^2)^{3/2}}$$

CHECK Point Now try Exercise 83.

| **Example 11**   **Rewriting a Difference Quotient**

The following expression from calculus is an example of a *difference quotient*.

$$\frac{\sqrt{x + h} - \sqrt{x}}{h}$$

Rewrite this expression by rationalizing its numerator.

Solution

$$\frac{\sqrt{x + h} - \sqrt{x}}{h} = \frac{\sqrt{x + h} - \sqrt{x}}{h} \cdot$$

$$= \frac{\left(\sqrt{x + h}\right)^2 - \left(\sqrt{x}\right)^2}{h\left(\sqrt{x + h} + \sqrt{x}\right)}$$

$$= \frac{h}{h\left(\sqrt{x + h} + \sqrt{x}\right)}$$

$$= \frac{1}{\sqrt{x + h} + \sqrt{x}}, \quad h \neq 0$$

Algebra Help

You can review the techniques for rationalizing a numerator in Section P.2.

Notice that the original expression is undefined when $h = 0$. So, you must exclude $h = 0$ from the domain of the simplified expression so that the expressions are equivalent.

CHECK Point   Now try Exercise 89.

Difference quotients, such as that in Example 11, occur frequently in calculus. Often, they need to be rewritten in an equivalent form that can be evaluated when $h = 0$. Note that the equivalent form is not simpler than the original form, but it has the advantage that it is defined when $h = 0$.

P.5   EXERCISES

VOCABULARY: Fill in the blanks.

1. The set of real numbers for which an algebraic expression is defined is the _____ of the expression.

2. The quotient of two algebraic expressions is a fractional expression and the quotient of two polynomials is a _____ _____.

3. Fractional expressions with separate fractions in the numerator, denominator, or both are called _____ fractions.

4. To simplify an expression with negative exponents, it is possible to begin by factoring out the common factor with the _____ exponent.

5. Two algebraic expressions that have the same domain and yield the same values for all numbers in their domains are called _____.

6. An important rational expression, such as $\dfrac{(x + h)^2 - x^2}{h}$, that occurs in calculus is called a _____ _____.

SKILLS AND APPLICATIONS

In Exercises 7–22, find the domain of the expression.

7. $3x^2 - 4x + 7$

8. $2x^2 + 5x - 2$

9. $4x^3 + 3, \quad x \geq 0$

10. $6x^2 - 9, \quad x > 0$

11. $\dfrac{1}{3 - x}$

12. $\dfrac{x + 6}{3x + 2}$

13. $\dfrac{x^2 - 1}{x^2 - 2x + 1}$

14. $\dfrac{x^2 - 5x + 6}{x^2 - 4}$

15. $\dfrac{x^2 - 2x - 3}{x^2 - 6x + 9}$

16. $\dfrac{x^2 - x - 12}{x^2 - 8x + 16}$

17. $\sqrt{x + 7}$

18. $\sqrt{4 - x}$

19. $\sqrt{2x - 5}$

20. $\sqrt{4x + 5}$

21. $\dfrac{1}{\sqrt{x - 3}}$

22. $\dfrac{1}{\sqrt{x + 2}}$

In Exercises 23 and 24, find the missing factor in the numerator such that the two fractions are equivalent.

23. $\dfrac{5}{2x} = \dfrac{5(\quad)}{6x^2}$

24. $\dfrac{3}{4} = \dfrac{3(\quad)}{4(x + 1)}$

In Exercises 25–42, write the rational expression in simplest form.

25. $\dfrac{15x^2}{10x}$

26. $\dfrac{18y^2}{60y^5}$

27. $\dfrac{3xy}{xy + x}$

28. $\dfrac{2x^2y}{xy - y}$

29. $\dfrac{4y - 8y^2}{10y - 5}$

30. $\dfrac{9x^2 + 9x}{2x + 2}$

31. $\dfrac{x - 5}{10 - 2x}$

32. $\dfrac{12 - 4x}{x - 3}$

33. $\dfrac{y^2 - 16}{y + 4}$

34. $\dfrac{x^2 - 25}{5 - x}$

35. $\dfrac{x^3 + 5x^2 + 6x}{x^2 - 4}$

36. $\dfrac{x^2 + 8x - 20}{x^2 + 11x + 10}$

37. $\dfrac{y^2 - 7y + 12}{y^2 + 3y - 18}$

38. $\dfrac{x^2 - 7x + 6}{x^2 + 11x + 10}$

39. $\dfrac{2 - x + 2x^2 - x^3}{x^2 - 4}$

40. $\dfrac{x^2 - 9}{x^3 + x^2 - 9x - 9}$

41. $\dfrac{z^3 - 8}{z^2 + 2z + 4}$

42. $\dfrac{y^3 - 2y^2 - 3y}{y^3 + 1}$

43. ERROR ANALYSIS Describe the error.

$$\dfrac{5x^3}{2x^3 + 4} = \dfrac{5x^3}{2x^3 + 4} = \dfrac{5}{2 + 4} = \dfrac{5}{6}$$

44. ERROR ANALYSIS Describe the error.

$$\dfrac{x^3 + 25x}{x^2 - 2x - 15} = \dfrac{x(x^2 + 25)}{(x - 5)(x + 3)}$$

$$= \dfrac{x(x + 5)(x - 5)}{(x - 5)(x + 3)} = \dfrac{x(x + 5)}{x + 3}$$

In Exercises 45 and 46, complete the table. What can you conclude?

45.

x	0	1	2	3	4	5	6
$\dfrac{x^2 - 2x - 3}{x - 3}$							
$x + 1$							

46.

x	0	1	2	3	4	5	6
$\dfrac{x - 3}{x^2 - x - 6}$							
$\dfrac{1}{x + 2}$							

GEOMETRY In Exercises 47 and 48, find the ratio of the area of the shaded portion of the figure to the total area of the figure.

47.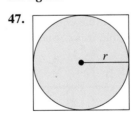

48.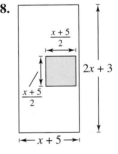

In Exercises 49–56, perform the multiplication or division and simplify.

49. $\dfrac{5}{x - 1} \cdot \dfrac{x - 1}{25(x - 2)}$

50. $\dfrac{x + 13}{x^3(3 - x)} \cdot \dfrac{x(x - 3)}{5}$

51. $\dfrac{r}{r - 1} \div \dfrac{r^2}{r^2 - 1}$

52. $\dfrac{4y - 16}{5y + 15} \div \dfrac{4 - y}{2y + 6}$

53. $\dfrac{t^2 - t - 6}{t^2 + 6t + 9} \cdot \dfrac{t + 3}{t^2 - 4}$

54. $\dfrac{x^2 + xy - 2y^2}{x^3 + x^2y} \cdot \dfrac{x}{x^2 + 3xy + 2y^2}$

55. $\dfrac{x^2 - 36}{x} \div \dfrac{x^3 - 6x^2}{x^2 + x}$

56. $\dfrac{x^2 - 14x + 49}{x^2 - 49} \div \dfrac{3x - 21}{x + 7}$

In Exercises 57–68, perform the addition or subtraction and simplify.

57. $6 - \dfrac{5}{x + 3}$

58. $\dfrac{3}{x - 1} - 5$

59. $\dfrac{5}{x - 1} + \dfrac{x}{x - 1}$

60. $\dfrac{2x - 1}{x + 3} + \dfrac{1 - x}{x + 3}$

61. $\dfrac{3}{x - 2} + \dfrac{5}{2 - x}$

62. $\dfrac{2x}{x - 5} - \dfrac{5}{5 - x}$

63. $\dfrac{4}{2x + 1} - \dfrac{x}{x + 2}$

64. $\dfrac{2}{x - 3} + \dfrac{5x}{3x + 4}$

65. $\dfrac{1}{x^2 - x - 2} - \dfrac{x}{x^2 - 5x + 6}$

66. $\dfrac{2}{x^2 - x - 2} + \dfrac{10}{x^2 + 2x - 8}$

67. $-\dfrac{1}{x} + \dfrac{2}{x^2 + 1} + \dfrac{1}{x^3 + x}$

68. $\dfrac{2}{x + 1} + \dfrac{2}{x - 1} + \dfrac{1}{x^2 - 1}$

ERROR ANALYSIS In Exercises 69 and 70, describe the error.

69. $\dfrac{x + 4}{x + 2} - \dfrac{3x - 8}{x + 2} = \dfrac{x + 4 - 3x - 8}{x + 2}$

$= \dfrac{-2x - 4}{x + 2} = \dfrac{-2(x + 2)}{x + 2} = -2$

70. $\dfrac{6 - x}{x(x + 2)} + \dfrac{x + 2}{x^2} + \dfrac{8}{x^2(x + 2)}$

$= \dfrac{x(6 - x) + (x + 2)^2 + 8}{x^2(x + 2)}$

$= \dfrac{6x - x^2 + x^2 + 4 + 8}{x^2(x + 2)}$

$= \dfrac{6(x + 2)}{x^2(x + 2)} = \dfrac{6}{x^2}$

In Exercises 71–76, simplify the complex fraction.

71. $\dfrac{\left(\dfrac{x}{2} - 1\right)}{(x - 2)}$

72. $\dfrac{(x - 4)}{\left(\dfrac{x}{4} - \dfrac{4}{x}\right)}$

73. $\dfrac{\left[\dfrac{x^2}{(x + 1)^2}\right]}{\left[\dfrac{x}{(x + 1)^3}\right]}$

74. $\dfrac{\left(\dfrac{x^2 - 1}{x}\right)}{\left[\dfrac{(x - 1)^2}{x}\right]}$

75. $\dfrac{\left(\sqrt{x} - \dfrac{1}{2\sqrt{x}}\right)}{\sqrt{x}}$

76. $\dfrac{\left(\dfrac{t^2}{\sqrt{t^2 + 1}} - \sqrt{t^2 + 1}\right)}{t^2}$

In Exercises 77–82, factor the expression by removing the common factor with the smaller exponent.

77. $x^5 - 2x^{-2}$

78. $x^5 - 5x^{-3}$

79. $x^2(x^2 + 1)^{-5} - (x^2 + 1)^{-4}$

80. $2x(x - 5)^{-3} - 4x^2(x - 5)^{-4}$

81. $2x^2(x - 1)^{1/2} - 5(x - 1)^{-1/2}$

82. $4x^3(2x - 1)^{3/2} - 2x(2x - 1)^{-1/2}$

In Exercises 83 and 84, simplify the expression.

83. $\dfrac{3x^{1/3} - x^{-2/3}}{3x^{-2/3}}$

84. $\dfrac{-x^3(1 - x^2)^{-1/2} - 2x(1 - x^2)^{1/2}}{x^4}$

In Exercises 85–88, simplify the difference quotient.

85. $\dfrac{\left(\dfrac{1}{x + h} - \dfrac{1}{x}\right)}{h}$

86. $\dfrac{\left[\dfrac{1}{(x + h)^2} - \dfrac{1}{x^2}\right]}{h}$

87. $\dfrac{\left(\dfrac{1}{x + h - 4} - \dfrac{1}{x - 4}\right)}{h}$

88. $\dfrac{\left(\dfrac{x + h}{x + h + 1} - \dfrac{x}{x + 1}\right)}{h}$

In Exercises 89–94, simplify the difference quotient by rationalizing the numerator.

89. $\dfrac{\sqrt{x + 2} - \sqrt{x}}{2}$

90. $\dfrac{\sqrt{z - 3} - \sqrt{z}}{3}$

91. $\dfrac{\sqrt{t + 3} - \sqrt{3}}{t}$

92. $\dfrac{\sqrt{x + 5} - \sqrt{5}}{x}$

93. $\dfrac{\sqrt{x + h + 1} - \sqrt{x + 1}}{h}$

94. $\dfrac{\sqrt{x + h - 2} - \sqrt{x - 2}}{h}$

PROBABILITY In Exercises 95 and 96, consider an experiment in which a marble is tossed into a box whose base is shown in the figure. The probability that the marble will come to rest in the shaded portion of the box is equal to the ratio of the shaded area to the total area of the figure. Find the probability.

95.

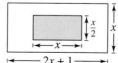

96.

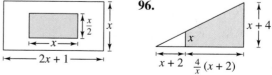

97. RATE A digital copier copies in color at a rate of 50 pages per minute.

(a) Find the time required to copy one page.

(b) Find the time required to copy x pages.

(c) Find the time required to copy 120 pages.

98. RATE After working together for t hours on a common task, two workers have done fractional parts of the job equal to $t/3$ and $t/5$, respectively. What fractional part of the task has been completed?

FINANCE In Exercises 99 and 100, the formula that approximates the annual interest rate r of a monthly installment loan is given by

$$r = \frac{\left[\dfrac{24(NM - P)}{N}\right]}{\left(P + \dfrac{NM}{12}\right)}$$

where N is the total number of payments, M is the monthly payment, and P is the amount financed.

99. (a) Approximate the annual interest rate for a four-year car loan of $20,000 that has monthly payments of $475.

(b) Simplify the expression for the annual interest rate r, and then rework part (a).

100. (a) Approximate the annual interest rate for a five-year car loan of $28,000 that has monthly payments of $525.

(b) Simplify the expression for the annual interest rate r, and then rework part (a).

101. REFRIGERATION When food (at room temperature) is placed in a refrigerator, the time required for the food to cool depends on the amount of food, the air circulation in the refrigerator, the original temperature of the food, and the temperature of the refrigerator. The model that gives the temperature of food that has an original temperature of 75°F and is placed in a 40°F refrigerator is

$$T = 10\left(\frac{4t^2 + 16t + 75}{t^2 + 4t + 10}\right)$$

where T is the temperature (in degrees Fahrenheit) and t is the time (in hours).

(a) Complete the table.

t	0	2	4	6	8	10	12
T							

t	14	16	18	20	22
T					

(b) What value of T does the mathematical model appear to be approaching?

102. INTERACTIVE MONEY MANAGEMENT The table shows the projected numbers of U.S. households (in millions) banking online and paying bills online from 2002 through 2007. (Source: eMarketer; Forrester Research)

Year	Banking	Paying Bills
2002	21.9	13.7
2003	26.8	17.4
2004	31.5	20.9
2005	35.0	23.9
2006	40.0	26.7
2007	45.0	29.1

Mathematical models for these data are

$$\text{Number banking online} = \frac{-0.728t^2 + 23.81t - 0.3}{-0.049t^2 + 0.61t + 1.0}$$

and

$$\text{Number paying bills online} = \frac{4.39t + 5.5}{0.002t^2 + 0.01t + 1.0}$$

where t represents the year, with $t = 2$ corresponding to 2002.

(a) Using the models, create a table to estimate the projected numbers of households banking online and the projected numbers of households paying bills online for the given years.

(b) Compare the values given by the models with the actual data.

(c) Determine a model for the ratio of the projected number of households paying bills online to the projected number of households banking online.

(d) Use the model from part (c) to find the ratios for the given years. Interpret your results.

EXPLORATION

TRUE OR FALSE? In Exercises 103 and 104, determine whether the statement is true or false. Justify your answer.

103. $\dfrac{x^{2n} - 1^{2n}}{x^n - 1^n} = x^n + 1^n$

104. $\dfrac{x^2 - 3x + 2}{x - 1} = x - 2$, for all values of x

105. THINK ABOUT IT How do you determine whether a rational expression is in simplest form?

106. CAPSTONE In your own words, explain how to divide rational expressions.

P.6 THE RECTANGULAR COORDINATE SYSTEM AND GRAPHS

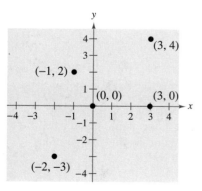

The Cartesian Plane

Just as you can represent real numbers by points on a real number line, you can represent ordered pairs of real numbers by points in a plane called the **rectangular coordinate system,** or the **Cartesian plane,** named after the French mathematician René Descartes (1596–1650).

The Cartesian plane is formed by using two real number lines intersecting at right angles, as shown in Figure P.14. The horizontal real number line is usually called the **x-axis,** and the vertical real number line is usually called the **y-axis.** The point of intersection of these two axes is the **origin,** and the two axes divide the plane into four parts called **quadrants.**

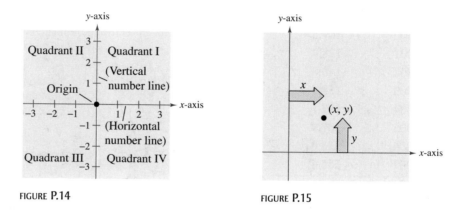

FIGURE P.14   FIGURE P.15

Each point in the plane corresponds to an **ordered pair** (x, y) of real numbers x and y, called **coordinates** of the point. The **x-coordinate** represents the directed distance from the y-axis to the point, and the **y-coordinate** represents the directed distance from the x-axis to the point, as shown in Figure P.15.

$$(x, y)$$

The notation (x, y) denotes both a point in the plane and an open interval on the real number line. The context will tell you which meaning is intended.

Example 1   Plotting Points in the Cartesian Plane

Plot the points $(-1, 2)$, $(3, 4)$, $(0, 0)$, $(3, 0)$, and $(-2, -3)$.

Solution

To plot the point $(-1, 2)$, imagine a vertical line through -1 on the x-axis and a horizontal line through 2 on the y-axis. The intersection of these two lines is the point $(-1, 2)$. The other four points can be plotted in a similar way, as shown in Figure P.16.

CHECK **Point**   Now try Exercise 7.

FIGURE P.16

The beauty of a rectangular coordinate system is that it allows you to *see* relationships between two variables. It would be difficult to overestimate the importance of Descartes's introduction of coordinates in the plane. Today, his ideas are in common use in virtually every scientific and business-related field.

Example 2   Sketching a Scatter Plot

From 1994 through 2007, the numbers N (in millions) of subscribers to a cellular telecommunication service in the United States are shown in the table, where t represents the year. Sketch a scatter plot of the data.   (Source: CTIA-The Wireless Association)

Solution

To sketch a *scatter plot* of the data shown in the table, you simply represent each pair of values by an ordered pair (t, N) and plot the resulting points, as shown in Figure P.17. For instance, the first pair of values is represented by the ordered pair (1994, 24.1). Note that the break in the t-axis indicates that the numbers between 0 and 1994 have been omitted.

Year, t	Subscribers, N
1994	24.1
1995	33.8
1996	44.0
1997	55.3
1998	69.2
1999	86.0
2000	109.5
2001	128.4
2002	140.8
2003	158.7
2004	182.1
2005	207.9
2006	233.0
2007	255.4

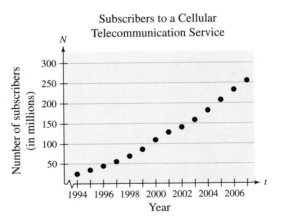

FIGURE **P.17**

CHECK*Point*   Now try Exercise 25.

In Example 2, you could have let $t = 1$ represent the year 1994. In that case, the horizontal axis would not have been broken, and the tick marks would have been labeled 1 through 14 (instead of 1994 through 2007).

TECHNOLOGY

The scatter plot in Example 2 is only one way to represent the data graphically. You could also represent the data using a bar graph or a line graph. If you have access to a graphing utility, try using it to represent graphically the data given in Example 2.

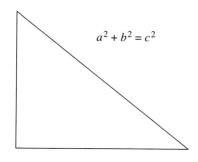

FIGURE **P.18**

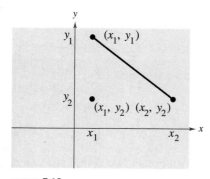

FIGURE **P.19**

The Distance Formula

Recall from the Pythagorean Theorem that, for a right triangle with hypotenuse of length c and sides of lengths a and b, you have

$$a^2 + b^2 = c^2$$

as shown in Figure P.18. (The converse is also true. That is, if $a^2 + b^2 = c^2$, then the triangle is a right triangle.)

Suppose you want to determine the distance d between two points (x_1, y_1) and (x_2, y_2) in the plane. With these two points, a right triangle can be formed, as shown in Figure P.19. The length of the vertical side of the triangle is $|y_2 - y_1|$, and the length of the horizontal side is $|x_2 - x_1|$. By the Pythagorean Theorem, you can write

$$d^2 = |x_2 - x_1|^2 + |y_2 - y_1|^2$$
$$d = \sqrt{|x_2 - x_1|^2 + |y_2 - y_1|^2} = \sqrt{(x_2 - x_1)^2 + (y_2 - y_1)^2}.$$

This result is the **Distance Formula.**

The Distance Formula

The distance d between the points (x_1, y_1) and (x_2, y_2) in the plane is

$$d = \sqrt{(x_2 - x_1)^2 + (y_2 - y_1)^2}.$$

Example 3   Finding a Distance

Find the distance between the points $(-2, 1)$ and $(3, 4)$.

Algebraic Solution

Let $(x_1, y_1) = (-2, 1)$ and $(x_2, y_2) = (3, 4)$. Then apply the Distance Formula.

$$d = \sqrt{(x_2 - x_1)^2 + (y_2 - y_1)^2}$$
$$= \sqrt{[\ -(\ )]^2 + (\ -\ )^2}$$
$$= \sqrt{(5)^2 + (3)^2}$$
$$= \sqrt{34}$$
$$\approx 5.83$$

So, the distance between the points is about 5.83 units. You can use the Pythagorean Theorem to check that the distance is correct.

$$d^2 \stackrel{?}{=} 3^2 + 5^2$$
$$(\quad)^2 \stackrel{?}{=} 3^2 + 5^2$$
$$34 = 34 \qquad \checkmark$$

Graphical Solution

Use centimeter graph paper to plot the points $A(-2, 1)$ and $B(3, 4)$. Carefully sketch the line segment from A to B. Then use a centimeter ruler to measure the length of the segment.

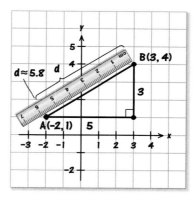

FIGURE **P.20**

The line segment measures about 5.8 centimeters, as shown in Figure P.20. So, the distance between the points is about 5.8 units.

CHECK*Point*   Now try Exercise 31.

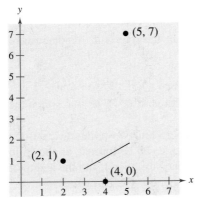

Example 4 Verifying a Right Triangle

Show that the points $(2, 1)$, $(4, 0)$, and $(5, 7)$ are vertices of a right triangle.

Solution

The three points are plotted in Figure P.21. Using the Distance Formula, you can find the lengths of the three sides as follows.

$$d_1 = \sqrt{(5 - 2)^2 + (7 - 1)^2} = \sqrt{9 + 36} = \sqrt{45}$$

$$d_2 = \sqrt{(4 - 2)^2 + (0 - 1)^2} = \sqrt{4 + 1} = \sqrt{5}$$

$$d_3 = \sqrt{(5 - 4)^2 + (7 - 0)^2} = \sqrt{1 + 49} = \sqrt{50}$$

Because

$$(d_1)^2 + (d_2)^2 = 45 + 5 = 50 = (d_3)^2$$

you can conclude by the Pythagorean Theorem that the triangle must be a right triangle.

CHECK*Point* Now try Exercise 43.

The Midpoint Formula

To find the **midpoint** of the line segment that joins two points in a coordinate plane, you can simply find the average values of the respective coordinates of the two endpoints using the **Midpoint Formula.**

> ### The Midpoint Formula
> The midpoint of the line segment joining the points (x_1, y_1) and (x_2, y_2) is given by the Midpoint Formula
> $$\text{Midpoint} = \left(\frac{x_1 + x_2}{2}, \frac{y_1 + y_2}{2} \right).$$

For a proof of the Midpoint Formula, see Proofs in Mathematics on page 72.

Example 5 Finding a Line Segment's Midpoint

Find the midpoint of the line segment joining the points $(-5, -3)$ and $(9, 3)$.

Solution

Let $(x_1, y_1) = (-5, -3)$ and $(x_2, y_2) = (9, 3)$.

$$\text{Midpoint} = \left(\frac{x_1 + x_2}{2}, \frac{y_1 + y_2}{2} \right)$$

$$= \left(\frac{+}{2}, \frac{+}{2} \right)$$

$$= (2, 0)$$

The midpoint of the line segment is $(2, 0)$, as shown in Figure P.22.

CHECK*Point* Now try Exercise 47(c).

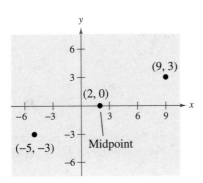

Applications

Example 6   Finding the Length of a Pass

A football quarterback throws a pass from the 28-yard line, 40 yards from the sideline. The pass is caught by a wide receiver on the 5-yard line, 20 yards from the same sideline, as shown in Figure P.23. How long is the pass?

Solution

You can find the length of the pass by finding the distance between the points $(40, 28)$ and $(20, 5)$.

$$d = \sqrt{(x_2 - x_1)^2 + (y_2 - y_1)^2}$$
$$= \sqrt{(\quad - \quad)^2 + (\quad - \quad)^2}$$
$$= \sqrt{400 + 529}$$
$$= \sqrt{929}$$
$$\approx 30$$

So, the pass is about 30 yards long.

CHECK**Point**   Now try Exercise 57.

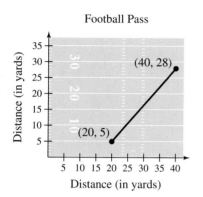

Football Pass

Distance (in yards)

$(40, 28)$

$(20, 5)$

Distance (in yards)

FIGURE **P.23**

In Example 6, the scale along the goal line does not normally appear on a football field. However, when you use coordinate geometry to solve real-life problems, you are free to place the coordinate system in any way that is convenient for the solution of the problem.

Example 7   Estimating Annual Revenue

Barnes & Noble had annual sales of approximately $5.1 billion in 2005, and $5.4 billion in 2007. Without knowing any additional information, what would you estimate the 2006 sales to have been?   (Source: Barnes & Noble, Inc.)

Solution

One solution to the problem is to assume that sales followed a linear pattern. With this assumption, you can estimate the 2006 sales by finding the midpoint of the line segment connecting the points $(2005, 5.1)$ and $(2007, 5.4)$.

$$\text{Midpoint} = \left(\frac{x_1 + x_2}{2}, \frac{y_1 + y_2}{2} \right)$$
$$= \left(\frac{\quad + \quad}{2}, \frac{\quad + \quad}{2} \right)$$
$$= (2006, 5.25)$$

So, you would estimate the 2006 sales to have been about $5.25 billion, as shown in Figure P.24. (The actual 2006 sales were about $5.26 billion.)

CHECK**Point**   Now try Exercise 59.

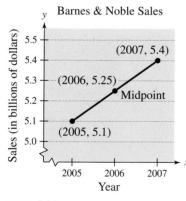

Barnes & Noble Sales

Sales (in billions of dollars)

$(2007, 5.4)$

$(2006, 5.25)$

Midpoint

$(2005, 5.1)$

Year

FIGURE **P.24**

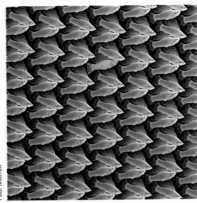

Much of computer graphics, including this computer-generated goldfish tessellation, consists of transformations of points in a coordinate plane. One type of transformation, a translation, is illustrated in Example 8. Other types include reflections, rotations, and stretches.

Example 8　Translating Points in the Plane

The triangle in Figure P.25 has vertices at the points $(-1, 2)$, $(1, -4)$, and $(2, 3)$. Shift the triangle three units to the right and two units upward and find the vertices of the shifted triangle, as shown in Figure P.26.

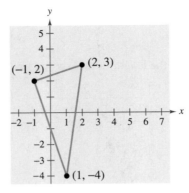

FIGURE **P.25**

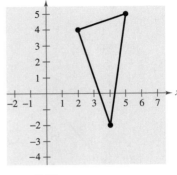

FIGURE **P.26**

Solution

To shift the vertices three units to the right, add 3 to each of the x-coordinates. To shift the vertices two units upward, add 2 to each of the y-coordinates.

Original Point	Translated Point
$(-1, 2)$	$(-1$ $, 2$ $) = (2, 4)$
$(1, -4)$	$(1$ $, -4$ $) = (4, -2)$
$(2, 3)$	$(2$ $, 3$ $) = (5, 5)$

CHECK*Point*　Now try Exercise 61.

The figures provided with Example 8 were not really essential to the solution. Nevertheless, it is strongly recommended that you develop the habit of including sketches with your solutions—even if they are not required.

CLASSROOM DISCUSSION

Extending the Example　Example 8 shows how to translate points in a coordinate plane. Write a short paragraph describing how each of the following transformed points is related to the original point.

Original Point	Transformed Point
(x, y)	$(-x, y)$
(x, y)	$(x, -y)$
(x, y)	$(-x, -y)$

P.6   EXERCISES

See www.CalcChat.com for worked-out solutions to odd-numbered exercises.

VOCABULARY

1. Match each term with its definition.

  (a) x-axis   (i) point of intersection of vertical axis and horizontal axis

  (b) y-axis   (ii) directed distance from the x-axis

  (c) origin   (iii) directed distance from the y-axis

  (d) quadrants   (iv) four regions of the coordinate plane

  (e) x-coordinate   (v) horizontal real number line

  (f) y-coordinate   (vi) vertical real number line

In Exercises 2–4, fill in the blanks.

2. An ordered pair of real numbers can be represented in a plane called the rectangular coordinate system or the _____ plane.

3. The _____ _____ is a result derived from the Pythagorean Theorem.

4. Finding the average values of the representative coordinates of the two endpoints of a line segment in a coordinate plane is also known as using the _____ _____.

SKILLS AND APPLICATIONS

In Exercises 5 and 6, approximate the coordinates of the points.

5.

6.
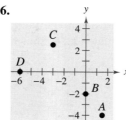

In Exercises 7–10, plot the points in the Cartesian plane.

7. $(-4, 2), (-3, -6), (0, 5), (1, -4)$

8. $(0, 0), (3, 1), (-2, 4), (1, -1)$

9. $(3, 8), (0.5, -1), (5, -6), (-2, 2.5)$

10. $\left(1, -\frac{1}{3}\right), \left(\frac{3}{4}, 3\right), (-3, 4), \left(-\frac{4}{3}, -\frac{3}{2}\right)$

In Exercises 11–14, find the coordinates of the point.

11. The point is located three units to the left of the y-axis and four units above the x-axis.

12. The point is located eight units below the x-axis and four units to the right of the y-axis.

13. The point is located five units below the x-axis and the coordinates of the point are equal.

14. The point is on the x-axis and 12 units to the left of the y-axis.

In Exercises 15–24, determine the quadrant(s) in which (x, y) is located so that the condition(s) is (are) satisfied.

15. $x > 0$ and $y < 0$ **16.** $x < 0$ and $y < 0$

17. $x = -4$ and $y > 0$ **18.** $x > 2$ and $y = 3$

19. $y < -5$ **20.** $x > 4$

21. $x < 0$ and $-y > 0$ **22.** $-x > 0$ and $y < 0$

23. $xy > 0$ **24.** $xy < 0$

In Exercises 25 and 26, sketch a scatter plot of the data shown in the table.

25. NUMBER OF STORES   The table shows the number y of Wal-Mart stores for each year x from 2000 through 2007. (Source: Wal-Mart Stores, Inc.)

Year, x	Number of stores, y
2000	4189
2001	4414
2002	4688
2003	4906
2004	5289
2005	6141
2006	6779
2007	7262

26. METEOROLOGY The table shows the lowest temperature on record y (in degrees Fahrenheit) in Duluth, Minnesota for each month x, where $x = 1$ represents January. (Source: NOAA)

Month, x	Temperature, y
1	-39
2	-39
3	-29
4	-5
5	17
6	27
7	35
8	32
9	22
10	8
11	-23
12	-34

In Exercises 27–38, find the distance between the points.

27. $(6, -3), (6, 5)$ **28.** $(1, 4), (8, 4)$

29. $(-3, -1), (2, -1)$ **30.** $(-3, -4), (-3, 6)$

31. $(-2, 6), (3, -6)$ **32.** $(8, 5), (0, 20)$

33. $(1, 4), (-5, -1)$ **34.** $(1, 3), (3, -2)$

35. $\left(\frac{1}{2}, \frac{4}{3}\right), (2, -1)$ **36.** $\left(-\frac{2}{3}, 3\right), \left(-1, \frac{5}{4}\right)$

37. $(-4.2, 3.1), (-12.5, 4.8)$

38. $(9.5, -2.6), (-3.9, 8.2)$

In Exercises 39–42, (a) find the length of each side of the right triangle, and (b) show that these lengths satisfy the Pythagorean Theorem.

39.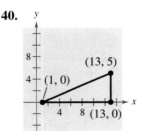

40.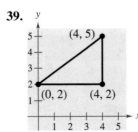

41.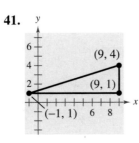

42.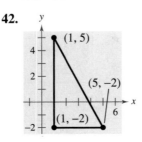

In Exercises 43–46, show that the points form the vertices of the indicated polygon.

43. Right triangle: $(4, 0), (2, 1), (-1, -5)$

44. Right triangle: $(-1, 3), (3, 5), (5, 1)$

45. Isosceles triangle: $(1, -3), (3, 2), (-2, 4)$

46. Isosceles triangle: $(2, 3), (4, 9), (-2, 7)$

In Exercises 47–56, (a) plot the points, (b) find the distance between the points, and (c) find the midpoint of the line segment joining the points.

47. $(1, 1), (9, 7)$ **48.** $(1, 12), (6, 0)$

49. $(-4, 10), (4, -5)$ **50.** $(-7, -4), (2, 8)$

51. $(-1, 2), (5, 4)$ **52.** $(2, 10), (10, 2)$

53. $\left(\frac{1}{2}, 1\right), \left(-\frac{5}{2}, \frac{4}{3}\right)$ **54.** $\left(-\frac{1}{3}, -\frac{1}{3}\right), \left(-\frac{1}{6}, -\frac{1}{2}\right)$

55. $(6.2, 5.4), (-3.7, 1.8)$ **56.** $(-16.8, 12.3), (5.6, 4.9)$

57. FLYING DISTANCE An airplane flies from Naples, Italy in a straight line to Rome, Italy, which is 120 kilometers north and 150 kilometers west of Naples. How far does the plane fly?

58. SPORTS A soccer player passes the ball from a point that is 18 yards from the endline and 12 yards from the sideline. The pass is received by a teammate who is 42 yards from the same endline and 50 yards from the same sideline, as shown in the figure. How long is the pass?

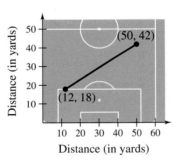

SALES In Exercises 59 and 60, use the Midpoint Formula to estimate the sales of Big Lots, Inc. and Dollar Tree Stores, Inc. in 2005, given the sales in 2003 and 2007. Assume that the sales followed a linear pattern. (Source: Big Lots, Inc.; Dollar Tree Stores, Inc.)

59. Big Lots

Year	Sales (in millions)
2003	$4174
2007	$4656

60. Dollar Tree

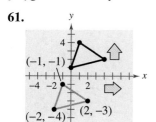

Year	Sales (in millions)
2003	$2800
2007	$4243

In Exercises 61–64, the polygon is shifted to a new position in the plane. Find the coordinates of the vertices of the polygon in its new position.

61.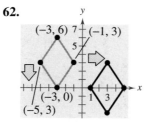

62.

63. Original coordinates of vertices: $(-7, -2)$, $(-2, 2)$, $(-2, -4)$, $(-7, -4)$

Shift: eight units upward, four units to the right

64. Original coordinates of vertices: $(5, 8)$, $(3, 6)$, $(7, 6)$, $(5, 2)$

Shift: 6 units downward, 10 units to the left

RETAIL PRICE In Exercises 65 and 66, use the graph, which shows the average retail prices of 1 gallon of whole milk from 1996 through 2007. (Source: U.S. Bureau of Labor Statistics)

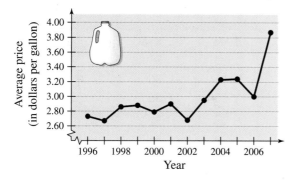

65. Approximate the highest price of a gallon of whole milk shown in the graph. When did this occur?

66. Approximate the percent change in the price of milk from the price in 1996 to the highest price shown in the graph.

67. **ADVERTISING** The graph shows the average costs of a 30-second television spot (in thousands of dollars) during the Super Bowl from 2000 through 2008. (Source: Nielson Media and TNS Media Intelligence)

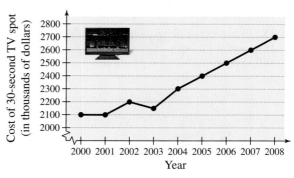

FIGURE FOR **67**

(a) Estimate the percent increase in the average cost of a 30-second spot from Super Bowl XXXIV in 2000 to Super Bowl XXXVIII in 2004.

(b) Estimate the percent increase in the average cost of a 30-second spot from Super Bowl XXXIV in 2000 to Super Bowl XLII in 2008.

68. **ADVERTISING** The graph shows the average costs of a 30-second television spot (in thousands of dollars) during the Academy Awards from 1995 through 2007. (Source: Nielson Monitor-Plus)

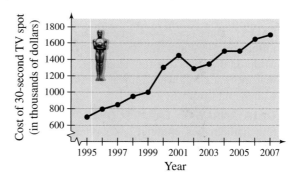

(a) Estimate the percent increase in the average cost of a 30-second spot in 1996 to the cost in 2002.

(b) Estimate the percent increase in the average cost of a 30-second spot in 1996 to the cost in 2007.

69. **MUSIC** The graph shows the numbers of performers who were elected to the Rock and Roll Hall of Fame from 1991 through 2008. Describe any trends in the data. From these trends, predict the number of performers elected in 2010. (Source: rockhall.com)

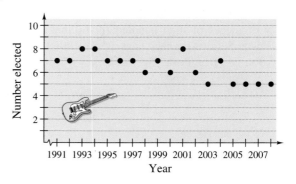

70. LABOR FORCE Use the graph below, which shows the minimum wage in the United States (in dollars) from 1950 through 2009. (Source: U.S. Department of Labor)

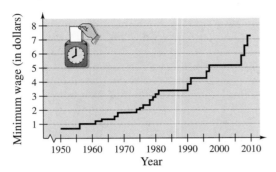

(a) Which decade shows the greatest increase in minimum wage?

(b) Approximate the percent increases in the minimum wage from 1990 to 1995 and from 1995 to 2009.

(c) Use the percent increase from 1995 to 2009 to predict the minimum wage in 2013.

(d) Do you believe that your prediction in part (c) is reasonable? Explain.

71. SALES The Coca-Cola Company had sales of $19,805 million in 1999 and $28,857 million in 2007. Use the Midpoint Formula to estimate the sales in 2003. Assume that the sales followed a linear pattern. (Source: The Coca-Cola Company)

72. DATA ANALYSIS: EXAM SCORES The table shows the mathematics entrance test scores x and the final examination scores y in an algebra course for a sample of 10 students.

x	22	29	35	40	44	48	53	58	65	76
y	53	74	57	66	79	90	76	93	83	99

(a) Sketch a scatter plot of the data.

(b) Find the entrance test score of any student with a final exam score in the 80s.

(c) Does a higher entrance test score imply a higher final exam score? Explain.

73. DATA ANALYSIS: MAIL The table shows the number y of pieces of mail handled (in billions) by the U.S. Postal Service for each year x from 1996 through 2008. (Source: U.S. Postal Service)

Year, x	Pieces of mail, y
1996	183
1997	191
1998	197
1999	202
2000	208
2001	207
2002	203
2003	202
2004	206
2005	212
2006	213
2007	212
2008	203

TABLE FOR 73

(a) Sketch a scatter plot of the data.

(b) Approximate the year in which there was the greatest decrease in the number of pieces of mail handled.

(c) Why do you think the number of pieces of mail handled decreased?

74. DATA ANALYSIS: ATHLETICS The table shows the numbers of men's M and women's W college basketball teams for each year x from 1994 through 2007. (Source: National Collegiate Athletic Association)

Year, x	Men's teams, M	Women's teams, W
1994	858	859
1995	868	864
1996	866	874
1997	865	879
1998	895	911
1999	926	940
2000	932	956
2001	937	958
2002	936	975
2003	967	1009
2004	981	1008
2005	983	1036
2006	984	1018
2007	982	1003

(a) Sketch scatter plots of these two sets of data on the same set of coordinate axes.

(b) Find the year in which the numbers of men's and women's teams were nearly equal.

(c) Find the year in which the difference between the numbers of men's and women's teams was the greatest. What was this difference?

EXPLORATION

75. A line segment has (x_1, y_1) as one endpoint and (x_m, y_m) as its midpoint. Find the other endpoint (x_2, y_2) of the line segment in terms of x_1, y_1, x_m, and y_m.

76. Use the result of Exercise 75 to find the coordinates of the endpoint of a line segment if the coordinates of the other endpoint and midpoint are, respectively,

(a) $(1, -2), (4, -1)$ and (b) $(-5, 11), (2, 4)$.

77. Use the Midpoint Formula three times to find the three points that divide the line segment joining (x_1, y_1) and (x_2, y_2) into four parts.

78. Use the result of Exercise 77 to find the points that divide the line segment joining the given points into four equal parts.

(a) $(1, -2), (4, -1)$ (b) $(-2, -3), (0, 0)$

79. MAKE A CONJECTURE Plot the points $(2, 1)$, $(-3, 5)$, and $(7, -3)$ on a rectangular coordinate system. Then change the sign of the x-coordinate of each point and plot the three new points on the same rectangular coordinate system. Make a conjecture about the location of a point when each of the following occurs.

(a) The sign of the x-coordinate is changed.

(b) The sign of the y-coordinate is changed.

(c) The signs of both the x- and y-coordinates are changed.

80. COLLINEAR POINTS Three or more points are *collinear* if they all lie on the same line. Use the steps below to determine if the set of points $\{A(2, 3), B(2, 6), C(6, 3)\}$ and the set of points $\{A(8, 3), B(5, 2), C(2, 1)\}$ are collinear.

(a) For each set of points, use the Distance Formula to find the distances from A to B, from B to C, and from A to C. What relationship exists among these distances for each set of points?

(b) Plot each set of points in the Cartesian plane. Do all the points of either set appear to lie on the same line?

(c) Compare your conclusions from part (a) with the conclusions you made from the graphs in part (b). Make a general statement about how to use the Distance Formula to determine collinearity.

TRUE OR FALSE? In Exercises 81 and 82, determine whether the statement is true or false. Justify your answer.

81. In order to divide a line segment into 16 equal parts, you would have to use the Midpoint Formula 16 times.

82. The points $(-8, 4)$, $(2, 11)$, and $(-5, 1)$ represent the vertices of an isosceles triangle.

83. THINK ABOUT IT When plotting points on the rectangular coordinate system, is it true that the scales on the x- and y-axes must be the same? Explain.

84. CAPSTONE Use the plot of the point (x_0, y_0) in the figure. Match the transformation of the point with the correct plot. Explain your reasoning. [The plots are labeled (i), (ii), (iii), and (iv).]

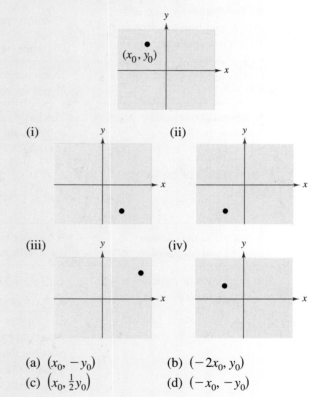

(a) $(x_0, -y_0)$ (b) $(-2x_0, y_0)$

(c) $\left(x_0, \frac{1}{2}y_0\right)$ (d) $(-x_0, -y_0)$

85. PROOF Prove that the diagonals of the parallelogram in the figure intersect at their midpoints.

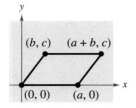

P CHAPTER SUMMARY

What Did You Learn?	Explanation/Examples	Review Exercises
Section P.1 Represent and classify real numbers *(p. 2)*.	**Real numbers:** set of all rational and irrational numbers **Rational numbers:** real numbers that can be written as the ratio of two integers **Irrational numbers:** real numbers that cannot be written as the ratio of two integers Real numbers can be represented on the real number line.	1, 2
Order real numbers and use inequalities *(p. 4)*.	$a < b$: a is less than b. $a > b$: a is greater than b. $a \leq b$: a is less than or equal to b. $a \geq b$: a is greater than or equal to b.	3–6
Find the absolute values of real numbers and find the distance between two real numbers *(p. 6)*.	**Absolute value of a:** $\|a\| = \begin{cases} a, & \text{if } a \geq 0 \\ -a, & \text{if } a < 0 \end{cases}$ **Distance between a and b:** $d(a, b) = \|b - a\| = \|a - b\|$	7–12
Evaluate algebraic expressions *(p. 8)*.	To evaluate an algebraic expression, substitute numerical values for each of the variables in the expression.	13–16
Use the basic rules and properties of algebra *(p. 9)*.	The basic rules of algebra, the properties of negation and equality, the properties of zero, and the properties and operations of fractions can be used to perform operations.	17–30
Section P.2 Use properties of exponents *(p. 15)*.	**1.** $a^m a^n = a^{m+n}$ **2.** $a^m/a^n = a^{m-n}$ **3.** $a^{-n} = (1/a)^n$ **4.** $a^0 = 1, a \neq 0$ **5.** $(ab)^m = a^m b^m$ **6.** $(a^m)^n = a^{mn}$ **7.** $(a/b)^m = a^m/b^m$ **8.** $\|a^2\| = a^2$	31–38
Use scientific notation to represent real numbers *(p. 17)*.	A number written in scientific notation has the form $\pm c \times 10^n$, where $1 \leq c < 10$ and n is an integer.	39–42
Use properties of radicals *(p. 19)* to simplify and combine radicals *(p. 21)*.	**1.** $\sqrt[n]{a^m} = \left(\sqrt[n]{a}\right)^m$ **2.** $\sqrt[n]{a} \cdot \sqrt[n]{b} = \sqrt[n]{ab}$ **3.** $\sqrt[n]{a}/\sqrt[n]{b} = \sqrt[n]{a/b}, b \neq 0$ **4.** $\sqrt[m]{\sqrt[n]{a}} = \sqrt[mn]{a}$ **5.** $\left(\sqrt[n]{a}\right)^n = a$ **6.** n **even:** $\sqrt[n]{a^n} = \|a\|$, n **odd:** $\sqrt[n]{a^n} = a$ A radical expression is in simplest form when (1) all possible factors have been removed from the radical, (2) all fractions have radical-free denominators, and (3) the index of the radical is reduced. Radical expressions can be combined if they are like radicals.	43–50
Rationalize denominators and numerators *(p. 22)*.	To rationalize a denominator or numerator of the form $a - b\sqrt{m}$ or $a + b\sqrt{m}$, multiply both numerator and denominator by a conjugate.	51–56
Use properties of rational exponents *(p. 23)*.	If a is a real number and n is a positive integer such that the principal nth root of a exists, then $a^{1/n}$ is defined as $a^{1/n} = \sqrt[n]{a}$, where $1/n$ is the rational exponent of a.	57–60
Section P.3 Write polynomials in standard form *(p. 28)*, and add, subtract, and multiply polynomials *(p. 29)*.	In standard form, a polynomial is written with descending powers of x. To add and subtract polynomials, add or subtract the like terms. To find the product of two polynomials, use the FOIL method.	61–72

	What Did You Learn?	Explanation/Examples	Review Exercises
Section P.3	Use special products to multiply polynomials *(p. 30)*.	**Sum and difference of same terms:** $$(u + v)(u - v) = u^2 - v^2$$ **Square of a binomial:** $(u + v)^2 = u^2 + 2uv + v^2$ $(u - v)^2 = u^2 - 2uv + v^2$ **Cube of a binomial:** $(u + v)^3 = u^3 + 3u^2v + 3uv^2 + v^3$ $(u - v)^3 = u^3 - 3u^2v + 3uv^2 - v^3$	73–76
	Use polynomials to solve real-life problems *(p. 32)*.	Polynomials can be used to find the volume of a box. (See Example 9.)	77–80
Section P.4	Remove common factors from polynomials *(p. 37)*.	The process of writing a polynomial as a product is called factoring. Removing (factoring out) any common factors is the first step in completely factoring a polynomial.	81, 82
	Factor special polynomial forms *(p. 38)*.	**Difference of two squares:** $u^2 - v^2 = (u + v)(u - v)$ **Perfect square trinomial:** $u^2 + 2uv + v^2 = (u + v)^2$ $u^2 - 2uv + v^2 = (u - v)^2$ **Sum or difference of two cubes:** $u^3 + v^3 = (u + v)(u^2 - uv + v^2)$ $u^3 - v^3 = (u - v)(u^2 + uv + v^2)$	83–86
	Factor trinomials as the product of two binomials *(p. 40)*.	$ax^2 + bx + c = (\quad x + \quad)(\quad x + \quad)$	87, 88
	Factor polynomials by grouping *(p. 41)*.	Polynomials with more than three terms can sometimes be factored by a method called factoring by grouping. (See Examples 9 and 10.)	89, 90
Section P.5	Find domains of algebraic expressions *(p. 45)*.	The set of real numbers for which an algebraic expression is defined is the domain of the expression.	91, 92
	Simplify rational expressions *(p. 45)*.	When simplifying rational expressions, be sure to factor each polynomial completely before concluding that the numerator and denominator have no factors in common.	93, 94
	Add, subtract, multiply, and divide rational expressions *(p. 47)*.	To add or subtract, use the LCD method or the basic definition $\dfrac{a}{b} \pm \dfrac{c}{d} = \dfrac{ad \pm bc}{bd}$, $b \neq 0$, $d \neq 0$. To multiply or divide, use the properties of fractions.	95–98
	Simplify complex fractions and rewrite difference quotients *(p. 49)*.	To simplify a complex fraction, combine the fractions in the numerator into a single fraction and then combine the fractions in the denominator into a single fraction. Then invert the denominator and multiply.	99–102
Section P.6	Plot points in the Cartesian plane *(p. 55)*.	For an ordered pair (x, y), the x-coordinate is the directed distance from the y-axis to the point, and the y-coordinate is the directed distance from the x-axis to the point.	103–106
	Use the Distance Formula *(p. 57)* and the Midpoint Formula *(p. 58)*.	**Distance Formula:** $d = \sqrt{(x_2 - x_1)^2 + (y_2 - y_1)^2}$ **Midpoint Formula:** Midpoint $= \left(\dfrac{x_1 + x_2}{2}, \dfrac{y_1 + y_2}{2}\right)$	107–110
	Use a coordinate plane to model and solve real-life problems *(p. 59)*.	The coordinate plane can be used to find the length of a football pass (See Example 6).	111–114

P REVIEW EXERCISES

See www.CalcChat.com for worked-out solutions to odd-numbered exercises.

P.1 In Exercises 1 and 2, determine which numbers in the set are (a) natural numbers, (b) whole numbers, (c) integers, (d) rational numbers, and (e) irrational numbers.

1. $\left\{11, -14, -\frac{8}{9}, \frac{5}{2}, \sqrt{6}, 0.4\right\}$

2. $\left\{\sqrt{15}, -22, -\frac{10}{3}, 0, 5.2, \frac{3}{7}\right\}$

In Exercises 3 and 4, use a calculator to find the decimal form of each rational number. If it is a nonterminating decimal, write the repeating pattern. Then plot the numbers on the real number line and place the appropriate inequality sign (< or >) between them.

3. (a) $\frac{5}{6}$ (b) $\frac{7}{8}$ **4.** (a) $\frac{9}{25}$ (b) $\frac{5}{7}$

In Exercises 5 and 6, give a verbal description of the subset of real numbers represented by the inequality, and sketch the subset on the real number line.

5. $x \le 7$ **6.** $x > 1$

In Exercises 7 and 8, find the distance between a and b.

7. $a = -74, b = 48$ **8.** $a = -112, b = -6$

In Exercises 9–12, use absolute value notation to describe the situation.

9. The distance between x and 7 is at least 4.

10. The distance between x and 25 is no more than 10.

11. The distance between y and -30 is less than 5.

12. The distance between z and -16 is greater than 8.

In Exercises 13–16, evaluate the expression for each value of x. (If not possible, state the reason.)

Expression	Values	
13. $12x - 7$	(a) $x = 0$	(b) $x = -1$
14. $x^2 - 6x + 5$	(a) $x = -2$	(b) $x = 2$
15. $-x^2 + x - 1$	(a) $x = 1$	(b) $x = -1$
16. $\dfrac{x}{x - 3}$	(a) $x = -3$	(b) $x = 3$

In Exercises 17–22, identify the rule of algebra illustrated by the statement.

17. $2x + (3x - 10) = (2x + 3x) - 10$

18. $4(t + 2) = 4 \cdot t + 4 \cdot 2$

19. $0 + (a - 5) = a - 5$

20. $\dfrac{2}{y + 4} \cdot \dfrac{y + 4}{2} = 1, \quad y \ne -4$

21. $(t^2 + 1) + 3 = 3 + (t^2 + 1)$

22. $1 \cdot (3x + 4) = 3x + 4$

In Exercises 23–30, perform the operation(s). (Write fractional answers in simplest form.)

23. $|-3| + 4(-2) - 6$ **24.** $\dfrac{|-10|}{-10}$

25. $\frac{5}{18} \div \frac{10}{3}$ **26.** $(16 - 8) \div 4$

27. $6[4 - 2(6 + 8)]$ **28.** $-4[16 - 3(7 - 10)]$

29. $\dfrac{x}{5} + \dfrac{7x}{12}$ **30.** $\dfrac{9}{x} \div \dfrac{1}{6}$

P.2 In Exercises 31–34, simplify each expression.

31. (a) $3x^2(4x^3)^3$ (b) $\dfrac{5y^6}{10y}$

32. (a) $(3a)^2(6a^3)$ (b) $\dfrac{36x^5}{9x^{10}}$

33. (a) $(-2z)^3$ (b) $\dfrac{(8y)^0}{y^2}$

34. (a) $[(x + 2)^2]^3$ (b) $\dfrac{40(b - 3)^5}{75(b - 3)^2}$

In Exercises 35–38, rewrite each expression with positive exponents and simplify.

35. (a) $\dfrac{a^2}{b^{-2}}$ (b) $(a^2b^4)(3ab^{-2})$

36. (a) $\dfrac{6^2u^3v^{-3}}{12u^{-2}v}$ (b) $\dfrac{3^{-4}m^{-1}n^{-3}}{9^{-2}mn^{-3}}$

37. (a) $\dfrac{(5a)^{-2}}{(5a)^2}$ (b) $\dfrac{4(x^{-1})^{-3}}{4^{-2}(x^{-1})^{-1}}$

38. (a) $(x + y^{-1})^{-1}$ (b) $\left(\dfrac{x^{-3}}{y}\right)\left(\dfrac{x}{y}\right)^{-1}$

In Exercises 39 and 40, write the number in scientific notation.

39. *Sales for Nautilus, Inc. in 2007:*
$501,500,000   (Source: Nautilus, Inc.)

40. *Number of meters in 1 foot:* 0.3048

In Exercises 41 and 42, write the number in decimal notation.

41. *Distance between the sun and Jupiter:* 4.84×10^8 miles

42. *Ratio of day to year:* 2.74×10^{-3}

In Exercises 43–46, simplify each expression.

43. (a) $\sqrt[3]{27^2}$ (b) $\sqrt{49^3}$

44. (a) $\sqrt[3]{\frac{64}{125}}$ (b) $\sqrt{\frac{81}{100}}$

45. (a) $\left(\sqrt[3]{216}\right)^3$ (b) $\sqrt[4]{32^4}$

46. (a) $\sqrt[3]{\frac{2x^3}{27}}$ (b) $\sqrt[5]{64x^6}$

In Exercises 47 and 48, simplify each expression.

47. (a) $\sqrt{50} - \sqrt{18}$ (b) $2\sqrt{32} + 3\sqrt{72}$

48. (a) $\sqrt{8x^3} + \sqrt{2x}$ (b) $\sqrt{18x^5} - \sqrt{8x^3}$

49. WRITING Explain why $\sqrt{5u} + \sqrt{3u} \neq 2\sqrt{2u}$.

50. ENGINEERING The rectangular cross section of a wooden beam cut from a log of diameter 24 inches (see figure) will have a maximum strength if its width w and height h are

$$w = 8\sqrt{3} \quad \text{and} \quad h = \sqrt{24^2 - \left(8\sqrt{3}\right)^2}.$$

Find the area of the rectangular cross section and write the answer in simplest form.

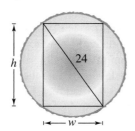

In Exercises 51–54, rationalize the denominator of the expression. Then, simplify your answer.

51. $\dfrac{3}{4\sqrt{3}}$ **52.** $\dfrac{12}{\sqrt[3]{4}}$

53. $\dfrac{1}{2 - \sqrt{3}}$ **54.** $\dfrac{1}{\sqrt{5} + 1}$

In Exercises 55 and 56, rationalize the numerator of the expression. Then, simplify your answer.

55. $\dfrac{\sqrt{7} + 1}{2}$ **56.** $\dfrac{\sqrt{2} - \sqrt{11}}{3}$

In Exercises 57–60, simplify the expression.

57. $16^{3/2}$ **58.** $64^{-2/3}$

59. $(3x^{2/5})(2x^{1/2})$ **60.** $(x - 1)^{1/3}(x - 1)^{-1/4}$

P.3 In Exercises 61–64, write the polynomial in standard form. Identify the degree and leading coefficient.

61. $3 - 11x^2$ **62.** $3x^3 - 5x^5 + x - 4$

63. $-4 - 12x^2$ **64.** $12x - 7x^2 + 6$

In Exercises 65–68, perform the operation and write the result in standard form.

65. $-(3x^2 + 2x) + (1 - 5x)$

66. $8y - [2y^2 - (3y - 8)]$

67. $2x(x^2 - 5x + 6)$

68. $(3x^3 - 1.5x^2 + 4)(-3x)$

In Exercises 69 and 70, perform the operation.

69. Add $2x^3 - 5x^2 + 10x - 7$ and $4x^2 - 7x - 2$.

70. Subtract $9x^4 - 11x^2 + 16$ from $6x^4 - 20x^2 - x + 3$.

In Exercises 71–76, find the product.

71. $(3x - 6)(5x + 1)$ **72.** $\left(x - \dfrac{1}{x}\right)(x + 2)$

73. $(2x - 3)^2$ **74.** $(6x + 5)(6x - 5)$

75. $\left(3\sqrt{5} + 2\right)\left(3\sqrt{5} - 2\right)$ **76.** $(x - 4)^3$

77. COMPOUND INTEREST After 2 years, an investment of \$2500 compounded annually at an interest rate r will yield an amount of $2500(1 + r)^2$. Write this polynomial in standard form.

78. SURFACE AREA The surface area S of a right circular cylinder is $S = 2\pi r^2 + 2\pi rh$.

(a) Draw a right circular cylinder of radius r and height h. Use the figure to explain how the surface area formula was obtained.

(b) Find the surface area when the radius is 6 inches and the height is 8 inches.

79. GEOMETRY Find a polynomial that represents the total number of square feet for the floor plan shown in the figure.

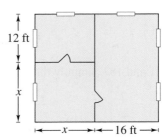

80. GEOMETRY Use the area model to write two different expressions for the area. Then equate the two expressions and name the algebraic property that is illustrated.

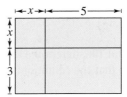

P.4 In Exercises 81–90, completely factor the expression.

81. $x^3 - x$

82. $x(x - 3) + 4(x - 3)$

83. $25x^2 - 49$

84. $x^2 - 12x + 36$

85. $x^3 - 64$

86. $8x^3 + 27$

87. $2x^2 + 21x + 10$

88. $3x^2 + 14x + 8$

89. $x^3 - x^2 + 2x - 2$

90. $x^3 - 4x^2 + 2x - 8$

P.5 In Exercises 91 and 92, find the domain of the expression.

91. $\dfrac{1}{x + 6}$

92. $\sqrt{x + 4}$

In Exercises 93 and 94, write the rational expression in simplest form.

93. $\dfrac{x^2 - 64}{5(3x + 24)}$

94. $\dfrac{x^3 + 27}{x^2 + x - 6}$

In Exercises 95–98, perform the indicated operation and simplify.

95. $\dfrac{x^2 - 4}{x^4 - 2x^2 - 8} \cdot \dfrac{x^2 + 2}{x^2}$

96. $\dfrac{4x - 6}{(x - 1)^2} \div \dfrac{2x^2 - 3x}{x^2 + 2x - 3}$

97. $\dfrac{1}{x - 1} + \dfrac{1 - x}{x^2 + x + 1}$

98. $\dfrac{3x}{x + 2} - \dfrac{4x^2 - 5}{2x^2 + 3x - 2}$

In Exercises 99 and 100, simplify the complex fraction.

99. $\dfrac{\left[\dfrac{3a}{(a^2/x) - 1}\right]}{\left(\dfrac{a}{x} - 1\right)}$

100. $\dfrac{\left(\dfrac{1}{2x - 3} - \dfrac{1}{2x + 3}\right)}{\left(\dfrac{1}{2x} - \dfrac{1}{2x + 3}\right)}$

In Exercises 101 and 102, simplify the difference quotient.

101. $\dfrac{\left[\dfrac{1}{2(x + h)} - \dfrac{1}{2x}\right]}{h}$

102. $\dfrac{\dfrac{1}{x + h - 3} - \dfrac{1}{x - 3}}{h}$

P.6 In Exercises 103 and 104, plot the points in the Cartesian plane.

103. $(5, 5), (-2, 0), (-3, 6), (-1, -7)$

104. $(0, 6), (8, 1), (4, -2), (-3, -3)$

In Exercises 105 and 106, determine the quadrant(s) in which (x, y) is located so that the condition(s) is (are) satisfied.

105. $x > 0$ and $y = -2$

106. $xy = 4$

In Exercises 107–110, (a) plot the points, (b) find the distance between the points, and (c) find the midpoint of the line segment joining the points.

107. $(-3, 8), (1, 5)$

108. $(-2, 6), (4, -3)$

109. $(5.6, 0), (0, 8.2)$

110. $(1.8, 7.4), (-0.6, -14.5)$

In Exercises 111 and 112, the polygon is shifted to a new position in the plane. Find the coordinates of the vertices of the polygon in its new position.

111. Original coordinates of vertices:

$(4, 8), (6, 8), (4, 3), (6, 3)$

Shift: eight units downward, four units to the left

112. Original coordinates of vertices:

$(0, 1), (3, 3), (0, 5), (-3, 3)$

Shift: three units upward, two units to the left

113. SALES Starbucks had annual sales of $2.17 billion in 2000 and $10.38 billion in 2008. Use the Midpoint Formula to estimate the sales in 2004. (Source: Starbucks Corp.)

114. METEOROLOGY The apparent temperature is a measure of relative discomfort to a person from heat and high humidity. The table shows the actual temperatures x (in degrees Fahrenheit) versus the apparent temperatures y (in degrees Fahrenheit) for a relative humidity of 75%.

x	70	75	80	85	90	95	100
y	70	77	85	95	109	130	150

(a) Sketch a scatter plot of the data shown in the table.

(b) Find the change in the apparent temperature when the actual temperature changes from 70°F to 100°F.

EXPLORATION

TRUE OR FALSE? In Exercises 115 and 116, determine whether the statement is true or false. Justify your answer.

115. A binomial sum squared is equal to the sum of the terms squared.

116. $x^n - y^n$ factors as conjugates for all values of n.

117. THINK ABOUT IT Is the following statement true for all nonzero real numbers a and b? Explain.

$$\dfrac{ax - b}{b - ax} = -1$$

P CHAPTER TEST

See www.CalcChat.com for worked-out solutions to odd-numbered exercises.

Take this test as you would take a test in class. When you are finished, check your work against the answers given in the back of the book.

1. Place $<$ or $>$ between the real numbers $-\frac{10}{3}$ and $-|-4|$.

2. Find the distance between the real numbers -5.4 and $3\frac{3}{4}$.

3. Identify the rule of algebra illustrated by $(5 - x) + 0 = 5 - x$.

In Exercises 4 and 5, evaluate each expression without using a calculator.

4. (a) $27\left(-\frac{2}{3}\right)$ (b) $\dfrac{5}{18} \div \dfrac{5}{8}$ (c) $\left(-\dfrac{3}{5}\right)^3$ (d) $\left(\dfrac{3^2}{2}\right)^{-3}$

5. (a) $\sqrt{5} \cdot \sqrt{125}$ (b) $\dfrac{\sqrt{27}}{\sqrt{2}}$ (c) $\dfrac{5.4 \times 10^8}{3 \times 10^3}$ (d) $(3 \times 10^4)^3$

In Exercises 6 and 7, simplify each expression.

6. (a) $3z^2(2z^3)^2$ (b) $(u - 2)^{-4}(u - 2)^{-3}$ (c) $\left(\dfrac{x^{-2}y^2}{3}\right)^{-1}$

7. (a) $9z\sqrt{8z} - 3\sqrt{2z^3}$ (b) $(4x^{3/5})(x^{1/3})$ (c) $\sqrt[3]{\dfrac{16}{v^5}}$

8. Write the polynomial $3 - 2x^5 + 3x^3 - x^4$ in standard form. Identify the degree and leading coefficient.

In Exercises 9–12, perform the operation and simplify.

9. $(x^2 + 3) - [3x + (8 - x^2)]$ **10.** $\left(x + \sqrt{5}\right)\left(x - \sqrt{5}\right)$

11. $\dfrac{5x}{x - 4} + \dfrac{20}{4 - x}$ **12.** $\dfrac{\left(\dfrac{2}{x} - \dfrac{2}{x + 1}\right)}{\left(\dfrac{4}{x^2 - 1}\right)}$

13. Factor (a) $2x^4 - 3x^3 - 2x^2$ and (b) $x^3 + 2x^2 - 4x - 8$ completely.

14. Rationalize each denominator.   (a) $\dfrac{16}{\sqrt[3]{16}}$ (b) $\dfrac{4}{1 - \sqrt{2}}$

15. Find the domain of $\dfrac{6 - x}{1 - x}$.

16. Multiply: $\dfrac{y^2 + 8y + 16}{2y - 4} \cdot \dfrac{8y - 16}{(y + 4)^3}$.

17. A T-shirt company can produce and sell x T-shirts per day. The total cost C (in dollars) for producing x T-shirts is $C = 1480 + 6x$, and the total revenue R (in dollars) is $R = 15x$. Find the profit obtained by selling 225 T-shirts per day.

18. Plot the points $(-2, 5)$ and $(6, 0)$. Find the coordinates of the midpoint of the line segment joining the points and the distance between the points.

19. Write an expression for the area of the shaded region in the figure at the left, and simplify the result.

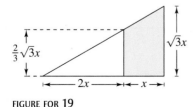

FIGURE FOR **19**

PROOFS IN MATHEMATICS

What does the word *proof* mean to you? In mathematics, the word *proof* is used to mean simply a valid argument. When you are proving a statement or theorem, you must use facts, definitions, and accepted properties in a logical order. You can also use previously proved theorems in your proof. For instance, the Distance Formula is used in the proof of the Midpoint Formula below. There are several different proof methods, which you will see in later chapters.

The Midpoint Formula   *(p. 58)*

The midpoint of the line segment joining the points (x_1, y_1) and (x_2, y_2) is given by the Midpoint Formula

$$\text{Midpoint} = \left(\frac{x_1 + x_2}{2}, \frac{y_1 + y_2}{2}\right).$$

Proof

Using the figure, you must show that $d_1 = d_2$ and $d_1 + d_2 = d_3$.

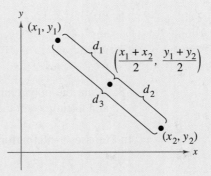

By the Distance Formula, you obtain

$$d_1 = \sqrt{\left(\frac{x_1 + x_2}{2} - x_1\right)^2 + \left(\frac{y_1 + y_2}{2} - y_1\right)^2}$$

$$= \frac{1}{2}\sqrt{(x_2 - x_1)^2 + (y_2 - y_1)^2}$$

$$d_2 = \sqrt{\left(x_2 - \frac{x_1 + x_2}{2}\right)^2 + \left(y_2 - \frac{y_1 + y_2}{2}\right)^2}$$

$$= \frac{1}{2}\sqrt{(x_2 - x_1)^2 + (y_2 - y_1)^2}$$

$$d_3 = \sqrt{(x_2 - x_1)^2 + (y_2 - y_1)^2}$$

So, it follows that $d_1 = d_2$ and $d_1 + d_2 = d_3$.

PROBLEM SOLVING

This collection of thought-provoking and challenging exercises further explores and expands upon concepts learned in this chapter.

1. The NCAA states that the men's and women's shots for track and field competition must comply with the following specifications. (Source: NCAA)

	Men's	Women's
Weight (minimum)	7.26 kg	4.0 kg
Diameter (minimum)	110 mm	95 mm
Diameter (maximum)	130 mm	110 mm

(a) Find the maximum and minimum volumes of both the men's and women's shots.

(b) The *density* of an object is an indication of how heavy the object is. To find the density of an object, divide its mass (weight) by its volume. Find the maximum and minimum densities of both the men's and women's shots.

(c) A shot is usually made out of iron. If a ball of cork has the same volume as an iron shot, do you think they would have the same density? Explain your reasoning.

2. Find an example for which

$$|a - b| > |a| - |b|,$$

and an example for which

$$|a - b| = |a| - |b|.$$

Then prove that

$$|a - b| \geq |a| - |b|$$

for all a, b.

3. A major feature of Epcot Center at Disney World is called Spaceship Earth. The building is shaped as a sphere and weighs 1.6×10^7 pounds, which is equal in weight to 1.58×10^8 golf balls. Use these values to find the approximate weight (in pounds) of one golf ball. Then convert the weight to ounces. (Source: Disney.com)

4. The average life expectancies at birth in 2005 for men and women were 75.2 years and 80.4 years, respectively. Assuming an average healthy heart rate of 70 beats per minute, find the numbers of beats in a lifetime for a man and for a woman. (Source: National Center for Health Statistics)

5. The accuracy of an approximation to a number is related to how many significant digits there are in the approximation. Write a definition of significant digits and illustrate the concept with examples.

6. The table shows the census population y (in millions) of the United States for each census year x from 1950 through 2000. (Source: U.S. Census Bureau)

Year, x	Population, y
1950	151.33
1960	179.32
1970	203.30
1980	226.54
1990	248.72
2000	281.42

(a) Sketch a scatter plot of the data. Describe any trends in the data.

(b) Find the increase in population from each census year to the next.

(c) Over which decade did the population increase the most? the least?

(d) Find the percent increase in population from each census year to the next.

(e) Over which decade was the percent increase the greatest? the least?

7. Find the annual depreciation rate r from the bar graph below. To find r by the declining balances method, use the formula

$$r = 1 - \left(\frac{S}{C}\right)^{1/n}$$

where n is the useful life of the item (in years), S is the salvage value (in dollars), and C is the original cost (in dollars).

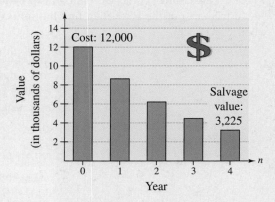

8. Johannes Kepler (1571–1630), a well-known German astronomer, discovered a relationship between the average distance of a planet from the sun and the time (or period) it takes the planet to orbit the sun. People then knew that planets that are closer to the sun take less time to complete an orbit than planets that are farther from the sun. Kepler discovered that the distance and period are related by an exact mathematical formula.

The table shows the average distances x (in astronomical units) and periods y (in years) for the five planets that are closest to the sun. By completing the table, can you rediscover Kepler's relationship? Write a paragraph that summarizes your conclusions.

Planet	Mercury	Venus	Earth	Mars	Jupiter
x	0.387	0.723	1.000	1.524	5.203
$\sqrt{x}$					
y	0.241	0.615	1.000	1.881	11.860
$\sqrt[3]{y}$					

9. A stained glass window is designed in the shape of a rectangle with a semicircular arch (see figure). The width of the window is 2 feet and the perimeter is approximately 13.14 feet. Find the smallest amount of glass required to construct the window.

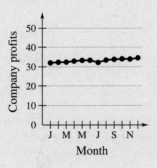

10. The volume V (in cubic inches) of the box shown in the figure is modeled by

$$V = 2x^3 + x^2 - 8x - 4$$

where x is measured in inches. Find an expression for the surface area of the box. Then find the surface area when $x = 6$ inches.

11. Verify that $y_1 \neq y_2$ by letting $x = 0$ and evaluating y_1 and y_2.

$$y_1 = 2x\sqrt{1 - x^2} - \frac{x^3}{\sqrt{1 - x^2}}$$

$$y_2 = \frac{2 - 3x^2}{\sqrt{1 - x^2}}$$

Change y_2 so that $y_1 = y_2$.

12. Prove that

$$\left(\frac{2x_1 + x_2}{3}, \frac{2y_1 + y_2}{3} \right)$$

is one of the points of trisection of the line segment joining (x_1, y_1) and (x_2, y_2). Find the midpoint of the line segment joining

$$\left(\frac{2x_1 + x_2}{3}, \frac{2y_1 + y_2}{3} \right)$$

and (x_2, y_2) to find the second point of trisection.

13. Use the results of Exercise 12 to find the points of trisection of the line segment joining each pair of points.
(a) $(1, -2), (4, 1)$ (b) $(-2, -3), (0, 0)$

14. Although graphs can help visualize relationships between two variables, they can also be used to mislead people. The graphs shown below represent the same data points.

(a) Which of the two graphs is misleading, and why? Discuss other ways in which graphs can be misleading.

(b) Why would it be beneficial for someone to use a misleading graph?

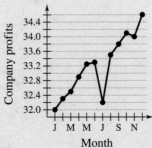

Equations, Inequalities, and Mathematical Modeling

1

In Mathematics

The methods used for solving equations are similar to the methods used for solving inequalities.

In Real Life

Real-life data can be modeled by many types of equations. These include linear, quadratic, radical, rational, and higher-order polynomial equations. Inequalities can also be used to model and solve real-life problems. For instance, inequalities can be used to represent the range of the target heart rates for a 20-year-old and a 40-year-old. (See Exercises 109 and 110, page 147.)

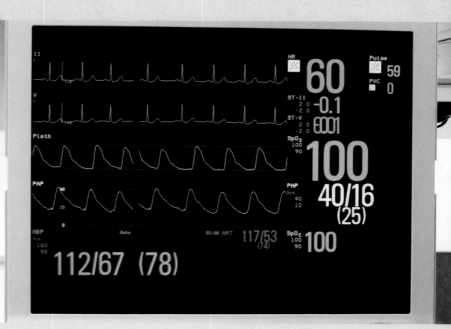

istockphoto.com

IN CAREERS

There are many careers that use equations and inequalities. Several are listed below.

- Electrician
  Exercise 80, page 86

- Anthropologist
  Exercise 107, page 94

- Physicist
  Exercises 93 and 94, page 106

- Physical Chemist
  Exercise 130, page 149

What you should learn

- Sketch graphs of equations.
- Find x- and y-intercepts of graphs of equations.
- Use symmetry to sketch graphs of equations.
- Find equations of and sketch graphs of circles.
- Use graphs of equations in solving real-life problems.

Why you should learn it

The graph of an equation can help you see relationships between real-life quantities. For example, in Exercise 79 on page 86, a graph can be used to estimate the life expectancies of children who are born in 2015.

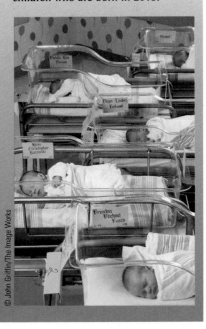

© John Griffin/The Image Works

Algebra Help

When evaluating an expression or an equation, remember to follow the Basic Rules of Algebra. To review these rules, see Section P.1.

1.1 GRAPHS OF EQUATIONS

The Graph of an Equation

In Section P.6, you used a coordinate system to represent graphically the relationship between two quantities. There, the graphical picture consisted of a collection of points in a coordinate plane.

Frequently, a relationship between two quantities is expressed as an **equation in two variables.** For instance, $y = 7 - 3x$ is an equation in x and y. An ordered pair (a, b) is a **solution** or **solution point** of an equation in x and y if the equation is true when a is substituted for x, and b is substituted for y. For instance, $(1, 4)$ is a solution of $y = 7 - 3x$ because $= 7 - 3(\ )$ is a true statement.

In this section you will review some basic procedures for sketching the graph of an equation in two variables. The **graph of an equation** is the set of all points that are solutions of the equation.

Example 1 Determining Solution Points

Determine whether (a) $(2, 13)$ and (b) $(-1, -3)$ lie on the graph of $y = 10x - 7$.

Solution

a. $y = 10x - 7$

$\overset{?}{=} 10(\ ) - 7$

$13 = 13$ ✓

The point $(2, 13)$ *does* lie on the graph of $y = 10x - 7$ because it is a solution point of the equation.

b. $y = 10x - 7$

$\overset{?}{=} 10(\ \ ) - 7$

$-3 \neq -17$

The point $(-1, -3)$ *does not* lie on the graph of $y = 10x - 7$ because it is *not* a solution point of the equation.

CHECK*Point* Now try Exercise 7.

The basic technique used for sketching the graph of an equation is the **point-plotting method.**

Sketching the Graph of an Equation by Point Plotting

1. If possible, rewrite the equation so that one of the variables is isolated on one side of the equation.

2. Make a table of values showing several solution points.

3. Plot these points on a rectangular coordinate system.

4. Connect the points with a smooth curve or line.

When making a table of solution points, be sure to use positive, zero, and negative values of x.

Example 2   Sketching the Graph of an Equation

Sketch the graph of

$$y = 7 - 3x.$$

Solution

Because the equation is already solved for y, construct a table of values that consists of several solution points of the equation. For instance, when $x = -1$,

$$y = 7 - 3(\quad)$$
$$= 10$$

which implies that $(-1, 10)$ is a solution point of the graph.

x	$y = 7 - 3x$	(x, y)
-1	10	$(-1, 10)$
0	7	$(0, 7)$
1	4	$(1, 4)$
2	1	$(2, 1)$
3	-2	$(3, -2)$
4	-5	$(4, -5)$

From the table, it follows that

$$(-1, 10), (0, 7), (1, 4), (2, 1), (3, -2), \text{ and } (4, -5)$$

are solution points of the equation. After plotting these points, you can see that they appear to lie on a line, as shown in Figure 1.1. The graph of the equation is the line that passes through the six plotted points.

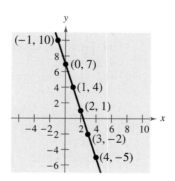

FIGURE **1.1**

CHECK *Point*   Now try Exercise 15.

Example 3 Sketching the Graph of an Equation

Sketch the graph of

$$y = x^2 - 2.$$

Solution

Because the equation is already solved for y, begin by constructing a table of values.

x	-2	-1	0	1	2	3
$y = x^2 - 2$	2	-1	-2	-1	2	7
(x, y)	$(-2, 2)$	$(-1, -1)$	$(0, -2)$	$(1, -1)$	$(2, 2)$	$(3, 7)$

Next, plot the points given in the table, as shown in Figure 1.2. Finally, connect the points with a smooth curve, as shown in Figure 1.3.

Study Tip

One of your goals in this course is to learn to classify the basic shape of a graph from its equation. For instance, you will learn that the *linear equation* in Example 2 has the form

$$y = mx + b$$

and its graph is a line. Similarly, the *quadratic equation* in Example 3 has the form

$$y = ax^2 + bx + c$$

and its graph is a parabola.

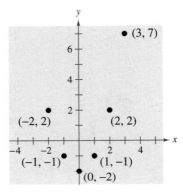

FIGURE 1.2

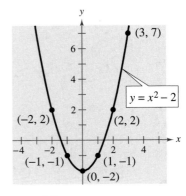

FIGURE 1.3

CHECK **Point** Now try Exercise 17.

The point-plotting method demonstrated in Examples 2 and 3 is easy to use, but it has some shortcomings. With too few solution points, you can misrepresent the graph of an equation. For instance, if only the four points

$$(-2, 2), (-1, -1), (1, -1), \text{ and } (2, 2)$$

in Figure 1.2 were plotted, any one of the three graphs in Figure 1.4 would be reasonable.

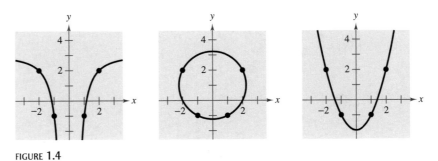

FIGURE 1.4

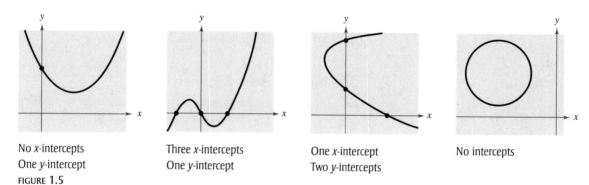

TECHNOLOGY

To graph an equation involving *x* and *y* on a graphing utility, use the following procedure.

1. Rewrite the equation so that *y* is isolated on the left side.

2. Enter the equation in the graphing utility.

3. Determine a *viewing window* that shows all important features of the graph.

4. Graph the equation.

Intercepts of a Graph

It is often easy to determine the solution points that have zero as either the *x*-coordinate or the *y*-coordinate. These points are called **intercepts** because they are the points at which the graph intersects or touches the *x*- or *y*-axis. It is possible for a graph to have no intercepts, one intercept, or several intercepts, as shown in Figure 1.5.

No *x*-intercepts
One *y*-intercept
FIGURE 1.5

Three *x*-intercepts
One *y*-intercept

One *x*-intercept
Two *y*-intercepts

No intercepts

Note that an *x*-intercept can be written as the ordered pair $(x, 0)$ and a *y*-intercept can be written as the ordered pair $(0, y)$. Some texts denote the *x*-intercept as the *x*-coordinate of the point $(a, 0)$ [and the *y*-intercept as the *y*-coordinate of the point $(0, b)$] rather than the point itself. Unless it is necessary to make a distinction, we will use the term *intercept* to mean either the point or the coordinate.

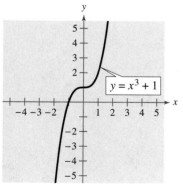

FIGURE 1.6

| **Example 4**   Identifying *x*- and *y*-Intercepts

Identify the *x*- and *y*-intercepts of the graph of

$$y = x^3 + 1$$

shown in Figure 1.6.

Solution

From the figure, you can see that the graph of the equation $y = x^3 + 1$ has an *x*-intercept (where *y* is zero) at $(-1, 0)$ and a *y*-intercept (where *x* is zero) at $(0, 1)$.

CHECK*Point*   Now try Exercise 19.

Symmetry

Graphs of equations can have **symmetry** with respect to one of the coordinate axes or with respect to the origin. Symmetry with respect to the x-axis means that if the Cartesian plane were folded along the x-axis, the portion of the graph above the x-axis would coincide with the portion below the x-axis. Symmetry with respect to the y-axis or the origin can be described in a similar manner, as shown in Figure 1.7.

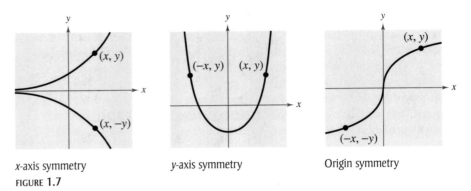

| x-axis symmetry | y-axis symmetry | Origin symmetry |

FIGURE 1.7

Knowing the symmetry of a graph *before* attempting to sketch it is helpful, because then you need only half as many solution points to sketch the graph. There are three basic types of symmetry, described as follows.

Graphical Tests for Symmetry

1. A graph is **symmetric with respect to the x-axis** if, whenever (x, y) is on the graph, $(x, -y)$ is also on the graph.

2. A graph is **symmetric with respect to the y-axis** if, whenever (x, y) is on the graph, $(-x, y)$ is also on the graph.

3. A graph is **symmetric with respect to the origin** if, whenever (x, y) is on the graph, $(-x, -y)$ is also on the graph.

You can conclude that the graph of $y = x^2 - 2$ is symmetric with respect to the y-axis because the point $(-x, y)$ is also on the graph of $y = x^2 - 2$. (See the table below and Figure 1.8.)

x	-3	-2	-1	1	2	3
y	7	2	-1	-1	2	7
(x, y)	$(-3, 7)$	$(-2, 2)$	$(-1, -1)$	$(1, -1)$	$(2, 2)$	$(3, 7)$

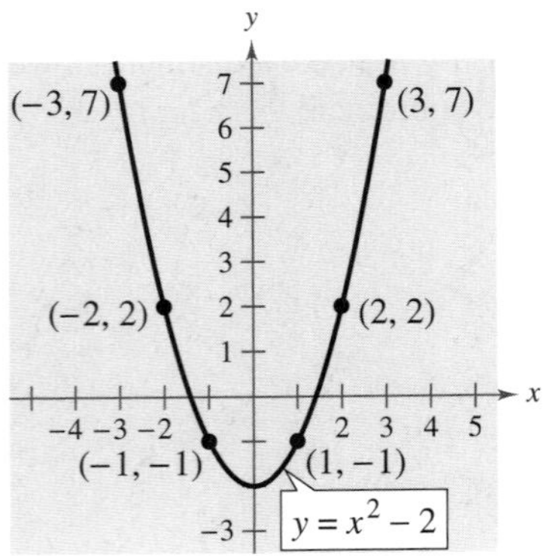

y-axis symmetry

FIGURE 1.8

Algebraic Tests for Symmetry

1. The graph of an equation is symmetric with respect to the x-axis if replacing y with $-y$ yields an equivalent equation.

2. The graph of an equation is symmetric with respect to the y-axis if replacing x with $-x$ yields an equivalent equation.

3. The graph of an equation is symmetric with respect to the origin if replacing x with $-x$ and y with $-y$ yields an equivalent equation.

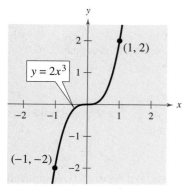

FIGURE **1.9**

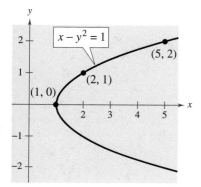

FIGURE **1.10**

Example 5   Testing for Symmetry

Test $y = 2x^3$ for symmetry with respect to both axes and the origin.

Solution

x-axis: $y = 2x^3$

$-y = 2x^3$

y-axis: $y = 2x^3$

$y = 2(-x)^3$

$y = -2x^3$

Origin: $y = 2x^3$

$-y = 2(-x)^3$

$-y = -2x^3$

$y = 2x^3$

Of the three tests for symmetry, the only one that is satisfied is the test for origin symmetry (see Figure 1.9).

CHECK*Point*   Now try Exercise 25.

Example 6   Using Symmetry as a Sketching Aid

Use symmetry to sketch the graph of $x - y^2 = 1$.

Solution

Of the three tests for symmetry, the only one that is satisfied is the test for x-axis symmetry because $x - (-y)^2 = 1$ is equivalent to $x - y^2 = 1$. So, the graph is symmetric with respect to the x-axis. Using symmetry, you only need to find the solution points above the x-axis and then reflect them to obtain the graph, as shown in Figure 1.10.

CHECK*Point*   Now try Exercise 41.

Example 7   Sketching the Graph of an Equation

Sketch the graph of $y = |x - 1|$.

Solution

This equation fails all three tests for symmetry and consequently its graph is not symmetric with respect to either axis or to the origin. The absolute value sign indicates that y is always nonnegative. Create a table of values and plot the points, as shown in Figure 1.11. From the table, you can see that $x = 0$ when $y = 1$. So, the y-intercept is $(0, 1)$. Similarly, $y = 0$ when $x = 1$. So, the x-intercept is $(1, 0)$.

x	-2	-1	0	1	2	3	4
$y = |x - 1|$	3	2	1	0	1	2	3
(x, y)	$(-2, 3)$	$(-1, 2)$	$(0, 1)$	$(1, 0)$	$(2, 1)$	$(3, 2)$	$(4, 3)$

CHECK*Point*   Now try Exercise 45.

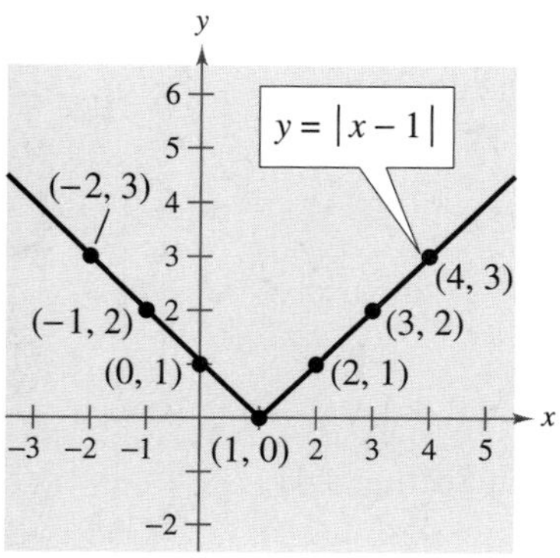

Algebra Help

In Example 7, $|x - 1|$ is an absolute value expression. You can review the techniques for evaluating an absolute value expression in Section P.1.

FIGURE **1.11**

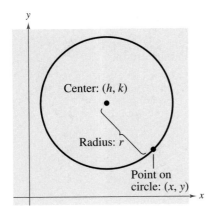

FIGURE **1.12**

Throughout this course, you will learn to recognize several types of graphs from their equations. For instance, you will learn to recognize that the graph of a second-degree equation of the form

$$y = ax^2 + bx + c$$

is a parabola (see Example 3). The graph of a **circle** is also easy to recognize.

Circles

Consider the circle shown in Figure 1.12. A point (x, y) is on the circle if and only if its distance from the center (h, k) is r. By the Distance Formula,

$$\sqrt{(x - h)^2 + (y - k)^2} = r.$$

By squaring each side of this equation, you obtain the **standard form of the equation of a circle.**

Standard Form of the Equation of a Circle

The point (x, y) lies on the circle of **radius** r and **center** (h, k) if and only if

$$(x - h)^2 + (y - k)^2 = r^2.$$

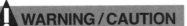

Be careful when you are finding h and k from the standard equation of a circle. For instance, to find the correct h and k from the equation of the circle in Example 8, rewrite the quantities $(x + 1)^2$ and $(y - 2)^2$ using subtraction.

$$(x + 1)^2 = [x - (-1)]^2,$$

$$(y - 2)^2 = [y - (2)]^2$$

So, $h = -1$ and $k = 2$.

From this result, you can see that the standard form of the equation of a circle *with its center at the origin*, $(h, k) = (0, 0)$, is simply

$$x^2 + y^2 = r^2.$$

Example 8 **Finding the Equation of a Circle**

The point $(3, 4)$ lies on a circle whose center is at $(-1, 2)$, as shown in Figure 1.13. Write the standard form of the equation of this circle.

Solution

The radius of the circle is the distance between $(-1, 2)$ and $(3, 4)$.

$$r = \sqrt{(x - h)^2 + (y - k)^2}$$
$$= \sqrt{[\quad - (\quad)]^2 + (\quad - \quad)^2}$$
$$= \sqrt{4^2 + 2^2}$$
$$= \sqrt{16 + 4}$$
$$= \sqrt{20}$$

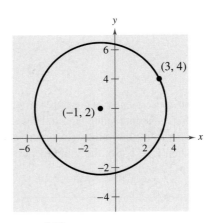

FIGURE **1.13**

Using $(h, k) = (-1, 2)$ and $r = \sqrt{20}$, the equation of the circle is

$$(x - h)^2 + (y - k)^2 = r^2$$
$$[x - (\quad)]^2 + (y - \quad)^2 = (\quad)^2$$
$$(x + 1)^2 + (y - 2)^2 = 20.$$

CHECK *Point* Now try Exercise 65.

You will learn more about writing equations of circles in Section 4.4.

Application

In this course, you will learn that there are many ways to approach a problem. Three common approaches are illustrated in Example 9.

A *Numerical Approach:* Construct and use a table.

A *Graphical Approach:* Draw and use a graph.

An *Algebraic Approach:* Use the rules of algebra.

Example 9   Recommended Weight

The median recommended weight y (in pounds) for men of medium frame who are 25 to 59 years old can be approximated by the mathematical model

$$y = 0.073x^2 - 6.99x + 289.0, \quad 62 \le x \le 76$$

where x is the man's height (in inches).   (Source: Metropolitan Life Insurance Company)

a. Construct a table of values that shows the median recommended weights for men with heights of 62, 64, 66, 68, 70, 72, 74, and 76 inches.

b. Use the table of values to sketch a graph of the model. Then use the graph to estimate *graphically* the median recommended weight for a man whose height is 71 inches.

c. Use the model to confirm *algebraically* the estimate you found in part (b).

Solution

a. You can use a calculator to complete the table, as shown at the left.

b. The table of values can be used to sketch the graph of the equation, as shown in Figure 1.14. From the graph, you can estimate that a height of 71 inches corresponds to a weight of about 161 pounds.

Height, x	Weight, y
62	136.2
64	140.6
66	145.6
68	151.2
70	157.4
72	164.2
74	171.5
76	179.4

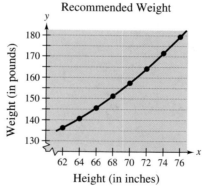

Recommended Weight

FIGURE **1.14**

c. To confirm algebraically the estimate found in part (b), you can substitute 71 for x in the model.

$$y = 0.073(\quad)^2 - 6.99(\quad) + 289.0 \approx 160.70$$

So, the graphical estimate of 161 pounds is fairly good.

CHECK *Point*   Now try Exercise 79.

1.1   EXERCISES

See www.CalcChat.com for worked-out solutions to odd-numbered exercises.

VOCABULARY: Fill in the blanks.

1. An ordered pair (a, b) is a _____ of an equation in x and y if the equation is true when a is substituted for x, and b is substituted for y.
2. The set of all solution points of an equation is the _____ of the equation.
3. The points at which a graph intersects or touches an axis are called the _____ of the graph.
4. A graph is symmetric with respect to the _____ if, whenever (x, y) is on the graph, $(-x, y)$ is also on the graph.
5. The equation $(x - h)^2 + (y - k)^2 = r^2$ is the standard form of the equation of a _____ with center _____ and radius _____.
6. When you construct and use a table to solve a problem, you are using a _____ approach.

SKILLS AND APPLICATIONS

In Exercises 7–14, determine whether each point lies on the graph of the equation.

Equation	Points	
7. $y = \sqrt{x + 4}$	(a) $(0, 2)$	(b) $(5, 3)$
8. $y = \sqrt{5 - x}$	(a) $(1, 2)$	(b) $(5, 0)$
9. $y = x^2 - 3x + 2$	(a) $(2, 0)$	(b) $(-2, 8)$
10. $y = 4 - |x - 2|$	(a) $(1, 5)$	(b) $(6, 0)$
11. $y = |x - 1| + 2$	(a) $(2, 3)$	(b) $(-1, 0)$
12. $2x - y - 3 = 0$	(a) $(1, 2)$	(b) $(1, -1)$
13. $x^2 + y^2 = 20$	(a) $(3, -2)$	(b) $(-4, 2)$
14. $y = \frac{1}{3}x^3 - 2x^2$	(a) $\left(2, -\frac{16}{3}\right)$	(b) $(-3, 9)$

In Exercises 15–18, complete the table. Use the resulting solution points to sketch the graph of the equation.

15. $y = -2x + 5$

x	-1	0	1	2	$\frac{5}{2}$
y					
(x, y)					

16. $y = \frac{3}{4}x - 1$

x	-2	0	1	$\frac{4}{3}$	2
y					
(x, y)					

17. $y = x^2 - 3x$

x	-1	0	1	2	3
y					
(x, y)					

18. $y = 5 - x^2$

x	-2	-1	0	1	2
y					
(x, y)					

In Exercises 19–24, graphically estimate the x- and y-intercepts of the graph.

19. $y = (x - 3)^2$

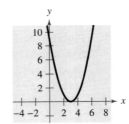

20. $y = 16 - 4x^2$

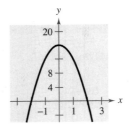

21. $y = |x + 2|$

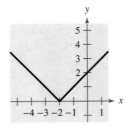

22. $y^2 = 4 - x$

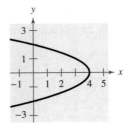

23. $y = 2 - 2x^3$

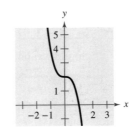

24. $y = x^3 - 4x$

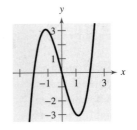

In Exercises 25–32, use the algebraic tests to check for symmetry with respect to both axes and the origin.

25. $x^2 - y = 0$

26. $x - y^2 = 0$

27. $y = x^3$

28. $y = x^4 - x^2 + 3$

29. $y = \dfrac{x}{x^2 + 1}$

30. $y = \dfrac{1}{x^2 + 1}$

31. $xy^2 + 10 = 0$

32. $xy = 4$

In Exercises 33–36, assume that the graph has the indicated type of symmetry. Sketch the complete graph of the equation. To print an enlarged copy of the graph, go to the website *www.mathgraphs.com*.

33.

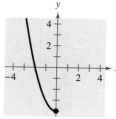

y-axis symmetry

34.

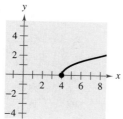

x-axis symmetry

35.

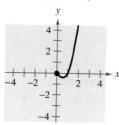

Origin symmetry

36.

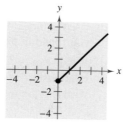

y-axis symmetry

In Exercises 37–48, identify any intercepts and test for symmetry. Then sketch the graph of the equation.

37. $y = -3x + 1$

38. $y = 2x - 3$

39. $y = x^2 - 2x$

40. $y = -x^2 - 2x$

41. $y = x^3 + 3$

42. $y = x^3 - 1$

43. $y = \sqrt{x - 3}$

44. $y = \sqrt{1 - x}$

45. $y = |x - 6|$

46. $y = 1 - |x|$

47. $x = y^2 - 1$

48. $x = y^2 - 5$

 In Exercises 49–60, use a graphing utility to graph the equation. Use a standard setting. Approximate any intercepts.

49. $y = 5 - \frac{1}{2}x$

50. $y = \frac{2}{3}x - 1$

51. $y = x^2 - 4x + 3$

52. $y = x^2 + x - 2$

53. $y = \dfrac{2x}{x - 1}$

54. $y = \dfrac{4}{x^2 + 1}$

55. $y = \sqrt[3]{x} + 2$

56. $y = \sqrt[3]{x + 1}$

57. $y = x\sqrt{x + 6}$

58. $y = (6 - x)\sqrt{x}$

59. $y = |x + 3|$

60. $y = 2 - |x|$

In Exercises 61–68, write the standard form of the equation of the circle with the given characteristics.

61. Center: $(0, 0)$; Radius: 4

62. Center: $(0, 0)$; Radius: 5

63. Center: $(2, -1)$; Radius: 4

64. Center: $(-7, -4)$; Radius: 7

65. Center: $(-1, 2)$; Solution point: $(0, 0)$

66. Center: $(3, -2)$; Solution point: $(-1, 1)$

67. Endpoints of a diameter: $(0, 0)$, $(6, 8)$

68. Endpoints of a diameter: $(-4, -1)$, $(4, 1)$

In Exercises 69–74, find the center and radius of the circle, and sketch its graph.

69. $x^2 + y^2 = 25$

70. $x^2 + y^2 = 36$

71. $(x - 1)^2 + (y + 3)^2 = 9$

72. $x^2 + (y - 1)^2 = 1$

73. $\left(x - \frac{1}{2}\right)^2 + \left(y - \frac{1}{2}\right)^2 = \frac{9}{4}$

74. $(x - 2)^2 + (y + 3)^2 = \frac{16}{9}$

75. DEPRECIATION   A hospital purchases a new magnetic resonance imaging (MRI) machine for $500,000. The depreciated value y (reduced value) after t years is given by $y = 500{,}000 - 40{,}000t, 0 \le t \le 8$. Sketch the graph of the equation.

76. CONSUMERISM   You purchase an all-terrain vehicle (ATV) for $8000. The depreciated value y after t years is given by $y = 8000 - 900t,\ 0 \le t \le 6$. Sketch the graph of the equation.

 77. GEOMETRY   A regulation NFL playing field (including the end zones) of length x and width y has a perimeter of $346\frac{2}{3}$ or $\frac{1040}{3}$ yards.

(a) Draw a rectangle that gives a visual representation of the problem. Use the specified variables to label the sides of the rectangle.

(b) Show that the width of the rectangle is $y = \dfrac{520}{3} - x$ and its area is $A = x\left(\dfrac{520}{3} - x\right)$.

(c) Use a graphing utility to graph the area equation. Be sure to adjust your window settings.

(d) From the graph in part (c), estimate the dimensions of the rectangle that yield a maximum area.

(e) Use your school's library, the Internet, or some other reference source to find the actual dimensions and area of a regulation NFL playing field and compare your findings with the results of part (d).

The symbol indicates an exercise or a part of an exercise in which you are instructed to use a graphing utility.

78. GEOMETRY A soccer playing field of length x and width y has a perimeter of 360 meters.

(a) Draw a rectangle that gives a visual representation of the problem. Use the specified variables to label the sides of the rectangle.

(b) Show that the width of the rectangle is $y = 180 - x$ and its area is $A = x(180 - x)$.

(c) Use a graphing utility to graph the area equation. Be sure to adjust your window settings.

(d) From the graph in part (c), estimate the dimensions of the rectangle that yield a maximum area.

(e) Use your school's library, the Internet, or some other reference source to find the actual dimensions and area of a regulation Major League Soccer field and compare your findings with the results of part (d).

79. POPULATION STATISTICS The table shows the life expectancies of a child (at birth) in the United States for selected years from 1920 to 2000. (Source: U.S. National Center for Health Statistics)

Year	Life Expectancy, y
1920	54.1
1930	59.7
1940	62.9
1950	68.2
1960	69.7
1970	70.8
1980	73.7
1990	75.4
2000	77.0

A model for the life expectancy during this period is

$$y = -0.0025t^2 + 0.574t + 44.25, \quad 20 \le t \le 100$$

where y represents the life expectancy and t is the time in years, with $t = 20$ corresponding to 1920.

 (a) Use a graphing utility to graph the data from the table and the model in the same viewing window. How well does the model fit the data? Explain.

(b) Determine the life expectancy in 1990 both graphically and algebraically.

(c) Use the graph to determine the year when life expectancy was approximately 76.0. Verify your answer algebraically.

(d) One projection for the life expectancy of a child born in 2015 is 78.9. How does this compare with the projection given by the model?

(e) Do you think this model can be used to predict the life expectancy of a child 50 years from now? Explain.

80. ELECTRONICS The resistance y (in ohms) of 1000 feet of solid copper wire at 68 degrees Fahrenheit can be approximated by the model

$$y = \frac{10{,}770}{x^2} - 0.37, \quad 5 \le x \le 100$$

where x is the diameter of the wire in mils (0.001 inch). (Source: American Wire Gage)

(a) Complete the table.

x	5	10	20	30	40	50
y						

x	60	70	80	90	100
y					

(b) Use the table of values in part (a) to sketch a graph of the model. Then use your graph to estimate the resistance when $x = 85.5$.

(c) Use the model to confirm algebraically the estimate you found in part (b).

(d) What can you conclude in general about the relationship between the diameter of the copper wire and the resistance?

EXPLORATION

81. THINK ABOUT IT Find a and b if the graph of $y = ax^2 + bx^3$ is symmetric with respect to (a) the y-axis and (b) the origin. (There are many correct answers.)

82. CAPSTONE Match the equation or equations with the given characteristic.

(i) $y = 3x^3 - 3x$ (ii) $y = (x + 3)^2$

(iii) $y = 3x - 3$ (iv) $y = \sqrt[3]{x}$

(v) $y = 3x^2 + 3$ (vi) $y = \sqrt{x + 3}$

(a) Symmetric with respect to the y-axis

(b) Three x-intercepts

(c) Symmetric with respect to the x-axis

(d) $(-2, 1)$ is a point on the graph

(e) Symmetric with respect to the origin

(f) Graph passes through the origin

1.2 LINEAR EQUATIONS IN ONE VARIABLE

What you should learn

- Identify different types of equations.
- Solve linear equations in one variable.
- Solve equations that lead to linear equations.
- Find *x*- and *y*-intercepts of graphs of equations algebraically.
- Use linear equations to model and solve real-life problems.

Why you should learn it

Linear equations are used in many real-life applications. For example, in Exercise 110 on page 95, linear equations can be used to model the number of women in the civilian work force over time.

© Andrew Douglas/Masterfile

Equations and Solutions of Equations

An **equation** in x is a statement that two algebraic expressions are equal. For example

$$3x - 5 = 7, \quad x^2 - x - 6 = 0, \quad \text{and} \quad \sqrt{2x} = 4$$

are equations. To **solve** an equation in x means to find all values of x for which the equation is true. Such values are **solutions.** For instance, $x = 4$ is a solution of the equation

$$3x - 5 = 7$$

because $3(\ ) - 5 = 7$ is a true statement.

The solutions of an equation depend on the kinds of numbers being considered. For instance, in the set of rational numbers, $x^2 = 10$ has no solution because there is no rational number whose square is 10. However, in the set of real numbers, the equation has the two solutions $x = \sqrt{10}$ and $x = -\sqrt{10}$.

An equation that is true for *every* real number in the domain of the variable is called an **identity.** For example

$$x^2 - 9 = (x + 3)(x - 3)$$

is an identity because it is a true statement for any real value of x. The equation

$$\frac{x}{3x^2} = \frac{1}{3x}$$

where $x \neq 0$, is an identity because it is true for any nonzero real value of x.

An equation that is true for just *some* (or even none) of the real numbers in the domain of the variable is called a **conditional equation.** For example, the equation

$$x^2 - 9 = 0$$

is conditional because $x = 3$ and $x = -3$ are the only values in the domain that satisfy the equation. The equation $2x - 4 = 2x + 1$ is conditional because there are no real values of x for which the equation is true. Learning to solve conditional equations is the primary focus of this chapter.

Linear Equations in One Variable

> **Definition of Linear Equation**
>
> A **linear equation in one variable** x is an equation that can be written in the standard form
>
> $$ax + b = 0$$
>
> where a and b are real numbers with $a \neq 0$.

HISTORICAL NOTE

This ancient Egyptian papyrus, discovered in 1858, contains one of the earliest examples of mathematical writing in existence. The papyrus itself dates back to around 1650 B.C., but it is actually a copy of writings from two centuries earlier. The algebraic equations on the papyrus were written in words. Diophantus, a Greek who lived around A.D. 250, is often called the Father of Algebra. He was the first to use abbreviated word forms in equations.

A linear equation has exactly one solution. To see this, consider the following steps. (Remember that $a \neq 0$.)

$$ax + b = 0$$

$$ax = -b$$

$$x = -\frac{b}{a}$$

To solve a conditional equation in x, isolate x on one side of the equation by a sequence of **equivalent** (and usually simpler) **equations,** each having the same solution(s) as the original equation. The operations that yield equivalent equations come from the Substitution Principle and the Properties of Equality studied in Chapter P.

Generating Equivalent Equations

An equation can be transformed into an *equivalent equation* by one or more of the following steps.

	Given Equation	*Equivalent Equation*
1. Remove symbols of grouping, combine like terms, or simplify fractions on one or both sides of the equation.	$2x - x = 4$	$x = 4$
2. Add (or subtract) the same quantity to (from) *each* side of the equation.	$x + 1 = 6$	$x = 5$
3. Multiply (or divide) *each* side of the equation by the same *nonzero* quantity.	$2x = 6$	$x = 3$
4. Interchange the two sides of the equation.	$2 = x$	$x = 2$

Example 1   Solving a Linear Equation

a. $3x - 6 = 0$

$\qquad 3x = 6$

$\qquad\ \ x = 2$

b. $5x + 4 = 3x - 8$

$\qquad 2x + 4 = -8$

$\qquad\ \ \ 2x = -12$

$\qquad\ \ \ \ x = -6$

CHECK *Point*   Now try Exercise 33.

After solving an equation, you should check each solution in the original equation. For instance, you can check the solution of Example 1(a) as follows.

$$3x - 6 = 0$$

$$3(\ ) - 6 \stackrel{?}{=} 0$$

$$0 = 0 \qquad \checkmark$$

Try checking the solution of Example 1(b).

Some equations have no solutions because all the x-terms sum to zero and a contradictory (false) statement such as $0 = 5$ or $12 = 7$ is obtained. For instance, the equation

$$x = x + 1$$

has no solution. Watch for this type of equation in the exercises.

Example 2   Solving a Linear Equation

Solve

$$6(x - 1) + 4 = 3(7x + 1).$$

Solution

$$6(x - 1) + 4 = 3(7x + 1)$$

$$6x - 6 + 4 = 21x + 3$$

$$6x - 2 = 21x + 3$$

$$-15x - 2 = 3$$

$$-15x = 5$$

$$x = -\frac{1}{3}$$

Check

Check this solution by substituting $-\frac{1}{3}$ for x in the original equation.

$$6(x - 1) + 4 = 3(7x + 1)$$

$$6(\quad - 1) + 4 \stackrel{?}{=} 3[7(\quad) + 1]$$

$$6\left(-\frac{4}{3}\right) + 4 \stackrel{?}{=} 3\left[-\frac{7}{3} + 1\right]$$

$$6\left(-\frac{4}{3}\right) + 4 \stackrel{?}{=} 3\left(-\frac{4}{3}\right)$$

$$-\frac{24}{3} + 4 \stackrel{?}{=} -\frac{12}{3}$$

$$-8 + 4 \stackrel{?}{=} -4$$

$$-4 = -4 \qquad \checkmark$$

So, the solution is $x = -\frac{1}{3}$. Note that if you subtracted $6x$ from each side of the equation and then subtracted 3 from each side of the equation, you would still obtain the solution $x = -\frac{1}{3}$.

CHECK*Point*   Now try Exercise 39.

TECHNOLOGY

You can use a graphing utility to check that a solution is reasonable. One way to do this is to graph the left side of the equation, then graph the right side of the equation, and determine the point of intersection. For instance, in Example 2, if you graph the equations

$$y_1 = 6(x - 1) + 4$$

$$y_2 = 3(7x + 1)$$

in the same viewing window, they should intersect at $x = -\frac{1}{3}$, as shown in the graph below.

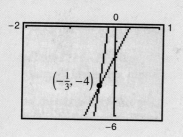

Equations That Lead to Linear Equations

To solve an equation involving fractional expressions, find the least common denominator (LCD) of all terms and multiply every term by the LCD. This process will clear the original equation of fractions and produce a simpler equation to work with.

Example 3   An Equation Involving Fractional Expressions

Solve $\dfrac{x}{3} + \dfrac{3x}{4} = 2$.

Solution

$$\frac{x}{3} + \frac{3x}{4} = 2$$

$$\frac{x}{3} + \frac{3x}{4} = 2$$

$$4x + 9x = 24$$

$$13x = 24$$

$$x = \frac{24}{13}$$

The solution is $x = \frac{24}{13}$. Check this in the original equation.

CHECK*Point* Now try Exercise 43.

When multiplying or dividing an equation by a *variable* quantity, it is possible to introduce an extraneous solution. An **extraneous solution** is one that does not satisfy the original equation. Therefore, it is essential that you check your solutions.

Example 4   An Equation with an Extraneous Solution

Solve $\dfrac{1}{x - 2} = \dfrac{3}{x + 2} - \dfrac{6x}{x^2 - 4}$.

Solution

The LCD is $x^2 - 4$, or $(x + 2)(x - 2)$. Multiply each term by this LCD.

$$\frac{1}{x - 2} = \frac{3}{x + 2} - \frac{6x}{x^2 - 4}$$

$$x + 2 = 3(x - 2) - 6x, \quad x \neq \pm 2$$

$$x + 2 = 3x - 6 - 6x$$

$$x + 2 = -3x - 6$$

$$4x = -8 \quad\Longrightarrow\quad x = -2$$

In the original equation, $x = -2$ yields a denominator of zero. So, $x = -2$ is an extraneous solution, and the original equation has *no solution*.

CHECK*Point* Now try Exercise 63.

Finding Intercepts Algebraically

In Section 1.1, you learned to find x- and y-intercepts using a graphical approach. Because all the points on the x-axis have a y-coordinate equal to zero, and all the points on the y-axis have an x-coordinate equal to zero, you can use an algebraic approach to find x- and y-intercepts, as follows.

> ### Finding Intercepts Algebraically
>
> **1.** To find x-intercepts, set y equal to zero and solve the equation for x.
>
> **2.** To find y-intercepts, set x equal to zero and solve the equation for y.

Here is an example.

$$y = 4x + 1 \Longrightarrow \quad = 4x + 1 \Longrightarrow -1 = 4x \Longrightarrow -\tfrac{1}{4} = x$$
$$y = 4x + 1 \Longrightarrow y = 4(\ ) + 1 \Longrightarrow y = 1$$

So, the x-intercept of $y = 4x + 1$ is $\left(-\tfrac{1}{4}, 0\right)$ and the y-intercept is $(0, 1)$.

Application

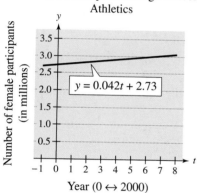

Female Participants in High School Athletics

$y = 0.042t + 2.73$

Year (0 ↔ 2000)

FIGURE 1.15

| **Example 5**   **Female Participants in Athletic Programs**

The number y (in millions) of female participants in high school athletic programs in the United States from 1999 through 2008 can be approximated by the linear model

$$y = 0.042t + 2.73, \quad -1 \le t \le 8$$

where $t = 0$ represents 2000. (a) Find algebraically the y-intercept of the graph of the linear model shown in Figure 1.15. (b) Assuming that this linear pattern continues, find the year in which there will be 3.36 million female participants. (Source: National Federation of State High School Associations)

Solution

a. To find the y-intercept, let $t = 0$ and solve for y, as follows.

$$y = 0.042t + 2.73$$
$$= 0.042(\ ) + 2.73$$
$$= 2.73$$

So, the y-intercept is $(0, 2.73)$.

b. Let $y = 3.36$ and solve the equation $3.36 = 0.042t + 2.73$ for t.

$$3.36 = 0.042t + 2.73$$
$$0.63 = 0.042t$$
$$15 = t$$

Because $t = 0$ represents 2000, $t = 15$ must represent 2015. So, from this model, there will be 3.36 million female participants in 2015.

CHECK**Point**   Now try Exercise 109.

1.2   EXERCISES

See www.CalcChat.com for worked-out solutions to odd-numbered exercises.

VOCABULARY: Fill in the blanks.

1. An _____ is a statement that equates two algebraic expressions.
2. To find all values that satisfy an equation is to _____ the equation.
3. There are two types of equations, _____ and _____ equations.
4. A linear equation in one variable is an equation that can be written in the standard form _____.
5. When solving an equation, it is possible to introduce an _____ solution, which is a value that does not satisfy the original equation.
6. To solve a conditional equation, isolate the variable on one side using transformations that produce _____ _____.

SKILLS AND APPLICATIONS

In Exercises 7–18, determine whether each value of x is a solution of the equation.

Equation	Values
7. $5x - 3 = 3x + 5$	(a) $x = 0$ (b) $x = -5$
	(c) $x = 4$ (d) $x = 10$
8. $7 - 3x = 5x - 17$	(a) $x = -3$ (b) $x = 0$
	(c) $x = 8$ (d) $x = 3$
9. $3x^2 + 2x - 5$	(a) $x = -3$ (b) $x = 1$
$= 2x^2 - 2$	(c) $x = 4$ (d) $x = -5$
10. $5x^3 + 2x - 3$	(a) $x = 2$ (b) $x = -2$
$= 4x^3 + 2x - 11$	(c) $x = 0$ (d) $x = 10$
11. $\dfrac{5}{2x} - \dfrac{4}{x} = 3$	(a) $x = -\tfrac{1}{2}$ (b) $x = 4$
	(c) $x = 0$ (d) $x = \tfrac{1}{4}$
12. $\dfrac{x}{2} + \dfrac{6x}{7} = \dfrac{19}{14}$	(a) $x = -2$ (b) $x = 1$
	(c) $x = \tfrac{1}{2}$ (d) $x = 7$
13. $3 + \dfrac{1}{x + 2} = 4$	(a) $x = -1$ (b) $x = -2$
	(c) $x = 0$ (d) $x = 5$
14. $\dfrac{(x + 5)(x - 3)}{2} = 24$	(a) $x = -3$ (b) $x = -2$
	(c) $x = 7$ (d) $x = 9$
15. $\sqrt{3x - 2} = 4$	(a) $x = 3$ (b) $x = 2$
	(c) $x = 9$ (d) $x = -6$
16. $\sqrt[3]{x - 8} = 3$	(a) $x = 2$ (b) $x = -5$
	(c) $x = 35$ (d) $x = 8$
17. $6x^2 - 11x - 35 = 0$	(a) $x = -\tfrac{5}{3}$ (b) $x = -\tfrac{2}{7}$
	(c) $x = \tfrac{7}{2}$ (d) $x = \tfrac{5}{3}$
18. $10x^2 + 21x - 10 = 0$	(a) $x = \tfrac{2}{5}$ (b) $x = -\tfrac{5}{2}$
	(c) $x = -\tfrac{1}{3}$ (d) $x = -2$

In Exercises 19–30, determine whether the equation is an identity or a conditional equation.

19. $2(x - 1) = 2x - 2$ **20.** $3(x + 2) = 5x + 4$

21. $-6(x - 3) + 5 = -2x + 10$

22. $3(x + 2) - 5 = 3x + 1$

23. $4(x + 1) - 2x = 2(x + 2)$

24. $-7(x - 3) + 4x = 3(7 - x)$

25. $x^2 - 8x + 5 = (x - 4)^2 - 11$

26. $x^2 + 2(3x - 2) = x^2 + 6x - 4$

27. $3 + \dfrac{1}{x + 1} = \dfrac{4x}{x + 1}$ **28.** $\dfrac{5}{x} + \dfrac{3}{x} = 24$

29. $2(x - 1) = 2x - 1$ **30.** $\tfrac{1}{4}(x - 4) = \tfrac{1}{4}x - 4$

In Exercises 31 and 32, justify each step of the solution.

31.
$$4x + 32 = 83$$
$$4x + 32 - 32 = 83 - 32$$
$$4x = 51$$
$$\frac{4x}{4} = \frac{51}{4}$$
$$x = \frac{51}{4}$$

32.
$$3(x - 4) + 10 = 7$$
$$3x - 12 + 10 = 7$$
$$3x - 2 = 7$$
$$3x - 2 + 2 = 7 + 2$$
$$3x = 9$$
$$\frac{3x}{3} = \frac{9}{3}$$
$$x = 3$$

In Exercises 33–48, solve the equation and check your solution.

33. $x + 11 = 15$ **34.** $7 - x = 19$

35. $7 - 2x = 25$ **36.** $7x + 2 = 23$

37. $3x - 5 = 2x + 7$ **38.** $5x + 3 = 6 - 2x$

39. $4y + 2 - 5y = 7 - 6y$ **40.** $5y + 1 = 8y - 5 + 6y$

41. $x - 3(2x + 3) = 8 - 5x$

42. $9x - 10 = 5x + 2(2x - 5)$

43. $\dfrac{5x}{4} + \dfrac{1}{2} = x - \dfrac{1}{2}$ **44.** $\dfrac{x}{5} - \dfrac{x}{2} = 3 + \dfrac{3x}{10}$

45. $\frac{3}{2}(z + 5) - \frac{1}{4}(z + 24) = 0$

46. $\dfrac{3x}{2} + \dfrac{1}{4}(x - 2) = 10$

47. $0.25x + 0.75(10 - x) = 3$

48. $0.60x + 0.40(100 - x) = 50$

In Exercises 49–52, solve the equation using two different methods. Then explain which method is easier.

49. $\dfrac{3x}{8} - \dfrac{4x}{3} = 4$ **50.** $\dfrac{3z}{8} - \dfrac{z}{10} = 6$

51. $\dfrac{2x}{5} + 5x = \dfrac{4}{3}$ **52.** $\dfrac{4y}{3} - 2y = \dfrac{16}{5}$

In Exercises 53–74, solve the equation and check your solution. (If not possible, explain why.)

53. $x + 8 = 2(x - 2) - x$

54. $8(x + 2) - 3(2x + 1) = 2(x + 5)$

55. $\dfrac{100 - 4x}{3} = \dfrac{5x + 6}{4} + 6$

56. $\dfrac{17 + y}{y} + \dfrac{32 + y}{y} = 100$

57. $\dfrac{5x - 4}{5x + 4} = \dfrac{2}{3}$ **58.** $\dfrac{10x + 3}{5x + 6} = \dfrac{1}{2}$

59. $10 - \dfrac{13}{x} = 4 + \dfrac{5}{x}$ **60.** $\dfrac{15}{x} - 4 = \dfrac{6}{x} + 3$

61. $3 = 2 + \dfrac{2}{z + 2}$ **62.** $\dfrac{1}{x} + \dfrac{2}{x - 5} = 0$

63. $\dfrac{x}{x + 4} + \dfrac{4}{x + 4} + 2 = 0$

64. $\dfrac{7}{2x + 1} - \dfrac{8x}{2x - 1} = -4$

65. $\dfrac{2}{(x - 4)(x - 2)} = \dfrac{1}{x - 4} + \dfrac{2}{x - 2}$

66. $\dfrac{4}{x - 1} + \dfrac{6}{3x + 1} = \dfrac{15}{3x + 1}$

67. $\dfrac{1}{x - 3} + \dfrac{1}{x + 3} = \dfrac{10}{x^2 - 9}$

68. $\dfrac{1}{x - 2} + \dfrac{3}{x + 3} = \dfrac{4}{x^2 + x - 6}$

69. $\dfrac{3}{x^2 - 3x} + \dfrac{4}{x} = \dfrac{1}{x - 3}$ **70.** $\dfrac{6}{x} - \dfrac{2}{x + 3} = \dfrac{3(x + 5)}{x^2 + 3x}$

71. $(x + 2)^2 + 5 = (x + 3)^2$

72. $4(x + 1) - 3x = x + 5$

73. $(x + 2)^2 - x^2 = 4(x + 1)$

74. $(2x - 1)^2 = 4(x^2 - x + 6)$

∿ **GRAPHICAL ANALYSIS**   In Exercises 75–80, use a graphing utility to graph the equation and approximate any x-intercepts. Set $y = 0$ and solve the resulting equation. Compare the results with the graph's x-intercept(s).

75. $y = 2(x - 1) - 4$ **76.** $y = \frac{4}{3}x + 2$

77. $y = 20 - (3x - 10)$ **78.** $y = 10 + 2(x - 2)$

79. $y = -38 + 5(9 - x)$ **80.** $y = 6x - 6\left(\frac{16}{11} + x\right)$

In Exercises 81–90, find the x- and y-intercepts of the graph of the equation algebraically.

81. $y = 12 - 5x$ **82.** $y = 16 - 3x$

83. $y = -3(2x + 1)$ **84.** $y = 5 - (6 - x)$

85. $2x + 3y = 10$ **86.** $4x - 5y = 12$

87. $\dfrac{2x}{5} + 8 - 3y = 0$ **88.** $\dfrac{8x}{3} + 5 - 2y = 0$

89. $4y - 0.75x + 1.2 = 0$ **90.** $3y + 2.5x - 3.4 = 0$

91. A student states that the solution of the equation

$$\dfrac{2}{x(x - 2)} + \dfrac{5}{x} = \dfrac{1}{x - 2}$$

is $x = 2$. Describe and correct the student's error.

92. A student states that the equation

$$-3(x + 2) = -3x + 6$$

is an identity. Describe and correct the student's error.

In Exercises 93–96, solve the equation for x. (Round your solution to three decimal places.)

93. $0.275x + 0.725(500 - x) = 300$

94. $2.763 - 4.5(2.1x - 5.1432) = 6.32x + 5$

95. $\dfrac{2}{7.398} - \dfrac{4.405}{x} = \dfrac{1}{x}$ **96.** $\dfrac{3}{6.350} - \dfrac{6}{x} = 18$

In Exercises 97–104, solve for x.

97. $4(x + 1) - ax = x + 5$

98. $4 - 2(x - 2b) = ax + 3$

99. $6x + ax = 2x + 5$ **100.** $5 + ax = 12 - bx$

101. $19x + \frac{1}{2}ax = x + 9$

102. $-5(3x - 6b) + 12 = 8 + 3ax$

103. $-2ax + 6(x + 3) = -4x + 1$

104. $\frac{4}{5}x - ax = 2\left(\frac{2}{5}x - 1\right) + 10$

105. GEOMETRY   The surface area S of the circular cylinder shown in the figure is

$$S = 2\pi(25) + 2\pi(5h).$$

Find the height h of the cylinder if the surface area is 471 square feet. Use 3.14 for π.

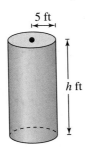

5 ft

h ft

106. GEOMETRY   The surface area S of the rectangular solid in the figure is $S = 2(24) + 2(4x) + 2(6x)$. Find the length x of the box if the surface area is 248 square centimeters.

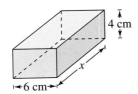

4 cm

x

6 cm

107. ANTHROPOLOGY   The relationship between the length of an adult's femur (thigh bone) and the height of the adult can be approximated by the linear equations

$$y = 0.432x - 10.44$$

$$y = 0.449x - 12.15$$

where y is the length of the femur in inches and x is the height of the adult in inches (see figure).

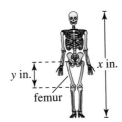

y in.

x in.

femur

(a) An anthropologist discovers a femur belonging to an adult human female. The bone is 16 inches long. Estimate the height of the female.

(b) From the foot bones of an adult human male, an anthropologist estimates that the person's height was 69 inches. A few feet away from the site where the foot bones were discovered, the anthropologist discovers a male adult femur that is 19 inches long. Is it likely that both the foot bones and the thigh bone came from the same person?

(c) Complete the table to determine if there is a height of an adult for which an anthropologist would not be able to determine whether the femur belonged to a male or a female.

Height, x	Female femur length, y	Male femur length, y
60		
70		
80		
90		
100		
110		

(d) Solve part (c) algebraically by setting the two equations equal to each other and solving for x. Compare your solutions. Do you believe an anthropologist would ever have the problem of not being able to determine whether a femur belonged to a male or a female? Why or why not?

108. TAX CREDITS   Use the following information about a possible tax credit for a family consisting of two adults and two children (see figure).

Earned income: E

Subsidy (a grant of money):

$$S = 10,000 - \tfrac{1}{2}E, \quad 0 \le E \le 20,000$$

Total income: $T = E + S$

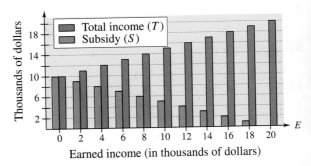

Earned income (in thousands of dollars)

(a) Write the total income T in terms of E.

(b) Find the earned income E if the subsidy is $6600.

(c) Find the earned income E if the total income is $13,800.

(d) Find the subsidy S if the total income is $12,500.

109. NEWSPAPERS The number of newspapers y in the United States from 1996 through 2007 can be approximated by the model $y = -7.69t + 1480.7$, $-4 \le t \le 7$, where t represents the year, with $t = 0$ corresponding to 2000. (Source: Editor & Publisher Co.)

(a) Sketch a graph of the model. Graphically estimate the y-intercept of the graph.

(b) Find the y-intercept of the graph algebraically.

(c) Assuming this linear pattern continues, find the year in which the number of newspapers will be 1373. Does your answer seem reasonable? Explain.

110. LABOR STATISTICS The number of women y (in millions) in the civilian work force in the United States from 2000 through 2007 (see figure) can be approximated by the model $y = 0.66t + 66.1$, $0 \le t \le 7$, where t represents the year, with $t = 0$ corresponding to 2000. (Source: U.S. Bureau of Labor Statistics)

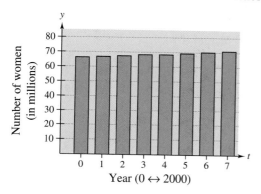

(a) According to this model, during which year did the number reach 70 million?

(b) Explain how you can solve part (a) graphically and algebraically.

111. OPERATING COST A delivery company has a fleet of vans. The annual operating cost C per van is $C = 0.32m + 2500$, where m is the number of miles traveled by a van in a year. What number of miles will yield an annual operating cost of $10,000?

112. FLOOD CONTROL A river has risen 8 feet above its flood stage. The water begins to recede at a rate of 3 inches per hour. Write a mathematical model that shows the number of feet above flood stage after t hours. If the water continually recedes at this rate, when will the river be 1 foot above its flood stage?

EXPLORATION

TRUE OR FALSE? In Exercises 113–117, determine whether the statement is true or false. Justify your answer.

113. The equation $x(3 - x) = 10$ is a linear equation.

114. The equation $x^2 + 9x - 5 = 4 - x^3$ has no real solution.

115. The equation $2(x - 3) + 1 = 2x - 5$ has no solution.

116. The equation $3(x - 1) - 2 = 3x - 6$ is an identity and therefore has all real number solutions.

117. The equation $2 - \dfrac{1}{x - 2} = \dfrac{3}{x - 2}$ has no solution because $x = 2$ is an extraneous solution.

118. THINK ABOUT IT What is meant by *equivalent equations*? Give an example of two equivalent equations.

119. THINK ABOUT IT

(a) Complete the table.

x	-1	0	1	2	3	4
$3.2x - 5.8$						

(b) Use the table in part (a) to determine the interval in which the solution of the equation $3.2x - 5.8 = 0$ is located. Explain your reasoning.

(c) Complete the table.

x	1.5	1.6	1.7	1.8	1.9	2.0
$3.2x - 5.8$						

(d) Use the table in part (c) to determine the interval in which the solution of the equation $3.2x - 5.8 = 0$ is located. Explain how this process can be used to approximate the solution to any desired degree of accuracy.

120. Use the procedure in Exercise 119 to approximate the solution of the equation $0.3(x - 1.5) - 2 = 0$, accurate to two decimal places.

121. GRAPHICAL REASONING

(a) Use a graphing utility to graph the equation $y = 3x - 6$.

(b) Use the result of part (a) to estimate the x-intercept of the graph.

(c) Explain how the x-intercept is related to the solution of the equation $3x - 6 = 0$, as shown in Example 1(a).

122. CAPSTONE

(a) Explain the difference between a conditional equation and an identity.

(b) Describe the steps used to transform an equation into an equivalent equation.

(c) What is meant by an equation having an extraneous solution?

1.3   MODELING WITH LINEAR EQUATIONS

What you should learn

- Use a verbal model in a problem-solving plan.
- Write and use mathematical models to solve real-life problems.
- Solve mixture problems.
- Use common formulas to solve real-life problems.

Why you should learn it

You can use linear equations to find the percent changes in the prices of various items or services. See Exercises 53–56 on page 104.

Introduction to Problem Solving

In this section, you will learn how algebra can be used to solve problems that occur in real-life situations. The process of translating phrases or sentences into algebraic expressions or equations is called **mathematical modeling.** A good approach to mathematical modeling is to use two stages. Begin by using the verbal description of the problem to form a *verbal model.* Then, after assigning labels to the quantities in the verbal model, form a *mathematical model* or *algebraic equation.*

$$\boxed{\text{Verbal Description}} \Rightarrow \boxed{\text{Verbal Model}} \Rightarrow \boxed{\text{Algebraic Equation}}$$

When you are constructing a verbal model, it is helpful to look for a *hidden equality.* For instance, in the following example the hidden equality equates your annual income to 24 paychecks and one bonus check.

Example 1   Using a Verbal Model

You have accepted a job for which your annual salary will be $32,300. This salary includes a year-end bonus of $500. You will be paid twice a month. What will your gross pay (pay before taxes) be for each paycheck?

Solution

Because there are 12 months in a year and you will be paid twice a month, it follows that you will receive 24 paychecks during the year.

Verbal Model: $\boxed{\text{Income for year}} = \boxed{24 \text{ paychecks}} + \boxed{\text{Bonus}}$

Labels: Income for year = 32,300   (dollars)
          Amount of each paycheck = x (dollars)
          Bonus = 500   (dollars)

Equation: $32,300 = 24x + 500$

The algebraic equation for this problem is a *linear equation* in the variable x, which you can solve as follows.

$$32,300 = 24x + 500$$
$$32,300 \qquad = 24x + 500$$
$$31,800 = 24x$$
$$\frac{31,800}{} = \frac{24x}{}$$
$$1325 = x$$

So, your gross pay for each paycheck will be $1325.

CHECK*Point* ▶ Now try Exercise 37.

A fundamental step in writing a mathematical model to represent a real-life problem is translating key words and phrases into algebraic expressions and equations. The following list gives several examples.

Translating Key Words and Phrases

Key Words and Phrases	*Verbal Description*	*Algebraic Expression or Equation*
Equality:		
Equals, equal to, is, are, was, will be, represents	• The sale price S is $10 less than the list price L.	$S = L - 10$
Addition:		
Sum, plus, greater than, increased by, more than, exceeds, total of	• The sum of 5 and x • Seven more than y	$5 + x$ or $x + 5$ $7 + y$ or $y + 7$
Subtraction:		
Difference, minus, less than, decreased by, subtracted from, reduced by, the remainder	• The difference of 4 and b • Three less than z	$4 - b$ $z - 3$
Multiplication:		
Product, multiplied by, twice, times, percent of	• Two times x • Three percent of t	$2x$ $0.03t$
Division:		
Quotient, divided by, ratio, per	• The ratio of x to 8	$\dfrac{x}{8}$

Using Mathematical Models

Example 2   Finding the Percent of a Raise

You have accepted a job that pays $10 an hour. You are told that after a two-month probationary period, your hourly wage will be increased to $11 an hour. What percent raise will you receive after the two-month period?

Solution

Verbal Model: Raise $=$ Percent $\cdot$ Old wage

Labels: Old wage $= 10$ (dollars per hour)
         New wage $= 11$ (dollars per hour)
         Raise $= 11 - 10 = 1$ (dollars per hour)
         Percent $= r$ (percent in decimal form)

Equation: $1 = r \cdot 10$

         $\frac{1}{10} = r$

         $0.1 = r$

You will receive a raise of 0.1 or 10%.

CHECK*Point*   Now try Exercise 49.

Study Tip

Writing the unit for each label in a real-life problem helps you determine the unit for the answer. This is called *unit analysis*. When the same unit of measure occurs in the numerator and denominator of an expression, you can divide out the unit. For instance, unit analysis verifies that the unit for time in the formula below is hours.

$$\text{Time} = \frac{\text{distance}}{\text{rate}}$$

$$= \frac{\text{miles}}{\dfrac{\text{miles}}{\text{hour}}}$$

$$= \text{miles} \cdot \frac{\text{hours}}{\text{miles}}$$

$$= \text{hours}$$

Example 3   Finding the Percent of Monthly Expenses

Your family has an annual income of $57,000 and the following monthly expenses: mortgage ($1100), car payment ($375), food ($295), utilities ($240), and credit cards ($220). The total value of the monthly expenses represents what percent of your family's annual income?

Solution

The total amount of your family's monthly expenses is $2230. The total monthly expenses for 1 year are $26,760.

Verbal Model: Monthly expenses = Percent · Income

Labels: Income = 57,000   (dollars)
             Monthly expenses = 26,760   (dollars)
             Percent = r (in decimal form)

Equation: $26,760 = r \cdot 57,000$

$$\frac{26,760}{57,000} = r$$

$$0.469 \approx r$$

Your family's monthly expenses are approximately 0.469 or 46.9% of your family's annual income.

CHECKPoint Now try Exercise 51.

Example 4   Finding the Dimensions of a Room

A rectangular kitchen is twice as long as it is wide, and its perimeter is 84 feet. Find the dimensions of the kitchen.

Solution

For this problem, it helps to sketch a diagram, as shown in Figure 1.16.

Verbal Model: 2 · Length + 2 · Width = Perimeter

Labels: Perimeter = 84   (feet)
             Width = w (feet)
             Length = $l = 2w$ (feet)

Equation: $2(2w) + 2w = 84$

$$6w = 84$$

$$w = 14$$

Because the length is twice the width, you have

$$l = 2w$$

$$= 2(\quad) = 28.$$

So, the dimensions of the room are 14 feet by 28 feet.

CHECKPoint Now try Exercise 57.

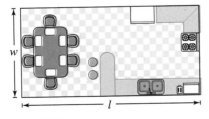

FIGURE 1.16

Example 5   A Distance Problem

A plane is flying nonstop from Atlanta to Portland, a distance of about 2700 miles, as shown in Figure 1.17. After 1.5 hours in the air, the plane flies over Kansas City (a distance of 820 miles from Atlanta). Estimate the time it will take the plane to fly from Atlanta to Portland.

Solution

Portland Kansas City Atlanta

FIGURE 1.17

Verbal Model: $\boxed{\text{Distance}} = \boxed{\text{Rate}} \cdot \boxed{\text{Time}}$

Labels: Distance = 2700   (miles)
Time = t (hours)
Rate = $\dfrac{\text{distance to Kansas City}}{\text{time to Kansas City}} = \dfrac{820}{1.5}$ (miles per hour)

Equation: $2700 = \dfrac{820}{1.5}t$

$4050 = 820t$

$\dfrac{4050}{820} = t$

$4.94 \approx t$

The trip will take about 4.94 hours, or about 4 hours and 56 minutes.

CHECK *Point*   Now try Exercise 61.

Example 6   An Application Involving Similar Triangles

To determine the height of the Aon Center Building (in Chicago), you measure the shadow cast by the building and find it to be 142 feet long, as shown in Figure 1.18. Then you measure the shadow cast by a four-foot post and find it to be 6 inches long. Estimate the building's height.

Solution

To solve this problem, you use a result from geometry that states that the ratios of corresponding sides of similar triangles are equal.

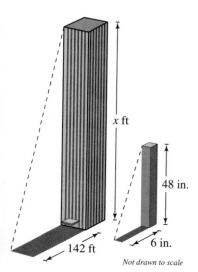

x ft

48 in.

142 ft   6 in.

Not drawn to scale

FIGURE 1.18

Verbal Model: $\dfrac{\boxed{\text{Height of building}}}{\boxed{\text{Length of building's shadow}}} = \dfrac{\boxed{\text{Height of post}}}{\boxed{\text{Length of post's shadow}}}$

Labels: Height of building = x (feet)
Length of building's shadow = 142   (feet)
Height of post = 4 feet = 48 inches   (inches)
Length of post's shadow = 6   (inches)

Equation: $\dfrac{x}{142} = \dfrac{48}{6}$

$x = 1136$

So, the Aon Center Building is about 1136 feet high.

CHECK *Point*   Now try Exercise 67.

Mixture Problems

Problems that involve two or more rates are called mixture problems. They are not limited to mixtures of chemical solutions, as shown in Examples 7 and 8.

Example 7   A Simple Interest Problem

You invested a total of $10,000 at $4\frac{1}{2}\%$ and $5\frac{1}{2}\%$ simple interest. During 1 year, the two accounts earned $508.75. How much did you invest in each account?

Solution

Verbal Model: Interest from $4\frac{1}{2}\%$ + Interest from $5\frac{1}{2}\%$ = Total interest

Labels: Amount invested at $4\frac{1}{2}\% = x$ (dollars)
Amount invested at $5\frac{1}{2}\% = 10,000 - x$ (dollars)
Interest from $4\frac{1}{2}\% = Prt = (x)(0.045)(1)$ (dollars)
Interest from $5\frac{1}{2}\% = Prt = (10,000 - x)(0.055)(1)$ (dollars)
Total interest $= 508.75$ (dollars)

Equation: $0.045x + 0.055(10,000 - x) = 508.75$

$$-0.01x = -41.25 \qquad x = 4125$$

So, $4125 was invested at $4\frac{1}{2}\%$ and $10,000 - x$ or $5875 was invested at $5\frac{1}{2}\%$.

CHECK*Point* Now try Exercise 71.

> **Study Tip**
>
> Example 7 uses the simple interest formula $I = Prt$, where I is the interest, P is the principal (original deposit), r is the annual interest rate (in decimal form), and t is the time in years. Notice that in this example the amount invested, $10,000, is separated into two parts, x and $10,000 - x$.

Example 8   An Inventory Problem

A store has $30,000 of inventory in single-disc DVD players and multi-disc DVD players. The profit on a single-disc player is 22% and the profit on a multi-disc player is 40%. The profit for the entire stock is 35%. How much was invested in each type of DVD player?

Solution

Verbal Model: Profit from single-disc players + Profit from multi-disc players = Total profit

Labels: Inventory of single-disc players $= x$ (dollars)
Inventory of multi-disc players $= 30,000 - x$ (dollars)
Profit from single-disc players $= 0.22x$ (dollars)
Profit from multi-disc players $= 0.40(30,000 - x)$ (dollars)
Total profit $= 0.35(30,000) = 10,500$ (dollars)

Equation: $0.22x + 0.40(30,000 - x) = 10,500$

$$-0.18x = -1500$$

$$x \approx 8333.33$$

So, $8333.33 was invested in single-disc DVD players and $30,000 - x$ or $21,666.67 was invested in multi-disc DVD players.

CHECK*Point* Now try Exercise 73.

Common Formulas

A **literal equation** is an equation that contains more than one variable. A *formula* is an example of a literal equation. Many common types of geometric, scientific, and investment problems use ready-made equations called **formulas.** Knowing these formulas will help you translate and solve a wide variety of real-life applications.

Common Formulas for Area A, Perimeter P, Circumference C, and Volume V

Square	Rectangle	Circle	Triangle
$A = s^2$	$A = lw$	$A = \pi r^2$	$A = \dfrac{1}{2}bh$
$P = 4s$	$P = 2l + 2w$	$C = 2\pi r$	$P = a + b + c$

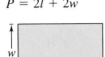

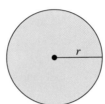

Cube	Rectangular Solid	Circular Cylinder	Sphere
$V = s^3$	$V = lwh$	$V = \pi r^2 h$	$V = \dfrac{4}{3}\pi r^3$

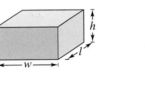

Miscellaneous Common Formulas

Temperature:

$$F = \frac{9}{5}C + 32 \qquad F = \text{degrees Fahrenheit}, \; C = \text{degrees Celsius}$$

$$C = \frac{5}{9}(F - 32)$$

Simple Interest:

$$I = Prt \qquad I = \text{interest}, \; P = \text{principal (original deposit)},$$
$$r = \text{annual interest rate (in decimal form)}, \; t = \text{time in years}$$

Compound Interest:

$$A = P\left(1 + \frac{r}{n}\right)^{nt} \qquad \begin{array}{l} n = \text{compoundings (number of times interest is calculated) per year}, \; t = \text{time in years}, \\ A = \text{balance}, \; P = \text{principal (original deposit)}, \; r = \text{annual interest rate (in decimal form)} \end{array}$$

Distance:

$$d = rt \qquad d = \text{distance traveled}, \; r = \text{rate}, \; t = \text{time}$$

When working with applied problems, you often need to rewrite a literal equation in terms of another variable. You can use the methods for solving linear equations to solve some literal equations for a specified variable. For instance, the formula for the perimeter of a rectangle, $P = 2l + 2w$, can be rewritten or solved for w as $w = \frac{1}{2}(P - 2l)$.

Example 9   Using a Formula

A cylindrical can has a volume of 200 cubic centimeters (cm^3) and a radius of 4 centimeters (cm), as shown in Figure 1.19. Find the height of the can.

Solution

The formula for the *volume of a cylinder* is $V = \pi r^2 h$. To find the height of the can, solve for h.

$$h = \frac{V}{\pi r^2}$$

Then, using $V = 200$ and $r = 4$, find the height.

$$h = \frac{}{\pi(\ )^2}$$

$$= \frac{200}{16\pi}$$

$$\approx 3.98$$

You can use unit analysis to check that your answer is reasonable.

$$\frac{200 \text{ cm}^3}{16\pi \text{ cm}^2} \approx 3.98 \text{ cm}$$

CHECKPoint ▶ Now try Exercise 95.

⊢—4 cm—⊣

Cat Food

h

FIGURE 1.19

CLASSROOM DISCUSSION

Translating Algebraic Formulas   Most people use algebraic formulas every day—sometimes without realizing it because they use a verbal form or think of an often-repeated calculation in steps. Translate each of the following verbal descriptions into an algebraic formula, and demonstrate the use of each formula.

a. *Designing Billboards*   "The letters on a sign or billboard are designed to be readable at a certain distance. Take half the letter height in inches and multiply by 100 to find the readable distance in feet."—Thos. Hodgson, Hodgson Signs (Source: *Rules of Thumb* by Tom Parker)

b. *Percent of Calories from Fat*   "To calculate percent of calories from fat, multiply grams of total fat per serving by 9, divide by the number of calories per serving," and then multiply by 100.   (Source: *Good Housekeeping*)

c. *Building Stairs*   "A set of steps will be comfortable to use if two times the height of one riser plus the width of one tread is equal to 26 inches." —Alice Lukens Bachelder, gardener   (Source: *Rules of Thumb* by Tom Parker)

1.3   EXERCISES

See www.CalcChat.com for worked-out solutions to odd-numbered exercises.

VOCABULARY

In Exercises 1 and 2, fill in the blanks.

1. The process of translating phrases or sentences into algebraic expressions or equations is called _____ _____.

2. A good approach to mathematical modeling is a two-stage approach, using a verbal description to form a _____ _____, and then, after assigning labels to the quantities, forming an _____ _____.

In Exercises 3–8, write the formula for the given quantity.

3. Area of a circle: _____

4. Perimeter of a rectangle: _____

5. Volume of a cube: _____

6. Volume of a circular cylinder: _____

7. Balance if P dollars is invested at $r\%$ compounded monthly for t years: _____

8. Simple interest if P dollars is invested at $r\%$ for t years: _____

SKILLS AND APPLICATIONS

In Exercises 9–18, write a verbal description of the algebraic expression without using the variable.

9. $x + 4$

10. $t - 10$

11. $\dfrac{u}{5}$

12. $\dfrac{2}{3}x$

13. $\dfrac{y - 4}{5}$

14. $\dfrac{z + 10}{7}$

15. $-3(b + 2)$

16. $12x(x - 5)$

17. $\dfrac{4(p - 1)}{p}$

18. $\dfrac{(q + 4)(3 - q)}{2q}$

In Exercises 19–30, write an algebraic expression for the verbal description.

19. The sum of two consecutive natural numbers

20. The product of two consecutive natural numbers

21. The product of two consecutive odd integers, the first of which is $2n - 1$

22. The sum of the squares of two consecutive even integers, the first of which is $2n$

23. The distance traveled in t hours by a car traveling at 55 miles per hour

24. The travel time for a plane traveling at a rate of r kilometers per hour for 900 kilometers

25. The amount of acid in x liters of a 20% acid solution

26. The sale price of an item that is discounted 33% of its list price L

27. The perimeter of a rectangle with a width x and a length that is twice the width

28. The area of a triangle with base 16 inches and height h inches

29. The total cost of producing x units for which the fixed costs are \$2500 and the cost per unit is \$40

30. The total revenue obtained by selling x units at \$12.99 per unit

In Exercises 31–34, translate the statement into an algebraic expression or equation.

31. Thirty percent of the list price L

32. The amount of water in q quarts of a liquid that is 28% water

33. The percent of 672 that is represented by the number N

34. The percent change in sales from one month to the next if the monthly sales are S_1 and S_2, respectively

In Exercises 35 and 36, write an expression for the area of the region in the figure.

35.

36.

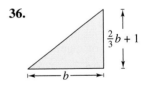

NUMBER PROBLEMS   In Exercises 37–42, write a mathematical model for the problem and solve.

37. The sum of two consecutive natural numbers is 525. Find the numbers.

38. The sum of three consecutive natural numbers is 804. Find the numbers.

39. One positive number is 5 times another number. The difference between the two numbers is 148. Find the numbers.

40. One positive number is $\frac{1}{5}$ of another number. The difference between the two numbers is 76. Find the numbers.

41. Find two consecutive integers whose product is 5 less than the square of the smaller number.

42. Find two consecutive natural numbers such that the difference of their reciprocals is $\frac{1}{4}$ the reciprocal of the smaller number.

In Exercises 43–48, solve the percent equation.

43. What is 30% of 45?   **44.** What is 175% of 360?

45. 432 is what percent of 1600?

46. 459 is what percent of 340?

47. 12 is $\frac{1}{2}$% of what number?

48. 70 is 40% of what number?

49. FINANCE A salesperson's weekly paycheck is 15% less than a second salesperson's paycheck. The two paychecks total $1125. Find the amount of each paycheck.

50. DISCOUNT The price of a swimming pool has been discounted 16.5%. The sale price is $1210.75. Find the original list price of the pool.

51. FINANCE A family has annual loan payments equaling 32% of their annual income. During the year, their loan payments total $15,125.50. What is their annual income?

52. FINANCE A family has a monthly mortgage payment of $500, which is 16% of their monthly income. What is their monthly income?

In Exercises 53–56, the prices of various items are given for 2000 and 2007. Find the percent change for each item. (Sources: U.S. Energy Information Association, SNL Kagan, U.S. Bureau of Labor Statistics, CTIA-The Wireless Association)

Item	2000	2007
53. Gallon of regular unleaded gasoline	$1.51	$2.80
54. Monthly cable rate	$30.37	$42.72
55. Pound of 100% ground beef	$1.63	$2.23
56. Monthly bill for cellular phone service	$45.27	$49.79

57. DIMENSIONS OF A ROOM A room is 1.5 times as long as it is wide, and its perimeter is 25 meters.

(a) Draw a diagram that represents the problem. Identify the length as l and the width as w.

(b) Write l in terms of w and write an equation for the perimeter in terms of w.

(c) Find the dimensions of the room.

58. DIMENSIONS OF A PICTURE FRAME A picture frame has a total perimeter of 3 meters. The height of the frame is $\frac{2}{3}$ times its width.

(a) Draw a diagram that represents the problem. Identify the width as w and the height as h.

(b) Write h in terms of w and write an equation for the perimeter in terms of w.

(c) Find the dimensions of the picture frame.

59. COURSE GRADE To get an A in a course, you must have an average of at least 90 on four tests of 100 points each. The scores on your first three tests were 87, 92, and 84. What must you score on the fourth test to get an A for the course?

60. COURSE GRADE You are taking a course that has four tests. The first three tests are 100 points each and the fourth test is 200 points. To get an A in the course, you must have an average of at least 90% on the four tests. Your scores on the first three tests were 87, 92, and 84. What must you score on the fourth test to get an A for the course?

61. TRAVEL TIME You are driving on a Canadian freeway to a town that is 500 kilometers from your home. After 30 minutes you pass a freeway exit that you know is 50 kilometers from your home. Assuming that you continue at the same constant speed, how long will it take for the entire trip?

62. TRAVEL TIME Students are traveling in two cars to a football game 135 miles away. The first car leaves on time and travels at an average speed of 45 miles per hour. The second car starts $\frac{1}{2}$ hour later and travels at an average speed of 55 miles per hour. How long will it take the second car to catch up to the first car? Will the second car catch up to the first car before the first car arrives at the game?

63. AVERAGE SPEED A truck driver traveled at an average speed of 55 miles per hour on a 200-mile trip to pick up a load of freight. On the return trip (with the truck fully loaded), the average speed was 40 miles per hour. What was the average speed for the round trip?

64. WIND SPEED An executive flew in the corporate jet to a meeting in a city 1500 kilometers away. After traveling the same amount of time on the return flight, the pilot mentioned that they still had 300 kilometers to go. The air speed of the plane was 600 kilometers per hour. How fast was the wind blowing? (Assume that the wind direction was parallel to the flight path and constant all day.)

65. PHYSICS Light travels at the speed of approximately 3.0×10^8 meters per second. Find the time in minutes required for light to travel from the sun to Earth (an approximate distance of 1.5×10^{11} meters).

66. RADIO WAVES Radio waves travel at the same speed as light, approximately 3.0×10^8 meters per second. Find the time required for a radio wave to travel from Mission Control in Houston to NASA astronauts on the surface of the moon 3.84×10^8 meters away.

67. HEIGHT OF A BUILDING To obtain the height of the Chrysler Building in New York, you measure the building's shadow and find that it is 87 feet long. You also measure the shadow of a four-foot stake and find that it is 4 inches long. How tall is the Chrysler Building?

68. HEIGHT OF A TREE To obtain the height of a tree (see figure), you measure the tree's shadow and find that it is 8 meters long. You also measure the shadow of a two-meter lamppost and find that it is 75 centimeters long. How tall is the tree?

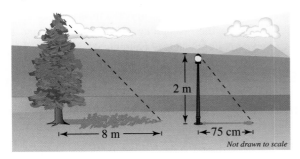

2 m

⊢— 8 m —⊣   ⊢75 cm⊣

Not drawn to scale

69. FLAGPOLE HEIGHT A person who is 6 feet tall walks away from a flagpole toward the tip of the shadow of the flagpole. When the person is 30 feet from the flagpole, the tips of the person's shadow and the shadow cast by the flagpole coincide at a point 5 feet in front of the person.

   (a) Draw a diagram that gives a visual representation of the problem. Let h represent the height of the flagpole.

   (b) Find the height of the flagpole.

70. SHADOW LENGTH A person who is 6 feet tall walks away from a 50-foot tower toward the tip of the tower's shadow. At a distance of 32 feet from the tower, the person's shadow begins to emerge beyond the tower's shadow. How much farther must the person walk to be completely out of the tower's shadow?

71. INVESTMENT You plan to invest $12,000 in two funds paying $4\frac{1}{2}\%$ and 5% simple interest. (There is more risk in the 5% fund.) Your goal is to obtain a total annual interest income of $580 from the investments. What is the smallest amount you can invest in the 5% fund and still meet your objective?

72. INVESTMENT You plan to invest $25,000 in two funds paying 3% and $4\frac{1}{2}\%$ simple interest. (There is more risk in the $4\frac{1}{2}\%$ fund.) Your goal is to obtain a total annual interest income of $1000 from the investments. What is the smallest amount you can invest in the $4\frac{1}{2}\%$ fund and still meet your objective?

73. INVENTORY A nursery has $40,000 of inventory in dogwood trees and red maple trees. The profit on a dogwood tree is 25% and the profit on a red maple tree is 17%. The profit for the entire stock is 20%. How much was invested in each type of tree?

74. INVENTORY An automobile dealer has $600,000 of inventory in minivans and alternative-fueled vehicles. The profit on a minivan is 24% and the profit on an alternative-fueled vehicle is 28%. The profit for the entire stock is 25%. How much was invested in each type of vehicle?

75. MIXTURE PROBLEM Using the values in the table, determine the amounts of solutions 1 and 2 needed to obtain the specified amount and concentration of the final mixture.

		Concentration		Amount of final solution
	Solution 1	Solution 2	Final solution	
(a)	10%	30%	25%	100 gal
(b)	25%	50%	30%	5 L
(c)	15%	45%	30%	10 qt
(d)	70%	90%	75%	25 gal

76. MIXTURE PROBLEM A 100% concentrate is to be mixed with a mixture having a concentration of 40% to obtain 55 gallons of a mixture with a concentration of 75%. How much of the 100% concentrate will be needed?

77. MIXTURE PROBLEM A forester mixes gasoline and oil to make 2 gallons of mixture for his two-cycle chainsaw engine. This mixture is 32 parts gasoline and 1 part two-cycle oil. How much gasoline must be added to bring the mixture to 40 parts gasoline and 1 part oil?

78. MIXTURE PROBLEM A grocer mixes peanuts that cost $1.49 per pound and walnuts that cost $2.69 per pound to make 100 pounds of a mixture that costs $2.21 per pound. How much of each kind of nut is put into the mixture?

79. COMPANY COSTS An outdoor furniture manufacturer has fixed costs of $14,000 per month and average variable costs of $12.75 per unit manufactured. The company has $110,000 available to cover the monthly costs. How many units can the company manufacture? (*Fixed costs* are those that occur regardless of the level of production. *Variable costs* depend on the level of production.)

80. COMPANY COSTS A plumbing supply company has fixed costs of $10,000 per month and average variable costs of $9.30 per unit manufactured. The company has $85,000 available to cover the monthly costs. How many units can the company manufacture? (*Fixed costs* are those that occur regardless of the level of production. *Variable costs* depend on the level of production.)

In Exercises 81–92, solve for the indicated variable.

81. AREA OF A TRIANGLE
Solve for h: $A = \frac{1}{2} bh$

82. AREA OF A TRAPEZOID
Solve for b: $A = \frac{1}{2}(a + b)h$

83. MARKUP   Solve for C: $S = C + RC$

84. INVESTMENT AT SIMPLE INTEREST
Solve for r: $A = P + Prt$

85. VOLUME OF AN OBLATE SPHEROID
Solve for b: $V = \frac{4}{3}\pi a^2 b$

86. VOLUME OF A SPHERICAL SEGMENT
Solve for r: $V = \frac{1}{3}\pi h^2(3r - h)$

87. FREE-FALLING BODY   Solve for a: $h = v_0 t + \frac{1}{2}at^2$

88. LENSMAKER'S EQUATION
Solve for R_1: $\dfrac{1}{f} = (n - 1)\left(\dfrac{1}{R_1} - \dfrac{1}{R_2}\right)$

89. CAPACITANCE IN SERIES CIRCUITS
Solve for C_1: $C = \dfrac{1}{\dfrac{1}{C_1} + \dfrac{1}{C_2}}$

90. ARITHMETIC PROGRESSION
Solve for a: $S = \dfrac{n}{2}[2a + (n - 1)d]$

91. ARITHMETIC PROGRESSION
Solve for n: $L = a + (n - 1)d$

92. GEOMETRIC PROGRESSION   Solve for r: $S = \dfrac{rL - a}{r - 1}$

PHYSICS   In Exercises 93 and 94, you have a uniform beam of length L with a fulcrum x feet from one end (see figure). Objects with weights W_1 and W_2 are placed at opposite ends of the beam. The beam will balance when $W_1 x = W_2(L - x)$. Find x such that the beam will balance.

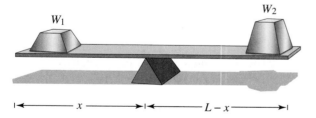

93. Two children weighing 50 pounds and 75 pounds are playing on a seesaw that is 10 feet long.

94. A person weighing 200 pounds is attempting to move a 550-pound rock with a bar that is 5 feet long.

95. VOLUME OF A BILLIARD BALL   A billiard ball has a volume of 5.96 cubic inches. Find the radius of a billiard ball.

96. LENGTH OF A TANK   The diameter of a cylindrical propane gas tank is 4 feet. The total volume of the tank is 603.2 cubic feet. Find the length of the tank.

97. TEMPERATURE   The average daily temperature in San Diego, California is 64.4°F. What is San Diego's average daily temperature in degrees Celsius?   (Source: NOAA)

98. TEMPERATURE   The average daily temperature in Duluth, Minnesota is 39.1°F. What is Duluth's average daily temperature in degrees Celsius?   (Source: NOAA)

99. TEMPERATURE   The highest temperature ever recorded in Phoenix, Arizona was 50°C. What is this temperature in degrees Fahrenheit?   (Source: NOAA)

100. TEMPERATURE   The lowest temperature ever recorded in Louisville, Kentucky was −30°C. What is this temperature in degrees Fahrenheit?   (Source: NOAA)

EXPLORATION

TRUE OR FALSE?   In Exercises 101 and 102, determine whether the statement is true or false. Justify your answer.

101. "8 less than z cubed divided by the difference of z squared and 9" can be written as $(z^3 - 8)/(z - 9)^2$.

102. The volume of a cube with a side of length 9.5 inches is greater than the volume of a sphere with a radius of 5.9 inches.

103. Consider the linear equation $ax + b = 0$.
(a) What is the sign of the solution if $ab > 0$?
(b) What is the sign of the solution if $ab < 0$?
In each case, explain your reasoning.

104. CAPSTONE   Arrange the following statements in the proper order to obtain a strategy for modeling and solving a real-life problem.

- Assign labels to each part of the verbal model—numbers to the known quantities and letters (or expressions) to the variable quantities.
- Answer the original question and check that your answer satisfies the original problem as stated.
- Solve the algebraic equation.
- Ask yourself what you need to know to solve the problem and then write a verbal model that includes arithmetic operations to describe the problem.
- Write an algebraic equation based on the verbal model.

105. Write a linear equation that has the solution $x = -3$. (There are many correct answers.)

What you should learn

- Solve quadratic equations by factoring.
- Solve quadratic equations by extracting square roots.
- Solve quadratic equations by completing the square.
- Use the Quadratic Formula to solve quadratic equations.
- Use quadratic equations to model and solve real-life problems.

Why you should learn it

Quadratic equations can be used to model and solve real-life problems. For instance, in Exercise 123 on page 119, you will use a quadratic equation to model average admission prices for movie theaters from 2001 through 2008.

1.4 QUADRATIC EQUATIONS AND APPLICATIONS

Factoring

A **quadratic equation** in x is an equation that can be written in the general form

$$ax^2 + bx + c = 0$$

where a, b, and c are real numbers with $a \neq 0$. A quadratic equation in x is also called a **second-degree polynomial equation** in x.

In this section, you will study four methods for solving quadratic equations: *factoring*, *extracting square roots*, *completing the square*, and the *Quadratic Formula*. The first method is based on the Zero-Factor Property from Section P.1.

If $ab = 0$, then $a = 0$ or $b = 0$.

To use this property, write the left side of the general form of a quadratic equation as the product of two linear factors. Then find the solutions of the quadratic equation by setting each linear factor equal to zero.

Example 1   Solving a Quadratic Equation by Factoring

a.
$$2x^2 + 9x + 7 = 3$$
$$2x^2 + 9x + 4 = 0$$
$$(2x + 1)(x + 4) = 0$$

$$2x + 1 = 0 \implies x = -\frac{1}{2}$$

$$x + 4 = 0 \implies x = -4$$

The solutions are $x = -\frac{1}{2}$ and $x = -4$. Check these in the original equation.

b.
$$6x^2 - 3x = 0$$
$$3x(2x - 1) = 0$$

$$3x = 0 \implies x = 0$$

$$2x - 1 = 0 \implies x = \frac{1}{2}$$

The solutions are $x = 0$ and $x = \frac{1}{2}$. Check these in the original equation.

CHECK*Point* Now try Exercise 15.

Be sure you see that the Zero-Factor Property works *only* for equations written in general form (in which the right side of the equation is zero). So, all terms must be collected on one side *before* factoring. For instance, in the equation $(x - 5)(x + 2) = 8$, it is *incorrect* to set each factor equal to 8. To solve this equation, you must multiply the binomials on the left side of the equation, and then subtract 8 from each side. After simplifying the left side of the equation, you can use the Zero-Factor Property to solve the equation. Try to solve this equation correctly.

TECHNOLOGY

You can use a graphing utility to check graphically the real solutions of a quadratic equation. Begin by writing the equation in general form. Then set y equal to the left side and graph the resulting equation. The x-intercepts of the equation represent the real solutions of the original equation. You can use the *zero* or *root* feature of a graphing utility to approximate the x-intercepts of the graph. For example, to check the solutions of $6x^2 - 3x = 0$, graph $y = 6x^2 - 3x$, and use the *zero* or *root* feature to approximate the x-intercepts to be $(0, 0)$ and $\left(\frac{1}{2}, 0\right)$, as shown below. These x-intercepts represent the solutions $x = 0$ and $x = \frac{1}{2}$, as found in Example 1(b).

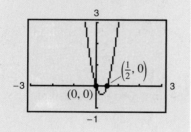

Extracting Square Roots

There is a nice shortcut for solving quadratic equations of the form $u^2 = d$, where $d > 0$ and u is an algebraic expression. By factoring, you can see that this equation has two solutions.

$$u^2 = d$$
$$u^2 - d = 0$$
$$\left(u + \sqrt{d}\right)\left(u - \sqrt{d}\right) = 0$$
$$u + \sqrt{d} = 0 \quad \Longrightarrow \quad u = -\sqrt{d}$$
$$u - \sqrt{d} = 0 \quad \Longrightarrow \quad u = \sqrt{d}$$

Because the two solutions differ only in sign, you can write the solutions together, using a "plus or minus sign," as

$$u = \pm\sqrt{d}.$$

This form of the solution is read as "u is equal to plus or minus the square root of d." Solving an equation of the form $u^2 = d$ without going through the steps of factoring is called **extracting square roots**.

Extracting Square Roots

The equation $u^2 = d$, where $d > 0$, has exactly two solutions:

$$u = \sqrt{d} \qquad \text{and} \qquad u = -\sqrt{d}.$$

These solutions can also be written as

$$u = \pm\sqrt{d}.$$

| Example 2   Extracting Square Roots

Solve each equation by extracting square roots.

a. $4x^2 = 12$ **b.** $(x - 3)^2 = 7$

Solution

a. $4x^2 = 12$

$$x^2 = 3$$
$$x = \pm\sqrt{3}$$

When you take the square root of a variable expression, you must account for both positive and negative solutions. So, the solutions are $x = \sqrt{3}$ and $x = -\sqrt{3}$. Check these in the original equation.

b. $(x - 3)^2 = 7$

$$x - 3 = \pm\sqrt{7}$$
$$x = 3 \pm \sqrt{7}$$

The solutions are $x = 3 \pm \sqrt{7}$. Check these in the original equation.

CHECK*Point*   Now try Exercise 33.

Completing the Square

The equation in Example 2(b) was given in the form $(x - 3)^2 = 7$ so that you could find the solution by extracting square roots. Suppose, however, that the equation had been given in the general form $x^2 - 6x + 2 = 0$. Because this equation is equivalent to the original, it has the same two solutions, $x = 3 \pm \sqrt{7}$. However, the left side of the equation is not factorable, and you cannot find its solutions unless you rewrite the equation by **completing the square.** Note that when you complete the square to solve a quadratic equation, you are just rewriting the equation so it can be solved by extracting square roots.

Completing the Square

To **complete the square** for the expression $x^2 + bx$, add $(b/2)^2$, which is the square of half the coefficient of x. Consequently,

$$x^2 + bx + \left(\frac{b}{2}\right)^2 = \left(x + \frac{b}{2}\right)^2.$$

Example 3   Completing the Square: Leading Coefficient Is 1

Solve $x^2 + 2x - 6 = 0$ by completing the square.

Solution

$$x^2 + 2x - 6 = 0$$
$$x^2 + 2x = 6$$
$$x^2 + 2x + \quad = 6 +$$

$$(x + 1)^2 = 7$$
$$x + 1 = \pm\sqrt{7}$$
$$x = -1 \pm \sqrt{7}$$

The solutions are $x = -1 \pm \sqrt{7}$. Check these in the original equation as follows.

Check

$$x^2 + 2x - 6 = 0$$
$$(\qquad)^2 + 2(\qquad) - 6 \stackrel{?}{=} 0$$
$$8 - 2\sqrt{7} - 2 + 2\sqrt{7} - 6 \stackrel{?}{=} 0$$
$$8 - 2 - 6 = 0 \qquad\qquad \checkmark$$

Check the second solution in the original equation.

CHECK*Point*▶   Now try Exercise 41.   ▌

When solving quadratic equations by completing the square, you must add $(b/2)^2$ to *each side* in order to maintain equality. If the leading coefficient is *not* 1, you must divide each side of the equation by the leading coefficient *before* completing the square, as shown in Example 4.

Example 4 Completing the Square: Leading Coefficient Is Not 1

Solve $2x^2 + 8x + 3 = 0$ by completing the square.

Solution

$$2x^2 + 8x + 3 = 0$$

$$2x^2 + 8x = -3$$

$$x^2 + 4x = -\frac{3}{2}$$

$$x^2 + 4x + \quad = -\frac{3}{2} +$$

$$(x + 2)^2 = \frac{5}{2}$$

$$x + 2 = \pm\sqrt{\frac{5}{2}}$$

$$x + 2 = \pm\frac{\sqrt{10}}{2}$$

$$x = -2 \pm \frac{\sqrt{10}}{2}$$

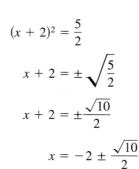

> **Algebra Help**
>
> You can review rationalizing denominators in Section P.2.

The solutions are $x = -2 \pm \dfrac{\sqrt{10}}{2}$. Check these in the original equation.

CHECK*Point* Now try Exercise 43.

Example 5 Completing the Square: Leading Coefficient Is Not 1

$$3x^2 - 4x - 5 = 0$$

$$3x^2 - 4x = 5$$

$$x^2 - \frac{4}{3}x = \frac{5}{3}$$

$$x^2 - \frac{4}{3}x + \quad = \frac{5}{3} +$$

$$x^2 - \frac{4}{3}x + \frac{4}{9} = \frac{19}{9}$$

$$\left(x - \frac{2}{3}\right)^2 = \frac{19}{9}$$

$$x - \frac{2}{3} = \pm\frac{\sqrt{19}}{3}$$

$$x = \frac{2}{3} \pm \frac{\sqrt{19}}{3}$$

CHECK*Point* Now try Exercise 47.

The Quadratic Formula

Often in mathematics you are taught the long way of solving a problem first. Then, the longer method is used to develop shorter techniques. The long way stresses understanding and the short way stresses efficiency.

For instance, you can think of completing the square as a "long way" of solving a quadratic equation. When you use completing the square to solve quadratic equations, you must complete the square for *each* equation separately. In the following derivation, you complete the square *once* in a general setting to obtain the **Quadratic Formula**— a shortcut for solving quadratic equations.

$$ax^2 + bx + c = 0$$

$$ax^2 + bx = -c$$

$$x^2 + \frac{b}{a}x = -\frac{c}{a}$$

$$x^2 + \frac{b}{a}x + \quad = -\frac{c}{a} +$$

$$\left(x + \frac{b}{2a}\right)^2 = \frac{b^2 - 4ac}{4a^2}$$

$$x + \frac{b}{2a} = \pm\sqrt{\frac{b^2 - 4ac}{4a^2}}$$

$$x = -\frac{b}{2a} \pm \frac{\sqrt{b^2 - 4ac}}{2|a|}$$

Note that because $\pm 2|a|$ represents the same numbers as $\pm 2a$, you can omit the absolute value sign. So, the formula simplifies to

$$x = \frac{-b \pm \sqrt{b^2 - 4ac}}{2a}.$$

Study Tip

You can solve every quadratic equation by completing the square or using the Quadratic Formula.

The Quadratic Formula

The solutions of a quadratic equation in the general form

$$ax^2 + bx + c = 0, \qquad a \neq 0$$

are given by the **Quadratic Formula**

$$x = \frac{-b \pm \sqrt{b^2 - 4ac}}{2a}.$$

The Quadratic Formula is one of the most important formulas in algebra. You should learn the verbal statement of the Quadratic Formula:

"Negative *b*, plus or minus the square root of *b* squared minus 4*ac*, all divided by 2*a*."

In the Quadratic Formula, the quantity under the radical sign, $b^2 - 4ac$, is called the **discriminant** of the quadratic expression $ax^2 + bx + c$. It can be used to determine the nature of the solutions of a quadratic equation.

Solutions of a Quadratic Equation

The solutions of a quadratic equation $ax^2 + bx + c = 0$, $a \neq 0$, can be classified as follows. If the discriminant $b^2 - 4ac$ is

1. *positive*, then the quadratic equation has *two* distinct real solutions and its graph has *two* x-intercepts.

2. *zero*, then the quadratic equation has *one* repeated real solution and its graph has *one* x-intercept.

3. *negative*, then the quadratic equation has *no* real solutions and its graph has *no* x-intercepts.

If the discriminant of a quadratic equation is negative, as in case 3 above, then its square root is imaginary (not a real number) and the Quadratic Formula yields two complex solutions. You will study complex solutions in Section 1.5.

When using the Quadratic Formula, remember that *before* the formula can be applied, you must first write the quadratic equation in general form.

Example 6   The Quadratic Formula: Two Distinct Solutions

Use the Quadratic Formula to solve $x^2 + 3x = 9$.

Solution

The general form of the equation is $x^2 + 3x - 9 = 0$. The discriminant is $b^2 - 4ac = 9 + 36 = 45$, which is positive. So, the equation has two real solutions. You can solve the equation as follows.

$$x^2 + 3x - 9 = 0$$

$$x = \frac{-b \pm \sqrt{b^2 - 4ac}}{2a}$$

$$x = \frac{-\quad \pm \sqrt{(\quad)^2 - 4(\quad)(\quad)}}{2(\quad)}$$

$$x = \frac{-3 \pm \sqrt{45}}{2}$$

$$x = \frac{-3 \pm 3\sqrt{5}}{2}$$

The two solutions are:

$$x = \frac{-3 + 3\sqrt{5}}{2} \qquad \text{and} \qquad x = \frac{-3 - 3\sqrt{5}}{2}.$$

Check these in the original equation.

CHECK*Point* ▶ Now try Exercise 81.

Applications

Quadratic equations often occur in problems dealing with area. Here is a simple example. "A square room has an area of 144 square feet. Find the dimensions of the room." To solve this problem, let x represent the length of each side of the room. Then, by solving the equation

$$x^2 = 144$$

you can conclude that each side of the room is 12 feet long. Note that although the equation $x^2 = 144$ has two solutions, $x = -12$ and $x = 12$, the negative solution does not make sense in the context of the problem, so you choose the positive solution.

| **Example 7**   **Finding the Dimensions of a Room**

A bedroom is 3 feet longer than it is wide (see Figure 1.20) and has an area of 154 square feet. Find the dimensions of the room.

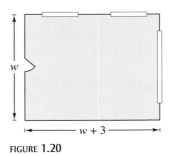

FIGURE **1.20**

Solution

Verbal Model: $\boxed{\text{Width of room}} \cdot \boxed{\text{Length of room}} = \boxed{\text{Area of room}}$

Labels: Width of room $= w$ (feet)
Length of room $= w + 3$ (feet)
Area of room $= 154$ (square feet)

Equation: $w(w + 3) = 154$

$$w^2 + 3w - 154 = 0$$

$$(w - 11)(w + 14) = 0$$

$$w - 11 = 0 \quad \Longrightarrow \quad w = 11$$

$$w + 14 = 0 \quad \Longrightarrow \quad w = -14$$

Choosing the positive value, you find that the width is 11 feet and the length is $w + 3$, or 14 feet. You can check this solution by observing that the length is 3 feet longer than the width *and* that the product of the length and width is 154 square feet.

CHECK*Point*   Now try Exercise 113.

Another common application of quadratic equations involves an object that is falling (or projected into the air). The general equation that gives the height of such an object is called a **position equation,** and on Earth's surface it has the form

$$s = -16t^2 + v_0 t + s_0.$$

In this equation, s represents the height of the object (in feet), v_0 represents the initial velocity of the object (in feet per second), s_0 represents the initial height of the object (in feet), and t represents the time (in seconds).

Example 8 Falling Time

A construction worker on the 24th floor of a building project (see Figure 1.21) accidentally drops a wrench and yells "Look out below!" Could a person at ground level hear this warning in time to get out of the way? (*Note:* The speed of sound is about 1100 feet per second.)

Solution

Assume that each floor of the building is 10 feet high, so that the wrench is dropped from a height of 235 feet (the construction worker's hand is 5 feet below the ceiling of the 24th floor). Because sound travels at about 1100 feet per second, it follows that a person at ground level hears the warning within 1 second of the time the wrench is dropped. To set up a mathematical model for the height of the wrench, use the position equation

$$s = -16t^2 + v_0 t + s_0.$$

Because the object is dropped rather than thrown, the initial velocity is $v_0 = 0$ feet per second. Moreover, because the initial height is $s_0 = 235$ feet, you have the following model.

$$s = -16t^2 + (0)t + 235 = -16t^2 + 235$$

After the wrench has fallen for 1 second, its height is $-16(\ )^2 + 235 = 219$ feet. After the wrench has fallen for 2 seconds, its height is $-16(\ )^2 + 235 = 171$ feet. To find the number of seconds it takes the wrench to hit the ground, let the height s be zero and solve the equation for t.

$$s = -16t^2 + 235$$
$$= -16t^2 + 235$$
$$16t^2 = 235$$
$$t^2 = \frac{235}{16}$$
$$t = \frac{\sqrt{235}}{4}$$
$$t \approx 3.83$$

The wrench will take about 3.83 seconds to hit the ground. If the person hears the warning 1 second after the wrench is dropped, the person still has almost 3 seconds to get out of the way.

CHECK**Point** Now try Exercise 119.

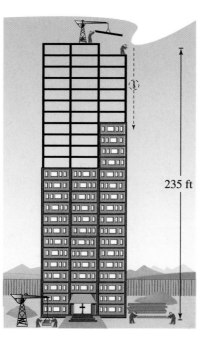

FIGURE **1.21**

235 ft

A third type of application of a quadratic equation is one in which a quantity is changing over time t according to a quadratic model.

Example 9   Quadratic Modeling: Internet Users

From 2000 through 2008, the estimated numbers of Internet users I (in millions) in the United States can be modeled by the quadratic equation

$$I = -1.446t^2 + 23.45t + 122.9, \quad 0 \le t \le 8$$

where t represents the year, with $t = 0$ corresponding to 2000. According to this model, in which year did the number of Internet users reach or surpass 200 million?   (Source: *International Telecommunication Union/The Nielsen Company*)

Algebraic Solution

To find the year in which the number of Internet users reached 200 million, you can solve the equation

$$-1.446t^2 + 23.45t + 122.9 = 200.$$

To begin, write the equation in general form.

$$-1.446t^2 + 23.45t - 77.1 = 0$$

Then apply the Quadratic Formula.

$$t = \frac{-b \pm \sqrt{b^2 - 4ac}}{2a}$$

$$t = \frac{-\quad \pm \sqrt{\quad^2 - 4(\quad)(\quad)}}{2(\quad)}$$

$$\approx \frac{-23.45 \pm \sqrt{103.96}}{-2.892}$$

$$\approx 4.6 \text{ or } 11.6$$

Choose the smaller value $t \approx 4.6$. Because $t = 0$ corresponds to 2000, it follows that $t \approx 4.6$ must correspond to 2004. So, the number of Internet users should have reached 200 million during the year 2004.

CHECK**Point**   Now try Exercise 123.

Numerical Solution

You can estimate the year in which the number of Internet users reached or surpassed 200 million by constructing a table of values. The table below shows the number of Internet users for each year from 2000 through 2008.

Year	t	I
2000	0	122.9
2001	1	144.9
2002	2	164.0
2003	3	180.2
2004	4	193.6
2005	5	204.0
2006	6	211.5
2007	7	216.2
2008	8	218.0

From the table, you can see that sometime during 2004 the number of Internet users reached 200 million.

TECHNOLOGY

You can also use a graphical approach to solve Example 9. Use a graphing utility to graph

$$y_1 = -1.446t^2 + 23.45t + 122.9 \quad \text{and} \quad y_2 = 200$$

in the same viewing window. Then use the *intersect* feature to find the point(s) of intersection of the two graphs. You should obtain $t \approx 4.6$, which verifies the answer obtained algebraically.

A fourth type of application that often involves a quadratic equation is one dealing with the hypotenuse of a right triangle. In these types of applications, the **Pythagorean Theorem** is often used. The Pythagorean Theorem states that

$$a^2 + b^2 = c^2$$

where a and b are the legs of a right triangle and c is the hypotenuse.

Example 10 An Application Involving the Pythagorean Theorem

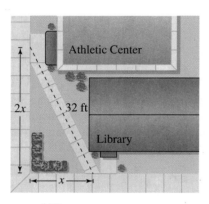

FIGURE **1.22**

An L-shaped sidewalk from the athletic center to the library on a college campus is shown in Figure 1.22. The sidewalk was constructed so that the length of one sidewalk forming the L was twice as long as the other. The length of the diagonal sidewalk that cuts across the grounds between the two buildings is 32 feet. How many feet does a person save by walking on the diagonal sidewalk?

Solution

Using the Pythagorean Theorem, you have the following.

$$x^2 + (2x)^2 = 32^2$$
$$5x^2 = 1024$$
$$x^2 = 204.8$$
$$x = \pm\sqrt{204.8}$$
$$x = \sqrt{204.8}$$

The total distance covered by walking on the L-shaped sidewalk is

$$x + 2x = 3x$$
$$= 3\sqrt{204.8}$$
$$\approx 42.9 \text{ feet.}$$

Walking on the diagonal sidewalk saves a person about $42.9 - 32 = 10.9$ feet.

CHECK**Point** Now try Exercise 125.

CLASSROOM DISCUSSION

Comparing Solution Methods In this section, you studied four algebraic methods for solving quadratic equations. Solve each of the quadratic equations below in several different ways. Write a short paragraph explaining which method(s) you prefer. Does your preferred method depend on the equation?

a. $x^2 - 4x - 5 = 0$

b. $x^2 - 4x = 0$

c. $x^2 - 4x - 3 = 0$

d. $x^2 - 4x - 6 = 0$

1.4   EXERCISES

See www.CalcChat.com for worked-out solutions to odd-numbered exercises.

VOCABULARY: Fill in the blanks.

1. A _____ _____ in x is an equation that can be written in the general form $ax^2 + bx + c = 0$, where a, b, and c are real numbers with $a \neq 0$.

2. A quadratic equation in x is also called a _____ _____ equation in x.

3. Four methods that can be used to solve a quadratic equation are _____, extracting _____ _____, _____ the _____, and the _____ _____.

4. The part of the Quadratic Formula, $b^2 - 4ac$, known as the _____, determines the type of solutions of a quadratic equation.

5. The general equation that gives the height of an object that is falling is called a _____ _____.

6. An important theorem that is sometimes used in applications that require solving quadratic equations is the _____ _____.

SKILLS AND APPLICATIONS

In Exercises 7–12, write the quadratic equation in general form.

7. $2x^2 = 3 - 5x$

8. $x^2 = 16x$

9. $(x - 3)^2 = 3$

10. $13 - 3(x + 7)^2 = 0$

11. $\frac{1}{5}(3x^2 - 10) = 12x$

12. $x(x + 2) = 5x^2 + 1$

In Exercises 13–24, solve the quadratic equation by factoring.

13. $6x^2 + 3x = 0$

14. $9x^2 - 1 = 0$

15. $x^2 - 2x - 8 = 0$

16. $x^2 - 10x + 9 = 0$

17. $x^2 + 10x + 25 = 0$

18. $4x^2 + 12x + 9 = 0$

19. $3 + 5x - 2x^2 = 0$

20. $2x^2 = 19x + 33$

21. $x^2 + 4x = 12$

22. $-x^2 + 8x = 12$

23. $\frac{3}{4}x^2 + 8x + 20 = 0$

24. $\frac{1}{8}x^2 - x - 16 = 0$

In Exercises 25–38, solve the equation by extracting square roots.

25. $x^2 = 49$

26. $x^2 = 144$

27. $x^2 = 11$

28. $x^2 = 32$

29. $3x^2 = 81$

30. $9x^2 = 36$

31. $(x - 12)^2 = 16$

32. $(x - 5)^2 = 25$

33. $(x + 2)^2 = 14$

34. $(x + 9)^2 = 24$

35. $(2x - 1)^2 = 18$

36. $(4x + 7)^2 = 44$

37. $(x - 7)^2 = (x + 3)^2$

38. $(x + 5)^2 = (x + 4)^2$

In Exercises 39–48, solve the quadratic equation by completing the square.

39. $x^2 + 4x - 32 = 0$

40. $x^2 - 2x - 3 = 0$

41. $x^2 + 6x + 2 = 0$

42. $x^2 + 8x + 14 = 0$

43. $9x^2 - 18x = -3$

44. $4x^2 - 4x = 1$

45. $7 + 2x - x^2 = 0$

46. $-x^2 + x - 1 = 0$

47. $2x^2 + 5x - 8 = 0$

48. $3x^2 - 4x - 7 = 0$

In Exercises 49–56, rewrite the quadratic portion of the algebraic expression as the sum or difference of two squares by completing the square.

49. $\dfrac{1}{x^2 + 2x + 5}$

50. $\dfrac{1}{x^2 - 12x + 19}$

51. $\dfrac{4}{x^2 + 4x - 3}$

52. $\dfrac{5}{x^2 + 25x + 11}$

53. $\dfrac{1}{4x^2 + 4x + 9}$

54. $\dfrac{1}{4x^2 - 4x + 25}$

55. $\dfrac{1}{\sqrt{6x - x^2}}$

56. $\dfrac{1}{\sqrt{16 - 6x - x^2}}$

GRAPHICAL ANALYSIS In Exercises 57–64, (a) use a graphing utility to graph the equation, (b) use the graph to approximate any x-intercepts of the graph, (c) set $y = 0$ and solve the resulting equation, and (d) compare the result of part (c) with the x-intercepts of the graph.

57. $y = (x + 3)^2 - 4$

58. $y = (x - 4)^2 - 1$

59. $y = 1 - (x - 2)^2$

60. $y = 9 - (x - 8)^2$

61. $y = -4x^2 + 4x + 3$

62. $y = 4x^2 - 1$

63. $y = x^2 + 3x - 4$

64. $y = x^2 - 5x - 24$

In Exercises 65–72, use the discriminant to determine the number of real solutions of the quadratic equation.

65. $2x^2 - 5x + 5 = 0$

66. $-5x^2 - 4x + 1 = 0$

67. $2x^2 - x - 1 = 0$

68. $x^2 - 4x + 4 = 0$

69. $\frac{1}{3}x^2 - 5x + 25 = 0$

70. $\frac{4}{7}x^2 - 8x + 28 = 0$

71. $0.2x^2 + 1.2x - 8 = 0$

72. $9 + 2.4x - 8.3x^2 = 0$

In Exercises 73–96, use the Quadratic Formula to solve the equation.

73. $2x^2 + x - 1 = 0$
74. $2x^2 - x - 1 = 0$
75. $16x^2 + 8x - 3 = 0$
76. $25x^2 - 20x + 3 = 0$
77. $2 + 2x - x^2 = 0$
78. $x^2 - 10x + 22 = 0$
79. $x^2 + 12x + 16 = 0$
80. $4x = 8 - x^2$
81. $x^2 + 8x - 4 = 0$
82. $2x^2 - 3x - 4 = 0$
83. $12x - 9x^2 = -3$
84. $9x^2 - 37 = 6x$
85. $9x^2 + 30x + 25 = 0$
86. $36x^2 + 24x - 7 = 0$
87. $4x^2 + 4x = 7$
88. $16x^2 - 40x + 5 = 0$
89. $28x - 49x^2 = 4$
90. $3x + x^2 - 1 = 0$
91. $8t = 5 + 2t^2$
92. $25h^2 + 80h + 61 = 0$
93. $(y - 5)^2 = 2y$
94. $(z + 6)^2 = -2z$
95. $\frac{1}{2}x^2 + \frac{3}{8}x = 2$
96. $\left(\frac{5}{7}x - 14\right)^2 = 8x$

In Exercises 97–104, use the Quadratic Formula to solve the equation. (Round your answer to three decimal places.)

97. $5.1x^2 - 1.7x - 3.2 = 0$
98. $2x^2 - 2.50x - 0.42 = 0$
99. $-0.067x^2 - 0.852x + 1.277 = 0$
100. $-0.005x^2 + 0.101x - 0.193 = 0$
101. $422x^2 - 506x - 347 = 0$
102. $1100x^2 + 326x - 715 = 0$
103. $12.67x^2 + 31.55x + 8.09 = 0$
104. $-3.22x^2 - 0.08x + 28.651 = 0$

In Exercises 105–112, solve the equation using any convenient method.

105. $x^2 - 2x - 1 = 0$
106. $11x^2 + 33x = 0$
107. $(x + 3)^2 = 81$
108. $x^2 - 14x + 49 = 0$
109. $x^2 - x - \frac{11}{4} = 0$
110. $x^2 + 3x - \frac{3}{4} = 0$
111. $(x + 1)^2 = x^2$
112. $3x + 4 = 2x^2 - 7$

113. FLOOR SPACE The floor of a one-story building is 14 feet longer than it is wide (see figure). The building has 1632 square feet of floor space.

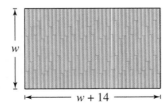

(a) Write a quadratic equation for the area of the floor in terms of w.

(b) Find the length and width of the floor.

114. DIMENSIONS OF A GARDEN A gardener has 100 meters of fencing to enclose two adjacent rectangular gardens (see figure). The gardener wants the enclosed area to be 350 square meters. What dimensions should the gardener use to obtain this area?

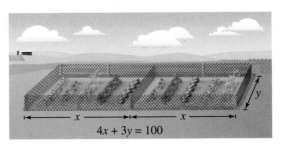

$4x + 3y = 100$

115. PACKAGING An open box with a square base (see figure) is to be constructed from 108 square inches of material. The height of the box is 3 inches. What are the dimensions of the box? (*Hint:* The surface area is $S = x^2 + 4xh$.)

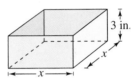

116. PACKAGING An open gift box is to be made from a square piece of material by cutting four-centimeter squares from the corners and turning up the sides (see figure). The volume of the finished box is to be 576 cubic centimeters. Find the size of the original piece of material.

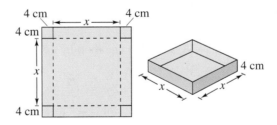

117. MOWING THE LAWN Two landscapers must mow a rectangular lawn that measures 100 feet by 200 feet. Each wants to mow no more than half of the lawn. The first starts by mowing around the outside of the lawn. The mower has a 24-inch cut. How wide a strip must the first landscaper mow on each of the four sides in order to mow no more than half of the lawn? Approximate the required number of trips around the lawn the first landscaper must take.

118. SEATING A rectangular classroom seats 72 students. If the seats were rearranged with three more seats in each row, the classroom would have two fewer rows. Find the original number of seats in each row.

In Exercises 119–122, use the position equation given in Example 8 as the model for the problem.

119. MILITARY   A C-141 Starlifter flying at 25,000 feet over level terrain drops a 500-pound supply package.

(a) How long will it take until the supply package strikes the ground?

(b) The plane is flying at 500 miles per hour. How far will the supply package travel horizontally during its descent?

120. EIFFEL TOWER   You drop a coin from the top of the Eiffel Tower in Paris. The building has a height of 984 feet.

(a) Use the position equation to write a mathematical model for the height of the coin.

(b) Find the height of the coin after 4 seconds.

(c) How long will it take before the coin strikes the ground?

121. SPORTS   Some Major League Baseball pitchers can throw a fastball at speeds of up to and over 100 miles per hour. Assume a Major League Baseball pitcher throws a baseball straight up into the air at 100 miles per hour from a height of 6 feet 3 inches.

(a) Use the position equation to write a mathematical model for the height of the baseball.

(b) Find the height of the baseball after 3 seconds, 4 seconds, and 5 seconds. What must have occurred sometime in the interval $3 \le t \le 5$? Explain.

(c) How many seconds is the baseball in the air?

122. CN TOWER   At 1815 feet tall, the CN Tower in Toronto, Ontario is the world's tallest self-supporting structure. An object is dropped from the top of the tower.

(a) Use the position equation to write a mathematical model for the height of the object.

(b) Complete the table.

Time, t	0	2	4	6	8	10	12
Height, s							

(c) From the table in part (b), determine the time interval during which the object reaches the ground. Numerically approximate the time it takes the object to reach the ground.

(d) Find the time it takes the object to reach the ground algebraically. How close was your numerical approximation?

 (e) Use a graphing utility with the appropriate viewing window to verify your answer(s) to parts (c) and (d).

123. DATA ANALYSIS: MOVIE TICKETS   The average admission prices P for movie theaters from 2001 through 2008 can be approximated by the model

$$P = 0.0103t^2 + 0.119t + 5.55, \quad 1 \le t \le 8$$

where t represents the year, with $t = 1$ corresponding to 2001. (Source: Motion Picture Association of America, Inc.)

(a) Use the model to complete the table to determine when the average admission price reached or surpassed $6.50.

t	1	2	3	4	5	6	7	8
P								

(b) Verify your result from part (a) algebraically.

(c) Use the model to predict the average admission price for movie theaters in 2014. Is this prediction reasonable? How does this value compare with the admission price where you live?

124. DATA ANALYSIS: MEDIAN INCOME   The median incomes I (in dollars) of U.S. households from 2000 through 2007 can be approximated by the model

$$I = 187.65t^2 - 119.1t + 42,013, \quad 0 \le t \le 7$$

where t represents the year, with $t = 0$ corresponding to 2000. (Source: U.S. Census Bureau)

 (a) Use a graphing utility to graph the model. Then use the graph to determine in which year the median income reached or surpassed $45,000.

(b) Verify your result from part (a) algebraically.

(c) Use the model to predict the median incomes of U.S. households in 2014 and 2018. Can this model be used to predict the median income of U.S. households after 2007? Before 2000? Explain.

125. BOATING   A winch is used to tow a boat to a dock. The rope is attached to the boat at a point 15 feet below the level of the winch (see figure).

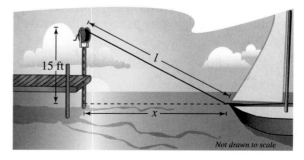

15 ft

l

x

Not drawn to scale

(a) Use the Pythagorean Theorem to write an equation giving the relationship between l and x.

(b) Find the distance from the boat to the dock when there is 75 feet of rope out.

126. FLYING SPEED Two planes leave simultaneously from Chicago's O'Hare Airport, one flying due north and the other due east (see figure). The northbound plane is flying 50 miles per hour faster than the eastbound plane. After 3 hours, the planes are 2440 miles apart. Find the speed of each plane.

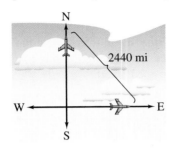

127. GEOMETRY The hypotenuse of an isosceles right triangle is 9 centimeters long. How long are its sides?

128. GEOMETRY An equilateral triangle has a height of 16 inches. How long is one of its sides? (*Hint:* Use the height of the triangle to partition the triangle into two congruent right triangles.)

129. REVENUE The demand equation for a product is $p = 20 - 0.0002x$, where p is the price per unit and x is the number of units sold. The total revenue for selling x units is

Revenue $= xp = x(20 - 0.0002x)$.

How many units must be sold to produce a revenue of $500,000?

130. REVENUE The demand equation for a product is $p = 60 - 0.0004x$, where p is the price per unit and x is the number of units sold. The total revenue for selling x units is

Revenue $= xp = x(60 - 0.0004x)$.

How many units must be sold to produce a revenue of $220,000?

COST In Exercises 131–134, use the cost equation to find the number of units x that a manufacturer can produce for the given cost C. Round your answer to the nearest positive integer.

131. $C = 0.125x^2 + 20x + 500$ $\qquad$ $C = \$14,000$

132. $C = 0.5x^2 + 15x + 5000$ $\qquad$ $C = \$11,500$

133. $C = 800 + 0.04x + 0.002x^2$ $\qquad$ $C = \$1680$

134. $C = 800 - 10x + \dfrac{x^2}{4}$ $\qquad$ $C = \$896$

135. PUBLIC DEBT The total public debt D (in trillions of dollars) in the United States at the beginning of each year from 2000 through 2008 can be approximated by the model

$D = 0.032t^2 + 0.21t + 5.6, \quad 0 \le t \le 8$

where t represents the year, with $t = 0$ corresponding to 2000. (Source: U.S. Department of the Treasury)

(a) Use the model to complete the table to determine when the total public debt reached or surpassed $7 trillion.

t	0	1	2	3	4	5	6	7	8
D									

(b) Verify your result from part (a) algebraically and graphically.

(c) Use the model to predict the public debt in 2014. Is this prediction reasonable? Explain.

136. BIOLOGY The metabolic rate of an ectothermic organism increases with increasing temperature within a certain range. Experimental data for the oxygen consumption C (in microliters per gram per hour) of a beetle at certain temperatures can be approximated by the model

$C = 0.45x^2 - 1.65x + 50.75, \quad 10 \le x \le 25$

where x is the air temperature in degrees Celsius.

(a) The oxygen consumption is 150 microliters per gram per hour. What is the air temperature?

(b) The temperature is increased from 10°C to 20°C. The oxygen consumption is increased by approximately what factor?

137. GEOMETRY An above ground swimming pool with the dimensions shown in the figure is to be constructed such that the volume of water in the pool is 1024 cubic feet. The height of the pool is to be 4 feet.

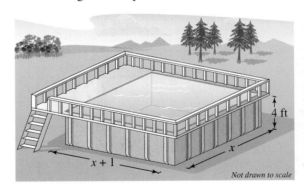

(a) What are the possible dimensions of the base?

(b) One cubic foot of water weighs approximately 62.4 pounds. Find the total weight of the water in the pool.

(c) A water pump is filling the pool at a rate of 5 gallons per minute. Find the time that will be required for the pump to fill the pool. (*Hint:* One gallon of water is approximately 0.13368 cubic foot.)

138. FLYING DISTANCE A commercial jet flies to three cities whose locations form the vertices of a right triangle (see figure). The total flight distance (from Oklahoma City to Austin to New Orleans and back to Oklahoma City) is approximately 1348 miles. It is 560 miles between Oklahoma City and New Orleans. Approximate the other two distances.

EXPLORATION

TRUE OR FALSE? In Exercises 139 and 140, determine whether the statement is true or false. Justify your answer.

139. The quadratic equation $-3x^2 - x = 10$ has two real solutions.

140. If $(2x - 3)(x + 5) = 8$, then either $2x - 3 = 8$ or $x + 5 = 8$.

141. To solve the equation $2x^2 + 3x = 15x$, a student divides each side by x and solves the equation $2x + 3 = 15$. The resulting solution ($x = 6$) satisfies the original equation. Is there an error? Explain.

142. The graphs show the solutions of equations plotted on the real number line. In each case, determine whether the solution(s) is (are) for a linear equation, a quadratic equation, both, or neither. Explain.

(a)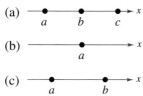

(b) ![] $\xrightarrow{\quad\bullet\quad} x$
a

(c) ![] $\xrightarrow{\bullet\quad\quad\bullet\quad} x$
ab

(d) ![] $\xrightarrow{\bullet\;\bullet\;\bullet\;\bullet} x$
$a\;b\;c\;d$

143. Solve $3(x + 4)^2 + (x + 4) - 2 = 0$ in two ways.

(a) Let $u = x + 4$, and solve the resulting equation for u. Then solve the u-solution for x.

(b) Expand and collect like terms in the equation, and solve the resulting equation for x.

(c) Which method is easier? Explain.

144. CAPSTONE Match the equation with a method you would use to solve it. Explain your reasoning. (Use each method once and do not solve the equations.)

(a) $3x^2 + 5x - 11 = 0$ (i) Factoring

(b) $x^2 + 10x = 3$ (ii) Extracting square roots

(c) $x^2 - 16x + 64 = 0$ (iii) Completing the square

(d) $x^2 - 15 = 0$ (iv) Quadratic Formula

THINK ABOUT IT In Exercises 145–150, write a quadratic equation that has the given solutions. (There are many correct answers.)

145. 3 and -5 **146.** -6 and -9

147. 8 and 14   **148.** $\frac{1}{6}$ and $-\frac{2}{5}$

149. $1 + \sqrt{2}$ and $1 - \sqrt{2}$

150. $-3 + \sqrt{5}$ and $-3 - \sqrt{5}$

151. From each graph, can you tell whether the discriminant is positive, zero, or negative? Explain your reasoning. Find each discriminant to verify your answers.

(a) $x^2 - 2x = 0$ (b) $x^2 - 2x + 1 = 0$

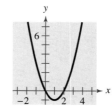   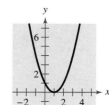

(c) $x^2 - 2x + 2 = 0$

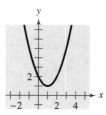

How many solutions would part (c) have if the linear term was $2x$? If the constant was -2?

152. THINK ABOUT IT Is it possible for a quadratic equation to have only one x-intercept? Explain.

153. PROOF Given that the solutions of a quadratic equation are $x = \left(-b \pm \sqrt{b^2 - 4ac}\right)/(2a)$, show that (a) the sum of the solutions is $S = -b/a$ and (b) the product of the solutions is $P = c/a$.

PROJECT: POPULATION To work an extended application analyzing the population of the United States, visit this text's website at *academic.cengage.com*. (Data Source: U.S. Census Bureau)

1.5 COMPLEX NUMBERS

What you should learn

- Use the imaginary unit *i* to write complex numbers.
- Add, subtract, and multiply complex numbers.
- Use complex conjugates to write the quotient of two complex numbers in standard form.
- Find complex solutions of quadratic equations.

Why you should learn it

You can use complex numbers to model and solve real-life problems in electronics. For instance, in Exercise 89 on page 128, you will learn how to use complex numbers to find the impedance of an electrical circuit.

The Imaginary Unit *i*

In Section 1.4, you learned that some quadratic equations have no real solutions. For instance, the quadratic equation $x^2 + 1 = 0$ has no real solution because there is no real number x that can be squared to produce -1. To overcome this deficiency, mathematicians created an expanded system of numbers using the **imaginary unit** *i*, defined as

$$i = \sqrt{-1}$$

where $i^2 = -1$. By adding real numbers to real multiples of this imaginary unit, the set of **complex numbers** is obtained. Each complex number can be written in the **standard form $a + bi$.** For instance, the standard form of the complex number $-5 + \sqrt{-9}$ is $-5 + 3i$ because

$$-5 + \sqrt{-9} = -5 + \sqrt{3^2(-1)} = -5 + 3\sqrt{-1} = -5 + 3i.$$

In the standard form $a + bi$, the real number a is called the **real part** of the **complex number $a + bi$,** and the number bi (where b is a real number) is called the **imaginary part** of the complex number.

Definition of a Complex Number

If a and b are real numbers, the number $a + bi$ is a **complex number,** and it is said to be written in **standard form.** If $b = 0$, the number $a + bi = a$ is a real number. If $b \neq 0$, the number $a + bi$ is called an **imaginary number.** A number of the form bi, where $b \neq 0$, is called a **pure imaginary number.**

The set of real numbers is a subset of the set of complex numbers, as shown in Figure 1.23. This is true because every real number a can be written as a complex number using $b = 0$. That is, for every real number a, you can write $a = a + 0i$.

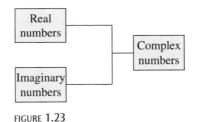

FIGURE 1.23

Equality of Complex Numbers

Two complex numbers $a + bi$ and $c + di$, written in standard form, are equal to each other

$$a + bi = c + di$$

if and only if $a = c$ and $b = d$.

Operations with Complex Numbers

To add (or subtract) two complex numbers, you add (or subtract) the real and imaginary parts of the numbers separately.

Addition and Subtraction of Complex Numbers

If $a + bi$ and $c + di$ are two complex numbers written in standard form, their sum and difference are defined as follows.

$\quad$ *Sum:* $(a + bi) + (c + di) = (a + c) + (b + d)i$

$\quad$ *Difference:* $(a + bi) - (c + di) = (a - c) + (b - d)i$

The **additive identity** in the complex number system is zero (the same as in the real number system). Furthermore, the **additive inverse** of the complex number $a + bi$ is

$$-(a + bi) = -a - bi.$$

So, you have

$$(a + bi) + (-a - bi) = 0 + 0i = 0.$$

▌Example 1   Adding and Subtracting Complex Numbers

a. $(4 + 7i) + (1 - 6i) = 4 + 7i + 1 - 6i$

$\qquad\qquad\qquad\quad = (4 + 1) + (7i - 6i)$

$\qquad\qquad\qquad\quad = 5 + i$

b. $(1 + 2i) - (4 + 2i) = 1 + 2i - 4 - 2i$

$\qquad\qquad\qquad\quad = (1 - 4) + (2i - 2i)$

$\qquad\qquad\qquad\quad = -3 + 0$

$\qquad\qquad\qquad\quad = -3$

c. $3i - (-2 + 3i) - (2 + 5i) = 3i + 2 - 3i - 2 - 5i$

$\qquad\qquad\qquad\qquad\qquad = (2 - 2) + (3i - 3i - 5i)$

$\qquad\qquad\qquad\qquad\qquad = 0 - 5i$

$\qquad\qquad\qquad\qquad\qquad = -5i$

d. $(3 + 2i) + (4 - i) - (7 + i) = 3 + 2i + 4 - i - 7 - i$

$\qquad\qquad\qquad\qquad\qquad\quad = (3 + 4 - 7) + (2i - i - i)$

$\qquad\qquad\qquad\qquad\qquad\quad = 0 + 0i$

$\qquad\qquad\qquad\qquad\qquad\quad = 0$

CHECK*Point* ▶ Now try Exercise 21. ▌

Note in Examples 1(b) and 1(d) that the sum of two complex numbers can be a real number.

Many of the properties of real numbers are valid for complex numbers as well. Here are some examples.

Associative Properties of Addition and Multiplication

Commutative Properties of Addition and Multiplication

Distributive Property of Multiplication Over Addition

Notice below how these properties are used when two complex numbers are multiplied.

$$(a + bi)(c + di) = a(c + di) + bi(c + di)$$
$$= ac + (ad)i + (bc)i + (bd)i^2$$
$$= ac + (ad)i + (bc)i + (bd)(-1)$$
$$= ac - bd + (ad)i + (bc)i$$
$$= (ac - bd) + (ad + bc)i$$

Rather than trying to memorize this multiplication rule, you should simply remember how the Distributive Property is used to multiply two complex numbers.

Example 2 Multiplying Complex Numbers

a. $4(-2 + 3i) = 4(-2) + 4(3i)$
$$= -8 + 12i$$

b. $(2 - i)(4 + 3i) = 2(4 + 3i) - i(4 + 3i)$
$$= 8 + 6i - 4i - 3i^2$$
$$= 8 + 6i - 4i - 3(-1)$$
$$= (8 + 3) + (6i - 4i)$$
$$= 11 + 2i$$

c. $(3 + 2i)(3 - 2i) = 3(3 - 2i) + 2i(3 - 2i)$
$$= 9 - 6i + 6i - 4i^2$$
$$= 9 - 6i + 6i - 4(-1)$$
$$= 9 + 4$$
$$= 13$$

d. $(3 + 2i)^2 = (3 + 2i)(3 + 2i)$
$$= 3(3 + 2i) + 2i(3 + 2i)$$
$$= 9 + 6i + 6i + 4i^2$$
$$= 9 + 6i + 6i + 4(-1)$$
$$= 9 + 12i - 4$$
$$= 5 + 12i$$

CHECK**Point** Now try Exercise 31.

Study Tip

The procedure described above is similar to multiplying two polynomials and combining like terms, as in the FOIL Method shown in Section P.3. For instance, you can use the FOIL Method to multiply the two complex numbers from Example 2(b).

$(2 - i)(4 + 3i) = 8 + 6i - 4i - 3i^2$

Complex Conjugates

Notice in Example 2(c) that the product of two complex numbers can be a real number. This occurs with pairs of complex numbers of the form $a + bi$ and $a - bi$, called **complex conjugates.**

$$(a + bi)(a - bi) = a^2 - abi + abi - b^2i^2$$
$$= a^2 - b^2(-1)$$
$$= a^2 + b^2$$

Algebra Help

You can compare complex conjugates with the method for rationalizing denominators in Section P.2.

Example 3   Multiplying Conjugates

Multiply each complex number by its complex conjugate.

a. $1 + i$ **b.** $4 - 3i$

Solution

a. The complex conjugate of $1 + i$ is $1 - i$.

$$(1 + i)(1 - i) = 1^2 - i^2 = 1 - (-1) = 2$$

b. The complex conjugate of $4 - 3i$ is $4 + 3i$.

$$(4 - 3i)(4 + 3i) = 4^2 - (3i)^2 = 16 - 9i^2 = 16 - 9(-1) = 25$$

CHECK**Point**   Now try Exercise 41.

To write the quotient of $a + bi$ and $c + di$ in standard form, where c and d are not both zero, multiply the numerator and denominator by the complex conjugate of the *denominator* to obtain

$$\frac{a + bi}{c + di} = \frac{a + bi}{c + di}$$
$$= \frac{(ac + bd) + (bc - ad)i}{c^2 + d^2}.$$

Study Tip

Note that when you multiply the numerator and denominator of a quotient of complex numbers by

$$\frac{c - di}{c - di}$$

you are actually multiplying the quotient by a form of 1. You are not changing the original expression, you are only creating an expression that is equivalent to the original expression.

Example 4   Writing a Quotient of Complex Numbers in Standard Form

$$\frac{2 + 3i}{4 - 2i} = \frac{2 + 3i}{4 - 2i}$$
$$= \frac{8 + 4i + 12i + 6i^2}{16 - 4i^2}$$
$$= \frac{8 - 6 + 16i}{16 + 4}$$
$$= \frac{2 + 16i}{20}$$
$$= \frac{1}{10} + \frac{4}{5}i$$

CHECK**Point**   Now try Exercise 53.

Complex Solutions of Quadratic Equations

When using the Quadratic Formula to solve a quadratic equation, you often obtain a result such as $\sqrt{-3}$, which you know is not a real number. By factoring out $i = \sqrt{-1}$, you can write this number in standard form.

$$\sqrt{-3} = \sqrt{3(-1)} = \sqrt{3}\sqrt{-1} = \sqrt{3}\,i$$

The number $\sqrt{3}\,i$ is called the *principal square root* of -3.

Algebra Help

You can review the techniques for using the Quadratic Formula in Section 1.4.

> **Principal Square Root of a Negative Number**
>
> If a is a positive number, the **principal square root** of the negative number $-a$ is defined as
>
> $$\sqrt{-a} = \sqrt{a}\,i.$$

⚠ **WARNING / CAUTION**

The definition of principal square root uses the rule

$$\sqrt{ab} = \sqrt{a}\sqrt{b}$$

for $a > 0$ and $b < 0$. This rule is not valid if *both* a and b are negative. For example,

$$\sqrt{-5}\sqrt{-5} = \sqrt{5(-1)}\sqrt{5(-1)}$$
$$= \sqrt{5}\,i\sqrt{5}\,i$$
$$= \sqrt{25}\,i^2$$
$$= 5i^2 = -5$$

whereas

$$\sqrt{(-5)(-5)} = \sqrt{25} = 5.$$

To avoid problems with square roots of negative numbers, be sure to convert complex numbers to standard form *before* multiplying.

Example 5 Writing Complex Numbers in Standard Form

a. $\sqrt{-3}\sqrt{-12} = \sqrt{3}\,i\sqrt{12}\,i = \sqrt{36}\,i^2 = 6(-1) = -6$

b. $\sqrt{-48} - \sqrt{-27} = \sqrt{48}\,i - \sqrt{27}\,i = 4\sqrt{3}\,i - 3\sqrt{3}\,i = \sqrt{3}\,i$

c. $\left(-1 + \sqrt{-3}\right)^2 = \left(-1 + \sqrt{3}\,i\right)^2$
$$= (-1)^2 - 2\sqrt{3}\,i + \left(\sqrt{3}\right)^2(i^2)$$
$$= 1 - 2\sqrt{3}\,i + 3(-1)$$
$$= -2 - 2\sqrt{3}\,i$$

CHECK Point Now try Exercise 63.

Example 6 Complex Solutions of a Quadratic Equation

Solve (a) $x^2 + 4 = 0$ and (b) $3x^2 - 2x + 5 = 0$.

Solution

a. $x^2 + 4 = 0$
$$x^2 = -4$$
$$x = \pm 2i$$

b. $3x^2 - 2x + 5 = 0$
$$x = \frac{-(\ ) \pm \sqrt{(\ )^2 - 4(\ )(\ )}}{2(\ )}$$
$$= \frac{2 \pm \sqrt{-56}}{6}$$
$$= \frac{2 \pm 2\sqrt{14}\,i}{6}$$
$$= \frac{1}{3} \pm \frac{\sqrt{14}}{3}\,i$$

CHECK Point Now try Exercise 69.

1.5   EXERCISES

See www.CalcChat.com for worked-out solutions to odd-numbered exercises.

VOCABULARY

1. Match the type of complex number with its definition.

(a) Real number   (i) $a + bi,\quad a \neq 0,\quad b \neq 0$

(b) Imaginary number   (ii) $a + bi,\quad a = 0,\quad b \neq 0$

(c) Pure imaginary number   (iii) $a + bi,\quad b = 0$

In Exercises 2–4, fill in the blanks.

2. The imaginary unit i is defined as $i =$ _____, where $i^2 =$ _____.

3. If a is a positive number, the _____ _____ root of the negative number $-a$ is defined as $\sqrt{-a} = \sqrt{a}\,i$.

4. The numbers $a + bi$ and $a - bi$ are called _____ _____, and their product is a real number $a^2 + b^2$.

SKILLS AND APPLICATIONS

In Exercises 5–8, find real numbers a and b such that the equation is true.

5. $a + bi = -12 + 7i$ **6.** $a + bi = 13 + 4i$

7. $(a - 1) + (b + 3)i = 5 + 8i$

8. $(a + 6) + 2bi = 6 - 5i$

In Exercises 9–20, write the complex number in standard form.

9. $8 + \sqrt{-25}$ **10.** $5 + \sqrt{-36}$

11. $2 - \sqrt{-27}$ **12.** $1 + \sqrt{-8}$

13. $\sqrt{-80}$ **14.** $\sqrt{-4}$

15. 14 **16.** 75

17. $-10i + i^2$ **18.** $-4i^2 + 2i$

19. $\sqrt{-0.09}$ **20.** $\sqrt{-0.0049}$

In Exercises 21–30, perform the addition or subtraction and write the result in standard form.

21. $(7 + i) + (3 - 4i)$ **22.** $(13 - 2i) + (-5 + 6i)$

23. $(9 - i) - (8 - i)$ **24.** $(3 + 2i) - (6 + 13i)$

25. $\left(-2 + \sqrt{-8}\right) + \left(5 - \sqrt{-50}\right)$

26. $\left(8 + \sqrt{-18}\right) - \left(4 + 3\sqrt{2}i\right)$

27. $13i - (14 - 7i)$ **28.** $25 + (-10 + 11i) + 15i$

29. $-\left(\frac{3}{2} + \frac{5}{2}i\right) + \left(\frac{5}{3} + \frac{11}{3}i\right)$

30. $(1.6 + 3.2i) + (-5.8 + 4.3i)$

In Exercises 31–40, perform the operation and write the result in standard form.

31. $(1 + i)(3 - 2i)$ **32.** $(7 - 2i)(3 - 5i)$

33. $12i(1 - 9i)$ **34.** $-8i(9 + 4i)$

35. $\left(\sqrt{14} + \sqrt{10}i\right)\left(\sqrt{14} - \sqrt{10}i\right)$

36. $\left(\sqrt{3} + \sqrt{15}i\right)\left(\sqrt{3} - \sqrt{15}i\right)$

37. $(6 + 7i)^2$ **38.** $(5 - 4i)^2$

39. $(2 + 3i)^2 + (2 - 3i)^2$ **40.** $(1 - 2i)^2 - (1 + 2i)^2$

In Exercises 41–48, write the complex conjugate of the complex number. Then multiply the number by its complex conjugate.

41. $9 + 2i$ **42.** $8 - 10i$

43. $-1 - \sqrt{5}i$ **44.** $-3 + \sqrt{2}i$

45. $\sqrt{-20}$ **46.** $\sqrt{-15}$

47. $\sqrt{6}$ **48.** $1 + \sqrt{8}$

In Exercises 49–58, write the quotient in standard form.

49. $\dfrac{3}{i}$ **50.** $-\dfrac{14}{2i}$

51. $\dfrac{2}{4 - 5i}$ **52.** $\dfrac{13}{1 - i}$

53. $\dfrac{5 + i}{5 - i}$ **54.** $\dfrac{6 - 7i}{1 - 2i}$

55. $\dfrac{9 - 4i}{i}$ **56.** $\dfrac{8 + 16i}{2i}$

57. $\dfrac{3i}{(4 - 5i)^2}$ **58.** $\dfrac{5i}{(2 + 3i)^2}$

In Exercises 59–62, perform the operation and write the result in standard form.

59. $\dfrac{2}{1 + i} - \dfrac{3}{1 - i}$ **60.** $\dfrac{2i}{2 + i} + \dfrac{5}{2 - i}$

61. $\dfrac{i}{3 - 2i} + \dfrac{2i}{3 + 8i}$ **62.** $\dfrac{1 + i}{i} - \dfrac{3}{4 - i}$

In Exercises 63–68, write the complex number in standard form.

63. $\sqrt{-6} \cdot \sqrt{-2}$ **64.** $\sqrt{-5} \cdot \sqrt{-10}$

65. $\left(\sqrt{-15}\right)^2$ **66.** $\left(\sqrt{-75}\right)^2$

67. $\left(3 + \sqrt{-5}\right)\left(7 - \sqrt{-10}\right)$ **68.** $\left(2 - \sqrt{-6}\right)^2$

In Exercises 69–78, use the Quadratic Formula to solve the quadratic equation.

69. $x^2 - 2x + 2 = 0$

70. $x^2 + 6x + 10 = 0$

71. $4x^2 + 16x + 17 = 0$

72. $9x^2 - 6x + 37 = 0$

73. $4x^2 + 16x + 15 = 0$

74. $16t^2 - 4t + 3 = 0$

75. $\frac{3}{2}x^2 - 6x + 9 = 0$

76. $\frac{7}{8}x^2 - \frac{3}{4}x + \frac{5}{16} = 0$

77. $1.4x^2 - 2x - 10 = 0$

78. $4.5x^2 - 3x + 12 = 0$

In Exercises 79–88, simplify the complex number and write it in standard form.

79. $-6i^3 + i^2$

80. $4i^2 - 2i^3$

81. $-14i^5$

82. $(-i)^3$

83. $\left(\sqrt{-72}\right)^3$

84. $\left(\sqrt{-2}\right)^6$

85. $\dfrac{1}{i^3}$

86. $\dfrac{1}{(2i)^3}$

87. $(3i)^4$

88. $(-i)^6$

89. IMPEDANCE The opposition to current in an electrical circuit is called its impedance. The impedance z in a parallel circuit with two pathways satisfies the equation

$$\frac{1}{z} = \frac{1}{z_1} + \frac{1}{z_2}$$

where z_1 is the impedance (in ohms) of pathway 1 and z_2 is the impedance of pathway 2.

(a) The impedance of each pathway in a parallel circuit is found by adding the impedances of all components in the pathway. Use the table to find z_1 and z_2.

(b) Find the impedance z.

	Resistor	Inductor	Capacitor
Symbol	⌇⌇⌇ $a\Omega$	⌒⌒⌒ $b\Omega$	⊣⊢ $c\Omega$
Impedance	a	bi	$-ci$

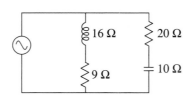

90. Cube each complex number.

(a) 2   (b) $-1 + \sqrt{3}i$ (c) $-1 - \sqrt{3}i$

91. Raise each complex number to the fourth power.

(a) 2   (b) -2 (c) $2i$ (d) $-2i$

92. Write each of the powers of i as i, $-i$, 1, or -1.

(a) i^{40} (b) i^{25} (c) i^{50} (d) i^{67}

EXPLORATION

TRUE OR FALSE? In Exercises 93–96, determine whether the statement is true or false. Justify your answer.

93. There is no complex number that is equal to its complex conjugate.

94. $-i\sqrt{6}$ is a solution of $x^4 - x^2 + 14 = 56$.

95. $i^{44} + i^{150} - i^{74} - i^{109} + i^{61} = -1$

96. The sum of two complex numbers is always a real number.

97. PATTERN RECOGNITION Complete the following.

$i^1 = i$ $i^2 = -1$ $i^3 = -i$ $i^4 = 1$

$i^5 = \rule{1cm}{0.15mm}$ $i^6 = \rule{1cm}{0.15mm}$ $i^7 = \rule{1cm}{0.15mm}$ $i^8 = \rule{1cm}{0.15mm}$

$i^9 = \rule{1cm}{0.15mm}$ $i^{10} = \rule{1cm}{0.15mm}$ $i^{11} = \rule{1cm}{0.15mm}$ $i^{12} = \rule{1cm}{0.15mm}$

What pattern do you see? Write a brief description of how you would find i raised to any positive integer power.

98. CAPSTONE Consider the binomials $x + 5$ and $2x - 1$ and the complex numbers $1 + 5i$ and $2 - i$.

(a) Find the sum of the binomials and the sum of the complex numbers.

(b) Find the difference of the binomials and the difference of the complex numbers.

(c) Describe the similarities and differences in your results for parts (a) and (b).

(d) Find the product of the binomials and the product of the complex numbers.

(e) Explain why the products you found in part (d) are not related in the same way as your results in parts (a) and (b).

(f) Write a brief paragraph that compares operations with binomials and operations with complex numbers.

99. ERROR ANALYSIS Describe the error.

$$\sqrt{-6}\sqrt{-6} = \sqrt{(-6)(-6)} = \sqrt{36} = 6$$

100. PROOF Prove that the complex conjugate of the product of two complex numbers $a_1 + b_1 i$ and $a_2 + b_2 i$ is the product of their complex conjugates.

101. PROOF Prove that the complex conjugate of the sum of two complex numbers $a_1 + b_1 i$ and $a_2 + b_2 i$ is the sum of their complex conjugates.

1.6   OTHER TYPES OF EQUATIONS

Polynomial Equations

In this section you will extend the techniques for solving equations to nonlinear and nonquadratic equations. At this point in the text, you have only four basic methods for solving nonlinear equations—*factoring*, *extracting square roots*, *completing the square*, and the *Quadratic Formula*. So the main goal of this section is to learn to *rewrite* nonlinear equations in a form to which you can apply one of these methods.

Example 1 shows how to use factoring to solve a **polynomial equation,** which is an equation that can be written in the general form

$$a_n x^n + a_{n-1} x^{n-1} + \cdots + a_2 x^2 + a_1 x + a_0 = 0.$$

Example 1   Solving a Polynomial Equation by Factoring

Solve $3x^4 = 48x^2$.

Solution

First write the polynomial equation in general form with zero on one side, factor the other side, and then set each factor equal to zero and solve.

$$3x^4 = 48x^2$$
$$3x^4 - 48x^2 = 0$$
$$3x^2(x^2 - 16) = 0$$
$$3x^2(x + 4)(x - 4) = 0$$

$3x^2 = 0 \implies x = 0$

$x + 4 = 0 \implies x = -4$

$x - 4 = 0 \implies x = 4$

You can check these solutions by substituting in the original equation, as follows.

Check

$3(\ )^4 = 48(\ )^2$ ✓

$3(\ \ )^4 = 48(\ \ )^2$ ✓

$3(\ )^4 = 48(\ )^2$ ✓

So, you can conclude that the solutions are $x = 0$, $x = -4$, and $x = 4$.

CHECK *Point* Now try Exercise 5.

A common mistake that is made in solving an equation like that in Example 1 is to divide each side of the equation by the variable factor x^2. This loses the solution $x = 0$. When solving an equation, always write the equation in general form, then factor the equation and set each factor equal to zero. Do not divide each side of an equation by a variable factor in an attempt to simplify the equation.

You can use a graphing utility to check graphically the solutions of the equation in Example 2. To do this, graph the equation

$$y = x^3 - 3x^2 + 3x - 9.$$

Then use the *zero* or *root* feature to approximate any *x*-intercepts. As shown below, the *x*-intercept of the graph occurs at (3, 0), confirming the *real* solution of $x = 3$ found in Example 2.

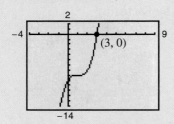

Try using a graphing utility to check the solutions found in Example 3.

For a review of factoring special polynomial forms, see Section P.4.

Example 2 Solving a Polynomial Equation by Factoring

Solve $x^3 - 3x^2 + 3x - 9 = 0$.

Solution

$$x^3 - 3x^2 + 3x - 9 = 0$$
$$x^2(x - 3) + 3(x - 3) = 0$$
$$(x - 3)(x^2 + 3) = 0$$
$$x - 3 = 0 \quad\Longrightarrow\quad x = 3$$
$$x^2 + 3 = 0 \quad\Longrightarrow\quad x = \pm\sqrt{3}i$$

The solutions are $x = 3$, $x = \sqrt{3}i$, and $x = -\sqrt{3}i$.

CHECK *Point*   Now try Exercise 13.

Occasionally, mathematical models involve equations that are of **quadratic type.** In general, an equation is of quadratic type if it can be written in the form

$$au^2 + bu + c = 0$$

where $a \neq 0$ and u is an algebraic expression.

Example 3 Solving an Equation of Quadratic Type

Solve $x^4 - 3x^2 + 2 = 0$.

Solution

This equation is of quadratic type with $u = x^2$.

$$(\quad)^2 - 3(\quad) + 2 = 0$$

To solve this equation, you can factor the left side of the equation as the product of two second-degree polynomials.

$$x^4 - 3x^2 + 2 = 0$$

$$(x^2)^2 - 3(x^2) + 2 = 0$$
$$(x^2 - 1)(x^2 - 2) = 0$$
$$(x + 1)(x - 1)(x^2 - 2) = 0$$
$$x + 1 = 0 \quad\Longrightarrow\quad x = -1$$
$$x - 1 = 0 \quad\Longrightarrow\quad x = 1$$
$$x^2 - 2 = 0 \quad\Longrightarrow\quad x = \pm\sqrt{2}$$

The solutions are $x = -1$, $x = 1$, $x = \sqrt{2}$, and $x = -\sqrt{2}$. Check these in the original equation.

CHECK *Point*   Now try Exercise 17.

Equations Involving Radicals

Operations such as squaring each side of an equation, raising each side of an equation to a rational power, and multiplying each side of an equation by a variable quantity all can introduce extraneous solutions. So, when you use any of these operations, checking your solutions is crucial.

> ### Example 4   Solving Equations Involving Radicals
>
> **a.** $\sqrt{2x + 7} - x = 2$
>
> $$\sqrt{2x + 7} = x + 2$$
>
> $$2x + 7 = x^2 + 4x + 4$$
>
> $$0 = x^2 + 2x - 3$$
>
> $$0 = (x + 3)(x - 1)$$
>
> $$x + 3 = 0 \implies x = -3$$
>
> $$x - 1 = 0 \implies x = 1$$
>
> By checking these values, you can determine that the only solution is $x = 1$.
>
> **b.** $\sqrt{2x - 5} - \sqrt{x - 3} = 1$
>
> $$\sqrt{2x - 5} = \sqrt{x - 3} + 1$$
>
> $$2x - 5 = x - 3 + 2\sqrt{x - 3} + 1$$
>
> $$2x - 5 = x - 2 + 2\sqrt{x - 3}$$
>
> $$x - 3 = 2\sqrt{x - 3}$$
>
> $$x^2 - 6x + 9 = 4(x - 3)$$
>
> $$x^2 - 10x + 21 = 0$$
>
> $$(x - 3)(x - 7) = 0$$
>
> $$x - 3 = 0 \implies x = 3$$
>
> $$x - 7 = 0 \implies x = 7$$
>
> The solutions are $x = 3$ and $x = 7$. Check these in the original equation.
>
> CHECK **Point**   Now try Exercise 37.

Study Tip

When an equation contains two radicals, it may not be possible to isolate both. In such cases, you may have to raise each side of the equation to a power at *two* different stages in the solution, as shown in Example 4(b).

> ### Example 5   Solving an Equation Involving a Rational Exponent
>
> $$(x - 4)^{2/3} = 25$$
>
> $$\sqrt[3]{(x - 4)^2} = 25$$
>
> $$(x - 4)^2 = 15{,}625$$
>
> $$x - 4 = \pm 125$$
>
> $$x = 129, \quad x = -121$$
>
> CHECK **Point**   Now try Exercise 51.

Equations with Fractions or Absolute Values

To solve an equation involving fractions, multiply each side of the equation by the least common denominator (LCD) of all terms in the equation. This procedure will "clear the equation of fractions." For instance, in the equation

$$\frac{2}{x^2 + 1} + \frac{1}{x} = \frac{2}{x}$$

you can multiply each side of the equation by $x(x^2 + 1)$. Try doing this and solve the resulting equation. You should obtain one solution: $x = 1$.

Example 6 Solving an Equation Involving Fractions

Solve $\frac{2}{x} = \frac{3}{x - 2} - 1$.

Solution

For this equation, the least common denominator of the three terms is $x(x - 2)$, so you begin by multiplying each term of the equation by this expression.

$$\frac{2}{x} = \frac{3}{x - 2} - 1$$

$$\frac{2}{x} = \qquad \frac{3}{x - 2} - \qquad (1)$$

$$2(x - 2) = 3x - x(x - 2)$$

$$2x - 4 = -x^2 + 5x$$

$$x^2 - 3x - 4 = 0$$

$$(x - 4)(x + 1) = 0$$

$$x - 4 = 0 \quad \Rightarrow \quad x = 4$$

$$x + 1 = 0 \quad \Rightarrow \quad x = -1$$

Check $x = 4$

$$\frac{2}{x} = \frac{3}{x - 2} - 1$$

$$\frac{2}{\ } \stackrel{?}{=} \frac{3}{\ - 2} - 1$$

$$\frac{1}{2} \stackrel{?}{=} \frac{3}{2} - 1$$

$$\frac{1}{2} = \frac{1}{2} \checkmark$$

Check $x = -1$

$$\frac{2}{x} = \frac{3}{x - 2} - 1$$

$$\frac{2}{\ } \stackrel{?}{=} \frac{3}{\ - 2} - 1$$

$$-2 \stackrel{?}{=} -1 - 1$$

$$-2 = -2 \checkmark$$

So, the solutions are $x = 4$ and $x = -1$.

CHECKPoint Now try Exercise 65.

Algebra Help

You can review the definition of absolute value in Section P.1.

To solve an equation involving an absolute value, remember that the expression inside the absolute value signs can be positive or negative. This results in *two* separate equations, each of which must be solved. For instance, the equation

$$|x - 2| = 3$$

results in the two equations $x - 2 = 3$ and $-(x - 2) = 3$, which implies that the equation has two solutions: $x = 5$ and $x = -1$.

Example 7   Solving an Equation Involving Absolute Value

Solve $|x^2 - 3x| = -4x + 6$.

Solution

Because the variable expression inside the absolute value signs can be positive or negative, you must solve the following two equations.

First Equation

$$x^2 - 3x = -4x + 6$$
$$x^2 + x - 6 = 0$$
$$(x + 3)(x - 2) = 0$$
$$x + 3 = 0 \quad \implies \quad x = -3$$
$$x - 2 = 0 \quad \implies \quad x = 2$$

Second Equation

$$-(x^2 - 3x) = -4x + 6$$
$$x^2 - 7x + 6 = 0$$
$$(x - 1)(x - 6) = 0$$
$$x - 1 = 0 \quad \implies \quad x = 1$$
$$x - 6 = 0 \quad \implies \quad x = 6$$

Check

$$|(\quad)^2 - 3(\quad)| \overset{?}{=} -4(\quad) + 6$$
$$18 = 18 \qquad\qquad \checkmark$$
$$|(\,)^2 - 3(\,)| \overset{?}{=} -4(\,) + 6$$
$$2 \neq -2$$
$$|(\,)^2 - 3(\,)| \overset{?}{=} -4(\,) + 6$$
$$2 = 2 \qquad\qquad \checkmark$$
$$|(\,)^2 - 3(\,)| \overset{?}{=} -4(\,) + 6$$
$$18 \neq -18$$

The solutions are $x = -3$ and $x = 1$.

CHECK*Point* Now try Exercise 73.

Applications

It would be impossible to categorize the many different types of applications that involve nonlinear and nonquadratic models. However, from the few examples and exercises that are given, you will gain some appreciation for the variety of applications that can occur.

Example 8   Reduced Rates

A ski club chartered a bus for a ski trip at a cost of $480. In an attempt to lower the bus fare per skier, the club invited nonmembers to go along. After five nonmembers joined the trip, the fare per skier decreased by $4.80. How many club members are going on the trip?

Solution

Begin the solution by creating a verbal model and assigning labels.

Verbal Model: Cost per skier · Number of skiers = Cost of trip

Labels:

Cost of trip = 480	(dollars)
Number of ski club members = x	(people)
Number of skiers = $x + 5$	(people)
Original cost per member = $\dfrac{480}{x}$	(dollars per person)
Cost per skier = $\dfrac{480}{x} - 4.80$	(dollars per person)

Equation:

$$\left(\frac{480}{x} - 4.80\right)(x + 5) = 480$$

$$\left(\frac{480 - 4.8x}{x}\right)(x + 5) = 480$$

$$(480 - 4.8x)(x + 5) = 480x$$

$$480x + 2400 - 4.8x^2 - 24x = 480x$$

$$-4.8x^2 - 24x + 2400 = 0$$

$$x^2 + 5x - 500 = 0$$

$$(x + 25)(x - 20) = 0$$

$$x + 25 = 0 \quad\Longrightarrow\quad x = -25$$

$$x - 20 = 0 \quad\Longrightarrow\quad x = 20$$

Choosing the positive value of x, you can conclude that 20 ski club members are going on the trip. Check this in the original statement of the problem, as follows.

$$\left(\frac{480}{} - 4.80\right)(+ 5) \stackrel{?}{=} 480$$

$$(24 - 4.80)25 \stackrel{?}{=} 480$$

$$480 = 480 \qquad\qquad \checkmark$$

CHECK*Point*   Now try Exercise 99.

Interest in a savings account is calculated by one of three basic methods: simple interest, interest compounded n times per year, and interest compounded continuously. The next example uses the formula for interest that is compounded n times per year.

$$A = P\left(1 + \frac{r}{n}\right)^{nt}$$

In this formula, A is the balance in the account, P is the principal (or original deposit), r is the annual interest rate (in decimal form), n is the number of compoundings per year, and t is the time in years. In Chapter 5, you will study a derivation of the formula above for interest compounded continuously.

Example 9   Compound Interest

When you were born, your grandparents deposited $5000 in a long-term investment in which the interest was compounded quarterly. Today, on your 25th birthday, the value of the investment is $25,062.59. What is the annual interest rate for this investment?

Solution

Formula: $A = P\left(1 + \dfrac{r}{n}\right)^{nt}$

Labels: Balance $= A = 25,062.59$ (dollars)
             Principal $= P = 5000$ (dollars)
             Time $= t = 25$ (years)
             Compoundings per year $= n = 4$ (compoundings per year)
             Annual interest rate $= r$ (percent in decimal form)

Equation: $25,062.59 = 5000\left(1 + \dfrac{r}{4}\right)^{4(25)}$

$\dfrac{25,062.59}{5000} = \left(1 + \dfrac{r}{4}\right)^{100}$

$5.0125 \approx \left(1 + \dfrac{r}{4}\right)^{100}$

$(5.0125)^{1/100} = 1 + \dfrac{r}{4}$

$1.01625 \approx 1 + \dfrac{r}{4}$

$0.01625 = \dfrac{r}{4}$

$0.065 = r$

The annual interest rate is about 0.065, or 6.5%. Check this in the original statement of the problem.

CHECK *Point*   Now try Exercise 103.

1.6   EXERCISES

See www.CalcChat.com for worked-out solutions to odd-numbered exercises.

VOCABULARY: Fill in the blanks.

1. The equation $a_n x^n + a_{n-1} x^{n-1} + \cdots + a_2 x^2 + a_1 x + a_0 = 0$ is a _____ equation in x written in general form.

2. Squaring each side of an equation, multiplying each side of an equation by a variable quantity, and raising each side of an equation to a rational power are all operations that can introduce _____ solutions to a given equation.

3. The equation $2x^4 + x^2 + 1 = 0$ is of _____ _____.

4. To clear the equation $\dfrac{4}{x} + 5 = \dfrac{6}{x-3}$ of fractions, multiply each side of the equation by the least common denominator _____.

SKILLS AND APPLICATIONS

In Exercises 5–30, find all solutions of the equation. Check your solutions in the original equation.

5. $6x^4 - 14x^2 = 0$
6. $36x^3 - 100x = 0$
7. $x^4 - 81 = 0$
8. $x^6 - 64 = 0$
9. $x^3 + 512 = 0$
10. $27x^3 - 343 = 0$
11. $5x^3 + 30x^2 + 45x = 0$
12. $9x^4 - 24x^3 + 16x^2 = 0$
13. $x^3 - 3x^2 - x + 3 = 0$
14. $x^3 + 2x^2 + 3x + 6 = 0$
15. $x^4 - x^3 + x - 1 = 0$
16. $x^4 + 2x^3 - 8x - 16 = 0$
17. $x^4 - 4x^2 + 3 = 0$
18. $x^4 + 5x^2 - 36 = 0$
19. $4x^4 - 65x^2 + 16 = 0$
20. $36t^4 + 29t^2 - 7 = 0$
21. $x^6 + 7x^3 - 8 = 0$
22. $x^6 + 3x^3 + 2 = 0$
23. $\dfrac{1}{x^2} + \dfrac{8}{x} + 15 = 0$
24. $1 + \dfrac{3}{x} = \dfrac{2}{x^2}$
25. $2\left(\dfrac{x}{x+2}\right)^2 - 3\left(\dfrac{x}{x+2}\right) - 2 = 0$
26. $6\left(\dfrac{x}{x+1}\right)^2 + 5\left(\dfrac{x}{x+1}\right) - 6 = 0$
27. $2x + 9\sqrt{x} = 5$
28. $6x - 7\sqrt{x} - 3 = 0$
29. $3x^{1/3} + 2x^{2/3} = 5$
30. $9t^{2/3} + 24t^{1/3} + 16 = 0$

GRAPHICAL ANALYSIS In Exercises 31–34, (a) use a graphing utility to graph the equation, (b) use the graph to approximate any x-intercepts of the graph, (c) set $y = 0$ and solve the resulting equation, and (d) compare the result of part (c) with the x-intercepts of the graph.

31. $y = x^3 - 2x^2 - 3x$
32. $y = 2x^4 - 15x^3 + 18x^2$
33. $y = x^4 - 10x^2 + 9$
34. $y = x^4 - 29x^2 + 100$

In Exercises 35–58, find all solutions of the equation. Check your solutions in the original equation.

35. $\sqrt{3x} - 12 = 0$
36. $7\sqrt{x} - 4 = 0$
37. $\sqrt{x - 10} - 4 = 0$
38. $\sqrt{5 - x} - 3 = 0$
39. $\sqrt[3]{2x + 5} + 3 = 0$
40. $\sqrt[3]{3x + 1} - 5 = 0$
41. $-\sqrt{26 - 11x} + 4 = x$
42. $x + \sqrt{31 - 9x} = 5$
43. $\sqrt{x + 1} = \sqrt{3x + 1}$
44. $\sqrt{x + 5} = \sqrt{x - 5}$
45. $\sqrt{x} - \sqrt{x - 5} = 1$
46. $\sqrt{x} + \sqrt{x - 20} = 10$
47. $\sqrt{x + 5} + \sqrt{x - 5} = 10$
48. $2\sqrt{x + 1} - \sqrt{2x + 3} = 1$
49. $\sqrt{x + 2} - \sqrt{2x - 3} = -1$
50. $4\sqrt{x - 3} - \sqrt{6x - 17} = 3$
51. $(x - 5)^{3/2} = 8$
52. $(x + 3)^{3/2} = 8$
53. $(x + 3)^{2/3} = 8$
54. $(x + 2)^{2/3} = 9$
55. $(x^2 - 5)^{3/2} = 27$
56. $(x^2 - x - 22)^{3/2} = 27$
57. $3x(x - 1)^{1/2} + 2(x - 1)^{3/2} = 0$
58. $4x^2(x - 1)^{1/3} + 6x(x - 1)^{4/3} = 0$

GRAPHICAL ANALYSIS In Exercises 59–62, (a) use a graphing utility to graph the equation, (b) use the graph to approximate any x-intercepts of the graph, (c) set $y = 0$ and solve the resulting equation, and (d) compare the result of part (c) with the x-intercepts of the graph.

59. $y = \sqrt{11x - 30} - x$
60. $y = 2x - \sqrt{15 - 4x}$
61. $y = \sqrt{7x + 36} - \sqrt{5x + 16} - 2$
62. $y = 3\sqrt{x} - \dfrac{4}{\sqrt{x}} - 4$

In Exercises 63–76, find all solutions of the equation. Check your solutions in the original equation.

63. $x = \dfrac{3}{x} + \dfrac{1}{2}$

64. $\dfrac{4}{x} - \dfrac{5}{3} = \dfrac{x}{6}$

65. $\dfrac{1}{x} - \dfrac{1}{x+1} = 3$

66. $\dfrac{4}{x+1} - \dfrac{3}{x+2} = 1$

67. $\dfrac{30 - x}{x} = x$

68. $4x + 1 = \dfrac{3}{x}$

69. $\dfrac{x}{x^2 - 4} + \dfrac{1}{x+2} = 3$

70. $\dfrac{x+1}{3} - \dfrac{x+1}{x+2} = 0$

71. $|2x - 5| = 11$

72. $|3x + 2| = 7$

73. $|x| = x^2 + x - 24$

74. $|x^2 + 6x| = 3x + 18$

75. $|x + 1| = x^2 - 5$

76. $|x - 15| = x^2 - 15x$

GRAPHICAL ANALYSIS In Exercises 77–80, (a) use a graphing utility to graph the equation, (b) use the graph to approximate any x-intercepts of the graph, (c) set $y = 0$ and solve the resulting equation, and (d) compare the result of part (c) with the x-intercepts of the graph.

77. $y = \dfrac{1}{x} - \dfrac{4}{x-1} - 1$

78. $y = x + \dfrac{9}{x+1} - 5$

79. $y = |x + 1| - 2$

80. $y = |x - 2| - 3$

In Exercises 81–88, find the real solutions of the equation algebraically. (Round your answers to three decimal places.)

81. $3.2x^4 - 1.5x^2 - 2.1 = 0$

82. $0.1x^4 - 2.4x^2 - 3.6 = 0$

83. $7.08x^6 + 4.15x^3 - 9.6 = 0$

84. $5.25x^6 - 0.2x^3 - 1.55 = 0$

85. $1.8x - 6\sqrt{x} - 5.6 = 0$

86. $2.4x - 12.4\sqrt{x} + 0.28 = 0$

87. $4x^{2/3} + 8x^{1/3} + 3.6 = 0$

88. $8.4x^{2/3} - 1.2x^{1/3} - 24 = 0$

THINK ABOUT IT In Exercises 89–98, find an equation that has the given solutions. (There are many correct answers.)

89. $-4, 7$

90. $0, 2, 9$

91. $-\dfrac{7}{3}, \dfrac{6}{7}$

92. $-\dfrac{1}{8}, -\dfrac{4}{5}$

93. $\sqrt{3}, -\sqrt{3}, 4$

94. $2\sqrt{7}, -\sqrt{7}$

95. $i, -i$

96. $2i, -2i$

97. $-1, 1, i, -i$

98. $4i, -4i, 6, -6$

99. CHARTERING A BUS A college charters a bus for $1700 to take a group to a museum. When six more students join the trip, the cost per student drops by $7.50. How many students were in the original group?

100. RENTING AN APARTMENT Three students are planning to rent an apartment for a year and share equally in the cost. By adding a fourth person, each person could save $75 a month. How much is the monthly rent?

101. AIRSPEED An airline runs a commuter flight between Portland, Oregon and Seattle, Washington, which are 145 miles apart. If the average speed of the plane could be increased by 40 miles per hour, the travel time would be decreased by 12 minutes. What airspeed is required to obtain this decrease in travel time?

102. AVERAGE SPEED A family drove 1080 miles to their vacation lodge. Because of increased traffic density, their average speed on the return trip was decreased by 6 miles per hour and the trip took $2\frac{1}{2}$ hours longer. Determine their average speed on the way to the lodge.

103. MUTUAL FUNDS A deposit of $2500 in a mutual fund reaches a balance of $3052.49 after 5 years. What annual interest rate on a certificate of deposit compounded monthly would yield an equivalent return?

104. MUTUAL FUNDS A sales representative for a mutual funds company describes a "guaranteed investment fund" that the company is offering to new investors. You are told that if you deposit $10,000 in the fund you will be guaranteed a return of at least $25,000 after 20 years. (Assume the interest is compounded quarterly.)

(a) What is the annual interest rate if the investment only meets the minimum guaranteed amount?

(b) After 20 years, you receive $32,000. What is the annual interest rate?

105. NUMBER OF DOCTORS The number of medical doctors D (in thousands) in the United States from 1998 through 2006 can be modeled by

$$D = 431.61 + 121.8\sqrt{t}, \quad 8 \le t \le 16$$

where t represents the year, with $t = 8$ corresponding to 1998. (Source: American Medical Association)

(a) In which year did the number of medical doctors reach 875,000?

(b) Use the model to predict when the number of medical doctors will reach 1,000,000. Is this prediction reasonable? Explain.

106. VOTING POPULATION The total voting-age population P (in millions) in the United States from 1990 through 2006 can be modeled by

$$P = \frac{182.17 - 1.552t}{1.00 - 0.018t}, \quad 0 \le t \le 16$$

where t represents the year, with $t = 0$ corresponding to 1990. (Source: U.S. Census Bureau)

(a) In which year did the total voting-age population reach 210 million?

(b) Use the model to predict when the total voting-age population will reach 245 million. Is this prediction reasonable? Explain.

107. SATURATED STEAM The temperature T (in degrees Fahrenheit) of saturated steam increases as pressure increases. This relationship is approximated by the model

$$T = 75.82 - 2.11x + 43.51\sqrt{x}, \quad 5 \le x \le 40$$

where x is the absolute pressure (in pounds per square inch).

(a) Use the model to complete the table.

Absolute pressure, x	5	10	15	20
Temperature, T				

Absolute pressure, x	25	30	35	40
Temperature, T				

(b) The temperature of steam at sea level is 212°F. Use the table in part (a) to approximate the absolute pressure at this temperature.

(c) Solve part (b) algebraically.

(d) Use a graphing utility to verify your solutions for parts (b) and (c).

108. AIRLINE PASSENGERS An airline offers daily flights between Chicago and Denver. The total monthly cost C (in millions of dollars) of these flights is

$$C = \sqrt{0.2x + 1}$$

where x is the number of passengers (in thousands). The total cost of the flights for June is 2.5 million dollars. How many passengers flew in June?

109. DEMAND The demand equation for a video game is modeled by

$$p = 40 - \sqrt{0.01x + 1}$$

where x is the number of units demanded per day and p is the price per unit. Approximate the demand when the price is $37.55.

110. DEMAND The demand equation for a high definition television set is modeled by

$$p = 800 - \sqrt{0.01x + 1}$$

where x is the number of units demanded per month and p is the price per unit. Approximate the demand when the price is $750.

111. BASEBALL A baseball diamond has the shape of a square in which the distance from home plate to second base is approximately $127\frac{1}{2}$ feet. Approximate the distance between the bases.

112. METEOROLOGY A meteorologist is positioned 100 feet from the point where a weather balloon is launched. When the balloon is at height h, the distance d (in feet) between the meteorologist and the balloon is $d = \sqrt{100^2 + h^2}$.

(a) Use a graphing utility to graph the equation. Use the *trace* feature to approximate the value of h when $d = 200$.

(b) Complete the table. Use the table to approximate the value of h when $d = 200$.

h	160	165	170	175	180	185
d						

(c) Find h algebraically when $d = 200$.

(d) Compare the results of the three methods. In each case, what information did you gain that was not apparent in another solution method?

113. GEOMETRY You construct a cone with a base radius of 8 inches. The lateral surface area S of the cone can be represented by the equation

$$S = 8\pi\sqrt{64 + h^2}$$

where h is the height of the cone.

 (a) Use a graphing utility to graph the equation. Use the *trace* feature to approximate the value of h when $S = 350$ square inches.

(b) Complete the table. Use the table to approximate the value of h when $S = 350$.

h	8	9	10	11	12	13
S						

(c) Find h algebraically when $S = 350$.

(d) Compare the results of the three methods. In each case, what information did you gain that was not apparent in another solution method?

114. LABOR Working together, two people can complete a task in 8 hours. Working alone, one person takes 2 hours longer than the other to complete the task. How long would it take for each person to complete the task?

115. LABOR Working together, two people can complete a task in 12 hours. Working alone, one person takes 3 hours longer than the other to complete the task. How long would it take for each person to complete the task?

 116. POWER LINE A power station is on one side of a river that is $\frac{3}{4}$ mile wide, and a factory is 8 miles downstream on the other side of the river, as shown in the figure. It costs $24 per foot to run power lines over land and $30 per foot to run them under water.

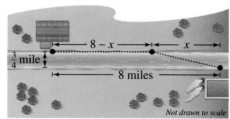

Not drawn to scale

(a) Write the total cost C of running power lines in terms of x (see figure).

(b) Find the total cost when $x = 3$.

(c) Find the length x when $C = \$1,098,662.40$.

(d) Use a graphing utility to graph the equation from part (a).

(e) Use your graph from part (d) to find the value of x that minimizes the cost.

In Exercises 117 and 118, solve for the indicated variable.

117. A PERSON'S TANGENTIAL SPEED IN A ROTOR

Solve for g: $v = \sqrt{\dfrac{gR}{\mu s}}$

118. INDUCTANCE

Solve for Q: $i = \pm\sqrt{\dfrac{1}{LC}}\sqrt{Q^2 - q}$

EXPLORATION

TRUE OR FALSE? In Exercises 119–121, determine whether the statement is true or false. Justify your answer.

119. An equation can never have more than one extraneous solution.

120. When solving an absolute value equation, you will always have to check more than one solution.

121. The equation $\sqrt{x + 10} - \sqrt{x - 10} = 0$ has no solution.

122. CAPSTONE When solving an equation, list three operations that can introduce an extraneous solution. Write an equation that has an extraneous solution.

In Exercises 123 and 124, find x such that the distance between the given points is 13. Explain your results.

123. $(1, 2), (x, -10)$ **124.** $(-8, 0), (x, 5)$

In Exercises 125 and 126, find y such that the distance between the given points is 17. Explain your results.

125. $(0, 0), (8, y)$

126. $(-8, 4), (7, y)$

In Exercises 127 and 128, consider an equation of the form $x + |x - a| = b$, where a and b are constants.

127. Find a and b when the solution of the equation is $x = 9$. (There are many correct answers.)

128. WRITING Write a short paragraph listing the steps required to solve this equation involving absolute values and explain why it is important to check your solutions.

In Exercises 129 and 130, consider an equation of the form $x + \sqrt{x - a} = b$, where a and b are constants.

129. Find a and b when the solution of the equation is $x = 20$. (There are many correct answers.)

130. WRITING Write a short paragraph listing the steps required to solve this equation involving radicals and explain why it is important to check your solutions.

1.7 LINEAR INEQUALITIES IN ONE VARIABLE

What you should learn

- Represent solutions of linear inequalities in one variable.
- Use properties of inequalities to create equivalent inequalities.
- Solve linear inequalities in one variable.
- Solve inequalities involving absolute values.
- Use inequalities to model and solve real-life problems.

Why you should learn it

Inequalities can be used to model and solve real-life problems. For instance, in Exercise 121 on page 148, you will use a linear inequality to analyze the average salary for elementary school teachers.

© Jose Luis Pelaez, Inc./Corbis

Introduction

Simple inequalities were discussed in Section P.1. There, you used the inequality symbols $<$, $\leq$, $>$, and $\geq$ to compare two numbers and to denote subsets of real numbers. For instance, the simple inequality

$$x \geq 3$$

denotes all real numbers x that are greater than or equal to 3.

Now, you will expand your work with inequalities to include more involved statements such as

$$5x - 7 < 3x + 9$$

and

$$-3 \leq 6x - 1 < 3.$$

As with an equation, you **solve an inequality** in the variable x by finding all values of x for which the inequality is true. Such values are **solutions** and are said to **satisfy** the inequality. The set of all real numbers that are solutions of an inequality is the **solution set** of the inequality. For instance, the solution set of

$$x + 1 < 4$$

is all real numbers that are less than 3.

The set of all points on the real number line that represents the solution set is the **graph of the inequality.** Graphs of many types of inequalities consist of intervals on the real number line. See Section P.1 to review the nine basic types of intervals on the real number line. Note that each type of interval can be classified as *bounded* or *unbounded*.

Example 1 Intervals and Inequalities

Write an inequality to represent each interval, and state whether the interval is bounded or unbounded.

a. $(-3, 5]$

b. $(-3, \infty)$

c. $[0, 2]$

d. $(-\infty, \infty)$

Solution

a. $(-3, 5]$ corresponds to $-3 < x \leq 5$.

b. $(-3, \infty)$ corresponds to $-3 < x$.

c. $[0, 2]$ corresponds to $0 \leq x \leq 2$.

d. $(-\infty, \infty)$ corresponds to $-\infty < x < \infty$.

CHECK**Point** Now try Exercise 9.

Properties of Inequalities

The procedures for solving linear inequalities in one variable are much like those for solving linear equations. To isolate the variable, you can make use of the **Properties of Inequalities.** These properties are similar to the properties of equality, but there are two important exceptions. When each side of an inequality is multiplied or divided by a negative number, the direction of the inequality symbol must be reversed. Here is an example.

$$-2 < 5$$
$$(-2) > \qquad (5)$$
$$6 > -15$$

Notice that if the inequality was not reversed, you would obtain the false statement $6 < -15$.

Two inequalities that have the same solution set are **equivalent.** For instance, the inequalities

$$x + 2 < 5$$

and

$$x < 3$$

are equivalent. To obtain the second inequality from the first, you can subtract 2 from each side of the inequality. The following list describes the operations that can be used to create equivalent inequalities.

Properties of Inequalities

Let a, b, c, and d be real numbers.

1. Transitive Property

$$a < b \text{ and } b < c \quad \Longrightarrow \quad a < c$$

2. Addition of Inequalities

$$a < b \text{ and } c < d \quad \Longrightarrow \quad a + c < b + d$$

3. Addition of a Constant

$$a < b \quad \Longrightarrow \quad a + c < b + c$$

4. Multiplication by a Constant

For $c > 0, a < b \quad \Longrightarrow \quad ac < bc$

For $c < 0, a < b \quad \Longrightarrow \quad ac > bc$

Each of the properties above is true if the symbol $<$ is replaced by $\leq$ and the symbol $>$ is replaced by $\geq$. For instance, another form of the multiplication property would be as follows.

For $c > 0, a \leq b \quad \Longrightarrow \quad ac \leq bc$

For $c < 0, a \leq b \quad \Longrightarrow \quad ac \geq bc$

Solving a Linear Inequality in One Variable

The simplest type of inequality is a **linear inequality** in one variable. For instance, $2x + 3 > 4$ is a linear inequality in x.

In the following examples, pay special attention to the steps in which the inequality symbol is reversed. Remember that when you multiply or divide by a negative number, you must reverse the inequality symbol.

Example 2   Solving a Linear Inequality

Solve $5x - 7 > 3x + 9$.

Solution

$$5x - 7 > 3x + 9$$

$$2x - 7 > 9$$

$$2x > 16$$

$$x > 8$$

The solution set is all real numbers that are greater than 8, which is denoted by $(8, \infty)$. The graph of this solution set is shown in Figure 1.24. Note that a parenthesis at 8 on the real number line indicates that 8 *is not* part of the solution set.

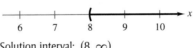

Solution interval: $(8, \infty)$

FIGURE **1.24**

CHECK*Point*   Now try Exercise 35.

> **Study Tip**
>
> Checking the solution set of an inequality is not as simple as checking the solutions of an equation. You can, however, get an indication of the validity of a solution set by substituting a few convenient values of x. For instance, in Example 2, try substituting $x = 5$ and $x = 10$ into the original inequality.

Example 3   Solving a Linear Inequality

Solve $1 - \frac{3}{2}x \geq x - 4$.

Algebraic Solution

$$1 - \frac{3x}{2} \geq x - 4$$

$$2 - 3x \geq 2x - 8$$

$$2 - 5x \geq -8$$

$$-5x \geq -10$$

$$x \leq 2$$

The solution set is all real numbers that are less than or equal to 2, which is denoted by $(-\infty, 2]$. The graph of this solution set is shown in Figure 1.25. Note that a bracket at 2 on the real number line indicates that 2 *is* part of the solution set.

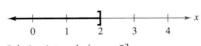

Solution interval: $(-\infty, 2]$

FIGURE **1.25**

CHECK*Point*   Now try Exercise 37.

Graphical Solution

Use a graphing utility to graph $y_1 = 1 - \frac{3}{2}x$ and $y_2 = x - 4$ in the same viewing window. In Figure 1.26, you can see that the graphs appear to intersect at the point $(2, -2)$. Use the *intersect* feature of the graphing utility to confirm this. The graph of y_1 lies above the graph of y_2 to the left of their point of intersection, which implies that $y_1 \geq y_2$ for all $x \leq 2$.

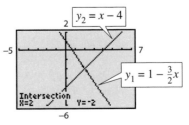

FIGURE **1.26**

Sometimes it is possible to write two inequalities as a **double inequality.** For instance, you can write the two inequalities $-4 \le 5x - 2$ and $5x - 2 < 7$ more simply as

$$-4 \le 5x - 2 < 7.$$

This form allows you to solve the two inequalities together, as demonstrated in Example 4.

Example 4   Solving a Double Inequality

To solve a double inequality, you can isolate x as the middle term.

$$-3 \le 6x - 1 < 3$$

$$-3 \quad \le 6x - 1 \quad < 3$$

$$-2 \le 6x < 4$$

$$\frac{-2}{-} \le \frac{6x}{-} < \frac{4}{-}$$

$$-\frac{1}{3} \le x < \frac{2}{3}$$

The solution set is all real numbers that are greater than or equal to $-\frac{1}{3}$ and less than $\frac{2}{3}$, which is denoted by $\left[-\frac{1}{3}, \frac{2}{3}\right)$. The graph of this solution set is shown in Figure 1.27.

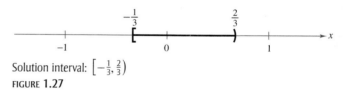

Solution interval: $\left[-\frac{1}{3}, \frac{2}{3}\right)$
FIGURE **1.27**

CHECK *Point*   Now try Exercise 47.

The double inequality in Example 4 could have been solved in two parts, as follows.

$$-3 \le 6x - 1 \qquad \text{and} \qquad 6x - 1 < 3$$

$$-2 \le 6x \qquad\qquad\qquad 6x < 4$$

$$-\frac{1}{3} \le x \qquad\qquad\qquad x < \frac{2}{3}$$

The solution set consists of all real numbers that satisfy *both* inequalities. In other words, the solution set is the set of all values of x for which

$$-\frac{1}{3} \le x < \frac{2}{3}.$$

When combining two inequalities to form a double inequality, be sure that the inequalities satisfy the Transitive Property. For instance, it is *incorrect* to combine the inequalities $3 < x$ and $x \le -1$ as $3 < x \le -1$. This "inequality" is wrong because 3 is not less than -1.

Inequalities Involving Absolute Values

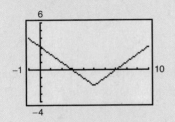

Solving an Absolute Value Inequality

Let *x* be a variable or an algebraic expression and let *a* be a real number such that $a \geq 0$.

1. The solutions of $|x| < a$ are all values of *x* that lie between $-a$ and *a*.

   $$|x| < a \quad \text{if and only if} \quad -a < x < a.$$

2. The solutions of $|x| > a$ are all values of *x* that are less than $-a$ or greater than *a*.

   $$|x| > a \quad \text{if and only if} \quad x < -a \quad \text{or} \quad x > a.$$

These rules are also valid if < is replaced by $\leq$ and > is replaced by $\geq$.

Example 5 Solving an Absolute Value Inequality

Solve each inequality.

a. $|x - 5| < 2$ **b.** $|x + 3| \geq 7$

Solution

a. $|x - 5| < 2$

$$-2 < x - 5 < 2$$
$$-2 \quad < x - 5 \quad < 2$$
$$3 < x < 7$$

The solution set is all real numbers that are greater than 3 and less than 7, which is denoted by (3, 7). The graph of this solution set is shown in Figure 1.28.

b. $|x + 3| \geq 7$

$$x + 3 \leq -7 \quad \text{or} \quad x + 3 \geq 7$$
$$x + 3 \quad \leq -7 \qquad x + 3 \quad \geq 7$$
$$x \leq -10 \qquad x \geq 4$$

The solution set is all real numbers that are less than or equal to -10 *or* greater than or equal to 4. The interval notation for this solution set is $(-\infty, -10] \cup [4, \infty)$. The symbol $\cup$ is called a *union* symbol and is used to denote the combining of two sets. The graph of this solution set is shown in Figure 1.29.

$|x - 5| < 2$: Solutions lie inside (3, 7).

FIGURE 1.28

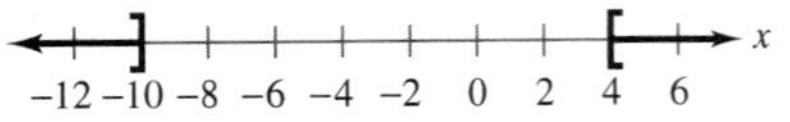

$|x + 3| \geq 7$: Solutions lie outside $(-10, 4)$.

FIGURE 1.29

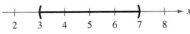 Now try Exercise 61.

Applications

A problem-solving plan can be used to model and solve real-life problems that involve inequalities, as illustrated in Example 6.

Example 6   Comparative Shopping

You are choosing between two different cell phone plans. Plan A costs $49.99 per month for 500 minutes plus $0.40 for each additional minute. Plan B costs $45.99 per month for 500 minutes plus $0.45 for each additional minute. How many *additional* minutes must you use in one month for plan B to cost more than plan A?

Solution

Verbal Model: | Monthly cost for plan B | > | Monthly cost for plan A |

Labels: Minutes used (over 500) in one month $= m$ (minutes)
Monthly cost for plan A $= 0.40m + 49.99$ (dollars)
Monthly cost for plan B $= 0.45m + 45.99$ (dollars)

Inequality: $0.45m + 45.99 > 0.40m + 49.99$

$$0.05m > 4$$

$$m > 80 \text{ minutes}$$

Plan B costs more if you use more than 80 additional minutes in one month.

CHECK Point   Now try Exercise 111.

Example 7   Accuracy of a Measurement

You go to a candy store to buy chocolates that cost $9.89 per pound. The scale that is used in the store has a state seal of approval that indicates the scale is accurate to within half an ounce (or $\frac{1}{32}$ of a pound). According to the scale, your purchase weighs one-half pound and costs $4.95. How much might you have been undercharged or overcharged as a result of inaccuracy in the scale?

Solution

Let x represent the *true* weight of the candy. Because the scale is accurate to within half an ounce (or $\frac{1}{32}$ of a pound), the difference between the exact weight (x) and the scale weight $\left(\frac{1}{2}\right)$ is less than or equal to $\frac{1}{32}$ of a pound. That is, $\left| x - \frac{1}{2} \right| \le \frac{1}{32}$. You can solve this inequality as follows.

$$-\tfrac{1}{32} \le x - \tfrac{1}{2} \le \tfrac{1}{32}$$
$$\tfrac{15}{32} \le x \le \tfrac{17}{32}$$

$$0.46875 \le x \le 0.53125$$

In other words, your "one-half pound" of candy could have weighed as little as 0.46875 pound (which would have cost $4.64) or as much as 0.53125 pound (which would have cost $5.25). So, you could have been overcharged by as much as $0.31 or undercharged by as much as $0.30.

CHECK Point   Now try Exercise 125.

1.7 EXERCISES

See www.CalcChat.com for worked-out solutions to odd-numbered exercises.

VOCABULARY: Fill in the blanks.

1. The set of all real numbers that are solutions of an inequality is the _____ _____ of the inequality.

2. The set of all points on the real number line that represents the solution set of an inequality is the _____ of the inequality.

3. To solve a linear inequality in one variable, you can use the properties of inequalities, which are identical to those used to solve equations, with the exception of multiplying or dividing each side by a _____ number.

4. Two inequalities that have the same solution set are _____.

5. It is sometimes possible to write two inequalities as one inequality, called a _____ inequality.

6. The symbol $\cup$ is called a _____ symbol and is used to denote the combining of two sets.

SKILLS AND APPLICATIONS

In Exercises 7–14, (a) write an inequality that represents the interval and (b) state whether the interval is bounded or unbounded.

7. $[0, 9)$
8. $(-7, 4)$
9. $[-1, 5]$
10. $(2, 10]$
11. $(11, \infty)$
12. $[-5, \infty)$
13. $(-\infty, -2)$
14. $(-\infty, 7]$

In Exercises 15–22, match the inequality with its graph. [The graphs are labeled (a)–(h).]

(a)

(b)
```
         )
  2    3      4      5      6
```

(c)
```
    (        )
-3 -2 -1  0  1  2  3  4  5  6
```

(d)
```
     [              ]
-1   0   1   2   3   4   5
```

(e)
```
(                  ]
-3 -2 -1  0  1  2  3  4  5  6
```

(f)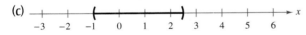
```
     (              )
-5 -4 -3 -2 -1  0  1  2  3  4  5
```

(g)
```
        [          ]
-3 -2 -1  0  1  2  3  4  5  6
```

(h)
```
        [
  4      5      6      7      8
```

15. $x < 3$
16. $x \geq 5$
17. $-3 < x \leq 4$
18. $0 \leq x \leq \frac{9}{2}$
19. $|x| < 3$
20. $|x| > 4$
21. $-1 \leq x \leq \frac{5}{2}$
22. $-1 < x < \frac{5}{2}$

In Exercises 23–28, determine whether each value of x is a solution of the inequality.

Inequality		*Values*	
23. $5x - 12 > 0$	(a) $x = 3$	(b) $x = -3$	
	(c) $x = \frac{5}{2}$	(d) $x = \frac{3}{2}$	
24. $2x + 1 < -3$	(a) $x = 0$	(b) $x = -\frac{1}{4}$	
	(c) $x = -4$	(d) $x = -\frac{3}{2}$	
25. $0 < \dfrac{x - 2}{4} < 2$	(a) $x = 4$	(b) $x = 10$	
	(c) $x = 0$	(d) $x = \frac{7}{2}$	
26. $-5 < 2x - 1 \leq 1$	(a) $x = -\frac{1}{2}$	(b) $x = -\frac{5}{2}$	
	(c) $x = \frac{4}{3}$	(d) $x = 0$	
27. $|x - 10| \geq 3$	(a) $x = 13$	(b) $x = -1$	
	(c) $x = 14$	(d) $x = 9$	
28. $|2x - 3| < 15$	(a) $x = -6$	(b) $x = 0$	
	(c) $x = 12$	(d) $x = 7$	

In Exercises 29–56, solve the inequality and sketch the solution on the real number line. (Some inequalities have no solutions.)

29. $4x < 12$
30. $10x < -40$
31. $-2x > -3$
32. $-6x > 15$
33. $x - 5 \geq 7$
34. $x + 7 \leq 12$
35. $2x + 7 < 3 + 4x$
36. $3x + 1 \geq 2 + x$
37. $2x - 1 \geq 1 - 5x$
38. $6x - 4 \leq 2 + 8x$
39. $4 - 2x < 3(3 - x)$
40. $4(x + 1) < 2x + 3$
41. $\frac{3}{4}x - 6 \leq x - 7$
42. $3 + \frac{2}{7}x > x - 2$
43. $\frac{1}{2}(8x + 1) \geq 3x + \frac{5}{2}$
44. $9x - 1 < \frac{3}{4}(16x - 2)$
45. $3.6x + 11 \geq -3.4$
46. $15.6 - 1.3x < -5.2$
47. $1 < 2x + 3 < 9$
48. $-8 \leq -(3x + 5) < 13$
49. $-8 \leq 1 - 3(x - 2) < 13$
50. $0 \leq 2 - 3(x + 1) < 20$

51. $-4 < \dfrac{2x - 3}{3} < 4$ **52.** $0 \le \dfrac{x + 3}{2} < 5$

53. $\dfrac{3}{4} > x + 1 > \dfrac{1}{4}$ **54.** $-1 < 2 - \dfrac{x}{3} < 1$

55. $3.2 \le 0.4x - 1 \le 4.4$ **56.** $4.5 > \dfrac{1.5x + 6}{2} > 10.5$

In Exercises 57–72, solve the inequality and sketch the solution on the real number line. (Some inequalities have no solution.)

57. $|x| < 5$ **58.** $|x| \ge 8$

59. $\left|\dfrac{x}{2}\right| > 1$ **60.** $\left|\dfrac{x}{5}\right| > 3$

61. $|x - 5| < -1$ **62.** $|x - 7| < -5$

63. $|x - 20| \le 6$ **64.** $|x - 8| \ge 0$

65. $|3 - 4x| \ge 9$ **66.** $|1 - 2x| < 5$

67. $\left|\dfrac{x - 3}{2}\right| \ge 4$ **68.** $\left|1 - \dfrac{2x}{3}\right| < 1$

69. $|9 - 2x| - 2 < -1$ **70.** $|x + 14| + 3 > 17$

71. $2|x + 10| \ge 9$ **72.** $3|4 - 5x| \le 9$

GRAPHICAL ANALYSIS In Exercises 73–82, use a graphing utility to graph the inequality and identify the solution set.

73. $6x > 12$ **74.** $3x - 1 \le 5$

75. $5 - 2x \ge 1$ **76.** $20 < 6x - 1$

77. $4(x - 3) \le 8 - x$ **78.** $3(x + 1) < x + 7$

79. $|x - 8| \le 14$ **80.** $|2x + 9| > 13$

81. $2|x + 7| \ge 13$ **82.** $\frac{1}{2}|x + 1| \le 3$

GRAPHICAL ANALYSIS In Exercises 83–88, use a graphing utility to graph the equation. Use the graph to approximate the values of x that satisfy each inequality.

Equation	*Inequalities*	
83. $y = 2x - 3$	(a) $y \ge 1$	(b) $y \le 0$
84. $y = \frac{2}{3}x + 1$	(a) $y \le 5$	(b) $y \ge 0$
85. $y = -\frac{1}{2}x + 2$	(a) $0 \le y \le 3$	(b) $y \ge 0$
86. $y = -3x + 8$	(a) $-1 \le y \le 3$	(b) $y \le 0$
87. $y = |x - 3|$	(a) $y \le 2$	(b) $y \ge 4$
88. $y = \left|\frac{1}{2}x + 1\right|$	(a) $y \le 4$	(b) $y \ge 1$

In Exercises 89–94, find the interval(s) on the real number line for which the radicand is nonnegative.

89. $\sqrt{x - 5}$ **90.** $\sqrt{x - 10}$

91. $\sqrt{x + 3}$ **92.** $\sqrt{3 - x}$

93. $\sqrt[4]{7 - 2x}$ **94.** $\sqrt[4]{6x + 15}$

95. THINK ABOUT IT   The graph of $|x - 5| < 3$ can be described as all real numbers within three units of 5. Give a similar description of $|x - 10| < 8$.

96. THINK ABOUT IT   The graph of $|x - 2| > 5$ can be described as all real numbers more than five units from 2. Give a similar description of $|x - 8| > 4$.

In Exercises 97–104, use absolute value notation to define the interval (or pair of intervals) on the real number line.

97.

98.

99.

100.

101. All real numbers within 10 units of 12

102. All real numbers at least five units from 8

103. All real numbers more than four units from -3

104. All real numbers no more than seven units from -6

In Exercises 105–108, use inequality notation to describe the subset of real numbers.

105. A company expects its earnings per share E for the next quarter to be no less than \$4.10 and no more than \$4.25.

106. The estimated daily oil production p at a refinery is greater than 2 million barrels but less than 2.4 million barrels.

107. According to a survey, the percent p of U.S. citizens that now conduct most of their banking transactions online is no more than 45%.

108. The net income I of a company is expected to be no less than \$239 million.

PHYSIOLOGY   In Exercises 109 and 110, use the following information. The maximum heart rate of a person in normal health is related to the person's age by the equation $r = 220 - A$, where r is the maximum heart rate in beats per minute and A is the person's age in years. Some physiologists recommend that during physical activity a sedentary person should strive to increase his or her heart rate to at least 50% of the maximum heart rate, and a highly fit person should strive to increase his or her heart rate to at most 85% of the maximum heart rate.   (Source: American Heart Association)

109. Express as an interval the range of the target heart rate for a 20-year-old.

110. Express as an interval the range of the target heart rate for a 40-year-old.

111. JOB OFFERS You are considering two job offers. The first job pays $13.50 per hour. The second job pays $9.00 per hour plus $0.75 per unit produced per hour. Write an inequality yielding the number of units x that must be produced per hour to make the second job pay the greater hourly wage. Solve the inequality.

112. JOB OFFERS You are considering two job offers. The first job pays $3000 per month. The second job pays $1000 per month plus a commission of 4% of your gross sales. Write an inequality yielding the gross sales x per month for which the second job will pay the greater monthly wage. Solve the inequality.

113. INVESTMENT In order for an investment of $1000 to grow to more than $1062.50 in 2 years, what must the annual interest rate be? $[A = P(1 + rt)]$

114. INVESTMENT In order for an investment of $750 to grow to more than $825 in 2 years, what must the annual interest rate be? $[A = P(1 + rt)]$

115. COST, REVENUE, AND PROFIT The revenue from selling x units of a product is $R = 115.95x$. The cost of producing x units is $C = 95x + 750$. To obtain a profit, the revenue must be greater than the cost. For what values of x will this product return a profit?

116. COST, REVENUE, AND PROFIT The revenue from selling x units of a product is $R = 24.55x$. The cost of producing x units is $C = 15.4x + 150,000$. To obtain a profit, the revenue must be greater than the cost. For what values of x will this product return a profit?

117. DAILY SALES A doughnut shop sells a dozen doughnuts for $4.50. Beyond the fixed costs (rent, utilities, and insurance) of $220 per day, it costs $2.75 for enough materials (flour, sugar, and so on) and labor to produce a dozen doughnuts. The daily profit from doughnut sales varies from $60 to $270. Between what levels (in dozens) do the daily sales vary?

118. WEIGHT LOSS PROGRAM A person enrolls in a diet and exercise program that guarantees a loss of at least $1\frac{1}{2}$ pounds per week. The person's weight at the beginning of the program is 164 pounds. Find the maximum number of weeks before the person attains a goal weight of 128 pounds.

 119. DATA ANALYSIS: IQ SCORES AND GPA The admissions office of a college wants to determine whether there is a relationship between IQ scores x and grade-point averages y after the first year of school. An equation that models the data the admissions office obtained is $y = 0.067x - 5.638$.

(a) Use a graphing utility to graph the model.

(b) Use the graph to estimate the values of x that predict a grade-point average of at least 3.0.

120. DATA ANALYSIS: WEIGHTLIFTING You want to determine whether there is a relationship between an athlete's weight x (in pounds) and the athlete's maximum bench-press weight y (in pounds). The table shows a sample of data from 12 athletes.

Athlete's weight, x	Bench-press weight, y
165	170
184	185
150	200
210	255
196	205
240	295
202	190
170	175
185	195
190	185
230	250
160	155

(a) Use a graphing utility to plot the data.

(b) A model for the data is $y = 1.3x - 36$. Use a graphing utility to graph the model in the same viewing window used in part (a).

(c) Use the graph to estimate the values of x that predict a maximum bench-press weight of at least 200 pounds.

(d) Verify your estimate from part (c) algebraically.

(e) Use the graph to write a statement about the accuracy of the model. If you think the graph indicates that an athlete's weight is not a particularly good indicator of the athlete's maximum bench-press weight, list other factors that might influence an individual's maximum bench-press weight.

121. TEACHERS' SALARIES The average salaries S (in thousands of dollars) for elementary school teachers in the United States from 1990 through 2005 are approximated by the model

$$S = 1.09t + 30.9, \quad 0 \le t \le 15$$

where t represents the year, with $t = 0$ corresponding to 1990. (Source: National Education Association)

(a) According to this model, when was the average salary at least $32,500, but not more than $42,000?

(b) According to this model, when will the average salary exceed $54,000?

122. EGG PRODUCTION The numbers of eggs E (in billions) produced in the United States from 1990 through 2006 can be modeled by

$$E = 1.52t + 68.0, \quad 0 \le t \le 16$$

where t represents the year, with $t = 0$ corresponding to 1990.   (Source: U.S. Department of Agriculture)

(a) According to this model, when was the annual egg production 70 billion, but no more than 80 billion?

(b) According to this model, when will the annual egg production exceed 100 billion?

123. GEOMETRY The side of a square is measured as 10.4 inches with a possible error of $\frac{1}{16}$ inch. Using these measurements, determine the interval containing the possible areas of the square.

124. GEOMETRY The side of a square is measured as 24.2 centimeters with a possible error of 0.25 centimeter. Using these measurements, determine the interval containing the possible areas of the square.

125. ACCURACY OF MEASUREMENT You stop at a self-service gas station to buy 15 gallons of 87-octane gasoline at $2.09 a gallon. The gas pump is accurate to within $\frac{1}{10}$ of a gallon. How much might you be undercharged or overcharged?

126. ACCURACY OF MEASUREMENT You buy six T-bone steaks that cost $14.99 per pound. The weight that is listed on the package is 5.72 pounds. The scale that weighed the package is accurate to within $\frac{1}{2}$ ounce. How much might you be undercharged or overcharged?

127. TIME STUDY A time study was conducted to determine the length of time required to perform a particular task in a manufacturing process. The times required by approximately two-thirds of the workers in the study satisfied the inequality

$$\left| \frac{t - 15.6}{1.9} \right| < 1$$

where t is time in minutes. Determine the interval on the real number line in which these times lie.

128. HEIGHT The heights h of two-thirds of the members of a population satisfy the inequality

$$\left| \frac{h - 68.5}{2.7} \right| \le 1$$

where h is measured in inches. Determine the interval on the real number line in which these heights lie.

129. METEOROLOGY An electronic device is to be operated in an environment with relative humidity h in the interval defined by $|h - 50| \le 30$. What are the minimum and maximum relative humidities for the operation of this device?

130. MUSIC Michael Kasha of Florida State University used physics and mathematics to design a new classical guitar. The model he used for the frequency of the vibrations on a circular plate was $v = (2.6t/d^2)\sqrt{E/\rho}$, where v is the frequency (in vibrations per second), t is the plate thickness (in millimeters), d is the diameter of the plate, E is the elasticity of the plate material, and ρ is the density of the plate material. For fixed values of d, E, and ρ, the graph of the equation is a line (see figure).

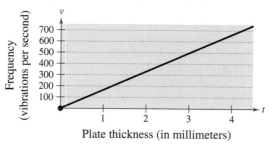

(a) Estimate the frequency when the plate thickness is 2 millimeters.

(b) Estimate the plate thickness when the frequency is 600 vibrations per second.

(c) Approximate the interval for the plate thickness when the frequency is between 200 and 400 vibrations per second.

(d) Approximate the interval for the frequency when the plate thickness is less than 3 millimeters.

EXPLORATION

TRUE OR FALSE? In Exercises 131 and 132, determine whether the statement is true or false. Justify your answer.

131. If a, b, and c are real numbers, and $a \le b$, then $ac \le bc$.

132. If $-10 \le x \le 8$, then $-10 \ge -x$ and $-x \ge -8$.

133. Identify the graph of the inequality $|x - a| \ge 2$.

(a) ![number line from $a-2$ to $a+2$]
(b) ![number line from $a-2$ to $a+2$]
(c) ![number line from $2-a$ to $2+a$]
(d) ![number line from $2-a$ to $2+a$]

134. Find sets of values of a, b, and c such that $0 \le x \le 10$ is a solution of the inequality $|ax - b| \le c$.

135. Give an example of an inequality with an unbounded solution set.

136. CAPSTONE Describe any differences between properties of equalities and properties of inequalities.

Spencer Grant / PhotoEdit

1.8   OTHER TYPES OF INEQUALITIES

What you should learn

- Solve polynomial inequalities.
- Solve rational inequalities.
- Use inequalities to model and solve real-life problems.

Why you should learn it

Inequalities can be used to model and solve real-life problems. For instance, in Exercise 77 on page 158, a polynomial inequality is used to model school enrollment in the United States.

Polynomial Inequalities

To solve a polynomial inequality such as $x^2 - 2x - 3 < 0$, you can use the fact that a polynomial can change signs only at its zeros (the x-values that make the polynomial equal to zero). Between two consecutive zeros, a polynomial must be entirely positive or entirely negative. This means that when the real zeros of a polynomial are put in order, they divide the real number line into intervals in which the polynomial has no sign changes. These zeros are the **key numbers** of the inequality, and the resulting intervals are the **test intervals** for the inequality. For instance, the polynomial above factors as

$$x^2 - 2x - 3 = (x + 1)(x - 3)$$

and has two zeros, $x = -1$ and $x = 3$. These zeros divide the real number line into three test intervals:

$$(-\infty, -1), \quad (-1, 3), \quad \text{and} \quad (3, \infty). \quad \text{(See Figure 1.30.)}$$

So, to solve the inequality $x^2 - 2x - 3 < 0$, you need only test one value from each of these test intervals to determine whether the value satisfies the original inequality. If so, you can conclude that the interval is a solution of the inequality.

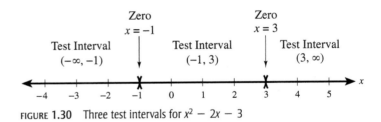

FIGURE 1.30   Three test intervals for $x^2 - 2x - 3$

You can use the same basic approach to determine the test intervals for any polynomial.

Finding Test Intervals for a Polynomial

To determine the intervals on which the values of a polynomial are entirely negative or entirely positive, use the following steps.

1. Find all real zeros of the polynomial, and arrange the zeros in increasing order (from smallest to largest). These zeros are the key numbers of the polynomial.

2. Use the key numbers of the polynomial to determine its test intervals.

3. Choose one representative x-value in each test interval and evaluate the polynomial at that value. If the value of the polynomial is negative, the polynomial will have negative values for every x-value in the interval. If the value of the polynomial is positive, the polynomial will have positive values for every x-value in the interval.

Example 1   Solving a Polynomial Inequality

Solve $x^2 - x - 6 < 0$.

Solution

By factoring the polynomial as

$$x^2 - x - 6 = (x + 2)(x - 3)$$

you can see that the key numbers are $x = -2$ and $x = 3$. So, the polynomial's test intervals are

$$(-\infty, -2), \quad (-2, 3), \quad \text{and} \quad (3, \infty).$$

In each test interval, choose a representative x-value and evaluate the polynomial.

Test Interval	x-Value	Polynomial Value	Conclusion
$(-\infty, -2)$	$x = -3$	$(\ )^2 - (\ ) - 6 = 6$	Positive
$(-2, 3)$	$x = 0$	$(\ )^2 - (\ ) - 6 = -6$	Negative
$(3, \infty)$	$x = 4$	$(\ )^2 - (\ ) - 6 = 6$	Positive

From this you can conclude that the inequality is satisfied for all x-values in $(-2, 3)$. This implies that the solution of the inequality $x^2 - x - 6 < 0$ is the interval $(-2, 3)$, as shown in Figure 1.31. Note that the original inequality contains a "less than" symbol. This means that the solution set does not contain the endpoints of the test interval $(-2, 3)$.

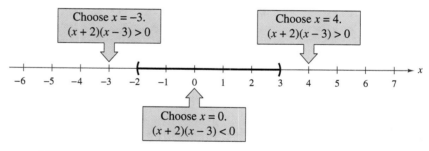

FIGURE **1.31**

CHECK*Point*   Now try Exercise 21.

As with linear inequalities, you can check the reasonableness of a solution by substituting x-values into the original inequality. For instance, to check the solution found in Example 1, try substituting several x-values from the interval $(-2, 3)$ into the inequality

$$x^2 - x - 6 < 0.$$

Regardless of which x-values you choose, the inequality should be satisfied.

You can also use a graph to check the result of Example 1. Sketch the graph of $y = x^2 - x - 6$, as shown in Figure 1.32. Notice that the graph is below the x-axis on the interval $(-2, 3)$.

In Example 1, the polynomial inequality was given in general form (with the polynomial on one side and zero on the other). Whenever this is not the case, you should begin the solution process by writing the inequality in general form.

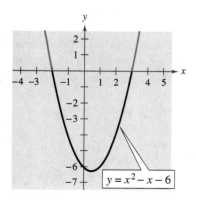

FIGURE **1.32**

Example 2 Solving a Polynomial Inequality

Solve $2x^3 - 3x^2 - 32x > -48$.

Solution

$$2x^3 - 3x^2 - 32x + 48 > 0$$

$$(x - 4)(x + 4)(2x - 3) > 0$$

The key numbers are $x = -4$, $x = \frac{3}{2}$, and $x = 4$, and the test intervals are $(-\infty, -4)$, $\left(-4, \frac{3}{2}\right)$, $\left(\frac{3}{2}, 4\right)$, and $(4, \infty)$.

Test Interval	x-Value	Polynomial Value	Conclusion
$(-\infty, -4)$	$x = -5$	$2(\quad)^3 - 3(\quad)^2 - 32(\quad) + 48$	Negative
$\left(-4, \frac{3}{2}\right)$	$x = 0$	$2(\quad)^3 - 3(\quad)^2 - 32(\quad) + 48$	Positive
$\left(\frac{3}{2}, 4\right)$	$x = 2$	$2(\quad)^3 - 3(\quad)^2 - 32(\quad) + 48$	Negative
$(4, \infty)$	$x = 5$	$2(\quad)^3 - 3(\quad)^2 - 32(\quad) + 48$	Positive

From this you can conclude that the inequality is satisfied on the open intervals $\left(-4, \frac{3}{2}\right)$ and $(4, \infty)$. So, the solution set is $\left(-4, \frac{3}{2}\right) \cup (4, \infty)$, as shown in Figure 1.33.

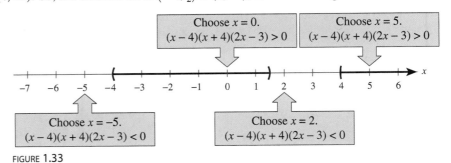

FIGURE **1.33**

CHECK Point Now try Exercise 27.

Example 3 Solving a Polynomial Inequality

Solve $4x^2 - 5x > 6$.

Algebraic Solution

$$4x^2 - 5x - 6 > 0$$

$$(x - 2)(4x + 3) > 0$$

Key Numbers: $x = -\frac{3}{4}$, $x = 2$

Test Intervals: $\left(-\infty, -\frac{3}{4}\right)$, $\left(-\frac{3}{4}, 2\right)$, $(2, \infty)$

Test: Is $(x - 2)(4x + 3) > 0$?

After testing these intervals, you can see that the polynomial $4x^2 - 5x - 6$ is positive on the open intervals $\left(-\infty, -\frac{3}{4}\right)$ and $(2, \infty)$. So, the solution set of the inequality is $\left(-\infty, -\frac{3}{4}\right) \cup (2, \infty)$.

CHECK Point Now try Exercise 23.

Graphical Solution

First write the polynomial inequality $4x^2 - 5x > 6$ as $4x^2 - 5x - 6 > 0$. Then use a graphing utility to graph $y = 4x^2 - 5x - 6$. In Figure 1.34, you can see that the graph is *above* the x-axis when x is less than $-\frac{3}{4}$ or when x is greater than 2. So, you can graphically approximate the solution set to be $\left(-\infty, -\frac{3}{4}\right) \cup (2, \infty)$.

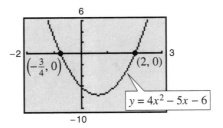

FIGURE **1.34**

Study Tip

You may find it easier to determine the sign of a polynomial from its *factored* form. For instance, in Example 3, if the test value $x = 1$ is substituted into the factored form

$$(x - 2)(4x + 3)$$

you can see that the sign pattern of the factors is

$$(-)(+)$$

which yields a negative result. Try using the factored forms of the polynomials to determine the signs of the polynomials in the test intervals of the other examples in this section.

When solving a polynomial inequality, be sure you have accounted for the particular type of inequality symbol given in the inequality. For instance, in Example 3, note that the original inequality contained a "greater than" symbol and the solution consisted of two open intervals. If the original inequality had been

$$4x^2 - 5x \geq 6$$

the solution would have consisted of the intervals $\left(-\infty, -\frac{3}{4}\right]$ and $[2, \infty)$.

Each of the polynomial inequalities in Examples 1, 2, and 3 has a solution set that consists of a single interval or the union of two intervals. When solving the exercises for this section, watch for unusual solution sets, as illustrated in Example 4.

Example 4   Unusual Solution Sets

a. The solution set of the following inequality consists of the entire set of real numbers, $(-\infty, \infty)$. In other words, the value of the quadratic $x^2 + 2x + 4$ is positive for every real value of x.

$$x^2 + 2x + 4 > 0$$

b. The solution set of the following inequality consists of the single real number $\{-1\}$, because the quadratic $x^2 + 2x + 1$ has only one key number, $x = -1$, and it is the only value that satisfies the inequality.

$$x^2 + 2x + 1 \leq 0$$

c. The solution set of the following inequality is empty. In other words, the quadratic $x^2 + 3x + 5$ is not less than zero for any value of x.

$$x^2 + 3x + 5 < 0$$

d. The solution set of the following inequality consists of all real numbers except $x = 2$. In interval notation, this solution set can be written as $(-\infty, 2) \cup (2, \infty)$.

$$x^2 - 4x + 4 > 0$$

CHECK*Point*   Now try Exercise 29.

Rational Inequalities

The concepts of key numbers and test intervals can be extended to rational inequalities. To do this, use the fact that the value of a rational expression can change sign only at its *zeros* (the x-values for which its numerator is zero) and its *undefined values* (the x-values for which its denominator is zero). These two types of numbers make up the *key numbers* of a rational inequality. When solving a rational inequality, begin by writing the inequality in general form with the rational expression on the left and zero on the right.

| **Example 5** | **Solving a Rational Inequality** |

Solve $\dfrac{2x - 7}{x - 5} \le 3$.

Solution

$$\frac{2x - 7}{x - 5} \le 3$$

$$\frac{2x - 7}{x - 5} - 3 \le 0$$

$$\frac{2x - 7 - 3x + 15}{x - 5} \le 0$$

$$\frac{-x + 8}{x - 5} \le 0$$

Key Numbers: $x = 5, x = 8$

Test Intervals: $(-\infty, 5), (5, 8), (8, \infty)$

Test: Is $\dfrac{-x + 8}{x - 5} \le 0$?

After testing these intervals, as shown in Figure 1.35, you can see that the inequality is satisfied on the open intervals $(-\infty, 5)$ and $(8, \infty)$. Moreover, because $\dfrac{-x + 8}{x - 5} = 0$ when $x = 8$, you can conclude that the solution set consists of all real numbers in the intervals $(-\infty, 5) \cup [8, \infty)$. (Be sure to use a closed interval to indicate that x can equal 8.)

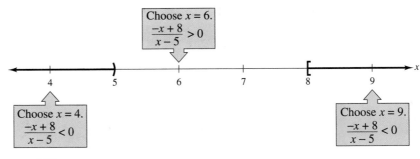

FIGURE **1.35**

CHECK*Point*   Now try Exercise 45.

Applications

One common application of inequalities comes from business and involves profit, revenue, and cost. The formula that relates these three quantities is

Profit $=$ Revenue $-$ Cost

$$P = R - C.$$

Example 6   Increasing the Profit for a Product

The marketing department of a calculator manufacturer has determined that the demand for a new model of calculator is

$$p = 100 - 0.00001x, \quad 0 \leq x \leq 10{,}000{,}000$$

where p is the price per calculator (in dollars) and x represents the number of calculators sold. (If this model is accurate, no one would be willing to pay $100 for the calculator. At the other extreme, the company couldn't sell more than 10 million calculators.) The revenue for selling x calculators is

$$R = xp = x(\qquad\qquad)$$

as shown in Figure 1.36. The total cost of producing x calculators is $10 per calculator plus a development cost of $2,500,000. So, the total cost is

$$C = 10x + 2{,}500{,}000.$$

What price should the company charge per calculator to obtain a profit of at least $190,000,000?

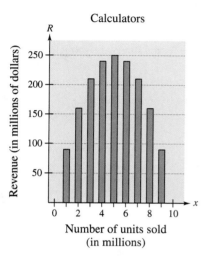

Calculators

Revenue (in millions of dollars)

Number of units sold
(in millions)

FIGURE 1.36

Solution

Verbal Model: Profit $=$ Revenue $-$ Cost

Equation: $P = R - C$

$$P = 100x - 0.00001x^2 - (10x + 2{,}500{,}000)$$

$$P = -0.00001x^2 + 90x - 2{,}500{,}000$$

To answer the question, solve the inequality

$$P \geq 190{,}000{,}000$$

$$\geq 190{,}000{,}000.$$

When you write the inequality in general form, find the key numbers and the test intervals, and then test a value in each test interval, you can find the solution to be

$$3{,}500{,}000 \leq x \leq 5{,}500{,}000$$

as shown in Figure 1.37. Substituting the x-values in the original price equation shows that prices of

$$\$45.00 \leq p \leq \$65.00$$

will yield a profit of at least $190,000,000.

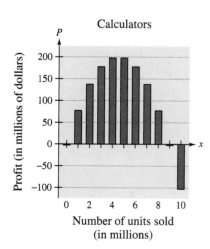

Calculators

Profit (in millions of dollars)

Number of units sold
(in millions)

FIGURE 1.37

CHECK*Point*   Now try Exercise 75.

Another common application of inequalities is finding the domain of an expression that involves a square root, as shown in Example 7.

Example 7 Finding the Domain of an Expression

Find the domain of $\sqrt{64 - 4x^2}$.

Algebraic Solution

Remember that the domain of an expression is the set of all x-values for which the expression is defined. Because $\sqrt{64 - 4x^2}$ is defined (has real values) only if $64 - 4x^2$ is nonnegative, the domain is given by $64 - 4x^2 \geq 0$.

$$64 - 4x^2 \geq 0$$

$$16 - x^2 \geq 0$$

$$(4 - x)(4 + x) \geq 0$$

So, the inequality has two key numbers: $x = -4$ and $x = 4$. You can use these two numbers to test the inequality, as follows.

Key numbers: $x = -4, x = 4$

Test intervals: $(-\infty, -4), (-4, 4), (4, \infty)$

Test: For what values of x is $\sqrt{64 - 4x^2} \geq 0$?

A test shows that the inequality is satisfied in the *closed interval* $[-4, 4]$. So, the domain of the expression $\sqrt{64 - 4x^2}$ is the interval $[-4, 4]$.

CHECK*Point* Now try Exercise 59.

Graphical Solution

Begin by sketching the graph of the equation $y = \sqrt{64 - 4x^2}$, as shown in Figure 1.38. From the graph, you can determine that the x-values extend from -4 to 4 (including -4 and 4). So, the domain of the expression $\sqrt{64 - 4x^2}$ is the interval $[-4, 4]$.

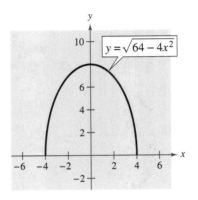

FIGURE **1.38**

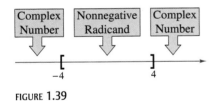

FIGURE **1.39**

To analyze a test interval, choose a representative x-value in the interval and evaluate the expression at that value. For instance, in Example 7, if you substitute any number from the interval $[-4, 4]$ into the expression $\sqrt{64 - 4x^2}$, you will obtain a nonnegative number under the radical symbol that simplifies to a real number. If you substitute any number from the intervals $(-\infty, -4)$ and $(4, \infty)$, you will obtain a complex number. It might be helpful to draw a visual representation of the intervals, as shown in Figure 1.39.

CLASSROOM DISCUSSION

Profit Analysis Consider the relationship

$$P = R - C$$

described on page 155. Write a paragraph discussing why it might be beneficial to solve $P < 0$ if you owned a business. Use the situation described in Example 6 to illustrate your reasoning.

1.8   EXERCISES

See www.CalcChat.com for worked-out solutions to odd-numbered exercises.

VOCABULARY: Fill in the blanks.

1. Between two consecutive zeros, a polynomial must be entirely _____ or entirely _____.
2. To solve a polynomial inequality, find the _____ numbers of the polynomial, and use these numbers to create _____ _____ for the inequality.
3. The key numbers of a rational expression are its _____ and its _____ _____.
4. The formula that relates cost, revenue, and profit is _____.

SKILLS AND APPLICATIONS

In Exercises 5–8, determine whether each value of x is a solution of the inequality.

	Inequality		*Values*	
5.	$x^2 - 3 < 0$	(a) $x = 3$	(b) $x = 0$
		(c) $x = \frac{3}{2}$	(d) $x = -5$
6.	$x^2 - x - 12 \geq 0$	(a) $x = 5$	(b) $x = 0$
		(c) $x = -4$	(d) $x = -3$
7.	$\dfrac{x + 2}{x - 4} \geq 3$	(a) $x = 5$	(b) $x = 4$
		(c) $x = -\frac{9}{2}$	(d) $x = \frac{9}{2}$
8.	$\dfrac{3x^2}{x^2 + 4} < 1$	(a) $x = -2$	(b) $x = -1$
		(c) $x = 0$	(d) $x = 3$

In Exercises 9–12, find the key numbers of the expression.

9. $3x^2 - x - 2$

10. $9x^3 - 25x^2$

11. $\dfrac{1}{x - 5} + 1$

12. $\dfrac{x}{x + 2} - \dfrac{2}{x - 1}$

In Exercises 13–30, solve the inequality and graph the solution on the real number line.

13. $x^2 < 9$

14. $x^2 \leq 16$

15. $(x + 2)^2 \leq 25$

16. $(x - 3)^2 \geq 1$

17. $x^2 + 4x + 4 \geq 9$

18. $x^2 - 6x + 9 < 16$

19. $x^2 + x < 6$

20. $x^2 + 2x > 3$

21. $x^2 + 2x - 3 < 0$

22. $x^2 > 2x + 8$

23. $3x^2 - 11x > 20$

24. $-2x^2 + 6x + 15 \leq 0$

25. $x^2 - 3x - 18 > 0$

26. $x^3 + 2x^2 - 4x - 8 \leq 0$

27. $x^3 - 3x^2 - x > -3$

28. $2x^3 + 13x^2 - 8x - 46 \geq 6$

29. $4x^2 - 4x + 1 \leq 0$

30. $x^2 + 3x + 8 > 0$

In Exercises 31–36, solve the inequality and write the solution set in interval notation.

31. $4x^3 - 6x^2 < 0$

32. $4x^3 - 12x^2 > 0$

33. $x^3 - 4x \geq 0$

34. $2x^3 - x^4 \leq 0$

35. $(x - 1)^2(x + 2)^3 \geq 0$

36. $x^4(x - 3) \leq 0$

GRAPHICAL ANALYSIS   In Exercises 37–40, use a graphing utility to graph the equation. Use the graph to approximate the values of x that satisfy each inequality.

	Equation	*Inequalities*	
37.	$y = -x^2 + 2x + 3$	(a) $y \leq 0$	(b) $y \geq 3$
38.	$y = \frac{1}{2}x^2 - 2x + 1$	(a) $y \leq 0$	(b) $y \geq 7$
39.	$y = \frac{1}{8}x^3 - \frac{1}{2}x$	(a) $y \geq 0$	(b) $y \leq 6$
40.	$y = x^3 - x^2 - 16x + 16$	(a) $y \leq 0$	(b) $y \geq 36$

In Exercises 41–54, solve the inequality and graph the solution on the real number line.

41. $\dfrac{4x - 1}{x} > 0$

42. $\dfrac{x^2 - 1}{x} < 0$

43. $\dfrac{3x - 5}{x - 5} \geq 0$

44. $\dfrac{5 + 7x}{1 + 2x} \leq 4$

45. $\dfrac{x + 6}{x + 1} - 2 < 0$

46. $\dfrac{x + 12}{x + 2} - 3 \geq 0$

47. $\dfrac{2}{x + 5} > \dfrac{1}{x - 3}$

48. $\dfrac{5}{x - 6} > \dfrac{3}{x + 2}$

49. $\dfrac{1}{x - 3} \leq \dfrac{9}{4x + 3}$

50. $\dfrac{1}{x} \geq \dfrac{1}{x + 3}$

51. $\dfrac{x^2 + 2x}{x^2 - 9} \leq 0$

52. $\dfrac{x^2 + x - 6}{x} \geq 0$

53. $\dfrac{3}{x - 1} + \dfrac{2x}{x + 1} > -1$

54. $\dfrac{3x}{x - 1} \leq \dfrac{x}{x + 4} + 3$

GRAPHICAL ANALYSIS In Exercises 55–58, use a graphing utility to graph the equation. Use the graph to approximate the values of x that satisfy each inequality.

	Equation		*Inequalities*	

55. $y = \dfrac{3x}{x-2}$ (a) $y \le 0$ (b) $y \ge 6$

56. $y = \dfrac{2(x-2)}{x+1}$ (a) $y \le 0$ (b) $y \ge 8$

57. $y = \dfrac{2x^2}{x^2+4}$ (a) $y \ge 1$ (b) $y \le 2$

58. $y = \dfrac{5x}{x^2+4}$ (a) $y \ge 1$ (b) $y \le 0$

In Exercises 59–64, find the domain of x in the expression. Use a graphing utility to verify your result.

59. $\sqrt{4 - x^2}$ **60.** $\sqrt{x^2 - 4}$

61. $\sqrt{x^2 - 9x + 20}$ **62.** $\sqrt{81 - 4x^2}$

63. $\sqrt{\dfrac{x}{x^2 - 2x - 35}}$ **64.** $\sqrt{\dfrac{x}{x^2 - 9}}$

In Exercises 65–70, solve the inequality. (Round your answers to two decimal places.)

65. $0.4x^2 + 5.26 < 10.2$

66. $-1.3x^2 + 3.78 > 2.12$

67. $-0.5x^2 + 12.5x + 1.6 > 0$

68. $1.2x^2 + 4.8x + 3.1 < 5.3$

69. $\dfrac{1}{2.3x - 5.2} > 3.4$ **70.** $\dfrac{2}{3.1x - 3.7} > 5.8$

HEIGHT OF A PROJECTILE In Exercises 71 and 72, use the position equation $s = -16t^2 + v_0 t + s_0$, where s represents the height of an object (in feet), v_0 represents the initial velocity of the object (in feet per second), s_0 represents the initial height of the object (in feet), and t represents the time (in seconds).

71. A projectile is fired straight upward from ground level ($s_0 = 0$) with an initial velocity of 160 feet per second.

    (a) At what instant will it be back at ground level?

    (b) When will the height exceed 384 feet?

72. A projectile is fired straight upward from ground level ($s_0 = 0$) with an initial velocity of 128 feet per second.

    (a) At what instant will it be back at ground level?

    (b) When will the height be less than 128 feet?

73. GEOMETRY A rectangular playing field with a perimeter of 100 meters is to have an area of at least 500 square meters. Within what bounds must the length of the rectangle lie?

74. GEOMETRY A rectangular parking lot with a perimeter of 440 feet is to have an area of at least 8000 square feet. Within what bounds must the length of the rectangle lie?

75. COST, REVENUE, AND PROFIT The revenue and cost equations for a product are $R = x(75 - 0.0005x)$ and $C = 30x + 250{,}000$, where R and C are measured in dollars and x represents the number of units sold. How many units must be sold to obtain a profit of at least \$750,000? What is the price per unit?

76. COST, REVENUE, AND PROFIT The revenue and cost equations for a product are

$$R = x(50 - 0.0002x) \quad \text{and} \quad C = 12x + 150{,}000$$

where R and C are measured in dollars and x represents the number of units sold. How many units must be sold to obtain a profit of at least \$1,650,000? What is the price per unit?

 77. SCHOOL ENROLLMENT The numbers N (in millions) of students enrolled in schools in the United States from 1995 through 2006 are shown in the table. (Source: U.S. Census Bureau)

Year	Number, N
1995	69.8
1996	70.3
1997	72.0
1998	72.1
1999	72.4
2000	72.2
2001	73.1
2002	74.0
2003	74.9
2004	75.5
2005	75.8
2006	75.2

(a) Use a graphing utility to create a scatter plot of the data. Let t represent the year, with $t = 5$ corresponding to 1995.

(b) Use the *regression* feature of a graphing utility to find a quartic model for the data.

(c) Graph the model and the scatter plot in the same viewing window. How well does the model fit the data?

(d) According to the model, during what range of years will the number of students enrolled in schools exceed 74 million?

(e) Is the model valid for long-term predictions of student enrollment in schools? Explain.

78. SAFE LOAD The maximum safe load uniformly distributed over a one-foot section of a two-inch-wide wooden beam is approximated by the model Load $= 168.5d^2 - 472.1$, where d is the depth of the beam.

(a) Evaluate the model for $d = 4$, $d = 6$, $d = 8$, $d = 10$, and $d = 12$. Use the results to create a bar graph.

(b) Determine the minimum depth of the beam that will safely support a load of 2000 pounds.

79. RESISTORS When two resistors of resistances R_1 and R_2 are connected in parallel (see figure), the total resistance R satisfies the equation

$$\frac{1}{R} = \frac{1}{R_1} + \frac{1}{R_2}.$$

Find R_1 for a parallel circuit in which $R_2 = 2$ ohms and R must be at least 1 ohm.

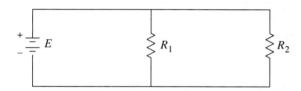

80. TEACHERS' SALARIES The mean salaries S (in thousands of dollars) of classroom teachers in the United States from 2000 through 2007 are shown in the table.

Year	Salary, S
2000	42.2
2001	43.7
2002	43.8
2003	45.0
2004	45.6
2005	45.9
2006	48.2
2007	49.3

A model that approximates these data is given by

$$S = \frac{42.6 - 1.95t}{1 - 0.06t}$$

where t represents the year, with $t = 0$ corresponding to 2000. (Source: Educational Research Service, Arlington, VA)

(a) Use a graphing utility to create a scatter plot of the data. Then graph the model in the same viewing window.

(b) How well does the model fit the data? Explain.

(c) According to the model, in what year will the salary for classroom teachers exceed $60,000?

(d) Is the model valid for long-term predictions of classroom teacher salaries? Explain.

EXPLORATION

TRUE OR FALSE? In Exercises 81 and 82, determine whether the statement is true or false. Justify your answer.

81. The zeros of the polynomial $x^3 - 2x^2 - 11x + 12 \geq 0$ divide the real number line into four test intervals.

82. The solution set of the inequality $\frac{3}{2}x^2 + 3x + 6 \geq 0$ is the entire set of real numbers.

In Exercises 83–86, (a) find the interval(s) for b such that the equation has at least one real solution and (b) write a conjecture about the interval(s) based on the values of the coefficients.

83. $x^2 + bx + 4 = 0$ **84.** $x^2 + bx - 4 = 0$

85. $3x^2 + bx + 10 = 0$ **86.** $2x^2 + bx + 5 = 0$

87. GRAPHICAL ANALYSIS You can use a graphing utility to verify the results in Example 4. For instance, the graph of $y = x^2 + 2x + 4$ is shown below. Notice that the y-values are greater than 0 for all values of x, as stated in Example 4(a). Use the graphing utility to graph $y = x^2 + 2x + 1$, $y = x^2 + 3x + 5$, and $y = x^2 - 4x + 4$. Explain how you can use the graphs to verify the results of parts (b), (c), and (d) of Example 4.

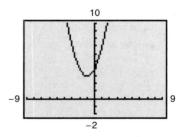

88. CAPSTONE Consider the polynomial

$$(x - a)(x - b)$$

and the real number line shown below.

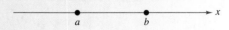

(a) Identify the points on the line at which the polynomial is zero.

(b) In each of the three subintervals of the line, write the sign of each factor and the sign of the product.

(c) At what x-values does the polynomial change signs?

1 CHAPTER SUMMARY

What Did You Learn?	Explanation/Examples	Review Exercises
Section 1.1		
Sketch graphs of equations (*p. 76*), and find *x*- and *y*-intercepts of graphs of equations (*p. 79*).	To graph an equation, make a table of values, plot the points, and connect the points with a smooth curve or line. The points at which a graph intersects or touches the *x*- or *y*-axis are called intercepts.	1–4
Use symmetry to sketch graphs of equations (*p. 80*).	Graphs can have symmetry with respect to one of the coordinate axes or with respect to the origin. You can test for symmetry algebraically and graphically.	5–12
Find equations of and sketch graphs of circles (*p. 82*).	The point (x, y) lies on the circle of radius r and center (h, k) if and only if $(x - h)^2 + (y - k)^2 = r^2$.	13–18
Use graphs of equations in solving real-life problems (*p. 83*).	The graph of an equation can be used to estimate the recommended weight for a man. (See Example 9.)	19, 20
Section 1.2		
Identify different types of equations (*p. 87*).	**Identity:** true for *every* real number in the domain **Conditional equation:** true for just *some* (or even none) of the real numbers in the domain	21–24
Solve linear equations in one variable (*p. 87*), and solve equations that lead to linear equations (*p. 90*).	**Linear equation in one variable:** An equation that can be written in the standard form $ax + b = 0$, where a and b are real numbers with $a \neq 0$. To solve an equation involving fractional expressions, find the LCD of all terms and multiply every term by the LCD.	25–30
Find *x*- and *y*-intercepts algebraically (*p. 91*).	To find *x*-intercepts, set *y* equal to zero and solve for *x*. To find *y*-intercepts, set *x* equal to zero and solve for *y*.	31–36
Use linear equations to model and solve real-life problems (*p. 91*).	A linear equation can be used to model the number of female participants in athletic programs. (See Example 5.)	37, 38
Section 1.3		
Use a verbal model in a problem-solving plan (*p. 96*).	Verbal Description $\Rightarrow$ Verbal Model $\Rightarrow$ Algebraic Equation	39, 40
Use mathematical models to solve real-life problems (*p. 97*).	Mathematical models can be used to find the percent of a raise, and a building's height. (See Examples 2 and 6.)	41, 42
Solve mixture problems (*p. 100*).	Mixture problems include simple interest problems and inventory problems. (See Examples 7 and 8.)	43, 44
Use common formulas to solve real-life problems (*p. 101*).	A literal equation contains more than one variable. A formula is an example of a literal equation. (See Example 9.)	45, 46
Section 1.4		
Solve quadratic equations by factoring (*p. 107*).	The method of factoring is based on the Zero-Factor Property, which states if $ab = 0$, then $a = 0$ or $b = 0$.	47, 48
Solve quadratic equations by extracting square roots (*p. 108*).	The equation $u^2 = d$, where $d > 0$, has exactly two solutions: $u = \sqrt{d}$ and $u = -\sqrt{d}$.	49–52
Solve quadratic equations by completing the square (*p. 109*) and using the Quadratic Formula (*p. 111*).	To complete the square for $x^2 + bx$, add $(b/2)^2$. **Quadratic Formula:** $x = \dfrac{-b \pm \sqrt{b^2 - 4ac}}{2a}$	53–56

	What Did You Learn?	Explanation/Examples	Review Exercises
Section 1.4	Use quadratic equations to model and solve real-life problems *(p. 113)*.	A quadratic equation can be used to model the number of Internet users in the United States from 2000 through 2008. (See Example 9.)	57, 58
Section 1.5	Use the imaginary unit i to write complex numbers *(p. 122)*, and add, subtract, and multiply complex numbers *(p. 123)*.	If a and b are real numbers, $a + bi$ is a complex number. **Sum:** $(a + bi) + (c + di) = (a + c) + (b + d)i$ **Difference:** $(a + bi) - (c + di) = (a - c) + (b - d)i$ The Distributive Property can be used to multiply.	59–66
	Use complex conjugates to write the quotient of two complex numbers in standard form *(p. 125)*.	To write $(a + bi)/(c + di)$ in standard form, multiply the numerator and denominator by the complex conjugate of the denominator, $c - di$.	67–70
	Find complex solutions of quadratic equations *(p. 126)*.	If a is a positive number, the principal square root of the negative number $-a$ is defined as $\sqrt{-a} = \sqrt{a}i$.	71–74
Section 1.6	Solve polynomial equations of degree three or greater *(p. 129)*.	Factoring is the most common method used to solve polynomial equations of degree three or greater.	75–78
	Solve equations involving radicals *(p. 131)*.	Solving equations involving radicals usually involves squaring or cubing each side of the equation.	79–82
	Solve equations involving fractions or absolute values *(p. 132)*.	To solve an equation involving fractions, multiply each side of the equation by the LCD of all terms in the equation. To solve an equation involving an absolute value, remember that the expression inside the absolute value signs can be positive or negative.	83–88
	Use polynomial equations and equations involving radicals to model and solve real-life problems *(p. 134)*.	Polynomial equations can be used to find the number of ski club members going on a ski trip, and the annual interest rate for an investment. (See Examples 8 and 9.)	89, 90
Section 1.7	Represent solutions of linear inequalities in one variable *(p. 140)*.	**Bounded** $[-1, 2) \rightarrow -1 \le x < 2$ $[-4, 5] \rightarrow -4 \le x \le 5$ **Unbounded** $(3, \infty) \rightarrow 3 < x$ $(-\infty, \infty) \rightarrow -\infty < x < \infty$	91–94
	Use properties of inequalities to create equivalent inequalities *(p. 141)* and solve linear inequalities in one variable *(p. 142)*.	Solving linear inequalities is similar to solving linear equations. Use the Properties of Inequalities to isolate the variable. Just remember to reverse the inequality symbol when you multiply or divide by a negative number.	95–98
	Solve inequalities involving absolute values *(p. 144)*.	Let x be a variable or an algebraic expression and let a be a real number such that $a \ge 0$. **1.** Solutions of $\lvert x \rvert < a$: All values of x that lie between $-a$ and a; $\lvert x \rvert < a$ if and only if $-a < x < a$. **2.** Solutions of $\lvert x \rvert > a$: All values of x that are less than $-a$ or greater than a; $\lvert x \rvert > a$ if and only if $x < -a$ or $x > a$.	99, 100
	Use inequalities to model and solve real-life problems *(p. 145)*.	An inequality can be used to determine the accuracy of a measurement. (See Example 7.)	101, 102
Section 1.8	Solve polynomial *(p. 150)* and rational inequalities *(p. 154)*.	Use the concepts of key numbers and test intervals to solve both polynomial and rational inequalities.	103–108
	Use inequalities to model and solve real-life problems *(p. 155)*.	A common application of inequalities involves profit P, revenue R, and cost C. (See Example 6.)	109, 110

1 REVIEW EXERCISES

See www.CalcChat.com for worked-out solutions to odd-numbered exercises.

1.1 In Exercises 1 and 2, complete a table of values. Use the resulting solution points to sketch the graph of the equation.

1. $y = -4x + 1$

2. $y = x^2 + 2x$

In Exercises 3 and 4, graphically estimate the *x*- and *y*-intercepts of the graph.

3. $y = (x - 3)^2 - 4$ **4.** $y = |x + 1| - 3$

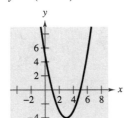   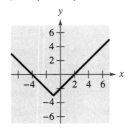

In Exercises 5–12, use the algebraic tests to check for symmetry with respect to both axes and the origin. Then sketch the graph of the equation.

5. $y = -4x + 1$ **6.** $y = 5x - 6$

7. $y = 7 - x^2$ **8.** $y = x^2 + 2$

9. $y = x^3 + 3$ **10.** $y = -6 - x^3$

11. $y = -|x| - 2$ **12.** $y = |x| + 9$

In Exercises 13–16, find the center and radius of the circle and sketch its graph.

13. $x^2 + y^2 = 9$ **14.** $x^2 + y^2 = 4$

15. $(x + 2)^2 + y^2 = 16$ **16.** $x^2 + (y - 8)^2 = 81$

17. Find the standard form of the equation of the circle for which the endpoints of a diameter are $(0, 0)$ and $(4, -6)$.

18. Find the standard form of the equation of the circle for which the endpoints of a diameter are $(-2, -3)$ and $(4, -10)$.

19. REVENUE The revenue R (in billions of dollars) for Target for the years 1998 through 2007 can be approximated by the model

$$R = 0.123t^2 + 0.43t + 20.0, \quad 8 \leq t \leq 17$$

where t represents the year, with $t = 8$ corresponding to 1998. (Source: Target Corp.)

(a) Sketch a graph of the model.

(b) Use the graph to estimate the year in which the revenue was 50 billion dollars.

20. PHYSICS The force F (in pounds) required to stretch a spring x inches from its natural length (see figure) is

$$F = \frac{5}{4}x, \quad 0 \leq x \leq 20.$$

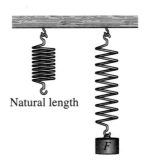

Natural length   *x* in.

(a) Use the model to complete the table.

x	0	4	8	12	16	20
Force, F						

(b) Sketch a graph of the model.

(c) Use the graph to estimate the force necessary to stretch the spring 10 inches.

1.2 In Exercises 21–24, determine whether the equation is an identity or a conditional equation.

21. $6 - (x - 2)^2 = 2 + 4x - x^2$

22. $3(x - 2) + 2x = 2(x + 3)$

23. $-x^3 + x(7 - x) + 3 = x(-x^2 - x) + 7(x + 1) - 4$

24. $3(x^2 - 4x + 8) = -10(x + 2) - 3x^2 + 6$

In Exercises 25–30, solve the equation (if possible) and check your solution.

25. $8x - 5 = 3x + 20$

26. $7x + 3 = 3x - 17$

27. $2(x + 5) - 7 = 3(x - 2)$

28. $3(x + 3) = 5(1 - x) - 1$

29. $\dfrac{x}{5} - 3 = \dfrac{x}{3} + 1$ **30.** $\dfrac{4x - 3}{6} + \dfrac{x}{4} = x - 2$

In Exercises 31–36, find the *x*- and *y*-intercepts of the graph of the equation algebraically.

31. $y = 3x - 1$ **32.** $y = -5x + 6$

33. $y = 2(x - 4)$ **34.** $y = 4(7x + 1)$

35. $y = -\frac{1}{2}x + \frac{2}{3}$ **36.** $y = \frac{3}{4}x - \frac{1}{4}$

37. GEOMETRY The surface area S of the cylinder shown in the figure is approximated by the equation $S = 2(3.14)(3)^2 + 2(3.14)(3)h$. The surface area is 244.92 square inches. Find the height h of the cylinder.

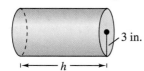

3 in.

$\overset{\longmapsto}{} h \overset{\longmapsto}{}$

38. TEMPERATURE The Fahrenheit and Celsius temperature scales are related by the equation

$$C = \frac{5}{9}F - \frac{160}{9}.$$

Find the Fahrenheit temperature that corresponds to $100°$ Celsius.

39. PROFIT In October, a greeting card company's total profit was 12% more than it was in September. The total profit for the two months was $689,000. Write a verbal model, assign labels, and write an algebraic equation to find the profit for each month.

40. DISCOUNT The price of a digital camera has been discounted $85. The sale price is $340. Write a verbal model, assign labels, and write an algebraic equation to find the percent discount.

41. BUSINESS VENTURE You are planning to start a small business that will require an investment of $90,000. You have found some people who are willing to share equally in the venture. If you can find three more people, each person's share will decrease by $2500. How many people have you found so far?

42. AVERAGE SPEED You commute 56 miles one way to work. The trip to work takes 10 minutes longer than the trip home. Your average speed on the trip home is 8 miles per hour faster. What is your average speed on the trip home?

43. MIXTURE PROBLEM A car radiator contains 10 liters of a 30% antifreeze solution. How many liters will have to be replaced with pure antifreeze if the resulting solution is to be 50% antifreeze?

44. INVESTMENT You invested $6000 at $4\frac{1}{2}\%$ and $5\frac{1}{2}\%$ simple interest. During the first year, the two accounts earned $305. How much did you invest in each fund? (*Note:* The $5\frac{1}{2}\%$ account is more risky.)

In Exercises 45 and 46, solve for the indicated variable.

45. Volume of a Cone
Solve for h: $V = \frac{1}{3}\pi r^2 h$

46. Kinetic Energy
Solve for m: $E = \frac{1}{2}mv^2$

1.4 In Exercises 47–56, use any method to solve the quadratic equation.

47. $15 + x - 2x^2 = 0$
48. $2x^2 - x - 28 = 0$
49. $6 = 3x^2$
50. $16x^2 = 25$
51. $(x + 13)^2 = 25$
52. $(x - 5)^2 = 30$
53. $x^2 + 12x = -25$
54. $9x^2 - 12x = 14$
55. $-2x^2 - 5x + 27 = 0$
56. $-20 - 3x + 3x^2 = 0$

 57. SIMPLY SUPPORTED BEAM A simply supported 20-foot beam supports a uniformly distributed load of 1000 pounds per foot. The bending moment M (in foot-pounds) x feet from one end of the beam is given by $M = 500x(20 - x)$.

(a) Where is the bending moment zero?

(b) Use a graphing utility to graph the equation.

(c) Use the graph to determine the point on the beam where the bending moment is the greatest.

58. SPORTS You throw a softball straight up into the air at a velocity of 30 feet per second. You release the softball at a height of 5.8 feet and catch it when it falls back to a height of 6.2 feet.

(a) Use the position equation to write a mathematical model for the height of the softball.

(b) What is the height of the softball after 1 second?

(c) How many seconds is the softball in the air?

1.5 In Exercises 59–62, write the complex number in standard form.

59. $4 + \sqrt{-9}$
60. $3 + \sqrt{-16}$
61. $i^2 + 3i$
62. $-5i + i^2$

In Exercises 63–66, perform the operation and write the result in standard form.

63. $(7 + 5i) + (-4 + 2i)$

64. $\left(\frac{\sqrt{2}}{2} - \frac{\sqrt{2}}{2}i\right) - \left(\frac{\sqrt{2}}{2} + \frac{\sqrt{2}}{2}i\right)$

65. $6i(5 - 2i)$
66. $(1 + 6i)(5 - 2i)$

In Exercises 67 and 68, write the quotient in standard form.

67. $\dfrac{6 - 5i}{i}$
68. $\dfrac{3 + 2i}{5 + i}$

In Exercises 69 and 70, perform the operation and write the result in standard form.

69. $\dfrac{4}{2 - 3i} + \dfrac{2}{1 + i}$
70. $\dfrac{1}{2 + i} - \dfrac{5}{1 + 4i}$

In Exercises 71–74, find all solutions of the equation.

71. $3x^2 + 1 = 0$
72. $2 + 8x^2 = 0$
73. $x^2 - 2x + 10 = 0$
74. $6x^2 + 3x + 27 = 0$

1.6 In Exercises 75–88, find all solutions of the equation. Check your solutions in the original equation.

75. $5x^4 - 12x^3 = 0$ **76.** $4x^3 - 6x^2 = 0$

77. $x^4 - 5x^2 + 6 = 0$

78. $9x^4 + 27x^3 - 4x^2 - 12x = 0$

79. $\sqrt{2x + 3} + \sqrt{x - 2} = 2$

80. $5\sqrt{x} - \sqrt{x - 1} = 6$

81. $(x - 1)^{2/3} - 25 = 0$ **82.** $(x + 2)^{3/4} = 27$

83. $\dfrac{5}{x} = 1 + \dfrac{3}{x + 2}$ **84.** $\dfrac{6}{x} + \dfrac{8}{x + 5} = 3$

85. $|x - 5| = 10$ **86.** $|2x + 3| = 7$

87. $|x^2 - 3| = 2x$ **88.** $|x^2 - 6| = x$

89. DEMAND The demand equation for a hair dryer is $p = 42 - \sqrt{0.001x + 2}$, where x is the number of units demanded per day and p is the price per unit. Find the demand if the price is set at $29.95.

90. DATA ANALYSIS: NEWSPAPERS The total numbers N of daily evening newspapers in the United States from 1970 through 2005 can be approximated by the model $N = 1465 - 4.2t^{3/2}$, $0 \le t \le 35$, where t represents the year, with $t = 0$ corresponding to 1970. The actual numbers of newspapers for selected years are shown in the table. (Source: Editor & Publisher Co.)

Year	Newspapers, N
1970	1429
1975	1436
1980	1388
1985	1220
1990	1084
1995	891
2000	727
2005	645

(a) Use a graphing utility to plot the data and graph the model in the same viewing window. How well does the model fit the data?

(b) Use the graph in part (a) to estimate the year in which there were 800 daily evening newspapers.

(c) Use the model to verify algebraically the estimate from part (b).

1.7 In Exercises 91–94, write an inequality that represents the interval and state whether the interval is bounded or unbounded.

91. $(-7, 2]$ **92.** $(4, \infty)$

93. $(-\infty, -10]$ **94.** $[-2, 2]$

In Exercises 95–100, solve the inequality.

95. $3(x + 2) + 7 < 2x - 5$

96. $2(x + 7) - 4 \ge 5(x - 3)$

97. $4(5 - 2x) \le \frac{1}{2}(8 - x)$ **98.** $\frac{1}{2}(3 - x) > \frac{1}{3}(2 - 3x)$

99. $|x - 3| > 4$ **100.** $\left|x - \frac{3}{2}\right| \ge \frac{3}{2}$

101. GEOMETRY The side of a square is measured as 19.3 centimeters with a possible error of 0.5 centimeter. Using these measurements, determine the interval containing the area of the square.

102. COST, REVENUE, AND PROFIT The revenue for selling x units of a product is $R = 125.33x$. The cost of producing x units is $C = 92x + 1200$. To obtain a profit, the revenue must be greater than the cost. Determine the smallest value of x for which this product returns a profit.

1.8 In Exercises 103–108, solve the inequality.

103. $x^2 - 6x - 27 < 0$ **104.** $x^2 - 2x \ge 3$

105. $6x^2 + 5x < 4$ **106.** $2x^2 + x \ge 15$

107. $\dfrac{2}{x + 1} \le \dfrac{3}{x - 1}$ **108.** $\dfrac{x - 5}{3 - x} < 0$

109. INVESTMENT P dollars invested at interest rate r compounded annually increases to an amount $A = P(1 + r)^2$ in 2 years. An investment of $5000 is to increase to an amount greater than $5500 in 2 years. The interest rate must be greater than what percent?

110. POPULATION OF A SPECIES A biologist introduces 200 ladybugs into a crop field. The population P of the ladybugs is approximated by the model $P = [1000(1 + 3t)]/(5 + t)$, where t is the time in days. Find the time required for the population to increase to at least 2000 ladybugs.

EXPLORATION

TRUE OR FALSE? In Exercises 111 and 112, determine whether the statement is true or false. Justify your answer.

111. $\sqrt{-18}\sqrt{-2} = \sqrt{(-18)(-2)}$

112. The equation $325x^2 - 717x + 398 = 0$ has no solution.

113. WRITING Explain why it is essential to check your solutions to radical, absolute value, and rational equations.

114. ERROR ANALYSIS What is wrong with the following solution?

$$|11x + 4| \ge 26$$
$$11x + 4 \le 26 \quad \text{or} \quad 11x + 4 \ge 26$$
$$11x \le 22 \qquad\qquad 11x \ge 22$$
$$x \le 2 \qquad\qquad\quad x \ge 2$$

1 CHAPTER TEST

See www.CalcChat.com for worked-out solutions to odd-numbered exercises.

Take this test as you would take a test in class. When you are finished, check your work against the answers given in the back of the book.

In Exercises 1–6, check for symmetry with respect to both axes and the origin. Then sketch the graph of the equation. Identify any x- and y-intercepts.

1. $y = 4 - \frac{3}{4}x$ **2.** $y = 4 - \frac{3}{4}|x|$ **3.** $y = 4 - (x - 2)^2$

4. $y = x - x^3$ **5.** $y = \sqrt{5 - x}$ **6.** $(x - 3)^2 + y^2 = 9$

In Exercises 7–12, solve the equation (if possible).

7. $\frac{2}{3}(x - 1) + \frac{1}{4}x = 10$ **8.** $(x - 4)(x + 2) = 7$

9. $\dfrac{x - 2}{x + 2} + \dfrac{4}{x + 2} + 4 = 0$ **10.** $x^4 + x^2 - 6 = 0$

11. $2\sqrt{x} - \sqrt{2x + 1} = 1$ **12.** $|3x - 1| = 7$

In Exercises 13–16, solve the inequality. Sketch the solution set on the real number line.

13. $-3 \le 2(x + 4) < 14$ **14.** $\dfrac{2}{x} > \dfrac{5}{x + 6}$

15. $2x^2 + 5x > 12$ **16.** $|3x + 5| \ge 10$

17. Perform each operation and write the result in standard form.

    (a) $10i - \left(3 + \sqrt{-25}\right)$ (b) $(-1 - 5i)(-1 + 5i)$

18. Write the quotient in standard form: $\dfrac{5}{2 + i}$.

19. The sales y (in billions of dollars) for Dell, Inc. from 1999 through 2008 can be approximated by the model

    $y = 4.41t - 14.6, \quad 9 \le t \le 18$

    where t represents the year, with $t = 9$ corresponding to 1999.   (Source: Dell, Inc.)

    (a) Sketch a graph of the model.

    (b) Assuming that the pattern continues, use the graph in part (a) to estimate the sales in 2013.

    (c) Use the model to verify algebraically the estimate from part (b).

20. A basketball has a volume of about 455.9 cubic inches. Find the radius of the basketball (accurate to three decimal places).

21. On the first part of a 350-kilometer trip, a salesperson travels 2 hours and 15 minutes at an average speed of 100 kilometers per hour. The salesperson needs to arrive at the destination in another hour and 20 minutes. Find the average speed required for the remainder of the trip.

22. The area of the ellipse in the figure at the left is $A = \pi ab$. If a and b satisfy the constraint $a + b = 100$, find a and b such that the area of the ellipse equals the area of the circle.

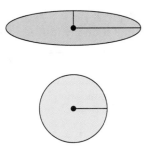

FIGURE FOR **22**

PROOFS IN MATHEMATICS

Conditional Statements

Many theorems are written in the **if-then form** "if p, then q," which is denoted by

$$p \rightarrow q$$

where p is the **hypothesis** and q is the **conclusion.** Here are some other ways to express the conditional statement $p \rightarrow q$.

p implies q. p, only if q. p is sufficient for q.

Conditional statements can be either true or false. The conditional statement $p \rightarrow q$ is false only when p is true and q is false. To show that a conditional statement is true, you must prove that the conclusion follows for all cases that fulfill the hypothesis. To show that a conditional statement is false, you need to describe only a single **counterexample** that shows that the statement is not always true.

For instance, $x = -4$ is a counterexample that shows that the following statement is false.

If $x^2 = 16$, then $x = 4$.

The hypothesis "$x^2 = 16$" is true because $(-4)^2 = 16$. However, the conclusion "$x = 4$" is false. This implies that the given conditional statement is false.

For the conditional statement $p \rightarrow q$, there are three important associated conditional statements.

1. The **converse** of $p \rightarrow q$: $q \rightarrow p$

2. The **inverse** of $p \rightarrow q$: $\sim p \rightarrow \sim q$

3. The **contrapositive** of $p \rightarrow q$: $\sim q \rightarrow \sim p$

The symbol $\sim$ means the **negation** of a statement. For instance, the negation of "The engine is running" is "The engine is not running."

Example   Writing the Converse, Inverse, and Contrapositive

Write the converse, inverse, and contrapositive of the conditional statement "If I get a B on my test, then I will pass the course."

Solution

a. *Converse:* If I pass the course, then I got a B on my test.

b. *Inverse:* If I do not get a B on my test, then I will not pass the course.

c. *Contrapositive:* If I do not pass the course, then I did not get a B on my test.

In the example above, notice that neither the converse nor the inverse is logically equivalent to the original conditional statement. On the other hand, the contrapositive *is* logically equivalent to the original conditional statement.

PROBLEM SOLVING

This collection of thought-provoking and challenging exercises further explores and expands upon concepts learned in this chapter.

1. Let x represent the time (in seconds) and let y represent the distance (in feet) between you and a tree. Sketch a possible graph that shows how x and y are related if you are walking toward the tree.

2. (a) Find the following sums

$$1 + 2 + 3 + 4 + 5 =$$

$$1 + 2 + 3 + 4 + 5 + 6 + 7 + 8 =$$

$$1 + 2 + 3 + 4 + 5 + 6$$
$$+ 7 + 8 + 9 + 10 =$$

(b) Use the following formula for the sum of the first n natural numbers to verify your answers to part (a).

$$1 + 2 + 3 + \cdots + n = \frac{1}{2}n(n + 1)$$

(c) Use the formula in part (b) to find n if the sum of the first n natural numbers is 210.

3. The area of an ellipse is given by $A = \pi ab$ (see figure). For a certain ellipse, it is required that $a + b = 20$.

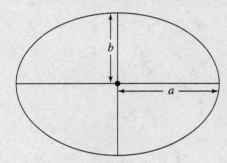

(a) Show that $A = \pi a(20 - a)$.

(b) Complete the table.

a	4	7	10	13	16
A					

(c) Find two values of a such that $A = 300$.

 (d) Use a graphing utility to graph the area equation.

(e) Find the a-intercepts of the graph of the area equation. What do these values represent?

(f) What is the maximum area? What values of a and b yield the maximum area?

4. A building code requires that a building be able to withstand a certain amount of wind pressure. The pressure P (in pounds per square foot) from wind blowing at s miles per hour is given by

$$P = 0.00256s^2.$$

(a) A two-story library is designed. Buildings this tall are often required to withstand wind pressure of 20 pounds per square foot. Under this requirement, how fast can the wind be blowing before it produces excessive stress on the building?

(b) To be safe, the library is designed so that it can withstand wind pressure of 40 pounds per square foot. Does this mean that the library can survive wind blowing at twice the speed you found in part (a)? Justify your answer.

(c) Use the pressure formula to explain why even a relatively small increase in the wind speed could have potentially serious effects on a building.

5. For a bathtub with a rectangular base, Toricelli's Law implies that the height h of water in the tub t seconds after it begins draining is given by

$$h = \left(\sqrt{h_0} - \frac{2\pi d^2 \sqrt{3}}{lw} t \right)^2$$

where l and w are the tub's length and width, d is the diameter of the drain, and h_0 is the water's initial height. (All measurements are in inches.) You completely fill a tub with water. The tub is 60 inches long by 30 inches wide by 25 inches high and has a drain with a two-inch diameter.

(a) Find the time it takes for the tub to go from being full to half-full.

(b) Find the time it takes for the tub to go from being half-full to empty.

(c) Based on your results in parts (a) and (b), what general statement can you make about the speed at which the water drains?

6. (a) Consider the sum of squares $x^2 + 9$. If the sum can be factored, then there are integers m and n such that $x^2 + 9 = (x + m)(x + n)$. Write two equations relating the sum and the product of m and n to the coefficients in $x^2 + 9$.

(b) Show that there are no integers m and n that satisfy both equations you wrote in part (a). What can you conclude?

7. A Pythagorean Triple is a group of three integers, such as 3, 4, and 5, that could be the lengths of the sides of a right triangle.

   (a) Find two other Pythagorean Triples.

   (b) Notice that $3 \cdot 4 \cdot 5 = 60$. Is the product of the three numbers in each Pythagorean Triple evenly divisible by 3? by 4? by 5?

   (c) Write a conjecture involving Pythagorean Triples and divisibility by 60.

8. Determine the solutions x_1 and x_2 of each quadratic equation. Use the values of x_1 and x_2 to fill in the boxes.

Equation	x_1, x_2	$x_1 + x_2$	$x_1 \cdot x_2$
(a) $x^2 - x - 6 = 0$			
(b) $2x^2 + 5x - 3 = 0$			
(c) $4x^2 - 9 = 0$			
(d) $x^2 - 10x + 34 = 0$			

9. Consider a general quadratic equation

   $$ax^2 + bx + c = 0$$

   whose solutions are x_1 and x_2. Use the results of Exercise 8 to determine a relationship among the coefficients a, b, and c and the sum $x_1 + x_2$ and the product $x_1 \cdot x_2$ of the solutions.

10. (a) The principal cube root of 125, $\sqrt[3]{125}$, is 5. Evaluate the expression x^3 for each value of x.

    (i) $x = \dfrac{-5 + 5\sqrt{3}i}{2}$

    (ii) $x = \dfrac{-5 - 5\sqrt{3}i}{2}$

    (b) The principal cube root of 27, $\sqrt[3]{27}$, is 3. Evaluate the expression x^3 for each value of x.

    (i) $x = \dfrac{-3 + 3\sqrt{3}i}{2}$

    (ii) $x = \dfrac{-3 - 3\sqrt{3}i}{2}$

    (c) Use the results of parts (a) and (b) to list possible cube roots of (i) 1, (ii) 8, and (iii) 64. Verify your results algebraically.

11. The multiplicative inverse of z is a complex number z_m such that $z \cdot z_m = 1$. Find the multiplicative inverse of each complex number.

    (a) $z = 1 + i$ (b) $z = 3 - i$ (c) $z = -2 + 8i$

12. Prove that the product of a complex number $a + bi$ and its complex conjugate is a real number.

13. A **fractal** is a geometric figure that consists of a pattern that is repeated infinitely on a smaller and smaller scale. The most famous fractal is called the **Mandelbrot Set,** named after the Polish-born mathematician Benoit Mandelbrot. To draw the Mandelbrot Set, consider the following sequence of numbers.

    $$c, c^2 + c, (c^2 + c)^2 + c, [(c^2 + c)^2 + c]^2 + c, \ldots$$

    The behavior of this sequence depends on the value of the complex number c. If the sequence is bounded (the absolute value of each number in the sequence, $|a + bi| = \sqrt{a^2 + b^2}$, is less than some fixed number N), the complex number c is in the Mandelbrot Set, and if the sequence is unbounded (the absolute value of the terms of the sequence become infinitely large), the complex number c is not in the Mandelbrot Set. Determine whether the complex number c is in the Mandelbrot Set.

    (a) $c = i$ (b) $c = 1 + i$ (c) $c = -2$

    The figure below shows a black and yellow photo of the Mandelbrot Set.

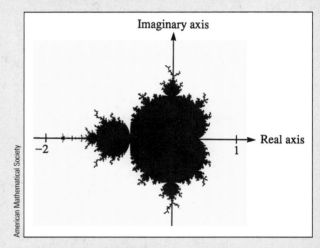

14. Use the equation $4\sqrt{x} = 2x + k$ to find three different values of k such that the equation has two solutions, one solution, and no solution. Describe the process you used to find the values.

15. Use the graph of $y = x^4 - x^3 - 6x^2 + 4x + 8$ to solve the inequality $x^4 - x^3 - 6x^2 + 4x + 8 > 0$.

16. When you buy a 16-ounce bag of chips, you expect to get precisely 16 ounces. The actual weight w (in ounces) of a "16-ounce" bag of chips is given by

    $$|w - 16| \le \frac{1}{2}.$$

    You buy four 16-ounce bags. What is the greatest amount you can expect to get? What is the smallest amount? Explain.

Functions and Their Graphs

2

In Mathematics

Functions show how one variable is related to another variable.

In Real Life

Functions are used to estimate values, stimulate processes, and discover relationships. You can model the enrollment rate of children in preschool and estimate the year in which the rate will reach a certain number. This estimate can be used to plan for future needs, such as adding teachers and buying books. (See Exercise 113, page 210.)

Jose Luis Pelaez/Getty Images

IN CAREERS

There are many careers that use functions. Several are listed below.

- Roofing Contractor
  Exercise 131, page 182

- Financial Analyst
  Exercise 95, page 197

- Sociologist
  Exercise 80, page 228

- Biologist
  Exercise 73, page 237

2.1 ## LINEAR EQUATIONS IN TWO VARIABLES

What you should learn

- Use slope to graph linear equations in two variables.
- Find the slope of a line given two points on the line.
- Write linear equations in two variables.
- Use slope to identify parallel and perpendicular lines.
- Use slope and linear equations in two variables to model and solve real-life problems.

Why you should learn it

Linear equations in two variables can be used to model and solve real-life problems. For instance, in Exercise 129 on page 182, you will use a linear equation to model student enrollment at the Pennsylvania State University.

Courtesy of Pennsylvania State University

Using Slope

The simplest mathematical model for relating two variables is the **linear equation in two variables** $y = mx + b$. The equation is called *linear* because its graph is a line. (In mathematics, the term *line* means *straight line*.) By letting $x = 0$, you obtain

$$y = m(0) + b$$
$$= b.$$

So, the line crosses the y-axis at $y = b$, as shown in Figure 2.1. In other words, the y-intercept is $(0, b)$. The steepness or slope of the line is m.

$$y = mx + b$$

The **slope** of a nonvertical line is the number of units the line rises (or falls) vertically for each unit of horizontal change from left to right, as shown in Figure 2.1 and Figure 2.2.

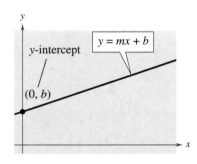

Positive slope, line rises.
FIGURE 2.1

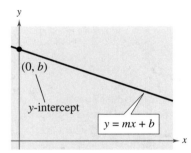

Negative slope, line falls.
FIGURE 2.2

A linear equation that is written in the form $y = mx + b$ is said to be written in **slope-intercept form.**

The Slope-Intercept Form of the Equation of a Line

The graph of the equation

$$y = mx + b$$

is a line whose slope is m and whose y-intercept is $(0, b)$.

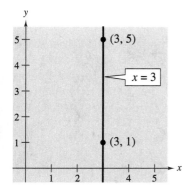

FIGURE **2.3**   Slope is undefined.

Once you have determined the slope and the y-intercept of a line, it is a relatively simple matter to sketch its graph. In the next example, note that none of the lines is vertical. A vertical line has an equation of the form

$$x = a.$$

The equation of a vertical line cannot be written in the form $y = mx + b$ because the slope of a vertical line is undefined, as indicated in Figure 2.3.

| **Example 1   Graphing a Linear Equation**

Sketch the graph of each linear equation.

a. $y = 2x + 1$

b. $y = 2$

c. $x + y = 2$

Solution

a. Because $b = 1$, the y-intercept is $(0, 1)$. Moreover, because the slope is $m = 2$, the line *rises* two units for each unit the line moves to the right, as shown in Figure 2.4.

b. By writing this equation in the form $y = (0)x + 2$, you can see that the y-intercept is $(0, 2)$ and the slope is zero. A zero slope implies that the line is horizontal—that is, it doesn't rise *or* fall, as shown in Figure 2.5.

c. By writing this equation in slope-intercept form

$$x + y = 2$$

$$y = -x + 2$$

$$y = (-1)x + 2$$

you can see that the y-intercept is $(0, 2)$. Moreover, because the slope is $m = -1$, the line *falls* one unit for each unit the line moves to the right, as shown in Figure 2.6.

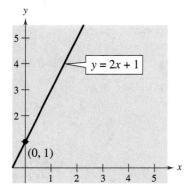

When *m* is positive, the line rises.
FIGURE **2.4**

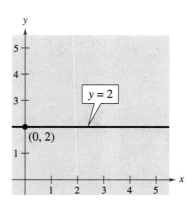

When *m* is 0, the line is horizontal.
FIGURE **2.5**

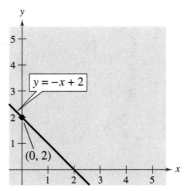

When *m* is negative, the line falls.
FIGURE **2.6**

CHECK*Point*   Now try Exercise 17.

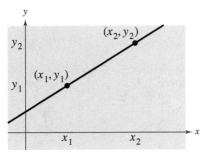

FIGURE **2.7**

Finding the Slope of a Line

Given an equation of a line, you can find its slope by writing the equation in slope-intercept form. If you are not given an equation, you can still find the slope of a line. For instance, suppose you want to find the slope of the line passing through the points (x_1, y_1) and (x_2, y_2), as shown in Figure 2.7. As you move from left to right along this line, a change of $(y_2 - y_1)$ units in the vertical direction corresponds to a change of $(x_2 - x_1)$ units in the horizontal direction.

$$y_2 - y_1 = \text{the change in } y = \text{rise}$$

and

$$x_2 - x_1 = \text{the change in } x = \text{run}$$

The ratio of $(y_2 - y_1)$ to $(x_2 - x_1)$ represents the slope of the line that passes through the points (x_1, y_1) and (x_2, y_2).

$$\text{Slope} = \frac{\text{change in } y}{\text{change in } x}$$

$$= \frac{\text{rise}}{\text{run}}$$

$$= \frac{y_2 - y_1}{x_2 - x_1}$$

The Slope of a Line Passing Through Two Points

The **slope** m of the nonvertical line through (x_1, y_1) and (x_2, y_2) is

$$m = \frac{y_2 - y_1}{x_2 - x_1}$$

where $x_1 \neq x_2$.

When this formula is used for slope, the *order of subtraction* is important. Given two points on a line, you are free to label either one of them as (x_1, y_1) and the other as (x_2, y_2). However, once you have done this, you must form the numerator and denominator using the same order of subtraction.

$$m = \frac{y_2 - y_1}{x_2 - x_1} \qquad m = \frac{y_1 - y_2}{x_1 - x_2} \qquad m = \frac{y_2 - y_1}{x_1 - x_2}$$

For instance, the slope of the line passing through the points $(3, 4)$ and $(5, 7)$ can be calculated as

$$m = \frac{7 - 4}{5 - 3} = \frac{3}{2}$$

or, reversing the subtraction order in both the numerator and denominator, as

$$m = \frac{4 - 7}{3 - 5} = \frac{-3}{-2} = \frac{3}{2}.$$

| Example 2 | **Finding the Slope of a Line Through Two Points** |

Find the slope of the line passing through each pair of points.

a. $(-2, 0)$ and $(3, 1)$ **b.** $(-1, 2)$ and $(2, 2)$

c. $(0, 4)$ and $(1, -1)$ **d.** $(3, 4)$ and $(3, 1)$

Solution

a. Letting $(x_1, y_1) = (-2, 0)$ and $(x_2, y_2) = (3, 1)$, you obtain a slope of

$$m = \frac{y_2 - y_1}{x_2 - x_1} = \frac{-}{-(\ )} = \frac{1}{5}.$$

b. The slope of the line passing through $(-1, 2)$ and $(2, 2)$ is

$$m = \frac{-}{-(\ )} = \frac{0}{3} = 0.$$

c. The slope of the line passing through $(0, 4)$ and $(1, -1)$ is

$$m = \frac{-}{-} = \frac{-5}{1} = -5.$$

d. The slope of the line passing through $(3, 4)$ and $(3, 1)$ is

$$m = \frac{1 - 4}{3 - 3} = \frac{-3}{0}.$$

Because division by 0 is undefined, the slope is undefined and the line is vertical.

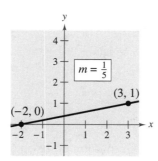

FIGURE 2.8

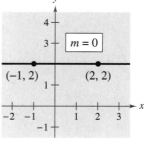

FIGURE 2.9

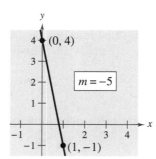

FIGURE 2.10

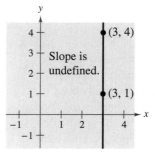

FIGURE 2.11

CHECK**Point**   Now try Exercise 31.

Writing Linear Equations in Two Variables

If (x_1, y_1) is a point on a line of slope m and (x, y) is *any other* point on the line, then

$$\frac{y - y_1}{x - x_1} = m.$$

This equation, involving the variables x and y, can be rewritten in the form

$$y - y_1 = m(x - x_1)$$

which is the **point-slope form** of the equation of a line.

Point-Slope Form of the Equation of a Line

The equation of the line with slope m passing through the point (x_1, y_1) is

$$y - y_1 = m(x - x_1).$$

The point-slope form is most useful for *finding* the equation of a line. You should remember this form.

Example 3   Using the Point-Slope Form

Find the slope-intercept form of the equation of the line that has a slope of 3 and passes through the point $(1, -2)$.

Solution

Use the point-slope form with $m = 3$ and $(x_1, y_1) = (1, -2)$.

$$y - y_1 = m(x - x_1)$$
$$y - (\quad) = \quad(x - \quad)$$
$$y + 2 = 3x - 3$$
$$y = 3x - 5$$

The slope-intercept form of the equation of the line is $y = 3x - 5$. The graph of this line is shown in Figure 2.12.

CHECK*Point*   Now try Exercise 51.

The point-slope form can be used to find an equation of the line passing through two points (x_1, y_1) and (x_2, y_2). To do this, first find the slope of the line

$$m = \frac{y_2 - y_1}{x_2 - x_1}, \quad x_1 \neq x_2$$

and then use the point-slope form to obtain the equation

$$y - y_1 = \frac{y_2 - y_1}{x_2 - x_1}(x - x_1).$$

This is sometimes called the **two-point form** of the equation of a line.

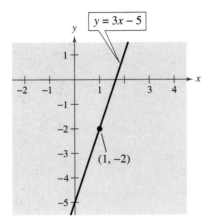

FIGURE **2.12**

Study Tip

When you find an equation of the line that passes through two given points, you only need to substitute the coordinates of one of the points in the point-slope form. It does not matter which point you choose because both points will yield the same result.

Parallel and Perpendicular Lines

Slope can be used to decide whether two nonvertical lines in a plane are parallel, perpendicular, or neither.

> ### Parallel and Perpendicular Lines
>
> 1. Two distinct nonvertical lines are **parallel** if and only if their slopes are equal. That is, $m_1 = m_2$.
>
> 2. Two nonvertical lines are **perpendicular** if and only if their slopes are negative reciprocals of each other. That is, $m_1 = -1/m_2$.

Example 4   Finding Parallel and Perpendicular Lines

Find the slope-intercept forms of the equations of the lines that pass through the point $(2, -1)$ and are (a) parallel to and (b) perpendicular to the line $2x - 3y = 5$.

Solution

By writing the equation of the given line in slope-intercept form

$$2x - 3y = 5$$
$$-3y = -2x + 5$$
$$y = \tfrac{2}{3}x - \tfrac{5}{3}$$

you can see that it has a slope of $m = \tfrac{2}{3}$, as shown in Figure 2.13.

a. Any line parallel to the given line must also have a slope of $\tfrac{2}{3}$. So, the line through $(2, -1)$ that is parallel to the given line has the following equation.

$$y - (\ ) = (x - \ )$$
$$3(y + 1) = 2(x - 2)$$
$$3y + 3 = 2x - 4$$
$$y = \tfrac{2}{3}x - \tfrac{7}{3}$$

b. Any line perpendicular to the given line must have a slope of $-\tfrac{3}{2}$ $\left(\text{because } -\tfrac{3}{2}\right.$ is the negative reciprocal of $\tfrac{2}{3}\right)$. So, the line through $(2, -1)$ that is perpendicular to the given line has the following equation.

$$y - (\ ) = \ (x - \ )$$
$$2(y + 1) = -3(x - 2)$$
$$2y + 2 = -3x + 6$$
$$y = -\tfrac{3}{2}x + 2$$

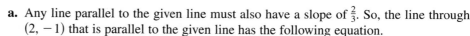 **CHECK*Point*   Now try Exercise 87.**

Notice in Example 4 how the slope-intercept form is used to obtain information about the graph of a line, whereas the point-slope form is used to write the equation of a line.

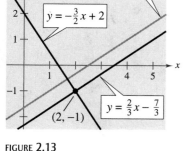

FIGURE 2.13

TECHNOLOGY

On a graphing utility, lines will not appear to have the correct slope unless you use a viewing window that has a square setting. For instance, try graphing the lines in Example 4 using the standard setting $-10 \le x \le 10$ and $-10 \le y \le 10$. Then reset the viewing window with the square setting $-9 \le x \le 9$ and $-6 \le y \le 6$. On which setting do the lines $y = \tfrac{2}{3}x - \tfrac{5}{3}$ and $y = -\tfrac{3}{2}x + 2$ appear to be perpendicular?

Applications

In real-life problems, the slope of a line can be interpreted as either a *ratio* or a *rate*. If the *x*-axis and *y*-axis have the same unit of measure, then the slope has no units and is a **ratio.** If the *x*-axis and *y*-axis have different units of measure, then the slope is a **rate** or **rate of change.**

Example 5 Using Slope as a Ratio

The maximum recommended slope of a wheelchair ramp is $\frac{1}{12}$. A business is installing a wheelchair ramp that rises 22 inches over a horizontal length of 24 feet. Is the ramp steeper than recommended?   (Source: *Americans with Disabilities Act Handbook*)

Solution

The horizontal length of the ramp is 24 feet or $12(24) = 288$ inches, as shown in Figure 2.14. So, the slope of the ramp is

$$\text{Slope} \; = \frac{\text{vertical change}}{\text{horizontal change}} = \text{———} \approx 0.076.$$

Because $\frac{1}{12} \approx 0.083$, the slope of the ramp is not steeper than recommended.

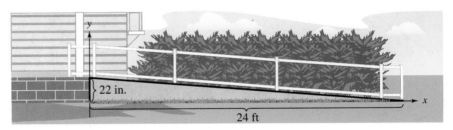

FIGURE 2.14

CHECK*Point* Now try Exercise 115.

Example 6 Using Slope as a Rate of Change

A kitchen appliance manufacturing company determines that the total cost in dollars of producing *x* units of a blender is

$$C = 25x + 3500.$$

Describe the practical significance of the *y*-intercept and slope of this line.

Solution

The *y*-intercept (0, 3500) tells you that the cost of producing zero units is $3500. This is the *fixed cost* of production—it includes costs that must be paid regardless of the number of units produced. The slope of $m = 25$ tells you that the cost of producing each unit is $25, as shown in Figure 2.15. Economists call the cost per unit the *marginal cost*. If the production increases by one unit, then the "margin," or extra amount of cost, is $25. So, the cost increases at a rate of $25 per unit.

CHECK*Point* Now try Exercise 119.

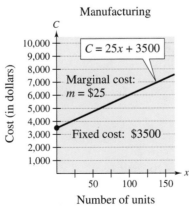

FIGURE 2.15 Production cost

Most business expenses can be deducted in the same year they occur. One exception is the cost of property that has a useful life of more than 1 year. Such costs must be *depreciated* (decreased in value) over the useful life of the property. If the *same amount* is depreciated each year, the procedure is called *linear* or *straight-line depreciation*. The *book value* is the difference between the original value and the total amount of depreciation accumulated to date.

Example 7   Straight-Line Depreciation

A college purchased exercise equipment worth $12,000 for the new campus fitness center. The equipment has a useful life of 8 years. The salvage value at the end of 8 years is $2000. Write a linear equation that describes the book value of the equipment each year.

Solution

Let V represent the value of the equipment at the end of year t. You can represent the initial value of the equipment by the data point $(0, 12,000)$ and the salvage value of the equipment by the data point $(8, 2000)$. The slope of the line is

$$m = \frac{2000 - 12,000}{8 - 0} = -\$1250$$

which represents the annual depreciation in *dollars per year*. Using the point-slope form, you can write the equation of the line as follows.

$$V - \quad = \quad (t - \ )$$
$$V = -1250t + 12,000$$

The table shows the book value at the end of each year, and the graph of the equation is shown in Figure 2.16.

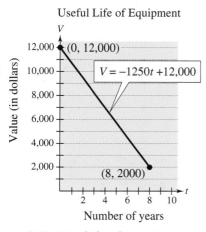

Useful Life of Equipment

$V = -1250t + 12,000$

(0, 12,000)

(8, 2000)

Number of years

FIGURE **2.16**   Straight-line depreciation

Year, t	Value, V
0	12,000
1	10,750
2	9500
3	8250
4	7000
5	5750
6	4500
7	3250
8	2000

CHECK*Point*   Now try Exercise 121.

In many real-life applications, the two data points that determine the line are often given in a disguised form. Note how the data points are described in Example 7.

Example 8 Predicting Sales

The sales for Best Buy were approximately \$35.9 billion in 2006 and \$40.0 billion in 2007. Using only this information, write a linear equation that gives the sales (in billions of dollars) in terms of the year. Then predict the sales for 2010. (Source: Best Buy Company, Inc.)

Solution

Let $t = 6$ represent 2006. Then the two given values are represented by the data points $(6, 35.9)$ and $(7, 40.0)$. The slope of the line through these points is

$$m = \frac{40.0 - 35.9}{7 - 6}$$

$$= 4.1.$$

Using the point-slope form, you can find the equation that relates the sales y and the year t to be

$$y - \quad = 4.1(t - \quad)$$

$$y = 4.1t + 11.3.$$

According to this equation, the sales for 2010 will be

$$y = 4.1(\quad) + 11.3 = 41 + 11.3 = \$52.3 \text{ billion. (See Figure 2.17.)}$$

CHECK**Point** Now try Exercise 129.

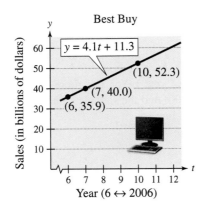

FIGURE **2.17**

The prediction method illustrated in Example 8 is called **linear extrapolation.** Note in Figure 2.18 that an extrapolated point does not lie between the given points. When the estimated point lies between two given points, as shown in Figure 2.19, the procedure is called **linear interpolation.**

Because the slope of a vertical line is not defined, its equation cannot be written in slope-intercept form. However, every line has an equation that can be written in the **general form**

$$Ax + By + C = 0$$

where A and B are not both zero. For instance, the vertical line given by $x = a$ can be represented by the general form $x - a = 0$.

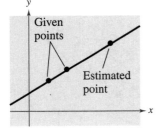

Linear extrapolation
FIGURE **2.18**

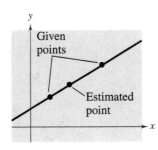

Linear interpolation
FIGURE **2.19**

Summary of Equations of Lines

1. General form: $Ax + By + C = 0$

2. Vertical line: $x = a$

3. Horizontal line: $y = b$

4. Slope-intercept form: $y = mx + b$

5. Point-slope form: $y - y_1 = m(x - x_1)$

6. Two-point form: $y - y_1 = \dfrac{y_2 - y_1}{x_2 - x_1}(x - x_1)$

2.1   EXERCISES

See www.CalcChat.com for worked-out solutions to odd-numbered exercises.

VOCABULARY

In Exercises 1–7, fill in the blanks.

1. The simplest mathematical model for relating two variables is the _____ equation in two variables $y = mx + b$.
2. For a line, the ratio of the change in y to the change in x is called the _____ of the line.
3. Two lines are _____ if and only if their slopes are equal.
4. Two lines are _____ if and only if their slopes are negative reciprocals of each other.
5. When the x-axis and y-axis have different units of measure, the slope can be interpreted as a _____.
6. The prediction method _____ _____ is the method used to estimate a point on a line when the point does not lie between the given points.
7. Every line has an equation that can be written in _____ form.
8. Match each equation of a line with its form.
   - (a) $Ax + By + C = 0$ (i) Vertical line
   - (b) $x = a$ (ii) Slope-intercept form
   - (c) $y = b$ (iii) General form
   - (d) $y = mx + b$ (iv) Point-slope form
   - (e) $y - y_1 = m(x - x_1)$ (v) Horizontal line

SKILLS AND APPLICATIONS

In Exercises 9 and 10, identify the line that has each slope.

9. (a) $m = \frac{2}{3}$
   (b) m is undefined.
   (c) $m = -2$

10. (a) $m = 0$
    (b) $m = -\frac{3}{4}$
    (c) $m = 1$

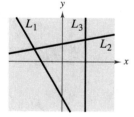

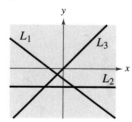

In Exercises 11 and 12, sketch the lines through the point with the indicated slopes on the same set of coordinate axes.

	Point		Slopes		
11.	$(2, 3)$	(a) 0 (b) 1	(c) 2	(d) -3	
12.	$(-4, 1)$	(a) 3 (b) -3	(c) $\frac{1}{2}$	(d) Undefined	

In Exercises 13–16, estimate the slope of the line.

13.

14.
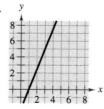

15.
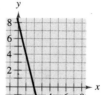

16.
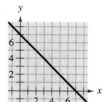

In Exercises 17–28, find the slope and y-intercept (if possible) of the equation of the line. Sketch the line.

17. $y = 5x + 3$
18. $y = x - 10$
19. $y = -\frac{1}{2}x + 4$
20. $y = -\frac{3}{2}x + 6$
21. $5x - 2 = 0$
22. $3y + 5 = 0$
23. $7x + 6y = 30$
24. $2x + 3y = 9$
25. $y - 3 = 0$ $m = 0$
26. $y + 4 = 0$ $m = 0$
27. $x + 5 = 0$ ND
28. $x - 2 = 0$ ND

In Exercises 29–40, plot the points and find the slope of the line passing through the pair of points.

29. $(0, 9), (6, 0)$
30. $(12, 0), (0, -8)$
31. $(-3, -2), (1, 6)$
32. $(2, 4), (4, -4)$
33. $(5, -7), (8, -7)$
34. $(-2, 1), (-4, -5)$
35. $(-6, -1), (-6, 4)$
36. $(0, -10), (-4, 0)$
37. $\left(\frac{11}{2}, -\frac{4}{3}\right), \left(-\frac{3}{2}, -\frac{1}{3}\right)$
38. $\left(\frac{7}{8}, \frac{3}{4}\right), \left(\frac{5}{4}, -\frac{1}{4}\right)$
39. $(4.8, 3.1), (-5.2, 1.6)$
40. $(-1.75, -8.3), (2.25, -2.6)$

In Exercises 41–50, use the point on the line and the slope m of the line to find three additional points through which the line passes. (There are many correct answers.)

41. $(2, 1)$, $m = 0$ **42.** $(3, -2)$, $m = 0$

43. $(5, -6)$, $m = 1$ **44.** $(10, -6)$, $m = -1$

45. $(-8, 1)$, m is undefined.

46. $(1, 5)$, m is undefined.

47. $(-5, 4)$, $m = 2$ **48.** $(0, -9)$, $m = -2$

49. $(7, -2)$, $m = \frac{1}{2}$ **50.** $(-1, -6)$, $m = -\frac{1}{2}$

In Exercises 51–64, find the slope-intercept form of the equation of the line that passes through the given point and has the indicated slope m. Sketch the line.

51. $(0, -2)$, $m = 3$ **52.** $(0, 10)$, $m = -1$

53. $(-3, 6)$, $m = -2$ **54.** $(0, 0)$, $m = 4$

55. $(4, 0)$, $m = -\frac{1}{3}$ **56.** $(8, 2)$, $m = \frac{1}{4}$

57. $(2, -3)$, $m = -\frac{1}{2}$ **58.** $(-2, -5)$, $m = \frac{3}{4}$

59. $(6, -1)$, m is undefined.

60. $(-10, 4)$, m is undefined.

61. $\left(4, \frac{5}{2}\right)$, $m = 0$ **62.** $\left(-\frac{1}{2}, \frac{3}{2}\right)$, $m = 0$

63. $(-5.1, 1.8)$, $m = 5$ **64.** $(2.3, -8.5)$, $m = -2.5$

In Exercises 65–78, find the slope-intercept form of the equation of the line passing through the points. Sketch the line.

65. $(5, -1)$, $(-5, 5)$ **66.** $(4, 3)$, $(-4, -4)$

67. $(-8, 1)$, $(-8, 7)$ **68.** $(-1, 4)$, $(6, 4)$

69. $\left(2, \frac{1}{2}\right)$, $\left(\frac{1}{2}, \frac{5}{4}\right)$ **70.** $(1, 1)$, $\left(6, -\frac{2}{3}\right)$

71. $\left(-\frac{1}{10}, -\frac{3}{5}\right)$, $\left(\frac{9}{10}, -\frac{9}{5}\right)$ **72.** $\left(\frac{3}{4}, \frac{3}{2}\right)$, $\left(-\frac{4}{3}, \frac{7}{4}\right)$

73. $(1, 0.6)$, $(-2, -0.6)$ **74.** $(-8, 0.6)$, $(2, -2.4)$

75. $(2, -1)$, $\left(\frac{1}{3}, -1\right)$ **76.** $\left(\frac{1}{5}, -2\right)$, $(-6, -2)$

77. $\left(\frac{7}{3}, -8\right)$, $\left(\frac{7}{3}, 1\right)$ **78.** $(1.5, -2)$, $(1.5, 0.2)$

In Exercises 79–82, determine whether the lines are parallel, perpendicular, or neither.

79. L_1: $y = \frac{1}{3}x - 2$

    L_2: $y = \frac{1}{3}x + 3$

80. L_1: $y = 4x - 1$

    L_2: $y = 4x + 7$

81. L_1: $y = \frac{1}{2}x - 3$

    L_2: $y = -\frac{1}{2}x + 1$

82. L_1: $y = -\frac{4}{5}x - 5$

    L_2: $y = \frac{5}{4}x + 1$

In Exercises 83–86, determine whether the lines L_1 and L_2 passing through the pairs of points are parallel, perpendicular, or neither.

83. L_1: $(0, -1)$, $(5, 9)$

    L_2: $(0, 3)$, $(4, 1)$

84. L_1: $(-2, -1)$, $(1, 5)$

    L_2: $(1, 3)$, $(5, -5)$

85. L_1: $(3, 6)$, $(-6, 0)$

    L_2: $(0, -1)$, $\left(5, \frac{7}{3}\right)$

86. L_1: $(4, 8)$, $(-4, 2)$

    L_2: $(3, -5)$, $\left(-1, \frac{1}{3}\right)$

In Exercises 87–96, write the slope-intercept forms of the equations of the lines through the given point (a) parallel to the given line and (b) perpendicular to the given line.

87. $4x - 2y = 3$, $(2, 1)$ **88.** $x + y = 7$, $(-3, 2)$

89. $3x + 4y = 7$, $\left(-\frac{2}{3}, \frac{7}{8}\right)$ **90.** $5x + 3y = 0$, $\left(\frac{7}{8}, \frac{3}{4}\right)$

91. $y + 3 = 0$, $(-1, 0)$ **92.** $y - 2 = 0$, $(-4, 1)$

93. $x - 4 = 0$, $(3, -2)$ **94.** $x + 2 = 0$, $(-5, 1)$

95. $x - y = 4$, $(2.5, 6.8)$

96. $6x + 2y = 9$, $(-3.9, -1.4)$

In Exercises 97–102, use the *intercept form* to find the equation of the line with the given intercepts. The intercept form of the equation of a line with intercepts $(a, 0)$ and $(0, b)$ is

$$\frac{x}{a} + \frac{y}{b} = 1, \quad a \neq 0, \ b \neq 0.$$

97. x-intercept: $(2, 0)$

    y-intercept: $(0, 3)$

98. x-intercept: $(-3, 0)$

    y-intercept: $(0, 4)$

99. x-intercept: $\left(-\frac{1}{6}, 0\right)$

    y-intercept: $\left(0, -\frac{2}{3}\right)$

100. x-intercept: $\left(\frac{2}{3}, 0\right)$

    y-intercept: $(0, -2)$

101. Point on line: $(1, 2)$

    x-intercept: $(c, 0)$

    y-intercept: $(0, c)$, $c \neq 0$

102. Point on line: $(-3, 4)$

    x-intercept: $(d, 0)$

    y-intercept: $(0, d)$, $d \neq 0$

GRAPHICAL ANALYSIS In Exercises 103–106, identify any relationships that exist among the lines, and then use a graphing utility to graph the three equations in the same viewing window. Adjust the viewing window so that the slope appears visually correct—that is, so that parallel lines appear parallel and perpendicular lines appear to intersect at right angles.

103. (a) $y = 2x$ (b) $y = -2x$ (c) $y = \frac{1}{2}x$

104. (a) $y = \frac{2}{3}x$ (b) $y = -\frac{3}{2}x$ (c) $y = \frac{2}{3}x + 2$

105. (a) $y = -\frac{1}{2}x$ (b) $y = -\frac{1}{2}x + 3$ (c) $y = 2x - 4$

106. (a) $y = x - 8$ (b) $y = x + 1$ (c) $y = -x + 3$

In Exercises 107–110, find a relationship between x and y such that (x, y) is equidistant (the same distance) from the two points.

107. $(4, -1)$, $(-2, 3)$ **108.** $(6, 5)$, $(1, -8)$

109. $\left(3, \frac{5}{2}\right)$, $(-7, 1)$ **110.** $\left(-\frac{1}{2}, -4\right)$, $\left(\frac{7}{2}, \frac{5}{4}\right)$

111. SALES The following are the slopes of lines representing annual sales y in terms of time x in years. Use the slopes to interpret any change in annual sales for a one-year increase in time.

(a) The line has a slope of $m = 135$.

(b) The line has a slope of $m = 0$.

(c) The line has a slope of $m = -40$.

112. REVENUE The following are the slopes of lines representing daily revenues y in terms of time x in days. Use the slopes to interpret any change in daily revenues for a one-day increase in time.

(a) The line has a slope of $m = 400$.

(b) The line has a slope of $m = 100$.

(c) The line has a slope of $m = 0$.

113. AVERAGE SALARY The graph shows the average salaries for senior high school principals from 1996 through 2008. (Source: Educational Research Service)

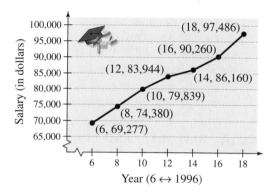

(a) Use the slopes of the line segments to determine the time periods in which the average salary increased the greatest and the least.

(b) Find the slope of the line segment connecting the points for the years 1996 and 2008.

(c) Interpret the meaning of the slope in part (b) in the context of the problem.

114. SALES The graph shows the sales (in billions of dollars) for Apple Inc. for the years 2001 through 2007. (Source: Apple Inc.)

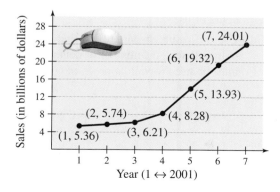

(a) Use the slopes of the line segments to determine the years in which the sales showed the greatest increase and the least increase.

(b) Find the slope of the line segment connecting the points for the years 2001 and 2007.

(c) Interpret the meaning of the slope in part (b) in the context of the problem.

115. ROAD GRADE You are driving on a road that has a 6% uphill grade (see figure). This means that the slope of the road is $\frac{6}{100}$. Approximate the amount of vertical change in your position if you drive 200 feet.

116. ROAD GRADE From the top of a mountain road, a surveyor takes several horizontal measurements x and several vertical measurements y, as shown in the table (x and y are measured in feet).

x	300	600	900	1200	1500	1800	2100
y	-25	-50	-75	-100	-125	-150	-175

(a) Sketch a scatter plot of the data.

(b) Use a straightedge to sketch the line that you think best fits the data.

(c) Find an equation for the line you sketched in part (b).

(d) Interpret the meaning of the slope of the line in part (c) in the context of the problem.

(e) The surveyor needs to put up a road sign that indicates the steepness of the road. For instance, a surveyor would put up a sign that states "8% grade" on a road with a downhill grade that has a slope of $-\frac{8}{100}$. What should the sign state for the road in this problem?

RATE OF CHANGE In Exercises 117 and 118, you are given the dollar value of a product in 2010 and the rate at which the value of the product is expected to change during the next 5 years. Use this information to write a linear equation that gives the dollar value V of the product in terms of the year t. (Let $t = 10$ represent 2010.)

	2010 Value	*Rate*
117.	$2540	$125 decrease per year
118.	$156	$4.50 increase per year

119. DEPRECIATION The value V of a molding machine t years after it is purchased is

$$V = -4000t + 58,500, \quad 0 \le t \le 5.$$

Explain what the V-intercept and the slope measure.

120. COST The cost C of producing n computer laptop bags is given by

$$C = 1.25n + 15,750, \quad 0 < n.$$

Explain what the C-intercept and the slope measure.

121. DEPRECIATION A sub shop purchases a used pizza oven for $875. After 5 years, the oven will have to be replaced. Write a linear equation giving the value V of the equipment during the 5 years it will be in use.

122. DEPRECIATION A school district purchases a high-volume printer, copier, and scanner for $25,000. After 10 years, the equipment will have to be replaced. Its value at that time is expected to be $2000. Write a linear equation giving the value V of the equipment during the 10 years it will be in use.

123. SALES A discount outlet is offering a 20% discount on all items. Write a linear equation giving the sale price S for an item with a list price L.

124. HOURLY WAGE A microchip manufacturer pays its assembly line workers $12.25 per hour. In addition, workers receive a piecework rate of $0.75 per unit produced. Write a linear equation for the hourly wage W in terms of the number of units x produced per hour.

125. MONTHLY SALARY A pharmaceutical salesperson receives a monthly salary of $2500 plus a commission of 7% of sales. Write a linear equation for the salesperson's monthly wage W in terms of monthly sales S.

126. BUSINESS COSTS A sales representative of a company using a personal car receives $120 per day for lodging and meals plus $0.55 per mile driven. Write a linear equation giving the daily cost C to the company in terms of x, the number of miles driven.

127. CASH FLOW PER SHARE The cash flow per share for the Timberland Co. was $1.21 in 1999 and $1.46 in 2007. Write a linear equation that gives the cash flow per share in terms of the year. Let $t = 9$ represent 1999. Then predict the cash flows for the years 2012 and 2014. (Source: The Timberland Co.)

128. NUMBER OF STORES In 2003 there were 1078 J.C. Penney stores and in 2007 there were 1067 stores. Write a linear equation that gives the number of stores in terms of the year. Let $t = 3$ represent 2003. Then predict the numbers of stores for the years 2012 and 2014. Are your answers reasonable? Explain. (Source: J.C. Penney Co.)

129. COLLEGE ENROLLMENT The Pennsylvania State University had enrollments of 40,571 students in 2000 and 44,112 students in 2008 at its main campus in University Park, Pennsylvania. (Source: Penn State Fact Book)

(a) Assuming the enrollment growth is linear, find a linear model that gives the enrollment in terms of the year t, where $t = 0$ corresponds to 2000.

(b) Use your model from part (a) to predict the enrollments in 2010 and 2015.

(c) What is the slope of your model? Explain its meaning in the context of the situation.

130. COLLEGE ENROLLMENT The University of Florida had enrollments of 46,107 students in 2000 and 51,413 students in 2008. (Source: University of Florida)

(a) What was the average annual change in enrollment from 2000 to 2008?

(b) Use the average annual change in enrollment to estimate the enrollments in 2002, 2004, and 2006.

(c) Write the equation of a line that represents the given data in terms of the year t, where $t = 0$ corresponds to 2000. What is its slope? Interpret the slope in the context of the problem.

(d) Using the results of parts (a)–(c), write a short paragraph discussing the concepts of *slope* and *average rate of change*.

131. COST, REVENUE, AND PROFIT A roofing contractor purchases a shingle delivery truck with a shingle elevator for $42,000. The vehicle requires an average expenditure of $6.50 per hour for fuel and maintenance, and the operator is paid $11.50 per hour.

(a) Write a linear equation giving the total cost C of operating this equipment for t hours. (Include the purchase cost of the equipment.)

(b) Assuming that customers are charged $30 per hour of machine use, write an equation for the revenue R derived from t hours of use.

(c) Use the formula for profit

$$P = R - C$$

to write an equation for the profit derived from t hours of use.

(d) Use the result of part (c) to find the break-even point—that is, the number of hours this equipment must be used to yield a profit of 0 dollars.

132. RENTAL DEMAND   A real estate office handles an apartment complex with 50 units. When the rent per unit is $580 per month, all 50 units are occupied. However, when the rent is $625 per month, the average number of occupied units drops to 47. Assume that the relationship between the monthly rent p and the demand x is linear.

(a) Write the equation of the line giving the demand x in terms of the rent p.

(b) Use this equation to predict the number of units occupied when the rent is $655.

(c) Predict the number of units occupied when the rent is $595.

133. GEOMETRY   The length and width of a rectangular garden are 15 meters and 10 meters, respectively. A walkway of width x surrounds the garden.

(a) Draw a diagram that gives a visual representation of the problem.

(b) Write the equation for the perimeter y of the walkway in terms of x.

(c) Use a graphing utility to graph the equation for the perimeter.

(d) Determine the slope of the graph in part (c). For each additional one-meter increase in the width of the walkway, determine the increase in its perimeter.

134. AVERAGE ANNUAL SALARY   The average salaries (in millions of dollars) of Major League Baseball players from 2000 through 2007 are shown in the scatter plot. Find the equation of the line that you think best fits these data. (Let y represent the average salary and let t represent the year, with $t = 0$ corresponding to 2000.)   (Source: Major League Baseball Players Association)

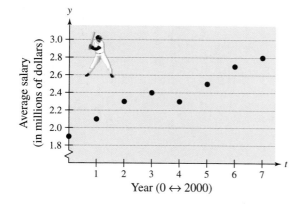

135. DATA ANALYSIS: NUMBER OF DOCTORS   The numbers of doctors of osteopathic medicine y (in thousands) in the United States from 2000 through 2008, where x is the year, are shown as data points (x, y).   (Source: American Osteopathic Association)

(2000, 44.9), (2001, 47.0), (2002, 49.2), (2003, 51.7), (2004, 54.1), (2005, 56.5), (2006, 58.9), (2007, 61.4), (2008, 64.0)

(a) Sketch a scatter plot of the data. Let $x = 0$ correspond to 2000.

(b) Use a straightedge to sketch the line that you think best fits the data.

(c) Find the equation of the line from part (b). Explain the procedure you used.

(d) Write a short paragraph explaining the meanings of the slope and y-intercept of the line in terms of the data.

(e) Compare the values obtained using your model with the actual values.

(f) Use your model to estimate the number of doctors of osteopathic medicine in 2012.

136. DATA ANALYSIS: AVERAGE SCORES   An instructor gives regular 20-point quizzes and 100-point exams in an algebra course. Average scores for six students, given as data points (x, y), where x is the average quiz score and y is the average test score, are (18, 87), (10, 55), (19, 96), (16, 79), (13, 76), and (15, 82). [*Note:* There are many correct answers for parts (b)–(d).]

(a) Sketch a scatter plot of the data.

(b) Use a straightedge to sketch the line that you think best fits the data.

(c) Find an equation for the line you sketched in part (b).

(d) Use the equation in part (c) to estimate the average test score for a person with an average quiz score of 17.

(e) The instructor adds 4 points to the average test score of each student in the class. Describe the changes in the positions of the plotted points and the change in the equation of the line.

EXPLORATION

TRUE OR FALSE? In Exercises 137 and 138, determine whether the statement is true or false. Justify your answer.

137. A line with a slope of $-\frac{5}{7}$ is steeper than a line with a slope of $-\frac{6}{7}$.

138. The line through $(-8, 2)$ and $(-1, 4)$ and the line through $(0, -4)$ and $(-7, 7)$ are parallel.

139. Explain how you could show that the points $A(2, 3)$, $B(2, 9)$, and $C(4, 3)$ are the vertices of a right triangle.

140. Explain why the slope of a vertical line is said to be undefined.

141. With the information shown in the graphs, is it possible to determine the slope of each line? Is it possible that the lines could have the same slope? Explain.

(a) 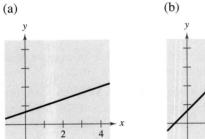   (b)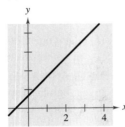

142. The slopes of two lines are -4 and $\frac{5}{2}$. Which is steeper? Explain.

143. Use a graphing utility to compare the slopes of the lines $y = mx$, where $m = 0.5, 1, 2,$ and 4. Which line rises most quickly? Now, let $m = -0.5, -1, -2,$ and -4. Which line falls most quickly? Use a square setting to obtain a true geometric perspective. What can you conclude about the slope and the "rate" at which the line rises or falls?

144. Find d_1 and d_2 in terms of m_1 and m_2, respectively (see figure). Then use the Pythagorean Theorem to find a relationship between m_1 and m_2.

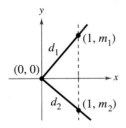

145. THINK ABOUT IT Is it possible for two lines with positive slopes to be perpendicular? Explain.

146. CAPSTONE Match the description of the situation with its graph. Also determine the slope and y-intercept of each graph and interpret the slope and y-intercept in the context of the situation. [The graphs are labeled (i), (ii), (iii), and (iv).]

(i) 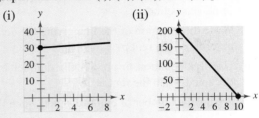   (ii)

(iii) 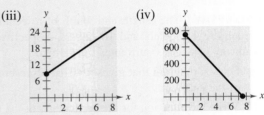   (iv)

(a) A person is paying $20 per week to a friend to repay a $200 loan.

(b) An employee is paid $8.50 per hour plus $2 for each unit produced per hour.

(c) A sales representative receives $30 per day for food plus $0.32 for each mile traveled.

(d) A computer that was purchased for $750 depreciates $100 per year.

PROJECT: BACHELOR'S DEGREES To work an extended application analyzing the numbers of bachelor's degrees earned by women in the United States from 1996 through 2007, visit this text's website at *academic.cengage.com*. (Data Source: U.S. National Center for Education Statistics)

2.2   FUNCTIONS

What you should learn

- Determine whether relations between two variables are functions.
- Use function notation and evaluate functions.
- Find the domains of functions.
- Use functions to model and solve real-life problems.
- Evaluate difference quotients.

Why you should learn it

Functions can be used to model and solve real-life problems. For instance, in Exercise 100 on page 198, you will use a function to model the force of water against the face of a dam.

Introduction to Functions

Many everyday phenomena involve two quantities that are related to each other by some rule of correspondence. The mathematical term for such a rule of correspondence is a **relation.** In mathematics, relations are often represented by mathematical equations and formulas. For instance, the simple interest I earned on \$1000 for 1 year is related to the annual interest rate r by the formula $I = 1000r$.

The formula $I = 1000r$ represents a special kind of relation that matches each item from one set with *exactly one* item from a different set. Such a relation is called a **function.**

Definition of Function

A **function** f from a set A to a set B is a relation that assigns to each element x in the set A exactly one element y in the set B. The set A is the **domain** (or set of inputs) of the function f, and the set B contains the **range** (or set of outputs).

To help understand this definition, look at the function that relates the time of day to the temperature in Figure 2.20.

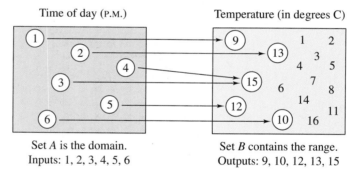

Set A is the domain.
Inputs: 1, 2, 3, 4, 5, 6

Set B contains the range.
Outputs: 9, 10, 12, 13, 15

FIGURE **2.20**

This function can be represented by the following ordered pairs, in which the first coordinate (x-value) is the input and the second coordinate (y-value) is the output.

$$\{(1, 9°), (2, 13°), (3, 15°), (4, 15°), (5, 12°), (6, 10°)\}$$

Characteristics of a Function from Set A to Set B

1. Each element in A must be matched with an element in B.

2. Some elements in B may not be matched with any element in A.

3. Two or more elements in A may be matched with the same element in B.

4. An element in A (the domain) cannot be matched with two different elements in B.

Functions are commonly represented in four ways.

Four Ways to Represent a Function

1. *Verbally* by a sentence that describes how the input variable is related to the output variable

2. *Numerically* by a table or a list of ordered pairs that matches input values with output values

3. *Graphically* by points on a graph in a coordinate plane in which the input values are represented by the horizontal axis and the output values are represented by the vertical axis

4. *Algebraically* by an equation in two variables

To determine whether or not a relation is a function, you must decide whether each input value is matched with exactly one output value. If any input value is matched with two or more output values, the relation is not a function.

Example 1   Testing for Functions

Determine whether the relation represents y as a function of x.

a. The input value x is the number of representatives from a state, and the output value y is the number of senators.

b.

Input, x	Output, y
2	11
2	10
3	8
4	5
5	1

c.

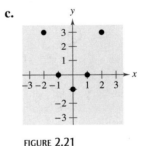

FIGURE **2.21**

Solution

a. This verbal description *does* describe y as a function of x. Regardless of the value of x, the value of y is always 2. Such functions are called *constant functions*.

b. This table *does not* describe y as a function of x. The input value 2 is matched with two different y-values.

c. The graph in Figure 2.21 *does* describe y as a function of x. Each input value is matched with exactly one output value.

CHECK Point   Now try Exercise 11.

Representing functions by sets of ordered pairs is common in *discrete mathematics*. In algebra, however, it is more common to represent functions by equations or formulas involving two variables. For instance, the equation

$$y = x^2$$

represents the variable y as a function of the variable x. In this equation, x is

HISTORICAL NOTE

Leonhard Euler (1707–1783), a Swiss mathematician, is considered to have been the most prolific and productive mathematician in history. One of his greatest influences on mathematics was his use of symbols, or notation. The function notation $y = f(x)$ was introduced by Euler.

the **independent variable** and y is the **dependent variable.** The domain of the function is the set of all values taken on by the independent variable x, and the range of the function is the set of all values taken on by the dependent variable y.

| **Example 2**   **Testing for Functions Represented Algebraically**

Which of the equations represent(s) y as a function of x?

a. $x^2 + y = 1$ **b.** $-x + y^2 = 1$

Solution

To determine whether y is a function of x, try to solve for y in terms of x.

a. Solving for y yields

$$x^2 + y = 1$$
$$y = 1 - x^2.$$

To each value of x there corresponds exactly one value of y. So, y is a function of x.

b. Solving for y yields

$$-x + y^2 = 1$$
$$y^2 = 1 + x$$
$$y = \pm\sqrt{1 + x}.$$

The $\pm$ indicates that to a given value of x there correspond two values of y. So, y is not a function of x.

CHECK Point   Now try Exercise 21.

Function Notation

When an equation is used to represent a function, it is convenient to name the function so that it can be referenced easily. For example, you know that the equation $y = 1 - x^2$ describes y as a function of x. Suppose you give this function the name "f." Then you can use the following **function notation.**

Input	*Output*	*Equation*
x	$f(x)$	$f(x) = 1 - x^2$

The symbol $f(x)$ is read as *the value of f at x* or simply *f of x.* The symbol $f(x)$ corresponds to the y-value for a given x. So, you can write $y = f(x)$. Keep in mind that f is the *name* of the function, whereas $f(x)$ is the *value* of the function at x. For instance, the function given by

$$f(x) = 3 - 2x$$

has *function values* denoted by $f(-1)$, $f(0)$, $f(2)$, and so on. To find these values, substitute the specified input values into the given equation.

For $x = -1$, $f(\ ) = 3 - 2(\ ) = 3 + 2 = 5.$

For $x = 0$, $f(\ ) = 3 - 2(\ ) = 3 - 0 = 3.$

For $x = 2$, $f(\ ) = 3 - 2(\ ) = 3 - 4 = -1.$

Although f is often used as a convenient function name and x is often used as the independent variable, you can use other letters. For instance,

$$f(x) = x^2 - 4x + 7, \quad f(t) = t^2 - 4t + 7, \quad \text{and} \quad g(s) = s^2 - 4s + 7$$

all define the same function. In fact, the role of the independent variable is that of a "placeholder." Consequently, the function could be described by

$$f(\;\;\;) = (\;\;\;)^2 - 4(\;\;\;) + 7.$$

> ⚠ **WARNING / CAUTION**
>
> In Example 3, note that $g(x + 2)$ is not equal to $g(x) + g(2)$. In general, $g(u + v) \neq g(u) + g(v)$.

Example 3 Evaluating a Function

Let $g(x) = -x^2 + 4x + 1$. Find each function value.

a. $g(2)$ **b.** $g(t)$ **c.** $g(x + 2)$

Solution

a. Replacing x with 2 in $g(x) = -x^2 + 4x + 1$ yields the following.

$$g(\;) = -(\;)^2 + 4(\;) + 1 = -4 + 8 + 1 = 5$$

b. Replacing x with t yields the following.

$$g(\;) = -(\;)^2 + 4(\;) + 1 = -t^2 + 4t + 1$$

c. Replacing x with $x + 2$ yields the following.

$$
\begin{aligned}
g(\quad) &= -(\quad)^2 + 4(\quad) + 1 \\
&= -(x^2 + 4x + 4) + 4x + 8 + 1 \\
&= -x^2 - 4x - 4 + 4x + 8 + 1 \\
&= -x^2 + 5
\end{aligned}
$$

CHECK*Point* Now try Exercise 41.

A function defined by two or more equations over a specified domain is called a **piecewise-defined function.**

Example 4 A Piecewise-Defined Function

Evaluate the function when $x = -1, 0,$ and 1.

$$f(x) = \begin{cases} x^2 + 1, & x < 0 \\ x - 1, & x \geq 0 \end{cases}$$

Solution

Because $x = -1$ is less than 0, use $f(x) = x^2 + 1$ to obtain

$$f(\;) = (\;)^2 + 1 = 2.$$

For $x = 0$, use $f(x) = x - 1$ to obtain

$$f(\;) = (\;) - 1 = -1.$$

For $x = 1$, use $f(x) = x - 1$ to obtain

$$f(\;) = (\;) - 1 = 0.$$

CHECK*Point* Now try Exercise 49.

| **Example 5**   **Finding Values for Which $f(x) = 0$**

Find all real values of x such that $f(x) = 0$.

a. $f(x) = -2x + 10$

b. $f(x) = x^2 - 5x + 6$

Solution

For each function, set $f(x) = 0$ and solve for x.

a. $-2x + 10 = 0$

$-2x = -10$

$x = 5$

So, $f(x) = 0$ when $x = 5$.

b. $x^2 - 5x + 6 = 0$

$(x - 2)(x - 3) = 0$

$x - 2 = 0$ ⟹ $x = 2$

$x - 3 = 0$ ⟹ $x = 3$

So, $f(x) = 0$ when $x = 2$ or $x = 3$.

CHECK Point   Now try Exercise 59.

| **Example 6**   **Finding Values for Which $f(x) = g(x)$**

Find the values of x for which $f(x) = g(x)$.

a. $f(x) = x^2 + 1$ and $g(x) = 3x - x^2$

b. $f(x) = x^2 - 1$ and $g(x) = -x^2 + x + 2$

Solution

a. $x^2 + 1 = 3x - x^2$

$2x^2 - 3x + 1 = 0$

$(2x - 1)(x - 1) = 0$

$2x - 1 = 0$ ⟹ $x = \frac{1}{2}$

$x - 1 = 0$ ⟹ $x = 1$

So, $f(x) = g(x)$ when $x = \dfrac{1}{2}$ or $x = 1$.

b. $x^2 - 1 = -x^2 + x + 2$

$2x^2 - x - 3 = 0$

$(2x - 3)(x + 1) = 0$

$2x - 3 = 0$ ⟹ $x = \frac{3}{2}$

$x + 1 = 0$ ⟹ $x = -1$

So, $f(x) = g(x)$ when $x = \dfrac{3}{2}$ or $x = -1$.

CHECK Point   Now try Exercise 67.

The Domain of a Function

The domain of a function can be described explicitly or it can be *implied* by the expression used to define the function. The **implied domain** is the set of all real numbers for which the expression is defined. For instance, the function given by

$$f(x) = \frac{1}{x^2 - 4}$$

has an implied domain that consists of all real x other than $x = \pm 2$. These two values are excluded from the domain because division by zero is undefined. Another common type of implied domain is that used to avoid even roots of negative numbers. For example, the function given by

$$f(x) = \sqrt{x}$$

is defined only for $x \geq 0$. So, its implied domain is the interval $[0, \infty)$. In general, the domain of a function *excludes* values that would cause division by zero *or* that would result in the even root of a negative number.

TECHNOLOGY

Use a graphing utility to graph the functions given by $y = \sqrt{4 - x^2}$ and $y = \sqrt{x^2 - 4}$. What is the domain of each function? Do the domains of these two functions overlap? If so, for what values do the domains overlap?

Example 7   Finding the Domain of a Function

Find the domain of each function.

a. f: $\{(-3, 0), (-1, 4), (0, 2), (2, 2), (4, -1)\}$ **b.** $g(x) = \dfrac{1}{x + 5}$

c. Volume of a sphere: $V = \frac{4}{3}\pi r^3$ **d.** $h(x) = \sqrt{4 - 3x}$

Solution

a. The domain of f consists of all first coordinates in the set of ordered pairs.

Domain $= \{-3, -1, 0, 2, 4\}$

b. Excluding x-values that yield zero in the denominator, the domain of g is the set of all real numbers x except $x = -5$.

c. Because this function represents the volume of a sphere, the values of the radius r must be positive. So, the domain is the set of all real numbers r such that $r > 0$.

d. This function is defined only for x-values for which

$$4 - 3x \geq 0.$$

Using the methods described in Section 1.8, you can conclude that $x \leq \frac{4}{3}$. So, the domain is the interval $\left(-\infty, \frac{4}{3}\right]$.

CHECK Point   Now try Exercise 73.

In Example 7(c), note that the domain of a function may be implied by the physical context. For instance, from the equation

$$V = \frac{4}{3}\pi r^3$$

you would have no reason to restrict r to positive values, but the physical context implies that a sphere cannot have a negative or zero radius.

FIGURE 2.22

Applications

Example 8   The Dimensions of a Container

You work in the marketing department of a soft-drink company and are experimenting with a new can for iced tea that is slightly narrower and taller than a standard can. For your experimental can, the ratio of the height to the radius is 4, as shown in Figure 2.22.

a. Write the volume of the can as a function of the radius r.

b. Write the volume of the can as a function of the height h.

Solution

a. $V(r) = \pi r^2 h = \pi r^2(4r) = 4\pi r^3$

b. $V(h) = \pi \left(\dfrac{h}{4}\right)^2 h = \dfrac{\pi h^3}{16}$

CHECK Point   Now try Exercise 87.

Example 9   The Path of a Baseball

A baseball is hit at a point 3 feet above ground at a velocity of 100 feet per second and an angle of 45°. The path of the baseball is given by the function

$$f(x) = -0.0032x^2 + x + 3$$

where x and $f(x)$ are measured in feet. Will the baseball clear a 10-foot fence located 300 feet from home plate?

Algebraic Solution

When $x = 300$, you can find the height of the baseball as follows.

$$f(x) = -0.0032x^2 + x + 3$$

$$f(\quad) = -0.0032(\quad)^2 + \quad + 3$$

$$= 15$$

When $x = 300$, the height of the baseball is 15 feet, so the baseball will clear a 10-foot fence.

Graphical Solution

Use a graphing utility to graph the function $y = -0.0032x^2 + x + 3$. Use the *value* feature or the *zoom* and *trace* features of the graphing utility to estimate that $y = 15$ when $x = 300$, as shown in Figure 2.23. So, the ball will clear a 10-foot fence.

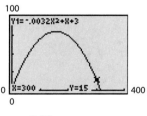

FIGURE 2.23

CHECK Point   Now try Exercise 93.

In the equation in Example 9, the height of the baseball is a function of the distance from home plate.

Number of Alternative-Fueled
Vehicles in the U.S.

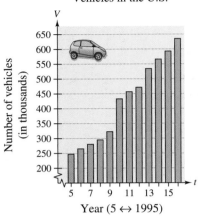

FIGURE **2.24**

Example 10 Alternative-Fueled Vehicles

The number V (in thousands) of alternative-fueled vehicles in the United States increased in a linear pattern from 1995 to 1999, as shown in Figure 2.24. Then, in 2000, the number of vehicles took a jump and, until 2006, increased in a different linear pattern. These two patterns can be approximated by the function

$$V(t) = \begin{cases} 18.08t + 155.3, & 5 \le t \le 9 \\ 34.75t + 74.9, & 10 \le t \le 16 \end{cases}$$

where t represents the year, with $t = 5$ corresponding to 1995. Use this function to approximate the number of alternative-fueled vehicles for each year from 1995 to 2006. (Source: Science Applications International Corporation; Energy Information Administration)

Solution

From 1995 to 1999, use $V(t) = 18.08t + 155.3$.

| 245.7 | 263.8 | 281.9 | 299.9 | 318.0 |

From 2000 to 2006, use $V(t) = 34.75t + 74.9$.

| 422.4 | 457.2 | 491.9 | 526.7 | 561.4 | 596.2 | 630.9 |

CHECK Point Now try Exercise 95.

Difference Quotients

One of the basic definitions in calculus employs the ratio

$$\frac{f(x + h) - f(x)}{h}, \quad h \ne 0.$$

This ratio is called a **difference quotient,** as illustrated in Example 11.

Example 11 Evaluating a Difference Quotient

For $f(x) = x^2 - 4x + 7$, find $\dfrac{f(x + h) - f(x)}{h}$.

Solution

$$\frac{f(\quad) - f()}{h} = \frac{[(\quad)^2 - 4(\quad) + 7] - (\ ^2 - 4\ + 7)}{h}$$

$$= \frac{x^2 + 2xh + h^2 - 4x - 4h + 7 - x^2 + 4x - 7}{h}$$

$$= \frac{2xh + h^2 - 4h}{h} = \frac{h(2x + h - 4)}{h} = 2x + h - 4, \quad h \ne 0$$

CHECK Point Now try Exercise 103.

The symbol **ʃ** indicates an example or exercise that highlights algebraic techniques specifically used in calculus.

You may find it easier to calculate the difference quotient in Example 11 by first finding $f(x + h)$, and then substituting the resulting expression into the difference quotient, as follows.

$$f(x + h) = \qquad ^2 - 4 \qquad + 7 = x^2 + 2xh + h^2 - 4x - 4h + 7$$

$$\frac{f(x + h) - f(x)}{h} = \frac{\qquad\qquad\qquad\qquad - (x^2 - 4x + 7)}{h}$$

$$= \frac{2xh + h^2 - 4h}{h} = \frac{h(2x + h - 4)}{h} = 2x + h - 4, \quad h \neq 0$$

Summary of Function Terminology

Function: A **function** is a relationship between two variables such that to each value of the independent variable there corresponds exactly one value of the dependent variable.

Function Notation: $y = f(x)$
    f is the *name* of the function.

    y is the **dependent variable.**

    x is the **independent variable.**

    $f(x)$ is the *value of the function at x.*

Domain: The **domain** of a function is the set of all values (inputs) of the independent variable for which the function is defined. If x is in the domain of f, f is said to be *defined* at x. If x is not in the domain of f, f is said to be *undefined* at x.

Range: The **range** of a function is the set of all values (outputs) assumed by the dependent variable (that is, the set of all function values).

Implied Domain: If f is defined by an algebraic expression and the domain is not specified, the **implied domain** consists of all real numbers for which the expression is defined.

CLASSROOM DISCUSSION

Everyday Functions   In groups of two or three, identify common real-life functions. Consider everyday activities, events, and expenses, such as long distance telephone calls and car insurance. Here are two examples.

a. The statement, "Your happiness is a function of the grade you receive in this course" is *not* a correct mathematical use of the word "function." The word "happiness" is ambiguous.

b. The statement, "Your federal income tax is a function of your adjusted gross income" *is* a correct mathematical use of the word "function." Once you have determined your adjusted gross income, your income tax can be determined.

Describe your functions in words. Avoid using ambiguous words. Can you find an example of a piecewise-defined function?

2.2 EXERCISES

See www.CalcChat.com for worked-out solutions to odd-numbered exercises.

VOCABULARY: Fill in the blanks.

1. A relation that assigns to each element x from a set of inputs, or _____, exactly one element y in a set of outputs, or _____, is called a _____.

2. Functions are commonly represented in four different ways, _____, _____, _____, and _____.

3. For an equation that represents y as a function of x, the set of all values taken on by the _____ variable x is the domain, and the set of all values taken on by the _____ variable y is the range.

4. The function given by

$$f(x) = \begin{cases} 2x - 1, & x < 0 \\ x^2 + 4, & x \geq 0 \end{cases}$$

is an example of a _____ function.

5. If the domain of the function f is not given, then the set of values of the independent variable for which the expression is defined is called the _____ _____.

6. In calculus, one of the basic definitions is that of a _____ _____, given by $\dfrac{f(x + h) - f(x)}{h}$, $h \neq 0$.

SKILLS AND APPLICATIONS

In Exercises 7–10, is the relationship a function?

7.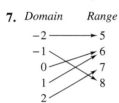

8.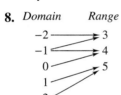

9. and **10.**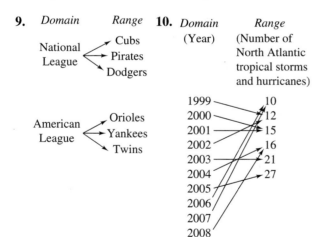

In Exercises 11–14, determine whether the relation represents y as a function of x.

11.

Input, x	-2	-1	0	1	2
Output, y	-8	-1	0	1	8

12.

Input, x	0	1	2	1	0
Output, y	-4	-2	0	2	4

13.

Input, x	10	7	4	7	10
Output, y	3	6	9	12	15

14.

Input, x	0	3	9	12	15
Output, y	3	3	3	3	3

In Exercises 15 and 16, which sets of ordered pairs represent functions from A to B? Explain.

15. $A = \{0, 1, 2, 3\}$ and $B = \{-2, -1, 0, 1, 2\}$
    (a) $\{(0, 1), (1, -2), (2, 0), (3, 2)\}$
    (b) $\{(0, -1), (2, 2), (1, -2), (3, 0), (1, 1)\}$
    (c) $\{(0, 0), (1, 0), (2, 0), (3, 0)\}$
    (d) $\{(0, 2), (3, 0), (1, 1)\}$

16. $A = \{a, b, c\}$ and $B = \{0, 1, 2, 3\}$
    (a) $\{(a, 1), (c, 2), (c, 3), (b, 3)\}$
    (b) $\{(a, 1), (b, 2), (c, 3)\}$
    (c) $\{(1, a), (0, a), (2, c), (3, b)\}$
    (d) $\{(c, 0), (b, 0), (a, 3)\}$

CIRCULATION OF NEWSPAPERS   In Exercises 17 and 18, use the graph, which shows the circulation (in millions) of daily newspapers in the United States. (Source: Editor & Publisher Company)

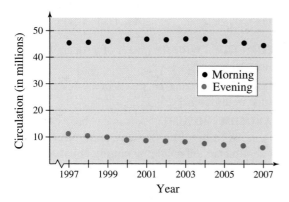

17. Is the circulation of morning newspapers a function of the year? Is the circulation of evening newspapers a function of the year? Explain.

18. Let $f(x)$ represent the circulation of evening newspapers in year x. Find $f(2002)$.

In Exercises 19–36, determine whether the equation represents y as a function of x.

19. $x^2 + y^2 = 4$ **20.** $x^2 - y^2 = 16$

21. $x^2 + y = 4$ **22.** $y - 4x^2 = 36$

23. $2x + 3y = 4$ **24.** $2x + 5y = 10$

25. $(x + 2)^2 + (y - 1)^2 = 25$

26. $(x - 2)^2 + y^2 = 4$

27. $y^2 = x^2 - 1$ **28.** $x + y^2 = 4$

29. $y = \sqrt{16 - x^2}$ **30.** $y = \sqrt{x + 5}$

31. $y = |4 - x|$ **32.** $|y| = 4 - x$

33. $x = 14$ **34.** $y = -75$

35. $y + 5 = 0$ **36.** $x - 1 = 0$

In Exercises 37–52, evaluate the function at each specified value of the independent variable and simplify.

37. $f(x) = 2x - 3$
  (a) $f(1)$ (b) $f(-3)$ (c) $f(x - 1)$

38. $g(y) = 7 - 3y$
  (a) $g(0)$ (b) $g\left(\frac{7}{3}\right)$ (c) $g(s + 2)$

39. $V(r) = \frac{4}{3}\pi r^3$
  (a) $V(3)$ (b) $V\left(\frac{3}{2}\right)$ (c) $V(2r)$

40. $S(r) = 4\pi r^2$
  (a) $S(2)$ (b) $S\left(\frac{1}{2}\right)$ (c) $S(3r)$

41. $g(t) = 4t^2 - 3t + 5$
  (a) $g(2)$ (b) $g(t - 2)$ (c) $g(t) - g(2)$

42. $h(t) = t^2 - 2t$
  (a) $h(2)$ (b) $h(1.5)$ (c) $h(x + 2)$

43. $f(y) = 3 - \sqrt{y}$
  (a) $f(4)$ (b) $f(0.25)$ (c) $f(4x^2)$

44. $f(x) = \sqrt{x + 8} + 2$
  (a) $f(-8)$ (b) $f(1)$ (c) $f(x - 8)$

45. $q(x) = 1/(x^2 - 9)$
  (a) $q(0)$ (b) $q(3)$ (c) $q(y + 3)$

46. $q(t) = (2t^2 + 3)/t^2$
  (a) $q(2)$ (b) $q(0)$ (c) $q(-x)$

47. $f(x) = |x|/x$
  (a) $f(2)$ (b) $f(-2)$ (c) $f(x - 1)$

48. $f(x) = |x| + 4$
  (a) $f(2)$ (b) $f(-2)$ (c) $f(x^2)$

49. $f(x) = \begin{cases} 2x + 1, & x < 0 \\ 2x + 2, & x \geq 0 \end{cases}$
  (a) $f(-1)$ (b) $f(0)$ (c) $f(2)$

50. $f(x) = \begin{cases} x^2 + 2, & x \leq 1 \\ 2x^2 + 2, & x > 1 \end{cases}$
  (a) $f(-2)$ (b) $f(1)$ (c) $f(2)$

51. $f(x) = \begin{cases} 3x - 1, & x < -1 \\ 4, & -1 \leq x \leq 1 \\ x^2, & x > 1 \end{cases}$
  (a) $f(-2)$ (b) $f\left(-\frac{1}{2}\right)$ (c) $f(3)$

52. $f(x) = \begin{cases} 4 - 5x, & x \leq -2 \\ 0, & -2 < x < 2 \\ x^2 + 1, & x \geq 2 \end{cases}$
  (a) $f(-3)$ (b) $f(4)$ (c) $f(-1)$

In Exercises 53–58, complete the table.

53. $f(x) = x^2 - 3$

x	-2	-1	0	1	2
$f(x)$					

54. $g(x) = \sqrt{x - 3}$

x	3	4	5	6	7
$g(x)$					

55. $h(t) = \frac{1}{2}|t + 3|$

t	-5	-4	-3	-2	-1
$h(t)$					

56. $f(s) = \dfrac{|s - 2|}{s - 2}$

s	0	1	$\frac{3}{2}$	$\frac{5}{2}$	4
$f(s)$					

57. $f(x) = \begin{cases} -\frac{1}{2}x + 4, & x \le 0 \\ (x - 2)^2, & x > 0 \end{cases}$

x	-2	-1	0	1	2
$f(x)$					

58. $f(x) = \begin{cases} 9 - x^2, & x < 3 \\ x - 3, & x \ge 3 \end{cases}$

x	1	2	3	4	5
$f(x)$					

In Exercises 59–66, find all real values of x such that $f(x) = 0$.

59. $f(x) = 15 - 3x$ **60.** $f(x) = 5x + 1$

61. $f(x) = \dfrac{3x - 4}{5}$ **62.** $f(x) = \dfrac{12 - x^2}{5}$

63. $f(x) = x^2 - 9$ **64.** $f(x) = x^2 - 8x + 15$

65. $f(x) = x^3 - x$

66. $f(x) = x^3 - x^2 - 4x + 4$

In Exercises 67–70, find the value(s) of x for which $f(x) = g(x)$.

67. $f(x) = x^2, \quad g(x) = x + 2$

68. $f(x) = x^2 + 2x + 1, \quad g(x) = 7x - 5$

69. $f(x) = x^4 - 2x^2, \quad g(x) = 2x^2$

70. $f(x) = \sqrt{x} - 4, \quad g(x) = 2 - x$

In Exercises 71–82, find the domain of the function.

71. $f(x) = 5x^2 + 2x - 1$ **72.** $g(x) = 1 - 2x^2$

73. $h(t) = \dfrac{4}{t}$ **74.** $s(y) = \dfrac{3y}{y + 5}$

75. $g(y) = \sqrt{y - 10}$ **76.** $f(t) = \sqrt[3]{t + 4}$

77. $g(x) = \dfrac{1}{x} - \dfrac{3}{x + 2}$ **78.** $h(x) = \dfrac{10}{x^2 - 2x}$

79. $f(s) = \dfrac{\sqrt{s - 1}}{s - 4}$ **80.** $f(x) = \dfrac{\sqrt{x + 6}}{6 + x}$

81. $f(x) = \dfrac{x - 4}{\sqrt{x}}$ **82.** $f(x) = \dfrac{x + 2}{\sqrt{x - 10}}$

In Exercises 83–86, assume that the domain of f is the set $A = \{-2, -1, 0, 1, 2\}$. Determine the set of ordered pairs that represents the function f.

83. $f(x) = x^2$ **84.** $f(x) = (x - 3)^2$

85. $f(x) = |x| + 2$ **86.** $f(x) = |x + 1|$

87. GEOMETRY Write the area A of a square as a function of its perimeter P.

88. GEOMETRY Write the area A of a circle as a function of its circumference C.

89. MAXIMUM VOLUME An open box of maximum volume is to be made from a square piece of material 24 centimeters on a side by cutting equal squares from the corners and turning up the sides (see figure).

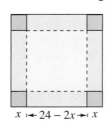 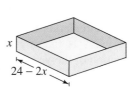

(a) The table shows the volumes V (in cubic centimeters) of the box for various heights x (in centimeters). Use the table to estimate the maximum volume.

Height, x	1	2	3	4	5	6
Volume, V	484	800	972	1024	980	864

(b) Plot the points (x, V) from the table in part (a). Does the relation defined by the ordered pairs represent V as a function of x?

(c) If V is a function of x, write the function and determine its domain.

90. MAXIMUM PROFIT The cost per unit in the production of an MP3 player is $60. The manufacturer charges $90 per unit for orders of 100 or less. To encourage large orders, the manufacturer reduces the charge by $0.15 per MP3 player for each unit ordered in excess of 100 (for example, there would be a charge of $87 per MP3 player for an order size of 120).

(a) The table shows the profits P (in dollars) for various numbers of units ordered, x. Use the table to estimate the maximum profit.

Units, x	110	120	130	140
Profit, P	3135	3240	3315	3360

Units, x	150	160	170
Profit, P	3375	3360	3315

(b) Plot the points (x, P) from the table in part (a). Does the relation defined by the ordered pairs represent P as a function of x?

(c) If P is a function of x, write the function and determine its domain.

91. GEOMETRY A right triangle is formed in the first quadrant by the x- and y-axes and a line through the point $(2, 1)$ (see figure). Write the area A of the triangle as a function of x, and determine the domain of the function.

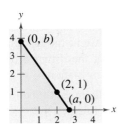

FIGURE FOR 91

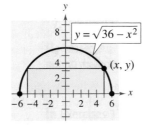

FIGURE FOR 92

92. GEOMETRY A rectangle is bounded by the x-axis and the semicircle $y = \sqrt{36 - x^2}$ (see figure). Write the area A of the rectangle as a function of x, and graphically determine the domain of the function.

93. PATH OF A BALL The height y (in feet) of a baseball thrown by a child is

$$y = -\frac{1}{10}x^2 + 3x + 6$$

where x is the horizontal distance (in feet) from where the ball was thrown. Will the ball fly over the head of another child 30 feet away trying to catch the ball? (Assume that the child who is trying to catch the ball holds a baseball glove at a height of 5 feet.)

94. PRESCRIPTION DRUGS The numbers d (in millions) of drug prescriptions filled by independent outlets in the United States from 2000 through 2007 (see figure) can be approximated by the model

$$d(t) = \begin{cases} 10.6t + 699, & 0 \le t \le 4 \\ 15.5t + 637, & 5 \le t \le 7 \end{cases}$$

where t represents the year, with $t = 0$ corresponding to 2000. Use this model to find the number of drug prescriptions filled by independent outlets in each year from 2000 through 2007. (Source: National Association of Chain Drug Stores)

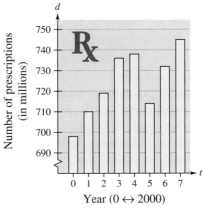

FIGURE FOR **94**

95. MEDIAN SALES PRICE The median sale prices p (in thousands of dollars) of an existing one-family home in the United States from 1998 through 2007 (see figure) can be approximated by the model

$$p(t) = \begin{cases} 1.011t^2 - 12.38t + 170.5, & 8 \le t \le 13 \\ -6.950t^2 + 222.55t - 1557.6, & 14 \le t \le 17 \end{cases}$$

where t represents the year, with $t = 8$ corresponding to 1998. Use this model to find the median sale price of an existing one-family home in each year from 1998 through 2007. (Source: National Association of Realtors)

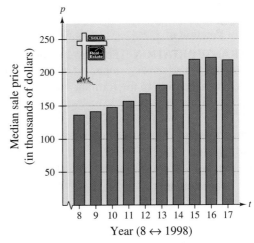

96. POSTAL REGULATIONS A rectangular package to be sent by the U.S. Postal Service can have a maximum combined length and girth (perimeter of a cross section) of 108 inches (see figure).

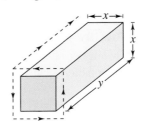

(a) Write the volume V of the package as a function of x. What is the domain of the function?

(b) Use a graphing utility to graph your function. Be sure to use an appropriate window setting.

(c) What dimensions will maximize the volume of the package? Explain your answer.

97. COST, REVENUE, AND PROFIT A company produces a product for which the variable cost is $12.30 per unit and the fixed costs are $98,000. The product sells for $17.98. Let x be the number of units produced and sold.

(a) The total cost for a business is the sum of the variable cost and the fixed costs. Write the total cost C as a function of the number of units produced.

(b) Write the revenue R as a function of the number of units sold.

(c) Write the profit P as a function of the number of units sold. (*Note:* $P = R - C$)

98. AVERAGE COST The inventor of a new game believes that the variable cost for producing the game is $0.95 per unit and the fixed costs are $6000. The inventor sells each game for $1.69. Let x be the number of games sold.

(a) The total cost for a business is the sum of the variable cost and the fixed costs. Write the total cost C as a function of the number of games sold.

(b) Write the average cost per unit $\overline{C} = C/x$ as a function of x.

99. TRANSPORTATION For groups of 80 or more people, a charter bus company determines the rate per person according to the formula

Rate $= 8 - 0.05(n - 80), \quad n \geq 80$

where the rate is given in dollars and n is the number of people.

(a) Write the revenue R for the bus company as a function of n.

(b) Use the function in part (a) to complete the table. What can you conclude?

n	90	100	110	120	130	140	150
$R(n)$							

100. PHYSICS The force F (in tons) of water against the face of a dam is estimated by the function $F(y) = 149.76\sqrt{10}y^{5/2}$, where y is the depth of the water (in feet).

(a) Complete the table. What can you conclude from the table?

y	5	10	20	30	40
$F(y)$					

(b) Use the table to approximate the depth at which the force against the dam is 1,000,000 tons.

(c) Find the depth at which the force against the dam is 1,000,000 tons algebraically.

101. HEIGHT OF A BALLOON A balloon carrying a transmitter ascends vertically from a point 3000 feet from the receiving station.

(a) Draw a diagram that gives a visual representation of the problem. Let h represent the height of the balloon and let d represent the distance between the balloon and the receiving station.

(b) Write the height of the balloon as a function of d. What is the domain of the function?

102. E-FILING The table shows the numbers of tax returns (in millions) made through e-file from 2000 through 2007. Let $f(t)$ represent the number of tax returns made through e-file in the year t. (Source: Internal Revenue Service)

Year	Number of tax returns made through e-file
2000	35.4
2001	40.2
2002	46.9
2003	52.9
2004	61.5
2005	68.5
2006	73.3
2007	80.0

(a) Find $\dfrac{f(2007) - f(2000)}{2007 - 2000}$ and interpret the result in the context of the problem.

(b) Make a scatter plot of the data.

(c) Find a linear model for the data algebraically. Let N represent the number of tax returns made through e-file and let $t = 0$ correspond to 2000.

(d) Use the model found in part (c) to complete the table.

t	0	1	2	3	4	5	6	7
N								

(e) Compare your results from part (d) with the actual data.

 (f) Use a graphing utility to find a linear model for the data. Let $x = 0$ correspond to 2000. How does the model you found in part (c) compare with the model given by the graphing utility?

114.

x	-4	-1	0	1	4
y	6	3	0	3	6

In Exercises 103–110, find the difference quotient and simplify your answer.

103. $f(x) = x^2 - x + 1, \quad \dfrac{f(2 + h) - f(2)}{h}, \quad h \neq 0$

104. $f(x) = 5x - x^2, \quad \dfrac{f(5 + h) - f(5)}{h}, \quad h \neq 0$

105. $f(x) = x^3 + 3x, \quad \dfrac{f(x + h) - f(x)}{h}, \quad h \neq 0$

106. $f(x) = 4x^2 - 2x, \quad \dfrac{f(x + h) - f(x)}{h}, \quad h \neq 0$

107. $g(x) = \dfrac{1}{x^2}, \quad \dfrac{g(x) - g(3)}{x - 3}, \quad x \neq 3$

108. $f(t) = \dfrac{1}{t - 2}, \quad \dfrac{f(t) - f(1)}{t - 1}, \quad t \neq 1$

109. $f(x) = \sqrt{5x}, \quad \dfrac{f(x) - f(5)}{x - 5}, \quad x \neq 5$

110. $f(x) = x^{2/3} + 1, \quad \dfrac{f(x) - f(8)}{x - 8}, \quad x \neq 8$

In Exercises 111–114, match the data with one of the following functions

$$f(x) = cx, \ g(x) = cx^2, \ h(x) = c\sqrt{|x|}, \ and \ r(x) = \dfrac{c}{x}$$

and determine the value of the constant c that will make the function fit the data in the table.

111.

x	-4	-1	0	1	4
y	-32	-2	0	-2	-32

112.

x	-4	-1	0	1	4
y	-1	$-\frac{1}{4}$	0	$\frac{1}{4}$	1

113.

x	-4	-1	0	1	4
y	-8	-32	Undefined	32	8

EXPLORATION

TRUE OR FALSE? In Exercises 115–118, determine whether the statement is true or false. Justify your answer.

115. Every relation is a function.

116. Every function is a relation.

117. The domain of the function given by $f(x) = x^4 - 1$ is $(-\infty, \infty)$, and the range of $f(x)$ is $(0, \infty)$.

118. The set of ordered pairs $\{(-8, -2), (-6, 0), (-4, 0), (-2, 2), (0, 4), (2, -2)\}$ represents a function.

119. THINK ABOUT IT Consider

$$f(x) = \sqrt{x - 1} \quad and \quad g(x) = \dfrac{1}{\sqrt{x - 1}}.$$

Why are the domains of f and g different?

120. THINK ABOUT IT Consider $f(x) = \sqrt{x - 2}$ and $g(x) = \sqrt[3]{x - 2}$. Why are the domains of f and g different?

121. THINK ABOUT IT Given $f(x) = x^2$, is f the independent variable? Why or why not?

122. CAPSTONE

(a) Describe any differences between a *relation* and a *function*.

(b) In your own words, explain the meanings of *domain* and *range*.

In Exercises 123 and 124, determine whether the statements use the word *function* in ways that are mathematically correct. Explain your reasoning.

123. (a) The sales tax on a purchased item is a function of the selling price.

(b) Your score on the next algebra exam is a function of the number of hours you study the night before the exam.

124. (a) The amount in your savings account is a function of your salary.

(b) The speed at which a free-falling baseball strikes the ground is a function of the height from which it was dropped.

The symbol  indicates an example or exercise that highlights algebraic techniques specifically used in calculus.

2.3 ANALYZING GRAPHS OF FUNCTIONS

What you should learn

- Use the Vertical Line Test for functions.
- Find the zeros of functions.
- Determine intervals on which functions are increasing or decreasing and determine relative maximum and relative minimum values of functions.
- Determine the average rate of change of a function.
- Identify even and odd functions.

Why you should learn it

Graphs of functions can help you visualize relationships between variables in real life. For instance, in Exercise 110 on page 210, you will use the graph of a function to represent visually the temperature of a city over a 24-hour period.

The Graph of a Function

In Section 2.2, you studied functions from an algebraic point of view. In this section, you will study functions from a graphical perspective.

The **graph of a function** f is the collection of ordered pairs $(x, f(x))$ such that x is in the domain of f. As you study this section, remember that

$x = $ the directed distance from the y-axis

$y = f(x) = $ the directed distance from the x-axis

as shown in Figure 2.25.

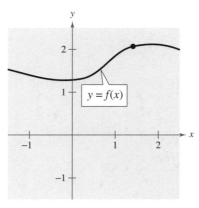

FIGURE **2.25**

Example 1   Finding the Domain and Range of a Function

Use the graph of the function f, shown in Figure 2.26, to find (a) the domain of f, (b) the function values $f(-1)$ and $f(2)$, and (c) the range of f.

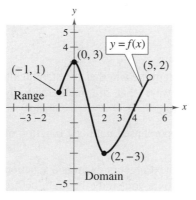

FIGURE **2.26**

Solution

a. The closed dot at $(-1, 1)$ indicates that $x = -1$ is in the domain of f, whereas the open dot at $(5, 2)$ indicates that $x = 5$ is not in the domain. So, the domain of f is all x in the interval $[-1, 5)$.

b. Because $(-1, 1)$ is a point on the graph of f, it follows that $f(-1) = 1$. Similarly, because $(2, -3)$ is a point on the graph of f, it follows that $f(2) = -3$.

c. Because the graph does not extend below $f(2) = -3$ or above $f(0) = 3$, the range of f is the interval $[-3, 3]$.

CHECK**Point**   Now try Exercise 9.

The use of dots (open or closed) at the extreme left and right points of a graph indicates that the graph does not extend beyond these points. If no such dots are shown, assume that the graph extends beyond these points.

By the definition of a function, at most one *y*-value corresponds to a given *x*-value. This means that the graph of a function cannot have two or more different points with the same *x*-coordinate, and no two points on the graph of a function can be vertically above or below each other. It follows, then, that a vertical line can intersect the graph of a function at most once. This observation provides a convenient visual test called the **Vertical Line Test** for functions.

Vertical Line Test for Functions

A set of points in a coordinate plane is the graph of *y* as a function of *x* if and only if no *vertical* line intersects the graph at more than one point.

Example 2   Vertical Line Test for Functions

Use the Vertical Line Test to decide whether the graphs in Figure 2.27 represent *y* as a function of *x*.

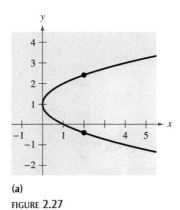

(a)

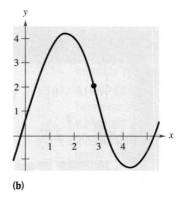

(b)

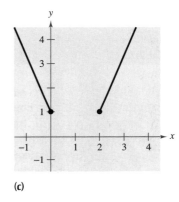

(c)

FIGURE 2.27

Solution

a. This *is not* a graph of *y* as a function of *x*, because you can find a vertical line that intersects the graph twice. That is, for a particular input *x*, there is more than one output *y*.

b. This *is* a graph of *y* as a function of *x*, because every vertical line intersects the graph at most once. That is, for a particular input *x*, there is at most one output *y*.

c. This *is* a graph of *y* as a function of *x*. (Note that if a vertical line does not intersect the graph, it simply means that the function is undefined for that particular value of *x*.) That is, for a particular input *x*, there is at most one output *y*.

CHECK*Point*   Now try Exercise 17.

TECHNOLOGY

Most graphing utilities are designed to graph functions of *x* more easily than other types of equations. For instance, the graph shown in Figure 2.27(a) represents the equation $x - (y - 1)^2 = 0$. To use a graphing utility to duplicate this graph, you must first solve the equation for *y* to obtain $y = 1 \pm \sqrt{x}$, and then graph the two equations $y_1 = 1 + \sqrt{x}$ and $y_2 = 1 - \sqrt{x}$ in the same viewing window.

Zeros of a Function

If the graph of a function of x has an x-intercept at $(a, 0)$, then a is a **zero** of the function.

> **Zeros of a Function**
>
> The **zeros of a function** f of x are the x-values for which $f(x) = 0$.

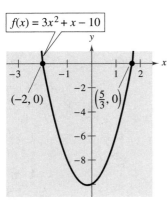

$f(x) = 3x^2 + x - 10$

$(-2, 0)$

$\left(\frac{5}{3}, 0\right)$

Zeros of f: $x = -2, x = \frac{5}{3}$

FIGURE **2.28**

Example 3   Finding the Zeros of a Function

Find the zeros of each function.

a. $f(x) = 3x^2 + x - 10$ **b.** $g(x) = \sqrt{10 - x^2}$ **c.** $h(t) = \dfrac{2t - 3}{t + 5}$

Solution

To find the zeros of a function, set the function equal to zero and solve for the independent variable.

a. $3x^2 + x - 10 = 0$

$(3x - 5)(x + 2) = 0$

$3x - 5 = 0 \quad\Longrightarrow\quad x = \frac{5}{3}$

$x + 2 = 0 \quad\Longrightarrow\quad x = -2$

The zeros of f are $x = \frac{5}{3}$ and $x = -2$. In Figure 2.28, note that the graph of f has $\left(\frac{5}{3}, 0\right)$ and $(-2, 0)$ as its x-intercepts.

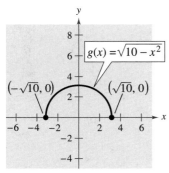

$g(x) = \sqrt{10 - x^2}$

$\left(-\sqrt{10}, 0\right)$

$\left(\sqrt{10}, 0\right)$

Zeros of g: $x = \pm\sqrt{10}$

FIGURE **2.29**

b. $\sqrt{10 - x^2} = 0$

$10 - x^2 = 0$

$10 = x^2$

$\pm\sqrt{10} = x$

The zeros of g are $x = -\sqrt{10}$ and $x = \sqrt{10}$. In Figure 2.29, note that the graph of g has $\left(-\sqrt{10}, 0\right)$ and $\left(\sqrt{10}, 0\right)$ as its x-intercepts.

c. $\dfrac{2t - 3}{t + 5} = 0$

$2t - 3 = 0$

$2t = 3$

$t = \dfrac{3}{2}$

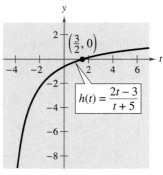

$\left(\frac{3}{2}, 0\right)$

$h(t) = \dfrac{2t - 3}{t + 5}$

Zero of h: $t = \frac{3}{2}$

FIGURE **2.30**

The zero of h is $t = \frac{3}{2}$. In Figure 2.30, note that the graph of h has $\left(\frac{3}{2}, 0\right)$ as its t-intercept.

CHECK **Point**   Now try Exercise 23.

Increasing and Decreasing Functions

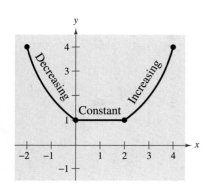

FIGURE **2.31**

The more you know about the graph of a function, the more you know about the function itself. Consider the graph shown in Figure 2.31. As you move from *left to right*, this graph falls from $x = -2$ to $x = 0$, is constant from $x = 0$ to $x = 2$, and rises from $x = 2$ to $x = 4$.

> ### Increasing, Decreasing, and Constant Functions
>
> A function f is **increasing** on an interval if, for any x_1 and x_2 in the interval, $x_1 < x_2$ implies $f(x_1) < f(x_2)$.
>
> A function f is **decreasing** on an interval if, for any x_1 and x_2 in the interval, $x_1 < x_2$ implies $f(x_1) > f(x_2)$.
>
> A function f is **constant** on an interval if, for any x_1 and x_2 in the interval, $f(x_1) = f(x_2)$.

Example 4   Increasing and Decreasing Functions

Use the graphs in Figure 2.32 to describe the increasing or decreasing behavior of each function.

Solution

a. This function is increasing over the entire real line.

b. This function is increasing on the interval $(-\infty, -1)$, decreasing on the interval $(-1, 1)$, and increasing on the interval $(1, \infty)$.

c. This function is increasing on the interval $(-\infty, 0)$, constant on the interval $(0, 2)$, and decreasing on the interval $(2, \infty)$.

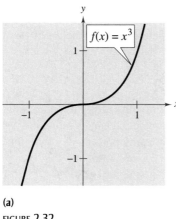

(a)

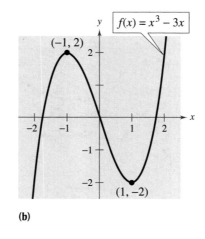

(b)

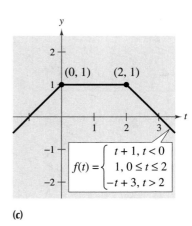

(c)

FIGURE **2.32**

CHECK***Point***   Now try Exercise 41.

To help you decide whether a function is increasing, decreasing, or constant on an interval, you can evaluate the function for several values of x. However, calculus is needed to determine, for certain, all intervals on which a function is increasing, decreasing, or constant.

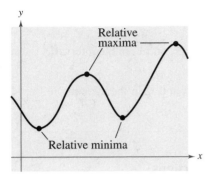

FIGURE 2.33

The points at which a function changes its increasing, decreasing, or constant behavior are helpful in determining the **relative minimum** or **relative maximum** values of the function.

Definitions of Relative Minimum and Relative Maximum

A function value $f(a)$ is called a **relative minimum** of f if there exists an interval (x_1, x_2) that contains a such that

$$x_1 < x < x_2 \quad \text{implies} \quad f(a) \leq f(x).$$

A function value $f(a)$ is called a **relative maximum** of f if there exists an interval (x_1, x_2) that contains a such that

$$x_1 < x < x_2 \quad \text{implies} \quad f(a) \geq f(x).$$

Figure 2.33 shows several different examples of relative minima and relative maxima. In Section 3.1, you will study a technique for finding the *exact point* at which a second-degree polynomial function has a relative minimum or relative maximum. For the time being, however, you can use a graphing utility to find reasonable approximations of these points.

Example 5   Approximating a Relative Minimum

Use a graphing utility to approximate the relative minimum of the function given by $f(x) = 3x^2 - 4x - 2$.

Solution

The graph of f is shown in Figure 2.34. By using the *zoom* and *trace* features or the *minimum* feature of a graphing utility, you can estimate that the function has a relative minimum at the point

$$(0.67, -3.33).$$

Later, in Section 3.1, you will be able to determine that the exact point at which the relative minimum occurs is $\left(\frac{2}{3}, -\frac{10}{3}\right)$.

CHECK*Point*   Now try Exercise 57.

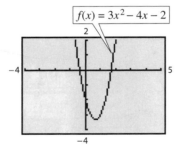

FIGURE 2.34

You can also use the *table* feature of a graphing utility to approximate numerically the relative minimum of the function in Example 5. Using a table that begins at 0.6 and increments the value of x by 0.01, you can approximate that the minimum of $f(x) = 3x^2 - 4x - 2$ occurs at the point $(0.67, -3.33)$.

TECHNOLOGY

If you use a graphing utility to estimate the x- and y-values of a relative minimum or relative maximum, the *zoom* feature will often produce graphs that are nearly flat. To overcome this problem, you can manually change the vertical setting of the viewing window. The graph will stretch vertically if the values of Ymin and Ymax are closer together.

Average Rate of Change

In Section 2.1, you learned that the slope of a line can be interpreted as a *rate of change*. For a nonlinear graph whose slope changes at each point, the **average rate of change** between any two points $(x_1, f(x_1))$ and $(x_2, f(x_2))$ is the slope of the line through the two points (see Figure 2.35). The line through the two points is called the **secant line,** and the slope of this line is denoted as m_{sec}.

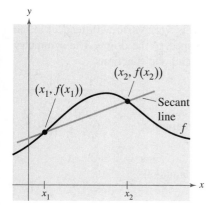

$$\text{Average rate of change of } f \text{ from } x_1 \text{ to } x_2 = \frac{f(x_2) - f(x_1)}{x_2 - x_1}$$

$$= \frac{\text{change in } y}{\text{change in } x}$$

$$= m_{sec}$$

FIGURE 2.35

Example 6   Average Rate of Change of a Function

Find the average rates of change of $f(x) = x^3 - 3x$ (a) from $x_1 = -2$ to $x_2 = 0$ and (b) from $x_1 = 0$ to $x_2 = 1$ (see Figure 2.36).

Solution

a. The average rate of change of f from $x_1 = -2$ to $x_2 = 0$ is

$$\frac{f(x_2) - f(x_1)}{x_2 - x_1} = \frac{f(\ ) - f(\ )}{-(\ )} = \frac{0 - (-2)}{2} = 1.$$

b. The average rate of change of f from $x_1 = 0$ to $x_2 = 1$ is

$$\frac{f(x_2) - f(x_1)}{x_2 - x_1} = \frac{f(\ ) - f(\ )}{-} = \frac{-2 - 0}{1} = -2.$$

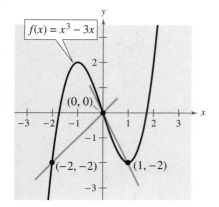

FIGURE 2.36

CHECK *Point* Now try Exercise 75.

Example 7   Finding Average Speed

The distance s (in feet) a moving car is from a stoplight is given by the function $s(t) = 20t^{3/2}$, where t is the time (in seconds). Find the average speed of the car (a) from $t_1 = 0$ to $t_2 = 4$ seconds and (b) from $t_1 = 4$ to $t_2 = 9$ seconds.

Solution

a. The average speed of the car from $t_1 = 0$ to $t_2 = 4$ seconds is

$$\frac{s(t_2) - s(t_1)}{t_2 - t_1} = \frac{s(\ ) - s(\ )}{-(\ )} = \frac{160 - 0}{4} = 40 \text{ feet per second.}$$

b. The average speed of the car from $t_1 = 4$ to $t_2 = 9$ seconds is

$$\frac{s(t_2) - s(t_1)}{t_2 - t_1} = \frac{s(\ ) - s(\ )}{-} = \frac{540 - 160}{5} = 76 \text{ feet per second.}$$

CHECK *Point* Now try Exercise 113.

Even and Odd Functions

In Section 1.1, you studied different types of symmetry of a graph. In the terminology of functions, a function is said to be **even** if its graph is symmetric with respect to the *y*-axis and to be **odd** if its graph is symmetric with respect to the origin. The symmetry tests in Section 1.1 yield the following tests for even and odd functions.

Tests for Even and Odd Functions

A function $y = f(x)$ is **even** if, for each x in the domain of f,

$$f(-x) = f(x).$$

A function $y = f(x)$ is **odd** if, for each x in the domain of f,

$$f(-x) = -f(x).$$

Example 8   Even and Odd Functions

a. The function $g(x) = x^3 - x$ is odd because $g(-x) = -g(x)$, as follows.

$$g(\quad) = (\quad)^3 - (\quad)$$
$$= -x^3 + x$$
$$= -(x^3 - x)$$
$$= -g(x)$$

b. The function $h(x) = x^2 + 1$ is even because $h(-x) = h(x)$, as follows.

$$h(\quad) = (\quad)^2 + 1$$
$$= x^2 + 1$$
$$= h(x)$$

The graphs and symmetry of these two functions are shown in Figure 2.37.

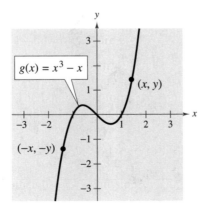

(a) Symmetric to origin: Odd Function

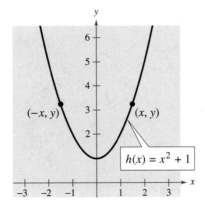

(b) Symmetric to *y*-axis: Even Function

FIGURE 2.37

CHECK *Point*   Now try Exercise 83.

2.3   EXERCISES

See www.CalcChat.com for worked-out solutions to odd-numbered exercises.

VOCABULARY: Fill in the blanks.

1. The graph of a function f is the collection of _____ _____ $(x, f(x))$ such that x is in the domain of f.

2. The _____ _____ _____ is used to determine whether the graph of an equation is a function of y in terms of x.

3. The _____ of a function f are the values of x for which $f(x) = 0$.

4. A function f is _____ on an interval if, for any x_1 and x_2 in the interval, $x_1 < x_2$ implies $f(x_1) > f(x_2)$.

5. A function value $f(a)$ is a relative _____ of f if there exists an interval (x_1, x_2) containing a such that $x_1 < x < x_2$ implies $f(a) \geq f(x)$.

6. The _____ _____ _____ _____ between any two points $(x_1, f(x_1))$ and $(x_2, f(x_2))$ is the slope of the line through the two points, and this line is called the _____ line.

7. A function f is _____ if, for each x in the domain of f, $f(-x) = -f(x)$.

8. A function f is _____ if its graph is symmetric with respect to the y-axis.

SKILLS AND APPLICATIONS

In Exercises 9–12, use the graph of the function to find the domain and range of f.

9.
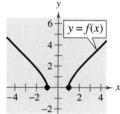

10.
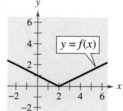

11.
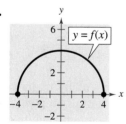

12.
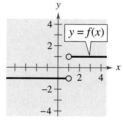

In Exercises 13–16, use the graph of the function to find the domain and range of f and the indicated function values.

13. (a) $f(-2)$ (b) $f(-1)$
      (c) $f\left(\frac{1}{2}\right)$ (d) $f(1)$

14. (a) $f(-1)$ (b) $f(2)$
      (c) $f(0)$ (d) $f(1)$

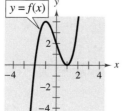
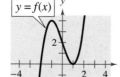

15. (a) $f(2)$ (b) $f(1)$
      (c) $f(3)$ (d) $f(-1)$

16. (a) $f(-2)$ (b) $f(1)$
      (c) $f(0)$ (d) $f(2)$

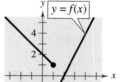

In Exercises 17–22, use the Vertical Line Test to determine whether y is a function of x. To print an enlarged copy of the graph, go to the website *www.mathgraphs.com*.

17. $y = \frac{1}{2}x^2$

18. $y = \frac{1}{4}x^3$

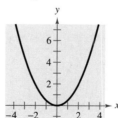
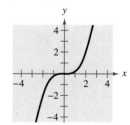

19. $x - y^2 = 1$

20. $x^2 + y^2 = 25$

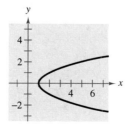
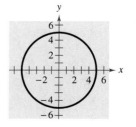

21. $x^2 = 2xy - 1$

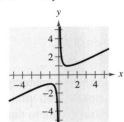

22. $x = |y + 2|$

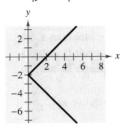

In Exercises 23–32, find the zeros of the function algebraically.

23. $f(x) = 2x^2 - 7x - 30$

24. $f(x) = 3x^2 + 22x - 16$

25. $f(x) = \dfrac{x}{9x^2 - 4}$

26. $f(x) = \dfrac{x^2 - 9x + 14}{4x}$

27. $f(x) = \frac{1}{2}x^3 - x$

28. $f(x) = x^3 - 4x^2 - 9x + 36$

29. $f(x) = 4x^3 - 24x^2 - x + 6$

30. $f(x) = 9x^4 - 25x^2$

31. $f(x) = \sqrt{2x} - 1$

32. $f(x) = \sqrt{3x + 2}$

 In Exercises 33–38, (a) use a graphing utility to graph the function and find the zeros of the function and (b) verify your results from part (a) algebraically.

33. $f(x) = 3 + \dfrac{5}{x}$

34. $f(x) = x(x - 7)$

35. $f(x) = \sqrt{2x + 11}$

36. $f(x) = \sqrt{3x - 14} - 8$

37. $f(x) = \dfrac{3x - 1}{x - 6}$

38. $f(x) = \dfrac{2x^2 - 9}{3 - x}$

In Exercises 39–46, determine the intervals over which the function is increasing, decreasing, or constant.

39. $f(x) = \frac{3}{2}x$

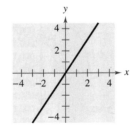

40. $f(x) = x^2 - 4x$

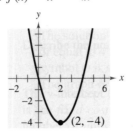

41. $f(x) = x^3 - 3x^2 + 2$

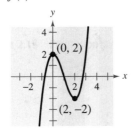

42. $f(x) = \sqrt{x^2 - 1}$

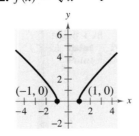

43. $f(x) = |x + 1| + |x - 1|$

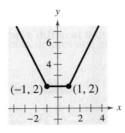

44. $f(x) = \dfrac{x^2 + x + 1}{x + 1}$

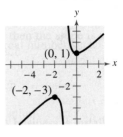

45. $f(x) = \begin{cases} x + 3, & x \le 0 \\ 3, & 0 < x \le 2 \\ 2x + 1, & x > 2 \end{cases}$

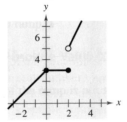

46. $f(x) = \begin{cases} 2x + 1, & x \le -1 \\ x^2 - 2, & x > -1 \end{cases}$

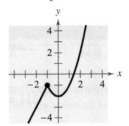

 In Exercises 47–56, (a) use a graphing utility to graph the function and visually determine the intervals over which the function is increasing, decreasing, or constant, and (b) make a table of values to verify whether the function is increasing, decreasing, or constant over the intervals you identified in part (a).

47. $f(x) = 3$

48. $g(x) = x$

49. $g(s) = \dfrac{s^2}{4}$

50. $h(x) = x^2 - 4$

51. $f(t) = -t^4$

52. $f(x) = 3x^4 - 6x^2$

53. $f(x) = \sqrt{1 - x}$

54. $f(x) = x\sqrt{x + 3}$

55. $f(x) = x^{3/2}$

56. $f(x) = x^{2/3}$

In Exercises 57–66, use a graphing utility to graph the function and approximate (to two decimal places) any relative minimum or relative maximum values.

57. $f(x) = (x - 4)(x + 2)$
58. $f(x) = 3x^2 - 2x - 5$
59. $f(x) = -x^2 + 3x - 2$
60. $f(x) = -2x^2 + 9x$
61. $f(x) = x(x - 2)(x + 3)$
62. $f(x) = x^3 - 3x^2 - x + 1$
63. $g(x) = 2x^3 + 3x^2 - 12x$
64. $h(x) = x^3 - 6x^2 + 15$
65. $h(x) = (x - 1)\sqrt{x}$
66. $g(x) = x\sqrt{4 - x}$

In Exercises 67–74, graph the function and determine the interval(s) for which $f(x) \geq 0$.

67. $f(x) = 4 - x$
68. $f(x) = 4x + 2$
69. $f(x) = 9 - x^2$
70. $f(x) = x^2 - 4x$
71. $f(x) = \sqrt{x - 1}$
72. $f(x) = \sqrt{x + 2}$
73. $f(x) = -(1 + |x|)$
74. $f(x) = \frac{1}{2}(2 + |x|)$

In Exercises 75–82, find the average rate of change of the function from x_1 to x_2.

	Function	*x-Values*
75.	$f(x) = -2x + 15$	$x_1 = 0, x_2 = 3$
76.	$f(x) = 3x + 8$	$x_1 = 0, x_2 = 3$
77.	$f(x) = x^2 + 12x - 4$	$x_1 = 1, x_2 = 5$
78.	$f(x) = x^2 - 2x + 8$	$x_1 = 1, x_2 = 5$
79.	$f(x) = x^3 - 3x^2 - x$	$x_1 = 1, x_2 = 3$
80.	$f(x) = -x^3 + 6x^2 + x$	$x_1 = 1, x_2 = 6$
81.	$f(x) = -\sqrt{x - 2} + 5$	$x_1 = 3, x_2 = 11$
82.	$f(x) = -\sqrt{x + 1} + 3$	$x_1 = 3, x_2 = 8$

In Exercises 83–90, determine whether the function is even, odd, or neither. Then describe the symmetry.

83. $f(x) = x^6 - 2x^2 + 3$
84. $h(x) = x^3 - 5$
85. $g(x) = x^3 - 5x$
86. $f(t) = t^2 + 2t - 3$
87. $h(x) = x\sqrt{x + 5}$
88. $f(x) = x\sqrt{1 - x^2}$
89. $f(s) = 4s^{3/2}$
90. $g(s) = 4s^{2/3}$

In Exercises 91–100, sketch a graph of the function and determine whether it is even, odd, or neither. Verify your answers algebraically.

91. $f(x) = 5$
92. $f(x) = -9$
93. $f(x) = 3x - 2$
94. $f(x) = 5 - 3x$
95. $h(x) = x^2 - 4$
96. $f(x) = -x^2 - 8$
97. $f(x) = \sqrt{1 - x}$
98. $g(t) = \sqrt[3]{t - 1}$
99. $f(x) = |x + 2|$
100. $f(x) = -|x - 5|$

In Exercises 101–104, write the height *h* of the rectangle as a function of *x*.

101.
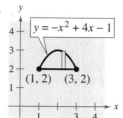

102.
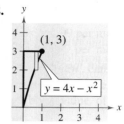

103.
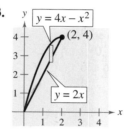

104.
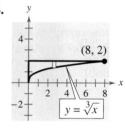

In Exercises 105–108, write the length *L* of the rectangle as a function of *y*.

105.
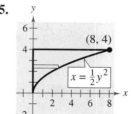

106.
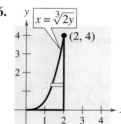

107.
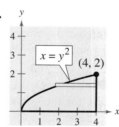

108.
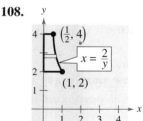

109. ELECTRONICS The number of lumens (time rate of flow of light) *L* from a fluorescent lamp can be approximated by the model

$$L = -0.294x^2 + 97.744x - 664.875, \quad 20 \leq x \leq 90$$

where *x* is the wattage of the lamp.

(a) Use a graphing utility to graph the function.

(b) Use the graph from part (a) to estimate the wattage necessary to obtain 2000 lumens.

110. DATA ANALYSIS: TEMPERATURE The table shows the temperatures y (in degrees Fahrenheit) in a certain city over a 24-hour period. Let x represent the time of day, where $x = 0$ corresponds to 6 A.M.

Time, x	Temperature, y
0	34
2	50
4	60
6	64
8	63
10	59
12	53
14	46
16	40
18	36
20	34
22	37
24	45

A model that represents these data is given by

$$y = 0.026x^3 - 1.03x^2 + 10.2x + 34, \quad 0 \le x \le 24.$$

(a) Use a graphing utility to create a scatter plot of the data. Then graph the model in the same viewing window.

(b) How well does the model fit the data?

(c) Use the graph to approximate the times when the temperature was increasing and decreasing.

(d) Use the graph to approximate the maximum and minimum temperatures during this 24-hour period.

(e) Could this model be used to predict the temperatures in the city during the next 24-hour period? Why or why not?

111. COORDINATE AXIS SCALE Each function described below models the specified data for the years 1998 through 2008, with $t = 8$ corresponding to 1998. Estimate a reasonable scale for the vertical axis (e.g., hundreds, thousands, millions, etc.) of the graph and justify your answer. (There are many correct answers.)

(a) $f(t)$ represents the average salary of college professors.

(b) $f(t)$ represents the U.S. population.

(c) $f(t)$ represents the percent of the civilian work force that is unemployed.

112. GEOMETRY Corners of equal size are cut from a square with sides of length 8 meters (see figure).

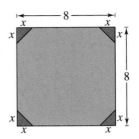

(a) Write the area A of the resulting figure as a function of x. Determine the domain of the function.

(b) Use a graphing utility to graph the area function over its domain. Use the graph to find the range of the function.

(c) Identify the figure that would result if x were chosen to be the maximum value in the domain of the function. What would be the length of each side of the figure?

113. ENROLLMENT RATE The enrollment rates r of children in preschool in the United States from 1970 through 2005 can be approximated by the model

$$r = -0.021t^2 + 1.44t + 39.3, \quad 0 \le t \le 35$$

where t represents the year, with $t = 0$ corresponding to 1970. (Source: U.S. Census Bureau)

(a) Use a graphing utility to graph the model.

(b) Find the average rate of change of the model from 1970 through 2005. Interpret your answer in the context of the problem.

114. VEHICLE TECHNOLOGY SALES The estimated revenues r (in millions of dollars) from sales of in-vehicle technologies in the United States from 2003 through 2008 can be approximated by the model

$$r = 157.30t^2 - 397.4t + 6114, \quad 3 \le t \le 8$$

where t represents the year, with $t = 3$ corresponding to 2003. (Source: Consumer Electronics Association)

(a) Use a graphing utility to graph the model.

(b) Find the average rate of change of the model from 2003 through 2008. Interpret your answer in the context of the problem.

PHYSICS In Exercises 115–120, (a) use the position equation $s = -16t^2 + v_0t + s_0$ to write a function that represents the situation, (b) use a graphing utility to graph the function, (c) find the average rate of change of the function from t_1 to t_2, (d) describe the slope of the secant line through t_1 and t_2, (e) find the equation of the secant line through t_1 and t_2, and (f) graph the secant line in the same viewing window as your position function.

115. An object is thrown upward from a height of 6 feet at a velocity of 64 feet per second.

$t_1 = 0, t_2 = 3$

116. An object is thrown upward from a height of 6.5 feet at a velocity of 72 feet per second.

$t_1 = 0, t_2 = 4$

117. An object is thrown upward from ground level at a velocity of 120 feet per second.

$t_1 = 3, t_2 = 5$

118. An object is thrown upward from ground level at a velocity of 96 feet per second.

$t_1 = 2, t_2 = 5$

119. An object is dropped from a height of 120 feet.

$t_1 = 0, t_2 = 2$

120. An object is dropped from a height of 80 feet.

$t_1 = 1, t_2 = 2$

EXPLORATION

TRUE OR FALSE? In Exercises 121 and 122, determine whether the statement is true or false. Justify your answer.

121. A function with a square root cannot have a domain that is the set of real numbers.

122. It is possible for an odd function to have the interval $[0, \infty)$ as its domain.

123. If f is an even function, determine whether g is even, odd, or neither. Explain.

(a) $g(x) = -f(x)$ (b) $g(x) = f(-x)$

(c) $g(x) = f(x) - 2$ (d) $g(x) = f(x - 2)$

124. THINK ABOUT IT Does the graph in Exercise 19 represent x as a function of y? Explain.

THINK ABOUT IT In Exercises 125–130, find the coordinates of a second point on the graph of a function f if the given point is on the graph and the function is (a) even and (b) odd.

125. $\left(-\frac{3}{2}, 4\right)$ **126.** $\left(-\frac{5}{3}, -7\right)$

127. $(4, 9)$ **128.** $(5, -1)$

129. $(x, -y)$ **130.** $(2a, 2c)$

131. WRITING Use a graphing utility to graph each function. Write a paragraph describing any similarities and differences you observe among the graphs.

(a) $y = x$ (b) $y = x^2$ (c) $y = x^3$

(d) $y = x^4$ (e) $y = x^5$ (f) $y = x^6$

132. CONJECTURE Use the results of Exercise 131 to make a conjecture about the graphs of $y = x^7$ and $y = x^8$. Use a graphing utility to graph the functions and compare the results with your conjecture.

133. Use the information in Example 7 to find the average speed of the car from $t_1 = 0$ to $t_2 = 9$ seconds. Explain why the result is less than the value obtained in part (b) of Example 7.

134. Graph each of the functions with a graphing utility. Determine whether the function is *even*, *odd*, or *neither*.

$f(x) = x^2 - x^4$

$g(x) = 2x^3 + 1$

$h(x) = x^5 - 2x^3 + x$

$j(x) = 2 - x^6 - x^8$

$k(x) = x^5 - 2x^4 + x - 2$

$p(x) = x^9 + 3x^5 - x^3 + x$

What do you notice about the equations of functions that are odd? What do you notice about the equations of functions that are even? Can you describe a way to identify a function as odd or even by inspecting the equation? Can you describe a way to identify a function as neither odd nor even by inspecting the equation?

135. WRITING Write a short paragraph describing three different functions that represent the behaviors of quantities between 1998 and 2009. Describe one quantity that decreased during this time, one that increased, and one that was constant. Present your results graphically.

136. CAPSTONE Use the graph of the function to answer (a)–(e).

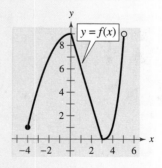

(a) Find the domain and range of f.

(b) Find the zero(s) of f.

(c) Determine the intervals over which f is increasing, decreasing, or constant.

(d) Approximate any relative minimum or relative maximum values of f.

(e) Is f even, odd, or neither?

What you should learn

- Identify and graph linear and squaring functions.
- Identify and graph cubic, square root, and reciprocal functions.
- Identify and graph step and other piecewise-defined functions.
- Recognize graphs of parent functions.

Why you should learn it

Step functions can be used to model real-life situations. For instance, in Exercise 69 on page 218, you will use a step function to model the cost of sending an overnight package from Los Angeles to Miami.

2.4 A LIBRARY OF PARENT FUNCTIONS

Linear and Squaring Functions

One of the goals of this text is to enable you to recognize the basic shapes of the graphs of different types of functions. For instance, you know that the graph of the **linear function** $f(x) = ax + b$ is a line with slope $m = a$ and y-intercept at $(0, b)$. The graph of the linear function has the following characteristics.

- The domain of the function is the set of all real numbers.
- The range of the function is the set of all real numbers.
- The graph has an x-intercept of $(-b/m, 0)$ and a y-intercept of $(0, b)$.
- The graph is increasing if $m > 0$, decreasing if $m < 0$, and constant if $m = 0$.

Example 1   Writing a Linear Function

Write the linear function f for which $f(1) = 3$ and $f(4) = 0$.

Solution

To find the equation of the line that passes through $(x_1, y_1) = (1, 3)$ and $(x_2, y_2) = (4, 0)$, first find the slope of the line.

$$m = \frac{y_2 - y_1}{x_2 - x_1} = \frac{0 - 3}{4 - 1} = \frac{-3}{3} = -1$$

Next, use the point-slope form of the equation of a line.

$$y - y_1 = m(x - x_1)$$
$$y - \quad = \quad (x - \quad)$$
$$y = -x + 4$$
$$f(x) = -x + 4$$

The graph of this function is shown in Figure 2.38.

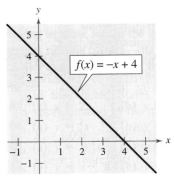

FIGURE 2.38

CHECK*Point* Now try Exercise 11.

There are two special types of linear functions, the **constant function** and the **identity function.** A constant function has the form

$$f(x) = c$$

and has the domain of all real numbers with a range consisting of a single real number c. The graph of a constant function is a horizontal line, as shown in Figure 2.39. The identity function has the form

$$f(x) = x.$$

Its domain and range are the set of all real numbers. The identity function has a slope of $m = 1$ and a y-intercept at $(0, 0)$. The graph of the identity function is a line for which each x-coordinate equals the corresponding y-coordinate. The graph is always increasing, as shown in Figure 2.40.

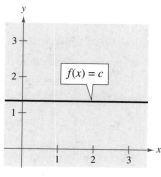

FIGURE 2.39

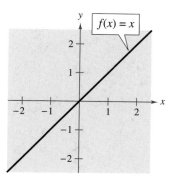

FIGURE 2.40

The graph of the **squaring function**

$$f(x) = x^2$$

is a U-shaped curve with the following characteristics.

- The domain of the function is the set of all real numbers.
- The range of the function is the set of all nonnegative real numbers.
- The function is even.
- The graph has an intercept at $(0, 0)$.
- The graph is decreasing on the interval $(-\infty, 0)$ and increasing on the interval $(0, \infty)$.
- The graph is symmetric with respect to the y-axis.
- The graph has a relative minimum at $(0, 0)$.

The graph of the squaring function is shown in Figure 2.41.

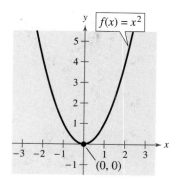

FIGURE 2.41

Cubic, Square Root, and Reciprocal Functions

The basic characteristics of the graphs of the **cubic, square root,** and **reciprocal functions** are summarized below.

1. The graph of the *cubic* function $f(x) = x^3$ has the following characteristics.
   - The domain of the function is the set of all real numbers.
   - The range of the function is the set of all real numbers.
   - The function is odd.
   - The graph has an intercept at $(0, 0)$.
   - The graph is increasing on the interval $(-\infty, \infty)$.
   - The graph is symmetric with respect to the origin.

   The graph of the cubic function is shown in Figure 2.42.

2. The graph of the *square root* function $f(x) = \sqrt{x}$ has the following characteristics.
   - The domain of the function is the set of all nonnegative real numbers.
   - The range of the function is the set of all nonnegative real numbers.
   - The graph has an intercept at $(0, 0)$.
   - The graph is increasing on the interval $(0, \infty)$.

   The graph of the square root function is shown in Figure 2.43.

3. The graph of the *reciprocal* function $f(x) = \dfrac{1}{x}$ has the following characteristics.
   - The domain of the function is $(-\infty, 0) \cup (0, \infty)$.
   - The range of the function is $(-\infty, 0) \cup (0, \infty)$.
   - The function is odd.
   - The graph does not have any intercepts.
   - The graph is decreasing on the intervals $(-\infty, 0)$ and $(0, \infty)$.
   - The graph is symmetric with respect to the origin.

   The graph of the reciprocal function is shown in Figure 2.44.

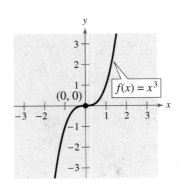

Cubic function
FIGURE 2.42

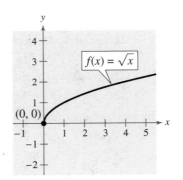

Square root function
FIGURE 2.43

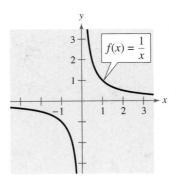

Reciprocal function
FIGURE 2.44

Step and Piecewise-Defined Functions

Functions whose graphs resemble sets of stairsteps are known as **step functions.** The most famous of the step functions is the **greatest integer function,** which is denoted by $[\![x]\!]$ and defined as

$$f(x) = [\![x]\!] = \text{the greatest integer less than or equal to } x.$$

Some values of the greatest integer function are as follows.

$$[\![-1]\!] = (\text{greatest integer} \le -1) = -1$$
$$[\![-\tfrac{1}{2}]\!] = (\text{greatest integer} \le -\tfrac{1}{2}) = -1$$
$$[\![\tfrac{1}{10}]\!] = (\text{greatest integer} \le \tfrac{1}{10}) = 0$$
$$[\![1.5]\!] = (\text{greatest integer} \le 1.5) = 1$$

The graph of the greatest integer function

$$f(x) = [\![x]\!]$$

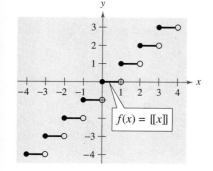

FIGURE 2.45

has the following characteristics, as shown in Figure 2.45.

- The domain of the function is the set of all real numbers.
- The range of the function is the set of all integers.
- The graph has a y-intercept at $(0, 0)$ and x-intercepts in the interval $[0, 1)$.
- The graph is constant between each pair of consecutive integers.
- The graph jumps vertically one unit at each integer value.

TECHNOLOGY

When graphing a step function, you should set your graphing utility to *dot* mode.

Example 2   Evaluating a Step Function

Evaluate the function when $x = -1, 2,$ and $\tfrac{3}{2}$.

$$f(x) = [\![x]\!] + 1$$

Solution

For $x = -1$, the greatest integer ≤ -1 is -1, so

$$f(\ ) = [\![\ \ ]\!] + 1 = -1 + 1 = 0.$$

For $x = 2$, the greatest integer ≤ 2 is 2, so

$$f(\ ) = [\![\ \ ]\!] + 1 = 2 + 1 = 3.$$

For $x = \tfrac{3}{2}$, the greatest integer $\le \tfrac{3}{2}$ is 1, so

$$f(\ ) = [\![\ \ ]\!] + 1 = 1 + 1 = 2.$$

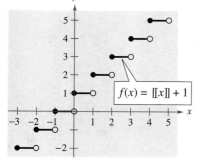

FIGURE 2.46

You can verify your answers by examining the graph of $f(x) = [\![x]\!] + 1$ shown in Figure 2.46.

CHECK**Point** Now try Exercise 43.

Recall from Section 2.2 that a piecewise-defined function is defined by two or more equations over a specified domain. To graph a piecewise-defined function, graph each equation separately over the specified domain, as shown in Example 3.

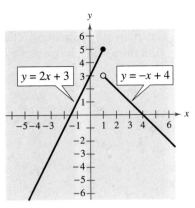

FIGURE **2.47**

Example 3 **Graphing a Piecewise-Defined Function**

Sketch the graph of

$$f(x) = \begin{cases} 2x + 3, & x \le 1 \\ -x + 4, & x > 1 \end{cases}.$$

Solution

This piecewise-defined function is composed of two linear functions. At $x = 1$ and to the left of $x = 1$ the graph is the line $y = 2x + 3$, and to the right of $x = 1$ the graph is the line $y = -x + 4$, as shown in Figure 2.47. Notice that the point $(1, 5)$ is a solid dot and the point $(1, 3)$ is an open dot. This is because $f(1) = 2(\ ) + 3 = 5$.

CHECK*Point* Now try Exercise 57.

Parent Functions

The eight graphs shown in Figure 2.48 represent the most commonly used functions in algebra. Familiarity with the basic characteristics of these simple graphs will help you analyze the shapes of more complicated graphs—in particular, graphs obtained from these graphs by the rigid and nonrigid transformations studied in the next section.

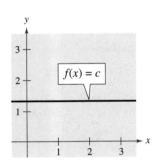

(a) Constant Function

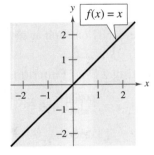

(b) Identity Function

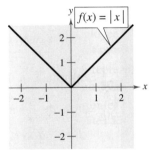

(c) Absolute Value Function

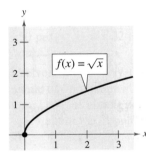

(d) Square Root Function

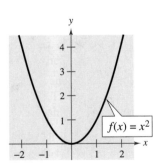

(e) Quadratic Function

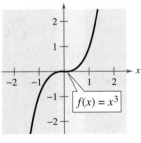

(f) Cubic Function

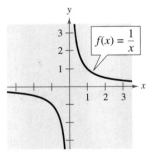

(g) Reciprocal Function

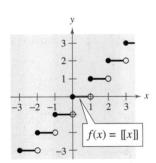

(h) Greatest Integer Function

FIGURE **2.48**

2.4   EXERCISES

See www.CalcChat.com for worked-out solutions to odd-numbered exercises.

VOCABULARY

In Exercises 1–9, match each function with its name.

1. $f(x) = [\![x]\!]$ 2. $f(x) = x$ 3. $f(x) = 1/x$
4. $f(x) = x^2$ 5. $f(x) = \sqrt{x}$ 6. $f(x) = c$
7. $f(x) = |x|$ 8. $f(x) = x^3$ 9. $f(x) = ax + b$

   (a) squaring function   (b) square root function   (c) cubic function
   (d) linear function   (e) constant function   (f) absolute value function
   (g) greatest integer function   (h) reciprocal function   (i) identity function

10. Fill in the blank: The constant function and the identity function are two special types of _____ functions.

SKILLS AND APPLICATIONS

In Exercises 11–18, (a) write the linear function f such that it has the indicated function values and (b) sketch the graph of the function.

11. $f(1) = 4, f(0) = 6$ 12. $f(-3) = -8, f(1) = 2$
13. $f(5) = -4, f(-2) = 17$ 14. $f(3) = 9, f(-1) = -11$
15. $f(-5) = -1, f(5) = -1$
16. $f(-10) = 12, f(16) = -1$
17. $f\left(\frac{1}{2}\right) = -6, f(4) = -3$
18. $f\left(\frac{2}{3}\right) = -\frac{15}{2}, f(-4) = -11$

In Exercises 19–42, use a graphing utility to graph the function. Be sure to choose an appropriate viewing window.

19. $f(x) = 0.8 - x$ 20. $f(x) = 2.5x - 4.25$
21. $f(x) = -\frac{1}{6}x - \frac{5}{2}$ 22. $f(x) = \frac{5}{6} - \frac{2}{3}x$
23. $g(x) = -2x^2$ 24. $h(x) = 1.5 - x^2$
25. $f(x) = 3x^2 - 1.75$ 26. $f(x) = 0.5x^2 + 2$
27. $f(x) = x^3 - 1$ 28. $f(x) = 8 - x^3$
29. $f(x) = (x - 1)^3 + 2$ 30. $g(x) = 2(x + 3)^3 + 1$
31. $f(x) = 4\sqrt{x}$ 32. $f(x) = 4 - 2\sqrt{x}$
33. $g(x) = 2 - \sqrt{x + 4}$ 34. $h(x) = \sqrt{x + 2} + 3$
35. $f(x) = -1/x$ 36. $f(x) = 4 + (1/x)$
37. $h(x) = 1/(x + 2)$ 38. $k(x) = 1/(x - 3)$
39. $g(x) = |x| - 5$ 40. $h(x) = 3 - |x|$
41. $f(x) = |x + 4|$ 42. $f(x) = |x - 1|$

In Exercises 43–50, evaluate the function for the indicated values.

43. $f(x) = [\![x]\!]$
   (a) $f(2.1)$ (b) $f(2.9)$ (c) $f(-3.1)$ (d) $f\left(\frac{7}{2}\right)$
44. $g(x) = 2[\![x]\!]$
   (a) $g(-3)$ (b) $g(0.25)$ (c) $g(9.5)$ (d) $g\left(\frac{11}{3}\right)$

45. $h(x) = [\![x + 3]\!]$
   (a) $h(-2)$ (b) $h\left(\frac{1}{2}\right)$ (c) $h(4.2)$ (d) $h(-21.6)$
46. $f(x) = 4[\![x]\!] + 7$
   (a) $f(0)$ (b) $f(-1.5)$ (c) $f(6)$ (d) $f\left(\frac{5}{3}\right)$
47. $h(x) = [\![3x - 1]\!]$
   (a) $h(2.5)$ (b) $h(-3.2)$ (c) $h\left(\frac{7}{3}\right)$ (d) $h\left(-\frac{21}{3}\right)$
48. $k(x) = \left[\!\left[\frac{1}{2}x + 6\right]\!\right]$
   (a) $k(5)$ (b) $k(-6.1)$ (c) $k(0.1)$ (d) $k(15)$
49. $g(x) = 3[\![x - 2]\!] + 5$
   (a) $g(-2.7)$ (b) $g(-1)$ (c) $g(0.8)$ (d) $g(14.5)$
50. $g(x) = -7[\![x + 4]\!] + 6$
   (a) $g\left(\frac{1}{8}\right)$ (b) $g(9)$ (c) $g(-4)$ (d) $g\left(\frac{3}{2}\right)$

In Exercises 51–56, sketch the graph of the function.

51. $g(x) = -[\![x]\!]$ 52. $g(x) = 4[\![x]\!]$
53. $g(x) = [\![x]\!] - 2$
54. $g(x) = [\![x]\!] - 1$
55. $g(x) = [\![x + 1]\!]$
56. $g(x) = [\![x - 3]\!]$

In Exercises 57–64, graph the function.

57. $f(x) = \begin{cases} 2x + 3, & x < 0 \\ 3 - x, & x \geq 0 \end{cases}$

58. $g(x) = \begin{cases} x + 6, & x \leq -4 \\ \frac{1}{2}x - 4, & x > -4 \end{cases}$

59. $f(x) = \begin{cases} \sqrt{4 + x}, & x < 0 \\ \sqrt{4 - x}, & x \geq 0 \end{cases}$

60. $f(x) = \begin{cases} 1 - (x - 1)^2, & x \leq 2 \\ \sqrt{x - 2}, & x > 2 \end{cases}$

61. $f(x) = \begin{cases} x^2 + 5, & x \leq 1 \\ -x^2 + 4x + 3, & x > 1 \end{cases}$

62. $h(x) = \begin{cases} 3 - x^2, & x < 0 \\ x^2 + 2, & x \geq 0 \end{cases}$

63. $h(x) = \begin{cases} 4 - x^2, & x < -2 \\ 3 + x, & -2 \leq x < 0 \\ x^2 + 1, & x \geq 0 \end{cases}$

64. $k(x) = \begin{cases} 2x + 1, & x \leq -1 \\ 2x^2 - 1, & -1 < x \leq 1 \\ 1 - x^2, & x > 1 \end{cases}$

In Exercises 65–68, (a) use a graphing utility to graph the function, (b) state the domain and range of the function, and (c) describe the pattern of the graph.

65. $s(x) = 2\left(\frac{1}{4}x - \left[\!\left[\frac{1}{4}x\right]\!\right]\right)$ **66.** $g(x) = 2\left(\frac{1}{4}x - \left[\!\left[\frac{1}{4}x\right]\!\right]\right)^2$

67. $h(x) = 4\left(\frac{1}{2}x - \left[\!\left[\frac{1}{2}x\right]\!\right]\right)$ **68.** $k(x) = 4\left(\frac{1}{2}x - \left[\!\left[\frac{1}{2}x\right]\!\right]\right)^2$

69. DELIVERY CHARGES The cost of sending an overnight package from Los Angeles to Miami is $23.40 for a package weighing up to but not including 1 pound and $3.75 for each additional pound or portion of a pound. A model for the total cost C (in dollars) of sending the package is $C = 23.40 + 3.75[\![x]\!]$, $x > 0$, where x is the weight in pounds.

(a) Sketch a graph of the model.

(b) Determine the cost of sending a package that weighs 9.25 pounds.

70. DELIVERY CHARGES The cost of sending an overnight package from New York to Atlanta is $22.65 for a package weighing up to but not including 1 pound and $3.70 for each additional pound or portion of a pound.

(a) Use the greatest integer function to create a model for the cost C of overnight delivery of a package weighing x pounds, $x > 0$.

(b) Sketch the graph of the function.

71. WAGES A mechanic is paid $14.00 per hour for regular time and time-and-a-half for overtime. The weekly wage function is given by

$$W(h) = \begin{cases} 14h, & 0 < h \leq 40 \\ 21(h - 40) + 560, & h > 40 \end{cases}$$

where h is the number of hours worked in a week.

(a) Evaluate $W(30)$, $W(40)$, $W(45)$, and $W(50)$.

(b) The company increased the regular work week to 45 hours. What is the new weekly wage function?

72. SNOWSTORM During a nine-hour snowstorm, it snows at a rate of 1 inch per hour for the first 2 hours, at a rate of 2 inches per hour for the next 6 hours, and at a rate of 0.5 inch per hour for the final hour. Write and graph a piecewise-defined function that gives the depth of the snow during the snowstorm. How many inches of snow accumulated from the storm?

73. REVENUE The table shows the monthly revenue y (in thousands of dollars) of a landscaping business for each month of the year 2008, with $x = 1$ representing January.

Month, x	Revenue, y
1	5.2
2	5.6
3	6.6
4	8.3
5	11.5
6	15.8
7	12.8
8	10.1
9	8.6
10	6.9
11	4.5
12	2.7

A mathematical model that represents these data is

$$f(x) = \begin{cases} -1.97x + 26.3 \\ 0.505x^2 - 1.47x + 6.3 \end{cases}.$$

(a) Use a graphing utility to graph the model. What is the domain of each part of the piecewise-defined function? How can you tell? Explain your reasoning.

(b) Find $f(5)$ and $f(11)$, and interpret your results in the context of the problem.

(c) How do the values obtained from the model in part (a) compare with the actual data values?

EXPLORATION

TRUE OR FALSE? In Exercises 74 and 75, determine whether the statement is true or false. Justify your answer.

74. A piecewise-defined function will always have at least one x-intercept or at least one y-intercept.

75. A linear equation will always have an x-intercept and a y-intercept.

76. CAPSTONE For each graph of f shown in Figure 2.48, do the following.

(a) Find the domain and range of f.

(b) Find the x- and y-intercepts of the graph of f.

(c) Determine the intervals over which f is increasing, decreasing, or constant.

(d) Determine whether f is even, odd, or neither. Then describe the symmetry.

2.5 TRANSFORMATIONS OF FUNCTIONS

Transtock Inc./Alamy

Shifting Graphs

Many functions have graphs that are simple transformations of the parent graphs summarized in Section 2.4. For example, you can obtain the graph of

$$h(x) = x^2 + 2$$

by shifting the graph of $f(x) = x^2$ *upward* two units, as shown in Figure 2.49. In function notation, h and f are related as follows.

$$h(x) = x^2 + 2 = f(x) + 2$$

Similarly, you can obtain the graph of

$$g(x) = (x - 2)^2$$

by shifting the graph of $f(x) = x^2$ to the *right* two units, as shown in Figure 2.50. In this case, the functions g and f have the following relationship.

$$g(x) = (x - 2)^2 = f(x - 2)$$

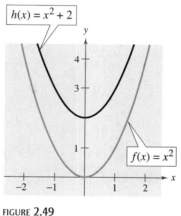

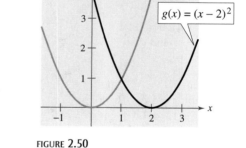

FIGURE **2.49**   FIGURE **2.50**

The following list summarizes this discussion about horizontal and vertical shifts.

Vertical and Horizontal Shifts

Let c be a positive real number. **Vertical and horizontal shifts** in the graph of $y = f(x)$ are represented as follows.

1. Vertical shift c units *upward:* $h(x) = f(x) + c$
2. Vertical shift c units *downward:* $h(x) = f(x) - c$
3. Horizontal shift c units to the *right:* $h(x) = f(x - c)$
4. Horizontal shift c units to the *left:* $h(x) = f(x + c)$

Some graphs can be obtained from combinations of vertical and horizontal shifts, as demonstrated in Example 1(b). Vertical and horizontal shifts generate a *family of functions*, each with the same shape but at different locations in the plane.

Example 1 Shifts in the Graphs of a Function

Use the graph of $f(x) = x^3$ to sketch the graph of each function.

a. $g(x) = x^3 - 1$ **b.** $h(x) = (x + 2)^3 + 1$

Solution

a. Relative to the graph of $f(x) = x^3$, the graph of

$$g(x) = x^3 - 1$$

is a downward shift of one unit, as shown in Figure 2.51.

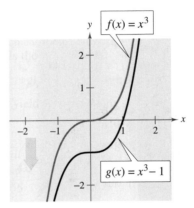

FIGURE 2.51

b. Relative to the graph of $f(x) = x^3$, the graph of

$$h(x) = (x + 2)^3 + 1$$

involves a left shift of two units and an upward shift of one unit, as shown in Figure 2.52.

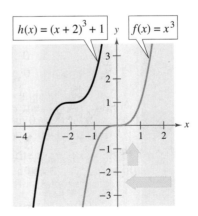

FIGURE 2.52

CHECK*Point* Now try Exercise 7.

In Figure 2.52, notice that the same result is obtained if the vertical shift precedes the horizontal shift *or* if the horizontal shift precedes the vertical shift.

Study Tip

In Example 1(a), note that $g(x) = f(x) - 1$ and that in Example 1(b), $h(x) = f(x + 2) + 1$.

Reflecting Graphs

The second common type of transformation is a **reflection.** For instance, if you consider the x-axis to be a mirror, the graph of

$$h(x) = -x^2$$

is the mirror image (or reflection) of the graph of

$$f(x) = x^2,$$

as shown in Figure 2.53.

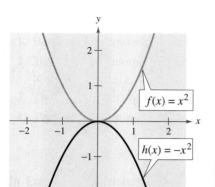

FIGURE 2.53

> ### Reflections in the Coordinate Axes
>
> **Reflections** in the coordinate axes of the graph of $y = f(x)$ are represented as follows.
>
> **1.** Reflection in the x-axis: $h(x) = -f(x)$
>
> **2.** Reflection in the y-axis: $h(x) = f(-x)$

| **Example 2** | **Finding Equations from Graphs** |

The graph of the function given by

$$f(x) = x^4$$

is shown in Figure 2.54. Each of the graphs in Figure 2.55 is a transformation of the graph of f. Find an equation for each of these functions.

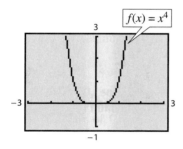

FIGURE 2.54

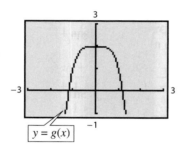

(a)

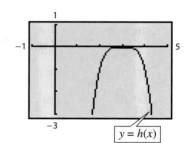

(b)

FIGURE 2.55

Solution

a. The graph of g is a reflection in the x-axis *followed by* an upward shift of two units of the graph of $f(x) = x^4$. So, the equation for g is

$$g(x) = -x^4 + 2.$$

b. The graph of h is a horizontal shift of three units to the right *followed by* a reflection in the x-axis of the graph of $f(x) = x^4$. So, the equation for h is

$$h(x) = -(x - 3)^4.$$

CHECK *Point* Now try Exercise 15.

Example 3 Reflections and Shifts

Compare the graph of each function with the graph of $f(x) = \sqrt{x}$.

a. $g(x) = -\sqrt{x}$ **b.** $h(x) = \sqrt{-x}$ **c.** $k(x) = -\sqrt{x+2}$

Algebraic Solution

a. The graph of g is a reflection of the graph of f in the x-axis because

$$g(x) = -\sqrt{x}$$
$$= -f(x).$$

b. The graph of h is a reflection of the graph of f in the y-axis because

$$h(x) = \sqrt{-x}$$
$$= f(-x).$$

c. The graph of k is a left shift of two units followed by a reflection in the x-axis because

$$k(x) = -\sqrt{x+2}$$
$$= -f(x+2).$$

Graphical Solution

a. Graph f and g on the same set of coordinate axes. From the graph in Figure 2.56, you can see that the graph of g is a reflection of the graph of f in the x-axis.

b. Graph f and h on the same set of coordinate axes. From the graph in Figure 2.57, you can see that the graph of h is a reflection of the graph of f in the y-axis.

c. Graph f and k on the same set of coordinate axes. From the graph in Figure 2.58, you can see that the graph of k is a left shift of two units of the graph of f, followed by a reflection in the x-axis.

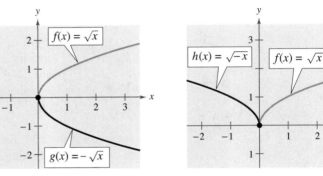

FIGURE 2.56   FIGURE 2.57

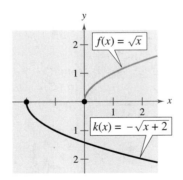

FIGURE 2.58

CHECK **Point** Now try Exercise 25.

When sketching the graphs of functions involving square roots, remember that the domain must be restricted to exclude negative numbers inside the radical. For instance, here are the domains of the functions in Example 3.

Domain of $g(x) = -\sqrt{x}$: $x \ge 0$

Domain of $h(x) = \sqrt{-x}$: $x \le 0$

Domain of $k(x) = -\sqrt{x+2}$: $x \ge -2$

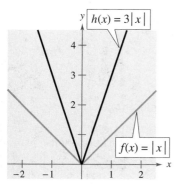

FIGURE **2.59**

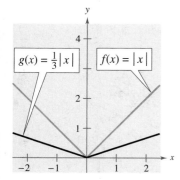

FIGURE **2.60**

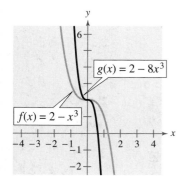

FIGURE **2.61**

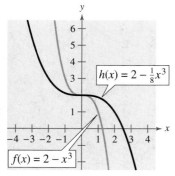

FIGURE **2.62**

Nonrigid Transformations

Horizontal shifts, vertical shifts, and reflections are **rigid transformations** because the basic shape of the graph is unchanged. These transformations change only the *position* of the graph in the coordinate plane. **Nonrigid transformations** are those that cause a *distortion*—a change in the shape of the original graph. For instance, a nonrigid transformation of the graph of $y = f(x)$ is represented by $g(x) = cf(x)$, where the transformation is a **vertical stretch** if $c > 1$ and a **vertical shrink** if $0 < c < 1$. Another nonrigid transformation of the graph of $y = f(x)$ is represented by $h(x) = f(cx)$, where the transformation is a **horizontal shrink** if $c > 1$ and a **horizontal stretch** if $0 < c < 1$.

Example 4   Nonrigid Transformations

Compare the graph of each function with the graph of $f(x) = |x|$.

a. $h(x) = 3|x|$ **b.** $g(x) = \frac{1}{3}|x|$

Solution

a. Relative to the graph of $f(x) = |x|$, the graph of

$$h(x) = 3|x| = 3f(x)$$

is a vertical stretch (each y-value is multiplied by 3) of the graph of f. (See Figure 2.59.)

b. Similarly, the graph of

$$g(x) = \frac{1}{3}|x| = \frac{1}{3}f(x)$$

is a vertical shrink $\left(\text{each } y\text{-value is multiplied by } \frac{1}{3}\right)$ of the graph of f.   (See Figure 2.60.)

CHECK*Point*   Now try Exercise 29.

Example 5   Nonrigid Transformations

Compare the graph of each function with the graph of $f(x) = 2 - x^3$.

a. $g(x) = f(2x)$ **b.** $h(x) = f\left(\frac{1}{2}x\right)$

Solution

a. Relative to the graph of $f(x) = 2 - x^3$, the graph of

$$g(x) = f(2x) = 2 - (2x)^3 = 2 - 8x^3$$

is a horizontal shrink $(c > 1)$ of the graph of f. (See Figure 2.61.)

b. Similarly, the graph of

$$h(x) = f\left(\frac{1}{2}x\right) = 2 - \left(\frac{1}{2}x\right)^3 = 2 - \frac{1}{8}x^3$$

is a horizontal stretch $(0 < c < 1)$ of the graph of f. (See Figure 2.62.)

CHECK*Point*   Now try Exercise 35.

2.5 EXERCISES

See www.CalcChat.com for worked-out solutions to odd-numbered exercises.

VOCABULARY

In Exercises 1–5, fill in the blanks.

1. Horizontal shifts, vertical shifts, and reflections are called _____ transformations.
2. A reflection in the x-axis of $y = f(x)$ is represented by $h(x) =$ _____, while a reflection in the y-axis of $y = f(x)$ is represented by $h(x) =$ _____.
3. Transformations that cause a distortion in the shape of the graph of $y = f(x)$ are called _____ transformations.
4. A nonrigid transformation of $y = f(x)$ represented by $h(x) = f(cx)$ is a _____ _____ if $c > 1$ and a _____ _____ if $0 < c < 1$.
5. A nonrigid transformation of $y = f(x)$ represented by $g(x) = cf(x)$ is a _____ _____ if $c > 1$ and a _____ _____ if $0 < c < 1$.

6. Match the rigid transformation of $y = f(x)$ with the correct representation of the graph of h, where $c > 0$.
   (a) $h(x) = f(x) + c$ (i) A horizontal shift of f, c units to the right
   (b) $h(x) = f(x) - c$ (ii) A vertical shift of f, c units downward
   (c) $h(x) = f(x + c)$ (iii) A horizontal shift of f, c units to the left
   (d) $h(x) = f(x - c)$ (iv) A vertical shift of f, c units upward

SKILLS AND APPLICATIONS

7. For each function, sketch (on the same set of coordinate axes) a graph of each function for $c = -1, 1$, and 3.
   (a) $f(x) = |x| + c$
   (b) $f(x) = |x - c|$
   (c) $f(x) = |x + 4| + c$

8. For each function, sketch (on the same set of coordinate axes) a graph of each function for $c = -3, -1, 1$, and 3.
   (a) $f(x) = \sqrt{x} + c$
   (b) $f(x) = \sqrt{x - c}$
   (c) $f(x) = \sqrt{x - 3} + c$

9. For each function, sketch (on the same set of coordinate axes) a graph of each function for $c = -2, 0$, and 2.
   (a) $f(x) = [\![x]\!] + c$
   (b) $f(x) = [\![x + c]\!]$
   (c) $f(x) = [\![x - 1]\!] + c$

10. For each function, sketch (on the same set of coordinate axes) a graph of each function for $c = -3, -1, 1$, and 3.
    (a) $f(x) = \begin{cases} x^2 + c, & x < 0 \\ -x^2 + c, & x \geq 0 \end{cases}$
    (b) $f(x) = \begin{cases} (x + c)^2, & x < 0 \\ -(x + c)^2, & x \geq 0 \end{cases}$

In Exercises 11–14, use the graph of f to sketch each graph. To print an enlarged copy of the graph, go to the website *www.mathgraphs.com*.

11. (a) $y = f(x) + 2$
    (b) $y = f(x - 2)$
    (c) $y = 2f(x)$
    (d) $y = -f(x)$
    (e) $y = f(x + 3)$
    (f) $y = f(-x)$
    (g) $y = f\left(\frac{1}{2}x\right)$

12. (a) $y = f(-x)$
    (b) $y = f(x) + 4$
    (c) $y = 2f(x)$
    (d) $y = -f(x - 4)$
    (e) $y = f(x) - 3$
    (f) $y = -f(x) - 1$
    (g) $y = f(2x)$

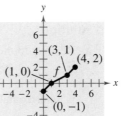

FIGURE FOR 11

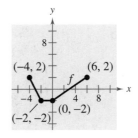

FIGURE FOR 12

13. (a) $y = f(x) - 1$
    (b) $y = f(x - 1)$
    (c) $y = f(-x)$
    (d) $y = f(x + 1)$
    (e) $y = -f(x - 2)$
    (f) $y = \frac{1}{2}f(x)$
    (g) $y = f(2x)$

14. (a) $y = f(x - 5)$
    (b) $y = -f(x) + 3$
    (c) $y = \frac{1}{3}f(x)$
    (d) $y = -f(x + 1)$
    (e) $y = f(-x)$
    (f) $y = f(x) - 10$
    (g) $y = f\left(\frac{1}{3}x\right)$

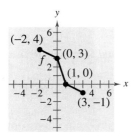

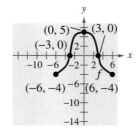

FIGURE FOR 13   FIGURE FOR 14

15. Use the graph of $f(x) = x^2$ to write an equation for each function whose graph is shown.

(a)

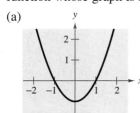

(b)

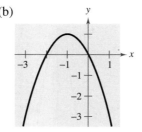

(c)

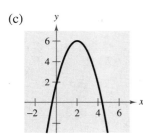

(d)

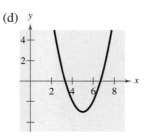

16. Use the graph of $f(x) = x^3$ to write an equation for each function whose graph is shown.

(a)

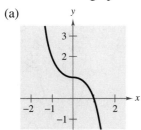

(b)

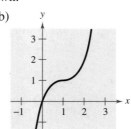

(c)

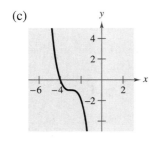

(d)

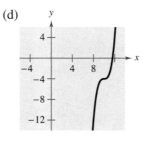

17. Use the graph of $f(x) = |x|$ to write an equation for each function whose graph is shown.

(a)

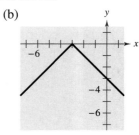

(b)

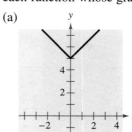

(c)

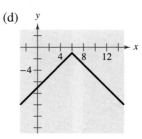

(d)

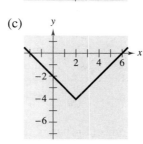

18. Use the graph of $f(x) = \sqrt{x}$ to write an equation for each function whose graph is shown.

(a)

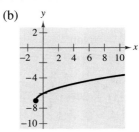

(b)

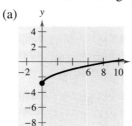

(c)

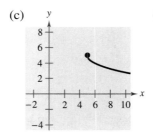

(d)

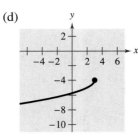

In Exercises 19–24, identify the parent function and the transformation shown in the graph. Write an equation for the function shown in the graph.

19.

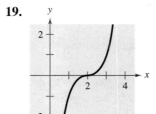

20.

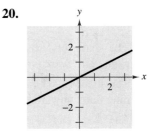

21.

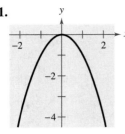

22.

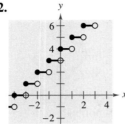

23.

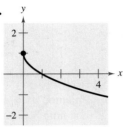

24.

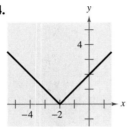

In Exercises 25–54, *g* is related to one of the parent functions described in Section 2.4. (a) Identify the parent function *f*. (b) Describe the sequence of transformations from *f* to *g*. (c) Sketch the graph of *g*. (d) Use function notation to write *g* in terms of *f*.

25. $g(x) = 12 - x^2$

26. $g(x) = (x - 8)^2$

27. $g(x) = x^3 + 7$

28. $g(x) = -x^3 - 1$

29. $g(x) = \frac{2}{3}x^2 + 4$

30. $g(x) = 2(x - 7)^2$

31. $g(x) = 2 - (x + 5)^2$

32. $g(x) = -(x + 10)^2 + 5$

33. $g(x) = 3 + 2(x - 4)^2$

34. $g(x) = -\frac{1}{4}(x + 2)^2 - 2$

35. $g(x) = \sqrt{3x}$

36. $g(x) = \sqrt{\frac{1}{4}x}$

37. $g(x) = (x - 1)^3 + 2$

38. $g(x) = (x + 3)^3 - 10$

39. $g(x) = 3(x - 2)^3$

40. $g(x) = -\frac{1}{2}(x + 1)^3$

41. $g(x) = -|x| - 2$

42. $g(x) = 6 - |x + 5|$

43. $g(x) = -|x + 4| + 8$

44. $g(x) = |-x + 3| + 9$

45. $g(x) = -2|x - 1| - 4$

46. $g(x) = \frac{1}{2}|x - 2| - 3$

47. $g(x) = 3 - [\![x]\!]$

48. $g(x) = 2[\![x + 5]\!]$

49. $g(x) = \sqrt{x - 9}$

50. $g(x) = \sqrt{x + 4} + 8$

51. $g(x) = \sqrt{7 - x} - 2$

52. $g(x) = -\frac{1}{2}\sqrt{x + 3} - 1$

53. $g(x) = \sqrt{\frac{1}{2}x} - 4$

54. $g(x) = \sqrt{3x} + 1$

In Exercises 55–62, write an equation for the function that is described by the given characteristics.

55. The shape of $f(x) = x^2$, but shifted three units to the right and seven units downward

56. The shape of $f(x) = x^2$, but shifted two units to the left, nine units upward, and reflected in the *x*-axis

57. The shape of $f(x) = x^3$, but shifted 13 units to the right

58. The shape of $f(x) = x^3$, but shifted six units to the left, six units downward, and reflected in the *y*-axis

59. The shape of $f(x) = |x|$, but shifted 12 units upward and reflected in the *x*-axis

60. The shape of $f(x) = |x|$, but shifted four units to the left and eight units downward

61. The shape of $f(x) = \sqrt{x}$, but shifted six units to the left and reflected in both the *x*-axis and the *y*-axis

62. The shape of $f(x) = \sqrt{x}$, but shifted nine units downward and reflected in both the *x*-axis and the *y*-axis

63. Use the graph of $f(x) = x^2$ to write an equation for each function whose graph is shown.

(a)

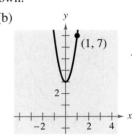

(b)

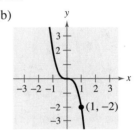

64. Use the graph of $f(x) = x^3$ to write an equation for each function whose graph is shown.

(a)

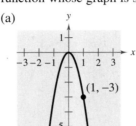

(b)

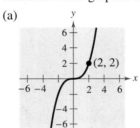

65. Use the graph of $f(x) = |x|$ to write an equation for each function whose graph is shown.

(a)

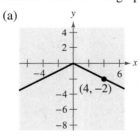

(b)

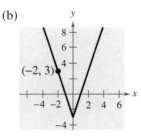

66. Use the graph of $f(x) = \sqrt{x}$ to write an equation for each function whose graph is shown.

(a)

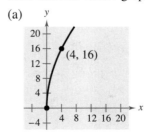

(b)

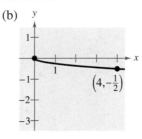

In Exercises 67–72, identify the parent function and the transformation shown in the graph. Write an equation for the function shown in the graph. Then use a graphing utility to verify your answer.

67.

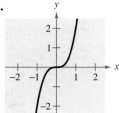

68.

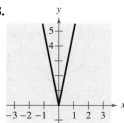

69.

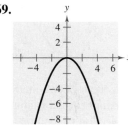

70.

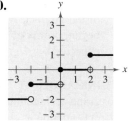

71.

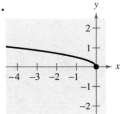

72.

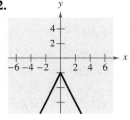

 GRAPHICAL ANALYSIS In Exercises 73–76, use the viewing window shown to write a possible equation for the transformation of the parent function.

73.

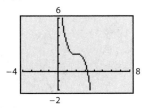

74.

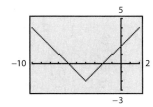

75.

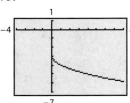

76.

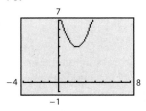

GRAPHICAL REASONING In Exercises 77 and 78, use the graph of f to sketch the graph of g. To print an enlarged copy of the graph, go to the website *www.mathgraphs.com*.

77.

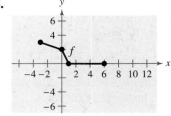

(a) $g(x) = f(x) + 2$ (b) $g(x) = f(x) - 1$

(c) $g(x) = f(-x)$ (d) $g(x) = -2f(x)$

(e) $g(x) = f(4x)$ (f) $g(x) = f\left(\frac{1}{2}x\right)$

78.

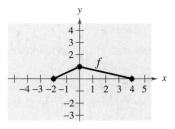

(a) $g(x) = f(x) - 5$ (b) $g(x) = f(x) + \frac{1}{2}$

(c) $g(x) = f(-x)$ (d) $g(x) = -4f(x)$

(e) $g(x) = f(2x) + 1$ (f) $g(x) = f\left(\frac{1}{4}x\right) - 2$

79. MILES DRIVEN The total numbers of miles M (in billions) driven by vans, pickups, and SUVs (sport utility vehicles) in the United States from 1990 through 2006 can be approximated by the function

$$M = 527 + 128.0\sqrt{t}, \quad 0 \le t \le 16$$

where t represents the year, with $t = 0$ corresponding to 1990. (Source: U.S. Federal Highway Administration)

(a) Describe the transformation of the parent function $f(x) = \sqrt{x}$. Then use a graphing utility to graph the function over the specified domain.

(b) Find the average rate of change of the function from 1990 to 2006. Interpret your answer in the context of the problem.

(c) Rewrite the function so that $t = 0$ represents 2000. Explain how you got your answer.

(d) Use the model from part (c) to predict the number of miles driven by vans, pickups, and SUVs in 2012. Does your answer seem reasonable? Explain.

80. MARRIED COUPLES The numbers N (in thousands) of married couples with stay-at-home mothers from 2000 through 2007 can be approximated by the function

$$N = -24.70(t - 5.99)^2 + 5617, \quad 0 \le t \le 7$$

where t represents the year, with $t = 0$ corresponding to 2000. (Source: U.S. Census Bureau)

(a) Describe the transformation of the parent function $f(x) = x^2$. Then use a graphing utility to graph the function over the specified domain.

(b) Find the average rate of the change of the function from 2000 to 2007. Interpret your answer in the context of the problem.

(c) Use the model to predict the number of married couples with stay-at-home mothers in 2015. Does your answer seem reasonable? Explain.

EXPLORATION

TRUE OR FALSE? In Exercises 81–84, determine whether the statement is true or false. Justify your answer.

81. The graph of $y = f(-x)$ is a reflection of the graph of $y = f(x)$ in the x-axis.

82. The graph of $y = -f(x)$ is a reflection of the graph of $y = f(x)$ in the y-axis.

83. The graphs of

$$f(x) = |x| + 6$$

and

$$f(x) = |-x| + 6$$

are identical.

84. If the graph of the parent function $f(x) = x^2$ is shifted six units to the right, three units upward, and reflected in the x-axis, then the point $(-2, 19)$ will lie on the graph of the transformation.

85. DESCRIBING PROFITS Management originally predicted that the profits from the sales of a new product would be approximated by the graph of the function f shown. The actual profits are shown by the function g along with a verbal description. Use the concepts of transformations of graphs to write g in terms of f.

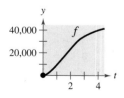

(a) The profits were only three-fourths as large as expected.

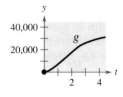

(b) The profits were consistently $10,000 greater than predicted.

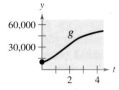

(c) There was a two-year delay in the introduction of the product. After sales began, profits grew as expected.

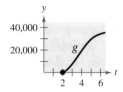

86. THINK ABOUT IT You can use either of two methods to graph a function: plotting points or translating a parent function as shown in this section. Which method of graphing do you prefer to use for each function? Explain.

(a) $f(x) = 3x^2 - 4x + 1$

(b) $f(x) = 2(x - 1)^2 - 6$

87. The graph of $y = f(x)$ passes through the points $(0, 1)$, $(1, 2)$, and $(2, 3)$. Find the corresponding points on the graph of $y = f(x + 2) - 1$.

88. Use a graphing utility to graph f, g, and h in the same viewing window. Before looking at the graphs, try to predict how the graphs of g and h relate to the graph of f.

(a) $f(x) = x^2$, $g(x) = (x - 4)^2$,
$\quad h(x) = (x - 4)^2 + 3$

(b) $f(x) = x^2$, $g(x) = (x + 1)^2$,
$\quad h(x) = (x + 1)^2 - 2$

(c) $f(x) = x^2$, $g(x) = (x + 4)^2$,
$\quad h(x) = (x + 4)^2 + 2$

89. Reverse the order of transformations in Example 2(a). Do you obtain the same graph? Do the same for Example 2(b). Do you obtain the same graph? Explain.

90. CAPSTONE Use the fact that the graph of $y = f(x)$ is increasing on the intervals $(-\infty, -1)$ and $(2, \infty)$ and decreasing on the interval $(-1, 2)$ to find the intervals on which the graph is increasing and decreasing. If not possible, state the reason.

(a) $y = f(-x)$ (b) $y = -f(x)$ (c) $y = \frac{1}{2}f(x)$

(d) $y = -f(x - 1)$ (e) $y = f(x - 2) + 1$

2.6 COMBINATIONS OF FUNCTIONS: COMPOSITE FUNCTIONS

What you should learn

- Add, subtract, multiply, and divide functions.
- Find the composition of one function with another function.
- Use combinations and compositions of functions to model and solve real-life problems.

Why you should learn it

Compositions of functions can be used to model and solve real-life problems. For instance, in Exercise 76 on page 237, compositions of functions are used to determine the price of a new hybrid car.

© Jim West/The Image Works

Arithmetic Combinations of Functions

Just as two real numbers can be combined by the operations of addition, subtraction, multiplication, and division to form other real numbers, two *functions* can be combined to create new functions. For example, the functions given by $f(x) = 2x - 3$ and $g(x) = x^2 - 1$ can be combined to form the sum, difference, product, and quotient of f and g.

$$f(x) + g(x) = (2x - 3) + (x^2 - 1)$$
$$= x^2 + 2x - 4$$
$$f(x) - g(x) = (2x - 3) - (x^2 - 1)$$
$$= -x^2 + 2x - 2$$
$$f(x)g(x) = (2x - 3)(x^2 - 1)$$
$$= 2x^3 - 3x^2 - 2x + 3$$
$$\frac{f(x)}{g(x)} = \frac{2x - 3}{x^2 - 1}, \quad x \neq \pm 1$$

The domain of an **arithmetic combination** of functions f and g consists of all real numbers that are common to the domains of f and g. In the case of the quotient $f(x)/g(x)$, there is the further restriction that $g(x) \neq 0$.

Sum, Difference, Product, and Quotient of Functions

Let f and g be two functions with overlapping domains. Then, for all x common to both domains, the *sum*, *difference*, *product*, and *quotient* of f and g are defined as follows.

1. *Sum:* $(f + g)(x) = f(x) + g(x)$

2. *Difference:* $(f - g)(x) = f(x) - g(x)$

3. *Product:* $(fg)(x) = f(x) \cdot g(x)$

4. *Quotient:* $\left(\dfrac{f}{g}\right)(x) = \dfrac{f(x)}{g(x)}, \quad g(x) \neq 0$

| **Example 1**   **Finding the Sum of Two Functions**

Given $f(x) = 2x + 1$ and $g(x) = x^2 + 2x - 1$, find $(f + g)(x)$. Then evaluate the sum when $x = 3$.

Solution

$$(f + g)(x) = f(x) + g(x) = \qquad + \qquad = x^2 + 4x$$

When $x = 3$, the value of this sum is

$$(f + g)(\ ) = \ ^2 + 4(\ ) = 21.$$

CHECK**Point**   Now try Exercise 9(a).

Example 2   Finding the Difference of Two Functions

Given $f(x) = 2x + 1$ and $g(x) = x^2 + 2x - 1$, find $(f - g)(x)$. Then evaluate the difference when $x = 2$.

Solution

The difference of f and g is

$$(f - g)(x) = f(x) - g(x) = \qquad - \qquad = -x^2 + 2.$$

When $x = 2$, the value of this difference is

$$(f - g)(\ ) = -(\ )^2 + 2 = -2.$$

CHECK Point   Now try Exercise 9(b).

Example 3   Finding the Product of Two Functions

Given $f(x) = x^2$ and $g(x) = x - 3$, find $(fg)(x)$. Then evaluate the product when $x = 4$.

Solution

$$(fg)(x) = f(x)g(x) = \qquad = x^3 - 3x^2$$

When $x = 4$, the value of this product is

$$(fg)(\ ) = {}^3 - 3(\ )^2 = 16.$$

CHECK Point   Now try Exercise 9(c).

In Examples 1–3, both f and g have domains that consist of all real numbers. So, the domains of $f + g, f - g$, and fg are also the set of all real numbers. Remember that any restrictions on the domains of f and g must be considered when forming the sum, difference, product, or quotient of f and g.

Example 4   Finding the Quotients of Two Functions

Find $(f/g)(x)$ and $(g/f)(x)$ for the functions given by $f(x) = \sqrt{x}$ and $g(x) = \sqrt{4 - x^2}$. Then find the domains of f/g and g/f.

Solution

The quotient of f and g is

$$\left(\frac{f}{g}\right)(x) = \frac{f(x)}{g(x)} = \frac{\sqrt{x}}{\sqrt{4 - x^2}}$$

and the quotient of g and f is

$$\left(\frac{g}{f}\right)(x) = \frac{g(x)}{f(x)} = \frac{\sqrt{4 - x^2}}{\sqrt{x}}.$$

The domain of f is $[0, \infty)$ and the domain of g is $[-2, 2]$. The intersection of these domains is $[0, 2]$. So, the domains of f/g and g/f are as follows.

Domain of f/g: $[0, 2)$ Domain of g/f: $(0, 2]$

CHECK Point   Now try Exercise 9(d).

Study Tip

Note that the domain of f/g includes $x = 0$, but not $x = 2$, because $x = 2$ yields a zero in the denominator, whereas the domain of g/f includes $x = 2$, but not $x = 0$, because $x = 0$ yields a zero in the denominator.

Composition of Functions

Another way of combining two functions is to form the **composition** of one with the other. For instance, if $f(x) = x^2$ and $g(x) = x + 1$, the composition of f with g is

$$f(\quad) = f(\quad)$$

$$= (x + 1)^2.$$

This composition is denoted as $f \circ g$ and reads as "f composed with g."

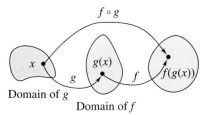

$f \circ g$

Domain of g

Domain of f

FIGURE **2.63**

> **Definition of Composition of Two Functions**
>
> The **composition** of the function f with the function g is
>
> $$(f \circ g)(x) = f(g(x)).$$
>
> The domain of $f \circ g$ is the set of all x in the domain of g such that $g(x)$ is in the domain of f. (See Figure 2.63.)

Example 5   Composition of Functions

Given $f(x) = x + 2$ and $g(x) = 4 - x^2$, find the following.

a. $(f \circ g)(x)$ **b.** $(g \circ f)(x)$ **c.** $(g \circ f)(-2)$

Solution

a. The composition of f with g is as follows.

$$(f \circ g)(x) = f(\quad)$$

$$= f(\qquad)$$

$$= (\qquad) + 2$$

$$= -x^2 + 6$$

b. The composition of g with f is as follows.

$$(g \circ f)(x) = g(\quad)$$

$$= g(\qquad)$$

$$= 4 - (\qquad)^2$$

$$= 4 - (x^2 + 4x + 4)$$

$$= -x^2 - 4x$$

Note that, in this case, $(f \circ g)(x) \neq (g \circ f)(x)$.

c. Using the result of part (b), you can write the following.

$$(g \circ f)(\quad) = -(\quad)^2 - 4(\quad)$$

$$= -4 + 8$$

$$= 4$$

CHECK*Point* Now try Exercise 37.

Study Tip

The following tables of values help illustrate the composition $(f \circ g)(x)$ given in Example 5.

x	0	1	2	3
$g(x)$	4	3	0	-5

$g(x)$	4	3	0	-5
$f(g(x))$	6	5	2	-3

x	0	1	2	3
$f(g(x))$	6	5	2	-3

Note that the first two tables can be combined (or "composed") to produce the values given in the third table.

Example 6   Finding the Domain of a Composite Function

Find the domain of $(f \circ g)(x)$ for the functions given by

$$f(x) = x^2 - 9 \quad \text{and} \quad g(x) = \sqrt{9 - x^2}.$$

Algebraic Solution

The composition of the functions is as follows.

$$(f \circ g)(x) = f(\quad)$$
$$= f(\qquad)$$
$$= (\qquad)^2 - 9$$
$$= 9 - x^2 - 9$$
$$= -x^2$$

From this, it might appear that the domain of the composition is the set of all real numbers. This, however, is not true. Because the domain of f is the set of all real numbers and the domain of g is $[-3, 3]$, the domain of $f \circ g$ is $[-3, 3]$.

Graphical Solution

You can use a graphing utility to graph the composition of the functions $(f \circ g)(x)$ as $y = \left(\sqrt{9 - x^2} \right)^2 - 9$. Enter the functions as follows.

$$y_1 = \sqrt{9 - x^2} \qquad y_2 = y_1^2 - 9$$

Graph y_2, as shown in Figure 2.64. Use the *trace* feature to determine that the x-coordinates of points on the graph extend from -3 to 3. So, you can graphically estimate the domain of $f \circ g$ to be $[-3, 3]$.

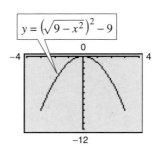

$$y = \left(\sqrt{9 - x^2} \right)^2 - 9$$

FIGURE 2.64

CHECK *Point*   Now try Exercise 41.

In Examples 5 and 6, you formed the composition of two given functions. In calculus, it is also important to be able to identify two functions that make up a given composite function. For instance, the function h given by $h(x) = (3x - 5)^3$ is the composition of f with g, where $f(x) = x^3$ and $g(x) = 3x - 5$. That is,

$$h(x) = (\quad)^3 = [\quad]^3 = f(\quad).$$

Basically, to "decompose" a composite function, look for an "inner" function and an "outer" function. In the function h above, $g(x) = 3x - 5$ is the inner function and $f(x) = x^3$ is the outer function.

Example 7   Decomposing a Composite Function

Write the function given by $h(x) = \dfrac{1}{(x - 2)^2}$ as a composition of two functions.

Solution

One way to write h as a composition of two functions is to take the inner function to be $g(x) = x - 2$ and the outer function to be

$$f(x) = \frac{1}{x^2} = x^{-2}.$$

Then you can write

$$h(x) = \frac{1}{(\quad)^2} = (\quad)^{-2} = f(\quad) = f(\quad).$$

CHECK *Point*   Now try Exercise 53.

Application

| Example 8 | **Bacteria Count**

The number N of bacteria in a refrigerated food is given by

$$N(T) = 20T^2 - 80T + 500, \quad 2 \le T \le 14$$

where T is the temperature of the food in degrees Celsius. When the food is removed from refrigeration, the temperature of the food is given by

$$T(t) = 4t + 2, \quad 0 \le t \le 3$$

where t is the time in hours. (a) Find the composition $N(T(t))$ and interpret its meaning in context. (b) Find the time when the bacteria count reaches 2000.

Solution

a. $N(\quad) = 20(\qquad)^2 - 80(\qquad) + 500$

$$= 20(16t^2 + 16t + 4) - 320t - 160 + 500$$

$$= 320t^2 + 320t + 80 - 320t - 160 + 500$$

$$= 320t^2 + 420$$

The composite function $N(T(t))$ represents the number of bacteria in the food as a function of the amount of time the food has been out of refrigeration.

b. The bacteria count will reach 2000 when $320t^2 + 420 = 2000$. Solve this equation to find that the count will reach 2000 when $t \approx 2.2$ hours. When you solve this equation, note that the negative value is rejected because it is not in the domain of the composite function.

CHECK*Point*   Now try Exercise 73.

CLASSROOM DISCUSSION

Analyzing Arithmetic Combinations of Functions

a. Use the graphs of f and $(f + g)$ in Figure 2.65 to make a table showing the values of $g(x)$ when $x = 1, 2, 3, 4, 5,$ and 6. Explain your reasoning.

b. Use the graphs of f and $(f - h)$ in Figure 2.65 to make a table showing the values of $h(x)$ when $x = 1, 2, 3, 4, 5,$ and 6. Explain your reasoning.

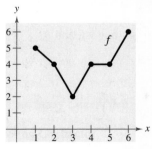
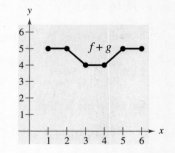
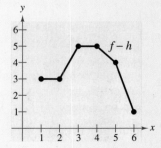

FIGURE **2.65**

2.6   EXERCISES

See www.CalcChat.com for worked-out solutions to odd-numbered exercises.

VOCABULARY: Fill in the blanks.

1. Two functions f and g can be combined by the arithmetic operations of _____, _____, _____, and _____ to create new functions.

2. The _____ of the function f with g is $(f \circ g)(x) = f(g(x))$.

3. The domain of $(f \circ g)$ is all x in the domain of g such that _____ is in the domain of f.

4. To decompose a composite function, look for an _____ function and an _____ function.

SKILLS AND APPLICATIONS

In Exercises 5–8, use the graphs of f and g to graph $h(x) = (f + g)(x)$. To print an enlarged copy of the graph, go to the website **www.mathgraphs.com**.

5.
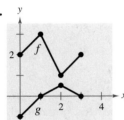

6.
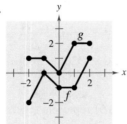

7.
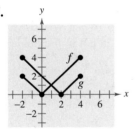

8.
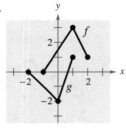

In Exercises 9–16, find (a) $(f + g)(x)$, (b) $(f - g)(x)$, (c) $(fg)(x)$, and (d) $(f/g)(x)$. What is the domain of f/g?

9. $f(x) = x + 2$, $g(x) = x - 2$

10. $f(x) = 2x - 5$, $g(x) = 2 - x$

11. $f(x) = x^2$, $g(x) = 4x - 5$

12. $f(x) = 3x + 1$, $g(x) = 5x - 4$

13. $f(x) = x^2 + 6$, $g(x) = \sqrt{1 - x}$

14. $f(x) = \sqrt{x^2 - 4}$, $g(x) = \dfrac{x^2}{x^2 + 1}$

15. $f(x) = \dfrac{1}{x}$, $g(x) = \dfrac{1}{x^2}$

16. $f(x) = \dfrac{x}{x + 1}$, $g(x) = x^3$

In Exercises 17–28, evaluate the indicated function for $f(x) = x^2 + 1$ and $g(x) = x - 4$.

17. $(f + g)(2)$

18. $(f - g)(-1)$

19. $(f - g)(0)$

20. $(f + g)(1)$

21. $(f - g)(3t)$

22. $(f + g)(t - 2)$

23. $(fg)(6)$

24. $(fg)(-6)$

25. $(f/g)(5)$

26. $(f/g)(0)$

27. $(f/g)(-1) - g(3)$

28. $(fg)(5) + f(4)$

In Exercises 29–32, graph the functions f, g, and $f + g$ on the same set of coordinate axes.

29. $f(x) = \frac{1}{2}x$, $g(x) = x - 1$

30. $f(x) = \frac{1}{3}x$, $g(x) = -x + 4$

31. $f(x) = x^2$, $g(x) = -2x$

32. $f(x) = 4 - x^2$, $g(x) = x$

GRAPHICAL REASONING In Exercises 33–36, use a graphing utility to graph f, g, and $f + g$ in the same viewing window. Which function contributes most to the magnitude of the sum when $0 \le x \le 2$? Which function contributes most to the magnitude of the sum when $x > 6$?

33. $f(x) = 3x$, $g(x) = -\dfrac{x^3}{10}$

34. $f(x) = \dfrac{x}{2}$, $g(x) = \sqrt{x}$

35. $f(x) = 3x + 2$, $g(x) = -\sqrt{x + 5}$

36. $f(x) = x^2 - \frac{1}{2}$, $g(x) = -3x^2 - 1$

In Exercises 37–40, find (a) $f \circ g$, (b) $g \circ f$, and (c) $g \circ g$.

37. $f(x) = x^2$, $g(x) = x - 1$

38. $f(x) = 3x + 5$, $g(x) = 5 - x$

39. $f(x) = \sqrt[3]{x - 1}$, $g(x) = x^3 + 1$

40. $f(x) = x^3$, $g(x) = \dfrac{1}{x}$

In Exercises 41–48, find (a) $f \circ g$ and (b) $g \circ f$. Find the domain of each function and each composite function.

41. $f(x) = \sqrt{x + 4}$, $g(x) = x^2$

42. $f(x) = \sqrt[3]{x - 5}$, $g(x) = x^3 + 1$

43. $f(x) = x^2 + 1, \quad g(x) = \sqrt{x}$

44. $f(x) = x^{2/3}, \quad g(x) = x^6$

45. $f(x) = |x|, \quad g(x) = x + 6$

46. $f(x) = |x - 4|, \quad g(x) = 3 - x$

47. $f(x) = \dfrac{1}{x}, \quad g(x) = x + 3$

48. $f(x) = \dfrac{3}{x^2 - 1}, \quad g(x) = x + 1$

In Exercises 49–52, use the graphs of f and g to evaluate the functions.

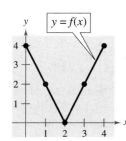 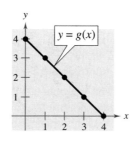

49. (a) $(f + g)(3)$ (b) $(f/g)(2)$

50. (a) $(f - g)(1)$ (b) $(fg)(4)$

51. (a) $(f \circ g)(2)$ (b) $(g \circ f)(2)$

52. (a) $(f \circ g)(1)$ (b) $(g \circ f)(3)$

In Exercises 53–60, find two functions f and g such that $(f \circ g)(x) = h(x)$. (There are many correct answers.)

53. $h(x) = (2x + 1)^2$ **54.** $h(x) = (1 - x)^3$

55. $h(x) = \sqrt[3]{x^2 - 4}$ **56.** $h(x) = \sqrt{9 - x}$

57. $h(x) = \dfrac{1}{x + 2}$ **58.** $h(x) = \dfrac{4}{(5x + 2)^2}$

59. $h(x) = \dfrac{-x^2 + 3}{4 - x^2}$ **60.** $h(x) = \dfrac{27x^3 + 6x}{10 - 27x^3}$

61. STOPPING DISTANCE The research and development department of an automobile manufacturer has determined that when a driver is required to stop quickly to avoid an accident, the distance (in feet) the car travels during the driver's reaction time is given by $R(x) = \frac{3}{4}x$, where x is the speed of the car in miles per hour. The distance (in feet) traveled while the driver is braking is given by $B(x) = \frac{1}{15}x^2$.

  (a) Find the function that represents the total stopping distance T.

  (b) Graph the functions R, B, and T on the same set of coordinate axes for $0 \le x \le 60$.

  (c) Which function contributes most to the magnitude of the sum at higher speeds? Explain.

62. SALES From 2003 through 2008, the sales R_1 (in thousands of dollars) for one of two restaurants owned by the same parent company can be modeled by

$$R_1 = 480 - 8t - 0.8t^2, \quad t = 3, 4, 5, 6, 7, 8$$

where $t = 3$ represents 2003. During the same six-year period, the sales R_2 (in thousands of dollars) for the second restaurant can be modeled by

$$R_2 = 254 + 0.78t, \quad t = 3, 4, 5, 6, 7, 8.$$

  (a) Write a function R_3 that represents the total sales of the two restaurants owned by the same parent company.

  (b) Use a graphing utility to graph R_1, R_2, and R_3 in the same viewing window.

63. VITAL STATISTICS Let $b(t)$ be the number of births in the United States in year t, and let $d(t)$ represent the number of deaths in the United States in year t, where $t = 0$ corresponds to 2000.

  (a) If $p(t)$ is the population of the United States in year t, find the function $c(t)$ that represents the percent change in the population of the United States.

  (b) Interpret the value of $c(5)$.

64. PETS Let $d(t)$ be the number of dogs in the United States in year t, and let $c(t)$ be the number of cats in the United States in year t, where $t = 0$ corresponds to 2000.

  (a) Find the function $p(t)$ that represents the total number of dogs and cats in the United States.

  (b) Interpret the value of $p(5)$.

  (c) Let $n(t)$ represent the population of the United States in year t, where $t = 0$ corresponds to 2000. Find and interpret

$$h(t) = \frac{p(t)}{n(t)}.$$

65. MILITARY PERSONNEL The total numbers of Navy personnel N (in thousands) and Marines personnel M (in thousands) from 2000 through 2007 can be approximated by the models

$$N(t) = 0.192t^3 - 3.88t^2 + 12.9t + 372$$

and

$$M(t) = 0.035t^3 - 0.23t^2 + 1.7t + 172$$

where t represents the year, with $t = 0$ corresponding to 2000. (Source: Department of Defense)

  (a) Find and interpret $(N + M)(t)$. Evaluate this function for $t = 0$, 6, and 12.

  (b) Find and interpret $(N - M)(t)$ Evaluate this function for $t = 0$, 6, and 12.

66. SPORTS The numbers of people playing tennis T (in millions) in the United States from 2000 through 2007 can be approximated by the function

$$T(t) = 0.0233t^4 - 0.3408t^3 + 1.556t^2 - 1.86t + 22.8$$

and the U.S. population P (in millions) from 2000 through 2007 can be approximated by the function $P(t) = 2.78t + 282.5$, where t represents the year, with $t = 0$ corresponding to 2000. (Source: Tennis Industry Association, U.S. Census Bureau)

(a) Find and interpret $h(t) = \dfrac{T(t)}{P(t)}$.

(b) Evaluate the function in part (a) for $t = 0$, 3, and 6.

BIRTHS AND DEATHS In Exercises 67 and 68, use the table, which shows the total numbers of births B (in thousands) and deaths D (in thousands) in the United States from 1990 through 2006. (Source: U.S. Census Bureau)

Year, t	Births, B	Deaths, D
1990	4158	2148
1991	4111	2170
1992	4065	2176
1993	4000	2269
1994	3953	2279
1995	3900	2312
1996	3891	2315
1997	3881	2314
1998	3942	2337
1999	3959	2391
2000	4059	2403
2001	4026	2416
2002	4022	2443
2003	4090	2448
2004	4112	2398
2005	4138	2448
2006	4266	2426

The models for these data are

$$B(t) = -0.197t^3 + 8.96t^2 - 90.0t + 4180$$

and

$$D(t) = -1.21t^2 + 38.0t + 2137$$

where t represents the year, with $t = 0$ corresponding to 1990.

67. Find and interpret $(B - D)(t)$.

68. Evaluate $B(t)$, $D(t)$, and $(B - D)(t)$ for the years 2010 and 2012. What does each function value represent?

69. GRAPHICAL REASONING An electronically controlled thermostat in a home is programmed to lower the temperature automatically during the night. The temperature in the house T (in degrees Fahrenheit) is given in terms of t, the time in hours on a 24-hour clock (see figure).

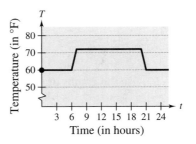

(a) Explain why T is a function of t.

(b) Approximate $T(4)$ and $T(15)$.

(c) The thermostat is reprogrammed to produce a temperature H for which $H(t) = T(t - 1)$. How does this change the temperature?

(d) The thermostat is reprogrammed to produce a temperature H for which $H(t) = T(t) - 1$. How does this change the temperature?

(e) Write a piecewise-defined function that represents the graph.

70. GEOMETRY A square concrete foundation is prepared as a base for a cylindrical tank (see figure).

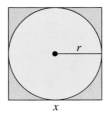

(a) Write the radius r of the tank as a function of the length x of the sides of the square.

(b) Write the area A of the circular base of the tank as a function of the radius r.

(c) Find and interpret $(A \circ r)(x)$.

71. RIPPLES A pebble is dropped into a calm pond, causing ripples in the form of concentric circles. The radius r (in feet) of the outer ripple is $r(t) = 0.6t$, where t is the time in seconds after the pebble strikes the water. The area A of the circle is given by the function $A(r) = \pi r^2$. Find and interpret $(A \circ r)(t)$.

72. POLLUTION The spread of a contaminant is increasing in a circular pattern on the surface of a lake. The radius of the contaminant can be modeled by $r(t) = 5.25\sqrt{t}$, where r is the radius in meters and t is the time in hours since contamination.

(a) Find a function that gives the area A of the circular leak in terms of the time t since the spread began.

(b) Find the size of the contaminated area after 36 hours.

(c) Find when the size of the contaminated area is 6250 square meters.

73. BACTERIA COUNT The number N of bacteria in a refrigerated food is given by

$$N(T) = 10T^2 - 20T + 600, \quad 1 \le T \le 20$$

where T is the temperature of the food in degrees Celsius. When the food is removed from refrigeration, the temperature of the food is given by

$$T(t) = 3t + 2, \quad 0 \le t \le 6$$

where t is the time in hours.

(a) Find the composition $N(T(t))$ and interpret its meaning in context.

(b) Find the bacteria count after 0.5 hour.

(c) Find the time when the bacteria count reaches 1500.

74. COST The weekly cost C of producing x units in a manufacturing process is given by $C(x) = 60x + 750$. The number of units x produced in t hours is given by $x(t) = 50t$.

(a) Find and interpret $(C \circ x)(t)$.

(b) Find the cost of the units produced in 4 hours.

(c) Find the time that must elapse in order for the cost to increase to $15,000.

75. SALARY You are a sales representative for a clothing manufacturer. You are paid an annual salary, plus a bonus of 3% of your sales over $500,000. Consider the two functions given by $f(x) = x - 500,000$ and $g(x) = 0.03x$. If x is greater than $500,000, which of the following represents your bonus? Explain your reasoning.

(a) $f(g(x))$ (b) $g(f(x))$

76. CONSUMER AWARENESS The suggested retail price of a new hybrid car is p dollars. The dealership advertises a factory rebate of $2000 and a 10% discount.

(a) Write a function R in terms of p giving the cost of the hybrid car after receiving the rebate from the factory.

(b) Write a function S in terms of p giving the cost of the hybrid car after receiving the dealership discount.

(c) Form the composite functions $(R \circ S)(p)$ and $(S \circ R)(p)$ and interpret each.

(d) Find $(R \circ S)(20,500)$ and $(S \circ R)(20,500)$. Which yields the lower cost for the hybrid car? Explain.

EXPLORATION

TRUE OR FALSE? In Exercises 77 and 78, determine whether the statement is true or false. Justify your answer.

77. If $f(x) = x + 1$ and $g(x) = 6x$, then

$$(f \circ g)(x) = (g \circ f)(x).$$

78. If you are given two functions $f(x)$ and $g(x)$, you can calculate $(f \circ g)(x)$ if and only if the range of g is a subset of the domain of f.

In Exercises 79 and 80, three siblings are of three different ages. The oldest is twice the age of the middle sibling, and the middle sibling is six years older than one-half the age of the youngest.

79. (a) Write a composite function that gives the oldest sibling's age in terms of the youngest. Explain how you arrived at your answer.

(b) If the oldest sibling is 16 years old, find the ages of the other two siblings.

80. (a) Write a composite function that gives the youngest sibling's age in terms of the oldest. Explain how you arrived at your answer.

(b) If the youngest sibling is two years old, find the ages of the other two siblings.

81. PROOF Prove that the product of two odd functions is an even function, and that the product of two even functions is an even function.

82. CONJECTURE Use examples to hypothesize whether the product of an odd function and an even function is even or odd. Then prove your hypothesis.

83. PROOF

(a) Given a function f, prove that $g(x)$ is even and $h(x)$ is odd, where $g(x) = \frac{1}{2}[f(x) + f(-x)]$ and $h(x) = \frac{1}{2}[f(x) - f(-x)]$.

(b) Use the result of part (a) to prove that any function can be written as a sum of even and odd functions. [*Hint:* Add the two equations in part (a).]

(c) Use the result of part (b) to write each function as a sum of even and odd functions.

$$f(x) = x^2 - 2x + 1, \quad k(x) = \frac{1}{x + 1}$$

84. CAPSTONE Consider the functions $f(x) = x^2$ and $g(x) = \sqrt{x}$.

(a) Find f/g and its domain.

(b) Find $f \circ g$ and $g \circ f$. Find the domain of each composite function. Are they the same? Explain.

Sean Gallup/Getty Images

2.7 INVERSE FUNCTIONS

What you should learn

- Find inverse functions informally and verify that two functions are inverse functions of each other.
- Use graphs of functions to determine whether functions have inverse functions.
- Use the Horizontal Line Test to determine if functions are one-to-one.
- Find inverse functions algebraically.

Why you should learn it

Inverse functions can be used to model and solve real-life problems. For instance, in Exercise 99 on page 246, an inverse function can be used to determine the year in which there was a given dollar amount of sales of LCD televisions in the United States.

Inverse Functions

Recall from Section 2.2 that a function can be represented by a set of ordered pairs. For instance, the function $f(x) = x + 4$ from the set $A = \{1, 2, 3, 4\}$ to the set $B = \{5, 6, 7, 8\}$ can be written as follows.

$$f(x) = x + 4: \{(1, 5), (2, 6), (3, 7), (4, 8)\}$$

In this case, by interchanging the first and second coordinates of each of these ordered pairs, you can form the **inverse function** of f, which is denoted by f^{-1}. It is a function from the set B to the set A, and can be written as follows.

$$f^{-1}(x) = x - 4: \{(5, 1), (6, 2), (7, 3), (8, 4)\}$$

Note that the domain of f is equal to the range of f^{-1}, and vice versa, as shown in Figure 2.66. Also note that the functions f and f^{-1} have the effect of "undoing" each other. In other words, when you form the composition of f with f^{-1} or the composition of f^{-1} with f, you obtain the identity function.

$$f(\quad) = f(\quad) = (\quad) + 4 = x$$

$$f^{-1}(\quad) = f^{-1}(\quad) = (\quad) - 4 = x$$

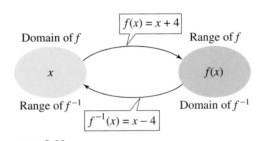

FIGURE 2.66

Example 1   Finding Inverse Functions Informally

Find the inverse function of $f(x) = 4x$. Then verify that both $f(f^{-1}(x))$ and $f^{-1}(f(x))$ are equal to the identity function.

Solution

The function f *multiplies* each input by 4. To "undo" this function, you need to *divide* each input by 4. So, the inverse function of $f(x) = 4x$ is

$$f^{-1}(x) = \frac{x}{4}.$$

You can verify that both $f(f^{-1}(x)) = x$ and $f^{-1}(f(x)) = x$ as follows.

$$f(\quad) = f\left(\quad\right) = 4\left(\quad\right) = x \qquad f^{-1}(\quad) = f^{-1}(\quad) = \frac{\quad}{4} = x$$

CHECK Point   Now try Exercise 7.

Definition of Inverse Function

Let f and g be two functions such that

$$f(g(x)) = x \qquad \text{for every } x \text{ in the domain of } g$$

and

$$g(f(x)) = x \qquad \text{for every } x \text{ in the domain of } f.$$

Under these conditions, the function g is the **inverse function** of the function f. The function g is denoted by f^{-1} (read "f-inverse"). So,

$$f(f^{-1}(x)) = x \qquad \text{and} \qquad f^{-1}(f(x)) = x.$$

The domain of f must be equal to the range of f^{-1}, and the range of f must be equal to the domain of f^{-1}.

Do not be confused by the use of -1 to denote the inverse function f^{-1}. In this text, whenever f^{-1} is written, it *always* refers to the inverse function of the function f and *not* to the reciprocal of $f(x)$.

If the function g is the inverse function of the function f, it must also be true that the function f is the inverse function of the function g. For this reason, you can say that the functions f and g are *inverse functions of each other*.

Example 2   Verifying Inverse Functions

Which of the functions is the inverse function of $f(x) = \dfrac{5}{x-2}$?

$$g(x) = \frac{x-2}{5} \qquad\qquad h(x) = \frac{5}{x} + 2$$

Solution

By forming the composition of f with g, you have

$$f(\quad) = f\left(\quad\right) = \cfrac{5}{\left(\quad\right) - 2} = \frac{25}{x - 12} \neq x.$$

Because this composition is not equal to the identity function x, it follows that g *is not* the inverse function of f. By forming the composition of f with h, you have

$$f(\quad) = f\left(\quad\right) = \cfrac{5}{\left(\quad\right) - 2} = \cfrac{5}{\left(\dfrac{5}{x}\right)} = x.$$

So, it appears that h *is* the inverse function of f. You can confirm this by showing that the composition of h with f is also equal to the identity function, as shown below.

$$h(\quad) = h\left(\quad\right) = \cfrac{5}{\left(\quad\right)} + 2 = x - 2 + 2 = x$$

CHECK*Point*   Now try Exercise 19.

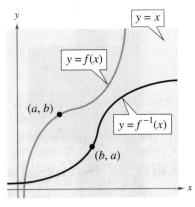

FIGURE 2.67

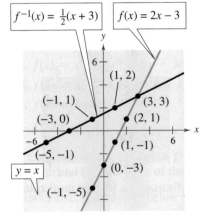

FIGURE 2.68

The Graph of an Inverse Function

The graphs of a function f and its inverse function f^{-1} are related to each other in the following way. If the point (a, b) lies on the graph of f, then the point (b, a) must lie on the graph of f^{-1}, and vice versa. This means that the graph of f^{-1} is a *reflection* of the graph of f in the line $y = x$, as shown in Figure 2.67.

Example 3 Finding Inverse Functions Graphically

Sketch the graphs of the inverse functions $f(x) = 2x - 3$ and $f^{-1}(x) = \frac{1}{2}(x + 3)$ on the same rectangular coordinate system and show that the graphs are reflections of each other in the line $y = x$.

Solution

The graphs of f and f^{-1} are shown in Figure 2.68. It appears that the graphs are reflections of each other in the line $y = x$. You can further verify this reflective property by testing a few points on each graph. Note in the following list that if the point (a, b) is on the graph of f, the point (b, a) is on the graph of f^{-1}.

Graph of $f(x) = 2x - 3$	*Graph of* $f^{-1}(x) = \frac{1}{2}(x + 3)$
$(-1, -5)$	$(-5, -1)$
$(0, -3)$	$(-3, 0)$
$(1, -1)$	$(-1, 1)$
$(2, 1)$	$(1, 2)$
$(3, 3)$	$(3, 3)$

CHECK *Point* Now try Exercise 25.

Example 4 Finding Inverse Functions Graphically

Sketch the graphs of the inverse functions $f(x) = x^2$ $(x \geq 0)$ and $f^{-1}(x) = \sqrt{x}$ on the same rectangular coordinate system and show that the graphs are reflections of each other in the line $y = x$.

Solution

The graphs of f and f^{-1} are shown in Figure 2.69. It appears that the graphs are reflections of each other in the line $y = x$. You can further verify this reflective property by testing a few points on each graph. Note in the following list that if the point (a, b) is on the graph of f, the point (b, a) is on the graph of f^{-1}.

Graph of $f(x) = x^2$, $x \geq 0$	*Graph of* $f^{-1}(x) = \sqrt{x}$
$(0, 0)$	$(0, 0)$
$(1, 1)$	$(1, 1)$
$(2, 4)$	$(4, 2)$
$(3, 9)$	$(9, 3)$

Try showing that $f(f^{-1}(x)) = x$ and $f^{-1}(f(x)) = x$.

CHECK *Point* Now try Exercise 27.

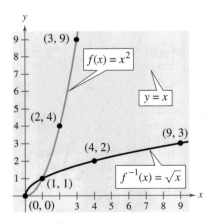

FIGURE 2.69

One-to-One Functions

The reflective property of the graphs of inverse functions gives you a nice *geometric* test for determining whether a function has an inverse function. This test is called the **Horizontal Line Test** for inverse functions.

> ### Horizontal Line Test for Inverse Functions
>
> A function f has an inverse function if and only if no *horizontal* line intersects the graph of f at more than one point.

If no horizontal line intersects the graph of f at more than one point, then no y-value is matched with more than one x-value. This is the essential characteristic of what are called **one-to-one functions.**

> ### One-to-One Functions
>
> A function f is **one-to-one** if each value of the dependent variable corresponds to exactly one value of the independent variable. A function f has an inverse function if and only if f is one-to-one.

Consider the function given by $f(x) = x^2$. The table on the left is a table of values for $f(x) = x^2$. The table of values on the right is made up by interchanging the columns of the first table. The table on the right does not represent a function because the input $x = 4$ is matched with two different outputs: $y = -2$ and $y = 2$. So, $f(x) = x^2$ is not one-to-one and does not have an inverse function.

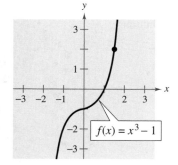

FIGURE 2.70

x	$f(x) = x^2$
-2	4
-1	1
0	0
1	1
2	4
3	9

x	y
1	-1
0	0
1	1
9	3

FIGURE 2.71

| **Example 5**   **Applying the Horizontal Line Test**

a. The graph of the function given by $f(x) = x^3 - 1$ is shown in Figure 2.70. Because no horizontal line intersects the graph of f at more than one point, you can conclude that f *is* a one-to-one function and *does* have an inverse function.

b. The graph of the function given by $f(x) = x^2 - 1$ is shown in Figure 2.71. Because it is possible to find a horizontal line that intersects the graph of f at more than one point, you can conclude that f *is not* a one-to-one function and *does not* have an inverse function.

CHECK *Point*   Now try Exercise 39.

Finding Inverse Functions Algebraically

For simple functions (such as the one in Example 1), you can find inverse functions by inspection. For more complicated functions, however, it is best to use the following guidelines. The key step in these guidelines is Step 3—interchanging the roles of x and y. This step corresponds to the fact that inverse functions have ordered pairs with the coordinates reversed.

! WARNING / CAUTION

Note what happens when you try to find the inverse function of a function that is not one-to-one.

$$f(x) = x^2 + 1$$

$$y = x^2 + 1$$

$$x = y^2 + 1$$

$$x - 1 = y^2$$

$$y = \pm\sqrt{x - 1}$$

You obtain two y-values for each x.

Finding an Inverse Function

1. Use the Horizontal Line Test to decide whether f has an inverse function.

2. In the equation for $f(x)$, replace $f(x)$ by y.

3. Interchange the roles of x and y, and solve for y.

4. Replace y by $f^{-1}(x)$ in the new equation.

5. Verify that f and f^{-1} are inverse functions of each other by showing that the domain of f is equal to the range of f^{-1}, the range of f is equal to the domain of f^{-1}, and $f(f^{-1}(x)) = x$ and $f^{-1}(f(x)) = x$.

| Example 6   Finding an Inverse Function Algebraically

Find the inverse function of

$$f(x) = \frac{5 - 3x}{2}.$$

Solution

The graph of f is a line, as shown in Figure 2.72. This graph passes the Horizontal Line Test. So, you know that f is one-to-one and has an inverse function.

$$f(x) = \frac{5 - 3x}{2}$$

$$y = \frac{5 - 3x}{2}$$

$$x = \frac{5 - 3y}{2}$$

$$2x = 5 - 3y$$

$$3y = 5 - 2x$$

$$y = \frac{5 - 2x}{3}$$

$$f^{-1}(x) = \frac{5 - 2x}{3}$$

Note that both f and f^{-1} have domains and ranges that consist of the entire set of real numbers. Check that $f(f^{-1}(x)) = x$ and $f^{-1}(f(x)) = x$.

CHECK*Point* ▶   Now try Exercise 63.

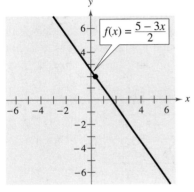

FIGURE **2.72**

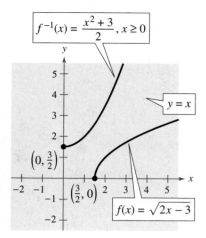

$$f^{-1}(x) = \frac{x^2 + 3}{2}, \; x \geq 0$$

$y = x$

$\left(0, \frac{3}{2}\right)$

$\left(\frac{3}{2}, 0\right)$

$f(x) = \sqrt{2x - 3}$

FIGURE 2.73

Example 7   Finding an Inverse Function

Find the inverse function of

$$f(x) = \sqrt{2x - 3}.$$

Solution

The graph of f is a curve, as shown in Figure 2.73. Because this graph passes the Horizontal Line Test, you know that f is one-to-one and has an inverse function.

$$f(x) = \sqrt{2x - 3}$$
$$= \sqrt{2x - 3}$$
$$x = \sqrt{2y - 3}$$
$$x^2 = 2y - 3$$
$$2y = x^2 + 3$$
$$y = \frac{x^2 + 3}{2}$$
$$= \frac{x^2 + 3}{2}, \quad x \geq 0$$

The graph of f^{-1} in Figure 2.73 is the reflection of the graph of f in the line $y = x$. Note that the range of f is the interval $[0, \infty)$, which implies that the domain of f^{-1} is the interval $[0, \infty)$. Moreover, the domain of f is the interval $\left[\frac{3}{2}, \infty\right)$, which implies that the range of f^{-1} is the interval $\left[\frac{3}{2}, \infty\right)$. Verify that $f(f^{-1}(x)) = x$ and $f^{-1}(f(x)) = x$.

CHECK **Point**   Now try Exercise 69.

CLASSROOM DISCUSSION

The Existence of an Inverse Function   Write a short paragraph describing why the following functions do or do not have inverse functions.

a. Let x represent the retail price of an item (in dollars), and let $f(x)$ represent the sales tax on the item. Assume that the sales tax is 6% of the retail price *and* that the sales tax is rounded to the nearest cent. Does this function have an inverse function? (*Hint:* Can you undo this function? For instance, if you know that the sales tax is $0.12, can you determine exactly what the retail price is?)

b. Let x represent the temperature in degrees Celsius, and let $f(x)$ represent the temperature in degrees Fahrenheit. Does this function have an inverse function? (*Hint:* The formula for converting from degrees Celsius to degrees Fahrenheit is $F = \frac{9}{5}C + 32$.)

2.7 EXERCISES

VOCABULARY: Fill in the blanks.

1. If the composite functions $f(g(x))$ and $g(f(x))$ both equal x, then the function g is the _____ function of f.
2. The inverse function of f is denoted by _____.
3. The domain of f is the _____ of f^{-1}, and the _____ of f^{-1} is the range of f.
4. The graphs of f and f^{-1} are reflections of each other in the line _____.
5. A function f is _____ if each value of the dependent variable corresponds to exactly one value of the independent variable.
6. A graphical test for the existence of an inverse function of f is called the _____ Line Test.

SKILLS AND APPLICATIONS

In Exercises 7–14, find the inverse function of f informally. Verify that $f(f^{-1}(x)) = x$ and $f^{-1}(f(x)) = x$.

7. $f(x) = 6x$　　　　**8.** $f(x) = \frac{1}{3}x$

9. $f(x) = x + 9$　　　**10.** $f(x) = x - 4$

11. $f(x) = 3x + 1$　　**12.** $f(x) = \dfrac{x - 1}{5}$

13. $f(x) = \sqrt[3]{x}$　　　**14.** $f(x) = x^5$

In Exercises 15–18, match the graph of the function with the graph of its inverse function. [The graphs of the inverse functions are labeled (a), (b), (c), and (d).]

(a)

(b)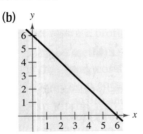

(c)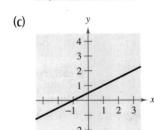

(d)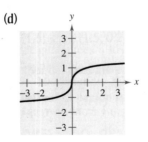

15.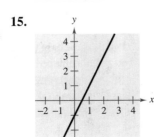

16.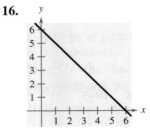

17.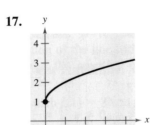

18.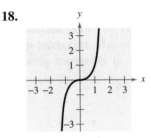

In Exercises 19–22, verify that f and g are inverse functions.

19. $f(x) = -\dfrac{7}{2}x - 3$,　$g(x) = -\dfrac{2x + 6}{7}$

20. $f(x) = \dfrac{x - 9}{4}$,　$g(x) = 4x + 9$

21. $f(x) = x^3 + 5$,　$g(x) = \sqrt[3]{x - 5}$

22. $f(x) = \dfrac{x^3}{2}$,　$g(x) = \sqrt[3]{2x}$

In Exercises 23–34, show that f and g are inverse functions (a) algebraically and (b) graphically.

23. $f(x) = 2x$,　$g(x) = \dfrac{x}{2}$

24. $f(x) = x - 5$,　$g(x) = x + 5$

25. $f(x) = 7x + 1$,　$g(x) = \dfrac{x - 1}{7}$

26. $f(x) = 3 - 4x$,　$g(x) = \dfrac{3 - x}{4}$

27. $f(x) = \dfrac{x^3}{8}$,　$g(x) = \sqrt[3]{8x}$

28. $f(x) = \dfrac{1}{x}$,　$g(x) = \dfrac{1}{x}$

29. $f(x) = \sqrt{x - 4}$,　$g(x) = x^2 + 4$,　$x \geq 0$

30. $f(x) = 1 - x^3$,　$g(x) = \sqrt[3]{1 - x}$

31. $f(x) = 9 - x^2$,　$x \geq 0$,　$g(x) = \sqrt{9 - x}$,　$x \leq 9$

32. $f(x) = \dfrac{1}{1 + x}$, $x \geq 0$, $g(x) = \dfrac{1 - x}{x}$, $0 < x \leq 1$

33. $f(x) = \dfrac{x - 1}{x + 5}$, $g(x) = -\dfrac{5x + 1}{x - 1}$

34. $f(x) = \dfrac{x + 3}{x - 2}$, $g(x) = \dfrac{2x + 3}{x - 1}$

In Exercises 35 and 36, does the function have an inverse function?

35.

x	-1	0	1	2	3	4
$f(x)$	-2	1	2	1	-2	-6

36.

x	-3	-2	-1	0	2	3
$f(x)$	10	6	4	1	-3	-10

In Exercises 37 and 38, use the table of values for $y = f(x)$ to complete a table for $y = f^{-1}(x)$.

37.

x	-2	-1	0	1	2	3
$f(x)$	-2	0	2	4	6	8

38.

x	-3	-2	-1	0	1	2
$f(x)$	-10	-7	-4	-1	2	5

In Exercises 39–42, does the function have an inverse function?

39.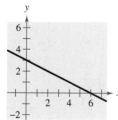

40.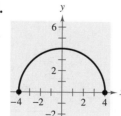

41.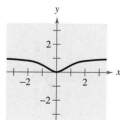

42.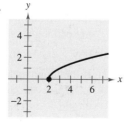

In Exercises 43–48, use a graphing utility to graph the function, and use the Horizontal Line Test to determine whether the function is one-to-one and so has an inverse function.

43. $g(x) = \dfrac{4 - x}{6}$

44. $f(x) = 10$

45. $h(x) = |x + 4| - |x - 4|$

46. $g(x) = (x + 5)^3$

47. $f(x) = -2x\sqrt{16 - x^2}$

48. $f(x) = \frac{1}{8}(x + 2)^2 - 1$

In Exercises 49–62, (a) find the inverse function of f, (b) graph both f and f^{-1} on the same set of coordinate axes, (c) describe the relationship between the graphs of f and f^{-1}, and (d) state the domain and range of f and f^{-1}.

49. $f(x) = 2x - 3$ **50.** $f(x) = 3x + 1$

51. $f(x) = x^5 - 2$ **52.** $f(x) = x^3 + 1$

53. $f(x) = \sqrt{4 - x^2}$, $0 \leq x \leq 2$

54. $f(x) = x^2 - 2$, $x \leq 0$

55. $f(x) = \dfrac{4}{x}$ **56.** $f(x) = -\dfrac{2}{x}$

57. $f(x) = \dfrac{x + 1}{x - 2}$ **58.** $f(x) = \dfrac{x - 3}{x + 2}$

59. $f(x) = \sqrt[3]{x - 1}$ **60.** $f(x) = x^{3/5}$

61. $f(x) = \dfrac{6x + 4}{4x + 5}$ **62.** $f(x) = \dfrac{8x - 4}{2x + 6}$

In Exercises 63–76, determine whether the function has an inverse function. If it does, find the inverse function.

63. $f(x) = x^4$ **64.** $f(x) = \dfrac{1}{x^2}$

65. $g(x) = \dfrac{x}{8}$ **66.** $f(x) = 3x + 5$

67. $p(x) = -4$ **68.** $f(x) = \dfrac{3x + 4}{5}$

69. $f(x) = (x + 3)^2$, $x \geq -3$

70. $q(x) = (x - 5)^2$

71. $f(x) = \begin{cases} x + 3, & x < 0 \\ 6 - x, & x \geq 0 \end{cases}$

72. $f(x) = \begin{cases} -x, & x \leq 0 \\ x^2 - 3x, & x > 0 \end{cases}$

73. $h(x) = -\dfrac{4}{x^2}$ **74.** $f(x) = |x - 2|$, $x \leq 2$

75. $f(x) = \sqrt{2x + 3}$ **76.** $f(x) = \sqrt{x - 2}$

THINK ABOUT IT In Exercises 77–86, restrict the domain of the function f so that the function is one-to-one and has an inverse function. Then find the inverse function f^{-1}. State the domains and ranges of f and f^{-1}. Explain your results. (There are many correct answers.)

77. $f(x) = (x - 2)^2$

78. $f(x) = 1 - x^4$

79. $f(x) = |x + 2|$

80. $f(x) = |x - 5|$

81. $f(x) = (x + 6)^2$

82. $f(x) = (x - 4)^2$

83. $f(x) = -2x^2 + 5$

84. $f(x) = \frac{1}{2}x^2 - 1$

85. $f(x) = |x - 4| + 1$

86. $f(x) = -|x - 1| - 2$

In Exercises 87–92, use the functions given by $f(x) = \frac{1}{8}x - 3$ and $g(x) = x^3$ to find the indicated value or function.

87. $(f^{-1} \circ g^{-1})(1)$

88. $(g^{-1} \circ f^{-1})(-3)$

89. $(f^{-1} \circ f^{-1})(6)$

90. $(g^{-1} \circ g^{-1})(-4)$

91. $(f \circ g)^{-1}$

92. $g^{-1} \circ f^{-1}$

In Exercises 93–96, use the functions given by $f(x) = x + 4$ and $g(x) = 2x - 5$ to find the specified function.

93. $g^{-1} \circ f^{-1}$

94. $f^{-1} \circ g^{-1}$

95. $(f \circ g)^{-1}$

96. $(g \circ f)^{-1}$

97. SHOE SIZES The table shows men's shoe sizes in the United States and the corresponding European shoe sizes. Let $y = f(x)$ represent the function that gives the men's European shoe size in terms of x, the men's U.S. size.

Men's U.S. shoe size	Men's European shoe size
8	41
9	42
10	43
11	45
12	46
13	47

(a) Is f one-to-one? Explain.

(b) Find $f(11)$.

(c) Find $f^{-1}(43)$, if possible.

(d) Find $f(f^{-1}(41))$.

(e) Find $f^{-1}(f(13))$.

98. SHOE SIZES The table shows women's shoe sizes in the United States and the corresponding European shoe sizes. Let $y = g(x)$ represent the function that gives the women's European shoe size in terms of x, the women's U.S. size.

Women's U.S. shoe size	Women's European shoe size
4	35
5	37
6	38
7	39
8	40
9	42

(a) Is g one-to-one? Explain.

(b) Find $g(6)$.

(c) Find $g^{-1}(42)$.

(d) Find $g(g^{-1}(39))$.

(e) Find $g^{-1}(g(5))$.

99. LCD TVS The sales S (in millions of dollars) of LCD televisions in the United States from 2001 through 2007 are shown in the table. The time (in years) is given by t, with $t = 1$ corresponding to 2001. (Source: Consumer Electronics Association)

Year, t	Sales, $S(t)$
1	62
2	246
3	664
4	1579
5	3258
6	8430
7	14,532

(a) Does S^{-1} exist?

(b) If S^{-1} exists, what does it represent in the context of the problem?

(c) If S^{-1} exists, find $S^{-1}(8430)$.

(d) If the table was extended to 2009 and if the sales of LCD televisions for that year was \$14,532 million, would S^{-1} exist? Explain.

100. POPULATION The projected populations P (in millions of people) in the United States for 2015 through 2040 are shown in the table. The time (in years) is given by t, with $t = 15$ corresponding to 2015. (Source: U.S. Census Bureau)

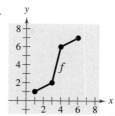

Year, t	Population, $P(t)$
15	325.5
20	341.4
25	357.5
30	373.5
35	389.5
40	405.7

(a) Does P^{-1} exist?

(b) If P^{-1} exists, what does it represent in the context of the problem?

(c) If P^{-1} exists, find $P^{-1}(357.5)$.

(d) If the table was extended to 2050 and if the projected population of the U.S. for that year was 373.5 million, would P^{-1} exist? Explain.

101. HOURLY WAGE Your wage is $10.00 per hour plus $0.75 for each unit produced per hour. So, your hourly wage y in terms of the number of units produced x is $y = 10 + 0.75x$.

(a) Find the inverse function. What does each variable represent in the inverse function?

(b) Determine the number of units produced when your hourly wage is $24.25.

102. DIESEL MECHANICS The function given by

$$y = 0.03x^2 + 245.50, \quad 0 < x < 100$$

approximates the exhaust temperature y in degrees Fahrenheit, where x is the percent load for a diesel engine.

(a) Find the inverse function. What does each variable represent in the inverse function?

(b) Use a graphing utility to graph the inverse function.

(c) The exhaust temperature of the engine must not exceed 500 degrees Fahrenheit. What is the percent load interval?

EXPLORATION

TRUE OR FALSE? In Exercises 103 and 104, determine whether the statement is true or false. Justify your answer.

103. If f is an even function, then f^{-1} exists.

104. If the inverse function of f exists and the graph of f has a y-intercept, then the y-intercept of f is an x-intercept of f^{-1}.

105. PROOF Prove that if f and g are one-to-one functions, then $(f \circ g)^{-1}(x) = (g^{-1} \circ f^{-1})(x)$.

106. PROOF Prove that if f is a one-to-one odd function, then f^{-1} is an odd function.

In Exercises 107 and 108, use the graph of the function f to create a table of values for the given points. Then create a second table that can be used to find f^{-1}, and sketch the graph of f^{-1} if possible.

107.

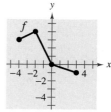

108.

In Exercises 109–112, determine if the situation could be represented by a one-to-one function. If so, write a statement that describes the inverse function.

109. The number of miles n a marathon runner has completed in terms of the time t in hours

110. The population p of South Carolina in terms of the year t from 1960 through 2008

111. The depth of the tide d at a beach in terms of the time t over a 24-hour period

112. The height h in inches of a human born in the year 2000 in terms of his or her age n in years.

113. THINK ABOUT IT The function given by $f(x) = k(2 - x - x^3)$ has an inverse function, and $f^{-1}(3) = -2$. Find k.

114. THINK ABOUT IT Consider the functions given by $f(x) = x + 2$ and $f^{-1}(x) = x - 2$. Evaluate $f(f^{-1}(x))$ and $f^{-1}(f(x))$ for the indicated values of x. What can you conclude about the functions?

x	-10	0	7	45
$f(f^{-1}(x))$				
$f^{-1}(f(x))$				

115. THINK ABOUT IT Restrict the domain of $f(x) = x^2 + 1$ to $x \geq 0$. Use a graphing utility to graph the function. Does the restricted function have an inverse function? Explain.

116. CAPSTONE Describe and correct the error.

Given $f(x) = \sqrt{x - 6}$, then $f^{-1}(x) = \dfrac{1}{\sqrt{x - 6}}$.

2 CHAPTER SUMMARY

What Did You Learn?	Explanation/Examples	Review Exercises
Section 2.1 Use slope to graph linear equations in two variables (p. 170).	**The Slope-Intercept Form of the Equation of a Line** The graph of the equation $y = mx + b$ is a line whose slope is m and whose y-intercept is $(0, b)$.	1–8
Find the slope of a line given two points on the line (p. 172).	The slope m of the nonvertical line through (x_1, y_1) and (x_2, y_2) is $m = (y_2 - y_1)/(x_2 - x_1)$, where $x_1 \neq x_2$.	9–12
Write linear equations in two variables (p. 174).	**Point-Slope Form of the Equation of a Line** The equation of the line with slope m passing through the point (x_1, y_1) is $y - y_1 = m(x - x_1)$.	13–20
Use slope to identify parallel and perpendicular lines (p. 175).	**Parallel lines:** Slopes are equal. **Perpendicular lines:** Slopes are negative reciprocals of each other.	21, 22
Use slope and linear equations in two variables to model and solve real-life problems (p. 176).	A linear equation in two variables can be used to describe the book value of exercise equipment in a given year. (See Example 7.)	23, 24
Section 2.2 Determine whether relations between two variables are functions (p. 185).	A function f from a set A (domain) to a set B (range) is a relation that assigns to each element x in the set A exactly one element y in the set B.	25–28
Use function notation, evaluate functions, and find domains (p. 187).	**Equation:** $f(x) = 5 - x^2$ $f(2)$: $f(2) = 5 - 2^2 = 1$ **Domain of $f(x) = 5 - x^2$:** All real numbers	29–36
Use functions to model and solve real-life problems (p. 191).	A function can be used to model the number of alternative-fueled vehicles in the United States. (See Example 10.)	37, 38
Evaluate difference quotients (p. 192).	**Difference quotient:** $[f(x + h) - f(x)]/h, h \neq 0$	39, 40
Section 2.3 Use the Vertical Line Test for functions (p. 201).	A graph represents a function if and only if no *vertical* line intersects the graph at more than one point.	41–44
Find the zeros of functions (p. 202).	**Zeros of $f(x)$:** x-values for which $f(x) = 0$	45–50
Determine intervals on which functions are increasing or decreasing (p. 203), find relative minimum and maximum values (p. 204), and find the average rate of change of a function (p. 205).	To determine whether a function is increasing, decreasing, or constant on an interval, evaluate the function for several values of x. The points at which the behavior of a function changes can help determine the relative minimum or relative maximum. The average rate of change between any two points is the slope of the line (secant line) through the two points.	51–60
Identify even and odd functions (p. 206).	**Even:** For each x in the domain of f, $f(-x) = f(x)$. **Odd:** For each x in the domain of f, $f(-x) = -f(x)$.	61–64
Section 2.4 Identify and graph linear (p. 212) and squaring functions (p. 213).	**Linear:** $f(x) = ax + b$ **Squaring:** $f(x) = x^2$ 	65–68

What Did You Learn?	Explanation/Examples	Review Exercises
Section 2.4 Identify and graph cubic, square root, reciprocal *(p. 214)*, step, and other piecewise-defined functions *(p. 215)*.	**Cubic:** $f(x) = x^3$ **Square Root:** $f(x) = \sqrt{x}$ $f(x) = x^3$ $(0,0)$ $f(x) = \sqrt{x}$ $(0,0)$ **Reciprocal:** $f(x) = 1/x$ **Step:** $f(x) = [\![x]\!]$ $f(x) = \dfrac{1}{x}$ $f(x) = [\![x]\!]$	69–78
Recognize graphs of parent functions *(p. 216)*.	Eight of the most commonly used functions in algebra are shown in Figure 2.48.	79, 80
Section 2.5 Use vertical and horizontal shifts *(p. 219)*, reflections *(p. 221)*, and nonrigid transformations *(p. 223)* to sketch graphs of functions.	**Vertical shifts:** $h(x) = f(x) + c$ or $h(x) = f(x) - c$ **Horizontal shifts:** $h(x) = f(x - c)$ or $h(x) = f(x + c)$ **Reflection in x-axis:** $h(x) = -f(x)$ **Reflection in y-axis:** $h(x) = f(-x)$ **Nonrigid transformations:** $h(x) = cf(x)$ or $h(x) = f(cx)$	81–94
Section 2.6 Add, subtract, multiply, and divide functions *(p. 229)*.	$(f + g)(x) = f(x) + g(x)$ $(f - g)(x) = f(x) - g(x)$ $(fg)(x) = f(x) \cdot g(x)$ $(f/g)(x) = f(x)/g(x), g(x) \neq 0$	95, 96
Find the composition of one function with another function *(p. 231)*.	The composition of the function f with the function g is $(f \circ g)(x) = f(g(x))$.	97–102
Use combinations and compositions of functions to model and solve real-life problems *(p. 233)*.	A composite function can be used to represent the number of bacteria in food as a function of the amount of time the food has been out of refrigeration. (See Example 8.)	103, 104
Section 2.7 Find inverse functions informally and verify that two functions are inverse functions of each other *(p. 238)*.	Let f and g be two functions such that $f(g(x)) = x$ for every x in the domain of g and $g(f(x)) = x$ for every x in the domain of f. Under these conditions, the function g is the inverse function of the function f.	105–108
Use graphs of functions to determine whether functions have inverse functions *(p. 240)*.	If the point (a, b) lies on the graph of f, then the point (b, a) must lie on the graph of f^{-1}, and vice versa. In short, f^{-1} is a reflection of f in the line $y = x$.	109, 110
Use the Horizontal Line Test to determine if functions are one-to-one *(p. 241)*.	**Horizontal Line Test for Inverse Functions** A function f has an inverse function if and only if no *horizontal* line intersects f at more than one point.	111–114
Find inverse functions algebraically *(p. 242)*.	To find inverse functions, replace $f(x)$ by y, interchange the roles of x and y, and solve for y. Replace y by $f^{-1}(x)$.	115–120

2 REVIEW EXERCISES

2.1 In Exercises 1–8, find the slope and *y*-intercept (if possible) of the equation of the line. Sketch the line.

1. $y = -2x - 7$

2. $y = 4x - 3$

3. $y = 6$

4. $x = -3$

5. $y = -\frac{5}{2}x - 1$

6. $y = \frac{5}{6}x + 5$

7. $-3x + y = 13$

8. $10x + 2y = 9$

In Exercises 9–12, plot the points and find the slope of the line passing through the pair of points.

9. $(6, 4), (-3, -4)$

10. $\left(\frac{3}{2}, 1\right), \left(5, \frac{5}{2}\right)$

11. $(-4.5, 6), (2.1, 3)$

12. $(-3, 2), (8, 2)$

In Exercises 13–16, find the slope-intercept form of the equation of the line that passes through the given point and has the indicated slope. Sketch the line.

Point	Slope
13. $(3, 0)$	$m = \frac{2}{3}$
14. $(-8, 5)$	$m = 0$
15. $(10, -3)$	$m = -\frac{1}{2}$
16. $(12, -6)$	m is undefined.

In Exercises 17–20, find the slope-intercept form of the equation of the line passing through the points.

17. $(0, 0), (0, 10)$

18. $(2, -1), (4, -1)$

19. $(-1, 0), (6, 2)$

20. $(11, -2), (6, -1)$

In Exercises 21 and 22, write the slope-intercept forms of the equations of the lines through the given point (a) parallel to the given line and (b) perpendicular to the given line.

Point	Line
21. $(3, -2)$	$5x - 4y = 8$
22. $(-8, 3)$	$2x + 3y = 5$

RATE OF CHANGE In Exercises 23 and 24, you are given the dollar value of a product in 2010 and the rate at which the value of the product is expected to change during the next 5 years. Use this information to write a linear equation that gives the dollar value *V* of the product in terms of the year *t*. (Let $t = 10$ represent 2010.)

2010 Value	Rate
23. $12,500	$850 decrease per year
24. $72.95	$5.15 increase per year

2.2 In Exercises 25–28, determine whether the equation represents *y* as a function of *x*.

25. $16x - y^4 = 0$

26. $2x - y - 3 = 0$

27. $y = \sqrt{1 - x}$

28. $|y| = x + 2$

In Exercises 29–32, evaluate the function at each specified value of the independent variable and simplify.

29. $f(x) = x^2 + 1$

    (a) $f(2)$ (b) $f(-4)$ (c) $f(t^2)$ (d) $f(t + 1)$

30. $g(x) = x^{4/3}$

    (a) $g(8)$ (b) $g(t + 1)$ (c) $g(-27)$ (d) $g(-x)$

31. $h(x) = \begin{cases} 2x + 1, & x \le -1 \\ x^2 + 2, & x > -1 \end{cases}$

    (a) $h(-2)$ (b) $h(-1)$ (c) $h(0)$ (d) $h(2)$

32. $f(x) = \dfrac{4}{x^2 + 1}$

    (a) $f(1)$ (b) $f(-5)$ (c) $f(-t)$ (d) $f(0)$

In Exercises 33–36, find the domain of the function. Verify your result with a graph.

33. $f(x) = \sqrt{25 - x^2}$

34. $g(s) = \dfrac{5s + 5}{3s - 9}$

35. $h(x) = \dfrac{x}{x^2 - x - 6}$

36. $h(t) = |t + 1|$

37. PHYSICS The velocity of a ball projected upward from ground level is given by $v(t) = -32t + 48$, where *t* is the time in seconds and *v* is the velocity in feet per second.

    (a) Find the velocity when $t = 1$.

    (b) Find the time when the ball reaches its maximum height. [*Hint:* Find the time when $v(t) = 0$.]

    (c) Find the velocity when $t = 2$.

38. MIXTURE PROBLEM From a full 50-liter container of a 40% concentration of acid, *x* liters is removed and replaced with 100% acid.

    (a) Write the amount of acid in the final mixture as a function of *x*.

    (b) Determine the domain and range of the function.

    (c) Determine *x* if the final mixture is 50% acid.

In Exercises 39 and 40, find the difference quotient and simplify your answer.

39. $f(x) = 2x^2 + 3x - 1$, $\dfrac{f(x + h) - f(x)}{h}$, $h \ne 0$

40. $f(x) = x^3 - 5x^2 + x$, $\dfrac{f(x + h) - f(x)}{h}$, $h \ne 0$

2.3 In Exercises 41–44, use the Vertical Line Test to determine whether y is a function of x. To print an enlarged copy of the graph, go to the website *www.mathgraphs.com*.

41. $y = (x - 3)^2$

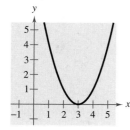

42. $y = -\frac{3}{5}x^3 - 2x + 1$

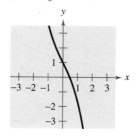

43. $x - 4 = y^2$

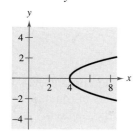

44. $x = -|4 - y|$

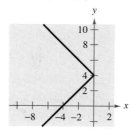

In Exercises 45–50, find the zeros of the function algebraically.

45. $f(x) = x^2 - 4x - 21$

46. $f(x) = 5x^2 + 4x - 1$

47. $f(x) = \dfrac{8x + 3}{11 - x}$

48. $f(x) = \sqrt{2x + 1}$

49. $f(x) = x^3 - x^2$

50. $f(x) = x^3 - x^2 - 25x + 25$

In Exercises 51 and 52, use a graphing utility to graph the function and visually determine the intervals over which the function is increasing, decreasing, or constant.

51. $f(x) = |x| + |x + 1|$

52. $f(x) = (x^2 - 4)^2$

In Exercises 53–56, use a graphing utility to graph the function and approximate any relative minimum or relative maximum values.

53. $f(x) = -x^2 + 2x + 1$

54. $f(x) = x^4 - 4x^2 - 2$

55. $f(x) = x^3 - 6x^4$

56. $f(x) = x^3 - 4x^2 - 1$

In Exercises 57–60, find the average rate of change of the function from x_1 to x_2.

Function	x-Values
57. $f(x) = -x^2 + 8x - 4$	$x_1 = 0, x_2 = 4$
58. $f(x) = x^3 + 12x - 2$	$x_1 = 0, x_2 = 4$
59. $f(x) = 2 - \sqrt{x + 1}$	$x_1 = 3, x_2 = 7$
60. $f(x) = 1 - \sqrt{x + 3}$	$x_1 = 1, x_2 = 6$

In Exercises 61–64, determine whether the function is even, odd, or neither.

61. $f(x) = x^5 + 4x - 7$

62. $f(x) = x^4 - 20x^2$

63. $f(x) = 2x\sqrt{x^2 + 3}$

64. $f(x) = \sqrt[5]{6x^2}$

2.4 In Exercises 65 and 66, write the linear function f such that it has the indicated function values. Then sketch the graph of the function.

65. $f(2) = -6, \quad f(-1) = 3$

66. $f(0) = -5, \quad f(4) = -8$

In Exercises 67–78, graph the function.

67. $f(x) = x^2 + 5$

68. $f(x) = 3 - x^2$

69. $g(x) = -3x^3$

70. $h(x) = x^3 - 2$

71. $f(x) = -\sqrt{x}$

72. $f(x) = \sqrt{x + 1}$

73. $g(x) = \dfrac{3}{x}$

74. $g(x) = \dfrac{1}{x + 5}$

75. $f(x) = [\![x]\!] + 2$

76. $g(x) = [\![x + 4]\!]$

77. $f(x) = \begin{cases} 5x - 3, & x \geq -1 \\ -4x + 5, & x < -1 \end{cases}$

78. $f(x) = \begin{cases} x^2 - 2, & x < -2 \\ 5, & -2 \leq x \leq 0 \\ 8x - 5, & x > 0 \end{cases}$

In Exercises 79 and 80, the figure shows the graph of a transformed parent function. Identify the parent function.

79.

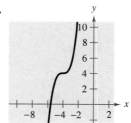

80.

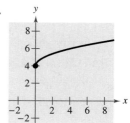

2.5 In Exercises 81–94, h is related to one of the parent functions described in this chapter. (a) Identify the parent function f. (b) Describe the sequence of transformations from f to h. (c) Sketch the graph of h. (d) Use function notation to write h in terms of f.

81. $h(x) = x^2 - 9$

82. $h(x) = (x - 2)^3 + 2$

83. $h(x) = -\sqrt{x} + 4$

84. $h(x) = |x + 3| - 5$

85. $h(x) = -(x + 2)^2 + 3$

86. $h(x) = \frac{1}{2}(x - 1)^2 - 2$

87. $h(x) = -[\![x]\!] + 6$

88. $h(x) = -\sqrt{x + 1} + 9$

89. $h(x) = -|-x + 4| + 6$

90. $h(x) = -(x + 1)^2 - 3$

91. $h(x) = 5[\![x - 9]\!]$

92. $h(x) = -\frac{1}{3}x^3$

93. $h(x) = -2\sqrt{x - 4}$

94. $h(x) = \frac{1}{2}|x| - 1$

2.6 In Exercises 95 and 96, find (a) $(f + g)(x)$, (b) $(f - g)(x)$, (c) $(fg)(x)$, and (d) $(f/g)(x)$. What is the domain of f/g?

95. $f(x) = x^2 + 3$, $g(x) = 2x - 1$
96. $f(x) = x^2 - 4$, $g(x) = \sqrt{3 - x}$

In Exercises 97–100, find (a) $f \circ g$ and (b) $g \circ f$. Find the domain of each function and each composite function.

97. $f(x) = \frac{1}{3}x - 3$, $g(x) = 3x + 1$

98. $f(x) = \frac{1}{x}$, $g(x) = 2x + 3$

99. $f(x) = x^3 - 4$, $g(x) = \sqrt[3]{x + 7}$
100. $f(x) = \sqrt{x + 1}$, $g(x) = x^2$

In Exercises 101 and 102, find two functions f and g such that $(f \circ g)(x) = h(x)$. (There are many correct answers.)

101. $h(x) = (1 - 2x)^3$ **102.** $h(x) = \sqrt[3]{x + 2}$

 103. PHONE EXPENDITURES The average annual expenditures (in dollars) for residential $r(t)$ and cellular $c(t)$ phone services from 2001 through 2006 can be approximated by the functions $r(t) = 27.5t + 705$ and $c(t) = 151.3t + 151$, where t represents the year, with $t = 1$ corresponding to 2001. (Source: Bureau of Labor Statistics)

(a) Find and interpret $(r + c)(t)$.

(b) Use a graphing utility to graph $r(t)$, $c(t)$, and $(r + c)(t)$ in the same viewing window.

(c) Find $(r + c)(13)$. Use the graph in part (b) to verify your result.

104. BACTERIA COUNT The number N of bacteria in a refrigerated food is given by

$$N(T) = 25T^2 - 50T + 300, \quad 2 \leq T \leq 20$$

where T is the temperature of the food in degrees Celsius. When the food is removed from refrigeration, the temperature of the food is given by

$$T(t) = 2t + 1, \quad 0 \leq t \leq 9$$

where t is the time in hours. (a) Find the composition $N(T(t))$ and interpret its meaning in context, and (b) find the time when the bacteria count reaches 750.

2.7 In Exercises 105–108, find the inverse function of f informally. Verify that $f(f^{-1}(x)) = x$ and $f^{-1}(f(x)) = x$.

105. $f(x) = 3x + 8$ **106.** $f(x) = \dfrac{x - 4}{5}$

107. $f(x) = x^3 - 1$ **108.** $f(x) = 2\sqrt[3]{x}$

In Exercises 109 and 110, determine whether the function has an inverse function.

109.
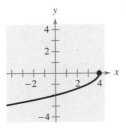

110.
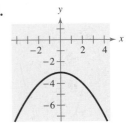

In Exercises 111–114, use a graphing utility to graph the function, and use the Horizontal Line Test to determine whether the function is one-to-one and so has an inverse function.

111. $f(x) = 4 - \frac{1}{3}x$ **112.** $f(x) = (x - 1)^2$

113. $h(t) = \dfrac{2}{t - 3}$ **114.** $g(x) = \sqrt{x + 6}$

In Exercises 115–118, (a) find the inverse function of f, (b) graph both f and f^{-1} on the same set of coordinate axes, (c) describe the relationship between the graphs of f and f^{-1}, and (d) state the domains and ranges of f and f^{-1}.

115. $f(x) = \frac{1}{2}x - 3$ **116.** $f(x) = 5x - 7$
117. $f(x) = \sqrt{x + 1}$ **118.** $f(x) = x^3 + 2$

In Exercises 119 and 120, restrict the domain of the function f to an interval over which the function is increasing and determine f^{-1} over that interval.

119. $f(x) = 2(x - 4)^2$ **120.** $f(x) = |x - 2|$

EXPLORATION

TRUE OR FALSE? In Exercises 121 and 122, determine whether the statement is true or false. Justify your answer.

121. Relative to the graph of $f(x) = \sqrt{x}$, the function given by $h(x) = -\sqrt{x + 9} - 13$ is shifted 9 units to the left and 13 units downward, then reflected in the x-axis.

122. If f and g are two inverse functions, then the domain of g is equal to the range of f.

123. WRITING Explain how to tell whether a relation between two variables is a function.

124. WRITING Explain the difference between the Vertical Line Test and the Horizontal Line Test.

125. WRITING Describe the basic characteristics of the cubic function. Describe the basic characteristics of $f(x) = x^3 + 1$.

2 CHAPTER TEST

See www.CalcChat.com for worked-out solutions to odd-numbered exercises.

Take this test as you would take a test in class. When you are finished, check your work against the answers given in the back of the book.

In Exercises 1 and 2, find the slope-intercept form of the equation of the line passing through the points. Then sketch the line.

1. $(4, -5), (-2, 7)$ **2.** $(3, 0.8), (7, -6)$

3. Find equations of the lines that pass through the point $(0, 4)$ and are (a) parallel to and (b) perpendicular to the line $5x + 2y = 3$.

In Exercises 4 and 5, evaluate the function at each specified value of the independent variable and simplify.

4. $f(x) = |x + 2| - 15$

    (a) $f(-8)$ (b) $f(14)$ (c) $f(x - 6)$

5. $f(x) = \dfrac{\sqrt{x + 9}}{x^2 - 81}$

    (a) $f(7)$ (b) $f(-5)$ (c) $f(x - 9)$

In Exercises 6 and 7, find the domain of the function.

6. $f(x) = |-x + 6| + 2$ **7.** $f(x) = 10 - \sqrt{3 - x}$

 In Exercises 8–10, (a) use a graphing utility to graph the function, (b) approximate the intervals over which the function is increasing, decreasing, or constant, and (c) determine whether the function is even, odd, or neither.

8. $f(x) = 2x^6 + 5x^4 - x^2$ **9.** $f(x) = 4x\sqrt{3 - x}$ **10.** $f(x) = |x + 5|$

11. Use a graphing utility to approximate any relative minimum or maximum values of $f(x) = -x^3 + 2x - 1$.

12. Find the average rate of change of $f(x) = -2x^2 + 5x - 3$ from $x_1 = 1$ to $x_2 = 3$.

13. Sketch the graph of $f(x) = \begin{cases} 3x + 7, & x \leq -3 \\ 4x^2 - 1, & x > -3 \end{cases}$.

In Exercises 14–16, (a) identify the parent function in the transformation, (b) describe the sequence of transformations from f to h, and (c) sketch the graph of h.

14. $h(x) = 3[\![x]\!]$ **15.** $h(x) = -\sqrt{x + 5} + 8$ **16.** $h(x) = -2(x - 5)^3 + 3$

In Exercises 17 and 18, find (a) $(f + g)(x)$, (b) $(f - g)(x)$, (c) $(fg)(x)$, (d) $(f/g)(x)$, (e) $(f \circ g)(x)$, and (f) $(g \circ f)(x)$.

17. $f(x) = 3x^2 - 7, \quad g(x) = -x^2 - 4x + 5$ **18.** $f(x) = \dfrac{1}{x}, \quad g(x) = 2\sqrt{x}$

In Exercises 19–21, determine whether the function has an inverse function, and if so, find the inverse function.

19. $f(x) = x^3 + 8$ **20.** $f(x) = |x^2 - 3| + 6$ **21.** $f(x) = 3x\sqrt{x}$

22. It costs a company \$58 to produce 6 units of a product and \$78 to produce 10 units. How much does it cost to produce 25 units, assuming that the cost function is linear?

2 CUMULATIVE TEST FOR CHAPTERS P–2

See www.CalcChat.com for worked-out solutions to odd-numbered exercises.

Take this test as you would take a test in class. When you are finished, check your work against the answers given in the back of the book.

In Exercises 1 and 2, simplify the expression.

1. $\dfrac{8x^2y^{-3}}{30x^{-1}y^2}$

2. $\sqrt{18x^3y^4}$

In Exercises 3–5, perform the operation and simplify the result.

3. $4x - [2x + 3(2 - x)]$

4. $(x - 2)(x^2 + x - 3)$

5. $\dfrac{2}{s + 3} - \dfrac{1}{s + 1}$

In Exercises 6–8, factor the expression completely.

6. $25 - (x - 2)^2$

7. $x - 5x^2 - 6x^3$

8. $54x^3 + 16$

In Exercises 9 and 10, write an expression for the area of the region.

9.

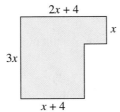

10.

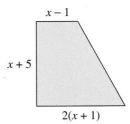

In Exercises 11–13, graph the equation without using a graphing utility.

11. $x - 3y + 12 = 0$

12. $y = x^2 - 9$

13. $y = \sqrt{4 - x}$

In Exercises 14–16, solve the equation and check your solution.

14. $3x - 5 = 6x + 8$

15. $-(x + 3) = 14(x - 6)$

16. $\dfrac{1}{x - 2} = \dfrac{10}{4x + 3}$

In Exercises 17–22, solve the equation using any convenient method and check your solutions. State the method you used.

17. $x^2 - 4x + 3 = 0$

18. $-2x^2 + 8x + 12 = 0$

19. $\frac{2}{3}x^2 = 24$

20. $3x^2 + 5x - 6 = 0$

21. $3x^2 + 9x + 1 = 0$

22. $\frac{1}{2}x^2 - 7 = 25$

In Exercises 23–28, solve the equation (if possible).

23. $x^4 + 12x^3 + 4x^2 + 48x = 0$

24. $8x^3 - 48x^2 + 72x = 0$

25. $x^{2/3} + 13 = 17$

26. $\sqrt{x + 10} = x - 2$

27. $|3(x - 4)| = 27$

28. $|x - 12| = -2$

In Exercises 29 and 30, determine whether each value of *x* is a solution of the inequality.

29. $4x + 2 > 7$

    (a) $x = -1$ (b) $x = \frac{1}{2}$

    (c) $x = \frac{3}{2}$ (d) $x = 2$

30. $|5x - 1| < 4$

    (a) $x = -1$ (b) $x = -\frac{1}{2}$

    (c) $x = 1$ (d) $x = 2$

In Exercises 31–34, solve the inequality and sketch the solution on the real number line.

31. $|x + 1| \le 6$

32. $|5 + 6x| > 3$

33. $5x^2 + 12x + 7 \ge 0$

34. $-x^2 + x + 4 < 0$

35. Find the slope-intercept form of the equation of the line passing through $\left(-\frac{1}{2}, 1\right)$ and $(3, 8)$.

36. Explain why the graph at the left does not represent *y* as a function of *x*.

37. Evaluate (if possible) the function given by $f(x) = \dfrac{x}{x - 2}$ for each value.

    (a) $f(6)$ (b) $f(2)$ (c) $f(s + 2)$

In Exercises 38–40, determine whether the function is even, odd, or neither.

38. $f(x) = 5 + \sqrt{4 - x}$ **39.** $f(x) = x^5 - x^3 + 2$ **40.** $f(x) = 2x^4 - 4$

41. Compare the graph of each function with the graph of $y = \sqrt[3]{x}$. (Note: It is not necessary to sketch the graphs.)

    (a) $r(x) = \frac{1}{2}\sqrt[3]{x}$ (b) $h(x) = \sqrt[3]{x} + 2$ (c) $g(x) = \sqrt[3]{x + 2}$

In Exercises 42 and 43, find (a) $(f + g)(x)$, (b) $(f - g)(x)$, (c) $(fg)(x)$, and (d) $(f/g)(x)$. What is the domain of *f/g*?

42. $f(x) = x - 4, \quad g(x) = 3x + 1$ **43.** $f(x) = \sqrt{x - 1}, \quad g(x) = x^2 + 1$

In Exercises 44 and 45, find (a) $f \circ g$ and (b) $g \circ f$. Find the domain of each composite function.

44. $f(x) = 2x^2, \quad g(x) = \sqrt{x + 6}$ **45.** $f(x) = x - 2, \quad g(x) = |x|$

46. Determine whether $h(x) = 3x - 4$ has an inverse function. If so, find the inverse function.

47. A group of *n* people decide to buy a $36,000 minibus. Each person will pay an equal share of the cost. If three additional people join the group, the cost per person will decrease by $1000. Find *n*.

48. For groups of 60 or more, a charter bus company determines the rate per person according to the formula

    Rate $= \$10.00 - \$0.05(n - 60), \quad n \ge 60$.

    (a) Write the revenue *R* as a function of *n*.

    (b) Use a graphing utility to graph the revenue function. Move the cursor along the function to estimate the number of passengers that will maximize the revenue.

49. The height of an object thrown vertically upward from a height of 8 feet at a velocity of 36 feet per second can be modeled by $s(t) = -16t^2 + 36t + 8$, where *s* is the height (in feet) and *t* is the time (in seconds). Find the average rate of change of the function from $t_1 = 0$ to $t_2 = 2$. Interpret your answer in the context of the problem.

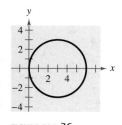

FIGURE FOR **36**

PROOFS IN MATHEMATICS

Biconditional Statements

Recall from the Proofs in Mathematics in Chapter 1 that a conditional statement is a statement of the form "if p, then q." A statement of the form "p if and only if q" is called a **biconditional statement.** A biconditional statement, denoted by

$$p \leftrightarrow q$$

is the conjunction of the conditional statement $p \rightarrow q$ and its converse $q \rightarrow p$.

A biconditional statement can be either true or false. To be true, *both* the conditional statement and its converse must be true.

Example 1   Analyzing a Biconditional Statement

Consider the statement $x = 3$ if and only if $x^2 = 9$.

a. Is the statement a biconditional statement?   **b.** Is the statement true?

Solution

a. The statement is a biconditional statement because it is of the form "p if and only if q."

b. The statement can be rewritten as the following conditional statement and its converse.

> *Conditional statement:* If $x = 3$, then $x^2 = 9$.
>
> *Converse:* If $x^2 = 9$, then $x = 3$.

The first of these statements is true, but the second is false because x could also equal -3. So, the biconditional statement is false.

Knowing how to use biconditional statements is an important tool for reasoning in mathematics.

Example 2   Analyzing a Biconditional Statement

Determine whether the biconditional statement is true or false. If it is false, provide a counterexample.

> A number is divisible by 5 if and only if it ends in 0.

Solution

The biconditional statement can be rewritten as the following conditional statement and its converse.

> *Conditional statement:* If a number is divisible by 5, then it ends in 0.
>
> *Converse:* If a number ends in 0, then it is divisible by 5.

The conditional statement is false. A counterexample is the number 15, which is divisible by 5 but does not end in 0.

PROBLEM SOLVING

This collection of thought-provoking and challenging exercises further explores and expands upon concepts learned in this chapter.

1. As a salesperson, you receive a monthly salary of $2000, plus a commission of 7% of sales. You are offered a new job at $2300 per month, plus a commission of 5% of sales.

   (a) Write a linear equation for your current monthly wage W_1 in terms of your monthly sales S.

   (b) Write a linear equation for the monthly wage W_2 of your new job offer in terms of the monthly sales S.

   (c) Use a graphing utility to graph both equations in the same viewing window. Find the point of intersection. What does it signify?

   (d) You think you can sell $20,000 per month. Should you change jobs? Explain.

2. For the numbers 2 through 9 on a telephone keypad (see figure), create two relations: one mapping numbers onto letters, and the other mapping letters onto numbers. Are both relations functions? Explain.

3. What can be said about the sum and difference of each of the following?

   (a) Two even functions   (b) Two odd functions

   (c) An odd function and an even function

4. The two functions given by

   $$f(x) = x \quad \text{and} \quad g(x) = -x$$

   are their own inverse functions. Graph each function and explain why this is true. Graph other linear functions that are their own inverse functions. Find a general formula for a family of linear functions that are their own inverse functions.

5. Prove that a function of the following form is even.

   $$y = a_{2n}x^{2n} + a_{2n-2}x^{2n-2} + \cdots + a_2 x^2 + a_0$$

6. A miniature golf professional is trying to make a hole-in-one on the miniature golf green shown. A coordinate plane is placed over the golf green. The golf ball is at the point $(2.5, 2)$ and the hole is at the point $(9.5, 2)$. The professional wants to bank the ball off the side wall of the green at the point (x, y). Find the coordinates of the point (x, y). Then write an equation for the path of the ball.

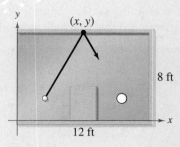

FIGURE FOR 6

7. At 2:00 P.M. on April 11, 1912, the *Titanic* left Cobh, Ireland, on her voyage to New York City. At 11:40 P.M. on April 14, the *Titanic* struck an iceberg and sank, having covered only about 2100 miles of the approximately 3400-mile trip.

   (a) What was the total duration of the voyage in hours?

   (b) What was the average speed in miles per hour?

   (c) Write a function relating the distance of the *Titanic* from New York City and the number of hours traveled. Find the domain and range of the function.

   (d) Graph the function from part (c).

8. Consider the function given by $f(x) = -x^2 + 4x - 3$. Find the average rate of change of the function from x_1 to x_2.

   (a) $x_1 = 1, x_2 = 2$ (b) $x_1 = 1, x_2 = 1.5$

   (c) $x_1 = 1, x_2 = 1.25$ (d) $x_1 = 1, x_2 = 1.125$

   (e) $x_1 = 1, x_2 = 1.0625$

   (f) Does the average rate of change seem to be approaching one value? If so, what value?

   (g) Find the equations of the secant lines through the points $(x_1, f(x_1))$ and $(x_2, f(x_2))$ for parts (a)–(e).

   (h) Find the equation of the line through the point $(1, f(1))$ using your answer from part (f) as the slope of the line.

9. Consider the functions given by $f(x) = 4x$ and $g(x) = x + 6$.

   (a) Find $(f \circ g)(x)$.   (b) Find $(f \circ g)^{-1}(x)$.

   (c) Find $f^{-1}(x)$ and $g^{-1}(x)$.

   (d) Find $(g^{-1} \circ f^{-1})(x)$ and compare the result with that of part (b).

   (e) Repeat parts (a) through (d) for $f(x) = x^3 + 1$ and $g(x) = 2x$.

   (f) Write two one-to-one functions f and g, and repeat parts (a) through (d) for these functions.

   (g) Make a conjecture about $(f \circ g)^{-1}(x)$ and $(g^{-1} \circ f^{-1})(x)$.

10. You are in a boat 2 miles from the nearest point on the coast. You are to travel to a point Q, 3 miles down the coast and 1 mile inland (see figure). You can row at 2 miles per hour and you can walk at 4 miles per hour.

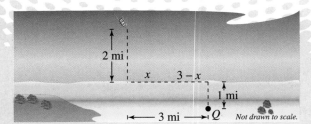

(a) Write the total time T of the trip as a function of x.

(b) Determine the domain of the function.

(c) Use a graphing utility to graph the function. Be sure to choose an appropriate viewing window.

(d) Use the *zoom* and *trace* features to find the value of x that minimizes T.

(e) Write a brief paragraph interpreting these values.

11. The **Heaviside function** $H(x)$ is widely used in engineering applications. (See figure.) To print an enlarged copy of the graph, go to the website *www.mathgraphs.com*.

$$H(x) = \begin{cases} 1, & x \geq 0 \\ 0, & x < 0 \end{cases}$$

Sketch the graph of each function by hand.

(a) $H(x) - 2$ (b) $H(x - 2)$ (c) $-H(x)$

(d) $H(-x)$ (e) $\frac{1}{2}H(x)$ (f) $-H(x - 2) + 2$

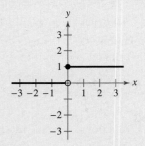

12. Let $f(x) = \dfrac{1}{1 - x}$.

(a) What are the domain and range of f?

(b) Find $f(f(x))$. What is the domain of this function?

(c) Find $f(f(f(x)))$. Is the graph a line? Why or why not?

13. Show that the Associative Property holds for compositions of functions—that is,

$$(f \circ (g \circ h))(x) = ((f \circ g) \circ h)(x).$$

14. Consider the graph of the function f shown in the figure. Use this graph to sketch the graph of each function. To print an enlarged copy of the graph, go to the website *www.mathgraphs.com*.

(a) $f(x + 1)$ (b) $f(x) + 1$ (c) $2f(x)$ (d) $f(-x)$

(e) $-f(x)$ (f) $|f(x)|$ (g) $f(|x|)$

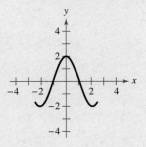

15. Use the graphs of f and f^{-1} to complete each table of function values.

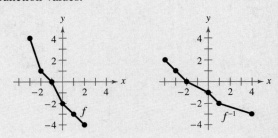

(a)

x	-4	-2	0	4
$(f(f^{-1}(x))$				

(b)

x	-3	-2	0	1
$(f + f^{-1})(x)$				

(c)

x	-3	-2	0	1
$(f \cdot f^{-1})(x)$				

(d)

x	-4	-3	0	4
$|f^{-1}(x)|$				

Polynomial Functions

3

In Mathematics

Functions defined by polynomial expressions are called polynomial functions.

In Real Life

Polynomial functions are used to model real-life situations, such as a company's revenue, the design of a propane tank, or the height of a thrown baseball. For instance, you can model the per capita cigarette consumption in the United States with a polynomial function. You can use the model to determine whether the addition of cigarette warnings affected consumption. (See Exercise 85, page 268.)

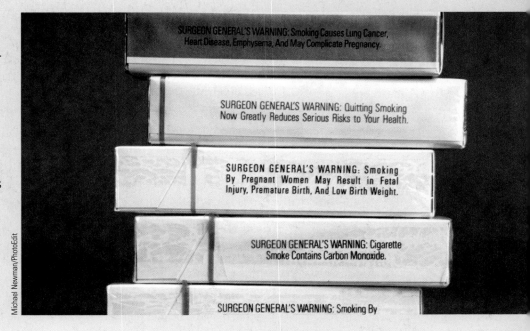

Michael Newman/PhotoEdit

IN CAREERS

There are many careers that use polynomial functions. Several are listed below.

3.1   QUADRATIC FUNCTIONS AND MODELS

What you should learn

- Analyze graphs of quadratic functions.
- Write quadratic functions in standard form and use the results to sketch graphs of functions.
- Find minimum and maximum values of quadratic functions in real-life applications.

Why you should learn it

Quadratic functions can be used to model data to analyze consumer behavior. For instance, in Exercise 79 on page 268, you will use a quadratic function to model the revenue earned from manufacturing handheld video games.

© John Henley/Corbis

The Graph of a Quadratic Function

In this and the next section, you will study the graphs of polynomial functions. In Section 2.4, you were introduced to the following basic functions.

$$f(x) = ax + b$$

$$f(x) = c$$

$$f(x) = x^2$$

These functions are examples of **polynomial functions.**

Definition of Polynomial Function

Let n be a nonnegative integer and let $a_n, a_{n-1}, \ldots, a_2, a_1, a_0$ be real numbers with $a_n \neq 0$. The function given by

$$f(x) = a_n x^n + a_{n-1} x^{n-1} + \cdots + a_2 x^2 + a_1 x + a_0$$

is called a **polynomial function of x with degree n.**

Polynomial functions are classified by degree. For instance, a constant function $f(x) = c$ with $c \neq 0$ has degree 0, and a linear function $f(x) = ax + b$ with $a \neq 0$ has degree 1. In this section, you will study second-degree polynomial functions, which are called **quadratic functions.**

For instance, each of the following functions is a quadratic function.

$$f(x) = x^2 + 6x + 2$$

$$g(x) = 2(x + 1)^2 - 3$$

$$h(x) = 9 + \tfrac{1}{4}x^2$$

$$k(x) = -3x^2 + 4$$

$$m(x) = (x - 2)(x + 1)$$

Note that the squaring function is a simple quadratic function that has degree 2.

Definition of Quadratic Function

Let a, b, and c be real numbers with $a \neq 0$. The function given by

$$f(x) = ax^2 + bx + c$$

is called a **quadratic function.**

The graph of a quadratic function is a special type of "U"-shaped curve called a **parabola.** Parabolas occur in many real-life applications—especially those involving reflective properties of satellite dishes and flashlight reflectors. You will study these properties in Section 4.3.

All parabolas are symmetric with respect to a line called the **axis of symmetry,** or simply the **axis** of the parabola. The point where the axis intersects the parabola is the **vertex** of the parabola, as shown in Figure 3.1. If the leading coefficient is positive, the graph of

$$f(x) = ax^2 + bx + c$$

is a parabola that opens upward. If the leading coefficient is negative, the graph of

$$f(x) = ax^2 + bx + c$$

is a parabola that opens downward.

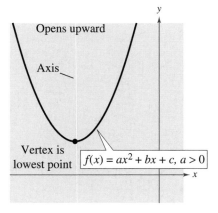

Leading coefficient is positive.
FIGURE 3.1

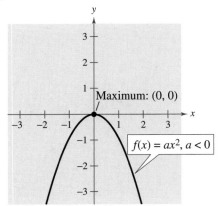

Leading coefficient is negative.

The simplest type of quadratic function is

$$f(x) = ax^2.$$

Its graph is a parabola whose vertex is $(0, 0)$. If $a > 0$, the vertex is the point with the *minimum* y-value on the graph, and if $a < 0$, the vertex is the point with the *maximum* y-value on the graph, as shown in Figure 3.2.

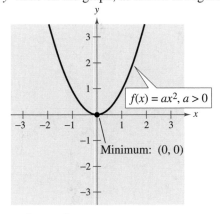

Leading coefficient is positive.
FIGURE 3.2

Leading coefficient is negative.

When sketching the graph of $f(x) = ax^2$, it is helpful to use the graph of $y = x^2$ as a reference, as discussed in Section 2.5.

Example 1 Sketching Graphs of Quadratic Functions

a. Compare the graphs of $y = x^2$ and $f(x) = \frac{1}{3}x^2$.

b. Compare the graphs of $y = x^2$ and $g(x) = 2x^2$.

Solution

a. Compared with $y = x^2$, each output of $f(x) = \frac{1}{3}x^2$ "shrinks" by a factor of $\frac{1}{3}$, creating the broader parabola shown in Figure 3.3.

b. Compared with $y = x^2$, each output of $g(x) = 2x^2$ "stretches" by a factor of 2, creating the narrower parabola shown in Figure 3.4.

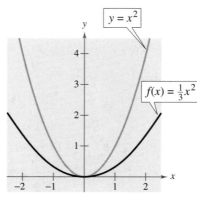

FIGURE **3.3**

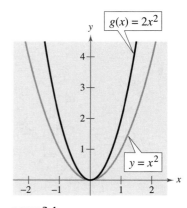

FIGURE **3.4**

CHECK Point Now try Exercise 13.

In Example 1, note that the coefficient a determines how widely the parabola given by $f(x) = ax^2$ opens. If $|a|$ is small, the parabola opens more widely than if $|a|$ is large.

Recall from Section 2.5 that the graphs of $y = f(x \pm c)$, $y = f(x) \pm c$, $y = f(-x)$, and $y = -f(x)$ are rigid transformations of the graph of $y = f(x)$. For instance, in Figure 3.5, notice how the graph of $y = x^2$ can be transformed to produce the graphs of $f(x) = -x^2 + 1$ and $g(x) = (x + 2)^2 - 3$.

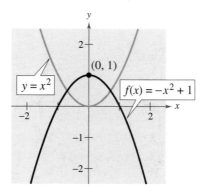

Reflection in *x*-axis followed by an upward shift of one unit

FIGURE **3.5**

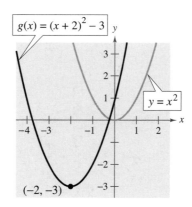

Left shift of two units followed by a downward shift of three units

Algebra Help

You can review the techniques for shifting, reflecting, and stretching graphs in Section 2.5.

The Standard Form of a Quadratic Function

The **standard form** of a quadratic function is $f(x) = a(x - h)^2 + k$. This form is especially convenient for sketching a parabola because it identifies the vertex of the parabola as (h, k).

> ### Standard Form of a Quadratic Function
>
> The quadratic function given by
>
> $$f(x) = a(x - h)^2 + k, \quad a \neq 0$$
>
> is in **standard form.** The graph of f is a parabola whose axis is the vertical line $x = h$ and whose vertex is the point (h, k). If $a > 0$, the parabola opens upward, and if $a < 0$, the parabola opens downward.

To graph a parabola, it is helpful to begin by writing the quadratic function in standard form using the process of completing the square, as illustrated in Example 2. In this example, notice that when completing the square, you *add and subtract* the square of half the coefficient of x within the parentheses instead of adding the value to each side of the equation as is done in Section 1.4.

Example 2   Graphing a Parabola in Standard Form

Sketch the graph of $f(x) = 2x^2 + 8x + 7$ and identify the vertex and the axis of the parabola.

Solution

Begin by writing the quadratic function in standard form. Notice that the first step in completing the square is to factor out any coefficient of x^2 that is not 1.

$$
\begin{aligned}
f(x) &= 2x^2 + 8x + 7 \\
&= 2(x^2 + 4x) + 7 \\
&= 2(x^2 + 4x \qquad\quad) + 7
\end{aligned}
$$

After adding and subtracting 4 within the parentheses, you must now regroup the terms to form a perfect square trinomial. The -4 can be removed from inside the parentheses; however, because of the 2 outside of the parentheses, you must multiply -4 by 2, as shown below.

$$
\begin{aligned}
f(x) &= 2(x^2 + 4x + 4) - 2(4) + 7 \\
&= 2(x^2 + 4x + 4) - 8 + 7 \\
&= 2(x + 2)^2 - 1
\end{aligned}
$$

From this form, you can see that the graph of f is a parabola that opens upward and has its vertex at $(-2, -1)$. This corresponds to a left shift of two units and a downward shift of one unit relative to the graph of $y = 2x^2$, as shown in Figure 3.6. In the figure, you can see that the axis of the parabola is the vertical line through the vertex, $x = -2$.

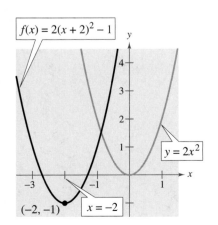

$f(x) = 2(x + 2)^2 - 1$

$y = 2x^2$

$(-2, -1)$ $x = -2$

FIGURE 3.6

CHECK Point Now try Exercise 19.

To find the x-intercepts of the graph of $f(x) = ax^2 + bx + c$, you must solve the equation $ax^2 + bx + c = 0$. If $ax^2 + bx + c$ does not factor, you can use the Quadratic Formula to find the x-intercepts. Remember, however, that a parabola may not have x-intercepts.

Example 3 Finding the Vertex and x-Intercepts of a Parabola

Sketch the graph of $f(x) = -x^2 + 6x - 8$ and identify the vertex and x-intercepts.

Solution

$$f(x) = -x^2 + 6x - 8$$
$$= -(x^2 - 6x) - 8$$
$$= -(x^2 - 6x \qquad) - 8$$

$$= -(x^2 - 6x + 9) - (-9) - 8$$
$$= -(x - 3)^2 + 1$$

From this form, you can see that f is a parabola that opens downward with vertex $(3, 1)$. The x-intercepts of the graph are determined as follows.

$$-(x^2 - 6x + 8) = 0$$
$$-(x - 2)(x - 4) = 0$$
$$x - 2 = 0 \implies x = 2$$
$$x - 4 = 0 \implies x = 4$$

So, the x-intercepts are $(2, 0)$ and $(4, 0)$, as shown in Figure 3.7.

CHECK Point Now try Exercise 25.

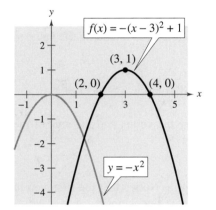

FIGURE **3.7**

Example 4 Writing the Equation of a Parabola

Write the standard form of the equation of the parabola whose vertex is $(1, 2)$ and that passes through the point $(3, -6)$.

Solution

Because the vertex of the parabola is at $(h, k) = (1, 2)$, the equation has the form

$$f(x) = a(x - \quad)^2 + \quad .$$

Because the parabola passes through the point $(3, -6)$, it follows that $f(3) = -6$. So,

$$f(x) = a(x - 1)^2 + 2$$
$$= a(\quad - 1)^2 + 2$$
$$-6 = 4a + 2$$
$$-8 = 4a$$
$$-2 = a.$$

The equation in standard form is $f(x) = -2(x - 1)^2 + 2$. The graph of f is shown in Figure 3.8.

CHECK Point Now try Exercise 47.

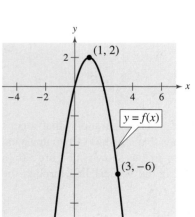

FIGURE **3.8**

Finding Minimum and Maximum Values

Many applications involve finding the maximum or minimum value of a quadratic function. By completing the square of the quadratic function $f(x) = ax^2 + bx + c$, you can rewrite the function in standard form (see Exercise 95).

$$f(x) = a\left(x + \frac{b}{2a}\right)^2 + \left(c - \frac{b^2}{4a}\right)$$

So, the vertex of the graph of f is $\left(-\dfrac{b}{2a}, f\left(-\dfrac{b}{2a}\right)\right)$, which implies the following.

Minimum and Maximum Values of Quadratic Functions

Consider the function $f(x) = ax^2 + bx + c$ with vertex $\left(-\dfrac{b}{2a},\ f\left(-\dfrac{b}{2a}\right)\right)$.

1. If $a > 0$, f has a *minimum* at $x = -\dfrac{b}{2a}$. The minimum value is $f\left(-\dfrac{b}{2a}\right)$.

2. If $a < 0$, f has a *maximum* at $x = -\dfrac{b}{2a}$. The maximum value is $f\left(-\dfrac{b}{2a}\right)$.

Example 5   The Maximum Height of a Baseball

A baseball is hit at a point 3 feet above the ground at a velocity of 100 feet per second and at an angle of 45° with respect to the ground. The path of the baseball is given by the function $f(x) = -0.0032x^2 + x + 3$, where $f(x)$ is the height of the baseball (in feet) and x is the horizontal distance from home plate (in feet). What is the maximum height reached by the baseball?

Algebraic Solution

For this quadratic function, you have

$$f(x) = ax^2 + bx + c$$
$$= -0.0032x^2 + x + 3$$

which implies that $a = -0.0032$ and $b = 1$. Because $a < 0$, the function has a maximum when $x = -b/(2a)$. So, you can conclude that the baseball reaches its maximum height when it is x feet from home plate, where x is

$$x = -\frac{b}{2a}$$

$$= -\frac{}{2(\quad)}$$

$$= 156.25 \text{ feet.}$$

At this distance, the maximum height is

$$f(\quad) = -0.0032(\quad) + \quad + 3$$
$$= 81.125 \text{ feet.}$$

CHECK**Point** Now try Exercise 75.

Graphical Solution

Use a graphing utility to graph

$$y = -0.0032x^2 + x + 3$$

so that you can see the important features of the parabola. Use the *maximum* feature (see Figure 3.9) or the *zoom* and *trace* features (see Figure 3.10) of the graphing utility to approximate the maximum height on the graph to be $y \approx 81.125$ feet at $x \approx 156.25$.

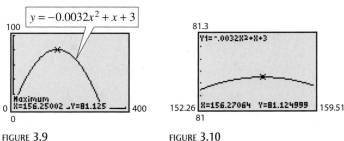

FIGURE 3.9   FIGURE 3.10

3.1   EXERCISES

See www.CalcChat.com for worked-out solutions to odd-numbered exercises.

VOCABULARY: Fill in the blanks.

1. Linear, constant, and squaring functions are examples of _____ functions.
2. A polynomial function of degree n and leading coefficient a_n is a function of the form
   $f(x) = a_n x^n + a_{n-1} x^{n-1} + \cdots + a_1 x + a_0$ $(a_n \neq 0)$ where n is a _____ _____ and $a_n, a_{n-1}, \ldots, a_1, a_0$
   are _____ numbers.
3. A _____ function is a second-degree polynomial function, and its graph is called a _____.
4. The graph of a quadratic function is symmetric about its _____.
5. If the graph of a quadratic function opens upward, then its leading coefficient is _____ and the vertex of the graph is a _____.
6. If the graph of a quadratic function opens downward, then its leading coefficient is _____ and the vertex of the graph is a _____.

SKILLS AND APPLICATIONS

In Exercises 7–12, match the quadratic function with its graph. [The graphs are labeled (a), (b), (c), (d), (e), and (f).]

(a)
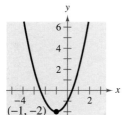

(b)
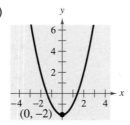

(c)
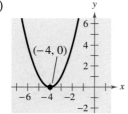

(d)
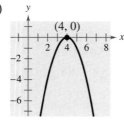

(e)
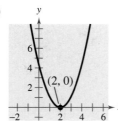

(f)
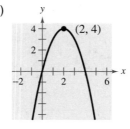

7. $f(x) = (x - 2)^2$
8. $f(x) = (x + 4)^2$
9. $f(x) = x^2 - 2$
10. $f(x) = (x + 1)^2 - 2$
11. $f(x) = 4 - (x - 2)^2$
12. $f(x) = -(x - 4)^2$

In Exercises 13–16, graph each function. Compare the graph of each function with the graph of $y = x^2$.

13. (a) $f(x) = \frac{1}{2} x^2$
    (b) $g(x) = -\frac{1}{8} x^2$
    (c) $h(x) = \frac{3}{2} x^2$
    (d) $k(x) = -3x^2$

14. (a) $f(x) = x^2 + 1$
    (b) $g(x) = x^2 - 1$
    (c) $h(x) = x^2 + 3$
    (d) $k(x) = x^2 - 3$
15. (a) $f(x) = (x - 1)^2$
    (b) $g(x) = (3x)^2 + 1$
    (c) $h(x) = \left(\frac{1}{3}x\right)^2 - 3$
    (d) $k(x) = (x + 3)^2$
16. (a) $f(x) = -\frac{1}{2}(x - 2)^2 + 1$
    (b) $g(x) = \left[\frac{1}{2}(x - 1)\right]^2 - 3$
    (c) $h(x) = -\frac{1}{2}(x + 2)^2 - 1$
    (d) $k(x) = [2(x + 1)]^2 + 4$

In Exercises 17–34, sketch the graph of the quadratic function without using a graphing utility. Identify the vertex, axis of symmetry, and x-intercept(s).

17. $f(x) = 1 - x^2$
18. $g(x) = x^2 - 8$
19. $f(x) = x^2 + 7$
20. $h(x) = 12 - x^2$
21. $f(x) = \frac{1}{2}x^2 - 4$
22. $f(x) = 16 - \frac{1}{4}x^2$
23. $f(x) = (x + 4)^2 - 3$
24. $f(x) = (x - 6)^2 + 8$
25. $h(x) = x^2 - 8x + 16$
26. $g(x) = x^2 + 2x + 1$
27. $f(x) = x^2 - x + \frac{5}{4}$
28. $f(x) = x^2 + 3x + \frac{1}{4}$
29. $f(x) = -x^2 + 2x + 5$
30. $f(x) = -x^2 - 4x + 1$
31. $h(x) = 4x^2 - 4x + 21$
32. $f(x) = 2x^2 - x + 1$
33. $f(x) = \frac{1}{4}x^2 - 2x - 12$
34. $f(x) = -\frac{1}{3}x^2 + 3x - 6$

In Exercises 35–42, use a graphing utility to graph the quadratic function. Identify the vertex, axis of symmetry, and x-intercepts. Then check your results algebraically by writing the quadratic function in standard form.

35. $f(x) = -(x^2 + 2x - 3)$
36. $f(x) = -(x^2 + x - 30)$
37. $g(x) = x^2 + 8x + 11$
38. $f(x) = x^2 + 10x + 14$
39. $f(x) = 2x^2 - 16x + 31$
40. $f(x) = -4x^2 + 24x - 41$
41. $g(x) = \frac{1}{2}(x^2 + 4x - 2)$
42. $f(x) = \frac{3}{5}(x^2 + 6x - 5)$

In Exercises 43–46, write an equation for the parabola in standard form.

43.

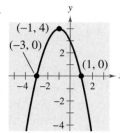

(−1, 4)
(−3, 0)
(1, 0)

44.

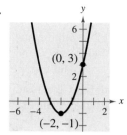

(0, 3)
(−2, −1)

45.

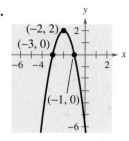

(−2, 2)
(−3, 0)
(−1, 0)

46.

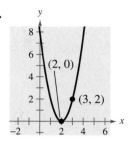

(2, 0)
(3, 2)

In Exercises 47–56, write the standard form of the equation of the parabola that has the indicated vertex and whose graph passes through the given point.

47. Vertex: $(-2, 5)$; point: $(0, 9)$

48. Vertex: $(4, -1)$; point: $(2, 3)$

49. Vertex: $(1, -2)$; point: $(-1, 14)$

50. Vertex: $(2, 3)$; point: $(0, 2)$

51. Vertex: $(5, 12)$; point: $(7, 15)$

52. Vertex: $(-2, -2)$; point: $(-1, 0)$

53. Vertex: $\left(-\frac{1}{4}, \frac{3}{2}\right)$; point: $(-2, 0)$

54. Vertex: $\left(\frac{5}{2}, -\frac{3}{4}\right)$; point: $(-2, 4)$

55. Vertex: $\left(-\frac{5}{2}, 0\right)$; point: $\left(-\frac{7}{2}, -\frac{16}{3}\right)$

56. Vertex: $(6, 6)$; point: $\left(\frac{61}{10}, \frac{3}{2}\right)$

GRAPHICAL REASONING In Exercises 57 and 58, determine the *x*-intercept(s) of the graph visually. Then find the *x*-intercept(s) algebraically to confirm your results.

57. $y = x^2 - 4x - 5$

58. $y = 2x^2 + 5x - 3$

In Exercises 59–64, use a graphing utility to graph the quadratic function. Find the *x*-intercepts of the graph and compare them with the solutions of the corresponding quadratic equation when $f(x) = 0$.

59. $f(x) = x^2 - 4x$

60. $f(x) = -2x^2 + 10x$

61. $f(x) = x^2 - 9x + 18$

62. $f(x) = x^2 - 8x - 20$

63. $f(x) = 2x^2 - 7x - 30$

64. $f(x) = \frac{7}{10}(x^2 + 12x - 45)$

In Exercises 65–70, find two quadratic functions, one that opens upward and one that opens downward, whose graphs have the given *x*-intercepts. (There are many correct answers.)

65. $(-1, 0), (3, 0)$

66. $(-5, 0), (5, 0)$

67. $(0, 0), (10, 0)$

68. $(4, 0), (8, 0)$

69. $(-3, 0), \left(-\frac{1}{2}, 0\right)$

70. $\left(-\frac{5}{2}, 0\right), (2, 0)$

In Exercises 71–74, find two positive real numbers whose product is a maximum.

71. The sum is 110.

72. The sum is S.

73. The sum of the first and twice the second is 24.

74. The sum of the first and three times the second is 42.

75. PATH OF A DIVER   The path of a diver is given by

$$y = -\frac{4}{9}x^2 + \frac{24}{9}x + 12$$

where y is the height (in feet) and x is the horizontal distance from the end of the diving board (in feet). What is the maximum height of the diver?

76. HEIGHT OF A BALL   The height y (in feet) of a punted football is given by

$$y = -\frac{16}{2025}x^2 + \frac{9}{5}x + 1.5$$

where x is the horizontal distance (in feet) from the point at which the ball is punted.

(a) How high is the ball when it is punted?

(b) What is the maximum height of the punt?

(c) How long is the punt?

77. MINIMUM COST   A manufacturer of lighting fixtures has daily production costs of $C = 800 - 10x + 0.25x^2$, where C is the total cost (in dollars) and x is the number of units produced. How many fixtures should be produced each day to yield a minimum cost?

78. MAXIMUM PROFIT   The profit P (in hundreds of dollars) that a company makes depends on the amount x (in hundreds of dollars) the company spends on advertising according to the model $P = 230 + 20x - 0.5x^2$. What expenditure for advertising will yield a maximum profit?

79. **MAXIMUM REVENUE** The total revenue R earned (in thousands of dollars) from manufacturing handheld video games is given by

$$R(p) = -25p^2 + 1200p$$

where p is the price per unit (in dollars).

(a) Find the revenues when the price per unit is $20, $25, and $30.

(b) Find the unit price that will yield a maximum revenue. What is the maximum revenue? Explain your results.

80. **MAXIMUM REVENUE** The total revenue R earned per day (in dollars) from a pet-sitting service is given by $R(p) = -12p^2 + 150p$, where p is the price charged per pet (in dollars).

(a) Find the revenues when the price per pet is $4, $6, and $8.

(b) Find the price that will yield a maximum revenue. What is the maximum revenue? Explain your results.

81. **NUMERICAL, GRAPHICAL, AND ANALYTICAL ANALYSIS** A rancher has 200 feet of fencing to enclose two adjacent rectangular corrals (see figure).

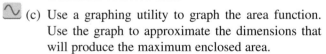

(a) Write the area A of the corrals as a function of x.

(b) Create a table showing possible values of x and the corresponding areas of the corral. Use the table to estimate the dimensions that will produce the maximum enclosed area.

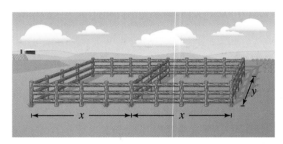 (c) Use a graphing utility to graph the area function. Use the graph to approximate the dimensions that will produce the maximum enclosed area.

(d) Write the area function in standard form to find analytically the dimensions that will produce the maximum area.

(e) Compare your results from parts (b), (c), and (d).

82. **GEOMETRY** An indoor physical fitness room consists of a rectangular region with a semicircle on each end. The perimeter of the room is to be a 200-meter single-lane running track.

(a) Draw a diagram that illustrates the problem. Let x and y represent the length and width of the rectangular region, respectively.

(b) Determine the radius of each semicircular end of the room. Determine the distance, in terms of y, around the inside edge of each semicircular part of the track.

(c) Use the result of part (b) to write an equation, in terms of x and y, for the distance traveled in one lap around the track. Solve for y.

(d) Use the result of part (c) to write the area A of the rectangular region as a function of x. What dimensions will produce a rectangle of maximum area?

83. **MAXIMUM REVENUE** A small theater has a seating capacity of 2000. When the ticket price is $20, attendance is 1500. For each $1 decrease in price, attendance increases by 100.

(a) Write the revenue R of the theater as a function of ticket price x.

(b) What ticket price will yield a maximum revenue? What is the maximum revenue?

84. **MAXIMUM AREA** A Norman window is constructed by adjoining a semicircle to the top of an ordinary rectangular window (see figure). The perimeter of the window is 16 feet.

(a) Write the area A of the window as a function of x.

(b) What dimensions will produce a window of maximum area?

85. **GRAPHICAL ANALYSIS** From 1950 through 2005, the per capita consumption C of cigarettes by Americans (age 18 and older) can be modeled by $C = 3565.0 + 60.30t - 1.783t^2$, $0 \le t \le 55$, where t is the year, with $t = 0$ corresponding to 1950. (Source: *Tobacco Outlook Report*)

(a) Use a graphing utility to graph the model.

(b) Use the graph of the model to approximate the maximum average annual consumption. Beginning in 1966, all cigarette packages were required by law to carry a health warning. Do you think the warning had any effect? Explain.

(c) In 2005, the U.S. population (age 18 and over) was 296,329,000. Of those, about 59,858,458 were smokers. What was the average annual cigarette consumption *per smoker* in 2005? What was the average daily cigarette consumption *per smoker*?

86. DATA ANALYSIS: SALES The sales y (in billions of dollars) for Harley-Davidson from 2000 through 2007 are shown in the table. (Source: U.S. Harley-Davidson, Inc.)

Year	Sales, y
2000	2.91
2001	3.36
2002	4.09
2003	4.62
2004	5.02
2005	5.34
2006	5.80
2007	5.73

(a) Use a graphing utility to create a scatter plot of the data. Let x represent the year, with $x = 0$ corresponding to 2000.

(b) Use the *regression* feature of the graphing utility to find a quadratic model for the data.

(c) Use the graphing utility to graph the model in the same viewing window as the scatter plot. How well does the model fit the data?

(d) Use the *trace* feature of the graphing utility to approximate the year in which the sales for Harley-Davidson were the greatest.

(e) Verify your answer to part (d) algebraically.

(f) Use the model to predict the sales for Harley-Davidson in 2010.

EXPLORATION

TRUE OR FALSE? In Exercises 87–90, determine whether the statement is true or false. Justify your answer.

87. The function given by $f(x) = -12x^2 - 1$ has no x-intercepts.

88. The graphs of $f(x) = -4x^2 - 10x + 7$ and $g(x) = 12x^2 + 30x + 1$ have the same axis of symmetry.

89. The graph of a quadratic function with a negative leading coefficient will have a maximum value at its vertex.

90. The graph of a quadratic function with a positive leading coefficient will have a minimum value at its vertex.

THINK ABOUT IT In Exercises 91–94, find the values of b such that the function has the given maximum or minimum value.

91. $f(x) = -x^2 + bx - 75$; Maximum value: 25

92. $f(x) = -x^2 + bx - 16$; Maximum value: 48

93. $f(x) = x^2 + bx + 26$; Minimum value: 10

94. $f(x) = x^2 + bx - 25$; Minimum value: -50

95. Write the quadratic function

$$f(x) = ax^2 + bx + c$$

in standard form to verify that the vertex occurs at

$$\left(-\frac{b}{2a}, \ f\left(-\frac{b}{2a}\right)\right).$$

96. CAPSTONE The profit P (in millions of dollars) for a recreational vehicle retailer is modeled by a quadratic function of the form

$$P = at^2 + bt + c$$

where t represents the year. If you were president of the company, which of the models below would you prefer? Explain your reasoning.

(a) a is positive and $-b/(2a) \le t$.

(b) a is positive and $t \le -b/(2a)$.

(c) a is negative and $-b/(2a) \le t$.

(d) a is negative and $t \le -b/(2a)$.

97. GRAPHICAL ANALYSIS

(a) Graph $y = ax^2$ for $a = -2, -1, -0.5, 0.5, 1$ and 2. How does changing the value of a affect the graph?

(b) Graph $y = (x - h)^2$ for $h = -4, -2, 2$, and 4. How does changing the value of h affect the graph?

(c) Graph $y = x^2 + k$ for $k = -4, -2, 2$, and 4. How does changing the value of k affect the graph?

98. Describe the sequence of transformation from f to g given that $f(x) = x^2$ and $g(x) = a(x - h)^2 + k$. (Assume a, h, and k are positive.)

99. Is it possible for a quadratic equation to have only one x-intercept? Explain.

100. Assume that the function given by

$$f(x) = ax^2 + bx + c, \quad a \ne 0$$

has two real zeros. Show that the x-coordinate of the vertex of the graph is the average of the zeros of f. (*Hint:* Use the Quadratic Formula.)

PROJECT: HEIGHT OF A BASKETBALL To work an extended application analyzing the height of a basketball after it has been dropped, visit this text's website at *academic.cengage.com*.

3.2 POLYNOMIAL FUNCTIONS OF HIGHER DEGREE

What you should learn

- Use transformations to sketch graphs of polynomial functions.
- Use the Leading Coefficient Test to determine the end behavior of graphs of polynomial functions.
- Find and use zeros of polynomial functions as sketching aids.
- Use the Intermediate Value Theorem to help locate zeros of polynomial functions.

Why you should learn it

You can use polynomial functions to analyze business situations such as how revenue is related to advertising expenses, as discussed in Exercise 104 on page 282.

Bill Aron/PhotoEdit, Inc.

Graphs of Polynomial Functions

In this section, you will study basic features of the graphs of polynomial functions. The first feature is that the graph of a polynomial function is **continuous.** Essentially, this means that the graph of a polynomial function has no breaks, holes, or gaps, as shown in Figure 3.11(a). The graph shown in Figure 3.11(b) is an example of a piecewise-defined function that is not continuous.

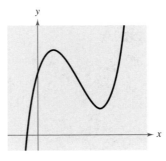

(a) Polynomial functions have continuous graphs.

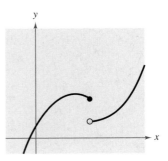

(b) Functions with graphs that are not continuous are not polynomial functions.

FIGURE **3.11**

The second feature is that the graph of a polynomial function has only smooth, rounded turns, as shown in Figure 3.12. A polynomial function cannot have a sharp turn. For instance, the function given by $f(x) = |x|$, which has a sharp turn at the point $(0, 0)$, as shown in Figure 3.13, is not a polynomial function.

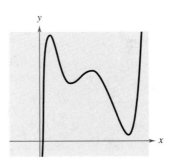

Polynomial functions have graphs with smooth, rounded turns.
FIGURE **3.12**

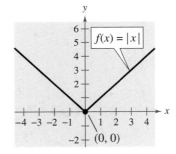

Graphs of polynomial functions cannot have sharp turns.
FIGURE **3.13**

The graphs of polynomial functions of degree greater than 2 are more difficult to analyze than the graphs of polynomials of degree 0, 1, or 2. However, using the features presented in this section, coupled with your knowledge of point plotting, intercepts, and symmetry, you should be able to make reasonably accurate sketches *by hand*.

For power functions given by $f(x) = x^n$, if n is even, then the graph of the function is symmetric with respect to the y-axis, and if n is odd, then the graph of the function is symmetric with respect to the origin.

The polynomial functions that have the simplest graphs are monomials of the form $f(x) = x^n$, where n is an integer greater than zero. From Figure 3.14, you can see that when n is *even*, the graph is similar to the graph of $f(x) = x^2$, and when n is *odd*, the graph is similar to the graph of $f(x) = x^3$. Moreover, the greater the value of n, the flatter the graph near the origin. Polynomial functions of the form $f(x) = x^n$ are often referred to as **power functions.**

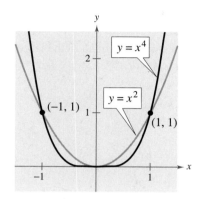

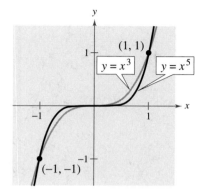

(a) If n is even, the graph of $y = x^n$ touches the axis at the x-intercept.

(b) If n is odd, the graph of $y = x^n$ crosses the axis at the x-intercept.

FIGURE **3.14**

| Example 1   Sketching Transformations of Polynomial Functions

Sketch the graph of each function.

a. $f(x) = -x^5$ **b.** $h(x) = (x + 1)^4$

Solution

a. Because the degree of $f(x) = -x^5$ is odd, its graph is similar to the graph of $y = x^3$. In Figure 3.15, note that the negative coefficient has the effect of reflecting the graph in the x-axis.

b. The graph of $h(x) = (x + 1)^4$, as shown in Figure 3.16, is a left shift by one unit of the graph of $y = x^4$.

You can review the techniques for shifting, reflecting, and stretching graphs in Section 2.5.

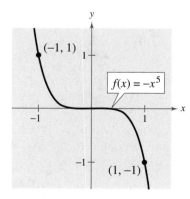

FIGURE **3.15**

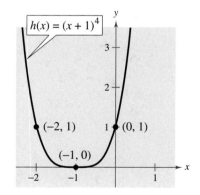

FIGURE **3.16**

CHECK*Point*   Now try Exercise 17.

The Leading Coefficient Test

In Example 1, note that both graphs eventually rise or fall without bound as x moves to the right. Whether the graph of a polynomial function eventually rises or falls can be determined by the function's degree (even or odd) and by its leading coefficient, as indicated in the **Leading Coefficient Test.**

Leading Coefficient Test

As x moves without bound to the left or to the right, the graph of the polynomial function $f(x) = a_n x^n + \cdots + a_1 x + a_0$ eventually rises or falls in the following manner.

1. When n is *odd:*

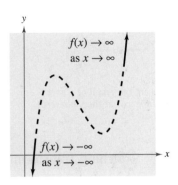

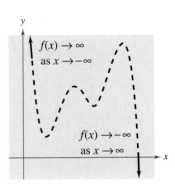

If the leading coefficient is positive $(a_n > 0)$, the graph falls to the left and rises to the right.

If the leading coefficient is negative $(a_n < 0)$, the graph rises to the left and falls to the right.

2. When n is *even:*

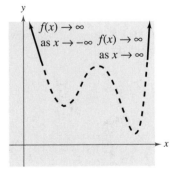

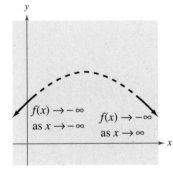

If the leading coefficient is positive $(a_n > 0)$, the graph rises to the left and right.

If the leading coefficient is negative $(a_n < 0)$, the graph falls to the left and right.

The dashed portions of the graphs indicate that the test determines *only* the right-hand and left-hand behavior of the graph.

Study Tip

The notation "$f(x) \to -\infty$ as $x \to -\infty$" indicates that the graph falls to the left. The notation "$f(x) \to \infty$ as $x \to \infty$" indicates that the graph rises to the right.

Example 2   **Applying the Leading Coefficient Test**

Describe the right-hand and left-hand behavior of the graph of each function.

a. $f(x) = -x^3 + 4x$ **b.** $f(x) = x^4 - 5x^2 + 4$ **c.** $f(x) = x^5 - x$

Solution

a. Because the degree is odd and the leading coefficient is negative, the graph rises to the left and falls to the right, as shown in Figure 3.17.

b. Because the degree is even and the leading coefficient is positive, the graph rises to the left and right, as shown in Figure 3.18.

c. Because the degree is odd and the leading coefficient is positive, the graph falls to the left and rises to the right, as shown in Figure 3.19.

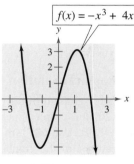
$f(x) = -x^3 + 4x$

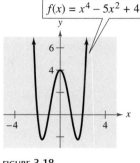
$f(x) = x^4 - 5x^2 + 4$

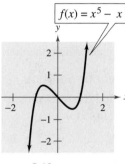
$f(x) = x^5 - x$

FIGURE **3.17**   FIGURE **3.18**   FIGURE **3.19**

CHECK Point   Now try Exercise 23.

In Example 2, note that the Leading Coefficient Test tells you only whether the graph *eventually* rises or falls to the right or left. Other characteristics of the graph, such as intercepts and minimum and maximum points, must be determined by other tests.

Zeros of Polynomial Functions

It can be shown that for a polynomial function f of degree n, the following statements are true.

1. The function f has, at most, n real zeros. (You will study this result in detail in the discussion of the Fundamental Theorem of Algebra in Section 3.4.)

2. The graph of f has, at most, $n - 1$ turning points. (Turning points, also called relative minima or relative maxima, are points at which the graph changes from increasing to decreasing or vice versa.)

Finding the zeros of polynomial functions is one of the most important problems in algebra. There is a strong interplay between graphical and algebraic approaches to this problem. Sometimes you can use information about the graph of a function to help find its zeros, and in other cases you can use information about the zeros of a function to help sketch its graph. Finding zeros of polynomial functions is closely related to factoring and finding x-intercepts.

Study Tip

Remember that the *zeros* of a function of x are the x-values for which the function is zero.

Algebra Help

To do Example 3 algebraically, you need to be able to completely factor polynomials. You can review the techniques for factoring in Section P.4.

Real Zeros of Polynomial Functions

If f is a polynomial function and a is a real number, the following statements are equivalent.

1. $x = a$ is a *zero* of the function f.

2. $x = a$ is a *solution* of the polynomial equation $f(x) = 0$.

3. $(x - a)$ is a *factor* of the polynomial $f(x)$.

4. $(a, 0)$ is an *x-intercept* of the graph of f.

Example 3   Finding the Zeros of a Polynomial Function

Find all real zeros of

$$f(x) = -2x^4 + 2x^2.$$

Then determine the number of turning points of the graph of the function.

Algebraic Solution

To find the real zeros of the function, set $f(x)$ equal to zero and solve for x.

$$-2x^4 + 2x^2 = 0$$

$$-2x^2(x^2 - 1) = 0$$

$$-2x^2(x - 1)(x + 1) = 0$$

So, the real zeros are $x = 0$, $x = 1$, and $x = -1$. Because the function is a fourth-degree polynomial, the graph of f can have at most $4 - 1 = 3$ turning points.

Graphical Solution

Use a graphing utility to graph $y = -2x^4 + 2x^2$. In Figure 3.20, the graph appears to have zeros at $(0, 0)$, $(1, 0)$, and $(-1, 0)$. Use the *zero* or *root* feature, or the *zoom* and *trace* features, of the graphing utility to verify these zeros. So, the real zeros are $x = 0$, $x = 1$, and $x = -1$. From the figure, you can see that the graph has three turning points. This is consistent with the fact that a fourth-degree polynomial can have at most three turning points.

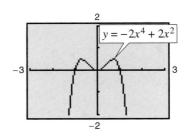

FIGURE 3.20

CHECK*Point*   Now try Exercise 35.

In Example 3, note that because the exponent is greater than 1, the factor $-2x^2$ yields the *repeated* zero $x = 0$. Because the exponent is even, the graph touches the x-axis at $x = 0$, as shown in Figure 3.20.

Repeated Zeros

A factor $(x - a)^k$, $k > 1$, yields a **repeated zero** $x = a$ of **multiplicity** k.

1. If k is odd, the graph *crosses* the x-axis at $x = a$.

2. If k is even, the graph *touches* the x-axis (but does not cross the x-axis) at $x = a$.

A polynomial function is written in **standard form** if its terms are written in descending order of exponents from left to right. Before applying the Leading Coefficient Test to a polynomial function, it is a good idea to check that the polynomial function is written in standard form.

TECHNOLOGY

Example 4 uses an *algebraic approach* to describe the graph of the function. A graphing utility is a complement to this approach. Remember that an important aspect of using a graphing utility is to find a viewing window that shows all significant features of the graph. For instance, the viewing window in part (a) illustrates all of the significant features of the function in Example 4 while the viewing window in part (b) does not.

a.

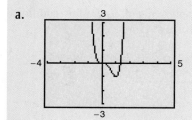

b.

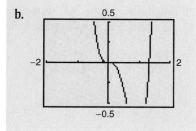

| **Example 4** | **Sketching the Graph of a Polynomial Function** |

Sketch the graph of $f(x) = 3x^4 - 4x^3$.

Solution

1. *Apply the Leading Coefficient Test.*   Because the leading coefficient is positive and the degree is even, you know that the graph eventually rises to the left and to the right (see Figure 3.21).

2. *Find the Zeros of the Polynomial.*   By factoring

$$f(x) = 3x^4 - 4x^3 = x^3(3x - 4)$$

   you can see that the zeros of f are $x = 0$ and $x = \frac{4}{3}$ (both of odd multiplicity). So, the x-intercepts occur at $(0, 0)$ and $\left(\frac{4}{3}, 0\right)$. Add these points to your graph, as shown in Figure 3.21.

3. *Plot a Few Additional Points.*   To sketch the graph by hand, find a few additional points, as shown in the table. Then plot the points (see Figure 3.22).

x	-1	0.5	1	1.5
$f(x)$	7	-0.3125	-1	1.6875

4. *Draw the Graph.*   Draw a continuous curve through the points, as shown in Figure 3.22. Because both zeros are of odd multiplicity, you know that the graph should cross the x-axis at $x = 0$ and $x = \frac{4}{3}$. If you are unsure of the shape of that portion of the graph, plot some additional points.

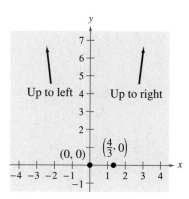

FIGURE **3.21**

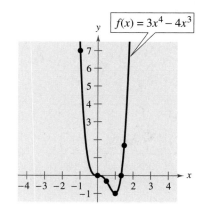

FIGURE **3.22**

CHECK*Point*   Now try Exercise 75.

| Example 5 | Sketching the Graph of a Polynomial Function |

Sketch the graph of $f(x) = -2x^3 + 6x^2 - \frac{9}{2}x$.

Solution

1. *Apply the Leading Coefficient Test.* Because the leading coefficient is negative and the degree is odd, you know that the graph eventually rises to the left and falls to the right (see Figure 3.23).

2. *Find the Zeros of the Polynomial.* By factoring

$$f(x) = -2x^3 + 6x^2 - \frac{9}{2}x$$
$$= -\frac{1}{2}x(4x^2 - 12x + 9)$$
$$= -\frac{1}{2}x(2x - 3)^2$$

you can see that the zeros of f are $x = 0$ (odd multiplicity) and $x = \frac{3}{2}$ (even multiplicity). So, the x-intercepts occur at $(0, 0)$ and $\left(\frac{3}{2}, 0\right)$. Add these points to your graph, as shown in Figure 3.23.

3. *Plot a Few Additional Points.* To sketch the graph by hand, find a few additional points, as shown in the table. Then plot the points (see Figure 3.24).

x	-0.5	0.5	1	2
$f(x)$	4	-1	-0.5	-1

4. *Draw the Graph.* Draw a continuous curve through the points, as shown in Figure 3.24. As indicated by the multiplicities of the zeros, the graph crosses the x-axis at $(0, 0)$ but does not cross the x-axis at $\left(\frac{3}{2}, 0\right)$.

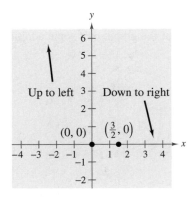

FIGURE **3.23**

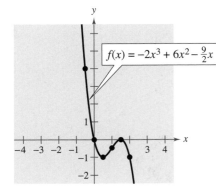

FIGURE **3.24**

 CHECK*Point* Now try Exercise 77.

Study Tip

Observe in Example 5 that the sign of $f(x)$ is positive to the left of and negative to the right of the zero $x = 0$. Similarly, the sign of $f(x)$ is negative to the left and to the right of the zero $x = \frac{3}{2}$. This suggests that if the zero of a polynomial function is of *odd* multiplicity, then the sign of $f(x)$ changes from one side to the other side of the zero. If the zero is of *even* multiplicity, then the sign of $f(x)$ does not change from one side of the zero to the other side. The following table helps to illustrate this concept.

x	-0.5	0	0.5
$f(x)$	4	0	-1
Sign	$+$		$-$

x	1	$\frac{3}{2}$	2
$f(x)$	-0.5	0	-1
Sign	$-$		$-$

This sign analysis may be helpful in graphing polynomial functions.

The Intermediate Value Theorem

The next theorem, called the **Intermediate Value Theorem,** illustrates the existence of real zeros of polynomial functions. This theorem implies that if $(a, f(a))$ and $(b, f(b))$ are two points on the graph of a polynomial function such that $f(a) \neq f(b)$, then for any number d between $f(a)$ and $f(b)$ there must be a number c between a and b such that $f(c) = d$. (See Figure 3.25.)

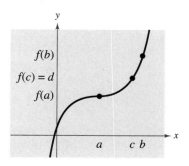

FIGURE **3.25**

Intermediate Value Theorem

Let a and b be real numbers such that $a < b$. If f is a polynomial function such that $f(a) \neq f(b)$, then, in the interval $[a, b]$, f takes on every value between $f(a)$ and $f(b)$.

The Intermediate Value Theorem helps you locate the real zeros of a polynomial function in the following way. If you can find a value $x = a$ at which a polynomial function is positive, and another value $x = b$ at which it is negative, you can conclude that the function has at least one real zero between these two values. For example, the function given by $f(x) = x^3 + x^2 + 1$ is negative when $x = -2$ and positive when $x = -1$. Therefore, it follows from the Intermediate Value Theorem that f must have a real zero somewhere between -2 and -1, as shown in Figure 3.26.

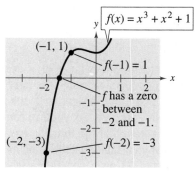

FIGURE **3.26**

By continuing this line of reasoning, you can approximate any real zeros of a polynomial function to any desired accuracy. This concept is further demonstrated in Example 6.

Example 6 Approximating a Zero of a Polynomial Function

Use the Intermediate Value Theorem to approximate the real zero of

$$f(x) = x^3 - x^2 + 1.$$

Solution

Begin by computing a few function values, as follows.

x	$f(x)$
-2	-11
-1	-1
0	1
1	1

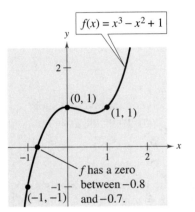

$f(x) = x^3 - x^2 + 1$

$(0, 1)$

$(1, 1)$

f has a zero between -0.8 and -0.7.

$(-1, -1)$

FIGURE 3.27

Because $f(-1)$ is negative and $f(0)$ is positive, you can apply the Intermediate Value Theorem to conclude that the function has a zero between -1 and 0. To pinpoint this zero more closely, divide the interval $[-1, 0]$ into tenths and evaluate the function at each point. When you do this, you will find that

$$f(-0.8) = -0.152 \quad \text{and} \quad f(-0.7) = 0.167.$$

So, f must have a zero between -0.8 and -0.7, as shown in Figure 3.27. For a more accurate approximation, compute function values between $f(-0.8)$ and $f(-0.7)$ and apply the Intermediate Value Theorem again. By continuing this process, you can approximate this zero to any desired accuracy.

CHECK*Point* Now try Exercise 93.

TECHNOLOGY

You can use the *table* feature of a graphing utility to approximate the zeros of a polynomial function. For instance, for the function given by

$$f(x) = -2x^3 - 3x^2 + 3$$

create a table that shows the function values for $-20 \leq x \leq 20$, as shown in the first table at the right. Scroll through the table looking for consecutive function values that differ in sign. From the table, you can see that $f(0)$ and $f(1)$ differ in sign. So, you can conclude from the Intermediate Value Theorem that the function has a zero between 0 and 1. You can adjust your table to show function values for $0 \leq x \leq 1$ using increments of 0.1, as shown in the second table at the right. By scrolling through the table you can see that $f(0.8)$ and $f(0.9)$ differ in sign. So, the function has a zero between 0.8 and 0.9. If you repeat this process several times, you should obtain $x \approx 0.806$ as the zero of the function. Use the *zero* or *root* feature of a graphing utility to confirm this result.

X	Y1
-2	7
-1	2
0	3
1	-2
2	-25
3	-78
4	-173

X=1

X	Y1
.4	2.392
.5	2
.6	1.488
.7	.844
.8	.056
.9	-.888
1	-2

X=.9

3.2   EXERCISES

See www.CalcChat.com for worked-out solutions to odd-numbered exercises.

VOCABULARY: Fill in the blanks.

1. The graphs of all polynomial functions are _____, which means that the graphs have no breaks, holes, or gaps.

2. The _____ _____ _____ is used to determine the left-hand and right-hand behavior of the graph of a polynomial function.

3. Polynomial functions of the form $f(x) =$ _____ are often referred to as power functions.

4. A polynomial function of degree n has at most _____ real zeros and at most _____ turning points.

5. If $x = a$ is a zero of a polynomial function f, then the following three statements are true.
    (a) $x = a$ is a _____ of the polynomial equation $f(x) = 0$.
    (b) _____ is a factor of the polynomial $f(x)$.
    (c) $(a, 0)$ is an _____ of the graph of f.

6. If a real zero of a polynomial function is of even multiplicity, then the graph of f _____ the x-axis at $x = a$, and if it is of odd multiplicity, then the graph of f _____ the x-axis at $x = a$.

7. A polynomial function is written in _____ form if its terms are written in descending order of exponents from left to right.

8. The _____ _____ Theorem states that if f is a polynomial function such that $f(a) \neq f(b)$, then, in the interval $[a, b]$, f takes on every value between $f(a)$ and $f(b)$.

SKILLS AND APPLICATIONS

In Exercises 9–16, match the polynomial function with its graph. [The graphs are labeled (a), (b), (c), (d), (e), (f), (g), and (h).]

(a)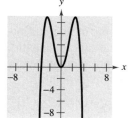

(b)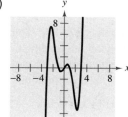

(g)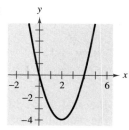

(h)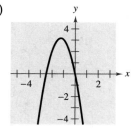

(c)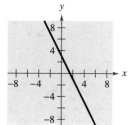

(d)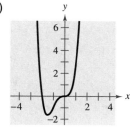

9. $f(x) = -2x + 3$

10. $f(x) = x^2 - 4x$

11. $f(x) = -2x^2 - 5x$

12. $f(x) = 2x^3 - 3x + 1$

13. $f(x) = -\frac{1}{4}x^4 + 3x^2$

14. $f(x) = -\frac{1}{3}x^3 + x^2 - \frac{4}{3}$

15. $f(x) = x^4 + 2x^3$

16. $f(x) = \frac{1}{5}x^5 - 2x^3 + \frac{9}{5}x$

In Exercises 17–20, sketch the graph of $y = x^n$ and each transformation.

17. $y = x^3$
    (a) $f(x) = (x - 4)^3$
    (b) $f(x) = x^3 - 4$
    (c) $f(x) = -\frac{1}{4}x^3$
    (d) $f(x) = (x - 4)^3 - 4$

18. $y = x^5$
    (a) $f(x) = (x + 1)^5$
    (b) $f(x) = x^5 + 1$
    (c) $f(x) = 1 - \frac{1}{2}x^5$
    (d) $f(x) = -\frac{1}{2}(x + 1)^5$

19. $y = x^4$
    (a) $f(x) = (x + 3)^4$
    (b) $f(x) = x^4 - 3$
    (c) $f(x) = 4 - x^4$
    (d) $f(x) = \frac{1}{2}(x - 1)^4$
    (e) $f(x) = (2x)^4 + 1$
    (f) $f(x) = \left(\frac{1}{2}x\right)^4 - 2$

(e)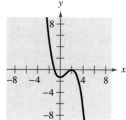

(f)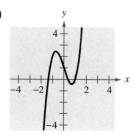

20. $y = x^6$

(a) $f(x) = -\frac{1}{8}x^6$ (b) $f(x) = (x + 2)^6 - 4$

(c) $f(x) = x^6 - 5$ (d) $f(x) = -\frac{1}{4}x^6 + 1$

(e) $f(x) = \left(\frac{1}{4}x\right)^6 - 2$ (f) $f(x) = (2x)^6 - 1$

In Exercises 21–30, describe the right-hand and left-hand behavior of the graph of the polynomial function.

21. $f(x) = \frac{1}{5}x^3 + 4x$ **22.** $f(x) = 2x^2 - 3x + 1$

23. $g(x) = 5 - \frac{7}{2}x - 3x^2$ **24.** $h(x) = 1 - x^6$

25. $f(x) = -2.1x^5 + 4x^3 - 2$

26. $f(x) = 4x^5 - 7x + 6.5$

27. $f(x) = 6 - 2x + 4x^2 - 5x^3$

28. $f(x) = (3x^4 - 2x + 5)/4$

29. $h(t) = -\frac{3}{4}(t^2 - 3t + 6)$

30. $f(s) = -\frac{7}{8}(s^3 + 5s^2 - 7s + 1)$

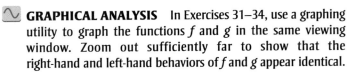 **GRAPHICAL ANALYSIS** In Exercises 31–34, use a graphing utility to graph the functions f and g in the same viewing window. Zoom out sufficiently far to show that the right-hand and left-hand behaviors of f and g appear identical.

31. $f(x) = 3x^3 - 9x + 1$, $g(x) = 3x^3$

32. $f(x) = -\frac{1}{3}(x^3 - 3x + 2)$, $g(x) = -\frac{1}{3}x^3$

33. $f(x) = -(x^4 - 4x^3 + 16x)$, $g(x) = -x^4$

34. $f(x) = 3x^4 - 6x^2$, $g(x) = 3x^4$

In Exercises 35–50, (a) find all the real zeros of the polynomial function, (b) determine the multiplicity of each zero and the number of turning points of the graph of the function, and (c) use a graphing utility to graph the function and verify your answers.

35. $f(x) = x^2 - 36$ **36.** $f(x) = 81 - x^2$

37. $h(t) = t^2 - 6t + 9$ **38.** $f(x) = x^2 + 10x + 25$

39. $f(x) = \frac{1}{3}x^2 + \frac{1}{3}x - \frac{2}{3}$ **40.** $f(x) = \frac{1}{2}x^2 + \frac{5}{2}x - \frac{3}{2}$

41. $f(x) = 3x^3 - 12x^2 + 3x$ **42.** $g(x) = 5x(x^2 - 2x - 1)$

43. $f(t) = t^3 - 8t^2 + 16t$ **44.** $f(x) = x^4 - x^3 - 30x^2$

45. $g(t) = t^5 - 6t^3 + 9t$ **46.** $f(x) = x^5 + x^3 - 6x$

47. $f(x) = 3x^4 + 9x^2 + 6$ **48.** $f(x) = 2x^4 - 2x^2 - 40$

49. $g(x) = x^3 + 3x^2 - 4x - 12$

50. $f(x) = x^3 - 4x^2 - 25x + 100$

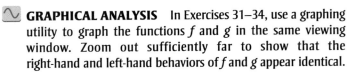 **GRAPHICAL ANALYSIS** In Exercises 51–54, (a) use a graphing utility to graph the function, (b) use the graph to approximate any x-intercepts of the graph, (c) set $y = 0$ and solve the resulting equation, and (d) compare the results of part (c) with any x-intercepts of the graph.

51. $y = 4x^3 - 20x^2 + 25x$

52. $y = 4x^3 + 4x^2 - 8x - 8$

53. $y = x^5 - 5x^3 + 4x$ **54.** $y = \frac{1}{4}x^3(x^2 - 9)$

In Exercises 55–64, find a polynomial function that has the given zeros. (There are many correct answers.)

55. $0, 8$ **56.** $0, -7$

57. $2, -6$ **58.** $-4, 5$

59. $0, -4, -5$ **60.** $0, 1, 10$

61. $4, -3, 3, 0$ **62.** $-2, -1, 0, 1, 2$

63. $1 + \sqrt{3}, 1 - \sqrt{3}$ **64.** $2, 4 + \sqrt{5}, 4 - \sqrt{5}$

In Exercises 65–74, find a polynomial of degree n that has the given zero(s). (There are many correct answers.)

	Zero(s)	Degree
65.	$x = -3$	$n = 2$
66.	$x = -12, -6$	$n = 2$
67.	$x = -5, 0, 1$	$n = 3$
68.	$x = -2, 4, 7$	$n = 3$
69.	$x = 0, \sqrt{3}, -\sqrt{3}$	$n = 3$
70.	$x = 9$	$n = 3$
71.	$x = -5, 1, 2$	$n = 4$
72.	$x = -4, -1, 3, 6$	$n = 4$
73.	$x = 0, -4$	$n = 5$
74.	$x = -1, 4, 7, 8$	$n = 5$

In Exercises 75–88, sketch the graph of the function by (a) applying the Leading Coefficient Test, (b) finding the zeros of the polynomial, (c) plotting sufficient solution points, and (d) drawing a continuous curve through the points.

75. $f(x) = x^3 - 25x$ **76.** $g(x) = x^4 - 9x^2$

77. $f(t) = \frac{1}{4}(t^2 - 2t + 15)$

78. $g(x) = -x^2 + 10x - 16$

79. $f(x) = x^3 - 2x^2$ **80.** $f(x) = 8 - x^3$

81. $f(x) = 3x^3 - 15x^2 + 18x$

82. $f(x) = -4x^3 + 4x^2 + 15x$

83. $f(x) = -5x^2 - x^3$ **84.** $f(x) = -48x^2 + 3x^4$

85. $f(x) = x^2(x - 4)$ **86.** $h(x) = \frac{1}{3}x^3(x - 4)^2$

87. $g(t) = -\frac{1}{4}(t - 2)^2(t + 2)^2$

88. $g(x) = \frac{1}{10}(x + 1)^2(x - 3)^3$

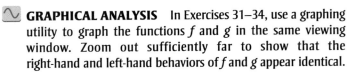 In Exercises 89–92, use a graphing utility to graph the function. Use the *zero* or *root* feature to approximate the real zeros of the function. Then determine the multiplicity of each zero.

89. $f(x) = x^3 - 16x$ **90.** $f(x) = \frac{1}{4}x^4 - 2x^2$

91. $g(x) = \frac{1}{5}(x + 1)^2(x - 3)(2x - 9)$

92. $h(x) = \frac{1}{5}(x + 2)^2(3x - 5)^2$

In Exercises 93–96, use the Intermediate Value Theorem and the *table* feature of a graphing utility to find intervals one unit in length in which the polynomial function is guaranteed to have a zero. Adjust the table to approximate the zeros of the function. Use the *zero* or *root* feature of the graphing utility to verify your results.

93. $f(x) = x^3 - 3x^2 + 3$

94. $f(x) = 0.11x^3 - 2.07x^2 + 9.81x - 6.88$

95. $g(x) = 3x^4 + 4x^3 - 3$

96. $h(x) = x^4 - 10x^2 + 3$

97. NUMERICAL AND GRAPHICAL ANALYSIS An open box is to be made from a square piece of material, 36 inches on a side, by cutting equal squares with sides of length x from the corners and turning up the sides (see figure).

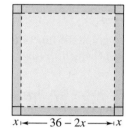
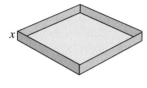

(a) Write a function $V(x)$ that represents the volume of the box.

(b) Determine the domain of the function.

(c) Use a graphing utility to create a table that shows box heights x and the corresponding volumes V. Use the table to estimate the dimensions that will produce a maximum volume.

(d) Use a graphing utility to graph V and use the graph to estimate the value of x for which $V(x)$ is maximum. Compare your result with that of part (c).

98. MAXIMUM VOLUME An open box with locking tabs is to be made from a square piece of material 24 inches on a side. This is to be done by cutting equal squares from the corners and folding along the dashed lines shown in the figure.

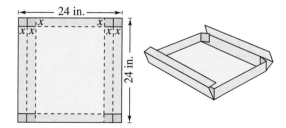

(a) Write a function $V(x)$ that represents the volume of the box.

(b) Determine the domain of the function V.

(c) Sketch a graph of the function and estimate the value of x for which $V(x)$ is maximum.

99. CONSTRUCTION A roofing contractor is fabricating gutters from 12-inch aluminum sheeting. The contractor plans to use an aluminum siding folding press to create the gutter by creasing equal lengths for the sidewalls (see figure).

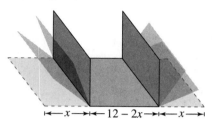

(a) Let x represent the height of the sidewall of the gutter. Write a function A that represents the cross-sectional area of the gutter.

(b) The length of the aluminum sheeting is 16 feet. Write a function V that represents the volume of one run of gutter in terms of x.

(c) Determine the domain of the function in part (b).

(d) Use a graphing utility to create a table that shows the sidewall heights x and the corresponding volumes V. Use the table to estimate the dimensions that will produce a maximum volume.

(e) Use a graphing utility to graph V. Use the graph to estimate the value of x for which $V(x)$ is a maximum. Compare your result with that of part (d).

(f) Would the value of x change if the aluminum sheeting were of different lengths? Explain.

100. CONSTRUCTION An industrial propane tank is formed by adjoining two hemispheres to the ends of a right circular cylinder. The length of the cylindrical portion of the tank is four times the radius of the hemispherical components (see figure).

(a) Write a function that represents the total volume V of the tank in terms of r.

(b) Find the domain of the function.

(c) Use a graphing utility to graph the function.

(d) The total volume of the tank is to be 120 cubic feet. Use the graph from part (c) to estimate the radius and length of the cylindrical portion of the tank.

101. REVENUE   The total revenues R (in millions of dollars) for Krispy Kreme from 2000 through 2007 are shown in the table.

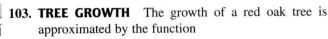

Year	Revenue, R
2000	300.7
2001	394.4
2002	491.5
2003	665.6
2004	707.8
2005	543.4
2006	461.2
2007	429.3

A model that represents these data is given by $R = 3.0711t^4 - 42.803t^3 + 160.59t^2 - 62.6t + 307$, $0 \le t \le 7$, where t represents the year, with $t = 0$ corresponding to 2000.   (Source: Krispy Kreme)

(a) Use a graphing utility to create a scatter plot of the data. Then graph the model in the same viewing window.

(b) How well does the model fit the data?

(c) Use a graphing utility to approximate any relative extrema of the model over its domain.

(d) Use a graphing utility to approximate the intervals over which the revenue for Krispy Kreme was increasing and decreasing over its domain.

(e) Use the results of parts (c) and (d) to write a short paragraph about Krispy Kreme's revenue during this time period.

102. REVENUE   The total revenues R (in millions of dollars) for Papa John's International from 2000 through 2007 are shown in the table.

Year	Revenue, R
2000	944.7
2001	971.2
2002	946.2
2003	917.4
2004	942.4
2005	968.8
2006	1001.6
2007	1063.6

A model that represents these data is given by $R = -0.5635t^4 + 9.019t^3 - 40.20t^2 + 49.0t + 947$, $0 \le t \le 7$, where t represents the year, with $t = 0$ corresponding to 2000.   (Source: Papa John's International)

(a) Use a graphing utility to create a scatter plot of the data. Then graph the model in the same viewing window.

(b) How well does the model fit the data?

(c) Use a graphing utility to approximate any relative extrema of the model over its domain.

(d) Use a graphing utility to approximate the intervals over which the revenue for Papa John's International was increasing and decreasing over its domain.

(e) Use the results of parts (c) and (d) to write a short paragraph about the revenue for Papa John's International during this time period.

103. TREE GROWTH   The growth of a red oak tree is approximated by the function

$$G = -0.003t^3 + 0.137t^2 + 0.458t - 0.839$$

where G is the height of the tree (in feet) and t $(2 \le t \le 34)$ is its age (in years).

(a) Use a graphing utility to graph the function. (*Hint:* Use a viewing window in which $-10 \le x \le 45$ and $-5 \le y \le 60$.)

(b) Estimate the age of the tree when it is growing most rapidly. This point is called the *point of diminishing returns* because the increase in size will be less with each additional year.

(c) Using calculus, the point of diminishing returns can also be found by finding the vertex of the parabola given by

$$y = -0.009t^2 + 0.274t + 0.458.$$

Find the vertex of this parabola.

(d) Compare your results from parts (b) and (c).

104. REVENUE   The total revenue R (in millions of dollars) for a company is related to its advertising expense by the function

$$R = \frac{1}{100,000}(-x^3 + 600x^2), \quad 0 \le x \le 400$$

where x is the amount spent on advertising (in tens of thousands of dollars). Use the graph of this function, shown in the figure on the next page, to estimate the point on the graph at which the function is increasing most rapidly. This point is called the *point of diminishing returns* because any expense above this amount will yield less return per dollar invested in advertising.

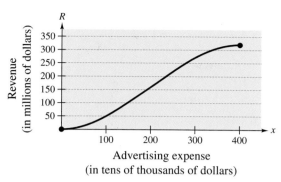

FIGURE FOR 104

EXPLORATION

TRUE OR FALSE? In Exercises 105–107, determine whether the statement is true or false. Justify your answer.

105. A fifth-degree polynomial can have five turning points in its graph.

106. It is possible for a sixth-degree polynomial to have only one solution.

107. The graph of the function given by

$$f(x) = 2 + x - x^2 + x^3 - x^4 + x^5 + x^6 - x^7$$

rises to the left and falls to the right.

108. CAPSTONE For each graph, describe a polynomial function that could represent the graph. (Indicate the degree of the function and the sign of its leading coefficient.)

(a)

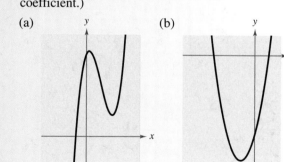

(b)

(c)

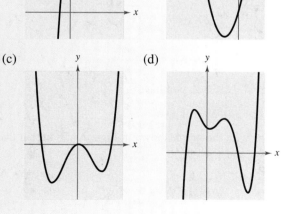

(d)

109. GRAPHICAL REASONING Sketch a graph of the function given by $f(x) = x^4$. Explain how the graph of each function g differs (if it does) from the graph of each function f. Determine whether g is odd, even, or neither.

(a) $g(x) = f(x) + 2$ (b) $g(x) = f(x + 2)$
(c) $g(x) = f(-x)$ (d) $g(x) = -f(x)$
(e) $g(x) = f\left(\frac{1}{2}x\right)$ (f) $g(x) = \frac{1}{2}f(x)$
(g) $g(x) = f\left(x^{3/4}\right)$ (h) $g(x) = (f \circ f)(x)$

110. THINK ABOUT IT For each function, identify the degree of the function and whether the degree of the function is even or odd. Identify the leading coefficient and whether the leading coefficient is positive or negative. Use a graphing utility to graph each function. Describe the relationship between the degree of the function and the sign of the leading coefficient of the function and the right-hand and left-hand behavior of the graph of the function.

(a) $f(x) = x^3 - 2x^2 - x + 1$
(b) $f(x) = 2x^5 + 2x^2 - 5x + 1$
(c) $f(x) = -2x^5 - x^2 + 5x + 3$
(d) $f(x) = -x^3 + 5x - 2$
(e) $f(x) = 2x^2 + 3x - 4$
(f) $f(x) = x^4 - 3x^2 + 2x - 1$
(g) $f(x) = x^2 + 3x + 2$

111. THINK ABOUT IT Sketch the graph of each polynomial function. Then count the number of zeros of the function and the numbers of relative minima and relative maxima. Compare these numbers with the degree of the polynomial. What do you observe?

(a) $f(x) = -x^3 + 9x$ (b) $f(x) = x^4 - 10x^2 + 9$
(c) $f(x) = x^5 - 16x$

112. Explore the transformations of the form

$$g(x) = a(x - h)^5 + k.$$

(a) Use a graphing utility to graph the functions $y_1 = -\frac{1}{3}(x - 2)^5 + 1$ and $y_2 = \frac{3}{5}(x + 2)^5 - 3$. Determine whether the graphs are increasing or decreasing. Explain.

(b) Will the graph of g always be increasing or decreasing? If so, is this behavior determined by a, h, or k? Explain.

(c) Use a graphing utility to graph the function given by $H(x) = x^5 - 3x^3 + 2x + 1$. Use the graph and the result of part (b) to determine whether H can be written in the form $H(x) = a(x - h)^5 + k$. Explain.

3.3 POLYNOMIAL AND SYNTHETIC DIVISION

What you should learn

- Use long division to divide polynomials by other polynomials.
- Use synthetic division to divide polynomials by binomials of the form $(x - k)$.
- Use the Remainder Theorem and the Factor Theorem.

Why you should learn it

Synthetic division can help you evaluate polynomial functions. For instance, in Exercise 85 on page 291, you will use synthetic division to determine the amount donated to support higher education in the United States in 2010.

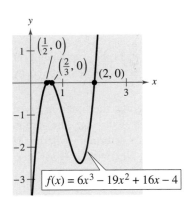

Long Division of Polynomials

In this section, you will study two procedures for *dividing* polynomials. These procedures are especially valuable in factoring and finding the zeros of polynomial functions. To begin, suppose you are given the graph of

$$f(x) = 6x^3 - 19x^2 + 16x - 4.$$

Notice that a zero of f occurs at $x = 2$, as shown in Figure 3.28. Because $x = 2$ is a zero of f, you know that $(x - 2)$ is a factor of $f(x)$. This means that there exists a second-degree polynomial $q(x)$ such that

$$f(x) = (x - 2) \cdot q(x).$$

To find $q(x)$, you can use **long division,** as illustrated in Example 1.

Example 1 Long Division of Polynomials

Divide $6x^3 - 19x^2 + 16x - 4$ by $x - 2$, and use the result to factor the polynomial completely.

Solution

$$
\begin{array}{r}
6x^2 - 7x + 2 \\
x - 2 \overline{)\, 6x^3 - 19x^2 + 16x - 4} \\
\underline{6x^3 - 12x^2} \\
-7x^2 + 16x \\
\underline{-7x^2 + 14x} \\
2x - 4 \\
\underline{2x - 4} \\
0
\end{array}
$$

From this division, you can conclude that

$$6x^3 - 19x^2 + 16x - 4 = (x - 2)(6x^2 - 7x + 2)$$

and by factoring the quadratic $6x^2 - 7x + 2$, you have

$$6x^3 - 19x^2 + 16x - 4 = (x - 2)(2x - 1)(3x - 2).$$

Note that this factorization agrees with the graph shown in Figure 3.28 in that the three x-intercepts occur at $x = 2$, $x = \frac{1}{2}$, and $x = \frac{2}{3}$.

CHECK**Point** Now try Exercise 11.

FIGURE 3.28

In Example 1, $x - 2$ is a factor of the polynomial $6x^3 - 19x^2 + 16x - 4$, and the long division process produces a remainder of zero. Often, long division will produce a nonzero remainder. For instance, if you divide $x^2 + 3x + 5$ by $x + 1$, you obtain the following.

$$
\begin{array}{r}
x + 2 \\
x + 1 \overline{)\, x^2 + 3x + 5} \\
\underline{x^2 + x} \\
2x + 5 \\
\underline{2x + 2} \\
3
\end{array}
$$

In fractional form, you can write this result as follows.

$$\frac{x^2 + 3x + 5}{x + 1} = x + 2 + \frac{3}{x + 1}$$

This implies that

$$x^2 + 3x + 5 = (x + 1)(x + 2) + 3$$

which illustrates the following theorem, called the **Division Algorithm.**

The Division Algorithm

If $f(x)$ and $d(x)$ are polynomials such that $d(x) \neq 0$, and the degree of $d(x)$ is less than or equal to the degree of $f(x)$, there exist unique polynomials $q(x)$ and $r(x)$ such that

$$f(x) = d(x)q(x) + r(x)$$

where $r(x) = 0$ *or* the degree of $r(x)$ is less than the degree of $d(x)$. If the remainder $r(x)$ is zero, $d(x)$ *divides evenly* into $f(x)$.

The Division Algorithm can also be written as

$$\frac{f(x)}{d(x)} = q(x) + \frac{r(x)}{d(x)}.$$

In the Division Algorithm, the rational expression $f(x)/d(x)$ is **improper** because the degree of $f(x)$ is greater than or equal to the degree of $d(x)$. On the other hand, the rational expression $r(x)/d(x)$ is **proper** because the degree of $r(x)$ is less than the degree of $d(x)$.

Before you apply the Division Algorithm, follow these steps.

1. Write the dividend and divisor in descending powers of the variable.

2. Insert placeholders with zero coefficients for missing powers of the variable.

Example 2   Long Division of Polynomials

Divide $x^3 - 1$ by $x - 1$.

Solution

Because there is no x^2-term or x-term in the dividend, you need to line up the subtraction by using zero coefficients (or leaving spaces) for the missing terms.

$$
\begin{array}{r}
x^2 + x + 1 \\
x - 1 \overline{)\, x^3 + 0x^2 + 0x - 1} \\
\underline{x^3 - x^2} \\
x^2 + 0x \\
\underline{x^2 - x} \\
x - 1 \\
\underline{x - 1} \\
0
\end{array}
$$

So, $x - 1$ divides evenly into $x^3 - 1$, and you can write

$$
\frac{x^3 - 1}{x - 1} = x^2 + x + 1, \quad x \neq 1.
$$

CHECK*Point* Now try Exercise 17.

You can check the result of Example 2 by multiplying.

$$(x - 1)(x^2 + x + 1) = x^3 + x^2 + x - x^2 - x - 1 = x^3 - 1$$

Algebra Help

You can check a long division problem by multiplying. You can review the techniques for multiplying polynomials in Section P.3.

Example 3   Long Division of Polynomials

Divide $-5x^2 - 2 + 3x + 2x^4 + 4x^3$ by $2x - 3 + x^2$.

Solution

Begin by writing the dividend and divisor in descending powers of x.

$$
\begin{array}{r}
2x^2 \qquad + 1 \\
x^2 + 2x - 3 \overline{)\, 2x^4 + 4x^3 - 5x^2 + 3x - 2} \\
\underline{2x^4 + 4x^3 - 6x^2} \\
x^2 + 3x - 2 \\
\underline{x^2 + 2x - 3} \\
x + 1
\end{array}
$$

Note that the first subtraction eliminated two terms from the dividend. When this happens, the quotient skips a term. You can write the result as

$$
\frac{2x^4 + 4x^3 - 5x^2 + 3x - 2}{x^2 + 2x - 3} = 2x^2 + 1 + \frac{x + 1}{x^2 + 2x - 3}.
$$

CHECK*Point* Now try Exercise 23.

Synthetic Division

There is a nice shortcut for long division of polynomials by divisors of the form $x - k$. This shortcut is called **synthetic division.** The pattern for synthetic division of a cubic polynomial is summarized as follows. (The pattern for higher-degree polynomials is similar.)

Synthetic Division (for a Cubic Polynomial)

To divide $ax^3 + bx^2 + cx + d$ by $x - k$, use the following pattern.

$$
\begin{array}{c|cccc}
k & a & b & c & d \\
  &   & ka &   &   \\
\hline
  & a &   &   & r \\
\end{array}
$$

Vertical pattern: Add terms.
Diagonal pattern: Multiply by k.

This algorithm for synthetic division works only for divisors of the form $x - k$. Remember that $x + k = x - (-k)$.

Example 4   Using Synthetic Division

Use synthetic division to divide $x^4 - 10x^2 - 2x + 4$ by $x + 3$.

Solution

You should set up the array as follows. Note that a zero is included for the missing x^3-term in the dividend.

$$
\begin{array}{c|ccccc}
-3 & 1 & 0 & -10 & -2 & 4 \\
   &   &   &   &   &   \\
\hline
\end{array}
$$

Then, use the synthetic division pattern by adding terms in columns and multiplying the results by -3.

$$
\begin{array}{c|ccccc}
-3 & 1 & 0 & -10 & -2 & 4 \\
   &   & -3 & 9 & 3 & -3 \\
\hline
   & 1 & -3 & -1 & 1 & 1 \\
\end{array}
$$

So, you have

$$
\frac{x^4 - 10x^2 - 2x + 4}{x + 3} = x^3 - 3x^2 - x + 1 + \frac{1}{x + 3}.
$$

CHECK*Point*   Now try Exercise 27.

The Remainder and Factor Theorems

The remainder obtained in the synthetic division process has an important interpretation, as described in the **Remainder Theorem.**

The Remainder Theorem

If a polynomial $f(x)$ is divided by $x - k$, the remainder is

$r = f(k)$.

For a proof of the Remainder Theorem, see Proofs in Mathematics on page 327.

The Remainder Theorem tells you that synthetic division can be used to evaluate a polynomial function. That is, to evaluate a polynomial function $f(x)$ when $x = k$, divide $f(x)$ by $x - k$. The remainder will be $f(k)$, as illustrated in Example 5.

Example 5 Using the Remainder Theorem

Use the Remainder Theorem to evaluate the following function at $x = -2$.

$f(x) = 3x^3 + 8x^2 + 5x - 7$

Solution

Using synthetic division, you obtain the following.

$$
\begin{array}{r|rrrr}
-2 & 3 & 8 & 5 & -7 \\
   &   & -6 & -4 & -2 \\
\hline
   & 3 & 2 & 1 & -9
\end{array}
$$

Because the remainder is $r = -9$, you can conclude that

$f(-2) = -9$.

This means that $(-2, -9)$ is a point on the graph of f. You can check this by substituting $x = -2$ in the original function.

Check

$f(\quad) = 3(\quad)^3 + 8(\quad)^2 + 5(\quad) - 7$

$= 3(-8) + 8(4) - 10 - 7 = -9$

CHECK *Point* Now try Exercise 55.

Another important theorem is the **Factor Theorem,** stated below. This theorem states that you can test to see whether a polynomial has $(x - k)$ as a factor by evaluating the polynomial at $x = k$. If the result is 0, $(x - k)$ is a factor.

The Factor Theorem

A polynomial $f(x)$ has a factor $(x - k)$ if and only if $f(k) = 0$.

For a proof of the Factor Theorem, see Proofs in Mathematics on page 327.

Example 6 Factoring a Polynomial: Repeated Division

Show that $(x - 2)$ and $(x + 3)$ are factors of $f(x) = 2x^4 + 7x^3 - 4x^2 - 27x - 18$. Then find the remaining factors of $f(x)$.

Algebraic Solution

Using synthetic division with the factor $(x - 2)$, you obtain the following.

$$
\begin{array}{r|rrrrr}
2 & 2 & 7 & -4 & -27 & -18 \\
  &   & 4 & 22 & 36 & 18 \\
\hline
  & 2 & 11 & 18 & 9 & 0
\end{array}
$$

Take the result of this division and perform synthetic division again using the factor $(x + 3)$.

$$
\begin{array}{r|rrrr}
-3 & 2 & 11 & 18 & 9 \\
   &   & -6 & -15 & -9 \\
\hline
   & 2 & 5 & 3 & 0
\end{array}
$$

Because the resulting quadratic expression factors as

$$2x^2 + 5x + 3 = (2x + 3)(x + 1)$$

the complete factorization of $f(x)$ is

$$f(x) = (x - 2)(x + 3)(2x + 3)(x + 1).$$

CHECK**Point** Now try Exercise 67.

Graphical Solution

From the graph of $f(x) = 2x^4 + 7x^3 - 4x^2 - 27x - 18$, you can see that there are four x-intercepts (see Figure 3.29). These occur at $x = -3$, $x = -\frac{3}{2}$, $x = -1$, and $x = 2$. (Check this algebraically.) This implies that $(x + 3)$, $\left(x + \frac{3}{2}\right)$, $(x + 1)$, and $(x - 2)$ are factors of $f(x)$. $\left[\text{Note that } \left(x + \frac{3}{2}\right) \text{ and } (2x + 3) \text{ are equivalent factors because they both yield the same zero, } x = -\frac{3}{2}.\right]$

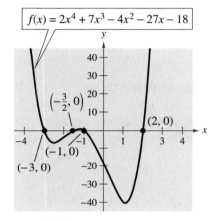

FIGURE **3.29**

Study Tip

Note in Example 6 that the complete factorization of $f(x)$ implies that f has four real zeros: $x = 2$, $x = -3$, $x = -\frac{3}{2}$, and $x = -1$. This is confirmed by the graph of f, which is shown in the Figure 3.29.

Uses of the Remainder in Synthetic Division

The remainder r, obtained in the synthetic division of $f(x)$ by $x - k$, provides the following information.

1. The remainder r gives the value of f at $x = k$. That is, $r = f(k)$.

2. If $r = 0$, $(x - k)$ is a factor of $f(x)$.

3. If $r = 0$, $(k, 0)$ is an x-intercept of the graph of f.

Throughout this text, the importance of developing several problem-solving strategies is emphasized. In the exercises for this section, try using more than one strategy to solve several of the exercises. For instance, if you find that $x - k$ divides evenly into $f(x)$ (with no remainder), try sketching the graph of f. You should find that $(k, 0)$ is an x-intercept of the graph.

3.3   EXERCISES

See www.CalcChat.com for worked-out solutions to odd-numbered exercises.

VOCABULARY

1. Two forms of the Division Algorithm are shown below. Identify and label each term or function.

$$f(x) = d(x)q(x) + r(x) \qquad \frac{f(x)}{d(x)} = q(x) + \frac{r(x)}{d(x)}$$

In Exercises 2–6, fill in the blanks.

2. The rational expression $p(x)/q(x)$ is called _____ if the degree of the numerator is greater than or equal to that of the denominator, and is called _____ if the degree of the numerator is less than that of the denominator.

3. In the Division Algorithm, the rational expression $f(x)/d(x)$ is _____ because the degree of $f(x)$ is greater than or equal to the degree of $d(x)$.

4. An alternative method to long division of polynomials is called _____ _____, in which the divisor must be of the form $x - k$.

5. The _____ Theorem states that a polynomial $f(x)$ has a factor $(x - k)$ if and only if $f(k) = 0$.

6. The _____ Theorem states that if a polynomial $f(x)$ is divided by $x - k$, the remainder is $r = f(k)$.

SKILLS AND APPLICATIONS

ANALYTICAL ANALYSIS In Exercises 7 and 8, use long division to verify that $y_1 = y_2$.

7. $y_1 = \dfrac{x^2}{x + 2}, \quad y_2 = x - 2 + \dfrac{4}{x + 2}$

8. $y_1 = \dfrac{x^4 - 3x^2 - 1}{x^2 + 5}, \quad y_2 = x^2 - 8 + \dfrac{39}{x^2 + 5}$

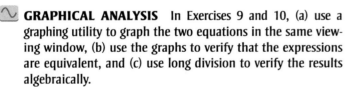 **GRAPHICAL ANALYSIS** In Exercises 9 and 10, (a) use a graphing utility to graph the two equations in the same viewing window, (b) use the graphs to verify that the expressions are equivalent, and (c) use long division to verify the results algebraically.

9. $y_1 = \dfrac{x^2 + 2x - 1}{x + 3}, \quad y_2 = x - 1 + \dfrac{2}{x + 3}$

10. $y_1 = \dfrac{x^4 + x^2 - 1}{x^2 + 1}, \quad y_2 = x^2 - \dfrac{1}{x^2 + 1}$

In Exercises 11–26, use long division to divide.

11. $(2x^2 + 10x + 12) \div (x + 3)$

12. $(5x^2 - 17x - 12) \div (x - 4)$

13. $(4x^3 - 7x^2 - 11x + 5) \div (4x + 5)$

14. $(6x^3 - 16x^2 + 17x - 6) \div (3x - 2)$

15. $(x^4 + 5x^3 + 6x^2 - x - 2) \div (x + 2)$

16. $(x^3 + 4x^2 - 3x - 12) \div (x - 3)$

17. $(x^3 - 27) \div (x - 3)$ **18.** $(x^3 + 125) \div (x + 5)$

19. $(7x + 3) \div (x + 2)$ **20.** $(8x - 5) \div (2x + 1)$

21. $(x^3 - 9) \div (x^2 + 1)$ **22.** $(x^5 + 7) \div (x^3 - 1)$

23. $(3x + 2x^3 - 9 - 8x^2) \div (x^2 + 1)$

24. $(5x^3 - 16 - 20x + x^4) \div (x^2 - x - 3)$

25. $\dfrac{x^4}{(x - 1)^3}$ **26.** $\dfrac{2x^3 - 4x^2 - 15x + 5}{(x - 1)^2}$

In Exercises 27–46, use synthetic division to divide.

27. $(3x^3 - 17x^2 + 15x - 25) \div (x - 5)$

28. $(5x^3 + 18x^2 + 7x - 6) \div (x + 3)$

29. $(6x^3 + 7x^2 - x + 26) \div (x - 3)$

30. $(2x^3 + 14x^2 - 20x + 7) \div (x + 6)$

31. $(4x^3 - 9x + 8x^2 - 18) \div (x + 2)$

32. $(9x^3 - 16x - 18x^2 + 32) \div (x - 2)$

33. $(-x^3 + 75x - 250) \div (x + 10)$

34. $(3x^3 - 16x^2 - 72) \div (x - 6)$

35. $(5x^3 - 6x^2 + 8) \div (x - 4)$

36. $(5x^3 + 6x + 8) \div (x + 2)$

37. $\dfrac{10x^4 - 50x^3 - 800}{x - 6}$ **38.** $\dfrac{x^5 - 13x^4 - 120x + 80}{x + 3}$

39. $\dfrac{x^3 + 512}{x + 8}$ **40.** $\dfrac{x^3 - 729}{x - 9}$

41. $\dfrac{-3x^4}{x - 2}$ **42.** $\dfrac{-3x^4}{x + 2}$

43. $\dfrac{180x - x^4}{x - 6}$ **44.** $\dfrac{5 - 3x + 2x^2 - x^3}{x + 1}$

45. $\dfrac{4x^3 + 16x^2 - 23x - 15}{x + \frac{1}{2}}$

46. $\dfrac{3x^3 - 4x^2 + 5}{x - \frac{3}{2}}$

In Exercises 47–54, write the function in the form $f(x) = (x - k)q(x) + r$ for the given value of k, and demonstrate that $f(k) = r$.

47. $f(x) = x^3 - x^2 - 14x + 11, \quad k = 4$

48. $f(x) = x^3 - 5x^2 - 11x + 8, \quad k = -2$

49. $f(x) = 15x^4 + 10x^3 - 6x^2 + 14, \quad k = -\frac{2}{3}$

50. $f(x) = 10x^3 - 22x^2 - 3x + 4, \quad k = \frac{1}{5}$

51. $f(x) = x^3 + 3x^2 - 2x - 14, \quad k = \sqrt{2}$

52. $f(x) = x^3 + 2x^2 - 5x - 4, \quad k = -\sqrt{5}$

53. $f(x) = -4x^3 + 6x^2 + 12x + 4, \quad k = 1 - \sqrt{3}$

54. $f(x) = -3x^3 + 8x^2 + 10x - 8, \quad k = 2 + \sqrt{2}$

In Exercises 55–58, use the Remainder Theorem and synthetic division to find each function value. Verify your answers using another method.

55. $f(x) = 2x^3 - 7x + 3$

    (a) $f(1)$ (b) $f(-2)$ (c) $f\!\left(\frac{1}{2}\right)$ (d) $f(2)$

56. $g(x) = 2x^6 + 3x^4 - x^2 + 3$

    (a) $g(2)$ (b) $g(1)$ (c) $g(3)$ (d) $g(-1)$

57. $h(x) = x^3 - 5x^2 - 7x + 4$

    (a) $h(3)$ (b) $h(2)$ (c) $h(-2)$ (d) $h(-5)$

58. $f(x) = 4x^4 - 16x^3 + 7x^2 + 20$

    (a) $f(1)$ (b) $f(-2)$ (c) $f(5)$ (d) $f(-10)$

In Exercises 59–66, use synthetic division to show that x is a solution of the third-degree polynomial equation, and use the result to factor the polynomial completely. List all real solutions of the equation.

59. $x^3 - 7x + 6 = 0, \quad x = 2$

60. $x^3 - 28x - 48 = 0, \quad x = -4$

61. $2x^3 - 15x^2 + 27x - 10 = 0, \quad x = \frac{1}{2}$

62. $48x^3 - 80x^2 + 41x - 6 = 0, \quad x = \frac{2}{3}$

63. $x^3 + 2x^2 - 3x - 6 = 0, \quad x = \sqrt{3}$

64. $x^3 + 2x^2 - 2x - 4 = 0, \quad x = \sqrt{2}$

65. $x^3 - 3x^2 + 2 = 0, \quad x = 1 + \sqrt{3}$

66. $x^3 - x^2 - 13x - 3 = 0, \quad x = 2 - \sqrt{5}$

In Exercises 67–74, (a) verify the given factors of the function f, (b) find the remaining factor(s) of f, (c) use your results to write the complete factorization of f, (d) list all real zeros of f, and (e) confirm your results by using a graphing utility to graph the function.

Function	Factors
67. $f(x) = 2x^3 + x^2 - 5x + 2$	$(x + 2), (x - 1)$
68. $f(x) = 3x^3 + 2x^2 - 19x + 6$	$(x + 3), (x - 2)$
69. $f(x) = x^4 - 4x^3 - 15x^2$ $\quad\quad + 58x - 40$	$(x - 5), (x + 4)$

Function	Factors
70. $f(x) = 8x^4 - 14x^3 - 71x^2$ $\quad\quad - 10x + 24$	$(x + 2), (x - 4)$
71. $f(x) = 6x^3 + 41x^2 - 9x - 14$	$(2x + 1), (3x - 2)$
72. $f(x) = 10x^3 - 11x^2 - 72x + 45$	$(2x + 5), (5x - 3)$
73. $f(x) = 2x^3 - x^2 - 10x + 5$	$(2x - 1), \left(x + \sqrt{5}\right)$
74. $f(x) = x^3 + 3x^2 - 48x - 144$	$\left(x + 4\sqrt{3}\right), (x + 3)$

 **GRAPHICAL ANALYSIS** In Exercises 75–80, (a) use the *zero* or *root* feature of a graphing utility to approximate the zeros of the function accurate to three decimal places, (b) determine one of the exact zeros, and (c) use synthetic division to verify your result from part (b), and then factor the polynomial completely.

75. $f(x) = x^3 - 2x^2 - 5x + 10$

76. $g(x) = x^3 - 4x^2 - 2x + 8$

77. $h(t) = t^3 - 2t^2 - 7t + 2$

78. $f(s) = s^3 - 12s^2 + 40s - 24$

79. $h(x) = x^5 - 7x^4 + 10x^3 + 14x^2 - 24x$

80. $g(x) = 6x^4 - 11x^3 - 51x^2 + 99x - 27$

In Exercises 81–84, simplify the rational expression by using long division or synthetic division.

81. $\dfrac{4x^3 - 8x^2 + x + 3}{2x - 3}$ **82.** $\dfrac{x^3 + x^2 - 64x - 64}{x + 8}$

83. $\dfrac{x^4 + 6x^3 + 11x^2 + 6x}{x^2 + 3x + 2}$

84. $\dfrac{x^4 + 9x^3 - 5x^2 - 36x + 4}{x^2 - 4}$

85. DATA ANALYSIS: HIGHER EDUCATION The amounts A (in billions of dollars) donated to support higher education in the United States from 2000 through 2007 are shown in the table, where t represents the year, with $t = 0$ corresponding to 2000.

Year, t	Amount, A
0	23.2
1	24.2
2	23.9
3	23.9
4	24.4
5	25.6
6	28.0
7	29.8

(a) Use a graphing utility to create a scatter plot of the data.

(b) Use the *regression* feature of the graphing utility to find a cubic model for the data. Graph the model in the same viewing window as the scatter plot.

(c) Use the model to create a table of estimated values of A. Compare the model with the original data.

(d) Use synthetic division to evaluate the model for the year 2010. Even though the model is relatively accurate for estimating the given data, would you use this model to predict the amount donated to higher education in the future? Explain.

86. DATA ANALYSIS: HEALTH CARE The amounts A (in billions of dollars) of national health care expenditures in the United States from 2000 through 2007 are shown in the table, where t represents the year, with $t = 0$ corresponding to 2000.

Year, t	Amount, A
0	30.5
1	32.2
2	34.2
3	38.0
4	42.7
5	47.9
6	52.7
7	57.6

(a) Use a graphing utility to create a scatter plot of the data.

(b) Use the *regression* feature of the graphing utility to find a cubic model for the data. Graph the model in the same viewing window as the scatter plot.

(c) Use the model to create a table of estimated values of A. Compare the model with the original data.

(d) Use synthetic division to evaluate the model for the year 2010.

EXPLORATION

TRUE OR FALSE? In Exercises 87–89, determine whether the statement is true or false. Justify your answer.

87. If $(7x + 4)$ is a factor of some polynomial function f, then $\frac{4}{7}$ is a zero of f.

88. $(2x - 1)$ is a factor of the polynomial
$$6x^6 + x^5 - 92x^4 + 45x^3 + 184x^2 + 4x - 48.$$

89. The rational expression
$$\frac{x^3 + 2x^2 - 13x + 10}{x^2 - 4x - 12}$$
is improper.

90. Use the form $f(x) = (x - k)q(x) + r$ to create a cubic function that (a) passes through the point $(2, 5)$ and rises to the right, and (b) passes through the point $(-3, 1)$ and falls to the right. (There are many correct answers.)

THINK ABOUT IT In Exercises 91 and 92, perform the division by assuming that n is a positive integer.

91. $\dfrac{x^{3n} + 9x^{2n} + 27x^n + 27}{x^n + 3}$ **92.** $\dfrac{x^{3n} - 3x^{2n} + 5x^n - 6}{x^n - 2}$

93. WRITING Briefly explain what it means for a divisor to divide evenly into a dividend.

94. WRITING Briefly explain how to check polynomial division, and justify your reasoning. Give an example.

EXPLORATION In Exercises 95 and 96, find the constant c such that the denominator will divide evenly into the numerator.

95. $\dfrac{x^3 + 4x^2 - 3x + c}{x - 5}$ **96.** $\dfrac{x^5 - 2x^2 + x + c}{x + 2}$

97. THINK ABOUT IT Find the value of k such that $x - 4$ is a factor of $x^3 - kx^2 + 2kx - 8$.

98. THINK ABOUT IT Find the value of k such that $x - 3$ is a factor of $x^3 - kx^2 + 2kx - 12$.

99. WRITING Complete each polynomial division. Write a brief description of the pattern that you obtain, and use your result to find a formula for the polynomial division $(x^n - 1)/(x - 1)$. Create a numerical example to test your formula.

(a) $\dfrac{x^2 - 1}{x - 1} = $ ▨ (b) $\dfrac{x^3 - 1}{x - 1} = $ ▨

(c) $\dfrac{x^4 - 1}{x - 1} = $ ▨

100. CAPSTONE Consider the division
$$f(x) \div (x - k)$$
where
$$f(x) = (x + 3)^2(x - 3)(x + 1)^3.$$

(a) What is the remainder when $k = -3$? Explain.

(b) If it is necessary to find $f(2)$, is it easier to evaluate the function directly or to use synthetic division? Explain.

What you should learn

- Use the Fundamental Theorem of Algebra to determine the number of zeros of polynomial functions.
- Find rational zeros of polynomial functions.
- Find conjugate pairs of complex zeros.
- Find zeros of polynomials by factoring.
- Use Descartes's Rule of Signs and the Upper and Lower Bound Rules to find zeros of polynomials.

Why you should learn it

Finding zeros of polynomial functions is an important part of solving real-life problems. For instance, in Exercise 120 on page 306, the zeros of a polynomial function can help you analyze the attendance at women's college basketball games.

3.4 ZEROS OF POLYNOMIAL FUNCTIONS

The Fundamental Theorem of Algebra

You know that an nth-degree polynomial can have at most n real zeros. In the complex number system, this statement can be improved. That is, in the complex number system, every nth-degree polynomial function has *precisely* n zeros. This important result is derived from the **Fundamental Theorem of Algebra,** first proved by the German mathematician Carl Friedrich Gauss (1777–1855).

> ### The Fundamental Theorem of Algebra
>
> If $f(x)$ is a polynomial of degree n, where $n > 0$, then f has at least one zero in the complex number system.

Using the Fundamental Theorem of Algebra and the equivalence of zeros and factors, you obtain the **Linear Factorization Theorem.**

> ### Linear Factorization Theorem
>
> If $f(x)$ is a polynomial of degree n, where $n > 0$, then f has precisely n linear factors
>
> $$f(x) = a_n(x - c_1)(x - c_2) \cdots (x - c_n)$$
>
> where $c_1, c_2, \ldots, c_n$ are complex numbers.

For a proof of the Linear Factorization Theorem, see Proofs in Mathematics on page 328.

Note that the Fundamental Theorem of Algebra and the Linear Factorization Theorem tell you only that the zeros or factors of a polynomial exist, not how to find them. Such theorems are called *existence theorems.* Remember that the n zeros of a polynomial function can be real or complex, and they may be repeated.

Study Tip

Recall that in order to find the zeros of a function $f(x)$, set $f(x)$ equal to 0 and solve the resulting equation for x. For instance, the function in Example 1(a) has a zero at $x = 2$ because

$$x - 2 = 0$$
$$x = 2.$$

Algebra Help

Examples 1(b), 1(c), and 1(d) involve factoring polynomials. You can review the techniques for factoring polynomials in Section P.4.

| Example 1   Zeros of Polynomial Functions

a. The first-degree polynomial $f(x) = x - 2$ has exactly *one* zero: $x = 2$.

b. Counting multiplicity, the second-degree polynomial function

$$f(x) = x^2 - 6x + 9 = (x - 3)(x - 3)$$

has exactly *two* zeros: $x = 3$ and $x = 3$. (This is called a *repeated zero.*)

c. The third-degree polynomial function

$$f(x) = x^3 + 4x = x(x^2 + 4) = x(x - 2i)(x + 2i)$$

has exactly *three* zeros: $x = 0$, $x = 2i$, and $x = -2i$.

d. The fourth-degree polynomial function

$$f(x) = x^4 - 1 = (x - 1)(x + 1)(x - i)(x + i)$$

has exactly *four* zeros: $x = 1$, $x = -1$, $x = i$, and $x = -i$.

CHECK*Point*   Now try Exercise 9. ∎

The Rational Zero Test

The **Rational Zero Test** relates the possible rational zeros of a polynomial (having integer coefficients) to the leading coefficient and to the constant term of the polynomial.

The Rational Zero Test

If the polynomial $f(x) = a_n x^n + a_{n-1} x^{n-1} + \cdots + a_2 x^2 + a_1 x + a_0$ has *integer* coefficients, every rational zero of f has the form

$$\text{Rational zero} = \frac{p}{q}$$

where p and q have no common factors other than 1, and

p = a factor of the constant term a_0

q = a factor of the leading coefficient a_n.

To use the Rational Zero Test, you should first list all rational numbers whose numerators are factors of the constant term and whose denominators are factors of the leading coefficient.

$$\text{Possible rational zeros} = \frac{\text{factors of constant term}}{\text{factors of leading coefficient}}$$

Having formed this list of *possible rational zeros*, use a trial-and-error method to determine which, if any, are actual zeros of the polynomial. Note that when the leading coefficient is 1, the possible rational zeros are simply the factors of the constant term.

Example 2 Rational Zero Test with Leading Coefficient of 1

Find the rational zeros of

$$f(x) = x^3 + x + 1.$$

Solution

Because the leading coefficient is 1, the possible rational zeros are ± 1, the factors of the constant term. By testing these possible zeros, you can see that neither works.

$$f(\ ) = (\ )^3 + \quad + 1$$
$$= 3$$
$$f(\ ) = (\ )^3 + (\ ) + 1$$
$$= -1$$

So, you can conclude that the given polynomial has *no* rational zeros. Note from the graph of f in Figure 3.30 that f does have one real zero between -1 and 0. However, by the Rational Zero Test, you know that this real zero is *not* a rational number.

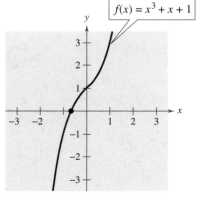

$f(x) = x^3 + x + 1$

FIGURE 3.30

CHECK **Point** Now try Exercise 15.

Example 3   Rational Zero Test with Leading Coefficient of 1

Find the rational zeros of $f(x) = x^4 - x^3 + x^2 - 3x - 6$.

Solution

Because the leading coefficient is 1, the possible rational zeros are the factors of the constant term.

Possible rational zeros: $\pm 1, \pm 2, \pm 3, \pm 6$

By applying synthetic division successively, you can determine that $x = -1$ and $x = 2$ are the only two rational zeros.

$$
\begin{array}{r|rrrrr}
-1 & 1 & -1 & 1 & -3 & -6 \\
   &   & -1 & 2 & -3 & 6 \\
\hline
   & 1 & -2 & 3 & -6 & 0
\end{array}
$$

$$
\begin{array}{r|rrrr}
2 & 1 & -2 & 3 & -6 \\
  &   & 2 & 0 & 6 \\
\hline
  & 1 & 0 & 3 & 0
\end{array}
$$

So, $f(x)$ factors as

$$f(x) = (x + 1)(x - 2)(x^2 + 3).$$

Because the factor $(x^2 + 3)$ produces no real zeros, you can conclude that $x = -1$ and $x = 2$ are the only *real* zeros of f, which is verified in Figure 3.31.

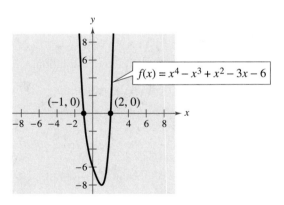

FIGURE **3.31**

CHECK*Point*   Now try Exercise 19.

If the leading coefficient of a polynomial is not 1, the list of possible rational zeros can increase dramatically. In such cases, the search can be shortened in several ways: (1) a programmable calculator can be used to speed up the calculations; (2) a graph, drawn either by hand or with a graphing utility, can give a good estimate of the locations of the zeros; (3) the Intermediate Value Theorem along with a table generated by a graphing utility can give approximations of zeros; and (4) synthetic division can be used to test the possible rational zeros.

Finding the first zero is often the most difficult part. After that, the search is simplified by working with the lower-degree polynomial obtained in synthetic division, as shown in Example 3.

Example 4 Using the Rational Zero Test

Find the rational zeros of $f(x) = 2x^3 + 3x^2 - 8x + 3$.

Solution

The leading coefficient is 2 and the constant term is 3.

$$Possible\ rational\ zeros: \frac{Factors\ of\ 3}{Factors\ of\ 2} = \frac{\pm 1, \pm 3}{\pm 1, \pm 2} = \pm 1, \pm 3, \pm\frac{1}{2}, \pm\frac{3}{2}$$

By synthetic division, you can determine that $x = 1$ is a rational zero.

$$\begin{array}{r|rrrr} 1 & 2 & 3 & -8 & 3 \\ & & 2 & 5 & -3 \\ \hline & 2 & 5 & -3 & 0 \end{array}$$

So, $f(x)$ factors as

$$f(x) = (x - 1)(2x^2 + 5x - 3)$$
$$= (x - 1)(2x - 1)(x + 3)$$

and you can conclude that the rational zeros of f are $x = 1$, $x = \frac{1}{2}$, and $x = -3$.

CHECK*Point* Now try Exercise 25.

Recall from Section 3.2 that if $x = a$ is a zero of the polynomial function f, then $x = a$ is a solution of the polynomial equation $f(x) = 0$.

Example 5 Solving a Polynomial Equation

Find all the real solutions of $-10x^3 + 15x^2 + 16x - 12 = 0$.

Solution

The leading coefficient is -10 and the constant term is -12.

$$Possible\ rational\ solutions: \frac{Factors\ of\ -12}{Factors\ of\ -10} = \frac{\pm 1, \pm 2, \pm 3, \pm 4, \pm 6, \pm 12}{\pm 1, \pm 2, \pm 5, \pm 10}$$

With so many possibilities (32, in fact), it is worth your time to stop and sketch a graph. From Figure 3.32, it looks like three reasonable solutions would be $x = -\frac{6}{5}$, $x = \frac{1}{2}$, and $x = 2$. Testing these by synthetic division shows that $x = 2$ is the only rational solution. So, you have

$$(x - 2)(-10x^2 - 5x + 6) = 0.$$

Using the Quadratic Formula for the second factor, you find that the two additional solutions are irrational numbers.

$$x = \frac{-5 - \sqrt{265}}{20} \approx -1.0639$$

and

$$x = \frac{-5 + \sqrt{265}}{20} \approx 0.5639$$

CHECK*Point* Now try Exercise 31.

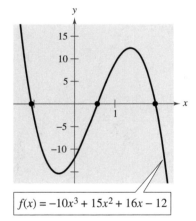

$f(x) = -10x^3 + 15x^2 + 16x - 12$

FIGURE 3.32

Conjugate Pairs

In Examples 1(c) and 1(d), note that the pairs of complex zeros are **conjugates.** That is, they are of the form $a + bi$ and $a - bi$.

Complex Zeros Occur in Conjugate Pairs

Let $f(x)$ be a polynomial function that has *real coefficients.* If $a + bi$, where $b \neq 0$, is a zero of the function, the conjugate $a - bi$ is also a zero of the function.

Be sure you see that this result is true only if the polynomial function has *real coefficients.* For instance, the result applies to the function given by $f(x) = x^2 + 1$ but not to the function given by $g(x) = x - i$.

Example 6   Finding a Polynomial with Given Zeros

Find a fourth-degree polynomial function with real coefficients that has -1, -1, and $3i$ as zeros.

Solution

Because $3i$ is a zero *and* the polynomial is stated to have real coefficients, you know that the conjugate $-3i$ must also be a zero. So, from the Linear Factorization Theorem, $f(x)$ can be written as

$$f(x) = a(x + 1)(x + 1)(x - 3i)(x + 3i).$$

For simplicity, let $a = 1$ to obtain

$$f(x) = (x^2 + 2x + 1)(x^2 + 9)$$

$$= x^4 + 2x^3 + 10x^2 + 18x + 9.$$

CHECK *Point*   Now try Exercise 45.

Factoring a Polynomial

The Linear Factorization Theorem shows that you can write any nth-degree polynomial as the product of n linear factors.

$$f(x) = a_n(x - c_1)(x - c_2)(x - c_3) \cdots (x - c_n)$$

However, this result includes the possibility that some of the values of c_i are complex. The following theorem says that even if you do not want to get involved with "complex factors," you can still write $f(x)$ as the product of linear and/or quadratic factors. For a proof of this theorem, see Proofs in Mathematics on page 328.

Factors of a Polynomial

Every polynomial of degree $n > 0$ with real coefficients can be written as the product of linear and quadratic factors with real coefficients, where the quadratic factors have no real zeros.

A quadratic factor with no real zeros is said to be *prime* or **irreducible over the reals.** Be sure you see that this is not the same as being *irreducible over the rationals*. For example, the quadratic $x^2 + 1 = (x - i)(x + i)$ is irreducible over the reals (and therefore over the rationals). On the other hand, the quadratic $x^2 - 2 = \left(x - \sqrt{2}\right)\left(x + \sqrt{2}\right)$ is irreducible over the rationals but *reducible* over the reals.

Example 7   Finding the Zeros of a Polynomial Function

Find all the zeros of $f(x) = x^4 - 3x^3 + 6x^2 + 2x - 60$ given that $1 + 3i$ is a zero of f.

Algebraic Solution

Because complex zeros occur in conjugate pairs, you know that $1 - 3i$ is also a zero of f. This means that both

$$[x - (1 + 3i)] \quad \text{and} \quad [x - (1 - 3i)]$$

are factors of f. Multiplying these two factors produces

$$[x - (1 + 3i)][x - (1 - 3i)] = [(x - 1) - 3i][(x - 1) + 3i]$$

$$= (x - 1)^2 - 9i^2$$

$$= x^2 - 2x + 10.$$

Using long division, you can divide $x^2 - 2x + 10$ into f to obtain the following.

$$
\require{enclose}
\begin{array}{r}
x^2 - x - 6 \\
x^2 - 2x + 10 \enclose{longdiv}{x^4 - 3x^3 + 6x^2 + 2x - 60} \\
\underline{x^4 - 2x^3 + 10x^2} \\
-x^3 - 4x^2 + 2x \\
\underline{-x^3 + 2x^2 - 10x} \\
-6x^2 + 12x - 60 \\
\underline{-6x^2 + 12x - 60} \\
0
\end{array}
$$

So, you have

$$f(x) = (x^2 - 2x + 10)(x^2 - x - 6)$$

$$= (x^2 - 2x + 10)(x - 3)(x + 2).$$

and you can conclude that the zeros of f are $x = 1 + 3i$, $x = 1 - 3i$, $x = 3$, and $x = -2$.

CHECK**Point**   Now try Exercise 55.

Graphical Solution

Because complex zeros always occur in conjugate pairs, you know that $1 - 3i$ is also a zero of f. Because the polynomial is a fourth-degree polynomial, you know that there are two other zeros of the function. Use a graphing utility to graph

$$y = x^4 - 3x^3 + 6x^2 + 2x - 60$$

as shown in Figure 3.33.

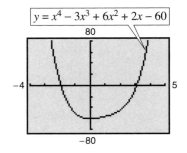

FIGURE **3.33**

You can see that -2 and 3 appear to be zeros of the graph of the function. Use the *zero* or *root* feature or the *zoom* and *trace* features of the graphing utility to confirm that $x = -2$ and $x = 3$ are zeros of the graph. So, you can conclude that the zeros of f are $x = 1 + 3i$, $x = 1 - 3i$, $x = 3$, and $x = -2$.

Algebra Help

You can review the techniques for polynomial long division in Section 3.3.

In Example 7, if you were not told that $1 + 3i$ is a zero of f, you could still find all zeros of the function by using synthetic division to find the real zeros -2 and 3. Then you could factor the polynomial as $(x + 2)(x - 3)(x^2 - 2x + 10)$. Finally, by using the Quadratic Formula, you could determine that the zeros are $x = -2$, $x = 3$, $x = 1 + 3i$, and $x = 1 - 3i$.

Example 8 shows how to find all the zeros of a polynomial function, including complex zeros.

Example 8   Finding the Zeros of a Polynomial Function

Write $f(x) = x^5 + x^3 + 2x^2 - 12x + 8$ as the product of linear factors, and list all of its zeros.

Solution

The possible rational zeros are $\pm 1, \pm 2, \pm 4$, and ± 8. Synthetic division produces the following.

$$
\begin{array}{r|rrrrrr}
1 & 1 & 0 & 1 & 2 & -12 & 8 \\
  &   & 1 & 1 & 2 & 4 & -8 \\
\hline
  & 1 & 1 & 2 & 4 & -8 & 0
\end{array}
$$

$$
\begin{array}{r|rrrrr}
-2 & 1 & 1 & 2 & 4 & -8 \\
   &   & -2 & 2 & -8 & 8 \\
\hline
   & 1 & -1 & 4 & -4 & 0
\end{array}
$$

So, you have

$$f(x) = x^5 + x^3 + 2x^2 - 12x + 8$$

$$= (x - 1)(x + 2)(x^3 - x^2 + 4x - 4).$$

You can factor $x^3 - x^2 + 4x - 4$ as $(x - 1)(x^2 + 4)$, and by factoring $x^2 + 4$ as

$$x^2 - (-4) = \left(x - \sqrt{-4}\right)\left(x + \sqrt{-4}\right)$$

$$= (x - 2i)(x + 2i)$$

you obtain

$$f(x) = (x - 1)(x - 1)(x + 2)(x - 2i)(x + 2i)$$

which gives the following five zeros of f.

$$x = 1, x = 1, x = -2, x = 2i, \quad \text{and} \quad x = -2i$$

From the graph of f shown in Figure 3.34, you can see that the *real* zeros are the only ones that appear as x-intercepts. Note that $x = 1$ is a repeated zero.

CHECK*Point*   Now try Exercise 77.

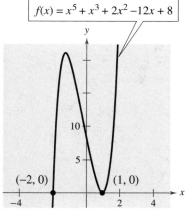

$f(x) = x^5 + x^3 + 2x^2 - 12x + 8$

(-2, 0)   (1, 0)

FIGURE **3.34**

TECHNOLOGY

You can use the *table* feature of a graphing utility to help you determine which of the possible rational zeros are zeros of the polynomial in Example 8. The table should be set to *ask* mode. Then enter each of the possible rational zeros in the table. When you do this, you will see that there are two rational zeros, −2 and 1, as shown at the right.

X	Y₁
-8	-33048
-4	-1000
-2	0
-1	20
1	0
2	32
4	1080

X=4

Other Tests for Zeros of Polynomials

You know that an nth-degree polynomial function can have *at most n* real zeros. Of course, many nth-degree polynomials do not have that many real zeros. For instance, $f(x) = x^2 + 1$ has no real zeros, and $f(x) = x^3 + 1$ has only one real zero. The following theorem, called **Descartes's Rule of Signs,** sheds more light on the number of real zeros of a polynomial.

Descartes's Rule of Signs

Let $f(x) = a_n x^n + a_{n-1} x^{n-1} + \cdots + a_2 x^2 + a_1 x + a_0$ be a polynomial with real coefficients and $a_0 \neq 0$.

1. The number of *positive real zeros* of f is either equal to the number of variations in sign of $f(x)$ or less than that number by an even integer.

2. The number of *negative real zeros* of f is either equal to the number of variations in sign of $f(-x)$ or less than that number by an even integer.

A **variation in sign** means that two consecutive coefficients have opposite signs.

When using Descartes's Rule of Signs, a zero of multiplicity k should be counted as k zeros. For instance, the polynomial $x^3 - 3x + 2$ has two variations in sign, and so has either two positive or no positive real zeros. Because

$$x^3 - 3x + 2 = (x - 1)(x - 1)(x + 2)$$

you can see that the two positive real zeros are $x = 1$ of multiplicity 2.

Example 9   Using Descartes's Rule of Signs

Describe the possible real zeros of

$$f(x) = 3x^3 - 5x^2 + 6x - 4.$$

Solution

The original polynomial has *three* variations in sign.

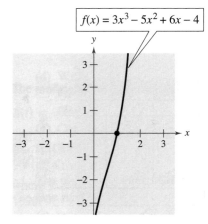

$$f(x) = 3x^3 - 5x^2 + 6x - 4$$

The polynomial

$$f(\quad) = 3(\quad)^3 - 5(\quad)^2 + 6(\quad) - 4$$
$$= -3x^3 - 5x^2 - 6x - 4$$

has no variations in sign. So, from Descartes's Rule of Signs, the polynomial $f(x) = 3x^3 - 5x^2 + 6x - 4$ has either three positive real zeros or one positive real zero, and has no negative real zeros. From the graph in Figure 3.35, you can see that the function has only one real zero, at $x = 1$.

FIGURE 3.35

CHECK*Point* Now try Exercise 87.

Another test for zeros of a polynomial function is related to the sign pattern in the last row of the synthetic division array. This test can give you an upper or lower bound of the real zeros of f. A real number b is an **upper bound** for the real zeros of f if no zeros are greater than b. Similarly, b is a **lower bound** if no real zeros of f are less than b.

Upper and Lower Bound Rules

Let $f(x)$ be a polynomial with real coefficients and a positive leading coefficient. Suppose $f(x)$ is divided by $x - c$, using synthetic division.

1. If $c > 0$ and each number in the last row is either positive or zero, c is an **upper bound** for the real zeros of f.

2. If $c < 0$ and the numbers in the last row are alternately positive and negative (zero entries count as positive or negative), c is a **lower bound** for the real zeros of f.

Example 10   Finding the Zeros of a Polynomial Function

Find the real zeros of $f(x) = 6x^3 - 4x^2 + 3x - 2$.

Solution

The possible real zeros are as follows.

$$\frac{\text{Factors of 2}}{\text{Factors of 6}} = \frac{\pm 1, \pm 2}{\pm 1, \pm 2, \pm 3, \pm 6} = \pm 1, \pm\frac{1}{2}, \pm\frac{1}{3}, \pm\frac{1}{6}, \pm\frac{2}{3}, \pm 2$$

The original polynomial $f(x)$ has three variations in sign. The polynomial

$$f(-x) = 6(-x)^3 - 4(-x)^2 + 3(-x) - 2$$
$$= -6x^3 - 4x^2 - 3x - 2$$

has no variations in sign. As a result of these two findings, you can apply Descartes's Rule of Signs to conclude that there are three positive real zeros or one positive real zero, and no negative zeros. Trying $x = 1$ produces the following.

```
1 | 6   -4    3   -2
  |       6    2    5
  ----------------------
    6    2    5    3
```

So, $x = 1$ is not a zero, but because the last row has all positive entries, you know that $x = 1$ is an upper bound for the real zeros. So, you can restrict the search to zeros between 0 and 1. By trial and error, you can determine that $x = \frac{2}{3}$ is a zero. So,

$$f(x) = \left(x - \frac{2}{3}\right)(6x^2 + 3).$$

Because $6x^2 + 3$ has no real zeros, it follows that $x = \frac{2}{3}$ is the only real zero.

CHECK*Point*   Now try Exercise 95.

Before concluding this section, here are two additional hints that can help you find the real zeros of a polynomial.

1. If the terms of $f(x)$ have a common monomial factor, it should be factored out before applying the tests in this section. For instance, by writing

$$f(x) = x^4 - 5x^3 + 3x^2 + x$$
$$= x(x^3 - 5x^2 + 3x + 1)$$

you can see that $x = 0$ is a zero of f and that the remaining zeros can be obtained by analyzing the cubic factor.

2. If you are able to find all but two zeros of $f(x)$, you can always use the Quadratic Formula on the remaining quadratic factor. For instance, if you succeeded in writing

$$f(x) = x^4 - 5x^3 + 3x^2 + x$$
$$= x(x - 1)(x^2 - 4x - 1)$$

you can apply the Quadratic Formula to $x^2 - 4x - 1$ to conclude that the two remaining zeros are $x = 2 + \sqrt{5}$ and $x = 2 - \sqrt{5}$.

Example 11   Using a Polynomial Model

You are designing candle-making kits. Each kit contains 25 cubic inches of candle wax and a mold for making a pyramid-shaped candle. You want the height of the candle to be 2 inches less than the length of each side of the candle's square base. What should the dimensions of your candle mold be?

Solution

The volume of a pyramid is $V = \frac{1}{3}Bh$, where B is the area of the base and h is the height. The area of the base is x^2 and the height is $(x - 2)$. So, the volume of the pyramid is $V = \frac{1}{3}x^2(x - 2)$. Substituting 25 for the volume yields the following.

$$= \frac{1}{3}x^2(x - 2)$$
$$75 = x^3 - 2x^2$$
$$0 = x^3 - 2x^2 - 75$$

The possible rational solutions are $x = \pm 1, \ \pm 3, \ \pm 5, \ \pm 15, \ \pm 25, \ \pm 75$. Use synthetic division to test some of the possible solutions. Note that in this case, it makes sense to test only positive x-values. Using synthetic division, you can determine that $x = 5$ is a solution.

$$
\begin{array}{r|rrrr}
5 & 1 & -2 & 0 & -75 \\
  &   & 5 & 15 & 75 \\
\hline
  & 1 & 3 & 15 & 0
\end{array}
$$

The other two solutions, which satisfy $x^2 + 3x + 15 = 0$, are imaginary and can be discarded. You can conclude that the base of the candle mold should be 5 inches by 5 inches and the height of the mold should be $5 - 2 = 3$ inches.

CHECK *Point*   Now try Exercise 115.

3.4 EXERCISES

See www.CalcChat.com for worked-out solutions to odd-numbered exercises.

VOCABULARY: Fill in the blanks.

1. The _____ _____ of _____ states that if $f(x)$ is a polynomial of degree n $(n > 0)$, then f has at least one zero in the complex number system.

2. The _____ _____ _____ states that if $f(x)$ is a polynomial of degree n $(n > 0)$, then f has precisely n linear factors, $f(x) = a_n(x - c_1)(x - c_2) \cdots (x - c_n)$, where $c_1, c_2, \ldots, c_n$ are complex numbers.

3. The test that gives a list of the possible rational zeros of a polynomial function is called the _____ _____ Test.

4. If $a + bi$ is a complex zero of a polynomial with real coefficients, then so is its _____, $a - bi$.

5. Every polynomial of degree $n > 0$ with real coefficients can be written as the product of _____ and _____ factors with real coefficients, where the _____ factors have no real zeros.

6. A quadratic factor that cannot be factored further as a product of linear factors containing real numbers is said to be _____ over the _____.

7. The theorem that can be used to determine the possible numbers of positive real zeros and negative real zeros of a function is called _____ _____ of _____.

8. A real number b is a(n) _____ bound for the real zeros of f if no real zeros are less than b, and is a(n) _____ bound if no real zeros are greater than b.

SKILLS AND APPLICATIONS

In Exercises 9–14, find all the zeros of the function.

9. $f(x) = x(x - 6)^2$

10. $f(x) = x^2(x + 3)(x^2 - 1)$

11. $g(x) = (x - 2)(x + 4)^3$

12. $f(x) = (x + 5)(x - 8)^2$

13. $f(x) = (x + 6)(x + i)(x - i)$

14. $h(t) = (t - 3)(t - 2)(t - 3i)(t + 3i)$

In Exercises 15–18, use the Rational Zero Test to list all possible rational zeros of f. Verify that the zeros of f shown on the graph are contained in the list.

15. $f(x) = x^3 + 2x^2 - x - 2$

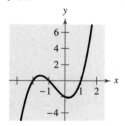

16. $f(x) = x^3 - 4x^2 - 4x + 16$

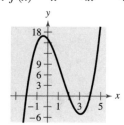

17. $f(x) = 2x^4 - 17x^3 + 35x^2 + 9x - 45$

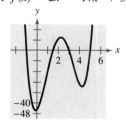

18. $f(x) = 4x^5 - 8x^4 - 5x^3 + 10x^2 + x - 2$

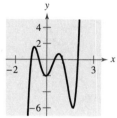

In Exercises 19–28, find all the rational zeros of the function.

19. $f(x) = x^3 - 6x^2 + 11x - 6$

20. $f(x) = x^3 - 7x - 6$

21. $g(x) = x^3 - 4x^2 - x + 4$

22. $h(x) = x^3 - 9x^2 + 20x - 12$

23. $h(t) = t^3 + 8t^2 + 13t + 6$

24. $p(x) = x^3 - 9x^2 + 27x - 27$

25. $C(x) = 2x^3 + 3x^2 - 1$

26. $f(x) = 3x^3 - 19x^2 + 33x - 9$

27. $f(x) = 9x^4 - 9x^3 - 58x^2 + 4x + 24$

28. $f(x) = 2x^4 - 15x^3 + 23x^2 + 15x - 25$

In Exercises 29–32, find all real solutions of the polynomial equation.

29. $z^4 + z^3 + z^2 + 3z - 6 = 0$

30. $x^4 - 13x^2 - 12x = 0$

31. $2y^4 + 3y^3 - 16y^2 + 15y - 4 = 0$

32. $x^5 - x^4 - 3x^3 + 5x^2 - 2x = 0$

In Exercises 33–36, (a) list the possible rational zeros of f, (b) sketch the graph of f so that some of the possible zeros in part (a) can be disregarded, and then (c) determine all real zeros of f.

33. $f(x) = x^3 + x^2 - 4x - 4$

34. $f(x) = -3x^3 + 20x^2 - 36x + 16$

35. $f(x) = -4x^3 + 15x^2 - 8x - 3$

36. $f(x) = 4x^3 - 12x^2 - x + 15$

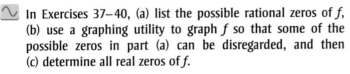 In Exercises 37–40, (a) list the possible rational zeros of f, (b) use a graphing utility to graph f so that some of the possible zeros in part (a) can be disregarded, and then (c) determine all real zeros of f.

37. $f(x) = -2x^4 + 13x^3 - 21x^2 + 2x + 8$

38. $f(x) = 4x^4 - 17x^2 + 4$

39. $f(x) = 32x^3 - 52x^2 + 17x + 3$

40. $f(x) = 4x^3 + 7x^2 - 11x - 18$

GRAPHICAL ANALYSIS In Exercises 41–44, (a) use the *zero* or *root* feature of a graphing utility to approximate the zeros of the function accurate to three decimal places, (b) determine one of the exact zeros (use synthetic division to verify your result), and (c) factor the polynomial completely.

41. $f(x) = x^4 - 3x^2 + 2$ **42.** $P(t) = t^4 - 7t^2 + 12$

43. $h(x) = x^5 - 7x^4 + 10x^3 + 14x^2 - 24x$

44. $g(x) = 6x^4 - 11x^3 - 51x^2 + 99x - 27$

In Exercises 45–50, find a polynomial function with real coefficients that has the given zeros. (There are many correct answers.)

45. $1, 5i$ **46.** $4, -3i$

47. $2, 5 + i$ **48.** $5, 3 - 2i$

49. $\frac{2}{3}, -1, 3 + \sqrt{2}i$ **50.** $-5, -5, 1 + \sqrt{3}i$

In Exercises 51–54, write the polynomial (a) as the product of factors that are irreducible over the *rationals*, (b) as the product of linear and quadratic factors that are irreducible over the *reals*, and (c) in completely factored form.

51. $f(x) = x^4 + 6x^2 - 27$

52. $f(x) = x^4 - 2x^3 - 3x^2 + 12x - 18$
   (*Hint:* One factor is $x^2 - 6$.)

53. $f(x) = x^4 - 4x^3 + 5x^2 - 2x - 6$
   (*Hint:* One factor is $x^2 - 2x - 2$.)

54. $f(x) = x^4 - 3x^3 - x^2 - 12x - 20$
   (*Hint:* One factor is $x^2 + 4$.)

In Exercises 55–62, use the given zero to find all the zeros of the function.

Function	Zero
55. $f(x) = x^3 - x^2 + 4x - 4$	$2i$
56. $f(x) = 2x^3 + 3x^2 + 18x + 27$	$3i$
57. $f(x) = 2x^4 - x^3 + 49x^2 - 25x - 25$	$5i$
58. $g(x) = x^3 - 7x^2 - x + 87$	$5 + 2i$
59. $g(x) = 4x^3 + 23x^2 + 34x - 10$	$-3 + i$
60. $h(x) = 3x^3 - 4x^2 + 8x + 8$	$1 - \sqrt{3}i$
61. $f(x) = x^4 + 3x^3 - 5x^2 - 21x + 22$	$-3 + \sqrt{2}i$
62. $f(x) = x^3 + 4x^2 + 14x + 20$	$-1 - 3i$

In Exercises 63–80, find all the zeros of the function and write the polynomial as a product of linear factors.

63. $f(x) = x^2 + 36$ **64.** $f(x) = x^2 - x + 56$

65. $h(x) = x^2 - 2x + 17$ **66.** $g(x) = x^2 + 10x + 17$

67. $f(x) = x^4 - 16$ **68.** $f(y) = y^4 - 256$

69. $f(z) = z^2 - 2z + 2$

70. $h(x) = x^3 - 3x^2 + 4x - 2$

71. $g(x) = x^3 - 3x^2 + x + 5$

72. $f(x) = x^3 - x^2 + x + 39$

73. $h(x) = x^3 - x + 6$

74. $h(x) = x^3 + 9x^2 + 27x + 35$

75. $f(x) = 5x^3 - 9x^2 + 28x + 6$

76. $g(x) = 2x^3 - x^2 + 8x + 21$

77. $g(x) = x^4 - 4x^3 + 8x^2 - 16x + 16$

78. $h(x) = x^4 + 6x^3 + 10x^2 + 6x + 9$

79. $f(x) = x^4 + 10x^2 + 9$

80. $f(x) = x^4 + 29x^2 + 100$

In Exercises 81–86, find all the zeros of the function. When there is an extended list of possible rational zeros, use a graphing utility to graph the function in order to discard any rational zeros that are obviously not zeros of the function.

81. $f(x) = x^3 + 24x^2 + 214x + 740$

82. $f(s) = 2s^3 - 5s^2 + 12s - 5$

83. $f(x) = 16x^3 - 20x^2 - 4x + 15$

84. $f(x) = 9x^3 - 15x^2 + 11x - 5$

85. $f(x) = 2x^4 + 5x^3 + 4x^2 + 5x + 2$

86. $g(x) = x^5 - 8x^4 + 28x^3 - 56x^2 + 64x - 32$

In Exercises 87–94, use Descartes's Rule of Signs to determine the possible numbers of positive and negative zeros of the function.

87. $g(x) = 2x^3 - 3x^2 - 3$ **88.** $h(x) = 4x^2 - 8x + 3$

89. $h(x) = 2x^3 + 3x^2 + 1$ **90.** $h(x) = 2x^4 - 3x + 2$

91. $g(x) = 5x^5 - 10x$

92. $f(x) = 4x^3 - 3x^2 + 2x - 1$

93. $f(x) = -5x^3 + x^2 - x + 5$

94. $f(x) = 3x^3 + 2x^2 + x + 3$

In Exercises 95–98, use synthetic division to verify the upper and lower bounds of the real zeros of f.

95. $f(x) = x^3 + 3x^2 - 2x + 1$
    (a) Upper: $x = 1$ (b) Lower: $x = -4$

96. $f(x) = x^3 - 4x^2 + 1$
    (a) Upper: $x = 4$ (b) Lower: $x = -1$

97. $f(x) = x^4 - 4x^3 + 16x - 16$
    (a) Upper: $x = 5$ (b) Lower: $x = -3$

98. $f(x) = 2x^4 - 8x + 3$
    (a) Upper: $x = 3$ (b) Lower: $x = -4$

In Exercises 99–102, find all the real zeros of the function.

99. $f(x) = 4x^3 - 3x - 1$

100. $f(z) = 12z^3 - 4z^2 - 27z + 9$

101. $f(y) = 4y^3 + 3y^2 + 8y + 6$

102. $g(x) = 3x^3 - 2x^2 + 15x - 10$

In Exercises 103–106, find all the rational zeros of the polynomial function.

103. $P(x) = x^4 - \frac{25}{4}x^2 + 9 = \frac{1}{4}(4x^4 - 25x^2 + 36)$

104. $f(x) = x^3 - \frac{3}{2}x^2 - \frac{23}{2}x + 6 = \frac{1}{2}(2x^3 - 3x^2 - 23x + 12)$

105. $f(x) = x^3 - \frac{1}{4}x^2 - x + \frac{1}{4} = \frac{1}{4}(4x^3 - x^2 - 4x + 1)$

106. $f(z) = z^3 + \frac{11}{6}z^2 - \frac{1}{2}z - \frac{1}{3} = \frac{1}{6}(6z^3 + 11z^2 - 3z - 2)$

In Exercises 107–110, match the cubic function with the numbers of rational and irrational zeros.

(a) Rational zeros: 0; irrational zeros: 1

(b) Rational zeros: 3; irrational zeros: 0

(c) Rational zeros: 1; irrational zeros: 2

(d) Rational zeros: 1; irrational zeros: 0

107. $f(x) = x^3 - 1$ **108.** $f(x) = x^3 - 2$

109. $f(x) = x^3 - x$ **110.** $f(x) = x^3 - 2x$

111. GEOMETRY An open box is to be made from a rectangular piece of material, 15 centimeters by 9 centimeters, by cutting equal squares from the corners and turning up the sides.

(a) Let x represent the length of the sides of the squares removed. Draw a diagram showing the squares removed from the original piece of material and the resulting dimensions of the open box.

(b) Use the diagram to write the volume V of the box as a function of x. Determine the domain of the function.

(c) Sketch the graph of the function and approximate the dimensions of the box that will yield a maximum volume.

(d) Find values of x such that $V = 56$. Which of these values is a physical impossibility in the construction of the box? Explain.

112. GEOMETRY A rectangular package to be sent by a delivery service (see figure) can have a maximum combined length and girth (perimeter of a cross section) of 120 inches.

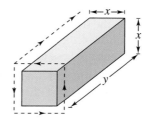

(a) Write a function $V(x)$ that represents the volume of the package.

(b) Use a graphing utility to graph the function and approximate the dimensions of the package that will yield a maximum volume.

(c) Find values of x such that $V = 13,500$. Which of these values is a physical impossibility in the construction of the package? Explain.

113. ADVERTISING COST A company that produces MP3 players estimates that the profit P (in dollars) for selling a particular model is given by

$$P = -76x^3 + 4830x^2 - 320,000, \quad 0 \le x \le 60$$

where x is the advertising expense (in tens of thousands of dollars). Using this model, find the smaller of two advertising amounts that will yield a profit of $2,500,000.

114. ADVERTISING COST A company that manufactures bicycles estimates that the profit P (in dollars) for selling a particular model is given by

$$P = -45x^3 + 2500x^2 - 275,000, \quad 0 \le x \le 50$$

where x is the advertising expense (in tens of thousands of dollars). Using this model, find the smaller of two advertising amounts that will yield a profit of $800,000.

115. GEOMETRY A bulk food storage bin with dimensions 2 feet by 3 feet by 4 feet needs to be increased in size to hold five times as much food as the current bin. (Assume each dimension is increased by the same amount.)

(a) Write a function that represents the volume V of the new bin.

(b) Find the dimensions of the new bin.

116. GEOMETRY A manufacturer wants to enlarge an existing manufacturing facility such that the total floor area is 1.5 times that of the current facility. The floor area of the current facility is rectangular and measures 250 feet by 160 feet. The manufacturer wants to increase each dimension by the same amount.

(a) Write a function that represents the new floor area A.

(b) Find the dimensions of the new floor.

(c) Another alternative is to increase the current floor's length by an amount that is twice an increase in the floor's width. The total floor area is 1.5 times that of the current facility. Repeat parts (a) and (b) using these criteria.

 117. COST The ordering and transportation cost C (in thousands of dollars) for the components used in manufacturing a product is given by

$$C = 100\left(\frac{200}{x^2} + \frac{x}{x + 30}\right), \quad x \geq 1$$

where x is the order size (in hundreds). In calculus, it can be shown that the cost is a minimum when

$$3x^3 - 40x^2 - 2400x - 36{,}000 = 0.$$

Use a calculator to approximate the optimal order size to the nearest hundred units.

118. HEIGHT OF A BASEBALL A baseball is thrown upward from a height of 6 feet with an initial velocity of 48 feet per second, and its height h (in feet) is

$$h(t) = -16t^2 + 48t + 6, \quad 0 \leq t \leq 3$$

where t is the time (in seconds). You are told the ball reaches a height of 64 feet. Is this possible?

119. PROFIT The demand equation for a certain product is $p = 140 - 0.0001x$, where p is the unit price (in dollars) of the product and x is the number of units produced and sold. The cost equation for the product is $C = 80x + 150{,}000$, where C is the total cost (in dollars) and x is the number of units produced. The total profit obtained by producing and selling x units is $P = R - C = xp - C$. You are working in the marketing department of the company that produces this product, and you are asked to determine a price p that will yield a profit of 9 million dollars. Is this possible? Explain.

 120. ATHLETICS The attendance A (in millions) at NCAA women's college basketball games for the years 2000 through 2007 is shown in the table. (Source: National Collegiate Athletic Association, Indianapolis, IN)

Year	Attendance, A
2000	8.7
2001	8.8
2002	9.5
2003	10.2
2004	10.0
2005	9.9
2006	9.9
2007	10.9

(a) Use a graphing utility to create a scatter plot of the data. Let t represent the year, with $t = 0$ corresponding to 2000.

(b) Use the *regression* feature of the graphing utility to find a quartic model for the data.

(c) Graph the model and the scatter plot in the same viewing window. How well does the model fit the data?

(d) According to the model in part (b), in what year(s) was the attendance at least 10 million?

(e) According to the model, will the attendance continue to increase? Explain.

EXPLORATION

TRUE OR FALSE? In Exercises 121 and 122, decide whether the statement is true or false. Justify your answer.

121. It is possible for a third-degree polynomial function with integer coefficients to have no real zeros.

122. If $x = -i$ is a zero of the function given by

$$f(x) = x^3 + ix^2 + ix - 1$$

then $x = i$ must also be a zero of f.

THINK ABOUT IT In Exercises 123–128, determine (if possible) the zeros of the function g if the function f has zeros at $x = r_1$, $x = r_2$, and $x = r_3$.

123. $g(x) = -f(x)$

124. $g(x) = 3f(x)$

125. $g(x) = f(x - 5)$

126. $g(x) = f(2x)$

127. $g(x) = 3 + f(x)$

128. $g(x) = f(-x)$

129. THINK ABOUT IT A third-degree polynomial function f has real zeros -2, $\frac{1}{2}$, and 3, and its leading coefficient is negative. Write an equation for f. Sketch the graph of f. How many different polynomial functions are possible for f?

130. CAPSTONE Use a graphing utility to graph the function given by $f(x) = x^4 - 4x^2 + k$ for different values of k. Find values of k such that the zeros of f satisfy the specified characteristics. (Some parts do not have unique answers.)

(a) Four real zeros

(b) Two real zeros, each of multiplicity 2

(c) Two real zeros and two complex zeros

(d) Four complex zeros

(e) Will the answers to parts (a) through (d) change for the function g, where $g(x) = f(x - 2)$?

(f) Will the answers to parts (a) through (d) change for the function g, where $g(x) = f(2x)$?

131. THINK ABOUT IT Sketch the graph of a fifth-degree polynomial function whose leading coefficient is positive and that has a zero at $x = 3$ of multiplicity 2.

132. WRITING Compile a list of all the various techniques for factoring a polynomial that have been covered so far in the text. Give an example illustrating each technique, and write a paragraph discussing when the use of each technique is appropriate.

133. THINK ABOUT IT Let $y = f(x)$ be a quartic polynomial with leading coefficient $a = 1$ and $f(i) = f(2i) = 0$. Write an equation for f.

134. THINK ABOUT IT Let $y = f(x)$ be a cubic polynomial with leading coefficient $a = -1$ and $f(2) = f(i) = 0$. Write an equation for f.

In Exercises 135 and 136, the graph of a cubic polynomial function $y = f(x)$ is shown. It is known that one of the zeros is $1 + i$. Write an equation for f.

135.

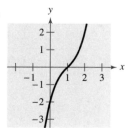

136.

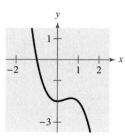

137. Use the information in the table to answer each question.

Interval	Value of $f(x)$
$(-\infty, -2)$	Positive
$(-2, 1)$	Negative
$(1, 4)$	Negative
$(4, \infty)$	Positive

(a) What are the three real zeros of the polynomial function f?

(b) What can be said about the behavior of the graph of f at $x = 1$?

(c) What is the least possible degree of f? Explain. Can the degree of f ever be odd? Explain.

(d) Is the leading coefficient of f positive or negative? Explain.

(e) Write an equation for f. (There are many correct answers.)

(f) Sketch a graph of the equation you wrote in part (e).

138. (a) Find a quadratic function f (with integer coefficients) that has $\pm \sqrt{b}\,i$ as zeros. Assume that b is a positive integer.

(b) Find a quadratic function f (with integer coefficients) that has $a \pm bi$ as zeros. Assume that b is a positive integer.

139. GRAPHICAL REASONING The graph of one of the following functions is shown below. Identify the function shown in the graph. Explain why each of the others is not the correct function. Use a graphing utility to verify your result.

(a) $f(x) = x^2(x + 2)(x - 3.5)$

(b) $g(x) = (x + 2)(x - 3.5)$

(c) $h(x) = (x + 2)(x - 3.5)(x^2 + 1)$

(d) $k(x) = (x + 1)(x + 2)(x - 3.5)$

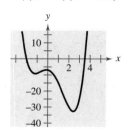

3.5   MATHEMATICAL MODELING AND VARIATION

Introduction

You have already studied some techniques for fitting models to data. For instance, in Section 2.1, you learned how to find the equation of a line that passes through two points. In this section, you will study other techniques for fitting models to data: *least squares regression* and *direct and inverse variation*. The resulting models are either polynomial functions or rational functions. (Rational functions will be studied in Chapter 4.)

Example 1   A Mathematical Model

The populations y (in millions) of the United States from 2000 through 2007 are shown in the table.   (Source: U.S. Census Bureau)

Year	Population, y
2000	282.4
2001	285.3
2002	288.2
2003	290.9
2004	293.6
2005	296.3
2006	299.2
2007	302.0

A linear model that approximates the data is $y = 2.78t + 282.5$ for $0 \le t \le 7$, where t is the year, with $t = 0$ corresponding to 2000. Plot the actual data *and* the model on the same graph. How closely does the model represent the data?

Solution

The actual data are plotted in Figure 3.36, along with the graph of the linear model. From the graph, it appears that the model is a "good fit" for the actual data. You can see how well the model fits by comparing the actual values of y with the values of y given by the model. The values given by the model are labeled $y*$ in the table below.

t	0	1	2	3	4	5	6	7
y	282.4	285.3	288.2	290.9	293.6	296.3	299.2	302.0
$y*$	282.5	285.3	288.1	290.8	293.6	296.4	299.2	302.0

CHECK*Point*   Now try Exercise 11.

Note in Example 1 that you could have chosen any two points to find a line that fits the data. However, the given linear model was found using the *regression* feature of a graphing utility and is the line that *best* fits the data. This concept of a "best-fitting" line is discussed on the next page.

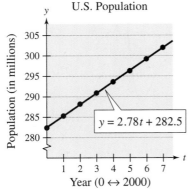

FIGURE 3.36

Least Squares Regression and Graphing Utilities

So far in this text, you have worked with many different types of mathematical models that approximate real-life data. In some instances the model was given (as in Example 1), whereas in other instances you were asked to find the model using simple algebraic techniques or a graphing utility.

To find a model that approximates the data most accurately, statisticians use a measure called the **sum of square differences,** which is the sum of the squares of the differences between actual data values and model values. The "best-fitting" linear model, called the **least squares regression line,** is the one with the least sum of square differences. Recall that you can approximate this line visually by plotting the data points and drawing the line that appears to fit best—or you can enter the data points into a calculator or computer and use the *linear regression* feature of the calculator or computer. When you use the *regression* feature of a graphing calculator or computer program, you will notice that the program may also output an "*r*-value." This *r*-value is the **correlation coefficient** of the data and gives a measure of how well the model fits the data. The closer the value of $|r|$ is to 1, the better the fit.

Example 2   Finding a Least Squares Regression Line

The data in the table show the outstanding household credit market debt D (in trillions of dollars) from 2000 through 2007. Construct a scatter plot that represents the data and find the least squares regression line for the data.   (Source: Board of Governors of the Federal Reserve System)

Household Credit Market Debt

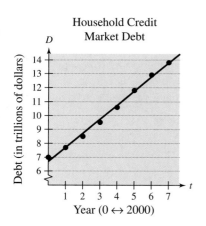

FIGURE 3.37

Year	Household credit market debt, D
2000	7.0
2001	7.7
2002	8.5
2003	9.5
2004	10.6
2005	11.8
2006	12.9
2007	13.8

t	D	$D*$
0	7.0	6.7
1	7.7	7.7
2	8.5	8.7
3	9.5	9.7
4	10.6	10.7
5	11.8	11.8
6	12.9	12.8
7	13.8	13.8

Solution

Let $t = 0$ represent 2000. The scatter plot for the points is shown in Figure 3.37. Using the *regression* feature of a graphing utility, you can determine that the equation of the least squares regression line is

$$D = 1.01t + 6.7.$$

To check this model, compare the actual D-values with the D-values given by the model, which are labeled $D*$ in the table at the left. The correlation coefficient for this model is $r \approx 0.997$, which implies that the model is a good fit.

CHECK**Point**   Now try Exercise 17.

Direct Variation

There are two basic types of linear models. The more general model has a y-intercept that is nonzero.

$$y = mx + b, \quad b \neq 0$$

The simpler model

$$y = kx$$

has a y-intercept that is zero. In the simpler model, y is said to **vary directly** as x, or to be **directly proportional** to x.

Direct Variation

The following statements are equivalent.

1. y **varies directly** as x.

2. y is **directly proportional** to x.

3. $y = kx$ for some nonzero constant k.

k is the **constant of variation** or the **constant of proportionality.**

Example 3 **Direct Variation**

In Pennsylvania, the state income tax is directly proportional to *gross income*. You are working in Pennsylvania and your state income tax deduction is $46.05 for a gross monthly income of $1500. Find a mathematical model that gives the Pennsylvania state income tax in terms of gross income.

Solution

Verbal Model:	State income tax $= k \cdot$ Gross income	

Labels:	State income tax $= y$	(dollars)
	Gross income $= x$	(dollars)
	Income tax rate $= k$	(percent in decimal form)

Equation: $y = kx$

To solve for k, substitute the given information into the equation $y = kx$, and then solve for k.

$$y = kx$$
$$= k(\quad)$$
$$0.0307 = k$$

So, the equation (or model) for state income tax in Pennsylvania is

$$y = 0.0307x.$$

In other words, Pennsylvania has a state income tax rate of 3.07% of gross income. The graph of this equation is shown in Figure 3.38.

CHECK *Point* Now try Exercise 43.

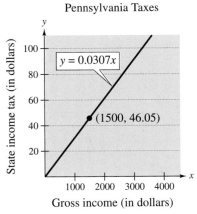

Pennsylvania Taxes

$y = 0.0307x$

$(1500, 46.05)$

State income tax (in dollars)

Gross income (in dollars)

FIGURE 3.38

Direct Variation as an *n*th Power

Another type of direct variation relates one variable to a *power* of another variable. For example, in the formula for the area of a circle

$$A = \pi r^2$$

the area A is directly proportional to the square of the radius r. Note that for this formula, π is the constant of proportionality.

Direct Variation as an *n*th Power

The following statements are equivalent.

1. y **varies directly as the *n*th power** of x.

2. y is **directly proportional to the *n*th power** of x.

3. $y = kx^n$ for some constant k.

| **Example 4**   **Direct Variation as *n*th Power**

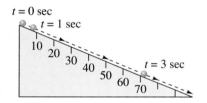

$t = 0$ sec
$t = 1$ sec
10 20 30 40 50 60 70
$t = 3$ sec

FIGURE 3.39

The distance a ball rolls down an inclined plane is directly proportional to the square of the time it rolls. During the first second, the ball rolls 8 feet. (See Figure 3.39.)

a. Write an equation relating the distance traveled to the time.

b. How far will the ball roll during the first 3 seconds?

Solution

a. Letting d be the distance (in feet) the ball rolls and letting t be the time (in seconds), you have

$$d = kt^2.$$

Now, because $d = 8$ when $t = 1$, you can see that $k = 8$, as follows.

$$d = kt^2$$
$$= k(\ )^2$$
$$8 = k$$

So, the equation relating distance to time is

$$d = 8t^2.$$

b. When $t = 3$, the distance traveled is $d = 8(\ )^2 = 8(9) = 72$ feet.

CHECK**Point**   Now try Exercise 75.

In Examples 3 and 4, the direct variations are such that an *increase* in one variable corresponds to an *increase* in the other variable. This is also true in the model $d = \frac{1}{5}F$, $F > 0$, where an increase in F results in an increase in d. You should not, however, assume that this always occurs with direct variation. For example, in the model $y = -3x$, an increase in x results in a *decrease* in y, and yet y is said to vary directly as x.

Inverse Variation

Inverse Variation

The following statements are equivalent.

1. *y* **varies inversely** as *x*.   **2.** *y* is **inversely proportional** to *x*.

3. $y = \dfrac{k}{x}$ for some constant *k*.

If *x* and *y* are related by an equation of the form $y = k/x^n$, then *y* varies inversely as the *n*th power of *x* (or *y* is inversely proportional to the *n*th power of *x*).

Some applications of variation involve problems with *both* direct and inverse variation in the same model. These types of models are said to have **combined variation.**

Example 5   **Direct and Inverse Variation**

A gas law states that the volume of an enclosed gas varies directly as the temperature *and* inversely as the pressure, as shown in Figure 3.40. The pressure of a gas is 0.75 kilogram per square centimeter when the temperature is 294 K and the volume is 8000 cubic centimeters. (a) Write an equation relating pressure, temperature, and volume. (b) Find the pressure when the temperature is 300 K and the volume is 7000 cubic centimeters.

Solution

a. Let *V* be volume (in cubic centimeters), let *P* be pressure (in kilograms per square centimeter), and let *T* be temperature (in Kelvin). Because *V* varies directly as *T* and inversely as *P*, you have

$$V = \frac{kT}{P}.$$

Now, because *P* = 0.75 when *T* = 294 and *V* = 8000, you have

$$= \frac{k(\quad)}{\underline{\qquad}}$$

$$k = \frac{6000}{294} = \frac{1000}{49}.$$

So, the equation relating pressure, temperature, and volume is

$$V = \frac{1000}{49}\left(\frac{T}{P}\right).$$

b. When *T* = 300 and *V* = 7000, the pressure is

$$P = \frac{1000}{49}\left(\underline{\quad}\right) = \frac{300}{343} \approx 0.87 \text{ kilogram per square centimeter.}$$

CHECK*Point*   Now try Exercise 77.

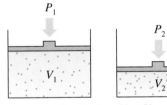

P_1

P_2

V_1

V_2

$P_2 > P_1$ then $V_2 < V_1$

FIGURE **3.40**   If the temperature is held constant and pressure increases, volume decreases.

Joint Variation

In Example 5, note that when a direct variation and an inverse variation occur in the same statement, they are coupled with the word "and." To describe two different *direct* variations in the same statement, the word **jointly** is used.

> ### Joint Variation
>
> The following statements are equivalent.
>
> **1.** z **varies jointly** as x and y.
>
> **2.** z is **jointly proportional** to x and y.
>
> **3.** $z = kxy$ for some constant k.

If x, y, and z are related by an equation of the form

$$z = kx^n y^m$$

then z varies jointly as the nth power of x and the mth power of y.

Example 6   Joint Variation

The *simple* interest for a certain savings account is jointly proportional to the time and the principal. After one quarter (3 months), the interest on a principal of $5000 is $43.75.

a. Write an equation relating the interest, principal, and time.

b. Find the interest after three quarters.

Solution

a. Let I = interest (in dollars), P = principal (in dollars), and t = time (in years). Because I is jointly proportional to P and t, you have

$$I = kPt.$$

For $I = 43.75$, $P = 5000$, and $t = \frac{1}{4}$, you have

$$= k\left(\quad\right)\left(\quad\right)$$

which implies that $k = 4(43.75)/5000 = 0.035$. So, the equation relating interest, principal, and time is

$$I = 0.035Pt$$

which is the familiar equation for simple interest where the constant of proportionality, 0.035, represents an annual interest rate of 3.5%.

b. When $P = \$5000$ and $t = \frac{3}{4}$, the interest is

$$I = (0.035)\left(\quad\right)\left(\quad\right)$$

$$= \$131.25.$$

CHECK*Point*   Now try Exercise 79.

3.5 EXERCISES

See www.CalcChat.com for worked-out solutions to odd-numbered exercises.

VOCABULARY: Fill in the blanks.

1. Two techniques for fitting models to data are called direct _____ and least squares _____.

2. Statisticians use a measure called _____ of_____ _____ to find a model that approximates a set of data most accurately.

3. The linear model with the least sum of square differences is called the _____ _____ _____ line.

4. An r-value of a set of data, also called a _____ _____, gives a measure of how well a model fits a set of data.

5. Direct variation models can be described as "y varies directly as x," or "y is _____ _____ to x."

6. In direct variation models of the form $y = kx$, k is called the _____ of _____.

7. The direct variation model $y = kx^n$ can be described as "y varies directly as the nth power of x," or "y is _____ _____ to the nth power of x."

8. The mathematical model $y = \dfrac{k}{x}$ is an example of _____ variation.

9. Mathematical models that involve both direct and inverse variation are said to have _____ variation.

10. The joint variation model $z = kxy$ can be described as "z varies jointly as x and y," or "z is _____ _____ to x and y."

SKILLS AND APPLICATIONS

11. **EMPLOYMENT** The total numbers of people (in thousands) in the U.S. civilian labor force from 1992 through 2007 are given by the following ordered pairs.

(1992, 128,105)	(2000, 142,583)
(1993, 129,200)	(2001, 143,734)
(1994, 131,056)	(2002, 144,863)
(1995, 132,304)	(2003, 146,510)
(1996, 133,943)	(2004, 147,401)
(1997, 136,297)	(2005, 149,320)
(1998, 137,673)	(2006, 151,428)
(1999, 139,368)	(2007, 153,124)

A linear model that approximates the data is $y = 1695.9t + 124{,}320$, where y represents the number of employees (in thousands) and $t = 2$ represents 1992. Plot the actual data and the model on the same set of coordinate axes. How closely does the model represent the data? (Source: U.S. Bureau of Labor Statistics)

12. **SPORTS** The winning times (in minutes) in the women's 400-meter freestyle swimming event in the Olympics from 1948 through 2008 are given by the following ordered pairs.

(1948, 5.30)	(1972, 4.32)	(1996, 4.12)
(1952, 5.20)	(1976, 4.16)	(2000, 4.10)
(1956, 4.91)	(1980, 4.15)	(2004, 4.09)
(1960, 4.84)	(1984, 4.12)	(2008, 4.05)
(1964, 4.72)	(1988, 4.06)	
(1968, 4.53)	(1992, 4.12)	

A linear model that approximates the data is $y = -0.020t + 5.00$, where y represents the winning time (in minutes) and $t = 0$ represents 1950. Plot the actual data and the model on the same set of coordinate axes. How closely does the model represent the data? Does it appear that another type of model may be a better fit? Explain. (Source: International Olympic Committee)

In Exercises 13–16, sketch the line that you think best approximates the data in the scatter plot. Then find an equation of the line. To print an enlarged copy of the graph, go to the website *www.mathgraphs.com*.

13.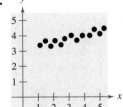

14.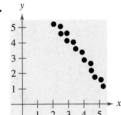

15.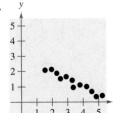

16.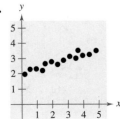

17. SPORTS The lengths (in feet) of the winning men's discus throws in the Olympics from 1920 through 2008 are listed below. (Source: International Olympic Committee)

1920	146.6	1956	184.9	1984	218.5
1924	151.3	1960	194.2	1988	225.8
1928	155.3	1964	200.1	1992	213.7
1932	162.3	1968	212.5	1996	227.7
1936	165.6	1972	211.3	2000	227.3
1948	173.2	1976	221.5	2004	229.3
1952	180.5	1980	218.7	2008	225.8

(a) Sketch a scatter plot of the data. Let y represent the length of the winning discus throw (in feet) and let $t = 20$ represent 1920.

(b) Use a straightedge to sketch the best-fitting line through the points and find an equation of the line.

(c) Use the *regression* feature of a graphing utility to find the least squares regression line that fits the data.

(d) Compare the linear model you found in part (b) with the linear model given by the graphing utility in part (c).

(e) Use the models from parts (b) and (c) to estimate the winning men's discus throw in the year 2012.

18. SALES The total sales (in billions of dollars) for Coca-Cola Enterprises from 2000 through 2007 are listed below. (Source: Coca-Cola Enterprises, Inc.)

2000	14.750	2004	18.185
2001	15.700	2005	18.706
2002	16.899	2006	19.804
2003	17.330	2007	20.936

(a) Sketch a scatter plot of the data. Let y represent the total revenue (in billions of dollars) and let $t = 0$ represent 2000.

(b) Use a straightedge to sketch the best-fitting line through the points and find an equation of the line.

(c) Use the *regression* feature of a graphing utility to find the least squares regression line that fits the data.

(d) Compare the linear model you found in part (b) with the linear model given by the graphing utility in part (c).

(e) Use the models from parts (b) and (c) to estimate the sales of Coca-Cola Enterprises in 2008.

(f) Use your school's library, the Internet, or some other reference source to analyze the accuracy of the estimate in part (e).

19. DATA ANALYSIS: BROADWAY SHOWS The table shows the annual gross ticket sales S (in millions of dollars) for Broadway shows in New York City from 1995 through 2006. (Source: The League of American Theatres and Producers, Inc.)

Year	Sales, S
1995	406
1996	436
1997	499
1998	558
1999	588
2000	603
2001	666
2002	643
2003	721
2004	771
2005	769
2006	862

(a) Use a graphing utility to create a scatter plot of the data. Let $t = 5$ represent 1995.

(b) Use the *regression* feature of a graphing utility to find the equation of the least squares regression line that fits the data.

(c) Use the graphing utility to graph the scatter plot you created in part (a) and the model you found in part (b) in the same viewing window. How closely does the model represent the data?

(d) Use the model to estimate the annual gross ticket sales in 2007 and 2009.

(e) Interpret the meaning of the slope of the linear model in the context of the problem.

20. DATA ANALYSIS: TELEVISION SETS The table shows the numbers N (in millions) of television sets in U.S. households from 2000 through 2006. (Source: Television Bureau of Advertising, Inc.)

Year	Television sets, N
2000	245
2001	248
2002	254
2003	260
2004	268
2005	287
2006	301

(a) Use the *regression* feature of a graphing utility to find the equation of the least squares regression line that fits the data. Let $t = 0$ represent 2000.

(b) Use the graphing utility to create a scatter plot of the data. Then graph the model you found in part (a) and the scatter plot in the same viewing window. How closely does the model represent the data?

(c) Use the model to estimate the number of television sets in U.S. households in 2008.

(d) Use your school's library, the Internet, or some other reference source to analyze the accuracy of the estimate in part (c).

THINK ABOUT IT In Exercises 21 and 22, use the graph to determine whether y varies directly as some power of x or inversely as some power of x. Explain.

21. 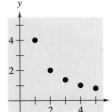 **22.**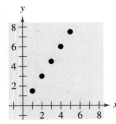

In Exercises 23–26, use the given value of k to complete the table for the direct variation model

$$y = kx^2.$$

Plot the points on a rectangular coordinate system.

x	2	4	6	8	10
$y = kx^2$					

23. $k = 1$ **24.** $k = 2$
25. $k = \frac{1}{2}$ **26.** $k = \frac{1}{4}$

In Exercises 27–30, use the given value of k to complete the table for the inverse variation model

$$y = \frac{k}{x^2}.$$

Plot the points on a rectangular coordinate system.

x	2	4	6	8	10
$y = \dfrac{k}{x^2}$					

27. $k = 2$ **28.** $k = 5$
29. $k = 10$ **30.** $k = 20$

In Exercises 31–34, determine whether the variation model is of the form $y = kx$ or $y = k/x$, and find k. Then write a model that relates y and x.

31.

x	5	10	15	20	25
y	1	$\frac{1}{2}$	$\frac{1}{3}$	$\frac{1}{4}$	$\frac{1}{5}$

32.

x	5	10	15	20	25
y	2	4	6	8	10

33.

x	5	10	15	20	25
y	-3.5	-7	-10.5	-14	-17.5

34.

x	5	10	15	20	25
y	24	12	8	6	$\frac{24}{5}$

DIRECT VARIATION In Exercises 35–38, assume that y is directly proportional to x. Use the given x-value and y-value to find a linear model that relates y and x.

35. $x = 5, y = 12$ **36.** $x = 2, y = 14$
37. $x = 10, y = 2050$ **38.** $x = 6, y = 580$

39. SIMPLE INTEREST The simple interest on an investment is directly proportional to the amount of the investment. By investing \$3250 in a certain bond issue, you obtained an interest payment of \$113.75 after 1 year. Find a mathematical model that gives the interest I for this bond issue after 1 year in terms of the amount invested P.

40. SIMPLE INTEREST The simple interest on an investment is directly proportional to the amount of the investment. By investing \$6500 in a municipal bond, you obtained an interest payment of \$211.25 after 1 year. Find a mathematical model that gives the interest I for this municipal bond after 1 year in terms of the amount invested P.

41. MEASUREMENT On a yardstick with scales in inches and centimeters, you notice that 13 inches is approximately the same length as 33 centimeters. Use this information to find a mathematical model that relates centimeters y to inches x. Then use the model to find the numbers of centimeters in 10 inches and 20 inches.

42. MEASUREMENT When buying gasoline, you notice that 14 gallons of gasoline is approximately the same amount of gasoline as 53 liters. Use this information to find a linear model that relates liters y to gallons x. Then use the model to find the numbers of liters in 5 gallons and 25 gallons.

43. TAXES Property tax is based on the assessed value of a property. A house that has an assessed value of $150,000 has a property tax of $5520. Find a mathematical model that gives the amount of property tax y in terms of the assessed value x of the property. Use the model to find the property tax on a house that has an assessed value of $225,000.

44. TAXES State sales tax is based on retail price. An item that sells for $189.99 has a sales tax of $11.40. Find a mathematical model that gives the amount of sales tax y in terms of the retail price x. Use the model to find the sales tax on a $639.99 purchase.

HOOKE'S LAW In Exercises 45–48, use Hooke's Law for springs, which states that the distance a spring is stretched (or compressed) varies directly as the force on the spring.

45. A force of 265 newtons stretches a spring 0.15 meter (see figure).

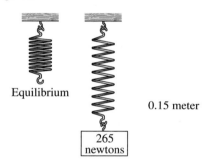

Equilibrium

0.15 meter

265 newtons

(a) How far will a force of 90 newtons stretch the spring?

(b) What force is required to stretch the spring 0.1 meter?

46. A force of 220 newtons stretches a spring 0.12 meter. What force is required to stretch the spring 0.16 meter?

47. The coiled spring of a toy supports the weight of a child. The spring is compressed a distance of 1.9 inches by the weight of a 25-pound child. The toy will not work properly if its spring is compressed more than 3 inches. What is the weight of the heaviest child who should be allowed to use the toy?

48. An overhead garage door has two springs, one on each side of the door (see figure). A force of 15 pounds is required to stretch each spring 1 foot. Because of a pulley system, the springs stretch only one-half the distance the door travels. The door moves a total of 8 feet, and the springs are at their natural length when the door is open. Find the combined lifting force applied to the door by the springs when the door is closed.

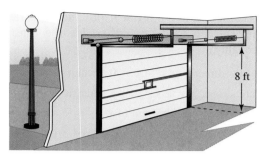

8 ft

FIGURE FOR **48**

In Exercises 49–58, find a mathematical model for the verbal statement.

49. A varies directly as the square of r.

50. V varies directly as the cube of e.

51. y varies inversely as the square of x.

52. h varies inversely as the square root of s.

53. F varies directly as g and inversely as r^2.

54. z is jointly proportional to the square of x and the cube of y.

55. BOYLE'S LAW: For a constant temperature, the pressure P of a gas is inversely proportional to the volume V of the gas.

56. NEWTON'S LAW OF COOLING: The rate of change R of the temperature of an object is proportional to the difference between the temperature T of the object and the temperature T_e of the environment in which the object is placed.

57. NEWTON'S LAW OF UNIVERSAL GRAVITATION: The gravitational attraction F between two objects of masses m_1 and m_2 is proportional to the product of the masses and inversely proportional to the square of the distance r between the objects.

58. LOGISTIC GROWTH: The rate of growth R of a population is jointly proportional to the size S of the population and the difference between S and the maximum population size L that the environment can support.

In Exercises 59–66, write a sentence using the variation terminology of this section to describe the formula.

59. *Area of a triangle:* $A = \frac{1}{2}bh$

60. *Area of a rectangle:* $A = lw$

61. *Area of an equilateral triangle:* $A = \left(\sqrt{3}s^2\right)/4$

62. *Surface area of a sphere:* $S = 4\pi r^2$

63. *Volume of a sphere:* $V = \frac{4}{3}\pi r^3$

64. *Volume of a right circular cylinder:* $V = \pi r^2 h$

65. *Average speed:* $r = d/t$

66. *Free vibrations:* $\omega = \sqrt{(kg)/W}$

In Exercises 67–74, find a mathematical model representing the statement. (In each case, determine the constant of proportionality.)

67. A varies directly as r^2. ($A = 9\pi$ when $r = 3$.)

68. y varies inversely as x. ($y = 3$ when $x = 25$.)

69. y is inversely proportional to x. ($y = 7$ when $x = 4$.)

70. z varies jointly as x and y. ($z = 64$ when $x = 4$ and $y = 8$.)

71. F is jointly proportional to r and the third power of s. ($F = 4158$ when $r = 11$ and $s = 3$.)

72. P varies directly as x and inversely as the square of y. ($P = \frac{28}{3}$ when $x = 42$ and $y = 9$.)

73. z varies directly as the square of x and inversely as y. ($z = 6$ when $x = 6$ and $y = 4$.)

74. v varies jointly as p and q and inversely as the square of s. ($v = 1.5$ when $p = 4.1$, $q = 6.3$, and $s = 1.2$.)

ECOLOGY In Exercises 75 and 76, use the fact that the diameter of the largest particle that can be moved by a stream varies approximately directly as the square of the velocity of the stream.

75. A stream with a velocity of $\frac{1}{4}$ mile per hour can move coarse sand particles about 0.02 inch in diameter. Approximate the velocity required to carry particles 0.12 inch in diameter.

76. A stream of velocity v can move particles of diameter d or less. By what factor does d increase when the velocity is doubled?

RESISTANCE In Exercises 77 and 78, use the fact that the resistance of a wire carrying an electrical current is directly proportional to its length and inversely proportional to its cross-sectional area.

77. If #28 copper wire (which has a diameter of 0.0126 inch) has a resistance of 66.17 ohms per thousand feet, what length of #28 copper wire will produce a resistance of 33.5 ohms?

78. A 14-foot piece of copper wire produces a resistance of 0.05 ohm. Use the constant of proportionality from Exercise 77 to find the diameter of the wire.

79. WORK The work W (in joules) done when lifting an object varies jointly with the mass m (in kilograms) of the object and the height h (in meters) that the object is lifted. The work done when a 120-kilogram object is lifted 1.8 meters is 2116.8 joules. How much work is done when lifting a 100-kilogram object 1.5 meters?

80. MUSIC The frequency of vibrations of a piano string varies directly as the square root of the tension on the string and inversely as the length of the string. The middle A string has a frequency of 440 vibrations per second. Find the frequency of a string that has 1.25 times as much tension and is 1.2 times as long.

81. FLUID FLOW The velocity v of a fluid flowing in a conduit is inversely proportional to the cross-sectional area of the conduit. (Assume that the volume of the flow per unit of time is held constant.) Determine the change in the velocity of water flowing from a hose when a person places a finger over the end of the hose to decrease its cross-sectional area by 25%.

82. BEAM LOAD The maximum load that can be safely supported by a horizontal beam varies jointly as the width of the beam and the square of its depth, and inversely as the length of the beam. Determine the changes in the maximum safe load under the following conditions.

(a) The width and length of the beam are doubled.

(b) The width and depth of the beam are doubled.

(c) All three of the dimensions are doubled.

(d) The depth of the beam is halved.

83. DATA ANALYSIS: OCEAN TEMPERATURES An oceanographer took readings of the water temperatures C (in degrees Celsius) at several depths d (in meters). The data collected are shown in the table.

Depth, d	Temperature, C
1000	4.2°
2000	1.9°
3000	1.4°
4000	1.2°
5000	0.9°

(a) Sketch a scatter plot of the data.

(b) Does it appear that the data can be modeled by the inverse variation model $C = k/d$? If so, find k for each pair of coordinates.

(c) Determine the mean value of k from part (b) to find the inverse variation model $C = k/d$.

(d) Use a graphing utility to plot the data points and the inverse model from part (c).

(e) Use the model to approximate the depth at which the water temperature is 3°C.

84. DATA ANALYSIS: PHYSICS EXPERIMENT An experiment in a physics lab requires a student to measure the compressed lengths y (in centimeters) of a spring when various forces of F pounds are applied. The data are shown in the table.

Force, F	Length, y
0	0
2	1.15
4	2.3
6	3.45
8	4.6
10	5.75
12	6.9

(a) Sketch a scatter plot of the data.

(b) Does it appear that the data can be modeled by Hooke's Law? If so, estimate k. (See Exercises 45–48.)

(c) Use the model in part (b) to approximate the force required to compress the spring 9 centimeters.

85. DATA ANALYSIS: LIGHT INTENSITY A light probe is located x centimeters from a light source, and the intensity y (in microwatts per square centimeter) of the light is measured. The results are shown as ordered pairs (x, y).

(30, 0.1881)   (34, 0.1543)   (38, 0.1172)

(42, 0.0998)   (46, 0.0775)   (50, 0.0645)

A model for the data is $y = 262.76/x^{2.12}$.

(a) Use a graphing utility to plot the data points and the model in the same viewing window.

(b) Use the model to approximate the light intensity 25 centimeters from the light source.

86. ILLUMINATION The illumination from a light source varies inversely as the square of the distance from the light source. When the distance from a light source is doubled, how does the illumination change? Discuss this model in terms of the data given in Exercise 85. Give a possible explanation of the difference.

EXPLORATION

TRUE OR FALSE? In Exercises 87 and 88, decide whether the statement is true or false. Justify your answer.

87. In the equation for kinetic energy, $E = \frac{1}{2}mv^2$, the amount of kinetic energy E is directly proportional to the mass m of an object and the square of its velocity v.

88. If the correlation coefficient for a least squares regression line is close to -1, the regression line cannot be used to describe the data.

89. Discuss how well the data shown in each scatter plot can be approximated by a linear model.

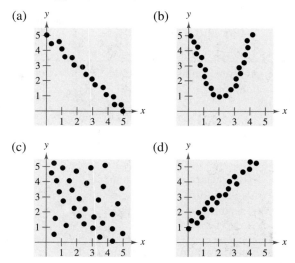

90. WRITING A linear model for predicting prize winnings at a race is based on data for 3 years. Write a paragraph discussing the potential accuracy or inaccuracy of such a model.

91. WRITING Suppose the constant of proportionality is positive and y varies directly as x. When one of the variables increases, how will the other change? Explain your reasoning.

92. WRITING Suppose the constant of proportionality is positive and y varies inversely as x. When one of the variables increases, how will the other change? Explain your reasoning.

93. WRITING

(a) Given that y varies inversely as the square of x and x is doubled, how will y change? Explain.

(b) Given that y varies directly as the square of x and x is doubled, how will y change? Explain.

94. CAPSTONE The prices of three sizes of pizza at a pizza shop are as follows.

9-inch: $8.78, 12-inch: $11.78, 15-inch: $14.18

You would expect that the price of a certain size of pizza would be directly proportional to its surface area. Is that the case for this pizza shop? If not, which size of pizza is the best buy?

PROJECT: FRAUD AND IDENTITY THEFT To work an extended application analyzing the numbers of fraud complaints and identity theft victims in the United States in 2007, visit this text's website at *academic.cengage.com*. (Data Source: U.S. Census Bureau)

3 Chapter Summary

What Did You Learn?	Explanation/Examples	Review Exercises
Section 3.1 Analyze graphs of quadratic functions (p. 260).	Let a, b, and c be real numbers with $a \neq 0$. The function given by $f(x) = ax^2 + bx + c$ is called a quadratic function. Its graph is a "U"-shaped curve called a parabola. All parabolas are symmetric with respect to a line called the axis of symmetry. The point where the axis of symmetry intersects the parabola is the vertex.	1, 2
Write quadratic functions in standard form and use the results to sketch graphs of functions (p. 263).	The quadratic function $f(x) = a(x - h)^2 + k$, $a \neq 0$, is in standard form. The graph of f is a parabola whose axis is the vertical line $x = h$ and whose vertex is (h, k). If $a > 0$, the parabola opens upward. If $a < 0$, the parabola opens downward.	3–20
Find minimum and maximum values of quadratic functions in real-life applications (p. 265).	Consider $f(x) = ax^2 + bx + c$ with vertex $\left(-\dfrac{b}{2a}, f\left(\dfrac{-b}{2a}\right)\right)$. If $a > 0$, then f has a *minimum* when $x = -b/(2a)$. If $a < 0$, then f has a *maximum* when $x = -b/(2a)$.	21–26
Section 3.2 Use transformations to sketch graphs of polynomial functions (p. 270).	The graph of a polynomial function is continuous (no breaks, holes, or gaps) and has only smooth, rounded turns.	27–32
Use the Leading Coefficient Test to determine the end behavior of graphs of polynomial functions (p. 272).	Consider the graph of $f(x) = a_n x^n + \cdots + a_1 x + a_0$. **When n is odd:** If $a_n > 0$, the graph falls to the left and rises to the right. If $a_n < 0$, the graph rises to the left and falls to the right. **When n is even:** If $a_n > 0$, the graph rises to the left and right. If $a_n < 0$, the graph falls to the left and right.	33–36
Find and use zeros of polynomial functions as sketching aids (p. 273).	If f is a polynomial function and a is a real number, the following are equivalent: (1) $x = a$ is a *zero* of f, (2) $x = a$ is a *solution* of the equation $f(x) = 0$, (3) $(x - a)$ is a *factor* of $f(x)$, and (4) $(a, 0)$ is an *x-intercept* of the graph of f.	37–46
Use the Intermediate Value Theorem to help locate zeros of polynomial functions (p. 277).	Let a and b be real numbers such that $a < b$. If f is a polynomial function such that $f(a) \neq f(b)$, then, in $[a, b]$, f takes on every value between $f(a)$ and $f(b)$.	47–50
Section 3.3 Use long division to divide polynomials by other polynomials (p. 284).	$$\dfrac{x^2 + 3x + 5}{x + 1} = x + 2 + \dfrac{3}{x + 1}$$	51–56
Use synthetic division to divide polynomials by binomials of the form $(x - k)$ (p. 287).	$$\begin{array}{r|rrrrr} -3 & 1 & 0 & -10 & -2 & 4 \\ & & -3 & 9 & 3 & -3 \\ \hline & 1 & -3 & -1 & 1 & 1 \end{array}$$	57–60
Use the Remainder Theorem and the Factor Theorem (p. 288).	**The Remainder Theorem:** If a polynomial $f(x)$ is divided by $x - k$, the remainder is $r = f(k)$. **The Factor Theorem:** A polynomial $f(x)$ has a factor $(x - k)$ if and only if $f(k) = 0$.	61–68

What Did You Learn?	Explanation/Examples	Review Exercises
Section 3.4 Use the Fundamental Theorem of Algebra to determine the number of zeros of polynomial functions (p. 293).	**The Fundamental Theorem of Algebra** If $f(x)$ is a polynomial of degree n, where $n > 0$, then f has at least one zero in the complex number system. **Linear Factorization Theorem** If $f(x)$ is a polynomial of degree n, where $n > 0$, then f has precisely n linear factors $f(x) = a_n(x - c_1)(x - c_2) \cdots (x - c_n)$ where $c_1, c_2, \ldots, c_n$ are complex numbers.	69–74
Find rational zeros of polynomial functions (p. 294).	The Rational Zero Test relates the possible rational zeros of a polynomial to the leading coefficient and to the constant term of the polynomial.	75–82
Find conjugate pairs of complex zeros (p. 297).	**Complex Zeros Occur in Conjugate Pairs** Let $f(x)$ be a polynomial function that has real coefficients. If $a + bi$ $(b \neq 0)$ is a zero of the function, the conjugate $a - bi$ is also a zero of the function.	83, 84
Find zeros of polynomials by factoring (p. 297).	Every polynomial of degree $n > 0$ with real coefficients can be written as the product of linear and quadratic factors with real coefficients, where the quadratic factors have no real zeros.	85–96
Use Descartes's Rule of Signs (p. 300) and the Upper and Lower Bound Rules (p. 301) to find zeros of polynomials.	**Descartes's Rule of Signs** Let $f(x) = a_n x^n + a_{n-1}x^{n-1} + \cdots + a_2 x^2 + a_1 x + a_0$ be a polynomial with real coefficients and $a_0 \neq 0$. **1.** The number of *positive real zeros* of f is either equal to the number of variations in sign of $f(x)$ or less than that number by an even integer. **2.** The number of *negative real zeros* of f is either equal to the number of variations in sign of $f(-x)$ or less than that number by an even integer.	97–100
Section 3.5 Use mathematical models to approximate sets of data points (p. 308).	To see how well a model fits a set of data, compare the actual values and model values of y. (see Example 1.)	101
Use the *regression* feature of a graphing utility to find the equation of a least squares regression line (p. 309).	The sum of square differences is the sum of the squares of the differences between actual data values and model values. The least squares regression line is the linear model with the least sum of square differences. The *regression* feature of a graphing utility can be used to find the least squares regression line. The correlation coefficient (r-value) of the data gives a measure of how well the model fits the data. The closer the value of $|r|$ is to 1, the better the fit.	102
Write mathematical models for direct variation (p. 310), direct variation as an *n*th power (p. 311), inverse variation (p. 312), and joint variation (p. 313).	**Direct variation:** $y = kx$ for some nonzero constant k **Direct variation as an *n*th power:** $y = kx^n$ for some constant k **Inverse variation:** $y = k/x$ for some constant k **Joint variation:** $z = kxy$ for some constant k	103–108

3 REVIEW EXERCISES

See www.CalcChat.com for worked-out solutions to odd-numbered exercises.

3.1 In Exercises 1 and 2, graph each function. Compare the graph of each function with the graph of $y = x^2$.

1. (a) $f(x) = 2x^2$
   (b) $g(x) = -2x^2$
   (c) $h(x) = x^2 + 2$
   (d) $k(x) = (x + 2)^2$

2. (a) $f(x) = x^2 - 4$
   (b) $g(x) = 4 - x^2$
   (c) $h(x) = (x - 3)^2$
   (d) $k(x) = \frac{1}{2}x^2 - 1$

In Exercises 3–14, write the quadratic function in standard form and sketch its graph. Identify the vertex, axis of symmetry, and x-intercept(s).

3. $g(x) = x^2 - 2x$
4. $f(x) = 6x - x^2$
5. $f(x) = x^2 + 8x + 10$
6. $h(x) = 3 + 4x - x^2$
7. $f(t) = -2t^2 + 4t + 1$
8. $f(x) = x^2 - 8x + 12$
9. $h(x) = 4x^2 + 4x + 13$
10. $f(x) = x^2 - 6x + 1$
11. $h(x) = x^2 + 5x - 4$
12. $f(x) = 4x^2 + 4x + 5$
13. $f(x) = \frac{1}{3}(x^2 + 5x - 4)$
14. $f(x) = \frac{1}{2}(6x^2 - 24x + 22)$

In Exercises 15–20, write the standard form of the equation of the parabola that has the indicated vertex and whose graph passes through the given point.

15.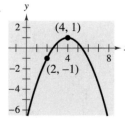

16.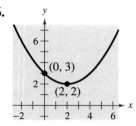

17. Vertex: $(1, -4)$; point: $(2, -3)$
18. Vertex: $(2, 3)$; point: $(-1, 6)$
19. Vertex: $\left(-\frac{3}{2}, 0\right)$; point: $\left(-\frac{9}{2}, -\frac{11}{4}\right)$
20. Vertex: $(3, 3)$; point: $\left(\frac{1}{4}, \frac{4}{5}\right)$

21. **NUMERICAL, GRAPHICAL, AND ANALYTICAL ANALYSIS** A rectangle is inscribed in the region bounded by the x-axis, the y-axis, and the graph of $x + 2y - 8 = 0$, as shown in the figure.

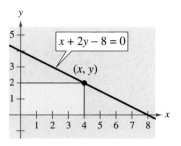

(a) Write the area A of the rectangle as a function of x.

(b) Determine the domain of the function in the context of the problem.

(c) Create a table showing possible values of x and the corresponding area of the rectangle.

(d) Use a graphing utility to graph the area function. Use the graph to approximate the dimensions that will produce the maximum area.

(e) Write the area function in standard form to find analytically the dimensions that will produce the maximum area.

22. **GEOMETRY** The perimeter of a rectangle is 200 meters.

(a) Draw a diagram that gives a visual representation of the problem. Label the length and width as x and y, respectively.

(b) Write y as a function of x. Use the result to write the area as a function of x.

(c) Of all possible rectangles with perimeters of 200 meters, find the dimensions of the one with the maximum area.

23. **MAXIMUM REVENUE** The total revenue R earned (in dollars) from producing a gift box of candles is given by

$$R(p) = -10p^2 + 800p$$

where p is the price per unit (in dollars).

(a) Find the revenues when the prices per box are $20, $25, and $30.

(b) Find the unit price that will yield a maximum revenue. What is the maximum revenue? Explain your results.

24. MAXIMUM PROFIT A real estate office handles an apartment building that has 50 units. When the rent is $540 per month, all units are occupied. However, for each $30 increase in rent, one unit becomes vacant. Each occupied unit requires an average of $18 per month for service and repairs. What rent should be charged to obtain the maximum profit?

25. MINIMUM COST A soft-drink manufacturer has daily production costs of

$$C = 70,000 - 120x + 0.055x^2$$

where C is the total cost (in dollars) and x is the number of units produced. How many units should be produced each day to yield a minimum cost?

26. SOCIOLOGY The average age of the groom at a first marriage for a given age of the bride can be approximated by the model

$$y = -0.107x^2 + 5.68x - 48.5, \quad 20 \le x \le 25$$

where y is the age of the groom and x is the age of the bride. Sketch a graph of the model. For what age of the bride is the average age of the groom 26? (Source: U.S. Census Bureau)

3.2 In Exercises 27–32, sketch the graphs of $y = x^n$ and the transformation.

27. $y = x^3, \quad f(x) = -(x - 2)^3$
28. $y = x^3, \quad f(x) = -4x^3$
29. $y = x^4, \quad f(x) = 6 - x^4$
30. $y = x^4, \quad f(x) = 2(x - 8)^4$
31. $y = x^5, \quad f(x) = (x - 5)^5$
32. $y = x^5, \quad f(x) = \frac{1}{2}x^5 + 3$

In Exercises 33–36, describe the right-hand and left-hand behavior of the graph of the polynomial function.

33. $f(x) = -2x^2 - 5x + 12$
34. $f(x) = \frac{1}{2}x^3 + 2x$
35. $g(x) = \frac{3}{4}(x^4 + 3x^2 + 2)$
36. $h(x) = -x^7 + 8x^2 - 8x$

In Exercises 37–42, find all the real zeros of the polynomial function. Determine the multiplicity of each zero and the number of turning points of the graph of the function. Use a graphing utility to verify your answers.

37. $f(x) = 3x^2 + 20x - 32$　　**38.** $f(x) = x(x + 3)^2$
39. $f(t) = t^3 - 3t$
40. $f(x) = x^3 - 8x^2$
41. $f(x) = -18x^3 + 12x^2$
42. $g(x) = x^4 + x^3 - 12x^2$

In Exercises 43–46, sketch the graph of the function by (a) applying the Leading Coefficient Test, (b) finding the zeros of the polynomial, (c) plotting sufficient solution points, and (d) drawing a continuous curve through the points.

43. $f(x) = -x^3 + x^2 - 2$
44. $g(x) = 2x^3 + 4x^2$
45. $f(x) = x(x^3 + x^2 - 5x + 3)$
46. $h(x) = 3x^2 - x^4$

In Exercises 47–50, (a) use the Intermediate Value Theorem and the *table* feature of a graphing utility to find intervals one unit in length in which the polynomial function is guaranteed to have a zero. (b) Adjust the table to approximate the zeros of the function. Use the *zero* or *root* feature of the graphing utility to verify your results.

47. $f(x) = 3x^3 - x^2 + 3$
48. $f(x) = 0.25x^3 - 3.65x + 6.12$
49. $f(x) = x^4 - 5x - 1$
50. $f(x) = 7x^4 + 3x^3 - 8x^2 + 2$

3.3 In Exercises 51–56, use long division to divide.

51. $\dfrac{30x^2 - 3x + 8}{5x - 3}$

52. $\dfrac{4x + 7}{3x - 2}$

53. $\dfrac{5x^3 - 21x^2 - 25x - 4}{x^2 - 5x - 1}$

54. $\dfrac{3x^4}{x^2 - 1}$

55. $\dfrac{x^4 - 3x^3 + 4x^2 - 6x + 3}{x^2 + 2}$

56. $\dfrac{6x^4 + 10x^3 + 13x^2 - 5x + 2}{2x^2 - 1}$

In Exercises 57–60, use synthetic division to divide.

57. $\dfrac{6x^4 - 4x^3 - 27x^2 + 18x}{x - 2}$

58. $\dfrac{0.1x^3 + 0.3x^2 - 0.5}{x - 5}$

59. $\dfrac{2x^3 - 25x^2 + 66x + 48}{x - 8}$

60. $\dfrac{5x^3 + 33x^2 + 50x - 8}{x + 4}$

In Exercises 61 and 62, use synthetic division to determine whether the given values of x are zeros of the function.

61. $f(x) = 20x^4 + 9x^3 - 14x^2 - 3x$

    (a) $x = -1$ (b) $x = \frac{3}{4}$ (c) $x = 0$ (d) $x = 1$

62. $f(x) = 3x^3 - 8x^2 - 20x + 16$

    (a) $x = 4$ (b) $x = -4$ (c) $x = \frac{2}{3}$ (d) $x = -1$

In Exercises 63 and 64, use the Remainder Theorem and synthetic division to find each function value.

63. $f(x) = x^4 + 10x^3 - 24x^2 + 20x + 44$

    (a) $f(-3)$ (b) $f(-1)$

64. $g(t) = 2t^5 - 5t^4 - 8t + 20$

    (a) $g(-4)$ (b) $g(\sqrt{2})$

In Exercises 65–68, (a) verify the given factor(s) of the function f, (b) find the remaining factors of f, (c) use your results to write the complete factorization of f, (d) list all real zeros of f, and (e) confirm your results by using a graphing utility to graph the function.

Function	Factor(s)
65. $f(x) = x^3 + 4x^2 - 25x - 28$	$(x - 4)$
66. $f(x) = 2x^3 + 11x^2 - 21x - 90$	$(x + 6)$
67. $f(x) = x^4 - 4x^3 - 7x^2 + 22x + 24$	$(x + 2)(x - 3)$
68. $f(x) = x^4 - 11x^3 + 41x^2 - 61x + 30$	$(x - 2)(x - 5)$

3.4 In Exercises 69–74, find all the zeros of the function.

69. $f(x) = 4x(x - 3)^2$

70. $f(x) = (x - 4)(x + 9)^2$

71. $f(x) = x^2 - 11x + 18$

72. $f(x) = x^3 + 10x$

73. $f(x) = (x + 4)(x - 6)(x - 2i)(x + 2i)$

74. $f(x) = (x - 8)(x - 5)^2(x - 3 + i)(x - 3 - i)$

In Exercises 75 and 76, use the Rational Zero Test to list all possible rational zeros of f.

75. $f(x) = -4x^3 + 8x^2 - 3x + 15$

76. $f(x) = 3x^4 + 4x^3 - 5x^2 - 8$

In Exercises 77–82, find all the rational zeros of the function.

77. $f(x) = x^3 + 3x^2 - 28x - 60$

78. $f(x) = 4x^3 - 27x^2 + 11x + 42$

79. $f(x) = x^3 - 10x^2 + 17x - 8$

80. $f(x) = x^3 + 9x^2 + 24x + 20$

81. $f(x) = x^4 + x^3 - 11x^2 + x - 12$

82. $f(x) = 25x^4 + 25x^3 - 154x^2 - 4x + 24$

In Exercises 83 and 84, find a polynomial function with real coefficients that has the given zeros. (There are many correct answers.)

83. $\frac{2}{3}, 4, \sqrt{3}i$

84. $2, -3, 1 - 2i$

In Exercises 85–88, use the given zero to find all the zeros of the function.

Function	Zero
85. $f(x) = x^3 - 4x^2 + x - 4$	i
86. $h(x) = -x^3 + 2x^2 - 16x + 32$	$-4i$
87. $g(x) = 2x^4 - 3x^3 - 13x^2 + 37x - 15$	$2 + i$
88. $f(x) = 4x^4 - 11x^3 + 14x^2 - 6x$	$1 - i$

In Exercises 89–92, find all the zeros of the function and write the polynomial as a product of linear factors.

89. $f(x) = x^3 + 4x^2 - 5x$

90. $g(x) = x^3 - 7x^2 + 36$

91. $g(x) = x^4 + 4x^3 - 3x^2 + 40x + 208$

92. $f(x) = x^4 + 8x^3 + 8x^2 - 72x - 153$

In Exercises 93–96, use a graphing utility to (a) graph the function, (b) determine the number of real zeros of the function, and (c) approximate the real zeros of the function to the nearest hundredth.

93. $f(x) = x^4 + 2x + 1$

94. $g(x) = x^3 - 3x^2 + 3x + 2$

95. $h(x) = x^3 - 6x^2 + 12x - 10$

96. $f(x) = x^5 + 2x^3 - 3x - 20$

In Exercises 97 and 98, use Descartes's Rule of Signs to determine the possible numbers of positive and negative zeros of the function.

97. $g(x) = 5x^3 + 3x^2 - 6x + 9$

98. $h(x) = -2x^5 + 4x^3 - 2x^2 + 5$

In Exercises 99 and 100, use synthetic division to verify the upper and lower bounds of the real zeros of f.

99. $f(x) = 4x^3 - 3x^2 + 4x - 3$

    (a) Upper: $x = 1$

    (b) Lower: $x = -\frac{1}{4}$

100. $f(x) = 2x^3 - 5x^2 - 14x + 8$

    (a) Upper: $x = 8$

    (b) Lower: $x = -4$

.5 101. COMPACT DISCS   The values V (in billions of dollars) of shipments of compact discs in the United States from 2000 through 2007 are shown in the table. A linear model that approximates these data is

$$V = -0.742t + 13.62$$

where t represents the year, with $t = 0$ corresponding to 2000.   (Source: Recording Industry Association of America)

Year	Value, V
2000	13.21
2001	12.91
2002	12.04
2003	11.23
2004	11.45
2005	10.52
2006	9.37
2007	7.45

(a) Plot the actual data and the model on the same set of coordinate axes.

(b) How closely does the model represent the data?

102. DATA ANALYSIS: TV USAGE   The table shows the projected numbers of hours H of television usage in the United States from 2003 through 2011.   (Source: Communications Industry Forecast and Report)

Year	Hours, H
2003	1615
2004	1620
2005	1659
2006	1673
2007	1686
2008	1704
2009	1714
2010	1728
2011	1742

(a) Use a graphing utility to create a scatter plot of the data. Let t represent the year, with $t = 3$ corresponding to 2003.

(b) Use the *regression* feature of the graphing utility to find the equation of the least squares regression line that fits the data. Then graph the model and the scatter plot you found in part (a) in the same viewing window. How closely does the model represent the data?

(c) Use the model to estimate the projected number of hours of television usage in 2020.

(d) Interpret the meaning of the slope of the linear model in the context of the problem.

103. MEASUREMENT   You notice a billboard indicating that it is 2.5 miles or 4 kilometers to the next restaurant of a national fast-food chain. Use this information to find a mathematical model that relates miles to kilometers. Then use the model to find the numbers of kilometers in 2 miles and 10 miles.

104. ENERGY   The power P produced by a wind turbine is proportional to the cube of the wind speed S. A wind speed of 27 miles per hour produces a power output of 750 kilowatts. Find the output for a wind speed of 40 miles per hour.

105. FRICTIONAL FORCE   The frictional force F between the tires and the road required to keep a car on a curved section of a highway is directly proportional to the square of the speed s of the car. If the speed of the car is doubled, the force will change by what factor?

106. DEMAND   A company has found that the daily demand x for its boxes of chocolates is inversely proportional to the price p. When the price is $5, the demand is 800 boxes. Approximate the demand when the price is increased to $6.

107. TRAVEL TIME   The travel time between two cities is inversely proportional to the average speed. A train travels between the cities in 3 hours at an average speed of 65 miles per hour. How long would it take to travel between the cities at an average speed of 80 miles per hour?

108. COST   The cost of constructing a wooden box with a square base varies jointly as the height of the box and the square of the width of the box. A box of height 16 inches and of width 6 inches costs $28.80. How much would a box of height 14 inches and of width 8 inches cost?

EXPLORATION

TRUE OR FALSE?   In Exercises 109 and 110, determine whether the statement is true or false. Justify your answer.

109. A fourth-degree polynomial with real coefficients can have -5, $-8i$, $4i$, and 5 as its zeros.

110. If y is directly proportional to x, then x is directly proportional to y.

111. WRITING   Explain how to determine the maximum or minimum value of a quadratic function.

112. WRITING   Explain the connections between factors of a polynomial, zeros of a polynomial function, and solutions of a polynomial equation.

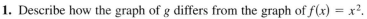

3 CHAPTER TEST

See www.CalcChat.com for worked-out solutions to odd-numbered exercises.

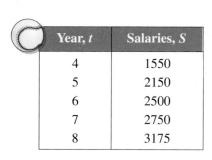

FIGURE FOR 3

Take this test as you would take a test in class. When you are finished, check your work against the answers given in the back of the book.

1. Describe how the graph of g differs from the graph of $f(x) = x^2$.
   (a) $g(x) = 2 - x^2$
   (b) $g(x) = \left(x - \frac{3}{2}\right)^2$

2. Identify the vertex and intercepts of the graph of $y = x^2 + 4x + 3$.

3. Find an equation of the parabola shown in the figure at the left.

4. The path of a ball is given by $y = -\frac{1}{20}x^2 + 3x + 5$, where y is the height (in feet) of the ball and x is the horizontal distance (in feet) from where the ball was thrown.
   (a) Find the maximum height of the ball.
   (b) Which number determines the height at which the ball was thrown? Does changing this value change the coordinates of the maximum height of the ball? Explain.

5. Determine the right-hand and left-hand behavior of the graph of the function $h(t) = -\frac{3}{4}t^5 + 2t^2$. Then sketch its graph.

6. Divide using long division.

$$\frac{3x^3 + 4x - 1}{x^2 + 1}$$

7. Divide using synthetic division.

$$\frac{2x^4 - 5x^2 - 3}{x - 2}$$

8. Use synthetic division to show that $x = \sqrt{3}$ is a zero of the function given by

$$f(x) = 2x^3 - 5x^2 - 6x + 15.$$

Use the result to factor the polynomial function completely and list all the real zeros of the function.

In Exercises 9 and 10, find all the rational zeros of the function.

9. $g(t) = 2t^4 - 3t^3 + 16t - 24$

10. $h(x) = 3x^5 + 2x^4 - 3x - 2$

In Exercises 11 and 12, find a polynomial function with real coefficients that has the given zeros. (There are many correct answers.)

11. $0, 3, 2 + i$

12. $1 - \sqrt{3}i, 2, 2$

In Exercises 13 and 14, find all the zeros of the function.

13. $f(x) = 3x^3 + 14x^2 - 7x - 10$

14. $f(x) = x^4 - 9x^2 - 22x - 24$

In Exercises 15–17, find a mathematical model that represents the statement. (In each case, determine the constant of proportionality.)

15. v varies directly as the square root of s. ($v = 24$ when $s = 16$.)

16. A varies jointly as x and y. ($A = 500$ when $x = 15$ and $y = 8$.)

17. b varies inversely as a. ($b = 32$ when $a = 1.5$.)

18. The table at the left shows the median salaries S (in thousands of dollars) for baseball players on the Chicago Cubs from 2004 through 2008, where $t = 4$ represents 2004. Use the *regression* feature of a graphing utility to find the equation of the least squares regression line that fits the data. How well does the model represent the data? (Source: USA Today)

Year, t	Salaries, S
4	1550
5	2150
6	2500
7	2750
8	3175

PROOFS IN MATHEMATICS

These two pages contain proofs of four important theorems about polynomial functions. The first two theorems are from Section 3.3, and the second two theorems are from Section 3.4.

The Remainder Theorem   *(p. 288)*

If a polynomial $f(x)$ is divided by $x - k$, the remainder is

$$r = f(k).$$

Proof

From the Division Algorithm, you have

$$f(x) = (x - k)q(x) + r(x)$$

and because either $r(x) = 0$ or the degree of $r(x)$ is less than the degree of $x - k$, you know that $r(x)$ must be a constant. That is, $r(x) = r$. Now, by evaluating $f(x)$ at $x = k$, you have

$$f(\ ) = (\ - k)q(\ ) + r$$
$$= (0)q(k) + r = r.$$

To be successful in algebra, it is important that you understand the connection among *factors* of a polynomial, *zeros* of a polynomial function, and *solutions* or *roots* of a polynomial equation. The Factor Theorem is the basis for this connection.

The Factor Theorem   *(p. 288)*

A polynomial $f(x)$ has a factor $(x - k)$ if and only if $f(k) = 0$.

Proof

Using the Division Algorithm with the factor $(x - k)$, you have

$$f(x) = (x - k)q(x) + r(x).$$

By the Remainder Theorem, $r(x) = r = f(k)$, and you have

$$f(x) = (x - k)q(x) + f(k)$$

where $q(x)$ is a polynomial of lesser degree than $f(x)$. If $f(k) = 0$, then

$$f(x) = (x - k)q(x)$$

and you see that $(x - k)$ is a factor of $f(x)$. Conversely, if $(x - k)$ is a factor of $f(x)$, division of $f(x)$ by $(x - k)$ yields a remainder of 0. So, by the Remainder Theorem, you have $f(k) = 0$.

PROOFS IN MATHEMATICS

Linear Factorization Theorem   *(p. 293)*

If $f(x)$ is a polynomial of degree n, where $n > 0$, then f has precisely n linear factors

$$f(x) = a_n(x - c_1)(x - c_2) \cdots (x - c_n)$$

where $c_1, c_2, \ldots, c_n$ are complex numbers.

The Fundamental Theorem of Algebra

The Linear Factorization Theorem is closely related to the Fundamental Theorem of Algebra. The Fundamental Theorem of Algebra has a long and interesting history. In the early work with polynomial equations, The Fundamental Theorem of Algebra was thought to have been not true, because imaginary solutions were not considered. In fact, in the very early work by mathematicians such as Abu al-Khwarizmi (c. 800 A.D.), negative solutions were also not considered.

Once imaginary numbers were accepted, several mathematicians attempted to give a general proof of the Fundamental Theorem of Algebra. These included Gottfried von Leibniz (1702), Jean d'Alembert (1746), Leonhard Euler (1749), Joseph-Louis Lagrange (1772), and Pierre Simon Laplace (1795). The mathematician usually credited with the first correct proof of the Fundamental Theorem of Algebra is Carl Friedrich Gauss, who published the proof in his doctoral thesis in 1799.

Proof

Using the Fundamental Theorem of Algebra, you know that f must have at least one zero, c_1. Consequently, $(x - c_1)$ is a factor of $f(x)$, and you have

$$f(x) = (x - c_1)f_1(x).$$

If the degree of $f_1(x)$ is greater than zero, you again apply the Fundamental Theorem to conclude that f_1 must have a zero c_2, which implies that

$$f(x) = (x - c_1)(x - c_2)f_2(x).$$

It is clear that the degree of $f_1(x)$ is $n - 1$, that the degree of $f_2(x)$ is $n - 2$, and that you can repeatedly apply the Fundamental Theorem n times until you obtain

$$f(x) = a_n(x - c_1)(x - c_2) \cdots (x - c_n)$$

where a_n is the leading coefficient of the polynomial $f(x)$.

Factors of a Polynomial   *(p. 297)*

Every polynomial of degree $n > 0$ with real coefficients can be written as the product of linear and quadratic factors with real coefficients, where the quadratic factors have no real zeros.

Proof

To begin, you use the Linear Factorization Theorem to conclude that $f(x)$ can be *completely* factored in the form

$$f(x) = d(x - c_1)(x - c_2)(x - c_3) \cdots (x - c_n).$$

If each c_i is real, there is nothing more to prove. If any c_i is complex ($c_i = a + bi$, $b \neq 0$), then, because the coefficients of $f(x)$ are real, you know that the conjugate $c_j = a - bi$ is also a zero. By multiplying the corresponding factors, you obtain

$$(x - c_i)(x - c_j) = [x - (a + bi)][x - (a - bi)]$$
$$= x^2 - 2ax + (a^2 + b^2)$$

where each coefficient is real.

PROBLEM SOLVING

This collection of thought-provoking and challenging exercises further explores and expands upon concepts learned in this chapter.

1. (a) Find the zeros of each quadratic function $g(x)$.

   (i) $g(x) = x^2 - 4x - 12$

   (ii) $g(x) = x^2 + 5x$

   (iii) $g(x) = x^2 + 3x - 10$

   (iv) $g(x) = x^2 - 4x + 4$

   (v) $g(x) = x^2 - 2x - 6$

   (vi) $g(x) = x^2 + 3x + 4$

   (b) For each function in part (a), use a graphing utility to graph $f(x) = (x - 2) \cdot g(x)$. Verify that $(2, 0)$ is an x-intercept of the graph of $f(x)$. Describe any similarities or differences in the behavior of the six functions at this x-intercept.

   (c) For each function in part (b), use the graph of $f(x)$ to approximate the other x-intercepts of the graph.

   (d) Describe the connections that you find among the results of parts (a), (b), and (c).

2. Quonset huts were developed during World War II. They were temporary housing structures that could be assembled quickly and easily. A Quonset hut is shaped like a half cylinder. A manufacturer has 600 square feet of material with which to build a Quonset hut.

   (a) The formula for the surface area of half a cylinder is $S = \pi r^2 + \pi r l$, where r is the radius and l is the length of the hut. Solve this equation for l when $S = 600$.

   (b) The formula for the volume of the hut is $V = \frac{1}{2}\pi r^2 l$. Write the volume V of the Quonset hut as a polynomial function of r.

   (c) Use the function you wrote in part (b) to find the maximum volume of a Quonset hut with a surface area of 600 square feet. What are the dimensions of the hut?

3. Show that if $f(x) = ax^3 + bx^2 + cx + d$ then $f(k) = r$, where $r = ak^3 + bk^2 + ck + d$ using long division. In other words, verify the Remainder Theorem for a third-degree polynomial function.

4. In 2000 B.C., the Babylonians solved polynomial equations by referring to tables of values. One such table gave the values of $y^3 + y^2$. To be able to use this table, the Babylonians sometimes had to manipulate the equation as shown below.

$$ax^3 + bx^2 = c$$

$$\frac{a^3 x^3}{b^3} + \frac{a^2 x^2}{b^2} = \frac{a^2 c}{b^3}$$

$$\left(\frac{ax}{b}\right)^3 + \left(\frac{ax}{b}\right)^2 = \frac{a^2 c}{b^3}$$

Then they would find $(a^2 c)/b^3$ in the $y^3 + y^2$ column of the table. Because they knew that the corresponding y-value was equal to $(ax)/b$, they could conclude that $x = (by)/a$.

   (a) Calculate $y^3 + y^2$ for $y = 1, 2, 3, \ldots, 10$. Record the values in a table.

   Use the table from part (a) and the method above to solve each equation.

   (b) $x^3 + x^2 = 252$

   (c) $x^3 + 2x^2 = 288$

   (d) $3x^3 + x^2 = 90$

   (e) $2x^3 + 5x^2 = 2500$

   (f) $7x^3 + 6x^2 = 1728$

   (g) $10x^3 + 3x^2 = 297$

   Using the methods from this chapter, verify your solution to each equation.

5. At a glassware factory, molten cobalt glass is poured into molds to make paperweights. Each mold is a rectangular prism whose height is 3 inches greater than the length of each side of the square base. A machine pours 20 cubic inches of liquid glass into each mold. What are the dimensions of the mold?

6. (a) Complete the table.

Function	Zeros	Sum of zeros	Product of zeros
$f_1(x) = x^2 - 5x + 6$			
$f_2(x) = x^3 - 7x + 6$			
$f_3(x) = x^4 + 2x^3 + x^2 + 8x - 12$			
$f_4(x) = x^5 - 3x^4 - 9x^3 + 25x^2 - 6x$			

   (b) Use the table to make a conjecture relating the sum of the zeros of a polynomial function to the coefficients of the polynomial function.

   (c) Use the table to make a conjecture relating the product of the zeros of a polynomial function to the coefficients of the polynomial function.

329

7. Determine whether the statement is true or false. If false, provide one or more reasons why the statement is false and correct the statement. Let

$$f(x) = ax^3 + bx^2 + cx + d, \quad a \neq 0$$

and let $f(2) = -1$. Then

$$\frac{f(x)}{x + 1} = q(x) + \frac{2}{x + 1}$$

where $q(x)$ is a second-degree polynomial.

8. The parabola shown in the figure has an equation of the form $y = ax^2 + bx + c$. Find the equation of this parabola by the following methods. (a) Find the equation analytically. (b) Use the *regression* feature of a graphing utility to find the equation.

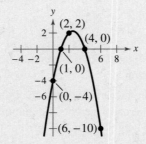

9. One of the fundamental themes of calculus is to find the slope of the tangent line to a curve at a point. To see how this can be done, consider the point $(2, 4)$ on the graph of the quadratic function $f(x) = x^2$, which is shown in the figure.

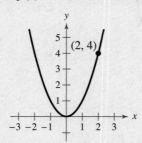

(a) Find the slope m_1 of the line joining $(2, 4)$ and $(3, 9)$. Is the slope of the tangent line at $(2, 4)$ greater than or less than the slope of the line through $(2, 4)$ and $(3, 9)$?

(b) Find the slope m_2 of the line joining $(2, 4)$ and $(1, 1)$. Is the slope of the tangent line at $(2, 4)$ greater than or less than the slope of the line through $(2, 4)$ and $(1, 1)$?

(c) Find the slope m_3 of the line joining $(2, 4)$ and $(2.1, 4.41)$. Is the slope of the tangent line at $(2, 4)$ greater than or less than the slope of the line through $(2, 4)$ and $(2.1, 4.41)$?

(d) Find the slope m_h of the line joining $(2, 4)$ and $(2 + h, f(2 + h))$ in terms of the nonzero number h.

(e) Evaluate the slope formula from part (d) for $h = -1$, 1, and 0.1. Compare these values with those in parts (a)–(c).

(f) What can you conclude the slope $m_{\tan}$ of the tangent line at $(2, 4)$ to be? Explain your answer.

10. A rancher plans to fence a rectangular pasture adjacent to a river (see figure). The rancher has 100 meters of fencing, and no fencing is needed along the river.

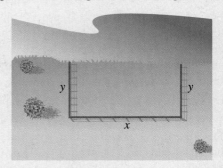

(a) Write the area A of the pasture as a function of x, the length of the side parallel to the river. What is the domain of $A(x)$?

(b) Graph the function $A(x)$ and estimate the dimensions that yield the maximum area of the pasture.

(c) Find the exact dimensions that yield the maximum area of the pasture by writing the quadratic function in standard form.

11. A wire 100 centimeters in length is cut into two pieces. One piece is bent to form a square and the other to form a circle. Let x equal the length of the wire used to form the square.

(a) Write the function that represents the combined area of the two figures.

(b) Determine the domain of the function.

(c) Find the value(s) of x that yield a maximum area and a minimum area.

(d) Explain your reasoning.

Rational Functions and Conics

4

In Mathematics

Functions defined by rational expressions are called rational functions. Conics are collections of points satisfying certain geometric properties.

In Real Life

Rational functions and conics are used to model real-life situations, such as the population growth of a deer herd, the concentration of a chemical in the bloodstream, or the path of a projectile. For instance, you can use a conic to model the path of a satellite as it escapes Earth's gravity. (See Exercise 42, page 368.)

Erik Simonsen/ Photographer's Choice/Getty Images

IN CAREERS

There are many careers that use rational functions and conics. Several are listed below.

What you should learn

- Find the domains of rational functions.
- Find the vertical and horizontal asymptotes of graphs of rational functions.
- Use rational functions to model and solve real-life problems.

Why you should learn it

Rational functions can be used to model and solve real-life problems relating to environmental scenarios. For instance, in Exercise 42 on page 338, a rational function shows how to determine the cost of supplying recycling bins in a pilot project.

4.1 RATIONAL FUNCTIONS AND ASYMPTOTES

Introduction

A **rational function** is a quotient of polynomial functions. It can be written in the form

$$f(x) = \frac{N(x)}{D(x)}$$

where $N(x)$ and $D(x)$ are polynomials and $D(x)$ is not the zero polynomial.

In general, the *domain* of a rational function of x includes all real numbers except x-values that make the denominator zero. Much of the discussion of rational functions will focus on their graphical behavior near the x-values excluded from the domain.

Example 1 Finding the Domain of a Rational Function

Find the domain of $f(x) = \dfrac{1}{x}$ and discuss the behavior of f near any excluded x-values.

Solution

Because the denominator is zero when $x = 0$, the domain of f is all real numbers except $x = 0$. To determine the behavior of f near this excluded value, evaluate $f(x)$ to the left and right of $x = 0$, as indicated in the following tables.

x	-1	-0.5	-0.1	-0.01	-0.001	$\longrightarrow 0$
$f(x)$	-1	-2	-10	-100	-1000	$\longrightarrow -\infty$

x	$0 \longleftarrow$	0.001	0.01	0.1	0.5	1
$f(x)$	$\infty \longleftarrow$	1000	100	10	2	1

Note that as x approaches 0 *from the left*, $f(x)$ decreases without bound. In contrast, as x approaches 0 *from the right*, $f(x)$ increases without bound. The graph of f is shown in Figure 4.1.

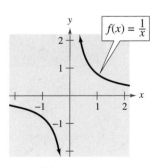

$f(x) = \frac{1}{x}$

FIGURE 4.1

Study Tip

Note that the rational function given by

$$f(x) = \frac{1}{x}$$

is also referred to as the reciprocal function discussed in Section 2.4.

CHECK **Point** Now try Exercise 5.

Vertical and Horizontal Asymptotes

In Example 1, the behavior of f near $x = 0$ is denoted as follows.

$$f(x) \longrightarrow -\infty \text{ as } x \longrightarrow 0^- \qquad f(x) \longrightarrow \infty \text{ as } x \longrightarrow 0^+$$

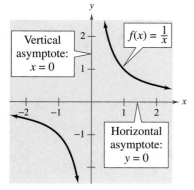

Vertical asymptote: $x = 0$

$f(x) = \frac{1}{x}$

Horizontal asymptote: $y = 0$

FIGURE 4.2

The line $x = 0$ is a **vertical asymptote** of the graph of f, as shown in Figure 4.2. From this figure, you can see that the graph of f also has a **horizontal asymptote**—the line $y = 0$. This means that the values of $f(x) = 1/x$ approach zero as x increases or decreases without bound.

$$f(x) \longrightarrow 0 \text{ as } x \longrightarrow -\infty \qquad f(x) \longrightarrow 0 \text{ as } x \longrightarrow \infty$$

Definitions of Vertical and Horizontal Asymptotes

1. The line $x = a$ is a **vertical asymptote** of the graph of f if

$$f(x) \longrightarrow \infty \quad \text{or} \quad f(x) \longrightarrow -\infty$$

as $x \longrightarrow a$, either from the right or from the left.

2. The line $y = b$ is a **horizontal asymptote** of the graph of f if

$$f(x) \longrightarrow b$$

as $x \longrightarrow \infty$ or $x \longrightarrow -\infty$.

Eventually (as $x \longrightarrow \infty$ or $x \longrightarrow -\infty$), the distance between the horizontal asymptote and the points on the graph must approach zero. Figure 4.3 shows the vertical and horizontal asymptotes of the graphs of three rational functions.

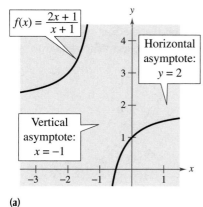

$f(x) = \frac{2x + 1}{x + 1}$

Horizontal asymptote: $y = 2$

Vertical asymptote: $x = -1$

(a)

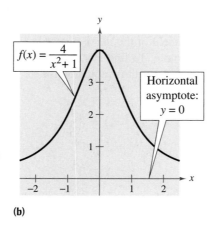

$f(x) = \frac{4}{x^2 + 1}$

Horizontal asymptote: $y = 0$

(b)

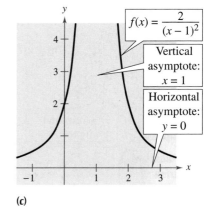

$f(x) = \frac{2}{(x - 1)^2}$

Vertical asymptote: $x = 1$

Horizontal asymptote: $y = 0$

(c)

FIGURE 4.3

The graphs of $f(x) = 1/x$ in Figure 4.2 and $f(x) = (2x + 1)/(x + 1)$ in Figure 4.3(a) are **hyperbolas.** You will study hyperbolas in Sections 4.3 and 4.4.

Vertical and Horizontal Asymptotes of a Rational Function

Let f be the rational function given by

$$f(x) = \frac{N(x)}{D(x)} = \frac{a_n x^n + a_{n-1}x^{n-1} + \cdots + a_1 x + a_0}{b_m x^m + b_{m-1}x^{m-1} + \cdots + b_1 x + b_0}$$

where $N(x)$ and $D(x)$ have no common factors.

1. The graph of f has *vertical* asymptotes at the zeros of $D(x)$.

2. The graph of f has one or no *horizontal* asymptote determined by comparing the degrees of $N(x)$ and $D(x)$.

    a. If $n < m$, the graph of f has the line $y = 0$ (the x-axis) as a horizontal asymptote.

    b. If $n = m$, the graph of f has the line $y = a_n/b_m$ (ratio of the leading coefficients) as a horizontal asymptote.

    c. If $n > m$, the graph of f has no horizontal asymptote.

Example 2   Finding Vertical and Horizontal Asymptotes

Find all vertical and horizontal asymptotes of the graph of each rational function.

a. $f(x) = \dfrac{2x}{3x^2 + 1}$ **b.** $f(x) = \dfrac{2x^2}{x^2 - 1}$

Solution

a. For this rational function, the degree of the numerator is *less than* the degree of the denominator, so the graph has the line $y = 0$ as a horizontal asymptote. To find any vertical asymptotes, set the denominator equal to zero and solve the resulting equation for x. Because the equation $3x^2 + 1 = 0$ has no real solutions, you can conclude that the graph has no vertical asymptote. The graph of the function is shown in Figure 4.4.

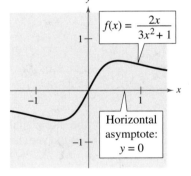

$f(x) = \dfrac{2x}{3x^2 + 1}$

Horizontal asymptote: $y = 0$

FIGURE **4.4**

b. For this rational function, the degree of the numerator is *equal to* the degree of the denominator. The leading coefficient of the numerator is 2 and the leading coefficient of the denominator is 1, so the graph has the line $y = 2$ as a horizontal asymptote. To find any vertical asymptotes, set the denominator equal to zero and solve the resulting equation for x.

$$x^2 - 1 = 0$$
$$(x + 1)(x - 1) = 0$$
$$x + 1 = 0 \quad\Longrightarrow\quad x = -1$$
$$x - 1 = 0 \quad\Longrightarrow\quad x = 1$$

This equation has two real solutions, $x = -1$ and $x = 1$, so the graph has the lines $x = -1$ and $x = 1$ as vertical asymptotes. The graph of the function is shown in Figure 4.5.

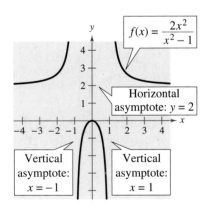

$f(x) = \dfrac{2x^2}{x^2 - 1}$

Horizontal asymptote: $y = 2$

Vertical asymptote: $x = -1$

Vertical asymptote: $x = 1$

FIGURE **4.5**

CHECK *Point*   Now try Exercise 13.

Example 3   Finding Vertical and Horizontal Asymptotes

Find all vertical and horizontal asymptotes of the graph of $f(x) = \dfrac{x^2 + x - 2}{x^2 - x - 6}$.

Solution

For this rational function, the degree of the numerator is *equal to* the degree of the denominator. The leading coefficient of both the numerator and denominator is 1, so the graph has the line $y = 1$ as a horizontal asymptote. To find any vertical asymptotes, first factor the numerator and denominator as follows.

$$f(x) = \frac{x^2 + x - 2}{x^2 - x - 6} = \frac{(x - 1)(x + 2)}{(x + 2)(x - 3)} = \frac{x - 1}{x - 3}, \quad x \neq -2$$

By setting the denominator $x - 3$ (of the simplified function) equal to zero, you can determine that the graph has the line $x = 3$ as a vertical asymptote.

CHECK *Point*   Now try Exercise 29.

Applications

There are many examples of asymptotic behavior in real life. For instance, Example 4 shows how a vertical asymptote can be used to analyze the cost of removing pollutants from smokestack emissions.

Example 4   Cost-Benefit Model

A utility company burns coal to generate electricity. The cost C (in dollars) of removing $p\%$ of the smokestack pollutants is given by $C = 80{,}000p/(100 - p)$ for $0 \leq p < 100$. Sketch the graph of this function. You are a member of a state legislature considering a law that would require utility companies to remove 90% of the pollutants from their smokestack emissions. The current law requires 85% removal. How much additional cost would the utility company incur as a result of the new law?

Solution

The graph of this function is shown in Figure 4.6. Note that the graph has a vertical asymptote at $p = 100$. Because the current law requires 85% removal, the current cost to the utility company is

$$C = \frac{80{,}000(\ )}{100 -} \approx \$453{,}333.$$

If the new law increases the percent removal to 90%, the cost will be

$$C = \frac{80{,}000(\ )}{100 -} = \$720{,}000.$$

So, the new law would require the utility company to spend an additional

$$720{,}000 - 453{,}333 = \$266{,}667.$$

CHECK *Point*   Now try Exercise 41.

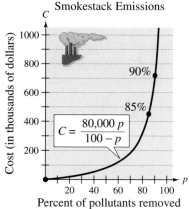

Smokestack Emissions

$C = \dfrac{80{,}000\,p}{100 - p}$

90%

85%

Cost (in thousands of dollars)

Percent of pollutants removed

FIGURE 4.6

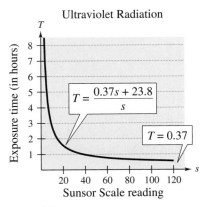

Ultraviolet Radiation

FIGURE 4.7

<hr>

Example 5 Ultraviolet Radiation

For a person with sensitive skin, the amount of time T (in hours) the person can be exposed to the sun with minimal burning can be modeled by

$$T = \frac{0.37s + 23.8}{s}, \quad 0 < s \leq 120$$

where s is the Sunsor Scale reading. The Sunsor Scale is based on the level of intensity of UVB rays. (Source: Sunsor, Inc.)

a. Find the amounts of time a person with sensitive skin can be exposed to the sun with minimal burning when $s = 10$, $s = 25$, and $s = 100$.

b. If the model were valid for all $s > 0$, what would be the horizontal asymptote of this function, and what would it represent?

Solution

a. When $s = 10$, $T = \dfrac{0.37(\quad) + 23.8}{}$

$$= 2.75 \text{ hours.}$$

When $s = 25$, $T = \dfrac{0.37(\quad) + 23.8}{}$

$$\approx 1.32 \text{ hours.}$$

When $s = 100$, $T = \dfrac{0.37(\quad) + 23.8}{}$

$$\approx 0.61 \text{ hour.}$$

b. As shown in Figure 4.7, the horizontal asymptote is the line $T = 0.37$. This line represents the shortest possible exposure time with minimal burning.

CHECK**Point** Now try Exercise 43.

CLASSROOM DISCUSSION

Asymptotes of Graphs of Rational Functions Do you think it is possible for the graph of a rational function to cross its horizontal asymptote? If so, how can you determine when the graph of a rational function will cross its horizontal asymptote? Use the graphs of the following functions to investigate these questions. Write a summary of your conclusions. Explain your reasoning.

a. $f(x) = \dfrac{x}{x^2 + 1}$

b. $g(x) = \dfrac{x}{x^2 - 3}$

c. $h(x) = \dfrac{x^2}{2x^3 - x}$

4.1   EXERCISES

See www.CalcChat.com for worked-out solutions to odd-numbered exercises.

VOCABULARY: Fill in the blanks.

1. Functions of the form $f(x) = N(x)/D(x)$, where $N(x)$ and $D(x)$ are polynomials and $D(x)$ is not the zero polynomial, are called _____ _____.

2. If $f(x) \to \pm\infty$ as $x \to a$ from the left or the right, then $x = a$ is a _____ _____ of the graph of f.

3. If $f(x) \to b$ as $x \to \pm\infty$, then $y = b$ is a _____ _____ of the graph of f.

4. The graph of $f(x) = 1/x$ is called a _____.

SKILLS AND APPLICATIONS

In Exercises 5–8, (a) find the domain of the function, (b) complete each table, and (c) discuss the behavior of f near any excluded x-values.

x	0.5	0.9	0.99	0.999
$f(x)$				

x	1.5	1.1	1.01	1.001
$f(x)$				

x	-1.5	-1.1	-1.01	-1.001
$f(x)$				

x	-0.5	-0.9	-0.99	-0.999
$f(x)$				

5. $f(x) = \dfrac{1}{x - 1}$

6. $f(x) = \dfrac{1}{(x - 1)^2}$

7. $f(x) = \dfrac{3x^2}{x^2 - 1}$

8. $f(x) = \dfrac{4x}{x^2 - 1}$

In Exercises 9–16, find the domain of the function and identify any vertical and horizontal asymptotes.

9. $f(x) = \dfrac{4}{x^2}$

10. $f(x) = \dfrac{1}{(x - 2)^3}$

11. $f(x) = \dfrac{5 + x}{5 - x}$

12. $f(x) = \dfrac{3 - 7x}{3 + 2x}$

13. $f(x) = \dfrac{x^3}{x^2 - 1}$

14. $f(x) = \dfrac{2x^2}{x + 1}$

15. $f(x) = \dfrac{3x^2 + 1}{x^2 + x + 9}$

16. $f(x) = \dfrac{3x^2 + x - 5}{x^2 + 1}$

In Exercises 17–20, match the rational function with its graph. [The graphs are labeled (a), (b), (c), and (d).]

(a)
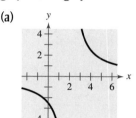

(b)
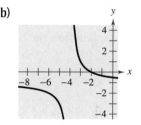

(c)
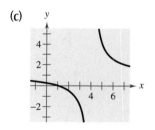

(d)
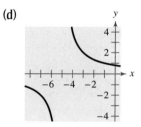

17. $f(x) = \dfrac{4}{x + 5}$

18. $f(x) = \dfrac{5}{x - 2}$

19. $f(x) = \dfrac{x - 1}{x - 4}$

20. $f(x) = -\dfrac{x + 2}{x + 4}$

In Exercises 21–28, find the zeros (if any) of the rational function.

21. $g(x) = \dfrac{x^2 - 1}{x + 1}$

22. $f(x) = \dfrac{x^2 - 2}{x - 3}$

23. $h(x) = 2 + \dfrac{5}{x^2 + 2}$

24. $f(x) = 1 + \dfrac{3}{x^2 - 4}$

25. $f(x) = 1 - \dfrac{3}{x - 3}$

26. $g(x) = 4 - \dfrac{2}{x + 5}$

27. $g(x) = \dfrac{x^3 - 8}{x^2 + 1}$

28. $f(x) = \dfrac{x^3 - 1}{x^2 + 6}$

In Exercises 29–36, find the domain of the function and identify any vertical and horizontal asymptotes.

29. $f(x) = \dfrac{x - 4}{x^2 - 16}$

30. $f(x) = \dfrac{x + 3}{x^2 - 9}$

31. $f(x) = \dfrac{x^2 - 1}{x^2 - 2x - 3}$

32. $f(x) = \dfrac{x^2 - 4}{x^2 - 3x + 2}$

33. $f(x) = \dfrac{x^2 - 3x - 4}{2x^2 + x - 1}$

34. $f(x) = \dfrac{x^2 + x - 2}{2x^2 + 5x + 2}$

35. $f(x) = \dfrac{6x^2 + 5x - 6}{3x^2 - 8x + 4}$

36. $f(x) = \dfrac{6x^2 - 11x + 3}{6x^2 - 7x - 3}$

ANALYTICAL AND NUMERICAL ANALYSIS In Exercises 37–40, (a) determine the domains of f and g, (b) simplify f and find any vertical asymptotes of f, (c) complete the table, and (d) explain how the two functions differ.

37. $f(x) = \dfrac{x^2 - 4}{x + 2}, \quad g(x) = x - 2$

x	-4	-3	-2.5	-2	-1.5	-1	0
$f(x)$							
$g(x)$							

38. $f(x) = \dfrac{x^2(x + 3)}{x^2 + 3x}, \quad g(x) = x$

x	-3	-2	-1	0	1	2	3
$f(x)$							
$g(x)$							

39. $f(x) = \dfrac{2x - 1}{2x^2 - x}, \quad g(x) = \dfrac{1}{x}$

x	-1	-0.5	0	0.5	2	3	4
$f(x)$							
$g(x)$							

40. $f(x) = \dfrac{2x - 8}{x^2 - 9x + 20}, \quad g(x) = \dfrac{2}{x - 5}$

x	0	1	2	3	4	5	6
$f(x)$							
$g(x)$							

41. POLLUTION The cost C (in millions of dollars) of removing $p\%$ of the industrial and municipal pollutants discharged into a river is given by

$$C = \frac{255p}{100 - p}, \quad 0 \le p < 100.$$

(a) Use a graphing utility to graph the cost function.

(b) Find the costs of removing 10%, 40%, and 75% of the pollutants.

(c) According to this model, would it be possible to remove 100% of the pollutants? Explain.

42. RECYCLING In a pilot project, a rural township is given recycling bins for separating and storing recyclable products. The cost C (in dollars) of supplying bins to $p\%$ of the population is given by

$$C = \frac{25{,}000p}{100 - p}, \quad 0 \le p < 100.$$

(a) Use a graphing utility to graph the cost function.

(b) Find the costs of supplying bins to 15%, 50%, and 90% of the population.

(c) According to this model, would it be possible to supply bins to 100% of the residents? Explain.

43. DATA ANALYSIS: PHYSICS EXPERIMENT Consider a physics laboratory experiment designed to determine an unknown mass. A flexible metal meter stick is clamped to a table with 50 centimeters overhanging the edge (see figure on next page). Known masses M ranging from 200 grams to 2000 grams are attached to the end of the meter stick. For each mass, the meter stick is displaced vertically and then allowed to oscillate. The average time t (in seconds) of one oscillation for each mass is recorded in the table.

Mass, M	Time, t
200	0.450
400	0.597
600	0.721
800	0.831
1000	0.906
1200	1.003
1400	1.008
1600	1.168
1800	1.218
2000	1.338

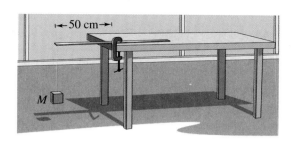

A model for the data that can be used to predict the time of one oscillation is

$$t = \frac{38M + 16{,}965}{10(M + 5000)}.$$

(a) Use this model to create a table showing the predicted time for each of the masses shown in the table.

(b) Compare the predicted times with the experimental times. What can you conclude?

(c) Use the model to approximate the mass of an object for which $t = 1.056$ seconds.

44. POPULATION GROWTH The game commission introduces 100 deer into newly acquired state game lands. The population N of the herd is modeled by

$$N = \frac{20(5 + 3t)}{1 + 0.04t}, \quad t \geq 0$$

where t is the time in years.

 (a) Use a graphing utility to graph this model.

(b) Find the populations when $t = 5$, $t = 10$, and $t = 25$.

(c) What is the limiting size of the herd as time increases?

45. FOOD CONSUMPTION A biology class performs an experiment comparing the quantity of food consumed by a certain kind of moth with the quantity supplied. The model for the experimental data is given by

$$y = \frac{1.568x - 0.001}{6.360x + 1}, \quad x > 0$$

where x is the quantity (in milligrams) of food supplied and y is the quantity (in milligrams) of food consumed.

 (a) Use a graphing utility to graph this model.

(b) At what level of consumption will the moth become satiated?

46. HUMAN MEMORY MODEL Psychologists have developed mathematical models to predict memory performance as a function of the number of trials n of a certain task. Consider the learning curve

$$P = \frac{0.5 + 0.9(n - 1)}{1 + 0.9(n - 1)}, \quad n > 0$$

where P is the fraction of correct responses after n trials.

(a) Complete the table for this model. What does it suggest?

n	1	2	3	4	5	6	7	8	9	10
P										

(b) According to this model, what is the limiting percent of correct responses as n increases?

EXPLORATION

TRUE OR FALSE? In Exercises 47 and 48, determine whether the statement is true or false. Justify your answer.

47. A polynomial function can have infinitely many vertical asymptotes.

48. $f(x) = x^3 - 2x^2 - 5x + 6$ is a rational function.

In Exercises 49–52, (a) determine the value that the function f approaches as the magnitude of x increases. Is $f(x)$ greater than or less than this functional value when (b) x is positive and large in magnitude and (c) x is negative and large in magnitude?

49. $f(x) = 4 - \dfrac{1}{x}$

50. $f(x) = 2 + \dfrac{1}{x - 3}$

51. $f(x) = \dfrac{2x - 1}{x - 3}$

52. $f(x) = \dfrac{2x - 1}{x^2 + 1}$

THINK ABOUT IT In Exercises 53 and 54, write a rational function f that has the specified characteristics. (There are many correct answers.)

53. Vertical asymptote: None

Horizontal asymptote: $y = 2$

54. Vertical asymptotes: $x = -2, x = 1$

Horizontal asymptote: None

55. THINK ABOUT IT Give an example of a rational function whose domain is the set of all real numbers. Give an example of a rational function whose domain is the set of all real numbers except $x = 15$.

56. CAPSTONE Given a polynomial $p(x)$, is it true that the graph of the function given by $f(x) = \dfrac{p(x)}{x^2 - 4}$ has a vertical asymptote at $x = 2$? Why or why not?

4.2 GRAPHS OF RATIONAL FUNCTIONS

What you should learn

- Analyze and sketch graphs of rational functions.
- Sketch graphs of rational functions that have slant asymptotes.
- Use graphs of rational functions to model and solve real-life problems.

Why you should learn it

You can use rational functions to model average speed over a distance. For instance, see Exercise 85 on page 348.

Analyzing Graphs of Rational Functions

To sketch the graph of a rational function, use the following guidelines.

> ### Guidelines for Analyzing Graphs of Rational Functions
>
> Let $f(x) = N(x)/D(x)$, where $N(x)$ and $D(x)$ are polynomials.
>
> 1. Simplify f, if possible.
>
> 2. Find and plot the y-intercept (if any) by evaluating $f(0)$.
>
> 3. Find the zeros of the numerator (if any) by solving the equation $N(x) = 0$. Then plot the corresponding x-intercepts.
>
> 4. Find the zeros of the denominator (if any) by solving the equation $D(x) = 0$. Then sketch the corresponding vertical asymptotes.
>
> 5. Find and sketch the horizontal asymptote (if any) by using the rule for finding the horizontal asymptote of a rational function.
>
> 6. Plot at least one point *between* and one point *beyond* each x-intercept and vertical asymptote.
>
> 7. Use smooth curves to complete the graph between and beyond the vertical asymptotes.

You may also want to test for symmetry when graphing rational functions, especially for simple rational functions. Recall from Section 2.4 that the graph of $f(x) = 1/x$ is symmetric with respect to the origin.

TECHNOLOGY

Some graphing utilities have difficulty graphing rational functions that have vertical asymptotes. Often, the utility will connect parts of the graph that are not supposed to be connected. For instance, the screen on the left below shows the graph of $f(x) = 1/(x - 2)$. Notice that the graph should consist of two unconnected portions—one to the left of $x = 2$ and the other to the right of $x = 2$. To eliminate this problem, you can try changing the mode of the graphing utility to *dot mode*. The problem with this is that the graph is then represented as a collection of dots (as shown in the screen on the right) rather than as a smooth curve.

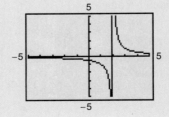

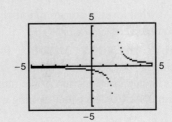

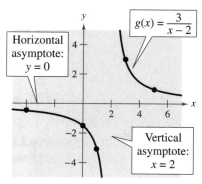

FIGURE 4.8

| Example 1 | **Sketching the Graph of a Rational Function** |

Sketch the graph of $g(x) = \dfrac{3}{x-2}$ and state its domain.

Solution

y-intercept:	$\left(0, -\frac{3}{2}\right)$, because $g(0) = -\frac{3}{2}$
x-intercept:	None, because $3 \neq 0$
Vertical asymptote:	$x = 2$, zero of denominator
Horizontal asymptote:	$y = 0$, because degree of $N(x) <$ degree of $D(x)$
Additional points:	

x	-4	1	2	3	5
$g(x)$	-0.5	-3	Undefined	3	1

By plotting the intercepts, asymptotes, and a few additional points, you can obtain the graph shown in Figure 4.8. The domain of g is all real numbers except $x = 2$.

CHECK Point   Now try Exercise 15.

The graph of g in Example 1 is a vertical stretch and a right shift of the graph of $f(x) = 1/x$, because

$$g(x) = \frac{3}{x-2}$$

$$= 3\left(\frac{1}{x-2}\right)$$

$$= 3f(x-2).$$

Study Tip

Note in the examples in this section that the vertical asymptotes are included in the table of additional points. This is done to emphasize numerically the behavior of the graph of the function.

| Example 2 | **Sketching the Graph of a Rational Function** |

Sketch the graph of $f(x) = \dfrac{2x-1}{x}$ and state its domain.

Solution

y-intercept:	None, because $x = 0$ is not in the domain
x-intercept:	$\left(\frac{1}{2}, 0\right)$, because $f\left(\frac{1}{2}\right) = 0$
Vertical asymptote:	$x = 0$, zero of denominator
Horizontal asymptote:	$y = 2$, because degree of $N(x) =$ degree of $D(x)$
Additional points:	

x	-4	-1	0	$\frac{1}{4}$	4
$f(x)$	2.25	3	Undefined	-2	1.75

By plotting the intercepts, asymptotes, and a few additional points, you can obtain the graph shown in Figure 4.9. The domain of f is all real numbers except $x = 0$.

CHECK Point   Now try Exercise 19.

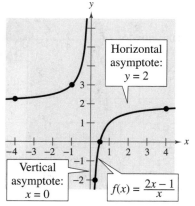

FIGURE 4.9

Example 3 Sketching the Graph of a Rational Function

Sketch the graph of $f(x) = \dfrac{x}{x^2 - x - 2}$.

Solution

Factor the denominator to determine more easily the zeros of the denominator.

$$f(x) = \frac{x}{x^2 - x - 2} = \frac{x}{(x + 1)(x - 2)}$$

y-intercept: (0, 0), because $f(0) = 0$

x-intercept: (0, 0), because $f(0) = 0$

Vertical asymptotes: $x = -1$, $x = 2$, zeros of denominator

Horizontal asymptote: $y = 0$, because degree of $N(x) <$ degree of $D(x)$

Additional points:

x	-3	-1	-0.5	1	2	3
$f(x)$	-0.3	Undefined	0.4	-0.5	Undefined	0.75

The graph is shown in Figure 4.10.

CHECK *Point* Now try Exercise 31.

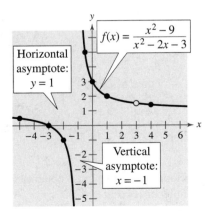

Vertical asymptote: $x = -1$

Vertical asymptote: $x = 2$

Horizontal asymptote: $y = 0$

$f(x) = \dfrac{x}{x^2 - x - 2}$

FIGURE **4.10**

Example 4 Sketching the Graph of a Rational Function

Sketch the graph of $f(x) = \dfrac{x^2 - 9}{x^2 - 2x - 3}$.

Solution

By factoring the numerator and denominator, you have

$$f(x) = \frac{x^2 - 9}{x^2 - 2x - 3} = \frac{(x - 3)(x + 3)}{(x - 3)(x + 1)} = \frac{x + 3}{x + 1}, \quad x \neq 3.$$

y-intercept: (0, 3), because $f(0) = 3$

x-intercept: (−3, 0), because $f(-3) = 0$

Vertical asymptote: $x = -1$, zero of (simplified) denominator

Horizontal asymptote: $y = 1$, because degree of $N(x) =$ degree of $D(x)$

Additional points:

x	-5	-2	-1	-0.5	1	3	4
$f(x)$	0.5	-1	Undefined	5	2	Undefined	1.4

The graph is shown in Figure 4.11. Notice that there is a hole in the graph at $x = 3$ because the function is not defined when $x = 3$.

CHECK *Point* Now try Exercise 39.

Horizontal asymptote: $y = 1$

$f(x) = \dfrac{x^2 - 9}{x^2 - 2x - 3}$

Vertical asymptote: $x = -1$

FIGURE **4.11** Hole at $x = 3$

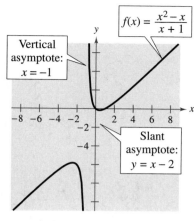

Vertical asymptote: $x = -1$

$f(x) = \dfrac{x^2 - x}{x + 1}$

Slant asymptote: $y = x - 2$

FIGURE **4.12**

Slant Asymptotes

Consider a rational function whose denominator is of degree 1 or greater. If the degree of the numerator is exactly *one more* than the degree of the denominator, the graph of the function has a **slant** (or **oblique**) **asymptote.** For example, the graph of

$$f(x) = \frac{x^2 - x}{x + 1}$$

has a slant asymptote, as shown in Figure 4.12. To find the equation of a slant asymptote, use long division. For instance, by dividing $x + 1$ into $x^2 - x$, you obtain

$$f(x) = \frac{x^2 - x}{x + 1} = x - 2 + \frac{2}{x + 1}.$$

As x increases or decreases without bound, the remainder term $2/(x + 1)$ approaches 0, so the graph of f approaches the line $y = x - 2$, as shown in Figure 4.12.

Example 5   A Rational Function with a Slant Asymptote

Sketch the graph of $f(x) = \dfrac{x^2 - x - 2}{x - 1}$.

Solution

First write $f(x)$ in two different ways. Factoring the numerator

$$f(x) = \frac{x^2 - x - 2}{x - 1} = \frac{(x - 2)(x + 1)}{x - 1}$$

allows you to recognize the *x*-intercepts. Long division

$$f(x) = \frac{x^2 - x - 2}{x - 1} = x - \frac{2}{x - 1}$$

allows you to recognize that the line $y = x$ is a slant asymptote of the graph.

y-intercept: $(0, 2)$, because $f(0) = 2$

x-intercepts: $(-1, 0)$ and $(2, 0)$, because $f(-1) = 0$ and $f(2) = 0$

Vertical asymptote: $x = 1$, zero of denominator

Slant asymptote: $y = x$

Additional points:

x	-2	0.5	1	1.5	3
$f(x)$	-1.33	4.5	Undefined	-2.5	2

The graph is shown in Figure 4.13.

CHECK *Point*   Now try Exercise 61.

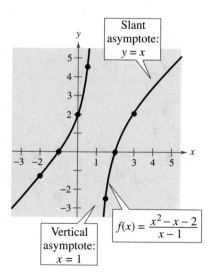

Slant asymptote: $y = x$

Vertical asymptote: $x = 1$

$f(x) = \dfrac{x^2 - x - 2}{x - 1}$

FIGURE **4.13**

Application

Example 6   Finding a Minimum Area

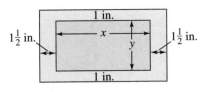

FIGURE **4.14**

A rectangular page is designed to contain 48 square inches of print. The margins at the top and bottom of the page are each 1 inch deep. The margins on each side are $1\frac{1}{2}$ inches wide. What should the dimensions of the page be so that the least amount of paper is used?

Graphical Solution

Let A be the area to be minimized. From Figure 4.14, you can write

$$A = (x + 3)(y + 2).$$

The printed area inside the margins is modeled by $48 = xy$ or $y = 48/x$. To find the minimum area, rewrite the equation for A in terms of just one variable by substituting $48/x$ for y.

$$A = (x + 3)\left(\quad + 2\right)$$

$$= \frac{(x + 3)(48 + 2x)}{x}, \quad x > 0$$

The graph of this rational function is shown in Figure 4.15. Because x represents the width of the printed area, you need consider only the portion of the graph for which x is positive. Using a graphing utility, you can approximate the minimum value of A to occur when $x \approx 8.5$ inches. The corresponding value of y is $48/8.5 \approx 5.6$ inches. So, the dimensions should be

$$x + 3 \approx 11.5 \text{ inches} \quad \text{by} \quad y + 2 \approx 7.6 \text{ inches.}$$

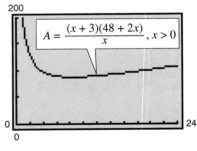

FIGURE **4.15**

CHECK**Point**   Now try Exercise 79.

Numerical Solution

Let A be the area to be minimized. From Figure 4.14, you can write

$$A = (x + 3)(y + 2).$$

The printed area inside the margins is modeled by $48 = xy$ or $y = 48/x$. To find the minimum area, rewrite the equation for A in terms of just one variable by substituting $48/x$ for y.

$$A = (x + 3)\left(\quad + 2\right) = \frac{(x + 3)(48 + 2x)}{x}, \quad x > 0$$

Use the *table* feature of a graphing utility to create a table of values for the function

$$y_1 = \frac{(x + 3)(48 + 2x)}{x}$$

beginning at $x = 1$. From the table, you can see that the minimum value of y_1 occurs when x is somewhere between 8 and 9, as shown in Figure 4.16. To approximate the minimum value of y_1 to one decimal place, change the table so that it starts at $x = 8$ and increases by 0.1. The minimum value of y_1 occurs when $x \approx 8.5$, as shown in Figure 4.17. The corresponding value of y is $48/8.5 \approx 5.6$ inches. So, the dimensions should be $x + 3 \approx 11.5$ inches by $y + 2 \approx 7.6$ inches.

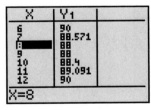

X	Y1
6	90
7	88.571
8	88
9	88
10	88.4
11	89.091
12	90

X=8

FIGURE **4.16**

X	Y1
8.2	87.961
8.3	87.949
8.4	87.943
8.5	87.941
8.6	87.944
8.7	87.952
8.8	87.964

X=8.5

FIGURE **4.17**

If you go on to take a course in calculus, you will learn an analytic technique for finding the exact value of x that produces a minimum area. In this case, that value is $x = 6\sqrt{2} \approx 8.485$.

4.2   EXERCISES

See www.CalcChat.com for worked-out solutions to odd-numbered exercises.

VOCABULARY: Fill in the blanks.

1. For the rational function given by $f(x) = N(x)/D(x)$, if the degree of $N(x)$ is exactly one more than the degree of $D(x)$, then the graph of f has a _____ (or oblique) _____.

2. The graph of $g(x) = 3/(x - 2)$ has a _____ asymptote at $x = 2$.

SKILLS AND APPLICATIONS

In Exercises 3–6, use the graph of $f(x) = 2/x$ to sketch the graph of g.

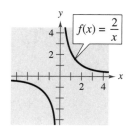

3. $g(x) = \dfrac{2}{x} + 4$

4. $g(x) = \dfrac{2}{x - 5}$

5. $g(x) = -\dfrac{2}{x}$

6. $g(x) = \dfrac{1}{x + 2}$

In Exercises 7–10, use the graph of $f(x) = 3/x^2$ to sketch the graph of g.

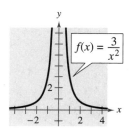

7. $g(x) = \dfrac{3}{x^2} - 1$

8. $g(x) = -\dfrac{3}{x^2}$

9. $g(x) = \dfrac{3}{(x - 1)^2}$

10. $g(x) = \dfrac{1}{x^2}$

In Exercises 11–14, use the graph of $f(x) = 4/x^3$ to sketch the graph of g.

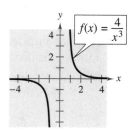

11. $g(x) = \dfrac{4}{(x + 2)^3}$

12. $g(x) = \dfrac{4}{x^3} + 2$

13. $g(x) = -\dfrac{4}{x^3}$

14. $g(x) = \dfrac{2}{x^3}$

In Exercises 15–44, (a) state the domain of the function, (b) identify all intercepts, (c) find any vertical and horizontal asymptotes, and (d) plot additional solution points as needed to sketch the graph of the rational function.

15. $f(x) = \dfrac{1}{x + 2}$

16. $f(x) = \dfrac{1}{x - 3}$

17. $h(x) = \dfrac{-1}{x + 4}$

18. $g(x) = \dfrac{1}{6 - x}$

19. $C(x) = \dfrac{7 + 2x}{2 + x}$

20. $P(x) = \dfrac{1 - 3x}{1 - x}$

21. $g(x) = \dfrac{1}{x + 2} + 2$

22. $f(x) = 2 - \dfrac{3}{x^2}$

23. $f(x) = \dfrac{x^2}{x^2 + 9}$

24. $f(t) = \dfrac{1 - 2t}{t}$

25. $h(x) = \dfrac{x^2}{x^2 - 9}$

26. $g(x) = \dfrac{x}{x^2 - 9}$

27. $g(s) = \dfrac{4s}{s^2 + 4}$

28. $f(x) = -\dfrac{1}{(x - 2)^2}$

29. $g(x) = \dfrac{4(x + 1)}{x(x - 4)}$

30. $h(x) = \dfrac{2}{x^2(x - 2)}$

31. $f(x) = \dfrac{2x}{x^2 - 3x - 4}$

32. $f(x) = \dfrac{3x}{x^2 + 2x - 3}$

33. $h(x) = \dfrac{x^2 - 5x + 4}{x^2 - 4}$

34. $g(x) = \dfrac{x^2 - 2x - 8}{x^2 - 9}$

35. $f(x) = \dfrac{6x}{x^2 - 5x - 14}$

36. $f(x) = \dfrac{3(x^2 + 1)}{x^2 + 2x - 15}$

37. $f(x) = \dfrac{2x^2 - 5x - 3}{x^3 - 2x^2 - x + 2}$

38. $f(x) = \dfrac{x^2 - x - 2}{x^3 - 2x^2 - 5x + 6}$

39. $f(x) = \dfrac{x^2 + 3x}{x^2 + x - 6}$ **40.** $f(x) = \dfrac{5(x + 4)}{x^2 + x - 12}$

41. $f(x) = \dfrac{2x^2 - 5x + 2}{2x^2 - x - 6}$ **42.** $f(x) = \dfrac{3x^2 - 8x + 4}{2x^2 - 3x - 2}$

43. $f(t) = \dfrac{t^2 - 1}{t - 1}$ **44.** $f(x) = \dfrac{x^2 - 36}{x + 6}$

〰 **ANALYTICAL, NUMERICAL, AND GRAPHICAL ANALYSIS**
In Exercises 45–48, do the following.

(a) Determine the domains of f and g.

(b) Simplify f and find any vertical asymptotes of the graph of f.

(c) Compare the functions by completing the table.

(d) Use a graphing utility to graph f and g in the same viewing window.

(e) Explain why the graphing utility may not show the difference in the domains of f and g.

45. $f(x) = \dfrac{x^2 - 1}{x + 1}$, $g(x) = x - 1$

x	-3	-2	-1.5	-1	-0.5	0	1
$f(x)$							
$g(x)$							

46. $f(x) = \dfrac{x^2(x - 2)}{x^2 - 2x}$, $g(x) = x$

x	-1	0	1	1.5	2	2.5	3
$f(x)$							
$g(x)$							

47. $f(x) = \dfrac{x - 2}{x^2 - 2x}$, $g(x) = \dfrac{1}{x}$

x	-0.5	0	0.5	1	1.5	2	3
$f(x)$							
$g(x)$							

48. $f(x) = \dfrac{2x - 6}{x^2 - 7x + 12}$, $g(x) = \dfrac{2}{x - 4}$

x	0	1	2	3	4	5	6
$f(x)$							
$g(x)$							

In Exercises 49–64, (a) state the domain of the function, (b) identify all intercepts, (c) identify any vertical and slant asymptotes, and (d) plot additional solution points as needed to sketch the graph of the rational function.

49. $h(x) = \dfrac{x^2 - 9}{x}$

50. $g(x) = \dfrac{x^2 + 5}{x}$

51. $f(x) = \dfrac{2x^2 + 1}{x}$

52. $f(x) = \dfrac{1 - x^2}{x}$

53. $g(x) = \dfrac{x^2 + 1}{x}$

54. $h(x) = \dfrac{x^2}{x - 1}$

55. $f(t) = -\dfrac{t^2 + 1}{t + 5}$

56. $f(x) = \dfrac{x^2}{3x + 1}$

57. $f(x) = \dfrac{x^3}{x^2 - 4}$

58. $g(x) = \dfrac{x^3}{2x^2 - 8}$

59. $f(x) = \dfrac{x^3 - 1}{x^2 - x}$

60. $f(x) = \dfrac{x^4 + x}{x^3}$

61. $f(x) = \dfrac{x^2 - x + 1}{x - 1}$

62. $f(x) = \dfrac{2x^2 - 5x + 5}{x - 2}$

63. $f(x) = \dfrac{2x^3 - x^2 - 2x + 1}{x^2 + 3x + 2}$

64. $f(x) = \dfrac{2x^3 + x^2 - 8x - 4}{x^2 - 3x + 2}$

〰 In Exercises 65–68, use a graphing utility to graph the rational function. Give the domain of the function and identify any asymptotes. Then zoom out sufficiently far so that the graph appears as a line. Identify the line.

65. $f(x) = \dfrac{x^2 + 5x + 8}{x + 3}$ **66.** $f(x) = \dfrac{2x^2 + x}{x + 1}$

67. $g(x) = \dfrac{1 + 3x^2 - x^3}{x^2}$ **68.** $h(x) = \dfrac{12 - 2x - x^2}{2(4 + x)}$

GRAPHICAL REASONING In Exercises 69–72, (a) use the graph to determine any *x*-intercepts of the graph of the rational function and (b) set $y = 0$ and solve the resulting equation to confirm your result in part (a).

69. $y = \dfrac{x + 1}{x - 3}$

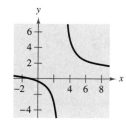

70. $y = \dfrac{2x}{x - 3}$

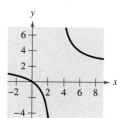

71. $y = \dfrac{1}{x} - x$

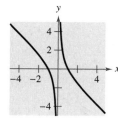

72. $y = x - 3 + \dfrac{2}{x}$

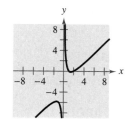

GRAPHICAL REASONING In Exercises 73–76, (a) use a graphing utility to graph the rational function and determine any *x*-intercepts of the graph and (b) set $y = 0$ and solve the resulting equation to confirm your result in part (a).

73. $y = \dfrac{1}{x + 5} + \dfrac{4}{x}$

74. $y = 20\left(\dfrac{2}{x + 1} - \dfrac{3}{x}\right)$

75. $y = x - \dfrac{6}{x - 1}$

76. $y = x - \dfrac{9}{x}$

77. CONCENTRATION OF A MIXTURE A 1000-liter tank contains 50 liters of a 25% brine solution. You add *x* liters of a 75% brine solution to the tank.

(a) Show that the concentration C, the proportion of brine to total solution, in the final mixture is
$$C = \dfrac{3x + 50}{4(x + 50)}.$$

(b) Determine the domain of the function based on the physical constraints of the problem.

(c) Sketch a graph of the concentration function.

(d) As the tank is filled, what happens to the rate at which the concentration of brine is increasing? What percent does the concentration of brine appear to approach?

78. GEOMETRY A rectangular region of length *x* and width *y* has an area of 500 square meters.

(a) Write the width *y* as a function of *x*.

(b) Determine the domain of the function based on the physical constraints of the problem.

(c) Sketch a graph of the function and determine the width of the rectangle when $x = 30$ meters.

79. PAGE DESIGN A page that is *x* inches wide and *y* inches high contains 30 square inches of print. The top and bottom margins are 1 inch deep and the margins on each side are 2 inches wide (see figure).

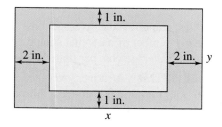

(a) Show that the total area *A* on the page is
$$A = \dfrac{2x(x + 11)}{x - 4}.$$

(b) Determine the domain of the function based on the physical constraints of the problem.

(c) Use a graphing utility to graph the area function and approximate the page size for which the least amount of paper will be used. Verify your answer numerically using the *table* feature of the graphing utility.

80. PAGE DESIGN A rectangular page is designed to contain 64 square inches of print. The margins at the top and bottom of the page are each 1 inch deep. The margins on each side are $1\frac{1}{2}$ inches wide. What should the dimensions of the page be so that the least amount of paper is used?

In Exercises 81 and 82, use a graphing utility to graph the function and locate any relative maximum or minimum points on the graph.

81. $f(x) = \dfrac{3(x + 1)}{x^2 + x + 1}$

82. $C(x) = x + \dfrac{32}{x}$

83. MINIMUM COST The ordering and transportation cost C (in thousands of dollars) for the components used in manufacturing a product is given by

$$C = 100\left(\frac{200}{x^2} + \frac{x}{x + 30}\right), \quad x \geq 1$$

where x is the order size (in hundreds). Use a graphing utility to graph the cost function. From the graph, estimate the order size that minimizes cost.

84. MINIMUM COST The cost C of producing x units of a product is given by

$$C = 0.2x^2 + 10x + 5$$

and the average cost per unit is given by

$$\overline{C} = \frac{C}{x} = \frac{0.2x^2 + 10x + 5}{x}, \quad x > 0.$$

Sketch the graph of the average cost function and estimate the number of units that should be produced to minimize the average cost per unit.

85. AVERAGE SPEED A driver averaged 50 miles per hour on the round trip between Akron, Ohio, and Columbus, Ohio, 100 miles away. The average speeds for going and returning were x and y miles per hour, respectively.

(a) Show that $y = \dfrac{25x}{x - 25}$.

(b) Determine the vertical and horizontal asymptotes of the graph of the function.

 (c) Use a graphing utility to graph the function.

(d) Complete the table.

x	30	35	40	45	50	55	60
y							

(e) Are the results in the table what you expected? Explain.

(f) Is it possible to average 20 miles per hour in one direction and still average 50 miles per hour on the round trip? Explain.

 86. MEDICINE The concentration C of a chemical in the bloodstream t hours after injection into muscle tissue is given by

$$C = \frac{3t^2 + t}{t^3 + 50}, \quad t > 0.$$

(a) Determine the horizontal asymptote of the graph of the function and interpret its meaning in the context of the problem.

(b) Use a graphing utility to graph the function and approximate the time when the bloodstream concentration is greatest.

(c) Use a graphing utility to determine when the concentration is less than 0.345.

EXPLORATION

TRUE OR FALSE? In Exercises 87–90, determine whether the statement is true or false. Justify your answer.

87. If the graph of a rational function f has a vertical asymptote at $x = 5$, it is possible to sketch the graph without lifting your pencil from the paper.

88. The graph of a rational function can never cross one of its asymptotes.

89. The graph of $f(x) = \dfrac{2x^3}{x + 1}$ has a slant asymptote.

90. Every rational function has a horizontal asymptote.

 THINK ABOUT IT In Exercises 91 and 92, use a graphing utility to graph the function. Explain why there is no vertical asymptote when a superficial examination of the function may indicate that there should be one.

91. $h(x) = \dfrac{6 - 2x}{3 - x}$

92. $g(x) = \dfrac{x^2 + x - 2}{x - 1}$

93. WRITING Given a rational function f, how can you determine whether f has a slant asymptote? If f has a slant asymptote, explain the process for finding it.

94. CAPSTONE Write a rational function satisfying the following criteria. Then sketch a graph of your function.

Vertical asymptote: $x = 2$

Slant asymptote: $y = x + 1$

Zero of the function: $x = -2$

PROJECT: DEPARTMENT OF DEFENSE To work an extended application analyzing the total numbers of Department of Defense personnel from 1980 through 2007, visit this text's website at *academic.cengage.com*. (Data Source: U.S. Department of Defense)

4.3 CONICS

What you should learn

- Recognize the four basic conics: circle, ellipse, parabola, and hyperbola.
- Recognize, graph, and write equations of parabolas (vertex at origin).
- Recognize, graph, and write equations of ellipses (center at origin).
- Recognize, graph, and write equations of hyperbolas (center at origin).

Why you should learn it

Conics have been used for hundreds of years to model and solve engineering problems. For instance, in Exercise 45 on page 359, a parabola can be used to model the cables of the Golden Gate Bridge.

Cosmo Condina/Getty Images

Introduction

Conic sections were discovered during the classical Greek period, 600 to 300 B.C. This early Greek study was largely concerned with the geometric properties of conics. It was not until the early 17th century that the broad applicability of conics became apparent and played a prominent role in the early development of calculus.

A **conic section** (or simply **conic**) is the intersection of a plane and a double-napped cone. Notice in Figure 4.18 that in the formation of the four basic conics, the intersecting plane does not pass through the vertex of the cone. When the plane does pass through the vertex, the resulting figure is a **degenerate conic,** as shown in Figure 4.19.

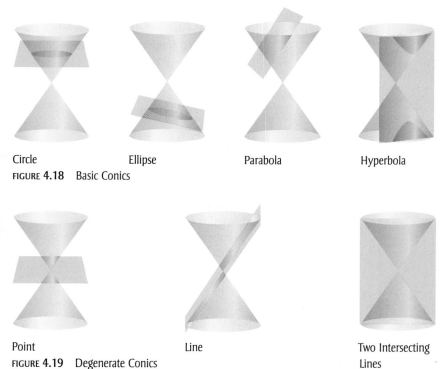

Circle   Ellipse   Parabola   Hyperbola

FIGURE **4.18**   Basic Conics

Point   Line   Two Intersecting
                                                             Lines
FIGURE **4.19**   Degenerate Conics

There are several ways to approach the study of conics. You could begin by defining conics in terms of the intersections of planes and cones, as the Greeks did, or you could define them algebraically, in terms of the general second-degree equation

$$Ax^2 + Bxy + Cy^2 + Dx + Ey + F = 0.$$

However, you will study a third approach, in which each of the conics is defined as a *locus* (collection) of points satisfying a certain geometric property. For example, in Section 1.1 you saw how the definition of a circle as *the collection of all points* (x, y) *that are equidistant from a fixed point* (h, k) led easily to the standard form of the equation of a circle

$$(x - h)^2 + (y - k)^2 = r^2.$$

Recall from Section 1.1 that the center of a circle is at (h, k) and that the radius of the circle is r.

Parabolas

In Section 3.1, you learned that the graph of the quadratic function

$$f(x) = ax^2 + bx + c$$

is a parabola that opens upward or downward. The following definition of a parabola is more general in the sense that it is independent of the orientation of the parabola.

Definition of a Parabola

A **parabola** is the set of all points (x, y) in a plane that are equidistant from a fixed line, the **directrix**, and a fixed point, the **focus**, not on the line. (See Figure 4.20.) The **vertex** is the midpoint between the focus and the directrix. The **axis** of the parabola is the line passing through the focus and the vertex.

Standard Equation of a Parabola (Vertex at Origin)

The **standard form of the equation of a parabola** with vertex at $(0, 0)$ and directrix $y = -p$ is

$$x^2 = 4py, \qquad p \neq 0.$$

For directrix $x = -p$, the equation is

$$y^2 = 4px, \qquad p \neq 0.$$

The focus is on the axis p units (directed distance) from the vertex.

FIGURE **4.20**   Parabola

For a proof of the standard form of the equation of a parabola, see Proofs in Mathematics on page 376.

Notice that a parabola can have a vertical or a horizontal axis and that a parabola is symmetric with respect to its axis. Examples of each are shown in Figure 4.21.

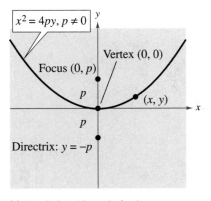

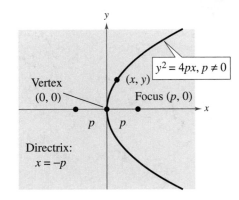

(a) Parabola with vertical axis

(b) Parabola with horizontal axis

FIGURE **4.21**

Example 1   Finding the Focus of a Parabola

Find the focus of the parabola whose equation is $y = -2x^2$.

Solution

Because the squared term in the equation involves x, you know that the axis is vertical, and the equation is of the form

$$x^2 = 4py.$$

You can write the original equation in this form as follows.

$$x^2 = -\frac{1}{2}y$$

$$x^2 = 4\left(-\frac{1}{8}\right)y$$

So, $p = -\frac{1}{8}$. Because p is negative, the parabola opens downward (see Figure 4.22), and the focus of the parabola is

$$(0, p) = \left(0, -\frac{1}{8}\right).$$

CHECK*Point*   Now try Exercise 21.

Example 2   A Parabola with a Horizontal Axis

Find the standard form of the equation of the parabola with vertex at the origin and focus at $(2, 0)$.

Solution

The axis of the parabola is horizontal, passing through $(0, 0)$ and $(2, 0)$, as shown in Figure 4.23. So, the standard form is

$$y^2 = 4px.$$

Because the focus is $p = 2$ units from the vertex, the equation is

$$y^2 = 4(\ )x$$

$$y^2 = 8x.$$

CHECK*Point*   Now try Exercise 27.

Parabolas occur in a wide variety of applications. For instance, a parabolic reflector can be formed by revolving a parabola about its axis. The resulting surface has the property that all incoming rays parallel to the axis are reflected through the focus of the parabola. This is the principle behind the construction of the parabolic mirrors used in reflecting telescopes. Conversely, the light rays emanating from the focus of a parabolic reflector used in a flashlight are all parallel to one another, as shown in Figure 4.24.

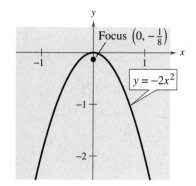

FIGURE **4.22**

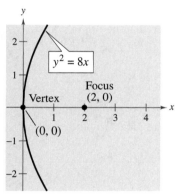

FIGURE **4.23**

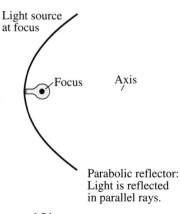

Parabolic reflector:
Light is reflected
in parallel rays.

FIGURE **4.24**

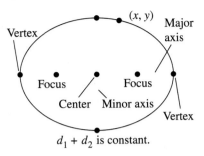

FIGURE **4.25**

$d_1 + d_2$ is constant.

Ellipses

Definition of an Ellipse

An **ellipse** is the set of all points (x, y) in a plane the sum of whose distances from two distinct fixed points (**foci**) is constant. See Figure 4.25.

The line through the foci intersects the ellipse at two points (**vertices**). The chord joining the vertices is the **major axis,** and its midpoint is the **center** of the ellipse. The chord perpendicular to the major axis at the center is the **minor axis.** (See Figure 4.25.)

You can visualize the definition of an ellipse by imagining two thumbtacks placed at the foci, as shown in Figure 4.26. If the ends of a fixed length of string are fastened to the thumbtacks and the string is *drawn taut* with a pencil, the path traced by the pencil will be an ellipse.

The standard form of the equation of an ellipse takes one of two forms, depending on whether the major axis is horizontal or vertical.

FIGURE **4.26**

Standard Equation of an Ellipse (Center at Origin)

The **standard form of the equation of an ellipse** centered at the origin with major and minor axes of lengths $2a$ and $2b$ (where $0 < b < a$) is

$$\frac{x^2}{a^2} + \frac{y^2}{b^2} = 1 \quad \text{or} \quad \frac{x^2}{b^2} + \frac{y^2}{a^2} = 1.$$

The vertices and foci lie on the major axis, a and c units, respectively, from the center, as shown in Figure 4.27. Moreover, a, b, and c are related by the equation $c^2 = a^2 - b^2$.

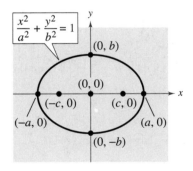

(a) **Major axis is horizontal; minor axis is vertical.**

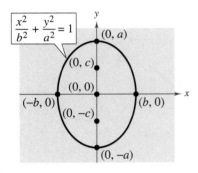

(b) **Major axis is vertical; minor axis is horizontal.**

FIGURE **4.27**

In Figure 4.27(a), note that because the sum of the distances from a point on the ellipse to the two foci is constant, it follows that

(Sum of distances from $(0, b)$ to foci) = (sum of distances from $(a, 0)$ to foci)

$$2\sqrt{b^2 + c^2} = (a + c) + (a - c)$$

$$\sqrt{b^2 + c^2} = a$$

$$c^2 = a^2 - b^2.$$

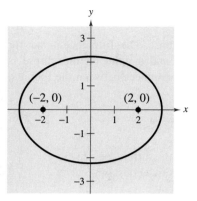

FIGURE **4.28**

TECHNOLOGY

Conics can be graphed using a graphing utility by first solving for *y*. You may have to graph the conic using two separate equations. For example, you can graph the ellipse from Example 4 by graphing both

$$y_1 = \sqrt{36 - 4x^2}$$

and

$$y_2 = -\sqrt{36 - 4x^2}$$

in the same viewing window.

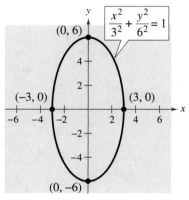

FIGURE **4.29**

Example 3   **Finding the Standard Equation of an Ellipse**

Find the standard form of the equation of the ellipse shown in Figure 4.28.

Solution

From Figure 4.28, the foci occur at $(-2, 0)$ and $(2, 0)$. So, the center of the ellipse is $(0, 0)$, the major axis is horizontal, and the ellipse has an equation of the form

$$\frac{x^2}{a^2} + \frac{y^2}{b^2} = 1.$$

Also from Figure 4.28, the length of the major axis is $2a = 6$. This implies that $a = 3$. Moreover, the distance from the center to either focus is $c = 2$. Finally,

$$b^2 = a^2 - c^2 = \quad^2 - \quad^2 = 9 - 4 = 5.$$

Substituting $a^2 = 3^2$ and $b^2 = \left(\sqrt{5}\right)^2$ yields the following equation in standard form.

$$\frac{x^2}{\quad} + \frac{y^2}{\quad} = 1$$

This equation simplifies to

$$\frac{x^2}{9} + \frac{y^2}{5} = 1.$$

CHECK Point   Now try Exercise 63.

Example 4   **Sketching an Ellipse**

Sketch the ellipse given by $4x^2 + y^2 = 36$, and identify the vertices.

Solution

$$4x^2 + y^2 = 36$$

$$\frac{4x^2}{36} + \frac{y^2}{36} = \frac{36}{36}$$

$$\frac{x^2}{9} + \frac{y^2}{36} = 1$$

$$\frac{x^2}{3^2} + \frac{y^2}{6^2} = 1$$

Because the denominator of the y^2-term is larger than the denominator of the x^2-term, you can conclude that the major axis is vertical. Moreover, because $a^2 = 6^2$, the endpoints of the major axis lie six units *up and down* from the center $(0, 0)$. So, the vertices of the ellipse are $(0, 6)$ and $(0, -6)$. Similarly, because the denominator of the x^2-term is $b^2 = 3^2$, the endpoints of the minor axis (or co-vertices) lie three units to the *right and left* of the center at $(3, 0)$ and $(-3, 0)$. The ellipse is shown in Figure 4.29.

CHECK Point   Now try Exercise 53.

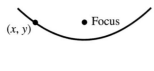

$d_2 - d_1$ is a positive constant.

(a)

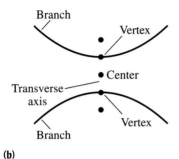

(b)

FIGURE **4.30**

Hyperbolas

The definition of a **hyperbola** is similar to that of an ellipse. The difference is that for an ellipse the *sum* of the distances between the foci and a point on the ellipse is constant, whereas for a hyperbola the *difference* of the distances between the foci and a point on the hyperbola is constant.

Definition of a Hyperbola

A **hyperbola** is the set of all points (x, y) in a plane the difference of whose distances from two distinct fixed points **(foci)** is a positive constant. See Figure 4.30(a).

The graph of a hyperbola has two disconnected parts **(branches).** The line through the two foci intersects the hyperbola at two points **(vertices).** The line segment connecting the vertices is the **transverse axis,** and the midpoint of the transverse axis is the **center** of the hyperbola. See Figure 4.30(b).

Standard Equation of a Hyperbola (Center at Origin)

The **standard form of the equation of a hyperbola** with center at the origin (where $a \neq 0$ and $b \neq 0$) is

$$\frac{x^2}{a^2} - \frac{y^2}{b^2} = 1$$

or

$$\frac{y^2}{a^2} - \frac{x^2}{b^2} = 1.$$

The vertices and foci are, respectively, a and c units from the center. Moreover, a, b, and c are related by the equation $b^2 = c^2 - a^2$. See Figure 4.31.

⚠ WARNING / CAUTION

Be careful when finding the foci of ellipses and hyperbolas. Notice that the relationships between a, b, and c differ slightly.

Finding the foci of an ellipse:

$$c^2 = a^2 - b^2$$

Finding the foci of a hyperbola:

$$c^2 = a^2 + b^2$$

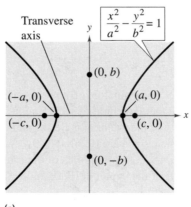

(a)

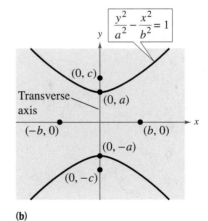

(b)

FIGURE **4.31**

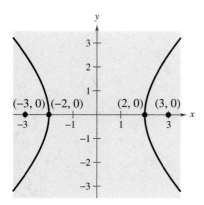

FIGURE 4.32

Example 5   Finding the Standard Equation of a Hyperbola

Find the standard form of the equation of the hyperbola with foci at $(-3, 0)$ and $(3, 0)$ and vertices at $(-2, 0)$ and $(2, 0)$, as shown in Figure 4.32.

Solution

From the graph, you can determine that $c = 3$, because the foci are three units from the center. Moreover, $a = 2$ because the vertices are two units from the center. So, it follows that

$$b^2 = c^2 - a^2$$

$$= 2^2 - 2^2$$

$$= 9 - 4$$

$$= 5.$$

Because the transverse axis is horizontal, the standard form of the equation is

$$\frac{x^2}{a^2} - \frac{y^2}{b^2} = 1.$$

Finally, substitute $a^2 = 2^2$ and $b^2 = \left(\sqrt{5}\right)^2$ to obtain

$$\frac{x^2}{} - \frac{y^2}{} = 1$$

$$\frac{x^2}{4} - \frac{y^2}{5} = 1.$$

CHECK *Point*   Now try Exercise 85.

An important aid in sketching the graph of a hyperbola is the determination of its *asymptotes*, as shown in Figure 4.33. Each hyperbola has two asymptotes that intersect at the center of the hyperbola. Furthermore, the asymptotes pass through the corners of a rectangle of dimensions $2a$ by $2b$. The line segment of length $2b$ joining $(0, b)$ and $(0, -b)$ [or $(-b, 0)$ and $(b, 0)$] is the **conjugate axis** of the hyperbola.

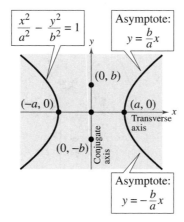

(a) **Transverse axis is horizontal;**
     conjugate axis is vertical.

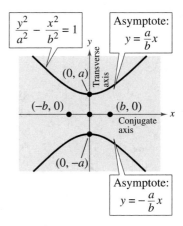

(b) **Transverse axis is vertical;**
     conjugate axis is horizontal.

FIGURE 4.33

> ### Asymptotes of a Hyperbola (Center at Origin)
>
> The **asymptotes of a hyperbola** with center at $(0, 0)$ are
>
> $$y = \frac{b}{a}x \quad \text{and} \quad y = -\frac{b}{a}x$$
>
> or
>
> $$y = \frac{a}{b}x \quad \text{and} \quad y = -\frac{a}{b}x.$$

Example 6   Sketching a Hyperbola

Sketch the hyperbola whose equation is

$$4x^2 - y^2 = 16.$$

Algebraic Solution

$$4x^2 - y^2 = 16$$

$$\frac{4x^2}{16} - \frac{y^2}{16} = \frac{16}{16}$$

$$\frac{x^2}{4} - \frac{y^2}{16} = 1$$

$$\frac{x^2}{2^2} - \frac{y^2}{4^2} = 1$$

Because the x^2-term is positive, you can conclude that the transverse axis is horizontal and the vertices occur at $(-2, 0)$ and $(2, 0)$. Moreover, the endpoints of the conjugate axis occur at $(0, -4)$ and $(0, 4)$, and you can sketch the rectangle shown in Figure 4.34. Finally, by drawing the asymptotes through the corners of this rectangle, you can complete the sketch shown in Figure 4.35. Note that the asymptotes are $y = 2x$ and $y = -2x$.

Graphical Solution

Solve the equation of the hyperbola for y as follows.

$$4x^2 - y^2 = 16$$

$$4x^2 - 16 = y^2$$

$$\pm\sqrt{4x^2 - 16} = y$$

Then use a graphing utility to graph

$$y_1 = \sqrt{4x^2 - 16}$$

and

$$y_2 = -\sqrt{4x^2 - 16}$$

in the same viewing window. Be sure to use a square setting. From the graph in Figure 4.36, you can see that the transverse axis is horizontal. You can use the *zoom* and *trace* features to approximate the vertices to be $(-2, 0)$ and $(2, 0)$.

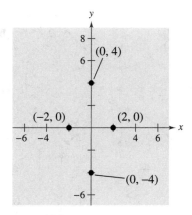

FIGURE 4.34

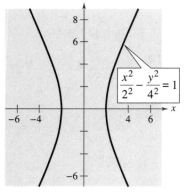

FIGURE 4.35

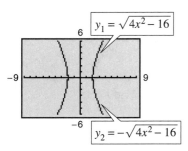

FIGURE 4.36

CHECK*Point*   Now try Exercise 81.

| **Example 7**   **Finding the Standard Equation of a Hyperbola**

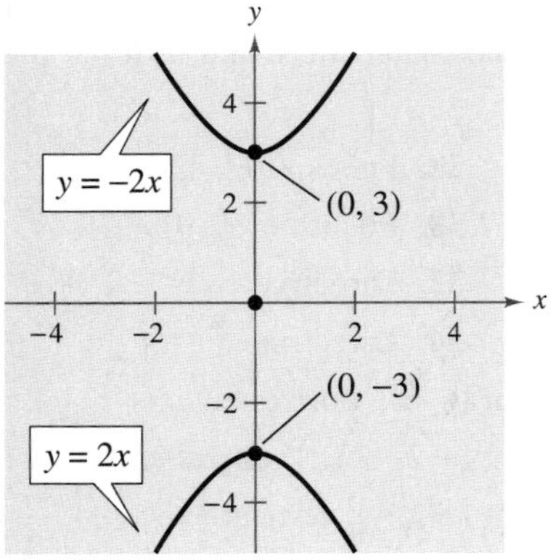

Find the standard form of the equation of the hyperbola that has vertices at $(0, -3)$ and $(0, 3)$ and asymptotes $y = -2x$ and $y = 2x$, as shown in Figure 4.37.

Solution

Because the transverse axis is vertical, the asymptotes are of the forms

$$y = \frac{a}{b}x \quad \text{and} \quad y = -\frac{a}{b}x.$$

Using the fact that $y = 2x$ and $y = -2x$, you can determine that

$$\frac{a}{b} = 2.$$

Because $a = 3$, you can determine that $b = \frac{3}{2}$. Finally, you can conclude that the hyperbola has the following equation.

$$\frac{y^2}{2} - \frac{x^2}{\left(\right)^2} = 1$$

$$\frac{y^2}{9} - \frac{x^2}{\frac{9}{4}} = 1$$

CHECK**Point**   Now try Exercise 87.

FIGURE **4.37**

4.3   EXERCISES

See www.CalcChat.com for worked-out solutions to odd-numbered exercises.

VOCABULARY: Fill in the blanks.

1. A _____ is the intersection of a plane and a double-napped cone.

2. The equation $(x - h)^2 + (y - k)^2 = r^2$ is the standard form of the equation of a _____ with center _____ and radius _____.

3. A _____ is the set of all points (x, y) in a plane that are equidistant from a fixed line, called the _____, and a fixed point, called the _____, not on the line.

4. The _____ of a parabola is the midpoint between the focus and the directrix.

5. The line that passes through the focus and the vertex of a parabola is called the _____ of the parabola.

6. An _____ is the set of all points (x, y) in a plane, the sum of whose distances from two distinct fixed points, called_____, is constant.

7. The chord joining the vertices of an ellipse is called the _____ _____, and its midpoint is the _____ of the ellipse.

8. The chord perpendicular to the major axis at the center of an ellipse is called the _____ _____ of the ellipse.

9. A _____ is the set of all points (x, y) in a plane, the difference of whose distances from two distinct fixed points, called _____, is a positive constant.

10. The line segment connecting the vertices of a hyperbola is called the _____ _____, and the midpoint of the line segment is the _____ of the hyperbola.

SKILLS AND APPLICATIONS

In Exercises 11–20, match the equation with its graph. If the graph of an equation is not shown, write "not shown." [The graphs are labeled (a), (b), (c), (d), (e), (f), (g), and (h).]

(a)
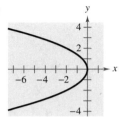

(b)
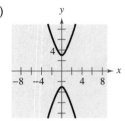

(c)
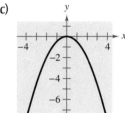

(d)
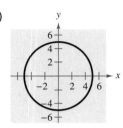

(e)
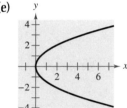

(f)
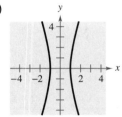

(g)
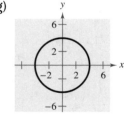

(h)
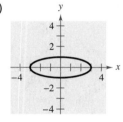

11. $x^2 = 2y$ **12.** $x^2 = -2y$

13. $y^2 = 2x$ **14.** $y^2 = -2x$

15. $9x^2 + y^2 = 9$ **16.** $x^2 + 9y^2 = 9$

17. $9x^2 - y^2 = 9$ **18.** $y^2 - 9x^2 = 9$

19. $x^2 + y^2 = 25$ **20.** $x^2 + y^2 = 16$

In Exercises 21–26, find the vertex and focus of the parabola and sketch its graph.

21. $y = \frac{1}{2}x^2$ **22.** $y = -4x^2$

23. $y^2 = -6x$ **24.** $y^2 = 3x$

25. $x^2 + 12y = 0$ **26.** $x + y^2 = 0$

In Exercises 27–38, find the standard form of the equation of the parabola with the given characteristic(s) and vertex at the origin.

27. Focus: $(-2, 0)$ **28.** Focus: $(0, -2)$

29. Focus: $\left(0, \frac{1}{2}\right)$ **30.** Focus: $\left(-\frac{3}{2}, 0\right)$

31. Directrix: $y = 1$ **32.** Directrix: $y = -2$

33. Directrix: $x = -1$ **34.** Directrix: $x = 4$

35. Passes through the point $(4, 6)$; horizontal axis

36. Passes through the point $(-2, -2)$; vertical axis

37. Passes through the point $\left(-2, \frac{1}{4}\right)$; vertical axis

38. Passes through the point $\left(\frac{1}{2}, -4\right)$; horizontal axis

In Exercises 39–42, find the standard form of the equation of the parabola and determine the coordinates of the focus.

39.
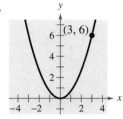

40.
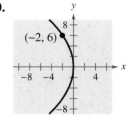

41.
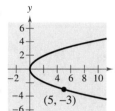

42.
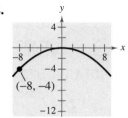

43. FLASHLIGHT   The light bulb in a flashlight is at the focus of the parabolic reflector, 1.5 centimeters from the vertex of the reflector (see figure). Write an equation for a cross section of the flashlight's reflector with its focus on the positive x-axis and its vertex at the origin.

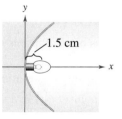

FIGURE FOR **43**

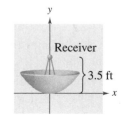

FIGURE FOR **44**

44. SATELLITE ANTENNA   Write an equation for a cross section of the parabolic satellite dish antenna shown in the figure.

45. SUSPENSION BRIDGE Each cable of the Golden Gate Bridge is suspended (in the shape of a parabola) between two towers that are 1280 meters apart. The top of each tower is 152 meters above the roadway. The cables touch the roadway at the midpoint between the towers.

(a) Draw a sketch of the bridge. Locate the origin of a rectangular coordinate system at the center of the roadway. Label the coordinates of the known points.

(b) Write an equation that models the cables.

(c) Complete the table by finding the height y of the suspension cables over the roadway at a distance of x meters from the center of the bridge.

Distance, x	0	200	400	500	600
Height, y					

46. BEAM DEFLECTION A simply supported beam (see figure) is 64 feet long and has a load at the center. The deflection of the beam at its center is 1 inch. The shape of the deflected beam is parabolic.

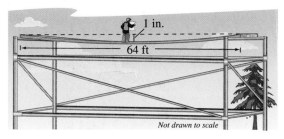

(a) Find an equation of the parabola. (Assume that the origin is at the center of the beam.)

(b) How far from the center of the beam is the deflection $\frac{1}{2}$ inch?

In Exercises 47–56, find the center and vertices of the ellipse and sketch its graph.

47. $\dfrac{x^2}{25} + \dfrac{y^2}{16} = 1$ **48.** $\dfrac{x^2}{121} + \dfrac{y^2}{144} = 1$

49. $\dfrac{x^2}{25/9} + \dfrac{y^2}{16/9} = 1$ **50.** $\dfrac{x^2}{4} + \dfrac{y^2}{1/4} = 1$

51. $\dfrac{x^2}{36} + \dfrac{y^2}{7} = 1$ **52.** $\dfrac{x^2}{28} + \dfrac{y^2}{64} = 1$

53. $4x^2 + y^2 = 1$ **54.** $4x^2 + 9y^2 = 36$

55. $\dfrac{1}{16}x^2 + \dfrac{1}{81}y^2 = 1$ **56.** $\dfrac{1}{100}x^2 + \dfrac{1}{49}y^2 = 1$

In Exercises 57–66, find the standard form of the equation of the ellipse with the given characteristics and center at the origin.

57. 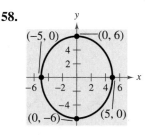 **58.**

59. 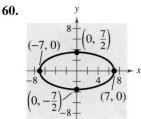 **60.**

61. Vertices: $(\pm 5, 0)$; foci: $(\pm 2, 0)$

62. Vertices: $(0, \pm 8)$; foci: $(0, \pm 4)$

63. Foci: $(\pm 5, 0)$; major axis of length 14

64. Foci: $(\pm 2, 0)$; major axis of length 10

65. Vertices: $(0, \pm 5)$; passes through the point $(4, 2)$

66. Vertical major axis; passes through the points $(0, 4)$ and $(2, 0)$

67. ARCHITECTURE A fireplace arch is to be constructed in the shape of a semiellipse. The opening is to have a height of 2 feet at the center and a width of 6 feet along the base (see figure). The contractor draws the outline of the ellipse on the wall by the method shown in Figure 4.26. Give the required positions of the tacks and the length of the string.

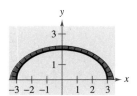

68. ARCHITECTURE A semielliptical arch over a tunnel for a one-way road through a mountain has a major axis of 50 feet and a height at the center of 10 feet.

(a) Sketch the arch of the tunnel on a rectangular coordinate system with the center of the road entering the tunnel at the origin. Identify the coordinates of the known points.

(b) Find an equation of the semielliptical arch over the tunnel.

(c) You are driving a moving truck that has a width of 8 feet and a height of 9 feet. Will the moving truck clear the opening of the arch?

69. ARCHITECTURE Repeat Exercise 68 for a semielliptical arch with a major axis of 40 feet and a height at the center of 15 feet. The dimensions of the truck are 10 feet wide by 14 feet high.

70. GEOMETRY A line segment through a focus of an ellipse with endpoints on the ellipse and perpendicular to the major axis is called a **latus rectum** of the ellipse. Therefore, an ellipse has two latera recta. Knowing the length of the latera recta is helpful in sketching an ellipse because it yields other points on the curve (see figure). Show that the length of each latus rectum is $2b^2/a$.

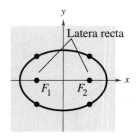

In Exercises 71–74, sketch the graph of the ellipse, using the latera recta (see Exercise 70).

71. $\dfrac{x^2}{4} + \dfrac{y^2}{1} = 1$ **72.** $\dfrac{x^2}{9} + \dfrac{y^2}{16} = 1$

73. $9x^2 + 4y^2 = 36$ **74.** $5x^2 + 3y^2 = 15$

In Exercises 75–84, find the center and vertices of the hyperbola and sketch its graph, using asymptotes as sketching aids.

75. $x^2 - y^2 = 1$ **76.** $\dfrac{x^2}{9} - \dfrac{y^2}{16} = 1$

77. $\dfrac{y^2}{1} - \dfrac{x^2}{4} = 1$ **78.** $\dfrac{y^2}{9} - \dfrac{x^2}{1} = 1$

79. $\dfrac{y^2}{49} - \dfrac{x^2}{196} = 1$ **80.** $\dfrac{x^2}{36} - \dfrac{y^2}{4} = 1$

81. $4y^2 - x^2 = 1$ **82.** $4y^2 - 9x^2 = 36$

83. $\dfrac{1}{36}y^2 - \dfrac{1}{100}x^2 = 1$ **84.** $\dfrac{1}{144}x^2 - \dfrac{1}{169}y^2 = 1$

In Exercises 85–92, find the standard form of the equation of the hyperbola with the given characteristics and center at the origin.

85. Vertices: $(0, \pm2)$; foci: $(0, \pm6)$

86. Vertices: $(\pm4, 0)$; foci: $(\pm5, 0)$

87. Vertices: $(\pm1, 0)$; asymptotes: $y = \pm3x$

88. Vertices: $(0, \pm3)$; asymptotes: $y = \pm3x$

89. Foci: $(0, \pm8)$; asymptotes: $y = \pm4x$

90. Foci: $(\pm10, 0)$; asymptotes: $y = \pm\frac{3}{4}x$

91. Vertices: $(0, \pm3)$; passes through the point $(-2, 5)$

92. Vertices: $(\pm2, 0)$; passes through the point $(3, \sqrt{3})$

93. ART A sculpture has a hyperbolic cross section (see figure).

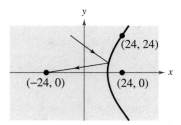

(a) Write an equation that models the curved sides of the sculpture.

(b) Each unit on the coordinate plane represents 1 foot. Find the width of the sculpture at a height of 5 feet.

94. OPTICS A hyperbolic mirror (used in some telescopes) has the property that a light ray directed at the focus will be reflected to the other focus. The focus of a hyperbolic mirror (see figure) has coordinates $(24, 0)$. Find the vertex of the mirror if its mount at the top edge of the mirror has coordinates $(24, 24)$.

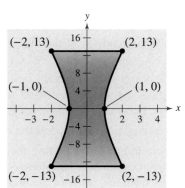

95. AERONAUTICS When an airplane travels faster than the speed of sound, the sound waves form a cone behind the airplane. If the airplane is flying parallel to the ground, the sound waves intersect the ground in a hyperbola with the airplane directly above its center (see figure). A sonic boom is heard along the hyperbola. You hear a sonic boom that is audible along a hyperbola with the equation

$$\frac{x^2}{100} - \frac{y^2}{4} = 1$$

where x and y are measured in miles. What is the shortest horizontal distance you could be from the airplane?

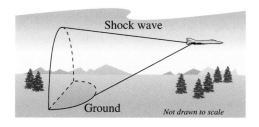

96. NAVIGATION Long distance radio navigation for aircraft and ships uses synchronized pulses transmitted by widely separated transmitting stations. These pulses travel at the speed of light (186,000 miles per second). The difference in the times of arrival of these pulses at an aircraft or ship is constant on a hyperbola having the transmitting stations as foci.

Assume that two stations 300 miles apart are positioned on a rectangular coordinate system at points with coordinates $(-150, 0)$ and $(150, 0)$ and that a ship is traveling on a path with coordinates $(x, 75)$, as shown in the figure. Find the x-coordinate of the position of the ship if the time difference between the pulses from the transmitting stations is 1000 micro-seconds (0.001 second).

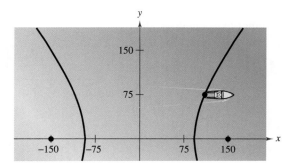

EXPLORATION

TRUE OR FALSE? In Exercises 97–100, determine whether the statement is true or false. Justify your answer.

97. The equation $x^2 - y^2 = 144$ represents a circle.

98. The major axis of the ellipse $y^2 + 16x^2 = 64$ is vertical.

99. It is possible for a parabola to intersect its directrix.

100. If the vertex and focus of a parabola are on a horizontal line, then the directrix of the parabola is vertical.

101. Consider the ellipse

$$\frac{x^2}{a^2} + \frac{y^2}{b^2} = 1, \quad a + b = 20.$$

(a) The area of the ellipse is given by $A = \pi ab$. Write the area of the ellipse as a function of a.

(b) Find the equation of an ellipse with an area of 264 square centimeters.

(c) Complete the table using your equation from part (a), and make a conjecture about the shape of the ellipse with maximum area.

a	8	9	10	11	12	13
A						

(d) Use a graphing utility to graph the area function and use the graph to support your conjecture in part (c).

102. CAPSTONE Identify the conic. Explain your reasoning.

(a) $4x^2 + 4y^2 - 16 = 0$

(b) $4y^2 - 5x^2 + 20 = 0$

(c) $3y^2 - 6x = 0$

(d) $2x^2 + 4y^2 - 12 = 0$

(e) $4x^2 + y^2 - 16 = 0$

(f) $2x^2 - 12y = 0$

103. THINK ABOUT IT How can you tell if an ellipse is a circle from the equation?

104. THINK ABOUT IT Is the graph of $x^2 + 4y^4 = 4$ an ellipse? Explain.

105. THINK ABOUT IT The graph of $x^2 - y^2 = 0$ is a degenerate conic. Sketch this graph and identify the degenerate conic.

106. THINK ABOUT IT Which part of the graph of the ellipse $4x^2 + 9y^2 = 36$ is represented by each equation? (Do not graph.)

(a) $x = -\frac{3}{2}\sqrt{4 - y^2}$

(b) $y = \frac{2}{3}\sqrt{9 - x^2}$

107. WRITING At the beginning of this section, you learned that each type of conic section can be formed by the intersection of a plane and a double-napped cone. Write a short paragraph describing examples of physical situations in which hyperbolas are formed.

108. WRITING Write a paragraph discussing the changes in the shape and orientation of the graph of the ellipse

$$\frac{x^2}{a^2} + \frac{y^2}{4^2} = 1$$

as a increases from 1 to 8.

109. Use the definition of an ellipse to derive the standard form of the equation of an ellipse.

110. Use the definition of a hyperbola to derive the standard form of the equation of a hyperbola.

111. An ellipse can be drawn using two thumbtacks placed at the foci of the ellipse, a string of fixed length (greater than the distance between the tacks), and a pencil, as shown in Figure 4.26. Try doing this. Vary the length of the string and the distance between the thumbtacks. Explain how to obtain ellipses that are almost circular. Explain how to obtain ellipses that are long and narrow.

4.4 TRANSLATIONS OF CONICS

Study Tip

Consider the equation of the ellipse

$$\frac{(x-h)^2}{a^2} + \frac{(y-k)^2}{b^2} = 1.$$

If you let $a = b$, then the equation can be rewritten as

$$(x-h)^2 + (y-k)^2 = a^2$$

which is the standard form of the equation of a circle with radius $r = a$ (see Section 1.1). Geometrically, when $a = b$ for an ellipse, the major and minor axes are of equal length, and so the graph is a circle [see Example 1(a)].

Vertical and Horizontal Shifts of Conics

In Section 4.3 you looked at conic sections whose graphs were in *standard position*. In this section you will study the equations of conic sections that have been shifted vertically or horizontally in the plane.

Standard Forms of Equations of Conics

Circle: Center = (h, k); radius = r

$$(x - h)^2 + (y - k)^2 = r^2$$

Ellipse: Center = (h, k)

Major axis length = $2a$; minor axis length = $2b$

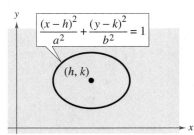
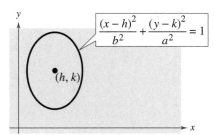

Hyperbola: Center = (h, k)

Transverse axis length = $2a$; conjugate axis length = $2b$

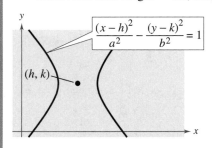
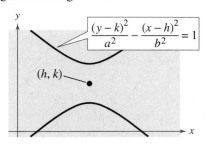

Parabola: Vertex = (h, k)

Directed distance from vertex to focus = p

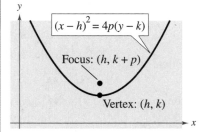
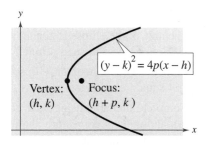

Example 1   Equations of Conic Sections

Identify each conic. Then describe the translation of the graph of the conic.

a. $(x - 1)^2 + (y + 2)^2 = 3^2$ **b.** $\dfrac{(x - 2)^2}{3^2} + \dfrac{(y - 1)^2}{2^2} = 1$

c. $\dfrac{(x - 3)^2}{1^2} - \dfrac{(y - 2)^2}{3^2} = 1$ **d.** $(x - 2)^2 = 4(-1)(y - 3)$

Solution

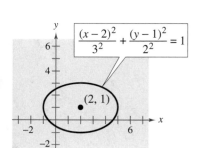

FIGURE **4.38**   Circle

a. The graph of $(x - 1)^2 + (y + 2)^2 = 3^2$ is a *circle* whose center is the point $(1, -2)$ and whose radius is 3, as shown in Figure 4.38. The graph of the circle has been shifted one unit to the right and two units downward from standard position.

b. The graph of

$$\frac{(x - 2)^2}{3^2} + \frac{(y - 1)^2}{2^2} = 1$$

is an *ellipse* whose center is the point $(2, 1)$. The major axis of the ellipse is horizontal and of length $2(\ ) = 6$, and the minor axis of the ellipse is vertical and of length $2(\ ) = 4$, as shown in Figure 4.39. The graph of the ellipse has been shifted two units to the right and one unit upward from standard position.

c. The graph of

$$\frac{(x - 3)^2}{1^2} - \frac{(y - 2)^2}{3^2} = 1$$

FIGURE **4.39**   Ellipse

is a *hyperbola* whose center is the point $(3, 2)$. The transverse axis is horizontal and of length $2(\ ) = 2$, and the conjugate axis is vertical and of length $2(\ ) = 6$, as shown in Figure 4.40. The graph of the hyperbola has been shifted three units to the right and two units upward from standard position.

d. The graph of

$$(x - 2)^2 = 4(-1)(y - 3)$$

is a *parabola* whose vertex is the point $(2, 3)$. The axis of the parabola is vertical. The focus is one unit above or below the vertex. Moreover, because $p = -1$, it follows that the focus lies *below* the vertex, as shown in Figure 4.41. The graph of the parabola has been reflected in the *x*-axis, shifted two units to the right and three units upward from standard position.

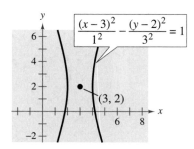

FIGURE **4.40**   Hyperbola

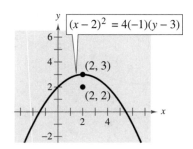

FIGURE **4.41**   Parabola

CHECK*Point*   Now try Exercise 11.

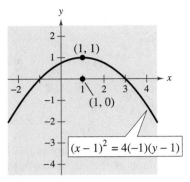

FIGURE **4.42**

Study Tip

Note in Example 2 that p is the directed distance from the vertex to the focus. Because the axis of the parabola is vertical and $p = -1$, the focus is one unit below the vertex, and the parabola opens downward.

Equations of Conics in Standard Form

Example 2   **Finding the Standard Equation of a Parabola**

Find the vertex and focus of the parabola $x^2 - 2x + 4y - 3 = 0$.

Solution

Complete the square to write the equation in standard form.

$$x^2 - 2x + 4y - 3 = 0$$
$$x^2 - 2x = -4y + 3$$
$$x^2 - 2x \quad = -4y + 3$$
$$(x - 1)^2 = -4y + 4$$
$$(x - 1)^2 = 4(-1)(y - 1)$$

From this standard form, it follows that $h = 1, k = 1$, and $p = -1$. Because the axis is vertical and p is negative, the parabola opens downward. The vertex is $(h, k) = (1, 1)$ and the focus is $(h, k + p) = (1, 0)$. (See Figure 4.42.)

CHECK Point   Now try Exercise 31.

Example 3   **Sketching an Ellipse**

Sketch the ellipse $x^2 + 4y^2 + 6x - 8y + 9 = 0$.

Solution

Complete the square to write the equation in standard form.

$$x^2 + 4y^2 + 6x - 8y + 9 = 0$$
$$(x^2 + 6x + \quad) + (4y^2 - 8y + \quad) = -9$$
$$(x^2 + 6x + \quad) + 4(y^2 - 2y + \quad) = -9$$
$$(x^2 + 6x \quad) + 4(y^2 - 2y \quad) = -9$$
$$(x + 3)^2 + 4(y - 1)^2 = 4$$
$$\frac{(x + 3)^2}{4} + \frac{4(y - 1)^2}{4} = 1$$
$$\frac{(x + 3)^2}{2^2} + \frac{(y - 1)^2}{1^2} = 1$$

From this standard form, it follows that the center is $(h, k) = (-3, 1)$. Because the denominator of the x-term is $a^2 = 2^2$, the endpoints of the major axis lie two units to the right and left of the center. Similarly, because the denominator of the y-term is $b^2 = 1^2$, the endpoints of the minor axis lie one unit up and down from the center. The ellipse is shown in Figure 4.43.

CHECK Point   Now try Exercise 47.

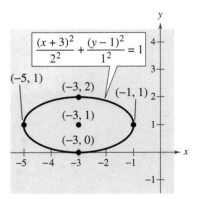

FIGURE **4.43**

| **Example 4** | **Sketching a Hyperbola** |

Sketch the hyperbola

$$y^2 - 4x^2 + 4y + 24x - 41 = 0.$$

Solution

Complete the square to write the equation in standard form.

$$y^2 - 4x^2 + 4y + 24x - 41 = 0$$

$$(y^2 + 4y + \quad) - (4x^2 - 24x + \quad) = 41$$

$$(y^2 + 4y + \quad) - 4(x^2 - 6x + \quad) = 41$$

$$(y^2 + 4y \quad) - 4(x^2 - 6x \quad) = 41$$

$$(y + 2)^2 - 4(x - 3)^2 = 9$$

$$\frac{(y + 2)^2}{9} - \frac{4(x - 3)^2}{9} = 1$$

$$\frac{(y + 2)^2}{9} - \frac{(x - 3)^2}{\frac{9}{4}} = 1$$

$$\frac{(y + 2)^2}{3^2} - \frac{(x - 3)^2}{\left(\frac{3}{2}\right)^2} = 1$$

From this standard form, it follows that the transverse axis is vertical and the center lies at $(h, k) = (3, -2)$. Because the denominator of the y-term is $a^2 = 3^2$, you know that the vertices occur three units above and below the center.

$$(3, 1) \qquad \text{and} \qquad (3, -5)$$

To sketch the hyperbola, draw a rectangle whose top and bottom pass through the vertices. Because the denominator of the x-term is $b^2 = \left(\frac{3}{2}\right)^2$, locate the sides of the rectangle $\frac{3}{2}$ units to the right and left of the center, as shown in Figure 4.44. Finally, sketch the asymptotes by drawing lines through the opposite corners of the rectangle. Using these asymptotes, you can complete the graph of the hyperbola, as shown in Figure 4.44.

CHECK*Point*   Now try Exercise 67.

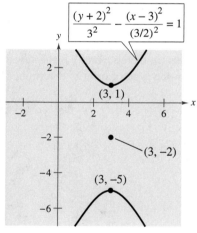

FIGURE 4.44

To find the foci in Example 4, first find c.

$$c^2 = a^2 + b^2$$

$$= \quad + \quad = \frac{45}{4} \quad \Longrightarrow \quad c = \frac{3\sqrt{5}}{2}$$

Because the transverse axis is vertical, the foci lie c units above and below the center.

$$\left(3, -2 + \tfrac{3}{2}\sqrt{5}\right) \qquad \text{and} \qquad \left(3, -2 - \tfrac{3}{2}\sqrt{5}\right)$$

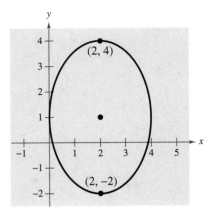

FIGURE **4.45**

Example 5 Writing the Equation of an Ellipse

Write the standard form of the equation of the ellipse whose vertices are $(2, -2)$ and $(2, 4)$. The length of the minor axis of the ellipse is 4, as shown in Figure 4.45.

Solution

The center of the ellipse lies at the midpoint of its vertices. So, the center is

$$(h, k) = (2, 1).$$

Because the vertices lie on a vertical line and are six units apart, it follows that the major axis is vertical and has a length of $2a = 6$. So, $a = 3$. Moreover, because the minor axis has a length of 4, it follows that $2b = 4$, which implies that $b = 2$. So, the standard form of the ellipse is as follows.

$$\frac{(x - h)^2}{b^2} + \frac{(y - k)^2}{a^2} = 1$$

$$\frac{(x - 2)^2}{2^2} + \frac{(y - 1)^2}{3^2} = 1$$

CHECK *Point* Now try Exercise 51.

An interesting application of conic sections involves the orbits of comets in our solar system. Of the 610 comets identified prior to 1970, 245 have elliptical orbits, 295 have parabolic orbits, and 70 have hyperbolic orbits. For example, Halley's comet has an elliptical orbit, and reappearance of this comet can be predicted every 76 years. The center of the sun is a focus of each of these orbits, and each orbit has a vertex at the point where the comet is closest to the sun, as shown in Figure 4.46.

If p is the distance between the vertex and the focus (in meters), and v is the speed of the comet at the vertex (in meters per second), then the type of orbit is determined as follows.

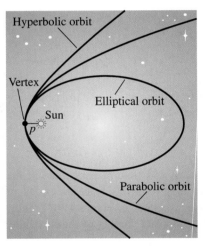

FIGURE **4.46**

1. Ellipse: $v < \sqrt{\dfrac{2GM}{p}}$

2. Parabola: $v = \sqrt{\dfrac{2GM}{p}}$

3. Hyperbola: $v > \sqrt{\dfrac{2GM}{p}}$

In each of these relations, $M = 1.989 \times 10^{30}$ kilograms (the mass of the sun) and $G \approx 6.67 \times 10^{-11}$ cubic meter per kilogram-second squared (the universal gravitational constant).

CLASSROOM DISCUSSION

Identifying Equations of Conics Use the Internet to research information about the orbits of comets in our solar system. What can you find about the orbits of comets that have been identified since 1970? Write a summary of your results. Identify your source. Does it seem reliable?

4.4   EXERCISES

See www.CalcChat.com for worked-out solutions to odd-numbered exercises.

VOCABULARY: Match the description of the conic with its standard equation. The equations are labeled (a), (b), (c), (d), (e), and (f).

(a) $\dfrac{(x - h)^2}{a^2} + \dfrac{(y - k)^2}{b^2} = 1$ (b) $\dfrac{(x - h)^2}{a^2} - \dfrac{(y - k)^2}{b^2} = 1$ (c) $\dfrac{(y - k)^2}{a^2} - \dfrac{(x - h)^2}{b^2} = 1$

(d) $\dfrac{(x - h)^2}{b^2} + \dfrac{(y - k)^2}{a^2} = 1$ (e) $(x - h)^2 = 4p(y - k)$ (f) $(y - k)^2 = 4p(x - h)$

1. Hyperbola with horizontal transverse axis   **2.** Ellipse with vertical major axis   **3.** Parabola with vertical axis

4. Hyperbola with vertical transverse axis   **5.** Ellipse with horizontal major axis   **6.** Parabola with horizontal axis

SKILLS AND APPLICATIONS

In Exercises 7–12, describe the translation of the graph of the conic.

7. $(x + 2)^2 + (y - 1)^2 = 4$ **8.** $(y - 1)^2 = 4(2)(x + 2)$

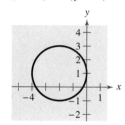   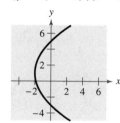

9. $\dfrac{(y + 3)^2}{4} - (x - 1)^2 = 1$ **10.** $\dfrac{(x - 2)^2}{9} + \dfrac{(y + 1)^2}{4} = 1$

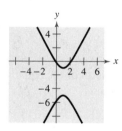   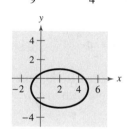

11. $\dfrac{(x + 4)^2}{9} + \dfrac{(y + 2)^2}{16} = 1$ **12.** $\dfrac{(x + 2)^2}{4} - \dfrac{(y - 3)^2}{9} = 1$

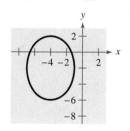   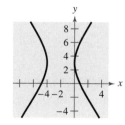

In Exercises 13–18, identify the center and radius of the circle.

13. $x^2 + y^2 = 49$ **14.** $x^2 + y^2 = 1$

15. $(x - 4)^2 + (y - 5)^2 = 36$

16. $(x + 8)^2 + (y + 1)^2 = 144$

17. $(x - 1)^2 + y^2 = 10$ **18.** $x^2 + (y + 12)^2 = 24$

In Exercises 19–24, write the equation of the circle in standard form, and then identify its center and radius.

19. $x^2 + y^2 - 2x + 6y + 9 = 0$

20. $x^2 + y^2 - 10x - 6y + 25 = 0$

21. $x^2 + y^2 - 8x = 0$

22. $2x^2 + 2y^2 - 2x - 2y - 7 = 0$

23. $4x^2 + 4y^2 + 12x - 24y + 41 = 0$

24. $9x^2 + 9y^2 + 54x - 36y + 17 = 0$

In Exercises 25–32, find the vertex, focus, and directrix of the parabola, and sketch its graph.

25. $(x - 1)^2 + 8(y + 2) = 0$

26. $(x + 2) + (y - 4)^2 = 0$

27. $\left(y + \tfrac{1}{2}\right)^2 = 2(x - 5)$ **28.** $\left(x + \tfrac{1}{2}\right)^2 = 4(y - 3)$

29. $y = \tfrac{1}{4}(x^2 - 2x + 5)$ **30.** $4x - y^2 - 2y - 33 = 0$

31. $y^2 + 6y + 8x + 25 = 0$

32. $y^2 - 4y - 4x = 0$

In Exercises 33–38, find the standard form of the equation of the parabola with the given characteristics.

33. Vertex: $(3, 2)$; focus: $(1, 2)$

34. Vertex: $(-1, 2)$; focus: $(-1, 0)$

35. Vertex: $(0, 4)$; directrix: $y = 2$

36. Vertex: $(-2, 1)$; directrix: $x = 1$

37. Focus: $(4, 4)$; directrix: $x = -4$

38. Focus: $(0, 0)$; directrix: $y = 4$

39. PROJECTILE MOTION   A cargo plane is flying at an altitude of 30,000 feet and a speed of 540 miles per hour (792 feet per second). How many feet will a supply crate dropped from the plane travel horizontally before it hits the ground if the path of the crate is modeled by

$x^2 = -39,204(y - 30,000)$?

40. PATH OF A PROJECTILE The path of a softball is modeled by $-12.5(y - 7.125) = (x - 6.25)^2$. The coordinates x and y are measured in feet, with $x = 0$ corresponding to the position from which the ball was thrown.

(a) Use a graphing utility to graph the trajectory of the softball.

(b) Use the *trace* feature of the graphing utility to approximate the highest point and the range of the trajectory.

41. SALES The sales S (in billions of dollars) for Texas Instruments, Inc. for the years 2002 through 2008 are shown in the table. (Source: Texas Instruments, Inc.)

Year	Sales, S
2002	8.4
2003	9.8
2004	12.6
2005	13.4
2006	14.3
2007	13.8
2008	12.5

(a) Use a graphing utility to find an equation of the parabola $y = at^2 + bt + c$ that models the data. Write the equation in standard form. Let t represent the year, with $t = 2$ corresponding to 2002.

(b) Find the coordinates of the vertex and interpret its meaning in the context of the problem.

(c) Use a graphing utility to graph the function.

(d) Use the *trace* feature of the graphing utility to approximate *graphically* the year in which sales were maximum.

(e) Use the *table* feature of the graphing utility to approximate *numerically* the year in which sales were maximum.

(f) Compare the results of parts (b), (d), and (e). What did you learn by using all three approaches?

42. SATELLITE ORBIT A satellite in a 100-mile-high circular orbit around Earth has a velocity of approximately 17,500 miles per hour (see figure). If this velocity is multiplied by $\sqrt{2}$, the satellite will have the minimum velocity necessary to escape Earth's gravity and it will follow a parabolic path with the center of Earth as the focus.

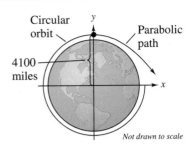

Not drawn to scale

FIGURE FOR **42**

(a) Find the escape velocity of the satellite.

(b) Find an equation of its path (assume that the radius of Earth is 4000 miles).

In Exercises 43–50, find the center, foci, and vertices of the ellipse, and sketch its graph.

43. $\dfrac{(x - 1)^2}{9} + \dfrac{(y - 5)^2}{25} = 1$

44. $\dfrac{(x - 6)^2}{4} + \dfrac{(y + 7)^2}{16} = 1$

45. $(x + 2)^2 + \dfrac{(y + 4)^2}{1/4} = 1$

46. $\dfrac{(x - 3)^2}{25/9} + (y - 8)^2 = 1$

47. $9x^2 + 25y^2 - 36x - 50y + 52 = 0$

48. $16x^2 + 25y^2 - 32x + 50y + 16 = 0$

49. $9x^2 + 4y^2 + 36x - 24y + 36 = 0$

50. $9x^2 + 4y^2 - 36x + 8y + 31 = 0$

In Exercises 51–58, find the standard form of the equation of the ellipse with the given characteristics.

51. Vertices: $(3, 3), (3, -3)$; minor axis of length 2

52. Vertices: $(-2, 3), (6, 3)$; minor axis of length 6

53. Foci: $(0, 0), (4, 0)$; major axis of length 8

54. Foci: $(0, 0), (0, 8)$; major axis of length 16

55. Center: $(0, 4)$; $a = 2c$; vertices: $(-4, 4), (4, 4)$

56. Center: $(3, 2)$; $a = 3c$; foci: $(1, 2), (5, 2)$

57. Vertices: $(0, 2), (4, 2)$;
endpoints of the minor axis: $(2, 3), (2, 1)$

58. Vertices: $(5, 0), (5, 12)$;
endpoints of the minor axis: $(0, 6), (10, 6)$

In Exercises 59 and 60, e is called the *eccentricity* of an ellipse, and is defined by $e = c/a$. It measures the flatness of the ellipse.

59. Find the standard form of the equation of the ellipse with vertices $(\pm 5, 0)$ and eccentricity $e = \frac{3}{5}$.

60. Find the standard form of the equation of the ellipse with vertices $(0, \pm 8)$ and eccentricity $e = \frac{1}{2}$.

61. PLANETARY MOTION The dwarf planet Pluto moves in an elliptical orbit with the sun at one of the foci, as shown in the figure. The length of half of the major axis, a, is 3.67×10^9 miles, and the eccentricity is 0.249. Find the smallest distance (*perihelion*) and the greatest distance (*aphelion*) of Pluto from the center of the sun.

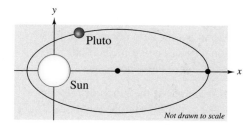

Not drawn to scale

62. AUSTRALIAN FOOTBALL In Australia, football by Australian Rules (or rugby) is played on elliptical fields. The field can be a maximum of 155 meters wide and a maximum of 185 meters long. Let the center of a field of maximum size be represented by the point $(0, 77.5)$. Write the standard form of the equation of the ellipse that represents this field. (Source: Australian Football League)

In Exercises 63–70, find the center, foci, and vertices of the hyperbola, and sketch its graph, using the asymptotes as an aid.

63. $\dfrac{(x-2)^2}{16} - \dfrac{(y+1)^2}{9} = 1$

64. $\dfrac{(x-1)^2}{144} - \dfrac{(y-4)^2}{25} = 1$

65. $(y+6)^2 - (x-2)^2 = 1$

66. $\dfrac{(y-1)^2}{1/4} - \dfrac{(x+3)^2}{1/9} = 1$

67. $x^2 - 9y^2 + 2x - 54y - 85 = 0$

68. $16y^2 - x^2 + 2x + 64y + 62 = 0$

69. $9x^2 - y^2 - 36x - 6y + 18 = 0$

70. $x^2 - 9y^2 + 36y - 72 = 0$

In Exercises 71–78, find the standard form of the equation of the hyperbola with the given characteristics.

71. Vertices: $(0, 2)$, $(0, 0)$; foci: $(0, 3)$, $(0, -1)$

72. Vertices: $(1, 2)$, $(5, 2)$; foci: $(0, 2)$, $(6, 2)$

73. Vertices: $(2, 0)$, $(6, 0)$; foci: $(0, 0)$, $(8, 0)$

74. Vertices: $(2, 3)$, $(2, -3)$; foci: $(2, 5)$, $(2, -5)$

75. Vertices: $(2, 3)$, $(2, -3)$; passes through the point $(0, 5)$

76. Vertices: $(-2, 1)$, $(2, 1)$; passes through the point $(4, 3)$

77. Vertices: $(0, 2)$, $(6, 2)$; asymptotes: $y = \frac{2}{3}x$, $y = 4 - \frac{2}{3}x$

78. Vertices: $(3, 0)$, $(3, 4)$; asymptotes: $y = \frac{2}{3}x$, $y = 4 - \frac{2}{3}x$

In Exercises 79–88, identify the conic by writing its equation in standard form. Then sketch its graph.

79. $x^2 + y^2 - 6x + 4y + 9 = 0$

80. $x^2 + 4y^2 - 6x + 16y + 21 = 0$

81. $y^2 - x^2 + 4y = 0$ **82.** $y^2 - 4y - 4x = 0$

83. $16y^2 + 128x + 8y - 7 = 0$

84. $4x^2 - y^2 - 4x - 3 = 0$

85. $9x^2 + 16y^2 + 36x + 128y + 148 = 0$

86. $25x^2 - 10x - 200y - 119 = 0$

87. $16x^2 + 16y^2 - 16x + 24y - 3 = 0$

88. $4x^2 + 3y^2 + 8x - 24y + 51 = 0$

EXPLORATION

TRUE OR FALSE? In Exercises 89 and 90, determine whether the statement is true or false. Justify your answer.

89. The conic represented by the equation
$3x^2 + 2y^2 - 18x - 16y + 58 = 0$ is an ellipse.

90. The graphs of $x^2 + 10y - 10x + 5 = 0$ and $x^2 + 16y^2 + 10x - 32y - 23 = 0$ do not intersect.

91. Consider the ellipse $\dfrac{x^2}{a^2} + \dfrac{y^2}{b^2} = 1$.

(a) Show that the equation of the ellipse can be written as

$$\dfrac{(x-h)^2}{a^2} + \dfrac{(y-k)^2}{a^2(1-e^2)} = 1$$

where e is the eccentricity (see Exercises 59 and 60).

(b) Use a graphing utility to graph the ellipse

$$\dfrac{(x-2)^2}{4} + \dfrac{(y-3)^2}{4(1-e^2)} = 1$$

for $e = 0.95, 0.75, 0.5, 0.25,$ and 0. Make a conjecture about the change in the shape of the ellipse as e approaches 0.

92. CAPSTONE Compare the graphs of the following equations.

(a) $\dfrac{(x-1)^2}{16} + \dfrac{(y+2)^2}{4} = 1$

(b) $\dfrac{(x-1)^2}{4} + \dfrac{(y+2)^2}{16} = 1$

(c) $\dfrac{(x-1)^2}{16} - \dfrac{(y+2)^2}{4} = 1$

(d) $\dfrac{(x-1)^2}{4} - \dfrac{(y+2)^2}{16} = 1$

(e) $\dfrac{(x-1)^2}{16} + \dfrac{(y+2)^2}{16} = 1$

4 CHAPTER SUMMARY

What Did You Learn?	Explanation/Examples	Review Exercises
Section 4.1 Find the domains of rational functions (*p. 332*).	A rational function is a quotient of polynomial functions. It can be written in the form $f(x) = N(x)/D(x)$, where $N(x)$ and $D(x)$ are polynomials and $D(x)$ is not the zero polynomial. In general, the domain of a rational function of x includes all real numbers except x-values that make the denominator zero.	1–4
Find the vertical and horizontal asymptotes of graphs of rational functions (*p. 333*).	The line $x = a$ is a vertical asymptote of the graph of f if $f(x) \to \infty$ or $f(x) \to -\infty$ as $x \to a$, either from the right or from the left. The line $y = b$ is a horizontal asymptote of the graph of f if $f(x) \to b$ as $x \to \infty$ or $x \to -\infty$.	5–10
Use rational functions to model and solve real-life problems (*p. 335*).	A rational function can be used to model the cost of removing a given percent of smokestack pollutants at a utility company that burns coal. (See Example 4.)	11, 12
Section 4.2 Analyze and sketch graphs of rational functions (*p. 340*).	Let $f(x) = N(x)/D(x)$, where $N(x)$ and $D(x)$ are polynomials. **1.** Simplify f, if possible. **2.** Find and plot the y-intercept (if any) by evaluating $f(0)$. **3.** Find the zeros of the numerator (if any) by solving the equation $N(x) = 0$. Then plot the corresponding x-intercepts. **4.** Find the zeros of the denominator (if any) by solving $D(x) = 0$. Then sketch the corresponding vertical asymptotes. **5.** Find and sketch the horizontal asymptote (if any). **6.** Plot at least one point *between* and one point *beyond* each x-intercept and vertical asymptote. **7.** Use smooth curves to complete the graph between and beyond the vertical asymptotes.	13–24
Sketch graphs of rational functions that have slant asymptotes (*p. 343*).	Consider a rational function whose denominator is of degree 1 or greater. If the degree of the numerator is exactly *one more* than the degree of the denominator, the graph of the function has a slant asymptote.	25–30
Use graphs of rational functions to model and solve real-life problems (*p. 344*).	The graph of a rational function can be used to model the printed area of a rectangular page that is to be minimized, and to find the page dimensions so that the least amount of paper is used. (See Example 6.)	31–34
Section 4.3 Recognize the four basic conics: circle, ellipse, parabola, and hyperbola (*p. 349*).	A conic section (or simply conic) is the intersection of a plane and a double-napped cone. Circle Ellipse Parabola Hyperbola	35–42

What Did You Learn?	Explanation/Examples	Review Exercises

Recognize, graph, and write equations of parabolas (vertex at origin) *(p. 350).*

The standard form of the equation of a parabola with vertex at $(0, 0)$ and directrix $y = -p$ is

$$x^2 = 4py, \quad p \neq 0.$$

For directrix $x = -p$, the equation is

$$y^2 = 4px, \quad p \neq 0.$$

The focus is on the axis p units from the vertex.

43–50

Recognize, graph, and write equations of ellipses (center at origin) *(p. 352).*

The standard form of the equation of an ellipse centered at the origin with major and minor axes of lengths $2a$ and $2b$ (where $0 < b < a$) is

$$\frac{x^2}{a^2} + \frac{y^2}{b^2} = 1$$

or

$$\frac{x^2}{b^2} + \frac{y^2}{a^2} = 1.$$

The vertices and foci lie on the major axis, a and c units, respectively, from the center. Moreover, a, b, and c are related by the equation $c^2 = a^2 - b^2$.

51–58

Recognize, graph, and write equations of hyperbolas (center at origin) *(p. 354).*

The standard form of the equation of a hyperbola with center at the origin (where $a \neq 0$ and $b \neq 0$) is

$$\frac{x^2}{a^2} - \frac{y^2}{b^2} = 1$$

or

$$\frac{y^2}{a^2} - \frac{x^2}{b^2} = 1.$$

The vertices and foci are, respectively, a and c units from the center. Moreover, a, b, and c are related by the equation $b^2 = c^2 - a^2$.

59–62

Recognize equations of conics that have been shifted vertically or horizontally in the plane *(p. 362)* **and write and graph equations of conics that have been shifted vertically or horizontally in the plane** *(p. 364).*

Circle: The graph of $(x - 2)^2 + (y + 1)^2 = 5^2$ is a circle whose center is the point $(2, -1)$ and whose radius is 5. The graph has been shifted two units to the right and one unit downward from standard position.

Ellipse: The graph of $\dfrac{(x - 1)^2}{4^2} + \dfrac{(y - 2)^2}{3^2} = 1$ is an ellipse whose center is the point $(1, 2)$. The major axis is horizontal and of length 8, and the minor axis is vertical and of length 6. The graph has been shifted one unit to the right and two units upward from standard position.

Hyperbola: The graph of $\dfrac{(x - 5)^2}{1^2} - \dfrac{(y - 4)^2}{2^2} = 1$ is a hyperbola whose center is the point $(5, 4)$. The transverse axis is horizontal and of length 2, and the conjugate axis is vertical and of length 4. The graph has been shifted five units to the right and four units upward from standard position.

Parabola: The graph of $(x - 1)^2 = 4(-1)(y - 6)$ is a parabola whose vertex is the point $(1, 6)$. The axis of the parabola is vertical. Because $p = -1$, the focus lies below the vertex. The graph has been reflected in the x-axis, shifted one unit to the right and six units upward from standard position.

63–87

4 REVIEW EXERCISES

See www.CalcChat.com for worked-out solutions to odd-numbered exercises.

4.1 In Exercises 1–4, find the domain of the rational function.

1. $f(x) = \dfrac{3x}{x + 10}$

2. $f(x) = \dfrac{4x^3}{2 + 5x}$

3. $f(x) = \dfrac{8}{x^2 - 10x + 24}$

4. $f(x) = \dfrac{x^2 + x - 2}{x^2 + 4}$

In Exercises 5–10, identify any vertical and horizontal asymptotes.

5. $f(x) = \dfrac{4}{x + 3}$

6. $f(x) = \dfrac{2x^2 + 5x - 3}{x^2 + 2}$

7. $g(x) = \dfrac{x^2}{x^2 - 4}$

8. $g(x) = \dfrac{1}{(x - 3)^2}$

9. $h(x) = \dfrac{5x + 20}{x^2 - 2x - 24}$

10. $h(x) = \dfrac{x^3 - 4x^2}{x^2 + 3x + 2}$

11. AVERAGE COST A business has a production cost of $C = 0.5x + 500$ for producing x units of a product. The average cost per unit, $\overline{C}$, is given by

$$\overline{C} = \frac{C}{x} = \frac{0.5x + 500}{x}, \quad x > 0.$$

Determine the average cost per unit as x increases without bound. (Find the horizontal asymptote.)

12. SEIZURE OF ILLEGAL DRUGS The cost C (in millions of dollars) for the federal government to seize $p\%$ of an illegal drug as it enters the country is given by

$$C = \frac{528p}{100 - p}, \quad 0 \le p < 100.$$

 (a) Use a graphing utility to graph the cost function.

(b) Find the costs of seizing 25%, 50%, and 75% of the drug.

(c) According to this model, would it be possible to seize 100% of the drug?

4.2 In Exercises 13–24, (a) state the domain of the function, (b) identify all intercepts, (c) find any vertical and horizontal asymptotes, and (d) plot additional solution points as needed to sketch the graph of the rational function.

13. $f(x) = \dfrac{-3}{2x^2}$

14. $f(x) = \dfrac{4}{x}$

15. $g(x) = \dfrac{2 + x}{1 - x}$

16. $h(x) = \dfrac{x - 4}{x - 7}$

17. $p(x) = \dfrac{5x^2}{4x^2 + 1}$

18. $f(x) = \dfrac{2x}{x^2 + 4}$

19. $f(x) = \dfrac{x}{x^2 + 1}$

20. $h(x) = \dfrac{9}{(x - 3)^2}$

21. $f(x) = \dfrac{-6x^2}{x^2 + 1}$

22. $y = \dfrac{2x^2}{x^2 - 4}$

23. $f(x) = \dfrac{6x^2 - 11x + 3}{3x^2 - x}$

24. $f(x) = \dfrac{6x^2 - 7x + 2}{4x^2 - 1}$

In Exercises 25–30, (a) state the domain of the function, (b) identify all intercepts, (c) identify any vertical and slant asymptotes, and (d) plot additional solution points as needed to sketch the graph of the rational function.

25. $f(x) = \dfrac{2x^3}{x^2 + 1}$

26. $f(x) = \dfrac{x^2 + 1}{x + 1}$

27. $f(x) = \dfrac{x^2 + 3x - 10}{x + 2}$

28. $f(x) = \dfrac{x^3}{x^2 - 25}$

29. $f(x) = \dfrac{3x^3 - 2x^2 - 3x + 2}{3x^2 - x - 4}$

30. $f(x) = \dfrac{3x^3 - 4x^2 - 12x + 16}{3x^2 + 5x - 2}$

31. AVERAGE COST The cost of producing x units of a product is C, and the average cost per unit $\overline{C}$ is given by

$$\overline{C} = \frac{C}{x} = \frac{100{,}000 + 0.9x}{x}, \quad x > 0.$$

(a) Graph the average cost function.

(b) Find the average costs of producing $x = 1000$, 10,000, and 100,000 units.

(c) By increasing the level of production, what is the smallest average cost per unit you can obtain? Explain your reasoning.

 32. PAGE DESIGN A page that is x inches wide and y inches high contains 30 square inches of print. The top and bottom margins are 2 inches deep and the margins on each side are 2 inches wide.

(a) Draw a diagram that gives a visual representation of the problem.

(b) Show that the total area A of the page is

$$A = \frac{2x(2x + 7)}{x - 4}.$$

(c) Determine the domain of the function based on the physical constraints of the problem.

(d) Use a graphing utility to graph the area function and approximate the page size for which the least amount of paper will be used. Verify your answer numerically using the *table* feature of the graphing utility.

33. PHOTOSYNTHESIS The amount y of CO_2 uptake (in milligrams per square decimeter per hour) at optimal temperatures and with the natural supply of CO_2 is approximated by the model

$$y = \frac{18.47x - 2.96}{0.23x + 1}, \quad x > 0$$

where x is the light intensity (in watts per square meter). Use a graphing utility to graph the function and determine the limiting amount of CO_2 uptake.

34. MEDICINE The concentration C of a medication in the bloodstream t hours after injection into muscle tissue is given by $C(t) = (2t + 1)/(t^2 + 4)$, $t > 0$.

(a) Determine the horizontal asymptote of the graph of the function and interpret its meaning in the context of the problem.

(b) Use a graphing utility to graph the function and approximate the time when the bloodstream concentration is greatest.

4.3 In Exercises 35–42, identify the conic.

35. $y^2 = -16x$ **36.** $16x^2 + y^2 = 16$

37. $\dfrac{x^2}{64} - \dfrac{y^2}{4} = 1$ **38.** $\dfrac{x^2}{1} + \dfrac{y^2}{36} = 1$

39. $x^2 + 20y = 0$ **40.** $x^2 + y^2 = 400$

41. $\dfrac{y^2}{49} - \dfrac{x^2}{144} = 1$ **42.** $\dfrac{x^2}{49} + \dfrac{y^2}{144} = 1$

In Exercises 43–48, find the standard form of the equation of the parabola with the given characteristic(s) and vertex at the origin.

43. Passes through the point $(3, 6)$; horizontal axis

44. Passes through the point $(4, -2)$; vertical axis

45. Focus: $(-6, 0)$ **46.** Focus: $(0, 7)$

47. Directrix: $y = -3$ **48.** Directrix: $x = 3$

49. SATELLITE ANTENNA A cross section of a large parabolic antenna (see figure) is modeled by $y = x^2/200$, $-100 \le x \le 100$. The receiving and transmitting equipment is positioned at the focus. Find the coordinates of the focus.

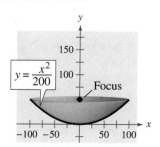

50. SUSPENSION BRIDGE Each cable of a suspension bridge is suspended (in the shape of a parabola) between two towers (see figure).

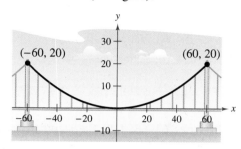

(a) Find the coordinates of the focus.

(b) Write an equation that models the cables.

In Exercises 51–56, find the standard form of the equation of the ellipse with the given characteristics and center at the origin.

51. Vertices: $(\pm 9, 0)$; minor axis of length 6

52. Vertices: $(0, \pm 10)$; minor axis of length 2

53. Vertices: $(0, \pm 6)$; passes through the point $(2, 2)$

54. Vertices: $(\pm 7, 0)$; foci: $(\pm 6, 0)$

55. Foci: $(\pm 14, 0)$; minor axis of length 10

56. Foci: $(\pm 3, 0)$; major axis of length 12

57. ARCHITECTURE A semielliptical archway is to be formed over the entrance to an estate (see figure). The arch is to be set on pillars that are 10 feet apart and is to have a height (atop the pillars) of 4 feet. Where should the foci be placed in order to sketch the arch?

58. WADING POOL You are building a wading pool that is in the shape of an ellipse. Your plans give an equation for the elliptical shape of the pool measured in feet as

$$\frac{x^2}{324} + \frac{y^2}{196} = 1.$$

Find the longest distance across the pool, the shortest distance, and the distance between the foci.

In Exercises 59–62, find the standard form of the equation of the hyperbola with the given characteristics and center at the origin.

59. Vertices: $(0, \pm1)$; foci: $(0, \pm5)$

60. Vertices: $(\pm4, 0)$; foci: $(\pm6, 0)$

61. Vertices: $(\pm1, 0)$; asymptotes: $y = \pm2x$

62. Vertices: $(0, \pm2)$; asymptotes: $y = \pm\dfrac{2}{\sqrt{5}}x$

4.4 In Exercises 63–66, find the standard form of the equation of the parabola with the given characteristics.

63. Vertex: $(-8, 8)$; directrix: $y = 1$

64. Focus: $(0, 5)$; directrix: $x = 6$

65. Vertex: $(4, 2)$; focus: $(4, 0)$

66. Vertex: $(2, 0)$; focus: $(0, 0)$

In Exercises 67–70, find the standard form of the equation of the ellipse with the given characteristics.

67. Vertices: $(0, 3)$, $(12, 3)$; passes through the point $(6, 0)$

68. Center: $(0, 4)$; vertices: $(0, 0)$, $(0, 8)$

69. Vertices: $(-3, 0)$, $(7, 0)$; foci: $(0, 0)$, $(4, 0)$

70. Vertices: $(2, 0)$, $(2, 4)$; foci: $(2, 1)$, $(2, 3)$

In Exercises 71–76, find the standard form of the equation of the hyperbola with the given characteristics.

71. Vertices: $(\pm6, 7)$;
asymptotes: $y = -\frac{1}{2}x + 7$, $y = \frac{1}{2}x + 7$

72. Vertices: $(0, 0)$, $(0, -4)$; passes through the point $\left(2, 2\left(\sqrt{5} - 1\right)\right)$

73. Vertices: $(-10, 3)$, $(6, 3)$; foci: $(-12, 3)$, $(8, 3)$

74. Vertices: $(2, 2)$, $(-2, 2)$; foci: $(4, 2)$, $(-4, 2)$

75. Foci: $(0, 0)$, $(8, 0)$; asymptotes: $y = \pm2(x - 4)$

76. Foci: $(3, \pm2)$; asymptotes: $y = \pm2(x - 3)$

In Exercises 77–84, identify the conic by writing its equation in standard form. Then sketch its graph and describe the translation.

77. $x^2 - 6x + 2y + 9 = 0$

78. $y^2 - 12y - 8x + 20 = 0$

79. $x^2 + y^2 - 2x - 4y + 5 = 0$

80. $16x^2 + 16y^2 - 16x + 24y - 3 = 0$

81. $x^2 + 9y^2 + 10x - 18y + 25 = 0$

82. $4x^2 + y^2 - 16x + 15 = 0$

83. $9x^2 - y^2 - 72x + 8y + 119 = 0$

84. $x^2 - 9y^2 + 10x + 18y + 7 = 0$

85. ARCHITECTURE A parabolic archway is 12 meters high at the vertex. At a height of 10 meters, the width of the archway is 8 meters (see figure). How wide is the archway at ground level?

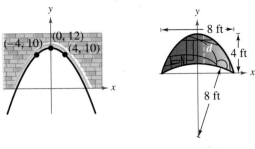

FIGURE FOR **85**   FIGURE FOR **86**

86. ARCHITECTURE A church window (see figure) is bounded above by a parabola and below by the arc of a circle.

(a) Find equations for the parabola and the circle.

(b) Complete the table by filling in the vertical distance d between the circle and the parabola for each given value of x.

x	0	1	2	3	4
d					

87. RUNNING PATH Let $(0, 0)$ represent a water fountain located in a city park. Each day you run through the park along a path given by

$$x^2 + y^2 - 200x - 52{,}500 = 0$$

where x and y are measured in meters.

(a) What type of conic is your path? Explain your reasoning.

(b) Write the equation of the path in standard form. Sketch a graph of the equation.

(c) After you run, you walk to the water fountain. If you stop running at $(-100, 150)$, how far must you walk for a drink of water?

EXPLORATION

TRUE OR FALSE? In Exercises 88 and 89, determine whether the statement is true or false. Justify your answer.

88. The domain of a rational function can never be the set of all real numbers.

89. The graph of the equation

$$Ax^2 + Bxy + Cy^2 + Dx + Ey + F = 0$$

can be a single point.

4 CHAPTER TEST

See www.CalcChat.com for worked-out solutions to odd-numbered exercises.

Take this test as you would take a test in class. When you are finished, check your work against the answers given in the back of the book.

In Exercises 1–3, find the domain of the function and identify any asymptotes.

1. $y = \dfrac{3x}{x+1}$

2. $f(x) = \dfrac{3-x^2}{3+x^2}$

3. $g(x) = \dfrac{x^2-7x+12}{x-3}$

In Exercises 4–9, identify any intercepts and asymptotes of the graph of the function. Then sketch a graph of the function.

4. $h(x) = \dfrac{4}{x^2} - 1$

5. $g(x) = \dfrac{x^2+2}{x-1}$

6. $f(x) = \dfrac{x+1}{x^2+x-12}$

7. $f(x) = \dfrac{2x^2-5x-12}{x^2-16}$

8. $f(x) = \dfrac{2x^2+9}{5x^2+9}$

9. $g(x) = \dfrac{2x^3-7x^2+4x+4}{x^2-x-2}$

10. A rectangular page is designed to contain 36 square inches of print. The margins at the top and bottom of the page are 2 inches deep. The margins on each side are 1 inch wide. What should the dimensions of the page be so that the least amount of paper is used?

11. A triangle is formed by the coordinate axes and a line through the point $(2, 1)$, as shown in the figure.

(a) Verify that $y = 1 + \dfrac{2}{x-2}$.

(b) Write the area A of the triangle as a function of x. Determine the domain of the function in the context of the problem.

(c) Graph the area function. Estimate the minimum area of the triangle from the graph.

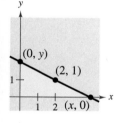

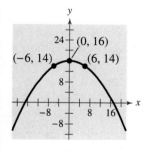

FIGURE FOR 11

In Exercises 12–17, graph the conic and identify the center, vertices, and foci, if applicable.

12. $y^2 - 4x = 0$

13. $x^2 + y^2 - 10x + 4y + 4 = 0$

14. $x^2 - 10x - 2y + 19 = 0$

15. $x^2 - \dfrac{y^2}{4} = 1$

16. $\dfrac{y^2}{4} - x^2 = 1$

17. $x^2 + 3y^2 - 2x + 36y + 100 = 0$

18. Find an equation of the ellipse with vertices $(0, 2)$ and $(8, 2)$ and minor axis of length 4.

19. Find an equation of the hyperbola with vertices $(0, \pm 3)$ and asymptotes $y = \pm\frac{3}{2}x$.

20. A parabolic archway is 16 meters high at the vertex. At a height of 14 meters, the width of the archway is 12 meters, as shown in the figure. How wide is the archway at ground level?

21. The moon orbits Earth in an elliptical path with the center of Earth at one focus, as shown in the figure. The major and minor axes of the orbit have lengths of 768,800 kilometers and 767,640 kilometers, respectively. Find the smallest distance (perigee) and the greatest distance (apogee) from the center of the moon to the center of Earth.

FIGURE FOR 20

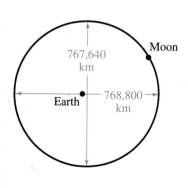

FIGURE FOR 21

PROOFS IN MATHEMATICS

You can use the definition of a parabola to derive the standard form of the equation of a parabola whose directrix is parallel to the x-axis or to the y-axis.

Parabolic Patterns

There are many natural occurrences of parabolas in real life. For instance, the famous astronomer Galileo discovered in the 17th century that an object that is projected upward and obliquely to the pull of gravity travels in a parabolic path. Examples of this are the center of gravity of a jumping dolphin and the path of water molecules of a drinking water fountain.

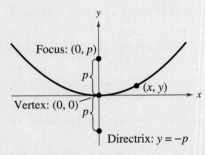

Parabola with vertical axis

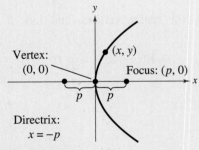

Parabola with horizontal axis

Standard Equation of a Parabola (Vertex at Origin) *(p. 350)*

The standard form of the equation of a parabola with vertex at $(0, 0)$ and directrix $y = -p$ is

$$x^2 = 4py, \quad p \neq 0.$$

For directrix $x = -p$, the equation is

$$y^2 = 4px, \quad p \neq 0.$$

The focus is on the axis p units (directed distance) from the vertex.

Proof

For the first case, suppose the directrix $(y = -p)$ is parallel to the x-axis. In the figure, you assume that $p > 0$, and because p is the directed distance from the vertex to the focus, the focus must lie above the vertex. Because the point (x, y) is equidistant from $(0, p)$ and $y = -p$, you can apply the Distance Formula to obtain

$$\sqrt{(x - 0)^2 + (y - p)^2} = y + p$$

$$x^2 + (y - p)^2 = (y + p)^2$$

$$x^2 + y^2 - 2py + p^2 = y^2 + 2py + p^2$$

$$x^2 = 4py.$$

A proof of the second case is similar to the proof of the first case. Suppose the directrix $(x = -p)$ is parallel to the y-axis. In the figure, you assume that $p > 0$, and because p is the directed distance from the vertex to the focus, the focus must lie to the right of the vertex. Because the point (x, y) is equidistant from $(p, 0)$ and $x = -p$, you can apply the Distance Formula as follows.

$$\sqrt{(x - p)^2 + (y - 0)^2} = x + p$$

$$(x - p)^2 + y^2 = (x + p)^2$$

$$x^2 - 2px + p^2 + y^2 = x^2 + 2px + p^2$$

$$y^2 = 4px$$

PROBLEM SOLVING

This collection of thought-provoking and challenging exercises further explores and expands upon concepts learned in this chapter.

1. Match the graph of the rational function given by

$$f(x) = \frac{ax + b}{cx + d}$$

with the given conditions.

(a)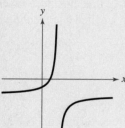

(b)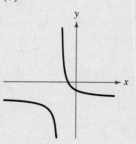

(c)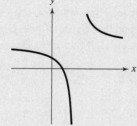

(d)

(i) $a > 0$ (ii) $a > 0$ (iii) $a < 0$ (iv) $a > 0$
    $b < 0$ $b > 0$ $b > 0$ $b < 0$
    $c > 0$ $c < 0$ $c > 0$ $c > 0$
    $d < 0$ $d < 0$ $d < 0$ $d > 0$

2. Consider the function given by

$$f(x) = \frac{ax}{(x - b)^2}.$$

   (a) Determine the effect on the graph of f if $b \neq 0$ and a is varied. Consider cases in which a is positive and a is negative.

   (b) Determine the effect on the graph of f if $a \neq 0$ and b is varied.

3. The endpoints of the interval over which distinct vision is possible is called the *near point* and *far point* of the eye (see figure). With increasing age, these points normally change. The table shows the approximate near points y (in inches) for various ages x (in years).

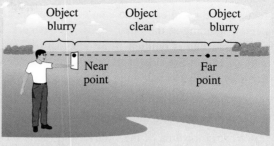

FIGURE FOR 3

Age, x	Near point, y
16	3.0
32	4.7
44	9.8
50	19.7
60	39.4

   (a) Use the *regression* feature of a graphing utility to find a quadratic model for the data. Use a graphing utility to plot the data and graph the model in the same viewing window.

   (b) Find a rational model for the data. Take the reciprocals of the near points to generate the points $(x, 1/y)$. Use the *regression* feature of a graphing utility to find a linear model for the data. The resulting line has the form

$$\frac{1}{y} = ax + b.$$

   Solve for y. Use a graphing utility to plot the data and graph the model in the same viewing window.

   (c) Use the *table* feature of a graphing utility to create a table showing the predicted near point based on each model for each of the ages in the original table. How well do the models fit the original data?

   (d) Use both models to estimate the near point for a person who is 25 years old. Which model is a better fit?

   (e) Do you think either model can be used to predict the near point for a person who is 70 years old? Explain.

377

4. Statuary Hall is an elliptical room in the United States Capitol in Washington D.C. The room is also called the Whispering Gallery because a person standing at one focus of the room can hear even a whisper spoken by a person standing at the other focus. This occurs because any sound that is emitted from one focus of an ellipse will reflect off the side of the ellipse to the other focus. Statuary Hall is 46 feet wide and 97 feet long.

   (a) Find an equation that models the shape of the room.

   (b) How far apart are the two foci?

   (c) What is the area of the floor of the room? (The area of an ellipse is $A = \pi ab$.)

5. Use the figure to show that $|d_2 - d_1| = 2a$.

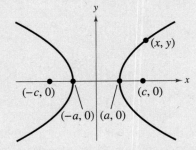

6. Find an equation of a hyperbola such that for any point on the hyperbola, the difference between its distances from the points $(2, 2)$ and $(10, 2)$ is 6.

7. The filament of a light bulb is a thin wire that glows when electricity passes through it. The filament of a car headlight is at the focus of a parabolic reflector, which sends light out in a straight beam. Given that the filament is 1.5 inches from the vertex, find an equation for the cross section of the reflector. A reflector is 7 inches wide. How deep is it?

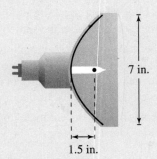

7 in.

1.5 in.

8. Consider the parabola $x^2 = 4py$.

   (a) Use a graphing utility to graph the parabola for $p = 1, p = 2, p = 3,$ and $p = 4$. Describe the effect on the graph when p increases.

   (b) Locate the focus for each parabola in part (a).

   (c) For each parabola in part (a), find the length of the chord passing through the focus and parallel to the directrix. How can the length of this chord be determined directly from the standard form of the equation of the parabola?

   (d) Explain how the result of part (c) can be used as a sketching aid when graphing parabolas.

9. Let (x_1, y_1) be the coordinates of a point on the parabola $x^2 = 4py$. The equation of the line that just touches the parabola at the point (x_1, y_1), called a *tangent line*, is given by

   $$y - y_1 = \frac{x_1}{2p}(x - x_1).$$

   (a) What is the slope of the tangent line?

   (b) For each parabola in Exercise 8, find the equations of the tangent lines at the endpoints of the chord. Use a graphing utility to graph the parabola and tangent lines.

10. A tour boat travels between two islands that are 12 miles apart (see figure). For each trip between the islands, there is enough fuel for a 20-mile trip.

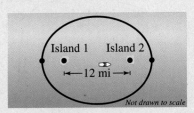

Island 1   Island 2

← 12 mi →

Not drawn to scale

   (a) Explain why the region in which the boat can travel is bounded by an ellipse.

   (b) Let $(0, 0)$ represent the center of the ellipse. Find the coordinates of the center of each island.

   (c) The boat travels from one island, straight past the other island to one vertex of the ellipse, and back to the second island. How many miles does the boat travel? Use your answer to find the coordinates of the vertex.

   (d) Use the results of parts (b) and (c) to write an equation of the ellipse that bounds the region in which the boat can travel.

11. Prove that the graph of the equation

   $$Ax^2 + Cy^2 + Dx + Ey + F = 0$$

   is one of the following (except in degenerate cases).

Conic	Condition
(a) Circle	$A = C$
(b) Parabola	$A = 0$ or $C = 0$ (but not both)
(c) Ellipse	$AC > 0$
(d) Hyperbola	$AC < 0$

Exponential and Logarithmic Functions

5

In Mathematics

Exponential functions involve a constant base and a variable exponent. The inverse of an exponential function is a logarithmic function.

In Real Life

Exponential and logarithmic functions are widely used in describing economic and physical phenomena such as compound interest, population growth, memory retention, and decay of radioactive material. For instance, a logarithmic function can be used to relate an animal's weight and its lowest galloping speed. (See Exercise 95, page 406.)

Juniors Bildarchiv / Alamy

IN CAREERS

There are many careers that use exponential and logarithmic functions. Several are listed below.

- Astronomer
  Example 7, page 404

- Psychologist
  Exercise 136, page 417

- Archeologist
  Example 3, page 422

- Forensic Scientist
  Exercise 75, page 430

What you should learn

- Recognize and evaluate exponential functions with base *a*.
- Graph exponential functions and use the One-to-One Property.
- Recognize, evaluate, and graph exponential functions with base *e*.
- Use exponential functions to model and solve real-life problems.

Why you should learn it

Exponential functions can be used to model and solve real-life problems. For instance, in Exercise 76 on page 390, an exponential function is used to model the concentration of a drug in the bloodstream.

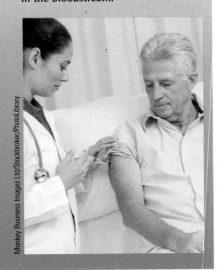

5.1 EXPONENTIAL FUNCTIONS AND THEIR GRAPHS

Exponential Functions

So far, this text has dealt mainly with **algebraic functions,** which include polynomial functions and rational functions. In this chapter, you will study two types of nonalgebraic functions—*exponential functions* and *logarithmic functions.* These functions are examples of **transcendental functions.**

Definition of Exponential Function

The **exponential function** *f* with base *a* is denoted by

$$f(x) = a^x$$

where $a > 0$, $a \neq 1$, and *x* is any real number.

The base $a = 1$ is excluded because it yields $f(x) = 1^x = 1$. This is a constant function, not an exponential function.

You have evaluated a^x for integer and rational values of *x*. For example, you know that $4^3 = 64$ and $4^{1/2} = 2$. However, to evaluate 4^x for any real number *x*, you need to interpret forms with *irrational* exponents. For the purposes of this text, it is sufficient to think of

$$a^{\sqrt{2}} \quad (\text{where } \sqrt{2} \approx 1.41421356)$$

as the number that has the successively closer approximations

$$a^{1.4}, a^{1.41}, a^{1.414}, a^{1.4142}, a^{1.41421}, \ldots$$

Example 1   Evaluating Exponential Functions

Use a calculator to evaluate each function at the indicated value of *x*.

Function	Value
a. $f(x) = 2^x$	$x = -3.1$
b. $f(x) = 2^{-x}$	$x = \pi$
c. $f(x) = 0.6^x$	$x = \frac{3}{2}$

Solution

Function Value	Graphing Calculator Keystrokes	Display
a. $f(\quad) = 2$	2 ^ (−) 3.1 ENTER	0.1166291
b. $f(\ ) = 2^-$	2 ^ (−) π ENTER	0.1133147
c. $f(\ ) = (0.6)$	.6 ^ (3 ÷ 2) ENTER	0.4647580

CHECK*Point* Now try Exercise 7.

When evaluating exponential functions with a calculator, remember to enclose fractional exponents in parentheses. Because the calculator follows the order of operations, parentheses are crucial in order to obtain the correct result.

Graphs of Exponential Functions

The graphs of all exponential functions have similar characteristics, as shown in Examples 2, 3, and 5.

Example 2   Graphs of $y = a^x$

In the same coordinate plane, sketch the graph of each function.

a. $f(x) = 2^x$ **b.** $g(x) = 4^x$

Solution

The table below lists some values for each function, and Figure 5.1 shows the graphs of the two functions. Note that both graphs are increasing. Moreover, the graph of $g(x) = 4^x$ is increasing more rapidly than the graph of $f(x) = 2^x$.

x	-3	-2	-1	0	1	2
2^x	$\frac{1}{8}$	$\frac{1}{4}$	$\frac{1}{2}$	1	2	4
4^x	$\frac{1}{64}$	$\frac{1}{16}$	$\frac{1}{4}$	1	4	16

CHECK **Point**   Now try Exercise 17.

The table in Example 2 was evaluated by hand. You could, of course, use a graphing utility to construct tables with even more values.

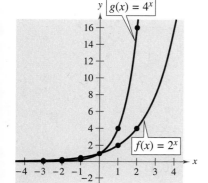

FIGURE **5.1**

Example 3   Graphs of $y = a^{-x}$

In the same coordinate plane, sketch the graph of each function.

a. $F(x) = 2^{-x}$ **b.** $G(x) = 4^{-x}$

Solution

The table below lists some values for each function, and Figure 5.2 shows the graphs of the two functions. Note that both graphs are decreasing. Moreover, the graph of $G(x) = 4^{-x}$ is decreasing more rapidly than the graph of $F(x) = 2^{-x}$.

x	-2	-1	0	1	2	3
2^{-x}	4	2	1	$\frac{1}{2}$	$\frac{1}{4}$	$\frac{1}{8}$
4^{-x}	16	4	1	$\frac{1}{4}$	$\frac{1}{16}$	$\frac{1}{64}$

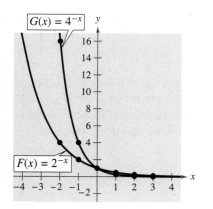

FIGURE **5.2**

CHECK **Point**   Now try Exercise 19.

In Example 3, note that by using one of the properties of exponents, the functions $F(x) = 2^{-x}$ and $G(x) = 4^{-x}$ can be rewritten with positive exponents.

$$F(x) = 2^{-x} = \frac{1}{2^x} = \left(\frac{1}{2}\right)^x \quad \text{and} \quad G(x) = 4^{-x} = \frac{1}{4^x} = \left(\frac{1}{4}\right)^x$$

Comparing the functions in Examples 2 and 3, observe that

$$F(x) = 2^{-x} = f(-x) \qquad \text{and} \qquad G(x) = 4^{-x} = g(-x).$$

Consequently, the graph of F is a reflection (in the y-axis) of the graph of f. The graphs of G and g have the same relationship. The graphs in Figures 5.1 and 5.2 are typical of the exponential functions $y = a^x$ and $y = a^{-x}$. They have one y-intercept and one horizontal asymptote (the x-axis), and they are continuous. The basic characteristics of these exponential functions are summarized in Figures 5.3 and 5.4.

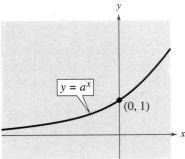

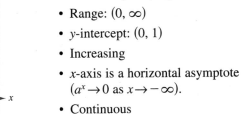

Graph of $y = a^x$, $a > 1$
- Domain: $(-\infty, \infty)$
- Range: $(0, \infty)$
- y-intercept: $(0, 1)$
- Increasing
- x-axis is a horizontal asymptote $(a^x \to 0$ as $x \to -\infty)$.
- Continuous

FIGURE **5.3**

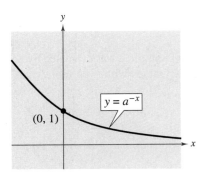

Graph of $y = a^{-x}$, $a > 1$
- Domain: $(-\infty, \infty)$
- Range: $(0, \infty)$
- y-intercept: $(0, 1)$
- Decreasing
- x-axis is a horizontal asymptote $(a^{-x} \to 0$ as $x \to \infty)$.
- Continuous

FIGURE **5.4**

From Figures 5.3 and 5.4, you can see that the graph of an exponential function is always increasing or always decreasing. As a result, the graphs pass the Horizontal Line Test, and therefore the functions are one-to-one functions. You can use the following **One-to-One Property** to solve simple exponential equations.

For $a > 0$ and $a \neq 1$, $a^x = a^y$ if and only if $x = y$.

Example 4   Using the One-to-One Property

a. $9 = 3^{x+1}$

$3^2 = 3^{x+1}$

$2 = x + 1$

$1 = x$

b. $\left(\frac{1}{2}\right)^x = 8 \Longrightarrow 2^{-x} = 2^3 \Longrightarrow x = -3$

CHECK *Point*   Now try Exercise 51.

In the following example, notice how the graph of $y = a^x$ can be used to sketch the graphs of functions of the form $f(x) = b \pm a^{x+c}$.

Algebra Help

You can review the techniques for transforming the graph of a function in Section 2.5.

Example 5   Transformations of Graphs of Exponential Functions

Each of the following graphs is a transformation of the graph of $f(x) = 3^x$.

a. Because $g(x) = 3^{x+1} = f(x + 1)$, the graph of g can be obtained by shifting the graph of f one unit to the *left*, as shown in Figure 5.5.

b. Because $h(x) = 3^x - 2 = f(x) - 2$, the graph of h can be obtained by shifting the graph of f *downward* two units, as shown in Figure 5.6.

c. Because $k(x) = -3^x = -f(x)$, the graph of k can be obtained by *reflecting* the graph of f in the *x*-axis, as shown in Figure 5.7.

d. Because $j(x) = 3^{-x} = f(-x)$, the graph of j can be obtained by *reflecting* the graph of f in the *y*-axis, as shown in Figure 5.8.

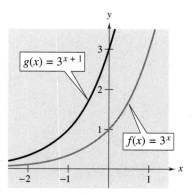

FIGURE 5.5   Horizontal shift

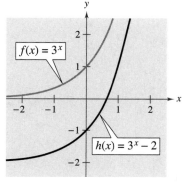

FIGURE 5.6   Vertical shift

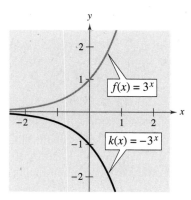

FIGURE 5.7   Reflection in *x*-axis

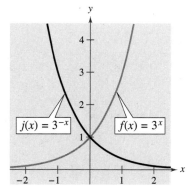

FIGURE 5.8   Reflection in *y*-axis

CHECK *Point*   Now try Exercise 23.

Notice that the transformations in Figures 5.5, 5.7, and 5.8 keep the *x*-axis as a horizontal asymptote, but the transformation in Figure 5.6 yields a new horizontal asymptote of $y = -2$. Also, be sure to note how the *y*-intercept is affected by each transformation.

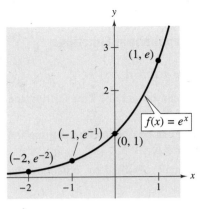

FIGURE 5.9

The Natural Base e

In many applications, the most convenient choice for a base is the irrational number

$$e \approx 2.718281828 \ldots.$$

This number is called the **natural base.** The function given by $f(x) = e^x$ is called the **natural exponential function.** Its graph is shown in Figure 5.9. Be sure you see that for the exponential function $f(x) = e^x$, e is the constant $2.718281828 \ldots$, whereas x is the variable.

Example 6   Evaluating the Natural Exponential Function

Use a calculator to evaluate the function given by $f(x) = e^x$ at each indicated value of x.

a. $x = -2$
b. $x = -1$
c. $x = 0.25$
d. $x = -0.3$

Solution

	Function Value	Graphing Calculator Keystrokes	Display
a.	$f(\quad) = e$	e^x $(-)$ 2 [ENTER]	0.1353353
b.	$f(\quad) = e$	e^x $(-)$ 1 [ENTER]	0.3678794
c.	$f(\quad) = e$	e^x 0.25 [ENTER]	1.2840254
d.	$f(\quad) = e$	e^x $(-)$ 0.3 [ENTER]	0.7408182

CHECK*Point*   Now try Exercise 33.

Example 7   Graphing Natural Exponential Functions

Sketch the graph of each natural exponential function.

a. $f(x) = 2e^{0.24x}$
b. $g(x) = \frac{1}{2}e^{-0.58x}$

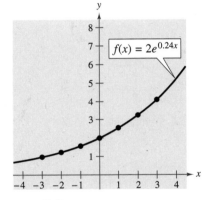

FIGURE 5.10

Solution

To sketch these two graphs, you can use a graphing utility to construct a table of values, as shown below. After constructing the table, plot the points and connect them with smooth curves, as shown in Figures 5.10 and 5.11. Note that the graph in Figure 5.10 is increasing, whereas the graph in Figure 5.11 is decreasing.

x	-3	-2	-1	0	1	2	3
$f(x)$	0.974	1.238	1.573	2.000	2.542	3.232	4.109
$g(x)$	2.849	1.595	0.893	0.500	0.280	0.157	0.088

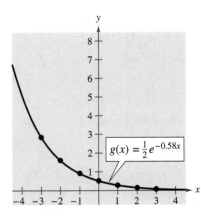

FIGURE 5.11

CHECK*Point*   Now try Exercise 41.

Applications

One of the most familiar examples of exponential growth is an investment earning *continuously compounded interest*. On page 135 in Section 1.6, you were introduced to the formula for the balance in an account that is compounded n times per year. Using exponential functions, you can now *develop* that formula and show how it leads to continuous compounding.

Suppose a principal P is invested at an annual interest rate r, compounded once per year. If the interest is added to the principal at the end of the year, the new balance P_1 is

$$P_1 = P + Pr$$
$$= P(1 + r).$$

This pattern of multiplying the previous principal by $1 + r$ is then repeated each successive year, as shown below.

Year	Balance After Each Compounding
0	$P = P$
1	$P_1 = P(1 + r)$
2	$P_2 = P_1(1 + r) = \qquad (1 + r) = P(1 + r)^2$
3	$P_3 = P_2(1 + r) = \qquad (1 + r) = P(1 + r)^3$
$\vdots$	$\vdots$
t	$P_t = P(1 + r)^t$

To accommodate more frequent (quarterly, monthly, or daily) compounding of interest, let n be the number of compoundings per year and let t be the number of years. Then the rate per compounding is r/n and the account balance after t years is

$$A = P\left(1 + \frac{r}{n}\right)^{nt}.$$

If you let the number of compoundings n increase without bound, the process approaches what is called **continuous compounding.** In the formula for n compoundings per year, let $m = n/r$. This produces

$$A = P\left(1 + \frac{r}{n}\right)^{nt}$$

$$= P\left(1 + \frac{r}{\quad}\right)^{\ t}$$

$$= P\left(1 + \frac{1}{m}\right)^{mrt}$$

$$= P\left[\left(1 + \frac{1}{m}\right)^m\right]^{rt}.$$

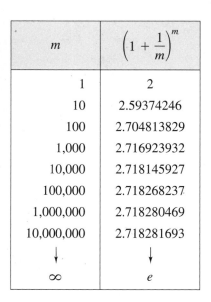

m	$\left(1 + \dfrac{1}{m}\right)^m$
1	2
10	2.59374246
100	2.704813829
1,000	2.716923932
10,000	2.718145927
100,000	2.718268237
1,000,000	2.718280469
10,000,000	2.718281693
$\downarrow$	$\downarrow$
∞	e

As m increases without bound, the table at the left shows that $[1 + (1/m)]^m \to e$ as $m \to \infty$. From this, you can conclude that the formula for continuous compounding is

$$A = P\ ^{rt}.$$

> ⚠️ **WARNING / CAUTION**
>
> Be sure you see that the annual interest rate must be written in decimal form. For instance, 6% should be written as 0.06.

Formulas for Compound Interest

After t years, the balance A in an account with principal P and annual interest rate r (in decimal form) is given by the following formulas.

1. For n compoundings per year: $A = P\left(1 + \dfrac{r}{n}\right)^{nt}$

2. For continuous compounding: $A = Pe^{rt}$

Example 8 Compound Interest

A total of $12,000 is invested at an annual interest rate of 9%. Find the balance after 5 years if it is compounded

a. quarterly.

b. monthly.

c. continuously.

Solution

a. For quarterly compounding, you have $n = 4$. So, in 5 years at 9%, the balance is

$$A = P\left(1 + \frac{r}{n}\right)^{nt}$$

$$= \left(1 + \frac{}{}\right)^{(\ )}$$

$$\approx \$18{,}726.11.$$

b. For monthly compounding, you have $n = 12$. So, in 5 years at 9%, the balance is

$$A = P\left(1 + \frac{r}{n}\right)^{nt}$$

$$= \left(1 + \frac{}{}\right)^{(\ )}$$

$$\approx \$18{,}788.17.$$

c. For continuous compounding, the balance is

$$A = Pe^{rt}$$

$$= e^{(\ )}$$

$$\approx \$18{,}819.75.$$

CHECK Point Now try Exercise 59.

In Example 8, note that continuous compounding yields more than quarterly or monthly compounding. This is typical of the two types of compounding. That is, for a given principal, interest rate, and time, continuous compounding will always yield a larger balance than compounding n times per year.

Example 9   Radioactive Decay

The *half-life* of radioactive radium (^{226}Ra) is about 1599 years. That is, for a given amount of radium, *half* of the original amount will remain after 1599 years. After another 1599 years, one-quarter of the original amount will remain, and so on. Let y represent the mass, in grams, of a quantity of radium. The quantity present after t years, then, is $y = 25\left(\frac{1}{2}\right)^{t/1599}$.

a. What is the initial mass (when $t = 0$)?

b. How much of the initial mass is present after 2500 years?

Algebraic Solution

a. $y = 25\left(\frac{1}{2}\right)^{t/1599}$

$= 25\left(\frac{1}{2}\right)^{/1599}$

$= 25$

So, the initial mass is 25 grams.

b. $y = 25\left(\frac{1}{2}\right)^{t/1599}$

$= 25\left(\frac{1}{2}\right)^{/1599}$

$\approx 25\left(\frac{1}{2}\right)^{1.563}$

≈ 8.46

So, about 8.46 grams is present after 2500 years.

CHECK *Point*   Now try Exercise 73.

Graphical Solution

Use a graphing utility to graph $y = 25\left(\frac{1}{2}\right)^{t/1599}$.

a. Use the *value* feature or the *zoom* and *trace* features of the graphing utility to determine that when $x = 0$, the value of y is 25, as shown in Figure 5.12. So, the initial mass is 25 grams.

b. Use the *value* feature or the *zoom* and *trace* features of the graphing utility to determine that when $x = 2500$, the value of y is about 8.46, as shown in Figure 5.13. So, about 8.46 grams is present after 2500 years.

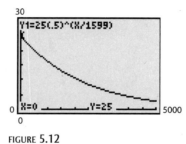

FIGURE 5.12

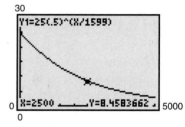

FIGURE 5.13

CLASSROOM DISCUSSION

Identifying Exponential Functions   Which of the following functions generated the two tables below? Discuss how you were able to decide. What do these functions have in common? Are any of them the same? If so, explain why.

a. $f_1(x) = 2^{(x+3)}$ **b.** $f_2(x) = 8\left(\frac{1}{2}\right)^x$ **c.** $f_3(x) = \left(\frac{1}{2}\right)^{(x-3)}$

d. $f_4(x) = \left(\frac{1}{2}\right)^x + 7$ **e.** $f_5(x) = 7 + 2^x$ **f.** $f_6(x) = 8(2^x)$

x	-1	0	1	2	3
$g(x)$	7.5	8	9	11	15

x	-2	-1	0	1	2
$h(x)$	32	16	8	4	2

Create two different exponential functions of the forms $y = a(b)^x$ and $y = c^x + d$ with y-intercepts of $(0, -3)$.

5.1 EXERCISES

See www.CalcChat.com for worked-out solutions to odd-numbered exercises.

VOCABULARY: Fill in the blanks.

1. Polynomial and rational functions are examples of _____ functions.

2. Exponential and logarithmic functions are examples of nonalgebraic functions, also called _____ functions.

3. You can use the _____ Property to solve simple exponential equations.

4. The exponential function given by $f(x) = e^x$ is called the _____ _____ function, and the base e is called the _____ base.

5. To find the amount A in an account after t years with principal P and an annual interest rate r compounded n times per year, you can use the formula _____.

6. To find the amount A in an account after t years with principal P and an annual interest rate r compounded continuously, you can use the formula _____.

SKILLS AND APPLICATIONS

In Exercises 7–12, evaluate the function at the indicated value of x. Round your result to three decimal places.

Function	Value
7. $f(x) = 0.9^x$	$x = 1.4$
8. $f(x) = 2.3^x$	$x = \frac{3}{2}$
9. $f(x) = 5^x$	$x = -\pi$
10. $f(x) = \left(\frac{2}{3}\right)^{5x}$	$x = \frac{3}{10}$
11. $g(x) = 5000(2^x)$	$x = -1.5$
12. $f(x) = 200(1.2)^{12x}$	$x = 24$

In Exercises 13–16, match the exponential function with its graph. [The graphs are labeled (a), (b), (c), and (d).]

(a)
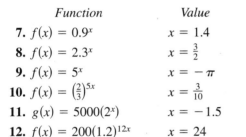

(b)
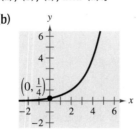

(c)
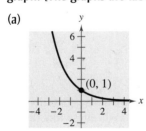

(d)
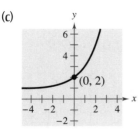

13. $f(x) = 2^x$

14. $f(x) = 2^x + 1$

15. $f(x) = 2^{-x}$

16. $f(x) = 2^{x-2}$

In Exercises 17–22, use a graphing utility to construct a table of values for the function. Then sketch the graph of the function.

17. $f(x) = \left(\frac{1}{2}\right)^x$ 18. $f(x) = \left(\frac{1}{2}\right)^{-x}$

19. $f(x) = 6^{-x}$ 20. $f(x) = 6^x$

21. $f(x) = 2^{x-1}$ 22. $f(x) = 4^{x-3} + 3$

In Exercises 23–28, use the graph of f to describe the transformation that yields the graph of g.

23. $f(x) = 3^x$, $g(x) = 3^x + 1$

24. $f(x) = 4^x$, $g(x) = 4^{x-3}$

25. $f(x) = 2^x$, $g(x) = 3 - 2^x$

26. $f(x) = 10^x$, $g(x) = 10^{-x+3}$

27. $f(x) = \left(\frac{7}{2}\right)^x$, $g(x) = -\left(\frac{7}{2}\right)^{-x}$

28. $f(x) = 0.3^x$, $g(x) = -0.3^x + 5$

In Exercises 29–32, use a graphing utility to graph the exponential function.

29. $y = 2^{-x^2}$ 30. $y = 3^{-|x|}$

31. $y = 3^{x-2} + 1$ 32. $y = 4^{x+1} - 2$

In Exercises 33–38, evaluate the function at the indicated value of x. Round your result to three decimal places.

Function	Value
33. $h(x) = e^{-x}$	$x = \frac{3}{4}$
34. $f(x) = e^x$	$x = 3.2$
35. $f(x) = 2e^{-5x}$	$x = 10$
36. $f(x) = 1.5e^{x/2}$	$x = 240$
37. $f(x) = 5000e^{0.06x}$	$x = 6$
38. $f(x) = 250e^{0.05x}$	$x = 20$

In Exercises 39–44, use a graphing utility to construct a table of values for the function. Then sketch the graph of the function.

39. $f(x) = e^x$

40. $f(x) = e^{-x}$

41. $f(x) = 3e^{x+4}$

42. $f(x) = 2e^{-0.5x}$

43. $f(x) = 2e^{x-2} + 4$

44. $f(x) = 2 + e^{x-5}$

In Exercises 45–50, use a graphing utility to graph the exponential function.

45. $y = 1.08^{-5x}$

46. $y = 1.08^{5x}$

47. $s(t) = 2e^{0.12t}$

48. $s(t) = 3e^{-0.2t}$

49. $g(x) = 1 + e^{-x}$

50. $h(x) = e^{x-2}$

In Exercises 51–58, use the One-to-One Property to solve the equation for x.

51. $3^{x+1} = 27$

52. $2^{x-3} = 16$

53. $\left(\frac{1}{2}\right)^x = 32$

54. $5^{x-2} = \frac{1}{125}$

55. $e^{3x+2} = e^3$

56. $e^{2x-1} = e^4$

57. $e^{x^2-3} = e^{2x}$

58. $e^{x^2+6} = e^{5x}$

COMPOUND INTEREST   In Exercises 59–62, complete the table to determine the balance A for P dollars invested at rate r for t years and compounded n times per year.

n	1	2	4	12	365	Continuous
A						

59. $P = \$1500, r = 2\%, t = 10$ years

60. $P = \$2500, r = 3.5\%, t = 10$ years

61. $P = \$2500, r = 4\%, t = 20$ years

62. $P = \$1000, r = 6\%, t = 40$ years

COMPOUND INTEREST   In Exercises 63–66, complete the table to determine the balance A for $\$12,000$ invested at rate r for t years, compounded continuously.

t	10	20	30	40	50
A					

63. $r = 4\%$

64. $r = 6\%$

65. $r = 6.5\%$

66. $r = 3.5\%$

67. TRUST FUND   On the day of a child's birth, a deposit of $30,000 is made in a trust fund that pays 5% interest, compounded continuously. Determine the balance in this account on the child's 25th birthday.

68. TRUST FUND   A deposit of $5000 is made in a trust fund that pays 7.5% interest, compounded continuously. It is specified that the balance will be given to the college from which the donor graduated after the money has earned interest for 50 years. How much will the college receive?

69. INFLATION   If the annual rate of inflation averages 4% over the next 10 years, the approximate costs C of goods or services during any year in that decade will be modeled by $C(t) = P(1.04)^t$, where t is the time in years and P is the present cost. The price of an oil change for your car is presently $23.95. Estimate the price 10 years from now.

70. COMPUTER VIRUS   The number V of computers infected by a computer virus increases according to the model $V(t) = 100e^{4.6052t}$, where t is the time in hours. Find the number of computers infected after (a) 1 hour, (b) 1.5 hours, and (c) 2 hours.

71. POPULATION GROWTH   The projected populations of California for the years 2015 through 2030 can be modeled by $P = 34.696e^{0.0098t}$, where P is the population (in millions) and t is the time (in years), with $t = 15$ corresponding to 2015.   (Source: U.S. Census Bureau)

(a) Use a graphing utility to graph the function for the years 2015 through 2030.

(b) Use the *table* feature of a graphing utility to create a table of values for the same time period as in part (a).

(c) According to the model, when will the population of California exceed 50 million?

72. POPULATION   The populations P (in millions) of Italy from 1990 through 2008 can be approximated by the model $P = 56.8e^{0.0015t}$, where t represents the year, with $t = 0$ corresponding to 1990.   (Source: U.S. Census Bureau, International Data Base)

(a) According to the model, is the population of Italy increasing or decreasing? Explain.

(b) Find the populations of Italy in 2000 and 2008.

(c) Use the model to predict the populations of Italy in 2015 and 2020.

73. RADIOACTIVE DECAY   Let Q represent a mass of radioactive plutonium (^{239}Pu) (in grams), whose half-life is 24,100 years. The quantity of plutonium present after t years is $Q = 16\left(\frac{1}{2}\right)^{t/24,100}$.

(a) Determine the initial quantity (when $t = 0$).

(b) Determine the quantity present after 75,000 years.

(c) Use a graphing utility to graph the function over the interval $t = 0$ to $t = 150,000$.

74. RADIOACTIVE DECAY   Let Q represent a mass of carbon 14 (^{14}C) (in grams), whose half-life is 5715 years. The quantity of carbon 14 present after t years is $Q = 10\left(\frac{1}{2}\right)^{t/5715}$.

(a) Determine the initial quantity (when $t = 0$).

(b) Determine the quantity present after 2000 years.

(c) Sketch the graph of this function over the interval $t = 0$ to $t = 10,000$.

75. DEPRECIATION   After t years, the value of a wheelchair conversion van that originally cost $30,500 depreciates so that each year it is worth $\frac{7}{8}$ of its value for the previous year.

(a) Find a model for $V(t)$, the value of the van after t years.

(b) Determine the value of the van 4 years after it was purchased.

76. DRUG CONCENTRATION   Immediately following an injection, the concentration of a drug in the bloodstream is 300 milligrams per milliliter. After t hours, the concentration is 75% of the level of the previous hour.

(a) Find a model for $C(t)$, the concentration of the drug after t hours.

(b) Determine the concentration of the drug after 8 hours.

EXPLORATION

TRUE OR FALSE?   In Exercises 77 and 78, determine whether the statement is true or false. Justify your answer.

77. The line $y = -2$ is an asymptote for the graph of $f(x) = 10^x - 2$.

78. $e = \dfrac{271{,}801}{99{,}990}$

THINK ABOUT IT   In Exercises 79–82, use properties of exponents to determine which functions (if any) are the same.

79. $f(x) = 3^{x-2}$
$g(x) = 3^x - 9$
$h(x) = \frac{1}{9}(3^x)$

80. $f(x) = 4^x + 12$
$g(x) = 2^{2x+6}$
$h(x) = 64(4^x)$

81. $f(x) = 16(4^{-x})$
$g(x) = \left(\frac{1}{4}\right)^{x-2}$
$h(x) = 16(2^{-2x})$

82. $f(x) = e^{-x} + 3$
$g(x) = e^{3-x}$
$h(x) = -e^{x-3}$

83. Graph the functions given by $y = 3^x$ and $y = 4^x$ and use the graphs to solve each inequality.

(a) $4^x < 3^x$ (b) $4^x > 3^x$

 84. Use a graphing utility to graph each function. Use the graph to find where the function is increasing and decreasing, and approximate any relative maximum or minimum values.

(a) $f(x) = x^2 e^{-x}$ (b) $g(x) = x2^{3-x}$

85. GRAPHICAL ANALYSIS   Use a graphing utility to graph $y_1 = (1 + 1/x)^x$ and $y_2 = e$ in the same viewing window. Using the *trace* feature, explain what happens to the graph of y_1 as x increases.

86. GRAPHICAL ANALYSIS   Use a graphing utility to graph

$$f(x) = \left(1 + \frac{0.5}{x}\right)^x \quad \text{and} \quad g(x) = e^{0.5}$$

in the same viewing window. What is the relationship between f and g as x increases and decreases without bound?

87. GRAPHICAL ANALYSIS   Use a graphing utility to graph each pair of functions in the same viewing window. Describe any similarities and differences in the graphs.

(a) $y_1 = 2^x,\ y_2 = x^2$ (b) $y_1 = 3^x,\ y_2 = x^3$

88. THINK ABOUT IT   Which functions are exponential?

(a) $3x$ (b) $3x^2$ (c) 3^x (d) 2^{-x}

89. COMPOUND INTEREST   Use the formula

$$A = P\left(1 + \frac{r}{n}\right)^{nt}$$

to calculate the balance of an account when $P = \$3000$, $r = 6\%$, and $t = 10$ years, and compounding is done (a) by the day, (b) by the hour, (c) by the minute, and (d) by the second. Does increasing the number of compoundings per year result in unlimited growth of the balance of the account? Explain.

90. CAPSTONE   The figure shows the graphs of $y = 2^x$, $y = e^x$, $y = 10^x$, $y = 2^{-x}$, $y = e^{-x}$, and $y = 10^{-x}$. Match each function with its graph. [The graphs are labeled (a) through (f).] Explain your reasoning.

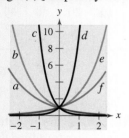

PROJECT: POPULATION PER SQUARE MILE   To work an extended application analyzing the population per square mile of the United States, visit this text's website at *academic.cengage.com*.   (Data Source: U.S. Census Bureau)

5.2 LOGARITHMIC FUNCTIONS AND THEIR GRAPHS

What you should learn

- Recognize and evaluate logarithmic functions with base a.
- Graph logarithmic functions.
- Recognize, evaluate, and graph natural logarithmic functions.
- Use logarithmic functions to model and solve real-life problems.

Why you should learn it

Logarithmic functions are often used to model scientific observations. For instance, in Exercise 97 on page 400, a logarithmic function is used to model human memory.

© Ariel Skelley/Corbis

Logarithmic Functions

In Section 2.7, you studied the concept of an inverse function. There, you learned that if a function is one-to-one—that is, if the function has the property that no horizontal line intersects the graph of the function more than once—the function must have an inverse function. By looking back at the graphs of the exponential functions introduced in Section 5.1, you will see that every function of the form $f(x) = a^x$ passes the Horizontal Line Test and therefore must have an inverse function. This inverse function is called the **logarithmic function with base** a.

> **Definition of Logarithmic Function with Base a**
>
> For $x > 0$, $a > 0$, and $a \neq 1$,
>
> $$y = \log_a x \text{ if and only if } x = a^y.$$
>
> The function given by
>
> $$f(x) = \log_a x$$
>
> is called the **logarithmic function with base** a.

The equations

$$y = \log_a x \qquad \text{and} \qquad x = a^y$$

are equivalent. The first equation is in logarithmic form and the second is in exponential form. For example, the logarithmic equation $2 = \log_3 9$ can be rewritten in exponential form as $9 = 3^2$. The exponential equation $5^3 = 125$ can be rewritten in logarithmic form as $\log_5 125 = 3$.

When evaluating logarithms, remember that *a logarithm is an exponent*. This means that $\log_a x$ is the exponent to which a must be raised to obtain x. For instance, $\log_2 8 = 3$ because 2 must be raised to the third power to get 8.

Example 1   Evaluating Logarithms

Use the definition of logarithmic function to evaluate each logarithm at the indicated value of x.

a. $f(x) = \log_2 x, \quad x = 32$ **b.** $f(x) = \log_3 x, \quad x = 1$
c. $f(x) = \log_4 x, \quad x = 2$ **d.** $f(x) = \log_{10} x, \quad x = \frac{1}{100}$

Solution

a. $f(\ ) = \log_2 \ = 5$ because $2^5 = 32$.
b. $f(\ ) = \log_3 \ = 0$ because $3^0 = 1$.
c. $f(\ ) = \log_4 \ = \frac{1}{2}$ because $4^{1/2} = \sqrt{4} = 2$.
d. $f(\ ) = \log_{10} \ = -2$ because $10^{-2} = \frac{1}{10^2} = \frac{1}{100}$.

CHECK **Point**   Now try Exercise 23.

The logarithmic function with base 10 is called the **common logarithmic function.** It is denoted by $\log_{10}$ or simply by log. On most calculators, this function is denoted by (LOG). Example 2 shows how to use a calculator to evaluate common logarithmic functions. You will learn how to use a calculator to calculate logarithms to any base in the next section.

| **Example 2** | **Evaluating Common Logarithms on a Calculator** |

Use a calculator to evaluate the function given by $f(x) = \log x$ at each value of x.

a. $x = 10$ **b.** $x = \frac{1}{3}$ **c.** $x = 2.5$ **d.** $x = -2$

Solution

Function Value	Graphing Calculator Keystrokes	Display
a. $f(\ ) = \log$	(LOG) 10 (ENTER)	1
b. $f(\ ) = \log$	(LOG) (() 1 (÷) 3 ()) (ENTER)	-0.4771213
c. $f(\ ) = \log$	(LOG) 2.5 (ENTER)	0.3979400
d. $f(\ ) = \log(\ )$	(LOG) (−) 2 (ENTER)	ERROR

Note that the calculator displays an error message (or a complex number) when you try to evaluate $\log(-2)$. The reason for this is that there is no real number power to which 10 can be raised to obtain -2.

CHECK*Point*   Now try Exercise 29.

The following properties follow directly from the definition of the logarithmic function with base a.

Properties of Logarithms

1. $\log_a 1 = 0$ because $a^0 = 1$.

2. $\log_a a = 1$ because $a^1 = a$.

3. $\log_a a^x = x$ and $a^{\log_a x} = x$

4. If $\log_a x = \log_a y$, then $x = y$.

| **Example 3** | **Using Properties of Logarithms** |

a. Simplify: $\log_4 1$ **b.** Simplify: $\log_{\sqrt{7}} \sqrt{7}$ **c.** Simplify: $6^{\log_6 20}$

Solution

a. Using Property 1, it follows that $\log_4 1 = 0$.

b. Using Property 2, you can conclude that $\log_{\sqrt{7}} \sqrt{7} = 1$.

c. Using the Inverse Property (Property 3), it follows that $6^{\log_6 20} = 20$.

CHECK*Point*   Now try Exercise 33.

You can use the One-to-One Property (Property 4) to solve simple logarithmic equations, as shown in Example 4.

Example 4   Using the One-to-One Property

a. $\log_3 x = \log_3 12$

$\qquad x = 12$

b. $\log(2x + 1) = \log 3x \implies 2x + 1 = 3x \implies 1 = x$

c. $\log_4(x^2 - 6) = \log_4 10 \implies x^2 - 6 = 10 \implies x^2 = 16 \implies x = \pm 4$

CHECK**Point**   Now try Exercise 85.

Graphs of Logarithmic Functions

To sketch the graph of $y = \log_a x$, you can use the fact that the graphs of inverse functions are reflections of each other in the line $y = x$.

Example 5   Graphs of Exponential and Logarithmic Functions

In the same coordinate plane, sketch the graph of each function.

a. $f(x) = 2^x$ **b.** $g(x) = \log_2 x$

Solution

a. For $f(x) = 2^x$, construct a table of values. By plotting these points and connecting them with a smooth curve, you obtain the graph shown in Figure 5.14.

x	-2	-1	0	1	2	3
$f(x) = 2^x$	$\frac{1}{4}$	$\frac{1}{2}$	1	2	4	8

b. Because $g(x) = \log_2 x$ is the inverse function of $f(x) = 2^x$, the graph of g is obtained by plotting the points $(f(x), x)$ and connecting them with a smooth curve. The graph of g is a reflection of the graph of f in the line $y = x$, as shown in Figure 5.14.

CHECK**Point**   Now try Exercise 37.

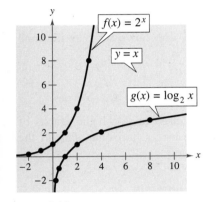

FIGURE 5.14

Example 6   Sketching the Graph of a Logarithmic Function

Sketch the graph of the common logarithmic function $f(x) = \log x$. Identify the vertical asymptote.

Solution

Begin by constructing a table of values. Note that some of the values can be obtained without a calculator by using the Inverse Property of Logarithms. Others require a calculator. Next, plot the points and connect them with a smooth curve, as shown in Figure 5.15. The vertical asymptote is $x = 0$ (y-axis).

FIGURE 5.15

x	Without calculator				With calculator		
	$\frac{1}{100}$	$\frac{1}{10}$	1	10	2	5	8
$f(x) = \log x$	-2	-1	0	1	0.301	0.699	0.903

CHECK**Point**   Now try Exercise 43.

The nature of the graph in Figure 5.15 is typical of functions of the form $f(x) = \log_a x, a > 1$. They have one x-intercept and one vertical asymptote. Notice how slowly the graph rises for $x > 1$. The basic characteristics of logarithmic graphs are summarized in Figure 5.16.

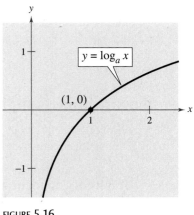

FIGURE 5.16

Graph of $y = \log_a x, a > 1$
- Domain: $(0, \infty)$
- Range: $(-\infty, \infty)$
- x-intercept: $(1, 0)$
- Increasing
- One-to-one, therefore has an inverse function
- y-axis is a vertical asymptote $(\log_a x \to -\infty$ as $x \to 0^+)$.
- Continuous
- Reflection of graph of $y = a^x$ about the line $y = x$

The basic characteristics of the graph of $f(x) = a^x$ are shown below to illustrate the inverse relation between $f(x) = a^x$ and $g(x) = \log_a x$.

- Domain: $(-\infty, \infty)$ • Range: $(0, \infty)$
- y-intercept: $(0, 1)$ • x-axis is a horizontal asymptote $(a^x \to 0$ as $x \to -\infty)$.

In the next example, the graph of $y = \log_a x$ is used to sketch the graphs of functions of the form $f(x) = b \pm \log_a(x + c)$. Notice how a horizontal shift of the graph results in a horizontal shift of the vertical asymptote.

Example 7 **Shifting Graphs of Logarithmic Functions**

The graph of each of the functions is similar to the graph of $f(x) = \log x$.

a. Because $g(x) = \log(x - 1) = f(x - 1)$, the graph of g can be obtained by shifting the graph of f one unit to the right, as shown in Figure 5.17.

b. Because $h(x) = 2 + \log x = 2 + f(x)$, the graph of h can be obtained by shifting the graph of f two units upward, as shown in Figure 5.18.

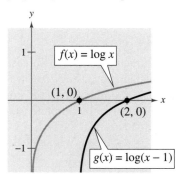

FIGURE 5.17

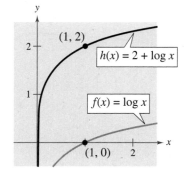

FIGURE 5.18

CHECK*Point* Now try Exercise 45.

The Natural Logarithmic Function

By looking back at the graph of the natural exponential function introduced on page 384 in Section 5.1, you will see that $f(x) = e^x$ is one-to-one and so has an inverse function. This inverse function is called the **natural logarithmic function** and is denoted by the special symbol $\ln x$, read as "the natural log of x" or "el en of x." Note that the natural logarithm is written without a base. The base is understood to be e.

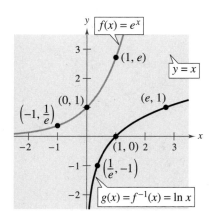

Reflection of graph of $f(x) = e^x$ about the line $y = x$

FIGURE 5.19

The Natural Logarithmic Function

The function defined by

$$f(x) = \log_e x = \ln x, \quad x > 0$$

is called the **natural logarithmic function.**

The definition above implies that the natural logarithmic function and the natural exponential function are inverse functions of each other. So, every logarithmic equation can be written in an equivalent exponential form, and every exponential equation can be written in logarithmic form. That is, $y = \ln x$ and $x = e^y$ are equivalent equations.

Because the functions given by $f(x) = e^x$ and $g(x) = \ln x$ are inverse functions of each other, their graphs are reflections of each other in the line $y = x$. This reflective property is illustrated in Figure 5.19.

On most calculators, the natural logarithm is denoted by $\boxed{\text{LN}}$, as illustrated in Example 8.

Example 8   Evaluating the Natural Logarithmic Function

Use a calculator to evaluate the function given by $f(x) = \ln x$ for each value of x.

a. $x = 2$

b. $x = 0.3$

c. $x = -1$

d. $x = 1 + \sqrt{2}$

Solution

	Function Value	Graphing Calculator Keystrokes	Display
a.	$f(\ ) = \ln$	$\boxed{\text{LN}}$ 2 $\boxed{\text{ENTER}}$	0.6931472
b.	$f(\ ) = \ln$	$\boxed{\text{LN}}$ .3 $\boxed{\text{ENTER}}$	−1.2039728
c.	$f(\ ) = \ln(\ )$	$\boxed{\text{LN}}$ $\boxed{(-)}$ 1 $\boxed{\text{ENTER}}$	ERROR
d.	$f(\quad) = \ln(\quad)$	$\boxed{\text{LN}}$ $\boxed{(}$ 1 $\boxed{+}$ $\boxed{\sqrt{\ }}$ 2 $\boxed{)}$ $\boxed{\text{ENTER}}$	0.8813736

CHECK **Point**   Now try Exercise 67.

In Example 8, be sure you see that $\ln(-1)$ gives an error message on most calculators. (Some calculators may display a complex number.) This occurs because the domain of $\ln x$ is the set of positive real numbers (see Figure 5.19). So, $\ln(-1)$ is undefined.

The four properties of logarithms listed on page 392 are also valid for natural logarithms.

⚠ **WARNING / CAUTION**

Notice that as with every other logarithmic function, the domain of the natural logarithmic function is the set of *positive real numbers*—be sure you see that $\ln x$ is not defined for zero or for negative numbers.

Properties of Natural Logarithms

1. $\ln 1 = 0$ because $e^0 = 1$.

2. $\ln e = 1$ because $e^1 = e$.

3. $\ln e^x = x$ and $e^{\ln x} = x$

4. If $\ln x = \ln y$, then $x = y$.

Example 9   Using Properties of Natural Logarithms

Use the properties of natural logarithms to simplify each expression.

a. $\ln \dfrac{1}{e}$ **b.** $e^{\ln 5}$ **c.** $\dfrac{\ln 1}{3}$ **d.** $2 \ln e$

Solution

a. $\ln \dfrac{1}{e} = \ln e^{-1} = -1$ **b.** $e^{\ln 5} = 5$

c. $\dfrac{\ln 1}{3} = \dfrac{0}{3} = 0$ **d.** $2 \ln e = 2(1) = 2$

✔**CHECK Point** Now try Exercise 71.

Example 10   Finding the Domains of Logarithmic Functions

Find the domain of each function.

a. $f(x) = \ln(x - 2)$ **b.** $g(x) = \ln(2 - x)$ **c.** $h(x) = \ln x^2$

Solution

a. Because $\ln(x - 2)$ is defined only if $x - 2 > 0$, it follows that the domain of f is $(2, \infty)$. The graph of f is shown in Figure 5.20.

b. Because $\ln(2 - x)$ is defined only if $2 - x > 0$, it follows that the domain of g is $(-\infty, 2)$. The graph of g is shown in Figure 5.21.

c. Because $\ln x^2$ is defined only if $x^2 > 0$, it follows that the domain of h is all real numbers except $x = 0$. The graph of h is shown in Figure 5.22.

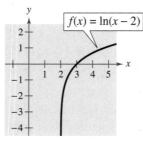

FIGURE 5.20

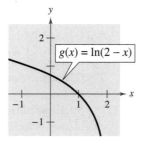

FIGURE 5.21

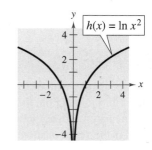

FIGURE 5.22

✔**CHECK Point** Now try Exercise 75.

Application

Example 11   Human Memory Model

Students participating in a psychology experiment attended several lectures on a subject and were given an exam. Every month for a year after the exam, the students were retested to see how much of the material they remembered. The average scores for the group are given by the *human memory model* $f(t) = 75 - 6\ln(t + 1)$, $0 \le t \le 12$, where t is the time in months.

a. What was the average score on the original ($t = 0$) exam?

b. What was the average score at the end of $t = 2$ months?

c. What was the average score at the end of $t = 6$ months?

Algebraic Solution

a. The original average score was

$$f(\ ) = 75 - 6\ln(\ + 1)$$

$$= 75 - 6\ln 1$$

$$= 75 - 6(0)$$

$$= 75.$$

b. After 2 months, the average score was

$$f(\ ) = 75 - 6\ln(\ + 1)$$

$$= 75 - 6\ln 3$$

$$\approx 75 - 6(1.0986)$$

$$\approx 68.4.$$

c. After 6 months, the average score was

$$f(\ ) = 75 - 6\ln(\ + 1)$$

$$= 75 - 6\ln 7$$

$$\approx 75 - 6(1.9459)$$

$$\approx 63.3.$$

Graphical Solution

Use a graphing utility to graph the model $y = 75 - 6\ln(x + 1)$. Then use the *value* or *trace* feature to approximate the following.

a. When $x = 0$, $y = 75$ (see Figure 5.23). So, the original average score was 75.

b. When $x = 2$, $y \approx 68.4$ (see Figure 5.24). So, the average score after 2 months was about 68.4.

c. When $x = 6$, $y \approx 63.3$ (see Figure 5.25). So, the average score after 6 months was about 63.3.

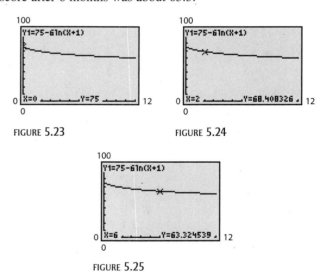

FIGURE **5.23**   FIGURE **5.24**

FIGURE **5.25**

CHECK*Point*   Now try Exercise 97.

CLASSROOM DISCUSSION

Analyzing a Human Memory Model   Use a graphing utility to determine the time in months when the average score in Example 11 was 60. Explain your method of solving the problem. Describe another way that you can use a graphing utility to determine the answer.

5.2 EXERCISES

See www.CalcChat.com for worked-out solutions to odd-numbered exercises.

VOCABULARY: Fill in the blanks.

1. The inverse function of the exponential function given by $f(x) = a^x$ is called the _____ function with base a.
2. The common logarithmic function has base _____ .
3. The logarithmic function given by $f(x) = \ln x$ is called the _____ logarithmic function and has base _____ .
4. The Inverse Properties of logarithms and exponentials state that $\log_a a^x = x$ and _____ .
5. The One-to-One Property of natural logarithms states that if $\ln x = \ln y$, then _____ .
6. The domain of the natural logarithmic function is the set of _____ _____ _____ .

SKILLS AND APPLICATIONS

In Exercises 7–14, write the logarithmic equation in exponential form. For example, the exponential form of $\log_5 25 = 2$ is $5^2 = 25$.

7. $\log_4 16 = 2$
8. $\log_7 343 = 3$
9. $\log_9 \frac{1}{81} = -2$
10. $\log \frac{1}{1000} = -3$
11. $\log_{32} 4 = \frac{2}{5}$
12. $\log_{16} 8 = \frac{3}{4}$
13. $\log_{64} 8 = \frac{1}{2}$
14. $\log_8 4 = \frac{2}{3}$

In Exercises 15–22, write the exponential equation in logarithmic form. For example, the logarithmic form of $2^3 = 8$ is $\log_2 8 = 3$.

15. $5^3 = 125$
16. $13^2 = 169$
17. $81^{1/4} = 3$
18. $9^{3/2} = 27$
19. $6^{-2} = \frac{1}{36}$
20. $4^{-3} = \frac{1}{64}$
21. $24^0 = 1$
22. $10^{-3} = 0.001$

In Exercises 23–28, evaluate the function at the indicated value of x without using a calculator.

Function	Value
23. $f(x) = \log_2 x$	$x = 64$
24. $f(x) = \log_{25} x$	$x = 5$
25. $f(x) = \log_8 x$	$x = 1$
26. $f(x) = \log x$	$x = 10$
27. $g(x) = \log_a x$	$x = a^2$
28. $g(x) = \log_b x$	$x = b^{-3}$

In Exercises 29–32, use a calculator to evaluate $f(x) = \log x$ at the indicated value of x. Round your result to three decimal places.

29. $x = \frac{7}{8}$
30. $x = \frac{1}{500}$
31. $x = 12.5$
32. $x = 96.75$

In Exercises 33–36, use the properties of logarithms to simplify the expression.

33. $\log_{11} 11^7$
34. $\log_{3.2} 1$
35. $\log_\pi \pi$
36. $9^{\log_9 15}$

In Exercises 37–44, find the domain, x-intercept, and vertical asymptote of the logarithmic function and sketch its graph.

37. $f(x) = \log_4 x$
38. $g(x) = \log_6 x$
39. $y = -\log_3 x + 2$
40. $h(x) = \log_4(x - 3)$
41. $f(x) = -\log_6(x + 2)$
42. $y = \log_5(x - 1) + 4$
43. $y = \log\left(\dfrac{x}{7}\right)$
44. $y = \log(-x)$

In Exercises 45–50, use the graph of $g(x) = \log_3 x$ to match the given function with its graph. Then describe the relationship between the graphs of f and g. [The graphs are labeled (a), (b), (c), (d), (e), and (f).]

(a)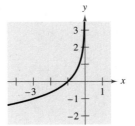

(b)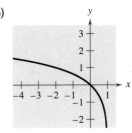

(c)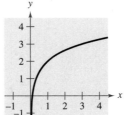

(d)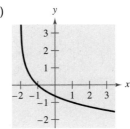

(e)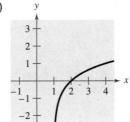

(f)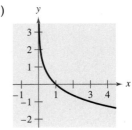

45. $f(x) = \log_3 x + 2$ **46.** $f(x) = -\log_3 x$

47. $f(x) = -\log_3(x + 2)$ **48.** $f(x) = \log_3(x - 1)$

49. $f(x) = \log_3(1 - x)$ **50.** $f(x) = -\log_3(-x)$

In Exercises 51–58, write the logarithmic equation in exponential form.

51. $\ln \frac{1}{2} = -0.693 \ldots$ **52.** $\ln \frac{2}{5} = -0.916 \ldots$

53. $\ln 7 = 1.945 \ldots$ **54.** $\ln 10 = 2.302 \ldots$

55. $\ln 250 = 5.521 \ldots$ **56.** $\ln 1084 = 6.988 \ldots$

57. $\ln 1 = 0$ **58.** $\ln e = 1$

In Exercises 59–66, write the exponential equation in logarithmic form.

59. $e^4 = 54.598 \ldots$ **60.** $e^2 = 7.3890 \ldots$

61. $e^{1/2} = 1.6487 \ldots$ **62.** $e^{1/3} = 1.3956 \ldots$

63. $e^{-0.9} = 0.406 \ldots$ **64.** $e^{-4.1} = 0.0165 \ldots$

65. $e^x = 4$ **66.** $e^{2x} = 3$

In Exercises 67–70, use a calculator to evaluate the function at the indicated value of x. Round your result to three decimal places.

Function	*Value*
67. $f(x) = \ln x$	$x = 18.42$
68. $f(x) = 3 \ln x$	$x = 0.74$
69. $g(x) = 8 \ln x$	$x = 0.05$
70. $g(x) = -\ln x$	$x = \frac{1}{2}$

In Exercises 71–74, evaluate $g(x) = \ln x$ at the indicated value of x without using a calculator.

71. $x = e^5$ **72.** $x = e^{-4}$

73. $x = e^{-5/6}$ **74.** $x = e^{-5/2}$

In Exercises 75–78, find the domain, x-intercept, and vertical asymptote of the logarithmic function and sketch its graph.

75. $f(x) = \ln(x - 4)$ **76.** $h(x) = \ln(x + 5)$

77. $g(x) = \ln(-x)$ **78.** $f(x) = \ln(3 - x)$

In Exercises 79–84, use a graphing utility to graph the function. Be sure to use an appropriate viewing window.

79. $f(x) = \log(x + 9)$ **80.** $f(x) = \log(x - 6)$

81. $f(x) = \ln(x - 1)$ **82.** $f(x) = \ln(x + 2)$

83. $f(x) = \ln x + 8$ **84.** $f(x) = 3 \ln x - 1$

In Exercises 85–92, use the One-to-One Property to solve the equation for x.

85. $\log_5(x + 1) = \log_5 6$ **86.** $\log_2(x - 3) = \log_2 9$

87. $\log(2x + 1) = \log 15$ **88.** $\log(5x + 3) = \log 12$

89. $\ln(x + 4) = \ln 12$ **90.** $\ln(x - 7) = \ln 7$

91. $\ln(x^2 - 2) = \ln 23$ **92.** $\ln(x^2 - x) = \ln 6$

93. MONTHLY PAYMENT   The model

$$t = 16.625 \ln\left(\frac{x}{x - 750}\right), \quad x > 750$$

approximates the length of a home mortgage of \$150,000 at 6% in terms of the monthly payment. In the model, t is the length of the mortgage in years and x is the monthly payment in dollars.

(a) Use the model to approximate the lengths of a \$150,000 mortgage at 6% when the monthly payment is \$897.72 and when the monthly payment is \$1659.24.

(b) Approximate the total amounts paid over the term of the mortgage with a monthly payment of \$897.72 and with a monthly payment of \$1659.24.

(c) Approximate the total interest charges for a monthly payment of \$897.72 and for a monthly payment of \$1659.24.

(d) What is the vertical asymptote for the model? Interpret its meaning in the context of the problem.

94. COMPOUND INTEREST   A principal P, invested at $5\frac{1}{2}\%$ and compounded continuously, increases to an amount K times the original principal after t years, where t is given by $t = (\ln K)/0.055$.

(a) Complete the table and interpret your results.

K	1	2	4	6	8	10	12
t							

(b) Sketch a graph of the function.

95. CABLE TELEVISION   The numbers of cable television systems C (in thousands) in the United States from 2001 through 2006 can be approximated by the model

$$C = 10.355 - 0.298t \ln t, \quad 1 \le t \le 6$$

where t represents the year, with $t = 1$ corresponding to 2001.   (Source: Warren Communication News)

(a) Complete the table.

t	1	2	3	4	5	6
C						

(b) Use a graphing utility to graph the function.

(c) Can the model be used to predict the numbers of cable television systems beyond 2006? Explain.

96. POPULATION The time t in years for the world population to double if it is increasing at a continuous rate of r is given by $t = (\ln 2)/r$.

(a) Complete the table and interpret your results.

r	0.005	0.010	0.015	0.020	0.025	0.030
t						

(b) Use a graphing utility to graph the function.

97. HUMAN MEMORY MODEL Students in a mathematics class were given an exam and then retested monthly with an equivalent exam. The average scores for the class are given by the human memory model $f(t) = 80 - 17\log(t + 1), 0 \le t \le 12$, where t is the time in months.

(a) Use a graphing utility to graph the model over the specified domain.

(b) What was the average score on the original exam $(t = 0)$?

(c) What was the average score after 4 months?

(d) What was the average score after 10 months?

98. SOUND INTENSITY The relationship between the number of decibels β and the intensity of a sound I in watts per square meter is

$$\beta = 10\log\left(\frac{I}{10^{-12}}\right).$$

(a) Determine the number of decibels of a sound with an intensity of 1 watt per square meter.

(b) Determine the number of decibels of a sound with an intensity of 10^{-2} watt per square meter.

(c) The intensity of the sound in part (a) is 100 times as great as that in part (b). Is the number of decibels 100 times as great? Explain.

EXPLORATION

TRUE OR FALSE? In Exercises 99 and 100, determine whether the statement is true or false. Justify your answer.

99. You can determine the graph of $f(x) = \log_6 x$ by graphing $g(x) = 6^x$ and reflecting it about the x-axis.

100. The graph of $f(x) = \log_3 x$ contains the point $(27, 3)$.

In Exercises 101–104, sketch the graphs of f and g and describe the relationship between the graphs of f and g. What is the relationship between the functions f and g?

101. $f(x) = 3^x, \quad g(x) = \log_3 x$

102. $f(x) = 5^x, \quad g(x) = \log_5 x$

103. $f(x) = e^x, \quad g(x) = \ln x$

104. $f(x) = 8^x, \quad g(x) = \log_8 x$

105. THINK ABOUT IT Complete the table for $f(x) = 10^x$.

x	-2	-1	0	1	2
$f(x)$					

Complete the table for $f(x) = \log x$.

x	$\frac{1}{100}$	$\frac{1}{10}$	1	10	100
$f(x)$					

Compare the two tables. What is the relationship between $f(x) = 10^x$ and $f(x) = \log x$?

106. GRAPHICAL ANALYSIS Use a graphing utility to graph f and g in the same viewing window and determine which is increasing at the greater rate as x approaches $+\infty$. What can you conclude about the rate of growth of the natural logarithmic function?

(a) $f(x) = \ln x, \quad g(x) = \sqrt{x}$

(b) $f(x) = \ln x, \quad g(x) = \sqrt[4]{x}$

107. (a) Complete the table for the function given by $f(x) = (\ln x)/x$.

x	1	5	10	10^2	10^4	10^6
$f(x)$						

(b) Use the table in part (a) to determine what value $f(x)$ approaches as x increases without bound.

(c) Use a graphing utility to confirm the result of part (b).

108. CAPSTONE The table of values was obtained by evaluating a function. Determine which of the statements may be true and which must be false.

x	y
1	0
2	1
8	3

(a) y is an exponential function of x.

(b) y is a logarithmic function of x.

(c) x is an exponential function of y.

(d) y is a linear function of x.

109. WRITING Explain why $\log_a x$ is defined only for $0 < a < 1$ and $a > 1$.

In Exercises 110 and 111, (a) use a graphing utility to graph the function, (b) use the graph to determine the intervals in which the function is increasing and decreasing, and (c) approximate any relative maximum or minimum values of the function.

110. $f(x) = |\ln x|$

111. $h(x) = \ln(x^2 + 1)$

5.3   PROPERTIES OF LOGARITHMS

Dynamic Graphics/ Jupiter Images

Change of Base

Most calculators have only two types of log keys, one for common logarithms (base 10) and one for natural logarithms (base e). Although common logarithms and natural logarithms are the most frequently used, you may occasionally need to evaluate logarithms with other bases. To do this, you can use the following **change-of-base formula.**

> **Change-of-Base Formula**
>
> Let a, b, and x be positive real numbers such that $a \neq 1$ and $b \neq 1$. Then $\log_a x$ can be converted to a different base as follows.
>
> | *Base b* | *Base 10* | *Base e* |
> |---|---|---|
> | $\log_a x = \dfrac{\log_b x}{\log_b a}$ | $\log_a x = \dfrac{\log x}{\log a}$ | $\log_a x = \dfrac{\ln x}{\ln a}$ |

One way to look at the change-of-base formula is that logarithms with base a are simply *constant multiples* of logarithms with base b. The constant multiplier is $1/(\log_b a)$.

Example 1   Changing Bases Using Common Logarithms

a. $\log_4 25 = \dfrac{\log 25}{\log 4}$

$\approx \dfrac{1.39794}{0.60206}$

≈ 2.3219

b. $\log_2 12 = \dfrac{\log 12}{\log 2} \approx \dfrac{1.07918}{0.30103} \approx 3.5850$

CHECK *Point*   Now try Exercise 7(a).

Example 2   Changing Bases Using Natural Logarithms

a. $\log_4 25 = \dfrac{\ln 25}{\ln 4}$

$\approx \dfrac{3.21888}{1.38629}$

≈ 2.3219

b. $\log_2 12 = \dfrac{\ln 12}{\ln 2} \approx \dfrac{2.48491}{0.69315} \approx 3.5850$

CHECK *Point*   Now try Exercise 7(b).

Properties of Logarithms

You know from the preceding section that the logarithmic function with base a is the *inverse function* of the exponential function with base a. So, it makes sense that the properties of exponents should have corresponding properties involving logarithms. For instance, the exponential property $a^0 = 1$ has the corresponding logarithmic property $\log_a 1 = 0$.

<table>
<tr><td colspan="3">

Properties of Logarithms

Let a be a positive number such that $a \neq 1$, and let n be a real number. If u and v are positive real numbers, the following properties are true.

</td></tr>
<tr><td></td><td align="center">*Logarithm with Base a*</td><td align="center">*Natural Logarithm*</td></tr>
<tr><td>**1. Product Property:**</td><td>$\log_a(uv) = \log_a u + \log_a v$</td><td>$\ln(uv) = \ln u + \ln v$</td></tr>
<tr><td>**2. Quotient Property:**</td><td>$\log_a \dfrac{u}{v} = \log_a u - \log_a v$</td><td>$\ln \dfrac{u}{v} = \ln u - \ln v$</td></tr>
<tr><td>**3. Power Property:**</td><td>$\log_a u^n = n \log_a u$</td><td>$\ln u^n = n \ln u$</td></tr>
</table>

For proofs of the properties listed above, see Proofs in Mathematics on page 440.

> ⚠️ **WARNING / CAUTION**
>
> There is no general property that can be used to rewrite $\log_a(u \pm v)$. Specifically, $\log_a(u + v)$ is *not* equal to $\log_a u + \log_a v$.

Example 3   Using Properties of Logarithms

Write each logarithm in terms of $\ln 2$ and $\ln 3$.

a. $\ln 6$ **b.** $\ln \dfrac{2}{27}$

Solution

a. $\ln 6 = \ln(2 \cdot 3)$

$\qquad = \ln 2 + \ln 3$

b. $\ln \dfrac{2}{27} = \ln 2 - \ln 27$

$\qquad\qquad = \ln 2 - \ln 3^3$

$\qquad\qquad = \ln 2 - 3 \ln 3$

CHECK**Point** Now try Exercise 27.

Example 4   Using Properties of Logarithms

Find the exact value of each expression without using a calculator.

a. $\log_5 \sqrt[3]{5}$ **b.** $\ln e^6 - \ln e^2$

Solution

a. $\log_5 \sqrt[3]{5} = \log_5 5^{1/3} = \frac{1}{3} \log_5 5 = \frac{1}{3}(1) = \frac{1}{3}$

b. $\ln e^6 - \ln e^2 = \ln \dfrac{e^6}{e^2} = \ln e^4 = 4 \ln e = 4(1) = 4$

CHECK**Point** Now try Exercise 29.

Rewriting Logarithmic Expressions

The properties of logarithms are useful for rewriting logarithmic expressions in forms that simplify the operations of algebra. This is true because these properties convert complicated products, quotients, and exponential forms into simpler sums, differences, and products, respectively.

Example 5   Expanding Logarithmic Expressions

Expand each logarithmic expression.

a. $\log_4 5x^3y$ **b.** $\ln \dfrac{\sqrt{3x-5}}{7}$

Solution

a. $\log_4 5x^3y = \log_4 5 + \log_4 x^3 + \log_4 y$

$\qquad\qquad = \log_4 5 + 3\log_4 x + \log_4 y$

b. $\ln \dfrac{\sqrt{3x-5}}{7} = \ln \dfrac{(3x-5)^{1/2}}{7}$

$\qquad\qquad\quad = \ln(3x-5)^{1/2} - \ln 7$

$\qquad\qquad\quad = \dfrac{1}{2}\ln(3x-5) - \ln 7$

CHECK**Point**   Now try Exercise 53.

In Example 5, the properties of logarithms were used to *expand* logarithmic expressions. In Example 6, this procedure is reversed and the properties of logarithms are used to *condense* logarithmic expressions.

Example 6   Condensing Logarithmic Expressions

Condense each logarithmic expression.

a. $\frac{1}{2}\log x + 3\log(x+1)$ **b.** $2\ln(x+2) - \ln x$

c. $\frac{1}{3}[\log_2 x + \log_2(x+1)]$

Solution

a. $\frac{1}{2}\log x + 3\log(x+1) = \log x^{1/2} + \log(x+1)^3$

$\qquad\qquad\qquad\qquad = \log\left[\sqrt{x}(x+1)^3\right]$

b. $2\ln(x+2) - \ln x = \ln(x+2)^2 - \ln x$

$\qquad\qquad\qquad = \ln\dfrac{(x+2)^2}{x}$

c. $\frac{1}{3}[\log_2 x + \log_2(x+1)] = \frac{1}{3}\{\log_2[x(x+1)]\}$

$\qquad\qquad\qquad\qquad = \log_2[x(x+1)]^{1/3}$

$\qquad\qquad\qquad\qquad = \log_2 \sqrt[3]{x(x+1)}$

CHECK**Point**   Now try Exercise 75.

Algebra Help

You can review rewriting radicals and rational exponents in Section P.2.

Application

One method of determining how the x- and y-values for a set of nonlinear data are related is to take the natural logarithm of each of the x- and y-values. If the points are graphed and fall on a line, then you can determine that the x- and y-values are related by the equation

$$\ln y = m \ln x$$

where m is the slope of the line.

Example 7   Finding a Mathematical Model

The table shows the mean distance from the sun x and the period y (the time it takes a planet to orbit the sun) for each of the six planets that are closest to the sun. In the table, the mean distance is given in terms of astronomical units (where Earth's mean distance is defined as 1.0), and the period is given in years. Find an equation that relates y and x.

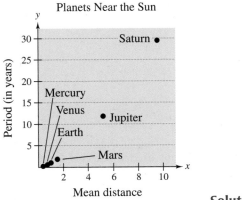

Planets Near the Sun

FIGURE 5.26

Planet	Mean distance, x	Period, y
Mercury	0.387	0.241
Venus	0.723	0.615
Earth	1.000	1.000
Mars	1.524	1.881
Jupiter	5.203	11.860
Saturn	9.537	29.460

Solution

The points in the table above are plotted in Figure 5.26. From this figure it is not clear how to find an equation that relates y and x. To solve this problem, take the natural logarithm of each of the x- and y-values in the table. This produces the following results.

Planet	Mercury	Venus	Earth	Mars	Jupiter	Saturn
$\ln x$	-0.949	-0.324	0.000	0.421	1.649	2.255
$\ln y$	-1.423	-0.486	0.000	0.632	2.473	3.383

Now, by plotting the points in the second table, you can see that all six of the points appear to lie in a line (see Figure 5.27). Choose any two points to determine the slope of the line. Using the two points $(0.421, 0.632)$ and $(0, 0)$, you can determine that the slope of the line is

$$m = \frac{0.632 - 0}{0.421 - 0} \approx 1.5 = \frac{3}{2}.$$

By the point-slope form, the equation of the line is $Y = \frac{3}{2}X$, where $Y = \ln y$ and $X = \ln x$. You can therefore conclude that $\ln y = \frac{3}{2}\ln x$.

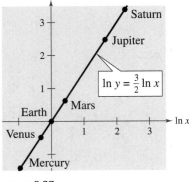

FIGURE 5.27

CHECK **Point**   Now try Exercise 91.

5.3   EXERCISES

See www.CalcChat.com for worked-out solutions to odd-numbered exercises.

VOCABULARY

In Exercises 1–3, fill in the blanks.

1. To evaluate a logarithm to any base, you can use the _____ formula.

2. The change-of-base formula for base e is given by $\log_a x =$ _____.

3. You can consider $\log_a x$ to be a constant multiple of $\log_b x$; the constant multiplier is _____.

In Exercises 4–6, match the property of logarithms with its name.

4. $\log_a(uv) = \log_a u + \log_a v$

5. $\ln u^n = n \ln u$

6. $\log_a \dfrac{u}{v} = \log_a u - \log_a v$

(a) Power Property

(b) Quotient Property

(c) Product Property

SKILLS AND APPLICATIONS

In Exercises 7–14, rewrite the logarithm as a ratio of (a) common logarithms and (b) natural logarithms.

7. $\log_5 16$

8. $\log_3 47$

9. $\log_{1/5} x$

10. $\log_{1/3} x$

11. $\log_x \frac{3}{10}$

12. $\log_x \frac{3}{4}$

13. $\log_{2.6} x$

14. $\log_{7.1} x$

In Exercises 15–22, evaluate the logarithm using the change-of-base formula. Round your result to three decimal places.

15. $\log_3 7$

16. $\log_7 4$

17. $\log_{1/2} 4$

18. $\log_{1/4} 5$

19. $\log_9 0.1$

20. $\log_{20} 0.25$

21. $\log_{15} 1250$

22. $\log_3 0.015$

In Exercises 23–28, use the properties of logarithms to rewrite and simplify the logarithmic expression.

23. $\log_4 8$

24. $\log_2(4^2 \cdot 3^4)$

25. $\log_5 \frac{1}{250}$

26. $\log \frac{9}{300}$

27. $\ln(5e^6)$

28. $\ln \frac{6}{e^2}$

In Exercises 29–44, find the exact value of the logarithmic expression without using a calculator. (If this is not possible, state the reason.)

29. $\log_3 9$

30. $\log_5 \frac{1}{125}$

31. $\log_2 \sqrt[4]{8}$

32. $\log_6 \sqrt[3]{6}$

33. $\log_4 16^2$

34. $\log_3 81^{-3}$

35. $\log_2(-2)$

36. $\log_3(-27)$

37. $\ln e^{4.5}$

38. $3 \ln e^4$

39. $\ln \dfrac{1}{\sqrt{e}}$

40. $\ln \sqrt[4]{e^3}$

41. $\ln e^2 + \ln e^5$

42. $2 \ln e^6 - \ln e^5$

43. $\log_5 75 - \log_5 3$

44. $\log_4 2 + \log_4 32$

In Exercises 45–66, use the properties of logarithms to expand the expression as a sum, difference, and/or constant multiple of logarithms. (Assume all variables are positive.)

45. $\ln 4x$

46. $\log_3 10z$

47. $\log_8 x^4$

48. $\log_{10} \dfrac{y}{2}$

49. $\log_5 \dfrac{5}{x}$

50. $\log_6 \dfrac{1}{z^3}$

51. $\ln \sqrt{z}$

52. $\ln \sqrt[3]{t}$

53. $\ln xyz^2$

54. $\log 4x^2 y$

55. $\ln z(z-1)^2, \ z > 1$

56. $\ln\left(\dfrac{x^2-1}{x^3}\right), \ x > 1$

57. $\log_2 \dfrac{\sqrt{a-1}}{9}, \ a > 1$

58. $\ln \dfrac{6}{\sqrt{x^2+1}}$

59. $\ln \sqrt[3]{\dfrac{x}{y}}$

60. $\ln \sqrt{\dfrac{x^2}{y^3}}$

61. $\ln x^2 \sqrt{\dfrac{y}{z}}$

62. $\log_2 x^4 \sqrt{\dfrac{y}{z^3}}$

63. $\log_5 \dfrac{x^2}{y^2 z^3}$

64. $\log_{10} \dfrac{xy^4}{z^5}$

65. $\ln \sqrt[4]{x^3(x^2+3)}$

66. $\ln \sqrt{x^2(x+2)}$

In Exercises 67–84, condense the expression to the logarithm of a single quantity.

67. $\ln 2 + \ln x$

68. $\ln y + \ln t$

69. $\log_4 z - \log_4 y$

70. $\log_5 8 - \log_5 t$

71. $2 \log_2 x + 4 \log_2 y$

72. $\frac{2}{3} \log_7(z - 2)$

73. $\frac{1}{4} \log_3 5x$

74. $-4 \log_6 2x$

75. $\log x - 2 \log(x + 1)$

76. $2 \ln 8 + 5 \ln(z - 4)$

77. $\log x - 2 \log y + 3 \log z$

78. $3 \log_3 x + 4 \log_3 y - 4 \log_3 z$

79. $\ln x - [\ln(x + 1) + \ln(x - 1)]$

80. $4[\ln z + \ln(z + 5)] - 2 \ln(z - 5)$

81. $\frac{1}{3}[2 \ln(x + 3) + \ln x - \ln(x^2 - 1)]$

82. $2[3 \ln x - \ln(x + 1) - \ln(x - 1)]$

83. $\frac{1}{3}[\log_8 y + 2 \log_8(y + 4)] - \log_8(y - 1)$

84. $\frac{1}{2}[\log_4(x + 1) + 2 \log_4(x - 1)] + 6 \log_4 x$

In Exercises 85 and 86, compare the logarithmic quantities. If two are equal, explain why.

85. $\dfrac{\log_2 32}{\log_2 4}$, $\log_2 \dfrac{32}{4}$, $\log_2 32 - \log_2 4$

86. $\log_7 \sqrt{70}$, $\log_7 35$, $\frac{1}{2} + \log_7 \sqrt{10}$

SOUND INTENSITY In Exercises 87–90, use the following information. The relationship between the number of decibels β and the intensity of a sound I in watts per square meter is given by

$$\beta = 10 \log\left(\frac{I}{10^{-12}}\right).$$

87. Use the properties of logarithms to write the formula in simpler form, and determine the number of decibels of a sound with an intensity of 10^{-6} watt per square meter.

88. Find the difference in loudness between an average office with an intensity of 1.26×10^{-7} watt per square meter and a broadcast studio with an intensity of 3.16×10^{-10} watt per square meter.

89. Find the difference in loudness between a vacuum cleaner with an intensity of 10^{-4} watt per square meter and rustling leaves with an intensity of 10^{-11} watt per square meter.

90. You and your roommate are playing your stereos at the same time and at the same intensity. How much louder is the music when both stereos are playing compared with just one stereo playing?

CURVE FITTING In Exercises 91–94, find a logarithmic equation that relates y and x. Explain the steps used to find the equation.

91.

x	1	2	3	4	5	6
y	1	1.189	1.316	1.414	1.495	1.565

92.

x	1	2	3	4	5	6
y	1	1.587	2.080	2.520	2.924	3.302

93.

x	1	2	3	4	5	6
y	2.5	2.102	1.9	1.768	1.672	1.597

94.

x	1	2	3	4	5	6
y	0.5	2.828	7.794	16	27.951	44.091

95. GALLOPING SPEEDS OF ANIMALS Four-legged animals run with two different types of motion: trotting and galloping. An animal that is trotting has at least one foot on the ground at all times, whereas an animal that is galloping has all four feet off the ground at some point in its stride. The number of strides per minute at which an animal breaks from a trot to a gallop depends on the weight of the animal. Use the table to find a logarithmic equation that relates an animal's weight x (in pounds) and its lowest galloping speed y (in strides per minute).

Weight, x	Galloping speed, y
25	191.5
35	182.7
50	173.8
75	164.2
500	125.9
1000	114.2

96. NAIL LENGTH The approximate lengths and diameters (in inches) of common nails are shown in the table. Find a logarithmic equation that relates the diameter y of a common nail to its length x.

Length, x	Diameter, y	Length, x	Diameter, y
1	0.072	4	0.203
2	0.120	5	0.238
3	0.148	6	0.284

97. COMPARING MODELS   A cup of water at an initial temperature of 78°C is placed in a room at a constant temperature of 21°C. The temperature of the water is measured every 5 minutes during a half-hour period. The results are recorded as ordered pairs of the form (t, T), where t is the time (in minutes) and T is the temperature (in degrees Celsius).

$(0, 78.0°)$, $(5, 66.0°)$, $(10, 57.5°)$, $(15, 51.2°)$,
$(20, 46.3°)$, $(25, 42.4°)$, $(30, 39.6°)$

(a) The graph of the model for the data should be asymptotic with the graph of the temperature of the room. Subtract the room temperature from each of the temperatures in the ordered pairs. Use a graphing utility to plot the data points (t, T) and $(t, T - 21)$.

(b) An exponential model for the data $(t, T - 21)$ is given by $T - 21 = 54.4(0.964)^t$. Solve for T and graph the model. Compare the result with the plot of the original data.

(c) Take the natural logarithms of the revised temperatures. Use a graphing utility to plot the points $(t, \ln(T - 21))$ and observe that the points appear to be linear. Use the *regression* feature of the graphing utility to fit a line to these data. This resulting line has the form $\ln(T - 21) = at + b$. Solve for T, and verify that the result is equivalent to the model in part (b).

(d) Fit a rational model to the data. Take the reciprocals of the y-coordinates of the revised data points to generate the points

$$\left(t, \frac{1}{T - 21}\right).$$

Use a graphing utility to graph these points and observe that they appear to be linear. Use the *regression* feature of a graphing utility to fit a line to these data. The resulting line has the form

$$\frac{1}{T - 21} = at + b.$$

Solve for T, and use a graphing utility to graph the rational function and the original data points.

(e) Why did taking the logarithms of the temperatures lead to a linear scatter plot? Why did taking the reciprocals of the temperatures lead to a linear scatter plot?

EXPLORATION

98. PROOF   Prove that $\log_b \dfrac{u}{v} = \log_b u - \log_b v$.

99. PROOF   Prove that $\log_b u^n = n \log_b u$.

100. CAPSTONE   A classmate claims that the following are true.

(a) $\ln(u + v) = \ln u + \ln v = \ln(uv)$

(b) $\ln(u - v) = \ln u - \ln v = \ln \dfrac{u}{v}$

(c) $(\ln u)^n = n(\ln u) = \ln u^n$

Discuss how you would demonstrate that these claims are not true.

TRUE OR FALSE?   In Exercises 101–106, determine whether the statement is true or false given that $f(x) = \ln x$. Justify your answer.

101. $f(0) = 0$

102. $f(ax) = f(a) + f(x), \quad a > 0, x > 0$

103. $f(x - 2) = f(x) - f(2), \quad x > 2$

104. $\sqrt{f(x)} = \frac{1}{2} f(x)$

105. If $f(u) = 2f(v)$, then $v = u^2$.

106. If $f(x) < 0$, then $0 < x < 1$.

In Exercises 107–112, use the change-of-base formula to rewrite the logarithm as a ratio of logarithms. Then use a graphing utility to graph the ratio.

107. $f(x) = \log_2 x$

108. $f(x) = \log_4 x$

109. $f(x) = \log_{1/2} x$

110. $f(x) = \log_{1/4} x$

111. $f(x) = \log_{11.8} x$

112. $f(x) = \log_{12.4} x$

113. THINK ABOUT IT   Consider the functions below.

$$f(x) = \ln \frac{x}{2}, \quad g(x) = \frac{\ln x}{\ln 2}, \quad h(x) = \ln x - \ln 2$$

Which two functions should have identical graphs? Verify your answer by sketching the graphs of all three functions on the same set of coordinate axes.

114. GRAPHICAL ANALYSIS   Use a graphing utility to graph the functions given by $y_1 = \ln x - \ln(x - 3)$ and $y_2 = \ln \dfrac{x}{x - 3}$ in the same viewing window. Does the graphing utility show the functions with the same domain? If so, should it? Explain your reasoning.

115. THINK ABOUT IT   For how many integers between 1 and 20 can the natural logarithms be approximated given the values $\ln 2 \approx 0.6931$, $\ln 3 \approx 1.0986$, and $\ln 5 \approx 1.6094$? Approximate these logarithms (do not use a calculator).

5.4 EXPONENTIAL AND LOGARITHMIC EQUATIONS

What you should learn

- Solve simple exponential and logarithmic equations.
- Solve more complicated exponential equations.
- Solve more complicated logarithmic equations.
- Use exponential and logarithmic equations to model and solve real-life problems.

Why you should learn it

Exponential and logarithmic equations are used to model and solve life science applications. For instance, in Exercise 132 on page 417, an exponential function is used to model the number of trees per acre given the average diameter of the trees.

Introduction

So far in this chapter, you have studied the definitions, graphs, and properties of exponential and logarithmic functions. In this section, you will study procedures for *solving equations* involving these exponential and logarithmic functions.

There are two basic strategies for solving exponential or logarithmic equations. The first is based on the One-to-One Properties and was used to solve simple exponential and logarithmic equations in Sections 5.1 and 5.2. The second is based on the Inverse Properties. For $a > 0$ and $a \neq 1$, the following properties are true for all x and y for which $\log_a x$ and $\log_a y$ are defined.

One-to-One Properties

$a^x = a^y$ if and only if $x = y$.

$\log_a x = \log_a y$ if and only if $x = y$.

Inverse Properties

$a^{\log_a x} = x$

$\log_a a^x = x$

Example 1 Solving Simple Equations

	Original Equation	Rewritten Equation	Solution	Property
a.	$2^x = 32$	$2^x = 2^5$	$x = 5$	One-to-One
b.	$\ln x - \ln 3 = 0$	$\ln x = \ln 3$	$x = 3$	One-to-One
c.	$\left(\frac{1}{3}\right)^x = 9$	$3^{-x} = 3^2$	$x = -2$	One-to-One
d.	$e^x = 7$	$\ln e^x = \ln 7$	$x = \ln 7$	Inverse
e.	$\ln x = -3$	$e^{\ln x} = e^{-3}$	$x = e^{-3}$	Inverse
f.	$\log x = -1$	$10^{\log x} = 10^{-1}$	$x = 10^{-1} = \frac{1}{10}$	Inverse
g.	$\log_3 x = 4$	$3^{\log_3 x} = 3^4$	$x = 81$	Inverse

CHECK Point Now try Exercise 17.

The strategies used in Example 1 are summarized as follows.

Strategies for Solving Exponential and Logarithmic Equations

1. Rewrite the original equation in a form that allows the use of the One-to-One Properties of exponential or logarithmic functions.

2. Rewrite an *exponential* equation in logarithmic form and apply the Inverse Property of logarithmic functions.

3. Rewrite a *logarithmic* equation in exponential form and apply the Inverse Property of exponential functions.

Solving Exponential Equations

Example 2   Solving Exponential Equations

Solve each equation and approximate the result to three decimal places, if necessary.

a. $e^{-x^2} = e^{-3x-4}$

b. $3(2^x) = 42$

Solution

a.
$$e^{-x^2} = e^{-3x-4}$$
$$-x^2 = -3x - 4$$
$$x^2 - 3x - 4 = 0$$
$$(x+1)(x-4) = 0$$
$$(x+1) = 0 \Rightarrow x = -1$$
$$(x-4) = 0 \Rightarrow x = 4$$

The solutions are $x = -1$ and $x = 4$. Check these in the original equation.

b.
$$3(2^x) = 42$$
$$2^x = 14$$
$$\log_2 2^x = \log_2 14$$
$$x = \log_2 14$$
$$x = \frac{\ln 14}{\ln 2} \approx 3.807$$

The solution is $x = \log_2 14 \approx 3.807$. Check this in the original equation.

CHECK*Point* Now try Exercise 29.

In Example 2(b), the exact solution is $x = \log_2 14$ and the approximate solution is $x \approx 3.807$. An exact answer is preferred when the solution is an intermediate step in a larger problem. For a final answer, an approximate solution is easier to comprehend.

Example 3   Solving an Exponential Equation

Solve $e^x + 5 = 60$ and approximate the result to three decimal places.

Solution
$$e^x + 5 = 60$$
$$e^x = 55$$
$$\ln e^x = \ln 55$$
$$x = \ln 55 \approx 4.007$$

The solution is $x = \ln 55 \approx 4.007$. Check this in the original equation.

CHECK*Point* Now try Exercise 55.

Study Tip

Another way to solve Example 2(b) is by taking the natural log of each side and then applying the Power Property, as follows.

$$3(2^x) = 42$$
$$2^x = 14$$
$$\ln 2^x = \ln 14$$
$$x \ln 2 = \ln 14$$
$$x = \frac{\ln 14}{\ln 2} \approx 3.807$$

As you can see, you obtain the same result as in Example 2(b).

Study Tip

Remember that the natural logarithmic function has a base of e.

Example 4 Solving an Exponential Equation

Solve $2(3^{2t-5}) - 4 = 11$ and approximate the result to three decimal places.

Solution

$$2(3^{2t-5}) - 4 = 11$$

$$2(3^{2t-5}) = 15$$

$$3^{2t-5} = \frac{15}{2}$$

$$\log_3 3^{2t-5} = \log_3 \frac{15}{2}$$

$$2t - 5 = \log_3 \frac{15}{2}$$

$$2t = 5 + \log_3 7.5$$

$$t = \frac{5}{2} + \frac{1}{2}\log_3 7.5$$

$$t \approx 3.417$$

The solution is $t = \frac{5}{2} + \frac{1}{2}\log_3 7.5 \approx 3.417$. Check this in the original equation.

CHECK*Point* Now try Exercise 57.

Study Tip

Remember that to evaluate a logarithm such as $\log_3 7.5$, you need to use the change-of-base formula.

$$\log_3 7.5 = \frac{\ln 7.5}{\ln 3} \approx 1.834$$

When an equation involves two or more exponential expressions, you can still use a procedure similar to that demonstrated in Examples 2, 3, and 4. However, the algebra is a bit more complicated.

Example 5 Solving an Exponential Equation of Quadratic Type

Solve $e^{2x} - 3e^x + 2 = 0$.

Algebraic Solution

$$e^{2x} - 3e^x + 2 = 0$$

$$(e^x)^2 - 3e^x + 2 = 0$$

$$(e^x - 2)(e^x - 1) = 0$$

$$e^x - 2 = 0$$

$$x = \ln 2$$

$$e^x - 1 = 0$$

$$x = 0$$

The solutions are $x = \ln 2 \approx 0.693$ and $x = 0$. Check these in the original equation.

CHECK*Point* Now try Exercise 59.

Graphical Solution

Use a graphing utility to graph $y = e^{2x} - 3e^x + 2$. Use the *zero* or *root* feature or the *zoom* and *trace* features of the graphing utility to approximate the values of x for which $y = 0$. In Figure 5.28, you can see that the zeros occur at $x = 0$ and at $x \approx 0.693$. So, the solutions are $x = 0$ and $x \approx 0.693$.

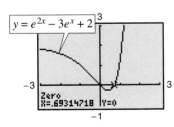

$y = e^{2x} - 3e^x + 2$

Zero
X=.69314718 Y=0

FIGURE **5.28**

Solving Logarithmic Equations

To solve a logarithmic equation, you can write it in exponential form.

$$\ln x = 3$$

$$e^{\ln x} = e^3$$

$$x = e^3$$

This procedure is called *exponentiating* each side of an equation.

Example 6   Solving Logarithmic Equations

a. $\ln x = 2$

$$e^{\ln x} = e^2$$

$$x = e^2$$

b. $\log_3(5x - 1) = \log_3(x + 7)$

$$5x - 1 = x + 7$$

$$4x = 8$$

$$x = 2$$

c. $\log_6(3x + 14) - \log_6 5 = \log_6 2x$

$$\log_6\left(\frac{3x + 14}{5}\right) = \log_6 2x$$

$$\frac{3x + 14}{5} = 2x$$

$$3x + 14 = 10x$$

$$-7x = -14$$

$$x = 2$$

CHECK Point Now try Exercise 83.

Example 7   Solving a Logarithmic Equation

Solve $5 + 2 \ln x = 4$ and approximate the result to three decimal places.

Algebraic Solution

$$5 + 2 \ln x = 4$$

$$2 \ln x = -1$$

$$\ln x = -\frac{1}{2}$$

$$e^{\ln x} = e^{-1/2}$$

$$x = e^{-1/2}$$

$$x \approx 0.607$$

Graphical Solution

Use a graphing utility to graph $y_1 = 5 + 2 \ln x$ and $y_2 = 4$ in the same viewing window. Use the *intersect* feature or the *zoom* and *trace* features to approximate the intersection point, as shown in Figure 5.29. So, the solution is $x \approx 0.607$.

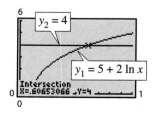

FIGURE 5.29

CHECK Point Now try Exercise 93.

Example 8 Solving a Logarithmic Equation

Solve $2 \log_5 3x = 4$.

Solution

$$2 \log_5 3x = 4$$

$$\log_5 3x = 2$$

$$5^{\log_5 3x} = 5^2$$

$$3x = 25$$

$$x = \frac{25}{3}$$

The solution is $x = \frac{25}{3}$. Check this in the original equation.

CHECK**Point** Now try Exercise 97.

Because the domain of a logarithmic function generally does not include all real numbers, you should be sure to check for extraneous solutions of logarithmic equations.

Example 9 Checking for Extraneous Solutions

Solve $\log 5x + \log(x - 1) = 2$.

Algebraic Solution

$$\log 5x + \log(x - 1) = 2$$

$$\log[5x(x - 1)] = 2$$

$$10^{\log(5x^2 - 5x)} = 10^2$$

$$5x^2 - 5x = 100$$

$$x^2 - x - 20 = 0$$

$$(x - 5)(x + 4) = 0$$

$$x - 5 = 0$$

$$x = 5$$

$$x + 4 = 0$$

$$x = -4$$

The solutions appear to be $x = 5$ and $x = -4$. However, when you check these in the original equation, you can see that $x = 5$ is the only solution.

CHECK**Point** Now try Exercise 109.

Graphical Solution

Use a graphing utility to graph $y_1 = \log 5x + \log(x - 1)$ and $y_2 = 2$ in the same viewing window. From the graph shown in Figure 5.30, it appears that the graphs intersect at one point. Use the *intersect* feature or the *zoom* and *trace* features to determine that the graphs intersect at approximately $(5, 2)$. So, the solution is $x = 5$. Verify that 5 is an exact solution algebraically.

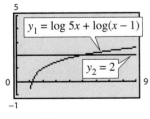

FIGURE 5.30

In Example 9, the domain of $\log 5x$ is $x > 0$ and the domain of $\log(x - 1)$ is $x > 1$, so the domain of the original equation is $x > 1$. Because the domain is all real numbers greater than 1, the solution $x = -4$ is extraneous. The graph in Figure 5.30 verifies this conclusion.

Applications

Example 10　Doubling an Investment

You have deposited $500 in an account that pays 6.75% interest, compounded continuously. How long will it take your money to double?

Solution

Using the formula for continuous compounding, you can find that the balance in the account is

$$A = Pe^{rt}$$

$$A = \quad e \quad {}^{t}.$$

To find the time required for the balance to double, let $A = 1000$ and solve the resulting equation for t.

$$500e^{0.0675t} = $$

$$e^{0.0675t} = 2$$

$$\ln e^{0.0675t} = \ln 2$$

$$0.0675t = \ln 2$$

$$t = \frac{\ln 2}{0.0675}$$

$$t \approx 10.27$$

The balance in the account will double after approximately 10.27 years. This result is demonstrated graphically in Figure 5.31.

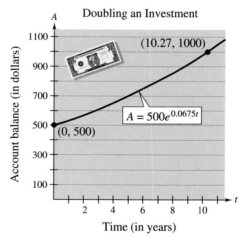

Doubling an Investment

(10.27, 1000)

$A = 500e^{0.0675t}$

(0, 500)

FIGURE **5.31**

CHECK *Point*　Now try Exercise 117.

In Example 10, an approximate answer of 10.27 years is given. Within the context of the problem, the exact solution, $(\ln 2)/0.0675$ years, does not make sense as an answer.

Retail Sales of e-Commerce
Companies

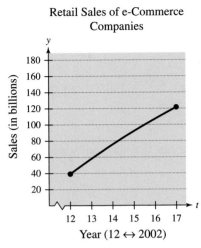

FIGURE 5.32

Example 11 Retail Sales

The retail sales y (in billions) of e-commerce companies in the United States from 2002 through 2007 can be modeled by

$$y = -549 + 236.7 \ln t, \quad 12 \le t \le 17$$

where t represents the year, with $t = 12$ corresponding to 2002 (see Figure 5.32). During which year did the sales reach \$108 billion? (Source: U.S. Census Bureau)

Solution

$$-549 + 236.7 \ln t = y$$

$$-549 + 236.7 \ln t =$$

$$236.7 \ln t = 657$$

$$\ln t = \frac{657}{236.7}$$

$$e^{\ln t} = e^{657/236.7}$$

$$t = e^{657/236.7}$$

$$t \approx 16$$

The solution is $t \approx 16$. Because $t = 12$ represents 2002, it follows that the sales reached \$108 billion in 2006.

CHECK**Point** Now try Exercise 133.

CLASSROOM DISCUSSION

Analyzing Relationships Numerically Use a calculator to fill in the table row-by-row. Discuss the resulting pattern. What can you conclude? Find two equations that summarize the relationships you discovered.

x	$\frac{1}{2}$	1	2	10	25	50
e^x						
$\ln(e^x)$						
$\ln x$						
$e^{\ln x}$						

5.4   EXERCISES

See www.CalcChat.com for worked-out solutions to odd-numbered exercises.

VOCABULARY: Fill in the blanks.

1. To _____ an equation in x means to find all values of x for which the equation is true.

2. To solve exponential and logarithmic equations, you can use the following One-to-One and Inverse Properties.
   (a) $a^x = a^y$ if and only if _____.
   (b) $\log_a x = \log_a y$ if and only if _____.
   (c) $a^{\log_a x} =$ _____
   (d) $\log_a a^x =$ _____

3. To solve exponential and logarithmic equations, you can use the following strategies.
   (a) Rewrite the original equation in a form that allows the use of the _____ Properties of exponential or logarithmic functions.
   (b) Rewrite an exponential equation in _____ form and apply the Inverse Property of _____ functions.
   (c) Rewrite a logarithmic equation in _____ form and apply the Inverse Property of _____ functions.

4. An _____ solution does not satisfy the original equation.

SKILLS AND APPLICATIONS

In Exercises 5–12, determine whether each x-value is a solution (or an approximate solution) of the equation.

5. $4^{2x-7} = 64$
   (a) $x = 5$
   (b) $x = 2$

6. $2^{3x+1} = 32$
   (a) $x = -1$
   (b) $x = 2$

7. $3e^{x+2} = 75$
   (a) $x = -2 + e^{25}$
   (b) $x = -2 + \ln 25$
   (c) $x \approx 1.219$

8. $4e^{x-1} = 60$
   (a) $x = 1 + \ln 15$
   (b) $x \approx 3.7081$
   (c) $x = \ln 16$

9. $\log_4(3x) = 3$
   (a) $x \approx 21.333$
   (b) $x = -4$
   (c) $x = \frac{64}{3}$

10. $\log_2(x + 3) = 10$
   (a) $x = 1021$
   (b) $x = 17$
   (c) $x = 10^2 - 3$

11. $\ln(2x + 3) = 5.8$
   (a) $x = \frac{1}{2}(-3 + \ln 5.8)$
   (b) $x = \frac{1}{2}(-3 + e^{5.8})$
   (c) $x \approx 163.650$

12. $\ln(x - 1) = 3.8$
   (a) $x = 1 + e^{3.8}$
   (b) $x \approx 45.701$
   (c) $x = 1 + \ln 3.8$

In Exercises 13–24, solve for x.

13. $4^x = 16$

14. $3^x = 243$

15. $\left(\frac{1}{2}\right)^x = 32$

16. $\left(\frac{1}{4}\right)^x = 64$

17. $\ln x - \ln 2 = 0$

18. $\ln x - \ln 5 = 0$

19. $e^x = 2$

20. $e^x = 4$

21. $\ln x = -1$

22. $\log x = -2$

23. $\log_4 x = 3$

24. $\log_5 x = \frac{1}{2}$

In Exercises 25–28, approximate the point of intersection of the graphs of f and g. Then solve the equation $f(x) = g(x)$ algebraically to verify your approximation.

25. $f(x) = 2^x$
   $g(x) = 8$

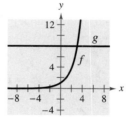

26. $f(x) = 27^x$
   $g(x) = 9$

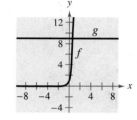

27. $f(x) = \log_3 x$
   $g(x) = 2$

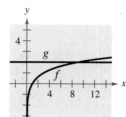

28. $f(x) = \ln(x - 4)$
   $g(x) = 0$

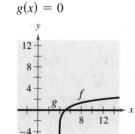

In Exercises 29–70, solve the exponential equation algebraically. Approximate the result to three decimal places.

29. $e^x = e^{x^2-2}$

30. $e^{2x} = e^{x^2-8}$

31. $e^{x^2-3} = e^{x-2}$

32. $e^{-x^2} = e^{x^2-2x}$

33. $4(3^x) = 20$

34. $2(5^x) = 32$

35. $2e^x = 10$

36. $4e^x = 91$

37. $e^x - 9 = 19$

38. $6^x + 10 = 47$

39. $3^{2x} = 80$

40. $6^{5x} = 3000$

41. $5^{-t/2} = 0.20$

42. $4^{-3t} = 0.10$

43. $3^{x-1} = 27$

44. $2^{x-3} = 32$

45. $2^{3-x} = 565$

46. $8^{-2-x} = 431$

47. $8(10^{3x}) = 12$

48. $5(10^{x-6}) = 7$

49. $3(5^{x-1}) = 21$

50. $8(3^{6-x}) = 40$

51. $e^{3x} = 12$

52. $e^{2x} = 50$

53. $500e^{-x} = 300$

54. $1000e^{-4x} = 75$

55. $7 - 2e^x = 5$

56. $-14 + 3e^x = 11$

57. $6(2^{3x-1}) - 7 = 9$

58. $8(4^{6-2x}) + 13 = 41$

59. $e^{2x} - 4e^x - 5 = 0$

60. $e^{2x} - 5e^x + 6 = 0$

61. $e^{2x} - 3e^x - 4 = 0$

62. $e^{2x} + 9e^x + 36 = 0$

63. $\dfrac{500}{100 - e^{x/2}} = 20$

64. $\dfrac{400}{1 + e^{-x}} = 350$

65. $\dfrac{3000}{2 + e^{2x}} = 2$

66. $\dfrac{119}{e^{6x} - 14} = 7$

67. $\left(1 + \dfrac{0.065}{365}\right)^{365t} = 4$

68. $\left(4 - \dfrac{2.471}{40}\right)^{9t} = 21$

69. $\left(1 + \dfrac{0.10}{12}\right)^{12t} = 2$

70. $\left(16 - \dfrac{0.878}{26}\right)^{3t} = 30$

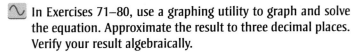 In Exercises 71–80, use a graphing utility to graph and solve the equation. Approximate the result to three decimal places. Verify your result algebraically.

71. $7 = 2^x$

72. $5^x = 212$

73. $6e^{1-x} = 25$

74. $-4e^{-x-1} + 15 = 0$

75. $3e^{3x/2} = 962$

76. $8e^{-2x/3} = 11$

77. $e^{0.09t} = 3$

78. $-e^{1.8x} + 7 = 0$

79. $e^{0.125t} - 8 = 0$

80. $e^{2.724x} = 29$

In Exercises 81–112, solve the logarithmic equation algebraically. Approximate the result to three decimal places.

81. $\ln x = -3$

82. $\ln x = 1.6$

83. $\ln x - 7 = 0$

84. $\ln x + 1 = 0$

85. $\ln 2x = 2.4$

86. $2.1 = \ln 6x$

87. $\log x = 6$

88. $\log 3z = 2$

89. $3 \ln 5x = 10$

90. $2 \ln x = 7$

91. $\ln\sqrt{x + 2} = 1$

92. $\ln\sqrt{x - 8} = 5$

93. $7 + 3 \ln x = 5$

94. $2 - 6 \ln x = 10$

95. $-2 + 2 \ln 3x = 17$

96. $2 + 3 \ln x = 12$

97. $6 \log_3(0.5x) = 11$

98. $4 \log(x - 6) = 11$

99. $\ln x - \ln(x + 1) = 2$

100. $\ln x + \ln(x + 1) = 1$

101. $\ln x + \ln(x - 2) = 1$

102. $\ln x + \ln(x + 3) = 1$

103. $\ln(x + 5) = \ln(x - 1) - \ln(x + 1)$

104. $\ln(x + 1) - \ln(x - 2) = \ln x$

105. $\log_2(2x - 3) = \log_2(x + 4)$

106. $\log(3x + 4) = \log(x - 10)$

107. $\log(x + 4) - \log x = \log(x + 2)$

108. $\log_2 x + \log_2(x + 2) = \log_2(x + 6)$

109. $\log_4 x - \log_4(x - 1) = \frac{1}{2}$

110. $\log_3 x + \log_3(x - 8) = 2$

111. $\log 8x - \log\left(1 + \sqrt{x}\right) = 2$

112. $\log 4x - \log\left(12 + \sqrt{x}\right) = 2$

In Exercises 113–116, use a graphing utility to graph and solve the equation. Approximate the result to three decimal places. Verify your result algebraically.

113. $3 - \ln x = 0$

114. $10 - 4 \ln(x - 2) = 0$

115. $2 \ln(x + 3) = 3$

116. $\ln(x + 1) = 2 - \ln x$

COMPOUND INTEREST In Exercises 117–120, $2500 is invested in an account at interest rate r, compounded continuously. Find the time required for the amount to (a) double and (b) triple.

117. $r = 0.05$

118. $r = 0.045$

119. $r = 0.025$

120. $r = 0.0375$

In Exercises 121–128, solve the equation algebraically. Round the result to three decimal places. Verify your answer using a graphing utility.

121. $2x^2e^{2x} + 2xe^{2x} = 0$

122. $-x^2e^{-x} + 2xe^{-x} = 0$

123. $-xe^{-x} + e^{-x} = 0$

124. $e^{-2x} - 2xe^{-2x} = 0$

125. $2x \ln x + x = 0$

126. $\dfrac{1 - \ln x}{x^2} = 0$

127. $\dfrac{1 + \ln x}{2} = 0$

128. $2x \ln\left(\dfrac{1}{x}\right) - x = 0$

129. DEMAND The demand equation for a limited edition coin set is

$$p = 1000\left(1 - \frac{5}{5 + e^{-0.001x}}\right).$$

Find the demand x for a price of (a) $p = \$139.50$ and (b) $p = \$99.99$.

130. DEMAND The demand equation for a hand-held electronic organizer is

$$p = 5000\left(1 - \frac{4}{4 + e^{-0.002x}}\right).$$

Find the demand x for a price of (a) $p = \$600$ and (b) $p = \$400$.

131. **FOREST YIELD**   The yield V (in millions of cubic feet per acre) for a forest at age t years is given by $V = 6.7e^{-48.1/t}$.

    (a) Use a graphing utility to graph the function.

    (b) Determine the horizontal asymptote of the function. Interpret its meaning in the context of the problem.

    (c) Find the time necessary to obtain a yield of 1.3 million cubic feet.

132. **TREES PER ACRE**   The number N of trees of a given species per acre is approximated by the model $N = 68(10^{-0.04x})$, $5 \le x \le 40$, where x is the average diameter of the trees (in inches) 3 feet above the ground. Use the model to approximate the average diameter of the trees in a test plot when $N = 21$.

133. **U.S. CURRENCY**   The values y (in billions of dollars) of U.S. currency in circulation in the years 2000 through 2007 can be modeled by $y = -451 + 444 \ln t$, $10 \le t \le 17$, where t represents the year, with $t = 10$ corresponding to 2000. During which year did the value of U.S. currency in circulation exceed \$690 billion?   (Source: Board of Governors of the Federal Reserve System)

134. **MEDICINE**   The numbers y of freestanding ambulatory care surgery centers in the United States from 2000 through 2007 can be modeled by

$$y = 2875 + \frac{2635.11}{1 + 14.215e^{-0.8038t}}, \quad 0 \le t \le 7$$

    where t represents the year, with $t = 0$ corresponding to 2000.   (Source: Verispan)

    (a) Use a graphing utility to graph the model.

    (b) Use the *trace* feature of the graphing utility to estimate the year in which the number of surgery centers exceeded 3600.

135. **AVERAGE HEIGHTS**   The percent m of American males between the ages of 18 and 24 who are no more than x inches tall is modeled by

$$m(x) = \frac{100}{1 + e^{-0.6114(x - 69.71)}}$$

    and the percent f of American females between the ages of 18 and 24 who are no more than x inches tall is modeled by

$$f(x) = \frac{100}{1 + e^{-0.66607(x - 64.51)}}.$$

    (Source: U.S. National Center for Health Statistics)

    (a) Use the graph to determine any horizontal asymptotes of the graphs of the functions. Interpret the meaning in the context of the problem.

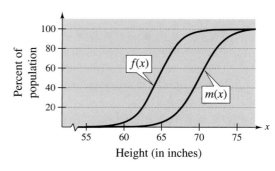

    (b) What is the average height of each sex?

136. **LEARNING CURVE**   In a group project in learning theory, a mathematical model for the proportion P of correct responses after n trials was found to be $P = 0.83/(1 + e^{-0.2n})$.

    (a) Use a graphing utility to graph the function.

    (b) Use the graph to determine any horizontal asymptotes of the graph of the function. Interpret the meaning of the upper asymptote in the context of this problem.

    (c) After how many trials will 60% of the responses be correct?

137. **AUTOMOBILES**   Automobiles are designed with crumple zones that help protect their occupants in crashes. The crumple zones allow the occupants to move short distances when the automobiles come to abrupt stops. The greater the distance moved, the fewer g's the crash victims experience. (One g is equal to the acceleration due to gravity. For very short periods of time, humans have withstood as much as 40 g's.) In crash tests with vehicles moving at 90 kilometers per hour, analysts measured the numbers of g's experienced during deceleration by crash dummies that were permitted to move x meters during impact. The data are shown in the table. A model for the data is given by $y = -3.00 + 11.88 \ln x + (36.94/x)$, where y is the number of g's.

x	g's
0.2	158
0.4	80
0.6	53
0.8	40
1.0	32

    (a) Complete the table using the model.

x	0.2	0.4	0.6	0.8	1.0
y					

(b) Use a graphing utility to graph the data points and the model in the same viewing window. How do they compare?

(c) Use the model to estimate the distance traveled during impact if the passenger deceleration must not exceed 30 g's.

(d) Do you think it is practical to lower the number of g's experienced during impact to fewer than 23? Explain your reasoning.

138. **DATA ANALYSIS** An object at a temperature of 160°C was removed from a furnace and placed in a room at 20°C. The temperature T of the object was measured each hour h and recorded in the table. A model for the data is given by $T = 20[1 + 7(2^{-h})]$. The graph of this model is shown in the figure.

Hour, h	Temperature, T
0	160°
1	90°
2	56°
3	38°
4	29°
5	24°

(a) Use the graph to identify the horizontal asymptote of the model and interpret the asymptote in the context of the problem.

(b) Use the model to approximate the time when the temperature of the object was 100°C.

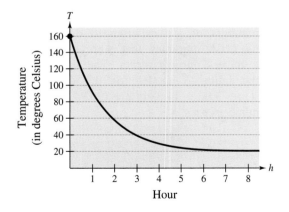

Hour

EXPLORATION

TRUE OR FALSE? In Exercises 139–142, rewrite each verbal statement as an equation. Then decide whether the statement is true or false. Justify your answer.

139. The logarithm of the product of two numbers is equal to the sum of the logarithms of the numbers.

140. The logarithm of the sum of two numbers is equal to the product of the logarithms of the numbers.

141. The logarithm of the difference of two numbers is equal to the difference of the logarithms of the numbers.

142. The logarithm of the quotient of two numbers is equal to the difference of the logarithms of the numbers.

143. **THINK ABOUT IT** Is it possible for a logarithmic equation to have more than one extraneous solution? Explain.

144. **FINANCE** You are investing P dollars at an annual interest rate of r, compounded continuously, for t years. Which of the following would result in the highest value of the investment? Explain your reasoning.

(a) Double the amount you invest.

(b) Double your interest rate.

(c) Double the number of years.

145. **THINK ABOUT IT** Are the times required for the investments in Exercises 117–120 to quadruple twice as long as the times for them to double? Give a reason for your answer and verify your answer algebraically.

146. The *effective yield* of a savings plan is the percent increase in the balance after 1 year. Find the effective yield for each savings plan when $1000 is deposited in a savings account. Which savings plan has the greatest effective yield? Which savings plan will have the highest balance after 5 years?

(a) 7% annual interest rate, compounded annually

(b) 7% annual interest rate, compounded continuously

(c) 7% annual interest rate, compounded quarterly

(d) 7.25% annual interest rate, compounded quarterly

147. **GRAPHICAL ANALYSIS** Let $f(x) = \log_a x$ and $g(x) = a^x$, where $a > 1$.

(a) Let $a = 1.2$ and use a graphing utility to graph the two functions in the same viewing window. What do you observe? Approximate any points of intersection of the two graphs.

(b) Determine the value(s) of a for which the two graphs have one point of intersection.

(c) Determine the value(s) of a for which the two graphs have two points of intersection.

148. **CAPSTONE** Write two or three sentences stating the general guidelines that you follow when solving (a) exponential equations and (b) logarithmic equations.

5.5 EXPONENTIAL AND LOGARITHMIC MODELS

Alan Becker/Stone/Getty Images

Introduction

The five most common types of mathematical models involving exponential functions and logarithmic functions are as follows.

1. **Exponential growth model:** $y = ae^{bx}, \quad b > 0$

2. **Exponential decay model:** $y = ae^{-bx}, \quad b > 0$

3. **Gaussian model:** $y = ae^{-(x-b)^2/c}$

4. **Logistic growth model:** $y = \dfrac{a}{1 + be^{-rx}}$

5. **Logarithmic models:** $y = a + b \ln x, \quad y = a + b \log x$

The basic shapes of the graphs of these functions are shown in Figure 5.33.

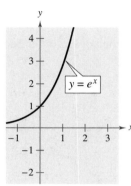

Exponential growth model

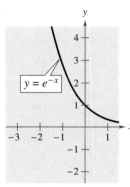

Exponential decay model

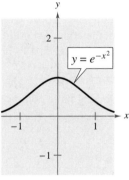

Gaussian model

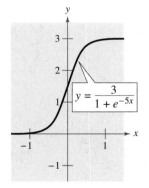

Logistic growth model
FIGURE 5.33

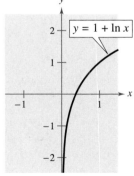

Natural logarithmic model

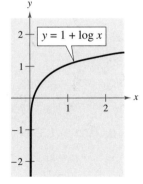

Common logarithmic model

You can often gain quite a bit of insight into a situation modeled by an exponential or logarithmic function by identifying and interpreting the function's asymptotes. Use the graphs in Figure 5.33 to identify the asymptotes of the graph of each function.

Exponential Growth and Decay

Example 1   Online Advertising

Estimates of the amounts (in billions of dollars) of U.S. online advertising spending from 2007 through 2011 are shown in the table. A scatter plot of the data is shown in Figure 5.34.   (Source: eMarketer)

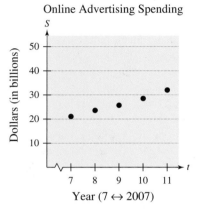

Year	Advertising spending
2007	21.1
2008	23.6
2009	25.7
2010	28.5
2011	32.0

Online Advertising Spending

FIGURE 5.34

An exponential growth model that approximates these data is given by $S = 10.33e^{0.1022t}$, $7 \leq t \leq 11$, where S is the amount of spending (in billions) and $t = 7$ represents 2007. Compare the values given by the model with the estimates shown in the table. According to this model, when will the amount of U.S. online advertising spending reach $40 billion?

Algebraic Solution

The following table compares the two sets of advertising spending figures.

Year	2007	2008	2009	2010	2011
Advertising spending	21.1	23.6	25.7	28.5	32.0
Model	21.1	23.4	25.9	28.7	31.8

To find when the amount of U.S. online advertising spending will reach $40 billion, let $S = 40$ in the model and solve for t.

$$10.33e^{0.1022t} = S$$

$$10.33e^{0.1022t} =$$

$$e^{0.1022t} \approx 3.8722$$

$$\ln e^{0.1022t} \approx \ln 3.8722$$

$$0.1022t \approx 1.3538$$

$$t \approx 13.2$$

According to the model, the amount of U.S. online advertising spending will reach $40 billion in 2013.

CHECK*Point*   Now try Exercise 43.

Graphical Solution

Use a graphing utility to graph the model $y = 10.33e^{0.1022x}$ and the data in the same viewing window. You can see in Figure 5.35 that the model appears to fit the data closely.

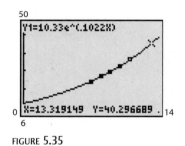

FIGURE 5.35

Use the *zoom* and *trace* features of the graphing utility to find that the approximate value of x for $y = 40$ is $x \approx 13.2$. So, according to the model, the amount of U.S. online advertising spending will reach $40 billion in 2013.

TECHNOLOGY

Some graphing utilities have an *exponential regression* feature that can be used to find exponential models that represent data. If you have such a graphing utility, try using it to find an exponential model for the data given in Example 1. How does your model compare with the model given in Example 1?

In Example 1, you were given the exponential growth model. But suppose this model were not given; how could you find such a model? One technique for doing this is demonstrated in Example 2.

Example 2   Modeling Population Growth

In a research experiment, a population of fruit flies is increasing according to the law of exponential growth. After 2 days there are 100 flies, and after 4 days there are 300 flies. How many flies will there be after 5 days?

Solution

Let y be the number of flies at time t. From the given information, you know that $y = 100$ when $t = 2$ and $y = 300$ when $t = 4$. Substituting this information into the model $y = ae^{bt}$ produces

$$= ae^{b} \quad \text{and} \quad = ae^{b}.$$

To solve for b, solve for a in the first equation.

$$100 = ae^{2b} \quad \Longrightarrow \quad a = \frac{100}{e^{2b}}$$

Then substitute the result into the second equation.

$$300 = ae^{4b}$$

$$300 = \left(\right)e^{4b}$$

$$\frac{300}{100} = e^{2b}$$

$$\ln 3 = 2b$$

$$\frac{1}{2}\ln 3 = b$$

Using $b = \frac{1}{2}\ln 3$ and the equation you found for a, you can determine that

$$a = \frac{100}{e^{2[]}}$$

$$= \frac{100}{e^{\ln 3}}$$

$$= \frac{100}{3}$$

$$\approx 33.33.$$

So, with $a \approx 33.33$ and $b = \frac{1}{2}\ln 3 \approx 0.5493$, the exponential growth model is

$$y = 33.33e^{0.5493t}$$

as shown in Figure 5.36. This implies that, after 5 days, the population will be

$$y = 33.33e^{0.5493()} \approx 520 \text{ flies.}$$

CHECK **Point**   Now try Exercise 49.

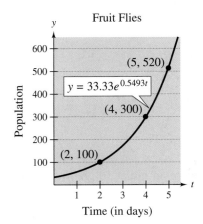

Fruit Flies

$y = 33.33e^{0.5493t}$

(5, 520)

(4, 300)

(2, 100)

Time (in days)

FIGURE 5.36

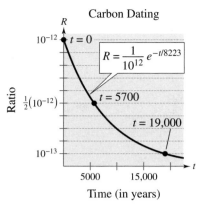

Carbon Dating

FIGURE 5.37

In living organic material, the ratio of the number of radioactive carbon isotopes (carbon 14) to the number of nonradioactive carbon isotopes (carbon 12) is about 1 to 10^{12}. When organic material dies, its carbon 12 content remains fixed, whereas its radioactive carbon 14 begins to decay with a half-life of about 5700 years. To estimate the age of dead organic material, scientists use the following formula, which denotes the ratio of carbon 14 to carbon 12 present at any time t (in years).

$$R = \frac{1}{10^{12}} e^{-t/8223}$$

The graph of R is shown in Figure 5.37. Note that R decreases as t increases.

Example 3   Carbon Dating

Estimate the age of a newly discovered fossil in which the ratio of carbon 14 to carbon 12 is

$R = 1/10^{13}$.

Algebraic Solution

In the carbon dating model, substitute the given value of R to obtain the following.

$$\frac{1}{10^{12}} e^{-t/8223} = R$$

$$\frac{e^{-t/8223}}{10^{12}} =$$

$$e^{-t/8223} = \frac{1}{10}$$

$$\ln e^{-t/8223} = \ln \frac{1}{10}$$

$$-\frac{t}{8223} \approx -2.3026$$

$$t \approx 18,934$$

So, to the nearest thousand years, the age of the fossil is about 19,000 years.

Graphical Solution

Use a graphing utility to graph the formula for the ratio of carbon 14 to carbon 12 at any time t as

$$y_1 = \frac{1}{10^{12}} e^{-x/8223}.$$

In the same viewing window, graph $y_2 = 1/(10^{13})$. Use the *intersect* feature or the *zoom* and *trace* features of the graphing utility to estimate that $x \approx 18,934$ when $y = 1/(10^{13})$, as shown in Figure 5.38.

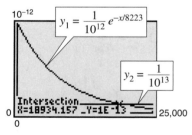

FIGURE 5.38

So, to the nearest thousand years, the age of the fossil is about 19,000 years.

CHECK *Point* Now try Exercise 51.

The value of b in the exponential decay model $y = ae^{-bt}$ determines the *decay* of radioactive isotopes. For instance, to find how much of an initial 10 grams of ^{226}Ra isotope with a half-life of 1599 years is left after 500 years, substitute this information into the model $y = ae^{-bt}$.

$$\frac{1}{2}(10) = 10e^{-b(1599)} \quad \Longrightarrow \quad \ln\frac{1}{2} = -1599b \quad \Longrightarrow \quad b = -\frac{\ln\frac{1}{2}}{1599}$$

Using the value of b found above and $a = 10$, the amount left is

$$y = 10e^{-[-\ln(1/2)/1599](500)} \approx 8.05 \text{ grams.}$$

Gaussian Models

As mentioned at the beginning of this section, Gaussian models are of the form

$$y = ae^{-(x-b)^2/c}.$$

This type of model is commonly used in probability and statistics to represent populations that are **normally distributed.** The graph of a Gaussian model is called a **bell-shaped curve.** Try graphing the normal distribution with a graphing utility. Can you see why it is called a bell-shaped curve?

For *standard* normal distributions, the model takes the form

$$y = \frac{1}{\sqrt{2\pi}}e^{-x^2/2}.$$

The **average value** of a population can be found from the bell-shaped curve by observing where the maximum y-value of the function occurs. The x-value corresponding to the maximum y-value of the function represents the average value of the independent variable—in this case, x.

Example 4   SAT Scores

In 2008, the Scholastic Aptitude Test (SAT) math scores for college-bound seniors roughly followed the normal distribution given by

$$y = 0.0034e^{-(x-515)^2/26,912}, \quad 200 \le x \le 800$$

where x is the SAT score for mathematics. Sketch the graph of this function. From the graph, estimate the average SAT score.   (Source: College Board)

Solution

The graph of the function is shown in Figure 5.39. On this bell-shaped curve, the maximum value of the curve represents the average score. From the graph, you can estimate that the average mathematics score for college-bound seniors in 2008 was 515.

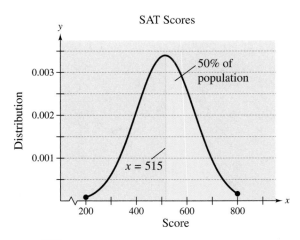

FIGURE 5.39

CHECK*Point* Now try Exercise 57.

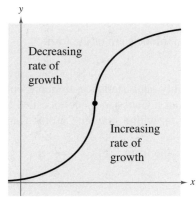

FIGURE 5.40

Logistic Growth Models

Some populations initially have rapid growth, followed by a declining rate of growth, as indicated by the graph in Figure 5.40. One model for describing this type of growth pattern is the **logistic curve** given by the function

$$y = \frac{a}{1 + be^{-rx}}$$

where y is the population size and x is the time. An example is a bacteria culture that is initially allowed to grow under ideal conditions, and then under less favorable conditions that inhibit growth. A logistic growth curve is also called a **sigmoidal curve.**

Example 5 Spread of a Virus

On a college campus of 5000 students, one student returns from vacation with a contagious and long-lasting flu virus. The spread of the virus is modeled by

$$y = \frac{5000}{1 + 4999e^{-0.8t}}, \quad t \geq 0$$

where y is the total number of students infected after t days. The college will cancel classes when 40% or more of the students are infected.

a. How many students are infected after 5 days?

b. After how many days will the college cancel classes?

Algebraic Solution

a. After 5 days, the number of students infected is

$$y = \frac{5000}{1 + 4999e^{-0.8(\ )}} = \frac{5000}{1 + 4999e^{-4}} \approx 54.$$

b. Classes are canceled when the number infected is $(0.40)(5000) = 2000$.

$$= \frac{5000}{1 + 4999e^{-0.8t}}$$

$$1 + 4999e^{-0.8t} = 2.5$$

$$e^{-0.8t} = \frac{1.5}{4999}$$

$$\ln e^{-0.8t} = \ln \frac{1.5}{4999}$$

$$-0.8t = \ln \frac{1.5}{4999}$$

$$t = -\frac{1}{0.8} \ln \frac{1.5}{4999}$$

$$t \approx 10.1$$

So, after about 10 days, at least 40% of the students will be infected, and the college will cancel classes.

CHECK*Point* Now try Exercise 59.

Graphical Solution

a. Use a graphing utility to graph $y = \dfrac{5000}{1 + 4999e^{-0.8x}}$. Use the *value* feature or the *zoom* and *trace* features of the graphing utility to estimate that $y \approx 54$ when $x = 5$. So, after 5 days, about 54 students will be infected.

b. Classes are canceled when the number of infected students is $(0.40)(5000) = 2000$. Use a graphing utility to graph

$$y_1 = \frac{5000}{1 + 4999e^{-0.8x}} \quad \text{and} \quad y_2 = 2000$$

in the same viewing window. Use the *intersect* feature or the *zoom* and *trace* features of the graphing utility to find the point of intersection of the graphs. In Figure 5.41, you can see that the point of intersection occurs near $x \approx 10.1$. So, after about 10 days, at least 40% of the students will be infected, and the college will cancel classes.

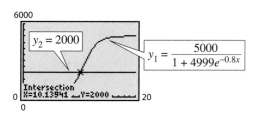

FIGURE 5.41

Logarithmic Models

Claro Cortes IV/Reuters /Landov

On May 12, 2008, an earthquake of magnitude 7.9 struck Eastern Sichuan Province, China. The total economic loss was estimated at 86 billion U.S. dollars.

Example 6   Magnitudes of Earthquakes

On the Richter scale, the magnitude R of an earthquake of intensity I is given by

$$R = \log \frac{I}{I_0}$$

where $I_0 = 1$ is the minimum intensity used for comparison. Find the intensity of each earthquake. (Intensity is a measure of the wave energy of an earthquake.)

a. Nevada in 2008: $R = 6.0$

b. Eastern Sichuan, China in 2008: $R = 7.9$

Solution

a. Because $I_0 = 1$ and $R = 6.0$, you have

$$= \log \frac{I}{-}$$

$$10^{6.0} = 10^{\log I}$$

$$I = 10^{6.0} = 1,000,000.$$

b. For $R = 7.9$, you have

$$= \log \frac{I}{-}$$

$$10^{7.9} = 10^{\log I}$$

$$I = 10^{7.9} \approx 79,400,000.$$

Note that an increase of 1.9 units on the Richter scale (from 6.0 to 7.9) represents an increase in intensity by a factor of

$$\frac{79,400,000}{1,000,000} = 79.4.$$

In other words, the intensity of the earthquake in Eastern Sichuan was about 79 times as great as that of the earthquake in Nevada.

CHECK**Point**   Now try Exercise 63.

CLASSROOM DISCUSSION

Comparing Population Models The populations P (in millions) of the United States for the census years from 1910 to 2000 are shown in the table at the left. Least squares regression analysis gives the best quadratic model for these data as $P = 1.0328t^2 + 9.607t + 81.82$, and the best exponential model for these data as $P = 82.677e^{0.124t}$. Which model better fits the data? Describe how you reached your conclusion.   (Source: U.S. Census Bureau)

t	Year	Population, P
1	1910	92.23
2	1920	106.02
3	1930	123.20
4	1940	132.16
5	1950	151.33
6	1960	179.32
7	1970	203.30
8	1980	226.54
9	1990	248.72
10	2000	281.42

5.5   EXERCISES

See www.CalcChat.com for worked-out solutions to odd-numbered exercises.

VOCABULARY: Fill in the blanks.

1. An exponential growth model has the form _____ and an exponential decay model has the form _____.
2. A logarithmic model has the form _____ or _____.
3. Gaussian models are commonly used in probability and statistics to represent populations that are _____ _____.
4. The graph of a Gaussian model is _____ shaped, where the _____ _____ is the maximum y-value of the graph.
5. A logistic growth model has the form _____.
6. A logistic curve is also called a _____ curve.

SKILLS AND APPLICATIONS

In Exercises 7–12, match the function with its graph. [The graphs are labeled (a), (b), (c), (d), (e), and (f).]

(a)
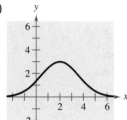

(b)
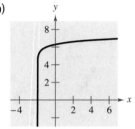

(c)
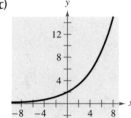

(d)
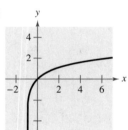

(e)
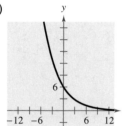

(f)
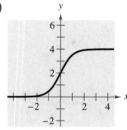

7. $y = 2e^{x/4}$

8. $y = 6e^{-x/4}$

9. $y = 6 + \log(x + 2)$

10. $y = 3e^{-(x-2)^2/5}$

11. $y = \ln(x + 1)$

12. $y = \dfrac{4}{1 + e^{-2x}}$

In Exercises 13 and 14, (a) solve for P and (b) solve for t.

13. $A = Pe^{rt}$

14. $A = P\left(1 + \dfrac{r}{n}\right)^{nt}$

COMPOUND INTEREST   In Exercises 15–22, complete the table for a savings account in which interest is compounded continuously.

	Initial Investment	Annual % Rate	Time to Double	Amount After 10 Years
15.	$1000	3.5%		
16.	$750	$10\frac{1}{2}$%		
17.	$750		$7\frac{3}{4}$ yr	
18.	$10,000		12 yr	
19.	$500			$1505.00
20.	$600			$19,205.00
21.		4.5%		$10,000.00
22.		2%		$2000.00

COMPOUND INTEREST   In Exercises 23 and 24, determine the principal P that must be invested at rate r, compounded monthly, so that $500,000 will be available for retirement in t years.

23. $r = 5\%, t = 10$

24. $r = 3\frac{1}{2}\%, t = 15$

COMPOUND INTEREST   In Exercises 25 and 26, determine the time necessary for $1000 to double if it is invested at interest rate r compounded (a) annually, (b) monthly, (c) daily, and (d) continuously.

25. $r = 10\%$

26. $r = 6.5\%$

27. **COMPOUND INTEREST**   Complete the table for the time t (in years) necessary for P dollars to triple if interest is compounded continuously at rate r.

r	2%	4%	6%	8%	10%	12%
t						

 28. **MODELING DATA**   Draw a scatter plot of the data in Exercise 27. Use the *regression* feature of a graphing utility to find a model for the data.

29. COMPOUND INTEREST Complete the table for the time t (in years) necessary for P dollars to triple if interest is compounded annually at rate r.

r	2%	4%	6%	8%	10%	12%
t						

30. MODELING DATA Draw a scatter plot of the data in Exercise 29. Use the *regression* feature of a graphing utility to find a model for the data.

31. COMPARING MODELS If $1 is invested in an account over a 10-year period, the amount in the account, where t represents the time in years, is given by $A = 1 + 0.075[\![t]\!]$ or $A = e^{0.07t}$ depending on whether the account pays simple interest at $7\frac{1}{2}$% or continuous compound interest at 7%. Graph each function on the same set of axes. Which grows at a higher rate? (Remember that $[\![t]\!]$ is the greatest integer function discussed in Section 2.4.)

32. COMPARING MODELS If $1 is invested in an account over a 10-year period, the amount in the account, where t represents the time in years, is given by $A = 1 + 0.06[\![t]\!]$ or $A = [1 + (0.055/365)]^{[\![365t]\!]}$ depending on whether the account pays simple interest at 6% or compound interest at $5\frac{1}{2}$% compounded daily. Use a graphing utility to graph each function in the same viewing window. Which grows at a higher rate?

RADIOACTIVE DECAY In Exercises 33–38, complete the table for the radioactive isotope.

Isotope	Half-life (years)	Initial Quantity	Amount After 1000 Years
33. ^{226}Ra	1599	10 g	
34. ^{14}C	5715	6.5 g	
35. ^{239}Pu	24,100	2.1 g	
36. ^{226}Ra	1599		2 g
37. ^{14}C	5715		2 g
38. ^{239}Pu	24,100		0.4 g

In Exercises 39–42, find the exponential model $y = ae^{bx}$ that fits the points shown in the graph or table.

39.
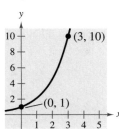

40.
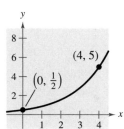

41.

x	0	4
y	5	1

42.

x	0	3
y	1	$\frac{1}{4}$

43. POPULATION The populations P (in thousands) of Horry County, South Carolina from 1970 through 2007 can be modeled by

$$P = -18.5 + 92.2e^{0.0282t}$$

where t represents the year, with $t = 0$ corresponding to 1970. (Source: U.S. Census Bureau)

(a) Use the model to complete the table.

Year	1970	1980	1990	2000	2007
Population					

(b) According to the model, when will the population of Horry County reach 300,000?

(c) Do you think the model is valid for long-term predictions of the population? Explain.

44. POPULATION The table shows the populations (in millions) of five countries in 2000 and the projected populations (in millions) for the year 2015. (Source: U.S. Census Bureau)

Country	2000	2015
Bulgaria	7.8	6.9
Canada	31.1	35.1
China	1268.9	1393.4
United Kingdom	59.5	62.2
United States	282.2	325.5

(a) Find the exponential growth or decay model $y = ae^{bt}$ or $y = ae^{-bt}$ for the population of each country by letting $t = 0$ correspond to 2000. Use the model to predict the population of each country in 2030.

(b) You can see that the populations of the United States and the United Kingdom are growing at different rates. What constant in the equation $y = ae^{bt}$ is determined by these different growth rates? Discuss the relationship between the different growth rates and the magnitude of the constant.

(c) You can see that the population of China is increasing while the population of Bulgaria is decreasing. What constant in the equation $y = ae^{bt}$ reflects this difference? Explain.

45. WEBSITE GROWTH The number y of hits a new search-engine website receives each month can be modeled by $y = 4080e^{kt}$, where t represents the number of months the website has been operating. In the website's third month, there were 10,000 hits. Find the value of k, and use this value to predict the number of hits the website will receive after 24 months.

46. VALUE OF A PAINTING The value V (in millions of dollars) of a famous painting can be modeled by $V = 10e^{kt}$, where t represents the year, with $t = 0$ corresponding to 2000. In 2008, the same painting was sold for $65 million. Find the value of k, and use this value to predict the value of the painting in 2014.

47. POPULATION The populations P (in thousands) of Reno, Nevada from 2000 through 2007 can be modeled by $P = 346.8e^{kt}$, where t represents the year, with $t = 0$ corresponding to 2000. In 2005, the population of Reno was about 395,000. (Source: U.S. Census Bureau)

(a) Find the value of k. Is the population increasing or decreasing? Explain.

(b) Use the model to find the populations of Reno in 2010 and 2015. Are the results reasonable? Explain.

(c) According to the model, during what year will the population reach 500,000?

48. POPULATION The populations P (in thousands) of Orlando, Florida from 2000 through 2007 can be modeled by $P = 1656.2e^{kt}$, where t represents the year, with $t = 0$ corresponding to 2000. In 2005, the population of Orlando was about 1,940,000. (Source: U.S. Census Bureau)

(a) Find the value of k. Is the population increasing or decreasing? Explain.

(b) Use the model to find the populations of Orlando in 2010 and 2015. Are the results reasonable? Explain.

(c) According to the model, during what year will the population reach 2.2 million?

49. BACTERIA GROWTH The number of bacteria in a culture is increasing according to the law of exponential growth. After 3 hours, there are 100 bacteria, and after 5 hours, there are 400 bacteria. How many bacteria will there be after 6 hours?

50. BACTERIA GROWTH The number of bacteria in a culture is increasing according to the law of exponential growth. The initial population is 250 bacteria, and the population after 10 hours is double the population after 1 hour. How many bacteria will there be after 6 hours?

51. CARBON DATING

(a) The ratio of carbon 14 to carbon 12 in a piece of wood discovered in a cave is $R = 1/8^{14}$. Estimate the age of the piece of wood.

(b) The ratio of carbon 14 to carbon 12 in a piece of paper buried in a tomb is $R = 1/13^{11}$. Estimate the age of the piece of paper.

52. RADIOACTIVE DECAY Carbon 14 dating assumes that the carbon dioxide on Earth today has the same radioactive content as it did centuries ago. If this is true, the amount of ^{14}C absorbed by a tree that grew several centuries ago should be the same as the amount of ^{14}C absorbed by a tree growing today. A piece of ancient charcoal contains only 15% as much radioactive carbon as a piece of modern charcoal. How long ago was the tree burned to make the ancient charcoal if the half-life of ^{14}C is 5715 years?

53. DEPRECIATION A sport utility vehicle that costs $23,300 new has a book value of $12,500 after 2 years.

(a) Find the linear model $V = mt + b$.

(b) Find the exponential model $V = ae^{kt}$.

(c) Use a graphing utility to graph the two models in the same viewing window. Which model depreciates faster in the first 2 years?

(d) Find the book values of the vehicle after 1 year and after 3 years using each model.

(e) Explain the advantages and disadvantages of using each model to a buyer and a seller.

54. DEPRECIATION A laptop computer that costs $1150 new has a book value of $550 after 2 years.

(a) Find the linear model $V = mt + b$.

(b) Find the exponential model $V = ae^{kt}$.

(c) Use a graphing utility to graph the two models in the same viewing window. Which model depreciates faster in the first 2 years?

(d) Find the book values of the computer after 1 year and after 3 years using each model.

(e) Explain the advantages and disadvantages of using each model to a buyer and a seller.

55. SALES The sales S (in thousands of units) of a new CD burner after it has been on the market for t years are modeled by $S(t) = 100(1 - e^{kt})$. Fifteen thousand units of the new product were sold the first year.

(a) Complete the model by solving for k.

(b) Sketch the graph of the model.

(c) Use the model to estimate the number of units sold after 5 years.

56. LEARNING CURVE The management at a plastics factory has found that the maximum number of units a worker can produce in a day is 30. The learning curve for the number N of units produced per day after a new employee has worked t days is modeled by $N = 30(1 - e^{kt})$. After 20 days on the job, a new employee produces 19 units.

(a) Find the learning curve for this employee (first, find the value of k).

(b) How many days should pass before this employee is producing 25 units per day?

57. IQ SCORES The IQ scores for a sample of a class of returning adult students at a small northeastern college roughly follow the normal distribution $y = 0.0266e^{-(x-100)^2/450}$, $70 \le x \le 115$, where x is the IQ score.

(a) Use a graphing utility to graph the function.

(b) From the graph in part (a), estimate the average IQ score of an adult student.

58. EDUCATION The amount of time (in hours per week) a student utilizes a math-tutoring center roughly follows the normal distribution $y = 0.7979e^{-(x-5.4)^2/0.5}$, $4 \le x \le 7$, where x is the number of hours.

(a) Use a graphing utility to graph the function.

(b) From the graph in part (a), estimate the average number of hours per week a student uses the tutoring center.

59. CELL SITES A cell site is a site where electronic communications equipment is placed in a cellular network for the use of mobile phones. The numbers y of cell sites from 1985 through 2008 can be modeled by

$$y = \frac{237{,}101}{1 + 1950e^{-0.355t}}$$

where t represents the year, with $t = 5$ corresponding to 1985. (Source: CTIA-The Wireless Association)

(a) Use the model to find the numbers of cell sites in the years 1985, 2000, and 2006.

(b) Use a graphing utility to graph the function.

(c) Use the graph to determine the year in which the number of cell sites will reach 235,000.

(d) Confirm your answer to part (c) algebraically.

60. POPULATION The populations P (in thousands) of Pittsburgh, Pennsylvania from 2000 through 2007 can be modeled by

$$P = \frac{2632}{1 + 0.083e^{0.0500t}}$$

where t represents the year, with $t = 0$ corresponding to 2000. (Source: U.S. Census Bureau)

(a) Use the model to find the populations of Pittsburgh in the years 2000, 2005, and 2007.

(b) Use a graphing utility to graph the function.

(c) Use the graph to determine the year in which the population will reach 2.2 million.

(d) Confirm your answer to part (c) algebraically.

61. POPULATION GROWTH A conservation organization releases 100 animals of an endangered species into a game preserve. The organization believes that the preserve has a carrying capacity of 1000 animals and that the growth of the pack will be modeled by the logistic curve

$$p(t) = \frac{1000}{1 + 9e^{-0.1656t}}$$

where t is measured in months (see figure).

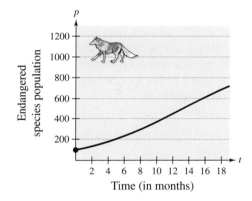

(a) Estimate the population after 5 months.

(b) After how many months will the population be 500?

(c) Use a graphing utility to graph the function. Use the graph to determine the horizontal asymptotes, and interpret the meaning of the asymptotes in the context of the problem.

62. SALES After discontinuing all advertising for a tool kit in 2004, the manufacturer noted that sales began to drop according to the model

$$S = \frac{500{,}000}{1 + 0.4e^{kt}}$$

where S represents the number of units sold and $t = 4$ represents 2004. In 2008, the company sold 300,000 units.

(a) Complete the model by solving for k.

(b) Estimate sales in 2012.

GEOLOGY In Exercises 63 and 64, use the Richter scale

$$R = \log \frac{I}{I_0}$$

for measuring the magnitudes of earthquakes.

63. Find the intensity I of an earthquake measuring R on the Richter scale (let $I_0 = 1$).

(a) Southern Sumatra, Indonesia in 2007, $R = 8.5$

(b) Illinois in 2008, $R = 5.4$

(c) Costa Rica in 2009, $R = 6.1$

64. Find the magnitude R of each earthquake of intensity I (let $I_0 = 1$).

(a) $I = 199{,}500{,}000$ (b) $I = 48{,}275{,}000$

(c) $I = 17{,}000$

INTENSITY OF SOUND In Exercises 65–68, use the following information for determining sound intensity. The level of sound β, in decibels, with an intensity of I, is given by $\beta = 10 \log(I/I_0)$, where I_0 is an intensity of 10^{-12} watt per square meter, corresponding roughly to the faintest sound that can be heard by the human ear. In Exercises 65 and 66, find the level of sound β.

65. (a) $I = 10^{-10}$ watt per m² (quiet room)

(b) $I = 10^{-5}$ watt per m² (busy street corner)

(c) $I = 10^{-8}$ watt per m² (quiet radio)

(d) $I = 10^0$ watt per m² (threshold of pain)

66. (a) $I = 10^{-11}$ watt per m² (rustle of leaves)

(b) $I = 10^2$ watt per m² (jet at 30 meters)

(c) $I = 10^{-4}$ watt per m² (door slamming)

(d) $I = 10^{-2}$ watt per m² (siren at 30 meters)

67. Due to the installation of noise suppression materials, the noise level in an auditorium was reduced from 93 to 80 decibels. Find the percent decrease in the intensity level of the noise as a result of the installation of these materials.

68. Due to the installation of a muffler, the noise level of an engine was reduced from 88 to 72 decibels. Find the percent decrease in the intensity level of the noise as a result of the installation of the muffler.

pH LEVELS In Exercises 69–74, use the acidity model given by pH $= -\log[\text{H}^+]$, where acidity (pH) is a measure of the hydrogen ion concentration $[\text{H}^+]$ (measured in moles of hydrogen per liter) of a solution.

69. Find the pH if $[\text{H}^+] = 2.3 \times 10^{-5}$.

70. Find the pH if $[\text{H}^+] = 1.13 \times 10^{-5}$.

71. Compute $[\text{H}^+]$ for a solution in which pH $= 5.8$.

72. Compute $[\text{H}^+]$ for a solution in which pH $= 3.2$.

73. Apple juice has a pH of 2.9 and drinking water has a pH of 8.0. The hydrogen ion concentration of the apple juice is how many times the concentration of drinking water?

74. The pH of a solution is decreased by one unit. The hydrogen ion concentration is increased by what factor?

75. FORENSICS At 8:30 A.M., a coroner was called to the home of a person who had died during the night. In order to estimate the time of death, the coroner took the person's temperature twice. At 9:00 A.M. the temperature was 85.7°F, and at 11:00 A.M. the temperature was 82.8°F. From these two temperatures, the coroner was able to determine that the time elapsed since death and the body temperature were related by the formula

$$t = -10 \ln \frac{T - 70}{98.6 - 70}$$

where t is the time in hours elapsed since the person died and T is the temperature (in degrees Fahrenheit) of the person's body. (This formula is derived from a general cooling principle called *Newton's Law of Cooling*. It uses the assumptions that the person had a normal body temperature of 98.6°F at death, and that the room temperature was a constant 70°F.) Use the formula to estimate the time of death of the person.

 76. HOME MORTGAGE A \$120,000 home mortgage for 30 years at $7\frac{1}{2}\%$ has a monthly payment of \$839.06. Part of the monthly payment is paid toward the interest charge on the unpaid balance, and the remainder of the payment is used to reduce the principal. The amount that is paid toward the interest is

$$u = M - \left(M - \frac{Pr}{12}\right)\left(1 + \frac{r}{12}\right)^{12t}$$

and the amount that is paid toward the reduction of the principal is

$$v = \left(M - \frac{Pr}{12}\right)\left(1 + \frac{r}{12}\right)^{12t}.$$

In these formulas, P is the size of the mortgage, r is the interest rate, M is the monthly payment, and t is the time (in years).

(a) Use a graphing utility to graph each function in the same viewing window. (The viewing window should show all 30 years of mortgage payments.)

(b) In the early years of the mortgage, is the larger part of the monthly payment paid toward the interest or the principal? Approximate the time when the monthly payment is evenly divided between interest and principal reduction.

(c) Repeat parts (a) and (b) for a repayment period of 20 years ($M = \$966.71$). What can you conclude?

77. HOME MORTGAGE The total interest u paid on a home mortgage of P dollars at interest rate r for t years is

$$u = P\left[\dfrac{rt}{1 - \left(\dfrac{1}{1 + r/12}\right)^{12t}} - 1\right].$$

Consider a \$120,000 home mortgage at $7\frac{1}{2}\%$.

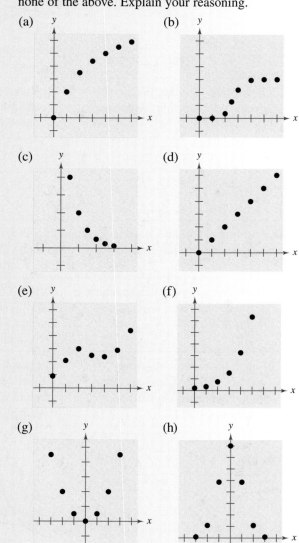

(a) Use a graphing utility to graph the total interest function.

(b) Approximate the length of the mortgage for which the total interest paid is the same as the size of the mortgage. Is it possible that some people are paying twice as much in interest charges as the size of the mortgage?

78. DATA ANALYSIS The table shows the time t (in seconds) required for a car to attain a speed of s miles per hour from a standing start.

Speed, s	Time, t
30	3.4
40	5.0
50	7.0
60	9.3
70	12.0
80	15.8
90	20.0

Two models for these data are as follows.

$$t_1 = 40.757 + 0.556s - 15.817 \ln s$$

$$t_2 = 1.2259 + 0.0023s^2$$

(a) Use the *regression* feature of a graphing utility to find a linear model t_3 and an exponential model t_4 for the data.

(b) Use a graphing utility to graph the data and each model in the same viewing window.

(c) Create a table comparing the data with estimates obtained from each model.

(d) Use the results of part (c) to find the sum of the absolute values of the differences between the data and the estimated values given by each model. Based on the four sums, which model do you think best fits the data? Explain.

EXPLORATION

TRUE OR FALSE? In Exercises 79–82, determine whether the statement is true or false. Justify your answer.

79. The domain of a logistic growth function cannot be the set of real numbers.

80. A logistic growth function will always have an x-intercept.

81. The graph of $f(x) = \dfrac{4}{1 + 6e^{-2x}} + 5$ is the graph of $g(x) = \dfrac{4}{1 + 6e^{-2x}}$ shifted to the right five units.

82. The graph of a Gaussian model will never have an x-intercept.

83. WRITING Use your school's library, the Internet, or some other reference source to write a paper describing John Napier's work with logarithms.

84. CAPSTONE Identify each model as exponential, Gaussian, linear, logarithmic, logistic, quadratic, or none of the above. Explain your reasoning.

PROJECT: SALES PER SHARE To work an extended application analyzing the sales per share for Kohl's Corporation from 1992 through 2007, visit this text's website at *academic.cengage.com*. (Data Source: Kohl's Corporation)

5 CHAPTER SUMMARY

What Did You Learn?	Explanation/Examples	Review Exercises
Section 5.1 Recognize and evaluate exponential functions with base a (p. 380).	The exponential function f with base a is denoted by $f(x) = a^x$ where $a > 0$, $a \neq 1$, and x is any real number.	1–6
Graph exponential functions and use the One-to-One Property (p. 381).	**One-to-One Property:** For $a > 0$ and $a \neq 1$, $a^x = a^y$ if and only if $x = y$.	7–24
Recognize, evaluate, and graph exponential functions with base e (p. 384).	The function $f(x) = e^x$ is called the natural exponential function.	25–32
Use exponential functions to model and solve real-life problems (p. 385).	Exponential functions are used in compound interest formulas (See Example 8.) and in radioactive decay models. (See Example 9.)	33–36
Section 5.2 Recognize and evaluate logarithmic functions with base a (p. 391).	For $x > 0$, $a > 0$, and $a \neq 1$, $y = \log_a x$ if and only if $x = a^y$. The function $f(x) = \log_a x$ is called the logarithmic function with base a. The logarithmic function with base 10 is the common logarithmic function. It is denoted by $\log_{10}$ or log.	37–48
Graph logarithmic functions (p. 393) and recognize, evaluate, and graph natural logarithmic functions (p. 395).	The graph of $y = \log_a x$ is a reflection of the graph of $y = a^x$ about the line $y = x$. The function defined by $f(x) = \ln x$, $x > 0$, is called the natural logarithmic function. Its graph is a reflection of the graph of $f(x) = e^x$ about the line $y = x$.	49–52
		53–58
Use logarithmic functions to model and solve real-life problems (p. 397).	A logarithmic function is used in the human memory model. (See Example 11.)	59, 60

What Did You Learn?	**Explanation/Examples**	**Review Exercises**
Section 5.3 Use the change-of-base formula to rewrite and evaluate logarithmic expressions *(p. 401)*.	Let a, b, and x be positive real numbers such that $a \neq 1$ and $b \neq 1$. Then $\log_a x$ can be converted to a different base as follows. *Base b* $\qquad\qquad$ *Base 10* $\qquad\qquad$ *Base e* $\log_a x = \dfrac{\log_b x}{\log_b a} \qquad \log_a x = \dfrac{\log x}{\log a} \qquad \log_a x = \dfrac{\ln x}{\ln a}$	61–64
Use properties of logarithms to evaluate, rewrite, expand, or condense logarithmic expressions *(p. 402)*.	Let a be a positive number $(a \neq 1)$, n be a real number, and u and v be positive real numbers. **1. Product Property:** $\log_a(uv) = \log_a u + \log_a v$ $\qquad\qquad\qquad\qquad\quad \ln(uv) = \ln u + \ln v$ **2. Quotient Property:** $\log_a(u/v) = \log_a u - \log_a v$ $\qquad\qquad\qquad\qquad\quad \ln(u/v) = \ln u - \ln v$ **3. Power Property:** $\qquad \log_a u^n = n \log_a u$, $\ln u^n = n \ln u$	65–80
Use logarithmic functions to model and solve real-life problems *(p. 404)*.	Logarithmic functions can be used to find an equation that relates the periods of several planets and their distances from the sun. (See Example 7.)	81, 82
Section 5.4 Solve simple exponential and logarithmic equations *(p. 408)*.	One-to-One Properties and Inverse Properties of exponential or logarithmic functions can be used to help solve exponential or logarithmic equations.	83–88
Solve more complicated exponential equations *(p. 409)* and logarithmic equations *(p. 411)*.	To solve more complicated equations, rewrite the equations so that the One-to-One Properties and Inverse Properties of exponential or logarithmic functions can be used. (See Examples 2–8.)	89–108
Use exponential and logarithmic equations to model and solve real-life problems *(p. 413)*.	Exponential and logarithmic equations can be used to find how long it will take to double an investment (see Example 10) and to find the year in which companies reached a given amount of sales. (See Example 11.)	109, 110
Section 5.5 Recognize the five most common types of models involving exponential and logarithmic functions *(p. 419)*.	**1. Exponential growth model:** $y = ae^{bx}$, $\quad b > 0$ **2. Exponential decay model:** $y = ae^{-bx}$, $\quad b > 0$ **3. Gaussian model:** $y = ae^{-(x-b)^2/c}$ **4. Logistic growth model:** $y = \dfrac{a}{1 + be^{-rx}}$ **5. Logarithmic models:** $y = a + b \ln x$, $y = a + b \log x$	111–116
Use exponential growth and decay functions to model and solve real-life problems *(p. 420)*.	An exponential growth function can be used to model a population of fruit flies (see Example 2) and an exponential decay function can be used to find the age of a fossil (see Example 3).	117–120
Use Gaussian functions *(p. 423)*, logistic growth functions *(p. 424)*, and logarithmic functions *(p. 425)* to model and solve real-life problems.	A Gaussian function can be used to model SAT math scores for college-bound seniors. (See Example 4.) A logistic growth function can be used to model the spread of a flu virus. (See Example 5.) A logarithmic function can be used to find the intensity of an earthquake using its magnitude. (See Example 6.)	121–123

5 REVIEW EXERCISES

See www.CalcChat.com for worked-out solutions to odd-numbered exercises.

5.1 In Exercises 1–6, evaluate the function at the indicated value of x. Round your result to three decimal places.

1. $f(x) = 0.3^x$, $x = 1.5$ **2.** $f(x) = 30^x$, $x = \sqrt{3}$

3. $f(x) = 2^{-0.5x}$, $x = \pi$ **4.** $f(x) = 1278^{x/5}$, $x = 1$

5. $f(x) = 7(0.2^x)$, $x = -\sqrt{11}$

6. $f(x) = -14(5^x)$, $x = -0.8$

In Exercises 7–14, use the graph of f to describe the transformation that yields the graph of g.

7. $f(x) = 2^x$, $g(x) = 2^x - 2$

8. $f(x) = 5^x$, $g(x) = 5^x + 1$

9. $f(x) = 4^x$, $g(x) = 4^{-x+2}$

10. $f(x) = 6^x$, $g(x) = 6^{x+1}$

11. $f(x) = 3^x$, $g(x) = 1 - 3^x$

12. $f(x) = 0.1^x$, $g(x) = -0.1^x$

13. $f(x) = \left(\frac{1}{2}\right)^x$, $g(x) = -\left(\frac{1}{2}\right)^{x+2}$

14. $f(x) = \left(\frac{2}{3}\right)^x$, $g(x) = 8 - \left(\frac{2}{3}\right)^x$

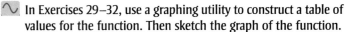 In Exercises 15–20, use a graphing utility to construct a table of values for the function. Then sketch the graph of the function.

15. $f(x) = 4^{-x} + 4$ **16.** $f(x) = 2.65^{x-1}$

17. $f(x) = 5^{x-2} + 4$ **18.** $f(x) = 2^{x-6} - 5$

19. $f(x) = \left(\frac{1}{2}\right)^{-x} + 3$ **20.** $f(x) = \left(\frac{1}{8}\right)^{x+2} - 5$

In Exercises 21–24, use the One-to-One Property to solve the equation for x.

21. $\left(\frac{1}{3}\right)^{x-3} = 9$ **22.** $3^{x+3} = \frac{1}{81}$

23. $e^{3x-5} = e^7$ **24.** $e^{8-2x} = e^{-3}$

In Exercises 25–28, evaluate $f(x) = e^x$ at the indicated value of x. Round your result to three decimal places.

25. $x = 8$ **26.** $x = \frac{5}{8}$

27. $x = -1.7$ **28.** $x = 0.278$

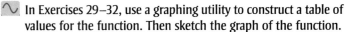 In Exercises 29–32, use a graphing utility to construct a table of values for the function. Then sketch the graph of the function.

29. $h(x) = e^{-x/2}$ **30.** $h(x) = 2 - e^{-x/2}$

31. $f(x) = e^{x+2}$ **32.** $s(t) = 4e^{-2/t}$, $t > 0$

COMPOUND INTEREST In Exercises 33 and 34, complete the table to determine the balance A for P dollars invested at rate r for t years and compounded n times per year.

n	1	2	4	12	365	Continuous
A						

TABLE FOR 33 AND 34

33. $P = \$5000$, $r = 3\%$, $t = 10$ years

34. $P = \$4500$, $r = 2.5\%$, $t = 30$ years

35. WAITING TIMES The average time between incoming calls at a switchboard is 3 minutes. The probability F of waiting less than t minutes until the next incoming call is approximated by the model $F(t) = 1 - e^{-t/3}$. A call has just come in. Find the probability that the next call will be within

(a) $\frac{1}{2}$ minute.   (b) 2 minutes.   (c) 5 minutes.

36. DEPRECIATION After t years, the value V of a car that originally cost $\$23,970$ is given by $V(t) = 23,970\left(\frac{3}{4}\right)^t$.

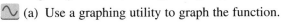 (a) Use a graphing utility to graph the function.

(b) Find the value of the car 2 years after it was purchased.

(c) According to the model, when does the car depreciate most rapidly? Is this realistic? Explain.

(d) According to the model, when will the car have no value?

5.2 In Exercises 37–40, write the exponential equation in logarithmic form. For example, the logarithmic form of $2^3 = 8$ is $\log_2 8 = 3$.

37. $3^3 = 27$ **38.** $25^{3/2} = 125$

39. $e^{0.8} = 2.2255\ldots$ **40.** $e^0 = 1$

In Exercises 41–44, evaluate the function at the indicated value of x without using a calculator.

41. $f(x) = \log x$, $x = 1000$ **42.** $g(x) = \log_9 x$, $x = 3$

43. $g(x) = \log_2 x$, $x = \frac{1}{4}$ **44.** $f(x) = \log_3 x$, $x = \frac{1}{81}$

In Exercises 45–48, use the One-to-One Property to solve the equation for x.

45. $\log_4(x + 7) = \log_4 14$ **46.** $\log_8(3x - 10) = \log_8 5$

47. $\ln(x + 9) = \ln 4$ **48.** $\ln(2x - 1) = \ln 11$

In Exercises 49–52, find the domain, x-intercept, and vertical asymptote of the logarithmic function and sketch its graph.

49. $g(x) = \log_7 x$ **50.** $f(x) = \log\left(\frac{x}{3}\right)$

51. $f(x) = 4 - \log(x + 5)$ **52.** $f(x) = \log(x - 3) + 1$

53. Use a calculator to evaluate $f(x) = \ln x$ at (a) $x = 22.6$ and (b) $x = 0.98$. Round your results to three decimal places if necessary.

54. Use a calculator to evaluate $f(x) = 5 \ln x$ at (a) $x = e^{-12}$ and (b) $x = \sqrt{3}$. Round your results to three decimal places if necessary.

In Exercises 55–58, find the domain, x-intercept, and vertical asymptote of the logarithmic function and sketch its graph.

55. $f(x) = \ln x + 3$ **56.** $f(x) = \ln(x - 3)$

57. $h(x) = \ln(x^2)$ **58.** $f(x) = \frac{1}{4} \ln x$

59. ANTLER SPREAD The antler spread a (in inches) and shoulder height h (in inches) of an adult male American elk are related by the model $h = 116 \log(a + 40) - 176$. Approximate the shoulder height of a male American elk with an antler spread of 55 inches.

60. SNOW REMOVAL The number of miles s of roads cleared of snow is approximated by the model

$$s = 25 - \frac{13 \ln(h/12)}{\ln 3}, \quad 2 \le h \le 15$$

where h is the depth of the snow in inches. Use this model to find s when $h = 10$ inches.

5.3 In Exercises 61–64, evaluate the logarithm using the change-of-base formula. Do each exercise twice, once with common logarithms and once with natural logarithms. Round the results to three decimal places.

61. $\log_2 6$ **62.** $\log_{12} 200$

63. $\log_{1/2} 5$ **64.** $\log_3 0.28$

In Exercises 65–68, use the properties of logarithms to rewrite and simplify the logarithmic expression.

65. $\log 18$ **66.** $\log_2\left(\frac{1}{12}\right)$

67. $\ln 20$ **68.** $\ln(3e^{-4})$

In Exercises 69–74, use the properties of logarithms to expand the expression as a sum, difference, and/or constant multiple of logarithms. (Assume all variables are positive.)

69. $\log_5 5x^2$ **70.** $\log 7x^4$

71. $\log_3 \dfrac{9}{\sqrt{x}}$ **72.** $\log_7 \dfrac{\sqrt[3]{x}}{14}$

73. $\ln x^2 y^2 z$ **74.** $\ln\left(\dfrac{y-1}{4}\right)^2, \quad y > 1$

In Exercises 75–80, condense the expression to the logarithm of a single quantity.

75. $\log_2 5 + \log_2 x$ **76.** $\log_6 y - 2 \log_6 z$

77. $\ln x - \frac{1}{4} \ln y$ **78.** $3 \ln x + 2 \ln(x + 1)$

79. $\frac{1}{2} \log_3 x - 2 \log_3(y + 8)$

80. $5 \ln(x - 2) - \ln(x + 2) - 3 \ln x$

81. CLIMB RATE The time t (in minutes) for a small plane to climb to an altitude of h feet is modeled by $t = 50 \log[18{,}000/(18{,}000 - h)]$, where 18,000 feet is the plane's absolute ceiling.

(a) Determine the domain of the function in the context of the problem.

(b) Use a graphing utility to graph the function and identify any asymptotes.

(c) As the plane approaches its absolute ceiling, what can be said about the time required to increase its altitude?

(d) Find the time for the plane to climb to an altitude of 4000 feet.

82. HUMAN MEMORY MODEL Students in a learning theory study were given an exam and then retested monthly for 6 months with an equivalent exam. The data obtained in the study are given as the ordered pairs (t, s), where t is the time in months after the initial exam and s is the average score for the class. Use these data to find a logarithmic equation that relates t and s.

(1, 84.2), (2, 78.4), (3, 72.1),
(4, 68.5), (5, 67.1), (6, 65.3)

5.4 In Exercises 83–88, solve for x.

83. $5^x = 125$ **84.** $6^x = \frac{1}{216}$

85. $e^x = 3$ **86.** $\log_6 x = -1$

87. $\ln x = 4$ **88.** $\ln x = -1.6$

In Exercises 89–92, solve the exponential equation algebraically. Approximate your result to three decimal places.

89. $e^{4x} = e^{x^2 + 3}$ **90.** $e^{3x} = 25$

91. $2^x - 3 = 29$ **92.** $e^{2x} - 6e^x + 8 = 0$

In Exercises 93 and 94, use a graphing utility to graph and solve the equation. Approximate the result to three decimal places.

93. $25e^{-0.3x} = 12$ **94.** $2^x = 3 + x - e^x$

In Exercises 95–104, solve the logarithmic equation algebraically. Approximate the result to three decimal places.

95. $\ln 3x = 8.2$ **96.** $4 \ln 3x = 15$

97. $\ln x - \ln 3 = 2$ **98.** $\ln x - \ln 5 = 4$

99. $\ln \sqrt{x} = 4$ **100.** $\ln \sqrt{x + 8} = 3$

101. $\log_8(x - 1) = \log_8(x - 2) - \log_8(x + 2)$

102. $\log_6(x + 2) - \log_6 x = \log_6(x + 5)$

103. $\log(1 - x) = -1$ **104.** $\log(-x - 4) = 2$

In Exercises 105–108, use a graphing utility to graph and solve the equation. Approximate the result to three decimal places.

105. $2 \ln(x + 3) - 3 = 0$ **106.** $x - 2 \log(x + 4) = 0$

107. $6 \log(x^2 + 1) - x = 0$

108. $3 \ln x + 2 \log x = e^x - 25$

109. COMPOUND INTEREST You deposit $8500 in an account that pays 3.5% interest, compounded continuously. How long will it take for the money to triple?

110. METEOROLOGY The speed of the wind S (in miles per hour) near the center of a tornado and the distance d (in miles) the tornado travels are related by the model $S = 93 \log d + 65$. On March 18, 1925, a large tornado struck portions of Missouri, Illinois, and Indiana with a wind speed at the center of about 283 miles per hour. Approximate the distance traveled by this tornado.

5.5 In Exercises 111–116, match the function with its graph. [The graphs are labeled (a), (b), (c), (d), (e), and (f).]

(a)
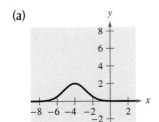

(b)
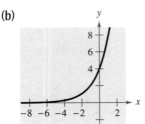

(c)
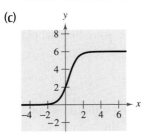

(d)
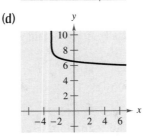

(e)
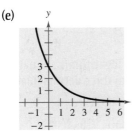

(f)
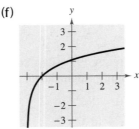

111. $y = 3e^{-2x/3}$ **112.** $y = 4e^{2x/3}$

113. $y = \ln(x + 3)$ **114.** $y = 7 - \log(x + 3)$

115. $y = 2e^{-(x+4)^2/3}$ **116.** $y = \dfrac{6}{1 + 2e^{-2x}}$

In Exercises 117 and 118, find the exponential model $y = ae^{bx}$ that passes through the points.

117. $(0, 2), (4, 3)$ **118.** $\left(0, \frac{1}{2}\right), (5, 5)$

119. POPULATION In 2007, the population of Florida residents aged 65 and over was about 3.10 million. In 2015 and 2020, the populations of Florida residents aged 65 and over are projected to be about 4.13 million and 5.11 million, respectively. An exponential growth model that approximates these data is given by $P = 2.36e^{0.0382t}$, $7 \le t \le 20$, where P is the population (in millions) and $t = 7$ represents 2007. (Source: U.S. Census Bureau)

(a) Use a graphing utility to graph the model and the data in the same viewing window. Is the model a good fit for the data? Explain.

(b) According to the model, when will the population of Florida residents aged 65 and over reach 5.5 million? Does your answer seem reasonable? Explain.

120. WILDLIFE POPULATION A species of bat is in danger of becoming extinct. Five years ago, the total population of the species was 2000. Two years ago, the total population of the species was 1400. What was the total population of the species one year ago?

121. TEST SCORES The test scores for a biology test follow a normal distribution modeled by $y = 0.0499e^{-(x-71)^2/128}$, $40 \le x \le 100$, where x is the test score. Use a graphing utility to graph the equation and estimate the average test score.

122. TYPING SPEED In a typing class, the average number N of words per minute typed after t weeks of lessons was found to be $N = 157/(1 + 5.4e^{-0.12t})$. Find the time necessary to type (a) 50 words per minute and (b) 75 words per minute.

123. SOUND INTENSITY The relationship between the number of decibels β and the intensity of a sound I in watts per square meter is $\beta = 10 \log(I/10^{-12})$. Find I for each decibel level β.

(a) $\beta = 60$ (b) $\beta = 135$ (c) $\beta = 1$

EXPLORATION

124. Consider the graph of $y = e^{kt}$. Describe the characteristics of the graph when k is positive and when k is negative.

TRUE OR FALSE? In Exercises 125 and 126, determine whether the equation is true or false. Justify your answer.

125. $\log_b b^{2x} = 2x$ **126.** $\ln(x + y) = \ln x + \ln y$

5 CHAPTER TEST

See www.CalcChat.com for worked-out solutions to odd-numbered exercises.

Take this test as you would take a test in class. When you are finished, check your work against the answers given in the back of the book.

In Exercises 1–4, evaluate the expression. Approximate your result to three decimal places.

1. $4.2^{0.6}$ **2.** $4^{3\pi/2}$ **3.** $e^{-7/10}$ **4.** $e^{3.1}$

In Exercises 5–7, construct a table of values. Then sketch the graph of the function.

5. $f(x) = 10^{-x}$ **6.** $f(x) = -6^{x-2}$ **7.** $f(x) = 1 - e^{2x}$

8. Evaluate (a) $\log_7 7^{-0.89}$ and (b) $4.6 \ln e^2$.

In Exercises 9–11, construct a table of values. Then sketch the graph of the function. Identify any asymptotes.

9. $f(x) = -\log x - 6$ **10.** $f(x) = \ln(x - 4)$ **11.** $f(x) = 1 + \ln(x + 6)$

In Exercises 12–14, evaluate the logarithm using the change-of-base formula. Round your result to three decimal places.

12. $\log_7 44$ **13.** $\log_{16} 0.63$ **14.** $\log_{3/4} 24$

In Exercises 15–17, use the properties of logarithms to expand the expression as a sum, difference, and/or constant multiple of logarithms.

15. $\log_2 3a^4$ **16.** $\ln \dfrac{5\sqrt{x}}{6}$ **17.** $\log \dfrac{(x-1)^3}{y^2 z}$

In Exercises 18–20, condense the expression to the logarithm of a single quantity.

18. $\log_3 13 + \log_3 y$ **19.** $4 \ln x - 4 \ln y$

20. $3 \ln x - \ln(x + 3) + 2 \ln y$

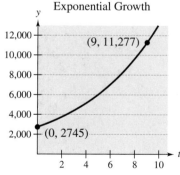

Exponential Growth

(9, 11,277)

(0, 2745)

FIGURE FOR **27**

In Exercises 21–26, solve the equation algebraically. Approximate your result to three decimal places.

21. $5^x = \dfrac{1}{25}$ **22.** $3e^{-5x} = 132$

23. $\dfrac{1025}{8 + e^{4x}} = 5$ **24.** $\ln x = \dfrac{1}{2}$

25. $18 + 4 \ln x = 7$ **26.** $\log x + \log(x - 15) = 2$

27. Find an exponential growth model for the graph shown in the figure.

28. The half-life of radioactive actinium (^{227}Ac) is 21.77 years. What percent of a present amount of radioactive actinium will remain after 19 years?

29. A model that can be used for predicting the height H (in centimeters) of a child based on his or her age is $H = 70.228 + 5.104x + 9.222 \ln x$, $\frac{1}{4} \le x \le 6$, where x is the age of the child in years. (Source: Snapshots of Applications in Mathematics)

(a) Construct a table of values. Then sketch the graph of the model.

(b) Use the graph from part (a) to estimate the height of a four-year-old child. Then calculate the actual height using the model.

5 CUMULATIVE TEST FOR CHAPTERS 3–5

See www.CalcChat.com for worked-out solutions to odd-numbered exercises.

Take this test as you would take a test in class. When you are finished, check your work against the answers given in the back of the book.

1. Find the quadratic function whose graph has a vertex at $(-8, 5)$ and passes through the point $(-4, -7)$.

In Exercises 2–4, sketch the graph of the function without the aid of a graphing utility.

2. $h(x) = -(x^2 + 4x)$ 3. $f(t) = \frac{1}{4}t(t - 2)^2$ 4. $g(s) = s^2 + 2s + 9$

In Exercises 5 and 6, find all the zeros of the function.

5. $f(x) = x^3 + 2x^2 + 4x + 8$ 6. $f(x) = x^4 + 4x^3 - 21x^2$

7. Divide: $\dfrac{6x^3 - 4x^2}{2x^2 + 1}$.

8. Use synthetic division to divide $3x^4 + 2x^2 - 5x + 3$ by $x - 2$.

 9. Use a graphing utility to approximate (to the nearest hundredth) the real zero of the function given by $g(x) = x^3 + 3x^2 - 6$.

10. Find a polynomial with real coefficients that has -5, -2, and $2 + \sqrt{3}i$ as its zeros.

In Exercises 11 and 12, find the domain of the function, and identify all asymptotes. Sketch the graph of the function.

11. $f(x) = \dfrac{2x}{x - 3}$ 12. $f(x) = \dfrac{4x^2}{x - 5}$

In Exercises 13–15, sketch the graph of the rational function by hand. Be sure to identify all intercepts and asymptotes.

13. $f(x) = \dfrac{2x}{x^2 + 2x - 3}$ 14. $f(x) = \dfrac{x^2 - 4}{x^2 + x - 2}$

15. $f(x) = \dfrac{x^3 - 2x^2 - 9x + 18}{x^2 + 4x + 3}$

In Exercises 16 and 17, sketch a graph of the conic.

16. $\dfrac{(x + 3)^2}{16} - \dfrac{(y + 4)^2}{25} = 1$ 17. $\dfrac{(x - 2)^2}{4} + \dfrac{(y + 1)^2}{9} = 1$

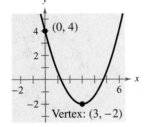

FIGURE FOR 18

18. Find an equation of the parabola shown in the figure.

19. Find an equation of the hyperbola with foci $(0, 0)$ and $(0, 4)$ and asymptotes $y = \pm\frac{1}{2}x + 2$.

 In Exercises 20 and 21, use the graph of f to describe the transformation that yields the graph of g. Use a graphing utility to graph both equations in the same viewing window.

20. $f(x) = \left(\frac{2}{5}\right)^x$, $g(x) = -\left(\frac{2}{5}\right)^{-x+3}$ 21. $f(x) = 2.2^x$, $g(x) = -2.2^x + 4$

In Exercises 22–25, use a calculator to evaluate each expression. Round your result to three decimal places.

22. $\log 98$ 23. $\log \frac{6}{7}$ 24. $\ln \sqrt{31}$ 25. $\ln\left(\sqrt{30} - 4\right)$

In Exercises 26–28, evaluate the logarithm using the change-of-base formula. Round your answer to three decimal places.

26. $\log_5 4.3$ **27.** $\log_3 0.149$ **28.** $\log_{1/2} 17$

29. Use the properties of logarithms to expand $\ln\left(\dfrac{x^2 - 16}{x^4}\right)$, where $x > 4$.

30. Write $2 \ln x - \frac{1}{2} \ln(x + 5)$ as a logarithm of a single quantity.

In Exercises 31–36, solve the equation algebraically. Approximate the result to three decimal places.

31. $6e^{2x} = 72$ **32.** $4^{x-5} + 21 = 30$

33. $e^{2x} - 13e^x + 42 = 0$ **34.** $\log_2 x + \log_2 5 = 6$

35. $\ln 4x - \ln 2 = 8$ **36.** $\ln \sqrt{x + 2} = 3$

 37. Use a graphing utility to graph

$$f(x) = \frac{1000}{1 + 4e^{-0.2x}}$$

and determine the horizontal asymptotes.

38. Let x be the amount (in hundreds of dollars) that an online stock-trading company spends on advertising, and let P be the profit (in thousands of dollars), where $P = 230 + 20x - \frac{1}{2}x^2$. What amount of advertising will yield a maximum profit?

 39. The sales S (in billions of dollars) of lottery tickets in the United States from 1997 through 2007 are shown in the table. (Source: TLF Publications, Inc.)

(a) Use a graphing utility to create a scatter plot of the data. Let t represent the year, with $t = 7$ corresponding to 1997.

(b) Use the *regression* feature of the graphing utility to find a cubic model for the data.

(c) Use the graphing utility to graph the model in the same viewing window used for the scatter plot. How well does the model fit the data?

(d) Use the model to predict the sales of lottery tickets in 2015. Does your answer seem reasonable? Explain.

40. On the day a grandchild is born, a grandparent deposits $2500 in a fund earning 7.5%, compounded continuously. Determine the balance in the account at the time of the grandchild's 25th birthday.

41. The number N of bacteria in a culture is given by the model $N = 175e^{kt}$, where t is the time in hours. If $N = 420$ when $t = 8$, estimate the time required for the population to double in size.

42. The population P of Texas (in thousands) from 2000 through 2007 can be modeled by $P = 20{,}879e^{0.0189t}$, where t represents the year, with $t = 0$ corresponding to 2000. According to this model, when will the population reach 28 million? (Source: U.S. Census Bureau)

43. The population p of a species of bird t years after it is introduced into a new habitat is given by

$$p = \frac{1200}{1 + 3e^{-t/5}}.$$

(a) Determine the population size that was introduced into the habitat.

(b) Determine the population after 5 years.

(c) After how many years will the population be 800?

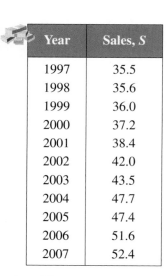

Year	Sales, S
1997	35.5
1998	35.6
1999	36.0
2000	37.2
2001	38.4
2002	42.0
2003	43.5
2004	47.7
2005	47.4
2006	51.6
2007	52.4

TABLE FOR **39**

PROOFS IN MATHEMATICS

Each of the following three properties of logarithms can be proved by using properties of exponential functions.

Slide Rules

The slide rule was invented by William Oughtred (1574–1660) in 1625. The slide rule is a computational device with a sliding portion and a fixed portion. A slide rule enables you to perform multiplication by using the Product Property of Logarithms. There are other slide rules that allow for the calculation of roots and trigonometric functions. Slide rules were used by mathematicians and engineers until the invention of the hand-held calculator in 1972.

Properties of Logarithms *(p. 402)*

Let a be a positive number such that $a \neq 1$, and let n be a real number. If u and v are positive real numbers, the following properties are true.

	Logarithm with Base a	*Natural Logarithm*
1. Product Property:	$\log_a(uv) = \log_a u + \log_a v$	$\ln(uv) = \ln u + \ln v$
2. Quotient Property:	$\log_a \dfrac{u}{v} = \log_a u - \log_a v$	$\ln \dfrac{u}{v} = \ln u - \ln v$
3. Power Property:	$\log_a u^n = n \log_a u$	$\ln u^n = n \ln u$

Proof

Let

$$x = \log_a u \quad \text{and} \quad y = \log_a v.$$

The corresponding exponential forms of these two equations are

$$a^x = u \quad \text{and} \quad a^y = v.$$

To prove the Product Property, multiply u and v to obtain

$$uv = a^x a^y = a^{x+y}.$$

The corresponding logarithmic form of $uv = a^{x+y}$ is $\log_a(uv) = x + y$. So,

$$\log_a(uv) = \log_a u + \log_a v.$$

To prove the Quotient Property, divide u by v to obtain

$$\frac{u}{v} = \frac{a^x}{a^y} = a^{x-y}.$$

The corresponding logarithmic form of $\dfrac{u}{v} = a^{x-y}$ is $\log_a \dfrac{u}{v} = x - y$. So,

$$\log_a \frac{u}{v} = \log_a u - \log_a v.$$

To prove the Power Property, substitute a^x for u in the expression $\log_a u^n$, as follows.

$$\log_a u^n = \log_a(\ )^n$$
$$= \log_a a^{nx}$$
$$= nx$$
$$= n$$

So, $\log_a u^n = n \log_a u.$

PROBLEM SOLVING

This collection of thought-provoking and challenging exercises further explores and expands upon concepts learned in this chapter.

1. Graph the exponential function given by $y = a^x$ for $a = 0.5, 1.2,$ and 2.0. Which of these curves intersects the line $y = x$? Determine all positive numbers a for which the curve $y = a^x$ intersects the line $y = x$.

2. Use a graphing utility to graph $y_1 = e^x$ and each of the functions $y_2 = x^2$, $y_3 = x^3$, $y_4 = \sqrt{x}$, and $y_5 = |x|$. Which function increases at the greatest rate as x approaches $+\infty$?

3. Use the result of Exercise 2 to make a conjecture about the rate of growth of $y_1 = e^x$ and $y = x^n$, where n is a natural number and x approaches $+\infty$.

4. Use the results of Exercises 2 and 3 to describe what is implied when it is stated that a quantity is growing exponentially.

5. Given the exponential function

   $f(x) = a^x$

   show that

   (a) $f(u + v) = f(u) \cdot f(v)$.   (b) $f(2x) = [f(x)]^2$.

6. Given that

   $f(x) = \dfrac{e^x + e^{-x}}{2}$ and $g(x) = \dfrac{e^x - e^{-x}}{2}$

   show that

   $[f(x)]^2 - [g(x)]^2 = 1$.

7. Use a graphing utility to compare the graph of the function given by $y = e^x$ with the graph of each given function. $[n!$ (read "n factorial") is defined as $n! = 1 \cdot 2 \cdot 3 \cdots (n - 1) \cdot n.]$

   (a) $y_1 = 1 + \dfrac{x}{1!}$

   (b) $y_2 = 1 + \dfrac{x}{1!} + \dfrac{x^2}{2!}$

   (c) $y_3 = 1 + \dfrac{x}{1!} + \dfrac{x^2}{2!} + \dfrac{x^3}{3!}$

8. Identify the pattern of successive polynomials given in Exercise 7. Extend the pattern one more term and compare the graph of the resulting polynomial function with the graph of $y = e^x$. What do you think this pattern implies?

9. Graph the function given by

   $f(x) = e^x - e^{-x}$.

   From the graph, the function appears to be one-to-one. Assuming that the function has an inverse function, find $f^{-1}(x)$.

10. Find a pattern for $f^{-1}(x)$ if

    $$f(x) = \dfrac{a^x + 1}{a^x - 1}$$

    where $a > 0$, $a \neq 1$.

11. By observation, identify the equation that corresponds to the graph. Explain your reasoning.

    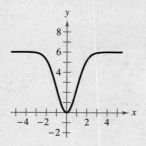

    (a) $y = 6e^{-x^2/2}$

    (b) $y = \dfrac{6}{1 + e^{-x/2}}$

    (c) $y = 6(1 - e^{-x^2/2})$

12. You have two options for investing $500. The first earns 7% compounded annually and the second earns 7% simple interest. The figure shows the growth of each investment over a 30-year period.

    (a) Identify which graph represents each type of investment. Explain your reasoning.

    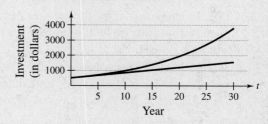

    (b) Verify your answer in part (a) by finding the equations that model the investment growth and graphing the models.

    (c) Which option would you choose? Explain your reasoning.

13. Two different samples of radioactive isotopes are decaying. The isotopes have initial amounts of c_1 and c_2, as well as half-lives of k_1 and k_2, respectively. Find the time t required for the samples to decay to equal amounts.

441

14. A lab culture initially contains 500 bacteria. Two hours later, the number of bacteria has decreased to 200. Find the exponential decay model of the form

$$B = B_0 a^{kt}$$

that can be used to approximate the number of bacteria after t hours.

15. The table shows the colonial population estimates of the American colonies from 1700 to 1780. (Source: U.S. Census Bureau)

Year	Population
1700	250,900
1710	331,700
1720	466,200
1730	629,400
1740	905,600
1750	1,170,800
1760	1,593,600
1770	2,148,100
1780	2,780,400

In each of the following, let y represent the population in the year t, with $t = 0$ corresponding to 1700.

(a) Use the *regression* feature of a graphing utility to find an exponential model for the data.

(b) Use the *regression* feature of the graphing utility to find a quadratic model for the data.

(c) Use the graphing utility to plot the data and the models from parts (a) and (b) in the same viewing window.

(d) Which model is a better fit for the data? Would you use this model to predict the population of the United States in 2015? Explain your reasoning.

16. Show that $\dfrac{\log_a x}{\log_{a/b} x} = 1 + \log_a \dfrac{1}{b}$.

17. Solve $(\ln x)^2 = \ln x^2$.

18. Use a graphing utility to compare the graph of the function $y = \ln x$ with the graph of each given function.

(a) $y_1 = x - 1$

(b) $y_2 = (x - 1) - \frac{1}{2}(x - 1)^2$

(c) $y_3 = (x - 1) - \frac{1}{2}(x - 1)^2 + \frac{1}{3}(x - 1)^3$

19. Identify the pattern of successive polynomials given in Exercise 18. Extend the pattern one more term and compare the graph of the resulting polynomial function with the graph of $y = \ln x$. What do you think the pattern implies?

20. Using

$$y = ab^x \quad \text{and} \quad y = ax^b$$

take the natural logarithm of each side of each equation. What are the slope and y-intercept of the line relating x and $\ln y$ for $y = ab^x$? What are the slope and y-intercept of the line relating $\ln x$ and $\ln y$ for $y = ax^b$?

In Exercises 21 and 22, use the model

$$y = 80.4 - 11 \ln x, \quad 100 \le x \le 1500$$

which approximates the minimum required ventilation rate in terms of the air space per child in a public school classroom. In the model, x is the air space per child in cubic feet and y is the ventilation rate per child in cubic feet per minute.

21. Use a graphing utility to graph the model and approximate the required ventilation rate if there is 300 cubic feet of air space per child.

22. A classroom is designed for 30 students. The air conditioning system in the room has the capacity of moving 450 cubic feet of air per minute.

(a) Determine the ventilation rate per child, assuming that the room is filled to capacity.

(b) Estimate the air space required per child.

(c) Determine the minimum number of square feet of floor space required for the room if the ceiling height is 30 feet.

 In Exercises 23–26, (a) use a graphing utility to create a scatter plot of the data, (b) decide whether the data could best be modeled by a linear model, an exponential model, or a logarithmic model, (c) explain why you chose the model you did in part (b), (d) use the *regression* feature of a graphing utility to find the model you chose in part (b) for the data and graph the model with the scatter plot, and (e) determine how well the model you chose fits the data.

23. $(1, 2.0), (1.5, 3.5), (2, 4.0), (4, 5.8), (6, 7.0), (8, 7.8)$

24. $(1, 4.4), (1.5, 4.7), (2, 5.5), (4, 9.9), (6, 18.1), (8, 33.0)$

25. $(1, 7.5), (1.5, 7.0), (2, 6.8), (4, 5.0), (6, 3.5), (8, 2.0)$

26. $(1, 5.0), (1.5, 6.0), (2, 6.4), (4, 7.8), (6, 8.6), (8, 9.0)$

Systems of Equations and Inequalities

In Mathematics

You can use a system of equations to solve a problem involving two or more equations.

In Real Life

Systems of equations and inequalities are used to determine the correct amounts to use in making an acid mixture, how much to invest in different funds, a break-even point for a business, and many other real-life applications. Systems of equations are also used to find least squares regression parabolas. For instance, a wildlife management team can use a system to model the reproduction rates of deer. (See Exercise 81, page 478.)

Krzysztof Wiktor/Shutterstock

IN CAREERS

There are many careers that use systems of equations and inequalities. Several are listed below.

- Economist
  Exercise 72, page 453

- Investor
  Exercises 53 and 54, page 465

- Dietitian
  Example 9, page 494

- Concert Promoter
  Exercise 78, page 496

6.1 LINEAR AND NONLINEAR SYSTEMS OF EQUATIONS

What you should learn

- Use the method of substitution to solve systems of linear equations in two variables.
- Use the method of substitution to solve systems of nonlinear equations in two variables.
- Use a graphical approach to solve systems of equations in two variables.
- Use systems of equations to model and solve real-life problems.

Why you should learn it

Graphs of systems of equations help you solve real-life problems. For instance, in Exercise 75 on page 453, you can use the graph of a system of equations to approximate when the consumption of wind energy surpassed the consumption of solar energy.

© M.L. Sinibaldi/Corbis

The Method of Substitution

Up to this point in the text, most problems have involved either a function of one variable or a single equation in two variables. However, many problems in science, business, and engineering involve two or more equations in two or more variables. To solve such problems, you need to find solutions of a **system of equations.** Here is an example of a system of two equations in two unknowns.

$$\begin{cases} 2x + y = 5 \\ 3x - 2y = 4 \end{cases}$$

A **solution** of this system is an ordered pair that satisfies each equation in the system. Finding the set of all solutions is called **solving the system of equations.** For instance, the ordered pair $(2, 1)$ is a solution of this system. To check this, you can substitute 2 for x and 1 for y in *each* equation.

Check (2, 1) in Equation 1 and Equation 2:

$$2x + y = 5$$
$$2(\ ) + \ \stackrel{?}{=} 5$$
$$4 + 1 = 5 \qquad \checkmark$$

$$3x - 2y = 4$$
$$3(\ ) - 2(\ ) \stackrel{?}{=} 4$$
$$6 - 2 = 4 \qquad \checkmark$$

In this chapter, you will study four ways to solve systems of equations, beginning with the **method of substitution.**

Method	Section	Type of System
1. Substitution	6.1	Linear or nonlinear, two variables
2. Graphical method	6.1	Linear or nonlinear, two variables
3. Elimination	6.2	Linear, two variables
4. Gaussian elimination	6.3	Linear, three or more variables

Method of Substitution

1. *Solve* one of the equations for one variable in terms of the other.

2. *Substitute* the expression found in Step 1 into the other equation to obtain an equation in one variable.

3. *Solve* the equation obtained in Step 2.

4. *Back-substitute* the value obtained in Step 3 into the expression obtained in Step 1 to find the value of the other variable.

5. *Check* that the solution satisfies *each* of the original equations.

Example 1   Solving a System of Equations by Substitution

Solve the system of equations.

$$\begin{cases} x + y = 4 \\ x - y = 2 \end{cases}$$

Solution

Begin by solving for y in Equation 1.

$$y = 4 - x$$

Next, substitute this expression for y into Equation 2 and solve the resulting single-variable equation for x.

$$x - y = 2$$
$$x - (\quad) = 2$$
$$x - 4 + x = 2$$
$$2x = 6$$
$$x = 3$$

Finally, you can solve for y by *back-substituting* $x = 3$ into the equation $y = 4 - x$, to obtain

$$y = 4 - x$$
$$y = 4 - $$
$$y = 1.$$

The solution is the ordered pair $(3, 1)$. You can check this solution as follows.

Check

Substitute $(3, 1)$ into Equation 1:

$$x + y = 4$$
$$+ \quad \stackrel{?}{=} 4$$
$$4 = 4 \qquad \checkmark$$

Substitute $(3, 1)$ into Equation 2:

$$x - y = 2$$
$$- \quad \stackrel{?}{=} 2$$
$$2 = 2 \qquad \checkmark$$

Because $(3, 1)$ satisfies both equations in the system, it is a solution of the system of equations.

CHECK*Point* Now try Exercise 11.

The term *back-substitution* implies that you work *backwards*. First you solve for one of the variables, and then you substitute that value *back* into one of the equations in the system to find the value of the other variable.

⚠ **WARNING/CAUTION**

Because many steps are required to solve a system of equations, it is very easy to make errors in arithmetic. So, you should always check your solution by substituting it into *each* equation in the original system.

| **Example 2** | **Solving a System by Substitution** |

A total of $12,000 is invested in two funds paying 5% and 3% simple interest. (Recall that the formula for simple interest is $I = Prt$, where P is the principal, r is the annual interest rate, and t is the time.) The yearly interest is $500. How much is invested at each rate?

Solution

Verbal Model: $\dfrac{5\%}{\text{fund}} + \dfrac{3\%}{\text{fund}} = \dfrac{\text{Total}}{\text{investment}}$

$\dfrac{5\%}{\text{interest}} + \dfrac{3\%}{\text{interest}} = \dfrac{\text{Total}}{\text{interest}}$

Labels:

Amount in 5% fund = x	(dollars)
Interest for 5% fund = $0.05x$	(dollars)
Amount in 3% fund = y	(dollars)
Interest for 3% fund = $0.03y$	(dollars)
Total investment = 12,000	(dollars)
Total interest = 500	(dollars)

System: $\begin{cases} x + y = 12{,}000 \\ 0.05x + 0.03y = 500 \end{cases}$

To begin, it is convenient to multiply each side of Equation 2 by 100. This eliminates the need to work with decimals.

$$100(0.05x + 0.03y) = 100(500)$$

$$5x + 3y = 50{,}000$$

To solve this system, you can solve for x in Equation 1.

$$x = 12{,}000 - y$$

Then, substitute this expression for x into revised Equation 2 and solve the resulting equation for y.

$$5x + 3y = 50{,}000$$

$$5(\qquad) + 3y = 50{,}000$$

$$60{,}000 - 5y + 3y = 50{,}000$$

$$-2y = -10{,}000$$

$$y = 5000$$

Next, back-substitute the value $y = 5000$ to solve for x.

$$x = 12{,}000 - y$$

$$x = 12{,}000 -$$

$$x = 7000$$

The solution is (7000, 5000). So, $7000 is invested at 5% and $5000 is invested at 3%. Check this in the original system.

CHECK**Point**   Now try Exercise 25.

Study Tip

When using the method of substitution, it does not matter which variable you choose to solve for first. Whether you solve for y first or x first, you will obtain the same solution. When making your choice, you should choose the variable and equation that are easier to work with. For instance, in Example 2, solving for x in Equation 1 is easier than solving for x in Equation 2.

TECHNOLOGY

One way to check the answers you obtain in this section is to use a graphing utility. For instance, enter the two equations in Example 2

$$y_1 = 12{,}000 - x$$

$$y_2 = \frac{500 - 0.05x}{0.03}$$

and find an appropriate viewing window that shows where the two lines intersect. Then use the *intersect* feature or the *zoom* and *trace* features to find the point of intersection. Does this point agree with the solution obtained at the right?

Nonlinear Systems of Equations

The equations in Examples 1 and 2 are linear. The method of substitution can also be used to solve systems in which one or both of the equations are nonlinear.

Example 3   Substitution: Two-Solution Case

Solve the system of equations.

$$\begin{cases} 3x^2 + 4x - y = 7 \\ 2x - y = -1 \end{cases}$$

Solution

Begin by solving for y in Equation 2 to obtain $y = 2x + 1$. Next, substitute this expression for y into Equation 1 and solve for x.

$$3x^2 + 4x - () = 7$$

$$3x^2 + 2x - 1 = 7$$

$$3x^2 + 2x - 8 = 0$$

$$(3x - 4)(x + 2) = 0$$

$$x = \frac{4}{3}, -2$$

Back-substituting these values of x to solve for the corresponding values of y produces the solutions $\left(\frac{4}{3}, \frac{11}{3}\right)$ and $(-2, -3)$. Check these in the original system.

CHECK **Point**   Now try Exercise 31.

> **Algebra Help**
>
> You can review the techniques for factoring in Section P.4.

When using the method of substitution, you may encounter an equation that has no solution, as shown in Example 4.

Example 4   Substitution: No-Real-Solution Case

Solve the system of equations.

$$\begin{cases} -x + y = 4 \\ x^2 + y = 3 \end{cases}$$

Solution

Begin by solving for y in Equation 1 to obtain $y = x + 4$. Next, substitute this expression for y into Equation 2 and solve for x.

$$x^2 + () = 3$$

$$x^2 + x + 1 = 0$$

$$x = \frac{-1 \pm \sqrt{-3}}{2}$$

Because the discriminant is negative, the equation $x^2 + x + 1 = 0$ has no (real) solution. So, the original system has no (real) solution.

CHECK **Point**   Now try Exercise 33.

Algebra Help

You can review the techniques for graphing equations in Section 1.1.

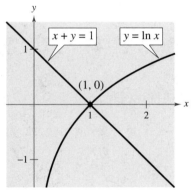

FIGURE **6.4**

Graphical Approach to Finding Solutions

From Examples 2, 3, and 4, you can see that a system of two equations in two unknowns can have exactly one solution, more than one solution, or no solution. By using a **graphical method,** you can gain insight about the number of solutions and the location(s) of the solution(s) of a system of equations by graphing each of the equations in the same coordinate plane. The solutions of the system correspond to the **points of intersection** of the graphs. For instance, the two equations in Figure 6.1 graph as two lines with a *single point* of intersection; the two equations in Figure 6.2 graph as a parabola and a line with *two points* of intersection; and the two equations in Figure 6.3 graph as a line and a parabola that have *no points* of intersection.

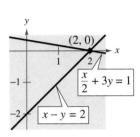

One intersection point
FIGURE **6.1**

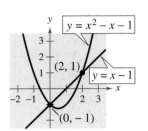

Two intersection points
FIGURE **6.2**

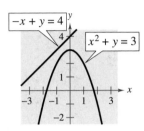

No intersection points
FIGURE **6.3**

Example 5 Solving a System of Equations Graphically

Solve the system of equations.

$$\begin{cases} y = \ln x \\ x + y = 1 \end{cases}$$

Solution

Sketch the graphs of the two equations. From the graphs of these equations, it is clear that there is only one point of intersection and that $(1, 0)$ is the solution point (see Figure 6.4). You can check this solution as follows.

Check (1, 0) in Equation 1:

$$y = \ln x$$
$$= \ln$$
$$0 = 0 \qquad \checkmark$$

Check (1, 0) in Equation 2:

$$x + y = 1$$
$$\quad + \quad = 1$$
$$1 = 1 \qquad \checkmark$$

CHECK*Point* Now try Exercise 39.

Example 5 shows the value of a graphical approach to solving systems of equations in two variables. Notice what would happen if you tried only the substitution method in Example 5. You would obtain the equation $x + \ln x = 1$. It would be difficult to solve this equation for x using standard algebraic techniques.

Applications

The total cost C of producing x units of a product typically has two components—the initial cost and the cost per unit. When enough units have been sold so that the total revenue R equals the total cost C, the sales are said to have reached the **break-even point.** You will find that the break-even point corresponds to the point of intersection of the cost and revenue curves.

Example 6   Break-Even Analysis

A shoe company invests $300,000 in equipment to produce a new line of athletic footwear. Each pair of shoes costs $5 to produce and is sold for $60. How many pairs of shoes must be sold before the business breaks even?

Algebraic Solution

The total cost of producing x units is

$$\text{Total cost} = \text{Cost per unit} \cdot \text{Number of units} + \text{Initial cost}$$

$$C = 5x + 300,000.$$

The revenue obtained by selling x units is

$$\text{Total revenue} = \text{Price per unit} \cdot \text{Number of units}$$

$$R = 60x.$$

Because the break-even point occurs when $R = C$, you have $C = 60x$, and the system of equations to solve is

$$\begin{cases} C = 5x + 300,000 \\ C = 60x \end{cases}.$$

Solve by substitution.

$$= 5x + 300,000$$

$$55x = 300,000$$

$$x \approx 5455$$

So, the company must sell about 5455 pairs of shoes to break even.

Graphical Solution

The total cost of producing x units is

$$\text{Total cost} = \text{Cost per unit} \cdot \text{Number of units} + \text{Initial cost}$$

$$C = 5x + 300,000.$$

The revenue obtained by selling x units is

$$\text{Total revenue} = \text{Price per unit} \cdot \text{Number of units}$$

$$R = 60x.$$

Because the break-even point occurs when $R = C$, you have $C = 60x$, and the system of equations to solve is

$$\begin{cases} C = 5x + 300,000 \\ C = 60x \end{cases}.$$

Use a graphing utility to graph $y_1 = 5x + 300,000$ and $y_2 = 60x$ in the same viewing window. Use the *intersect* feature or the *zoom* and *trace* features of the graphing utility to approximate the point of intersection of the graphs. The point of intersection (break-even point) occurs at $x \approx 5455$, as shown in Figure 6.5. So, the company must sell about 5455 pairs of shoes to break even.

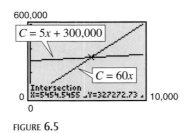

FIGURE **6.5**

CHECK Point   Now try Exercise 67.

Another way to view the solution in Example 6 is to consider the profit function

$$P = R - C.$$

The break-even point occurs when the profit is 0, which is the same as saying that $R = C$.

Example 7   Movie Ticket Sales

The weekly ticket sales for a new comedy movie decreased each week. At the same time, the weekly ticket sales for a new drama movie increased each week. Models that approximate the weekly ticket sales S (in millions of dollars) for each movie are

$$\begin{cases} S = 60 - 8x \\ S = 10 + 4.5x \end{cases}$$

where x represents the number of weeks each movie was in theaters, with $x = 0$ corresponding to the ticket sales during the opening weekend. After how many weeks will the ticket sales for the two movies be equal?

Algebraic Solution

Because the second equation has already been solved for S in terms of x, substitute this value into the first equation and solve for x, as follows.

$$= 60 - 8x$$
$$4.5x + 8x = 60 - 10$$
$$12.5x = 50$$
$$x = 4$$

So, the weekly ticket sales for the two movies will be equal after 4 weeks.

Numerical Solution

You can create a table of values for each model to determine when the ticket sales for the two movies will be equal.

Number of weeks, x	0	1	2	3	4	5	6
Sales, S (comedy)	60	52	44	36	28	20	12
Sales, S (drama)	10	14.5	19	23.5	28	32.5	37

So, from the table above, you can see that the weekly ticket sales for the two movies will be equal after 4 weeks.

CHECK Point   Now try Exercise 69.

CLASSROOM DISCUSSION

Interpreting Points of Intersection   You plan to rent a 14-foot truck for a two-day local move. At truck rental agency A, you can rent a truck for $29.95 per day plus $0.49 per mile. At agency B, you can rent a truck for $50 per day plus $0.25 per mile.

a. Write a total cost equation in terms of x and y for the total cost of renting the truck from each agency.

b. Use a graphing utility to graph the two equations in the same viewing window and find the point of intersection. Interpret the meaning of the point of intersection in the context of the problem.

c. Which agency should you choose if you plan to travel a total of 100 miles during the two-day move? Why?

d. How does the situation change if you plan to drive 200 miles during the two-day move?

6.1   EXERCISES

See www.CalcChat.com for worked-out solutions to odd-numbered exercises.

VOCABULARY: Fill in the blanks.

1. A set of two or more equations in two or more variables is called a _____ of _____.

2. A _____ of a system of equations is an ordered pair that satisfies each equation in the system.

3. Finding the set of all solutions to a system of equations is called _____ the system of equations.

4. The first step in solving a system of equations by the method of _____ is to solve one of the equations for one variable in terms of the other variable.

5. Graphically, the solution of a system of two equations is the _____ of _____ of the graphs of the two equations.

6. In business applications, the point at which the revenue equals costs is called the _____ point.

SKILLS AND APPLICATIONS

In Exercises 7–10, determine whether each ordered pair is a solution of the system of equations.

7. $\begin{cases} 2x - y = 4 \\ 8x + y = -9 \end{cases}$
   (a) $(0, -4)$ (b) $(-2, 7)$
   (c) $\left(\frac{3}{2}, -1\right)$ (d) $\left(-\frac{1}{2}, -5\right)$

8. $\begin{cases} 4x^2 + y = 3 \\ -x - y = 11 \end{cases}$
   (a) $(2, -13)$ (b) $(2, -9)$
   (c) $\left(-\frac{3}{2}, -\frac{31}{3}\right)$ (d) $\left(-\frac{7}{4}, -\frac{37}{4}\right)$

9. $\begin{cases} y = -4e^x \\ 7x - y = 4 \end{cases}$
   (a) $(-4, 0)$ (b) $(0, -4)$
   (c) $(0, -2)$ (d) $(-1, -3)$

10. $\begin{cases} -\log x + 3 = y \\ \frac{1}{9}x + y = \frac{28}{9} \end{cases}$
    (a) $\left(9, \frac{37}{9}\right)$ (b) $(10, 2)$
    (c) $(1, 3)$ (d) $(2, 4)$

In Exercises 11–20, solve the system by the method of substitution. Check your solution(s) graphically.

11. $\begin{cases} 2x + y = 6 \\ -x + y = 0 \end{cases}$

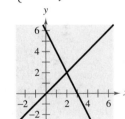

12. $\begin{cases} x - 4y = -11 \\ x + 3y = 3 \end{cases}$

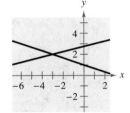

13. $\begin{cases} x - y = -4 \\ x^2 - y = -2 \end{cases}$

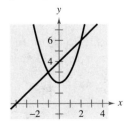

14. $\begin{cases} 3x + y = 2 \\ x^3 - 2 + y = 0 \end{cases}$

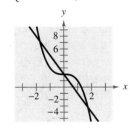

15. $\begin{cases} -\frac{1}{2}x + y = -\frac{5}{2} \\ x^2 + y^2 = 25 \end{cases}$

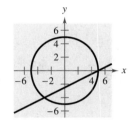

16. $\begin{cases} x + y = 0 \\ x^3 - 5x - y = 0 \end{cases}$

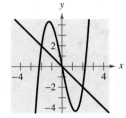

17. $\begin{cases} x^2 + y = 0 \\ x^2 - 4x - y = 0 \end{cases}$

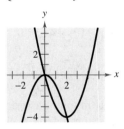

18. $\begin{cases} y = -2x^2 + 2 \\ y = 2(x^4 - 2x^2 + 1) \end{cases}$

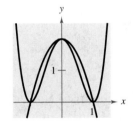

19. $\begin{cases} y = x^3 - 3x^2 + 1 \\ y = x^2 - 3x + 1 \end{cases}$

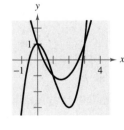

20. $\begin{cases} y = x^3 - 3x^2 + 4 \\ y = -2x + 4 \end{cases}$

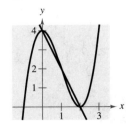

In Exercises 21–34, solve the system by the method of substitution.

21. $\begin{cases} x - y = 2 \\ 6x - 5y = 16 \end{cases}$

22. $\begin{cases} x + 4y = 3 \\ 2x - 7y = -24 \end{cases}$

23. $\begin{cases} 2x - y + 2 = 0 \\ 4x + y - 5 = 0 \end{cases}$

24. $\begin{cases} 6x - 3y - 4 = 0 \\ x + 2y - 4 = 0 \end{cases}$

25. $\begin{cases} 1.5x + 0.8y = 2.3 \\ 0.3x - 0.2y = 0.1 \end{cases}$ **26.** $\begin{cases} 0.5x + 3.2y = 9.0 \\ 0.2x - 1.6y = -3.6 \end{cases}$

27. $\begin{cases} \frac{1}{5}x + \frac{1}{2}y = 8 \\ x + y = 20 \end{cases}$ **28.** $\begin{cases} \frac{1}{2}x + \frac{3}{4}y = 10 \\ \frac{3}{4}x - y = 4 \end{cases}$

29. $\begin{cases} 6x + 5y = -3 \\ -x - \frac{5}{6}y = -7 \end{cases}$ **30.** $\begin{cases} -\frac{2}{3}x + y = 2 \\ 2x - 3y = 6 \end{cases}$

31. $\begin{cases} x^2 - y = 0 \\ 2x + y = 0 \end{cases}$ **32.** $\begin{cases} x - 2y = 0 \\ 3x - y^2 = 0 \end{cases}$

33. $\begin{cases} x - y = -1 \\ x^2 - y = -4 \end{cases}$ **34.** $\begin{cases} y = -x \\ y = x^3 + 3x^2 + 2x \end{cases}$

In Exercises 35–48, solve the system graphically.

35. $\begin{cases} -x + 2y = -2 \\ 3x + y = 20 \end{cases}$ **36.** $\begin{cases} x + y = 0 \\ 3x - 2y = 5 \end{cases}$

37. $\begin{cases} x - 3y = -3 \\ 5x + 3y = -6 \end{cases}$ **38.** $\begin{cases} -x + 2y = -7 \\ x - y = 2 \end{cases}$

39. $\begin{cases} x + y = 4 \\ x^2 + y^2 - 4x = 0 \end{cases}$ **40.** $\begin{cases} -x + y = 3 \\ x^2 - 6x - 27 + y^2 = 0 \end{cases}$

41. $\begin{cases} x - y + 3 = 0 \\ x^2 - 4x + 7 = y \end{cases}$ **42.** $\begin{cases} y^2 - 4x + 11 = 0 \\ -\frac{1}{2}x + y = -\frac{1}{2} \end{cases}$

43. $\begin{cases} 7x + 8y = 24 \\ x - 8y = 8 \end{cases}$ **44.** $\begin{cases} x - y = 0 \\ 5x - 2y = 6 \end{cases}$

45. $\begin{cases} 3x - 2y = 0 \\ x^2 - y^2 = 4 \end{cases}$ **46.** $\begin{cases} 2x - y + 3 = 0 \\ x^2 + y^2 - 4x = 0 \end{cases}$

47. $\begin{cases} x^2 + y^2 = 25 \\ 3x^2 - 16y = 0 \end{cases}$ **48.** $\begin{cases} x^2 + y^2 = 25 \\ (x - 8)^2 + y^2 = 41 \end{cases}$

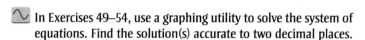 In Exercises 49–54, use a graphing utility to solve the system of equations. Find the solution(s) accurate to two decimal places.

49. $\begin{cases} y = e^x \\ x - y + 1 = 0 \end{cases}$ **50.** $\begin{cases} y = -4e^{-x} \\ y + 3x + 8 = 0 \end{cases}$

51. $\begin{cases} x + 2y = 8 \\ y = \log_2 x \end{cases}$ **52.** $\begin{cases} y + 2 = \ln(x - 1) \\ 3y + 2x = 9 \end{cases}$

53. $\begin{cases} x^2 + y^2 = 169 \\ x^2 - 8y = 104 \end{cases}$ **54.** $\begin{cases} x^2 + y^2 = 4 \\ 2x^2 - y = 2 \end{cases}$

In Exercises 55–64, solve the system graphically or algebraically. Explain your choice of method.

55. $\begin{cases} y = 2x \\ y = x^2 + 1 \end{cases}$ **56.** $\begin{cases} x^2 + y^2 = 25 \\ 2x + y = 10 \end{cases}$

57. $\begin{cases} x - 2y = 4 \\ x^2 - y = 0 \end{cases}$ **58.** $\begin{cases} y = (x + 1)^3 \\ y = \sqrt{x - 1} \end{cases}$

59. $\begin{cases} y - e^{-x} = 1 \\ y - \ln x = 3 \end{cases}$ **60.** $\begin{cases} x^2 + y = 4 \\ e^x - y = 0 \end{cases}$

61. $\begin{cases} y = x^4 - 2x^2 + 1 \\ y = 1 - x^2 \end{cases}$ **62.** $\begin{cases} y = x^3 - 2x^2 + x - 1 \\ y = -x^2 + 3x - 1 \end{cases}$

63. $\begin{cases} xy - 1 = 0 \\ 2x - 4y + 7 = 0 \end{cases}$ **64.** $\begin{cases} x - 2y = 1 \\ y = \sqrt{x - 1} \end{cases}$

BREAK-EVEN ANALYSIS In Exercises 65 and 66, find the sales necessary to break even ($R = C$) for the cost C of producing x units and the revenue R obtained by selling x units. (Round to the nearest whole unit.)

65. $C = 8650x + 250,000, \quad R = 9950x$

66. $C = 5.5\sqrt{x} + 10,000, \quad R = 3.29x$

67. BREAK-EVEN ANALYSIS A small software company invests $25,000 to produce a software package that will sell for $69.95. Each unit can be produced for $45.25.

(a) How many units must be sold to break even?

(b) How many units must be sold to make a profit of $100,000?

68. BREAK-EVEN ANALYSIS A small fast-food restaurant invests $10,000 to produce a new food item that will sell for $3.99. Each item can be produced for $1.90.

(a) How many items must be sold to break even?

(b) How many items must be sold to make a profit of $12,000?

69. DVD RENTALS The weekly rentals for a newly released DVD of an animated film at a local video store decreased each week. At the same time, the weekly rentals for a newly released DVD of a horror film increased each week. Models that approximate the weekly rentals R for each DVD are

$\begin{cases} R = 360 - 24x \\ R = 24 + 18x \end{cases}$

where x represents the number of weeks each DVD was in the store, with $x = 1$ corresponding to the first week.

(a) After how many weeks will the rentals for the two movies be equal?

(b) Use a table to solve the system of equations numerically. Compare your result with that of part (a).

70. SALES The total weekly sales for a newly released portable media player (PMP) increased each week. At the same time, the total weekly sales for another newly released PMP decreased each week. Models that approximate the total weekly sales S (in thousands of units) for each PMP are

$\begin{cases} S = 15x + 50 \\ S = -20x + 190 \end{cases}$

where x represents the number of weeks each PMP was in stores, with $x = 0$ corresponding to the PMP sales on the day each PMP was first released in stores.

(a) After how many weeks will the sales for the two PMPs be equal?

(b) Use a table to solve the system of equations numerically. Compare your result with that of part (a).

71. CHOICE OF TWO JOBS   You are offered two jobs selling dental supplies. One company offers a straight commission of 6% of sales. The other company offers a salary of $500 per week plus 3% of sales. How much would you have to sell in a week in order to make the straight commission offer better?

72. SUPPLY AND DEMAND   The supply and demand curves for a business dealing with wheat are

Supply: $p = 1.45 + 0.00014x^2$

Demand: $p = (2.388 - 0.007x)^2$

where p is the price in dollars per bushel and x is the quantity in bushels per day. Use a graphing utility to graph the supply and demand equations and find the market equilibrium. (The market equilibrium is the point of intersection of the graphs for $x > 0$.)

73. INVESTMENT PORTFOLIO   A total of $25,000 is invested in two funds paying 6% and 8.5% simple interest. (The 6% investment has a lower risk.) The investor wants a yearly interest income of $2000 from the two investments.

(a) Write a system of equations in which one equation represents the total amount invested and the other equation represents the $2000 required in interest. Let x and y represent the amounts invested at 6% and 8.5%, respectively.

(b) Use a graphing utility to graph the two equations in the same viewing window. As the amount invested at 6% increases, how does the amount invested at 8.5% change? How does the amount of interest income change? Explain.

(c) What amount should be invested at 6% to meet the requirement of $2000 per year in interest?

74. LOG VOLUME   You are offered two different rules for estimating the number of board feet in a 16-foot log. (A board foot is a unit of measure for lumber equal to a board 1 foot square and 1 inch thick.) The first rule is the *Doyle Log Rule* and is modeled by $V_1 = (D - 4)^2$, $5 \le D \le 40$, and the other is the *Scribner Log Rule* and is modeled by $V_2 = 0.79D^2 - 2D - 4$, $5 \le D \le 40$, where D is the diameter (in inches) of the log and V is its volume (in board feet).

(a) Use a graphing utility to graph the two log rules in the same viewing window.

(b) For what diameter do the two scales agree?

(c) You are selling large logs by the board foot. Which scale would you use? Explain your reasoning.

75. DATA ANALYSIS: RENEWABLE ENERGY   The table shows the consumption C (in trillions of Btus) of solar energy and wind energy in the United States from 1998 through 2006.   (Source: Energy Information Administration)

Year	Solar, C	Wind, C
1998	70	31
1999	69	46
2000	66	57
2001	65	70
2002	64	105
2003	64	115
2004	65	142
2005	66	178
2006	72	264

(a) Use the *regression* feature of a graphing utility to find a cubic model for the solar energy consumption data and a quadratic model for the wind energy consumption data. Let t represent the year, with $t = 8$ corresponding to 1998.

(b) Use a graphing utility to graph the data and the two models in the same viewing window.

(c) Use the graph from part (b) to approximate the point of intersection of the graphs of the models. Interpret your answer in the context of the problem.

(d) Describe the behavior of each model. Do you think the models can be used to predict consumption of solar energy and wind energy in the United States for future years? Explain.

(e) Use your school's library, the Internet, or some other reference source to research the advantages and disadvantages of using renewable energy.

76. DATA ANALYSIS: POPULATION   The table shows the populations P (in millions) of Georgia, New Jersey, and North Carolina from 2002 through 2007.   (Source: U.S. Census Bureau)

Year	Georgia, G	New Jersey, J	North Carolina, N
2002	8.59	8.56	8.32
2003	8.74	8.61	8.42
2004	8.92	8.64	8.54
2005	9.11	8.66	8.68
2006	9.34	8.67	8.87
2007	9.55	8.69	9.06

(a) Use the *regression* feature of a graphing utility to find linear models for each set of data. Let t represent the year, with $t = 2$ corresponding to 2002.

(b) Use a graphing utility to graph the data and the models in the same viewing window.

(c) Use the graph from part (b) to approximate any points of intersection of the graphs of the models. Interpret the points of intersection in the context of the problem.

(d) Verify your answers from part (c) algebraically.

77. DATA ANALYSIS: TUITION The table shows the average costs (in dollars) of one year's tuition for public and private universities in the United States from 2000 through 2006. (Source: U.S. National Center for Education Statistics)

Year	Public universities	Private universities
2000	2506	14,081
2001	2562	15,000
2002	2700	15,742
2003	2903	16,383
2004	3319	17,327
2005	3629	18,154
2006	3874	18,862

(a) Use the *regression* feature of a graphing utility to find a quadratic model T_1 for tuition at public universities and a linear model T_2 for tuition at private universities. Let t represent the year, with $t = 0$ corresponding to 2000.

(b) Use a graphing utility to graph the data and the two models in the same viewing window.

(c) Use the graph from part (b) to determine the year after 2006 in which tuition at public universities will exceed tuition at private universities.

(d) Verify your answer from part (c) algebraically.

GEOMETRY In Exercises 78–82, find the dimensions of the rectangle meeting the specified conditions.

78. The perimeter is 56 meters and the length is 4 meters greater than the width.

79. The perimeter is 280 centimeters and the width is 20 centimeters less than the length.

80. The perimeter is 42 inches and the width is three-fourths the length.

81. The perimeter is 484 feet and the length is $4\frac{1}{2}$ times the width.

82. The perimeter is 30.6 millimeters and the length is 2.4 times the width.

83. GEOMETRY What are the dimensions of a rectangular tract of land if its perimeter is 44 kilometers and its area is 120 square kilometers?

84. GEOMETRY What are the dimensions of an isosceles right triangle with a two-inch hypotenuse and an area of 1 square inch?

EXPLORATION

TRUE OR FALSE? In Exercises 85 and 86, determine whether the statement is true or false. Justify your answer.

85. In order to solve a system of equations by substitution, you must always solve for y in one of the two equations and then back-substitute.

86. If a system consists of a parabola and a circle, then the system can have at most two solutions.

87. GRAPHICAL REASONING Use a graphing utility to graph $y_1 = 4 - x$ and $y_2 = x - 2$ in the same viewing window. Use the *zoom* and *trace* features to find the coordinates of the point of intersection. What is the relationship between the point of intersection and the solution found in Example 1?

88. GRAPHICAL REASONING Use a graphing utility to graph the two equations in Example 3, $y_1 = 3x^2 + 4x - 7$ and $y_2 = 2x + 1$, in the same viewing window. How many solutions do you think this system has? Repeat this experiment for the equations in Example 4. How many solutions does this system have? Explain your reasoning.

89. THINK ABOUT IT When solving a system of equations by substitution, how do you recognize that the system has no solution?

90. CAPSTONE Consider the system of equations

$$\begin{cases} ax + by = c \\ dx + ey = f \end{cases}.$$

(a) Find values for a, b, c, d, e, and f so that the system has one distinct solution. (There is more than one correct answer.)

(b) Explain how to solve the system in part (a) by the method of substitution and graphically.

(c) Write a brief paragraph describing any advantages of the method of substitution over the graphical method of solving a system of equations.

91. Find equations of lines whose graphs intersect the graph of the parabola $y = x^2$ at (a) two points, (b) one point, and (c) no points. (There is more than one correct answer.) Use graphs to support your answers.

6.2 TWO-VARIABLE LINEAR SYSTEMS

What you should learn

- Use the method of elimination to solve systems of linear equations in two variables.
- Interpret graphically the numbers of solutions of systems of linear equations in two variables.
- Use systems of linear equations in two variables to model and solve real-life problems.

Why you should learn it

You can use systems of equations in two variables to model and solve real-life problems. For instance, in Exercise 61 on page 465, you will solve a system of equations to find a linear model that represents the relationship between wheat yield and amount of fertilizer applied.

© Bill Stormont/Corbis

The Method of Elimination

In Section 6.1, you studied two methods for solving a system of equations: substitution and graphing. Now you will study the **method of elimination.** The key step in this method is to obtain, for one of the variables, coefficients that differ only in sign so that *adding* the equations eliminates the variable.

$$
\begin{array}{rcl}
3x + 5y &=& 7 \\
\underline{-3x - 2y} &=& \underline{-1} \\
3y &=& 6
\end{array}
$$

Note that by adding the two equations, you eliminate the x-terms and obtain a single equation in y. Solving this equation for y produces $y = 2$, which you can then back-substitute into one of the original equations to solve for x.

| **Example 1** | **Solving a System of Equations by Elimination** |

Solve the system of linear equations.

$$
\begin{cases}
3x + 2y = 4 \\
5x - 2y = 12
\end{cases}
$$

Solution

Because the coefficients of y differ only in sign, you can eliminate the y-terms by adding the two equations.

$$
\begin{array}{rcl}
3x + 2y &=& 4 \\
\underline{5x - 2y} &=& \underline{12} \\
8x &=& 16 \\
x &=& 2
\end{array}
$$

By back-substituting $x = 2$ into Equation 1, you can solve for y.

$$
\begin{array}{rcl}
3x + 2y &=& 4 \\
3(\ ) + 2y &=& 4 \\
6 + 2y &=& 4 \\
y &=& -1
\end{array}
$$

The solution is $(2, -1)$. Check this in the original system, as follows.

Check

$$
\begin{array}{rcl}
3(\ ) + 2(\ ) &\overset{?}{=}& 4 \\
6 - 2 &=& 4 \qquad \checkmark \\
5(\ ) - 2(\ ) &\overset{?}{=}& 12 \\
10 + 2 &=& 12 \qquad \checkmark
\end{array}
$$

CHECK*Point*   Now try Exercise 13.

Method of Elimination

To use the **method of elimination** to solve a system of two linear equations in x and y, perform the following steps.

1. *Obtain coefficients* for x (or y) that differ only in sign by multiplying all terms of one or both equations by suitably chosen constants.

2. *Add* the equations to eliminate one variable.

3. *Solve* the equation obtained in Step 2.

4. *Back-substitute* the value obtained in Step 3 into either of the original equations and solve for the other variable.

5. *Check* that the solution satisfies *each* of the original equations.

Example 2   Solving a System of Equations by Elimination

Solve the system of linear equations.

$$\begin{cases} 2x - 4y = -7 \\ 5x + y = -1 \end{cases}$$

Solution

For this system, you can obtain coefficients that differ only in sign by multiplying Equation 2 by 4.

$$
\begin{array}{lll}
2x - 4y = -7 & \Longrightarrow & 2x - 4y = -7 \\
5x + y = -1 & \Longrightarrow & \underline{20x + 4y = -4} \\
 & & 22x = -11 \\
 & & x = -\tfrac{1}{2}
\end{array}
$$

By back-substituting $x = -\tfrac{1}{2}$ into Equation 1, you can solve for y.

$$2x - 4y = -7$$
$$2() - 4y = -7$$
$$-4y = -6$$
$$y = \tfrac{3}{2}$$

The solution is $\left(-\tfrac{1}{2}, \tfrac{3}{2}\right)$. Check this in the original system, as follows.

Check

$$2x - 4y = -7$$
$$2() - 4() \overset{?}{=} -7$$
$$-1 - 6 = -7 \qquad \checkmark$$
$$5x + y = -1$$
$$5() + \overset{?}{=} -1$$
$$-\tfrac{5}{2} + \tfrac{3}{2} = -1 \qquad \checkmark$$

CHECK*Point* Now try Exercise 15.

In Example 2, the two systems of linear equations (the original system and the system obtained by multiplying by constants)

$$\begin{cases} 2x - 4y = -7 \\ 5x + \ y = -1 \end{cases} \quad \text{and} \quad \begin{cases} 2x - 4y = -7 \\ 20x + 4y = -4 \end{cases}$$

are called **equivalent systems** because they have precisely the same solution set. The operations that can be performed on a system of linear equations to produce an equivalent system are (1) interchanging any two equations, (2) multiplying an equation by a nonzero constant, and (3) adding a multiple of one equation to any other equation in the system.

Example 3   Solving the System of Equations by Elimination

Solve the system of linear equations.

$$\begin{cases} 5x + 3y = \ 9 \\ 2x - 4y = 14 \end{cases}$$

Algebraic Solution

You can obtain coefficients that differ only in sign by multiplying Equation 1 by 4 and multiplying Equation 2 by 3.

$$5x + 3y = \ 9 \implies \ 20x + 12y = 36$$
$$\underline{2x - 4y = 14} \implies \ \underline{6x - 12y = 42}$$
$$26x \qquad = 78$$
$$x \qquad = \ 3$$

By back-substituting $x = 3$ into Equation 2, you can solve for y.

$$2x - 4y = 14$$
$$2(\ ) - 4y = 14$$
$$-4y = 8$$
$$y = -2$$

The solution is $(3, -2)$. Check this in the original system.

Graphical Solution

Solve each equation for y. Then use a graphing utility to graph $y_1 = -\frac{5}{3}x + 3$ and $y_2 = \frac{1}{2}x - \frac{7}{2}$ in the same viewing window. Use the *intersect* feature or the *zoom* and *trace* features to approximate the point of intersection of the graphs. From the graph in Figure 6.6, you can see that the point of intersection is $(3, -2)$. You can determine that this is the exact solution by checking $(3, -2)$ in both equations.

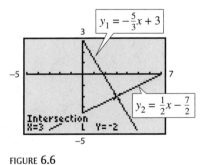

FIGURE 6.6

CHECK Point   Now try Exercise 17.

You can check the solution from Example 3 as follows.

$$5(\ ) + 3(\ ) \overset{?}{=} 9$$
$$15 - 6 = 9 \qquad \checkmark$$
$$2(\ ) - 4(\ ) \overset{?}{=} 14$$
$$6 + 8 = 14 \qquad \checkmark$$

Keep in mind that the terminology and methods discussed in this section apply only to systems of *linear* equations.

Graphical Interpretation of Solutions

It is possible for a *general* system of equations to have exactly one solution, two or more solutions, or no solution. If a system of *linear* equations has two different solutions, it must have an *infinite* number of solutions.

Graphical Interpretations of Solutions

For a system of two linear equations in two variables, the number of solutions is one of the following.

Number of Solutions	*Graphical Interpretation*	*Slopes of Lines*
1. Exactly one solution	The two lines intersect at one point.	The slopes of the two lines are not equal.
2. Infinitely many solutions	The two lines coincide (are identical).	The slopes of the two lines are equal.
3. No solution	The two lines are parallel.	The slopes of the two lines are equal.

A system of linear equations is **consistent** if it has at least one solution. A consistent system with exactly one solution is *independent*, whereas a consistent system with infinitely many solutions is *dependent*. A system is **inconsistent** if it has no solution.

Example 4   Recognizing Graphs of Linear Systems

Match each system of linear equations with its graph in Figure 6.7. Describe the number of solutions and state whether the system is consistent or inconsistent.

a. $\begin{cases} 2x - 3y = 3 \\ -4x + 6y = 6 \end{cases}$ **b.** $\begin{cases} 2x - 3y = 3 \\ x + 2y = 5 \end{cases}$ **c.** $\begin{cases} 2x - 3y = 3 \\ -4x + 6y = -6 \end{cases}$

i.

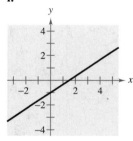

ii.

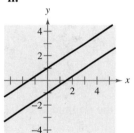

iii.

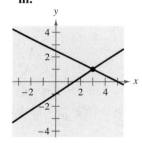

FIGURE 6.7

Solution

a. The graph of system (a) is a pair of parallel lines (ii). The lines have no point of intersection, so the system has no solution. The system is inconsistent.

b. The graph of system (b) is a pair of intersecting lines (iii). The lines have one point of intersection, so the system has exactly one solution. The system is consistent.

c. The graph of system (c) is a pair of lines that coincide (i). The lines have infinitely many points of intersection, so the system has infinitely many solutions. The system is consistent.

CHECK *Point*   Now try Exercises 31–34.

Study Tip

A comparison of the slopes of two lines gives useful information about the number of solutions of the corresponding system of equations. To solve a system of equations graphically, it helps to begin by writing the equations in slope-intercept form. Try doing this for the systems in Example 4.

In Examples 5 and 6, note how you can use the method of elimination to determine that a system of linear equations has no solution or infinitely many solutions.

Example 5   No-Solution Case: Method of Elimination

Solve the system of linear equations.

$$\begin{cases} x - 2y = 3 \\ -2x + 4y = 1 \end{cases}$$

Solution

To obtain coefficients that differ only in sign, you can multiply Equation 1 by 2.

$$
\begin{array}{lcl}
x - 2y = 3 & \Longrightarrow & 2x - 4y = 6 \\
\underline{-2x + 4y = 1} & \Longrightarrow & \underline{-2x + 4y = 1} \\
& & 0 = 7
\end{array}
$$

Because there are no values of x and y for which $0 = 7$, you can conclude that the system is inconsistent and has no solution. The lines corresponding to the two equations in this system are shown in Figure 6.8. Note that the two lines are parallel and therefore have no point of intersection.

CHECK*Point* ▶ Now try Exercise 21.   ∎

In Example 5, note that the occurrence of a false statement, such as $0 = 7$, indicates that the system has no solution. In the next example, note that the occurrence of a statement that is true for all values of the variables, such as $0 = 0$, indicates that the system has infinitely many solutions.

Example 6   Many-Solution Case: Method of Elimination

Solve the system of linear equations.

$$\begin{cases} 2x - y = 1 \\ 4x - 2y = 2 \end{cases}$$

Solution

To obtain coefficients that differ only in sign, you can multiply Equation 1 by -2.

$$
\begin{array}{lcl}
2x - y = 1 & \Longrightarrow & -4x + 2y = -2 \\
\underline{4x - 2y = 2} & \Longrightarrow & \underline{4x - 2y = 2} \\
& & 0 = 0
\end{array}
$$

Because the two equations are equivalent (have the same solution set), you can conclude that the system has infinitely many solutions. The solution set consists of all points (x, y) lying on the line $2x - y = 1$, as shown in Figure 6.9. Letting $x = a$, where a is any real number, you can see that the solutions of the system are $(a, 2a - 1)$.

CHECK*Point* ▶ Now try Exercise 23.   ∎

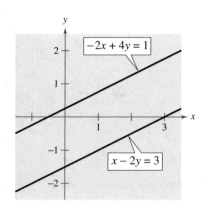

FIGURE **6.8**

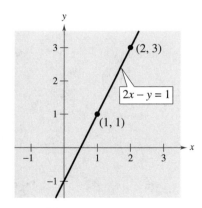

FIGURE **6.9**

The general solution of the linear system

$$\begin{cases} ax + by = c \\ dx + ey = f \end{cases}$$

is

$$x = \frac{ce - bf}{ae - bd}$$

and

$$y = \frac{af - cd}{ae - bd}.$$

If $ae - bd = 0$, the system does not have a unique solution. A graphing utility program (called Systems of Linear Equations) for solving such a system can be found at the website for this text at *academic.cengage.com*. Try using the program for your graphing utility to solve the system in Example 7.

Example 7 illustrates a strategy for solving a system of linear equations that has decimal coefficients.

Example 7   A Linear System Having Decimal Coefficients

Solve the system of linear equations.

$$\begin{cases} 0.02x - 0.05y = -0.38 \\ 0.03x + 0.04y = 1.04 \end{cases}$$

Solution

Because the coefficients in this system have two decimal places, you can begin by multiplying each equation by 100. This produces a system in which the coefficients are all integers.

$$\begin{cases} 2x - 5y = -38 \\ 3x + 4y = 104 \end{cases}$$

Now, to obtain coefficients that differ only in sign, multiply Equation 1 by 3 and multiply Equation 2 by -2.

$$2x - 5y = -38 \quad \Longrightarrow \quad 6x - 15y = -114$$

$$\underline{3x + 4y = 104} \quad \Longrightarrow \quad \underline{-6x - 8y = -208}$$

$$- 23y = -322$$

So, you can conclude that

$$y = \frac{-322}{-23}$$

$$= 14.$$

Back-substituting $y = 14$ into revised Equation 2 produces the following.

$$3x + 4y = 104$$

$$3x + 4() = 104$$

$$3x = 48$$

$$x = 16$$

The solution is $(16, 14)$. Check this in the original system, as follows.

Check

$$0.02x - 0.05y = -0.38$$

$$0.02() - 0.05() \overset{?}{=} -0.38$$

$$0.32 - 0.70 = -0.38 \qquad \checkmark$$

$$0.03x + 0.04y = 1.04$$

$$0.03() + 0.04() \overset{?}{=} 1.04$$

$$0.48 + 0.56 = 1.04 \qquad \checkmark$$

CHECK *Point*   Now try Exercise 25.

Applications

At this point, you may be asking the question "How can I tell which application problems can be solved using a system of linear equations?" The answer comes from the following considerations.

1. Does the problem involve more than one unknown quantity?

2. Are there two (or more) equations or conditions to be satisfied?

If one or both of these situations occur, the appropriate mathematical model for the problem may be a system of linear equations.

| Example 8   An Application of a Linear System

An airplane flying into a headwind travels the 2000-mile flying distance between Chicopee, Massachusetts and Salt Lake City, Utah in 4 hours and 24 minutes. On the return flight, the same distance is traveled in 4 hours. Find the airspeed of the plane and the speed of the wind, assuming that both remain constant.

Solution

The two unknown quantities are the speeds of the wind and the plane. If r_1 is the speed of the plane and r_2 is the speed of the wind, then

$$r_1 - r_2 = \text{speed of the plane against the wind}$$

$$r_1 + r_2 = \text{speed of the plane with the wind}$$

as shown in Figure 6.10. Using the formula distance $=$ (rate)(time) for these two speeds, you obtain the following equations.

$$2000 = (r_1 - r_2)\left(4 + \frac{24}{60}\right)$$

$$2000 = (r_1 + r_2)(4)$$

These two equations simplify as follows.

$$\begin{cases} 5000 = 11r_1 - 11r_2 \\ 500 = r_1 + r_2 \end{cases}$$

To solve this system by elimination, multiply Equation 2 by 11.

$$5000 = 11r_1 - 11r_2 \quad \Longrightarrow \quad 5000 = 11r_1 - 11r_2$$

$$\underline{500 = r_1 + r_2} \quad \Longrightarrow \quad \underline{5500 = 11r_1 + 11r_2}$$

$$10{,}500 = 22r_1$$

So,

$$r_1 = \frac{10{,}500}{22} = \frac{5250}{11} \approx 477.27 \text{ miles per hour}$$

and

$$r_2 = 500 - \frac{5250}{11} = \frac{250}{11} \approx 22.73 \text{ miles per hour.}$$

Check this solution in the original statement of the problem.

CHECK *Point*   Now try Exercise 43.

Original flight

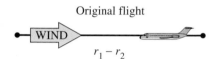

$$r_1 - r_2$$

Return flight

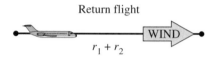

$$r_1 + r_2$$

FIGURE **6.10**

In a free market, the demands for many products are related to the prices of the products. As the prices decrease, the demands by consumers increase and the amounts that producers are able or willing to supply decrease.

Example 9   Finding the Equilibrium Point

The demand and supply equations for a new type of personal digital assistant are

$$\begin{cases} p = 150 - 0.00001x \\ p = 60 + 0.00002x \end{cases}$$

where p is the price in dollars and x represents the number of units. Find the equilibrium point for this market. The **equilibrium point** is the price p and number of units x that satisfy both the demand and supply equations.

Solution

Because p is written in terms of x, begin by substituting the value of p given in the supply equation into the demand equation.

$$p = 150 - 0.00001x$$

$$= 150 - 0.00001x$$

$$0.00003x = 90$$

$$x = 3{,}000{,}000$$

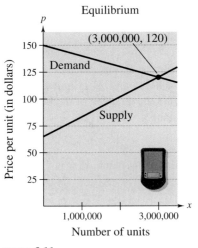

Equilibrium

(3,000,000, 120)

Demand

Supply

Price per unit (in dollars)

Number of units

1,000,000   3,000,000

FIGURE **6.11**

So, the equilibrium point occurs when the demand and supply are each 3 million units. (See Figure 6.11.) The price that corresponds to this x-value is obtained by back-substituting $x = 3{,}000{,}000$ into either of the original equations. For instance, back-substituting into the demand equation produces

$$p = 150 - 0.00001(\phantom{3{,}000{,}000})$$

$$= 150 - 30$$

$$= \$120.$$

The solution is $(3{,}000{,}000, 120)$. You can check this as follows.

Check

Substitute $(3{,}000{,}000, 120)$ into the demand equation.

$$p = 150 - 0.00001x$$

$$\overset{?}{=} 150 - 0.00001(\phantom{3{,}000{,}000})$$

$$120 = 120 \qquad\qquad ✓$$

Substitute $(3{,}000{,}000, 120)$ into the supply equation.

$$p = 60 + 0.00002x$$

$$\overset{?}{=} 60 + 0.00002(\phantom{3{,}000{,}000})$$

$$120 = 120 \qquad\qquad ✓$$

CHECK *Point* Now try Exercise 45.

6.2   EXERCISES

See www.CalcChat.com for worked-out solutions to odd-numbered exercises.

VOCABULARY: Fill in the blanks.

1. The first step in solving a system of equations by the method of _____ is to obtain coefficients for x (or y) that differ only in sign.

2. Two systems of equations that have the same solution set are called _____ systems.

3. A system of linear equations that has at least one solution is called _____, whereas a system of linear equations that has no solution is called _____.

4. In business applications, the _____ _____ is defined as the price p and the number of units x that satisfy both the demand and supply equations.

SKILLS AND APPLICATIONS

In Exercises 5–12, solve the system by the method of elimination. Label each line with its equation. To print an enlarged copy of the graph, go to the website *www.mathgraphs.com*.

5. $\begin{cases} 2x + y = 5 \\ x - y = 1 \end{cases}$

6. $\begin{cases} x + 3y = 1 \\ -x + 2y = 4 \end{cases}$

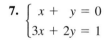
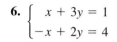

7. $\begin{cases} x + y = 0 \\ 3x + 2y = 1 \end{cases}$

8. $\begin{cases} 2x - y = -3 \\ 4x + 3y = -21 \end{cases}$

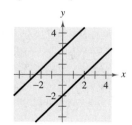
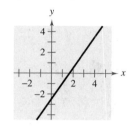

9. $\begin{cases} x - y = 2 \\ -2x + 2y = 5 \end{cases}$

10. $\begin{cases} 3x + 2y = 3 \\ 6x + 4y = 14 \end{cases}$

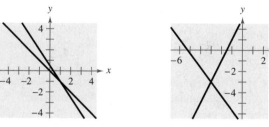
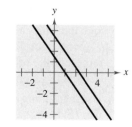

11. $\begin{cases} 3x - 2y = 5 \\ -6x + 4y = -10 \end{cases}$

12. $\begin{cases} 9x - 3y = -15 \\ -3x + y = 5 \end{cases}$

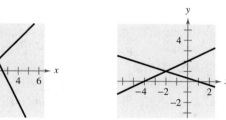
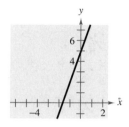

In Exercises 13–30, solve the system by the method of elimination and check any solutions algebraically.

13. $\begin{cases} x + 2y = 6 \\ x - 2y = 2 \end{cases}$

14. $\begin{cases} 3x - 5y = 8 \\ 2x + 5y = 22 \end{cases}$

15. $\begin{cases} 5x + 3y = 6 \\ 3x - y = 5 \end{cases}$

16. $\begin{cases} x + 5y = 10 \\ 3x - 10y = -5 \end{cases}$

17. $\begin{cases} 3x + 2y = 10 \\ 2x + 5y = 3 \end{cases}$

18. $\begin{cases} 2r + 4s = 5 \\ 16r + 50s = 55 \end{cases}$

19. $\begin{cases} 5u + 6v = 24 \\ 3u + 5v = 18 \end{cases}$

20. $\begin{cases} 3x + 11y = 4 \\ -2x - 5y = 9 \end{cases}$

21. $\begin{cases} \frac{9}{5}x + \frac{6}{5}y = 4 \\ 9x + 6y = 3 \end{cases}$

22. $\begin{cases} \frac{3}{4}x + y = \frac{1}{8} \\ \frac{9}{4}x + 3y = \frac{3}{8} \end{cases}$

23. $\begin{cases} -5x + 6y = -3 \\ 20x - 24y = 12 \end{cases}$

24. $\begin{cases} 7x + 8y = 6 \\ -14x - 16y = -12 \end{cases}$

25. $\begin{cases} 0.2x - 0.5y = -27.8 \\ 0.3x + 0.4y = 68.7 \end{cases}$

26. $\begin{cases} 0.05x - 0.03y = 0.21 \\ 0.07x + 0.02y = 0.16 \end{cases}$

27. $\begin{cases} 4b + 3m = 3 \\ 3b + 11m = 13 \end{cases}$

28. $\begin{cases} 2x + 5y = 8 \\ 5x + 8y = 10 \end{cases}$

29. $\begin{cases} \dfrac{x + 3}{4} + \dfrac{y - 1}{3} = 1 \\ 2x - y = 12 \end{cases}$

30. $\begin{cases} \dfrac{x - 1}{2} + \dfrac{y + 2}{3} = 4 \\ x - 2y = 5 \end{cases}$

In Exercises 31–34, match the system of linear equations with its graph. Describe the number of solutions and state whether the system is consistent or inconsistent. [The graphs are labeled (a), (b), (c) and (d).]

(a)
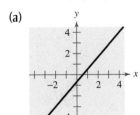

(b)
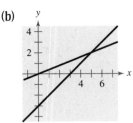

(c)
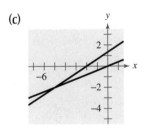

(d)
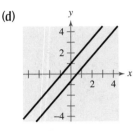

31. $\begin{cases} 2x - 5y = 0 \\ x - y = 3 \end{cases}$ **32.** $\begin{cases} 2x - 5y = 0 \\ 2x - 3y = -4 \end{cases}$

33. $\begin{cases} -7x + 6y = -4 \\ 14x - 12y = 8 \end{cases}$ **34.** $\begin{cases} 7x - 6y = -6 \\ -7x + 6y = -4 \end{cases}$

In Exercises 35–42, use any method to solve the system.

35. $\begin{cases} 3x - 5y = 7 \\ 2x + y = 9 \end{cases}$ **36.** $\begin{cases} -x + 3y = 17 \\ 4x + 3y = 7 \end{cases}$

37. $\begin{cases} y = 2x - 5 \\ y = 5x - 11 \end{cases}$ **38.** $\begin{cases} 7x + 3y = 16 \\ y = x + 2 \end{cases}$

39. $\begin{cases} x - 5y = 21 \\ 6x + 5y = 21 \end{cases}$ **40.** $\begin{cases} y = -2x - 17 \\ y = 2 - 3x \end{cases}$

41. $\begin{cases} -5x + 9y = 13 \\ y = x - 4 \end{cases}$ **42.** $\begin{cases} 4x - 3y = 6 \\ -5x + 7y = -1 \end{cases}$

43. AIRPLANE SPEED An airplane flying into a headwind travels the 1800-mile flying distance between Pittsburgh, Pennsylvania and Phoenix, Arizona in 3 hours and 36 minutes. On the return flight, the distance is traveled in 3 hours. Find the airspeed of the plane and the speed of the wind, assuming that both remain constant.

44. AIRPLANE SPEED Two planes start from Los Angeles International Airport and fly in opposite directions. The second plane starts $\frac{1}{2}$ hour after the first plane, but its speed is 80 kilometers per hour faster. Find the airspeed of each plane if 2 hours after the first plane departs the planes are 3200 kilometers apart.

SUPPLY AND DEMAND In Exercises 45–48, find the equilibrium point of the demand and supply equations. The equilibrium point is the price p and number of units x that satisfy both the demand and supply equations.

	Demand	*Supply*
45.	$p = 500 - 0.4x$	$p = 380 + 0.1x$
46.	$p = 100 - 0.05x$	$p = 25 + 0.1x$
47.	$p = 140 - 0.00002x$	$p = 80 + 0.00001x$
48.	$p = 400 - 0.0002x$	$p = 225 + 0.0005x$

49. NUTRITION Two cheeseburgers and one small order of French fries from a fast-food restaurant contain a total of 830 calories. Three cheeseburgers and two small orders of French fries contain a total of 1360 calories. Find the caloric content of each item.

50. NUTRITION One eight-ounce glass of apple juice and one eight-ounce glass of orange juice contain a total of 177.4 milligrams of vitamin C. Two eight-ounce glasses of apple juice and three eight-ounce glasses of orange juice contain a total of 436.7 milligrams of vitamin C. How much vitamin C is in an eight-ounce glass of each type of juice?

51. ACID MIXTURE Thirty liters of a 40% acid solution is obtained by mixing a 25% solution with a 50% solution.

(a) Write a system of equations in which one equation represents the amount of final mixture required and the other represents the percent of acid in the final mixture. Let x and y represent the amounts of the 25% and 50% solutions, respectively.

(b) Use a graphing utility to graph the two equations in part (a) in the same viewing window. As the amount of the 25% solution increases, how does the amount of the 50% solution change?

(c) How much of each solution is required to obtain the specified concentration of the final mixture?

52. FUEL MIXTURE Five hundred gallons of 89-octane gasoline is obtained by mixing 87-octane gasoline with 92-octane gasoline.

(a) Write a system of equations in which one equation represents the amount of final mixture required and the other represents the amounts of 87- and 92-octane gasolines in the final mixture. Let x and y represent the numbers of gallons of 87-octane and 92-octane gasolines, respectively.

(b) Use a graphing utility to graph the two equations in part (a) in the same viewing window. As the amount of 87-octane gasoline increases, how does the amount of 92-octane gasoline change?

(c) How much of each type of gasoline is required to obtain the 500 gallons of 89-octane gasoline?

53. INVESTMENT PORTFOLIO A total of $24,000 is invested in two corporate bonds that pay 3.5% and 5% simple interest. The investor wants an annual interest income of $930 from the investments. What amount should be invested in the 3.5% bond?

54. INVESTMENT PORTFOLIO A total of $32,000 is invested in two municipal bonds that pay 5.75% and 6.25% simple interest. The investor wants an annual interest income of $1900 from the investments. What amount should be invested in the 5.75% bond?

55. PRESCRIPTIONS The numbers of prescriptions P (in thousands) filled at two pharmacies from 2006 through 2010 are shown in the table.

Year	Pharmacy A	Pharmacy B
2006	19.2	20.4
2007	19.6	20.8
2008	20.0	21.1
2009	20.6	21.5
2010	21.3	22.0

(a) Use a graphing utility to create a scatter plot of the data for pharmacy A and use the *regression* feature to find a linear model. Let t represent the year, with $t = 6$ corresponding to 2006. Repeat the procedure for pharmacy B.

(b) Assuming the numbers for the given five years are representative of future years, will the number of prescriptions filled at pharmacy A ever exceed the number of prescriptions filled at pharmacy B? If so, when?

56. DATA ANALYSIS A store manager wants to know the demand for a product as a function of the price. The daily sales for different prices of the product are shown in the table.

Price, x	Demand, y
$1.00	45
$1.20	37
$1.50	23

(a) Find the least squares regression line $y = ax + b$ for the data by solving the system for a and b.

$$\begin{cases} 3.00b + 3.70a = 105.00 \\ 3.70b + 4.69a = 123.90 \end{cases}$$

(b) Use the regression feature of a graphing utility to confirm the result in part (a).

(c) Use the graphing utility to plot the data and graph the linear model from part (a) in the same viewing window.

(d) Use the linear model from part (a) to predict the demand when the price is $1.75.

FITTING A LINE TO DATA In Exercises 57–60, find the least squares regression line $y = ax + b$ for the points

$$(x_1, y_1), (x_2, y_2), \ldots, (x_n, y_n)$$

by solving the system for a and b.

$$nb + \left(\sum_{i=1}^{n} x_i\right)a = \left(\sum_{i=1}^{n} y_i\right)$$

$$\left(\sum_{i=1}^{n} x_i\right)b + \left(\sum_{i=1}^{n} x_i^2\right)a = \left(\sum_{i=1}^{n} x_i y_i\right)$$

Then use a graphing utility to confirm the result. (If you are unfamiliar with summation notation, look at the discussion in Section 8.1 or in Appendix B at the website for this text at *academic.cengage.com*.)

57.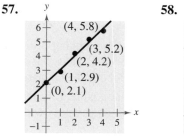

58.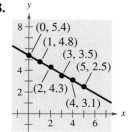

59. $(0, 8), (1, 6), (2, 4), (3, 2)$

60. $(1, 0.0), (2, 1.1), (3, 2.3), (4, 3.8),$

$(5, 4.0), (6, 5.5), (7, 6.7), (8, 6.9)$

61. DATA ANALYSIS An agricultural scientist used four test plots to determine the relationship between wheat yield y (in bushels per acre) and the amount of fertilizer x (in hundreds of pounds per acre). The results are shown in the table.

Fertilizer, x	Yield, y
1.0	32
1.5	41
2.0	48
2.5	53

(a) Use the technique demonstrated in Exercises 57–60 to set up a system of equations for the data and to find the least squares regression line $y = ax + b$.

(b) Use the linear model to predict the yield for a fertilizer application of 160 pounds per acre.

62. DEFENSE DEPARTMENT OUTLAYS The table shows the total national outlays y for defense functions (in billions of dollars) for the years 2000 through 2007. (Source: U.S. Office of Management and Budget)

Year	Outlays, y
2000	294.4
2001	304.8
2002	348.5
2003	404.8
2004	455.8
2005	495.3
2006	521.8
2007	552.6

(a) Use the technique demonstrated in Exercises 57–60 to set up a system of equations for the data and to find the least squares regression line $y = at + b$. Let t represent the year, with $t = 0$ corresponding to 2000.

(b) Use the *regression* feature of a graphing utility to find a linear model for the data. How does this model compare with the model obtained in part (a)?

(c) Use the linear model to create a table of estimated values of y. Compare the estimated values with the actual data.

(d) Use the linear model to estimate the total national outlay for 2008.

(e) Use the Internet, your school's library, or some other reference source to find the total national outlay for 2008. How does this value compare with your answer in part (d)?

(f) Is the linear model valid for long-term predictions of total national outlays? Explain.

EXPLORATION

TRUE OR FALSE? In Exercises 63 and 64, determine whether the statement is true or false. Justify your answer.

63. If two lines do not have exactly one point of intersection, then they must be parallel.

64. Solving a system of equations graphically will always give an exact solution.

65. WRITING Briefly explain whether or not it is possible for a consistent system of linear equations to have exactly two solutions.

66. THINK ABOUT IT Give examples of a system of linear equations that has (a) no solution and (b) an infinite number of solutions.

67. COMPARING METHODS Use the method of substitution to solve the system in Example 1. Is the method of substitution or the method of elimination easier? Explain.

68. CAPSTONE Rewrite each system of equations in slope-intercept form and sketch the graph of each system. What is the relationship among the slopes of the two lines, the number of points of intersection, and the number of solutions?

(a) $\begin{cases} 5x - y = -1 \\ -x + y = -5 \end{cases}$ (b) $\begin{cases} 4x - 3y = 1 \\ -8x + 6y = -2 \end{cases}$

(c) $\begin{cases} x + 2y = 3 \\ x + 2y = -8 \end{cases}$

THINK ABOUT IT In Exercises 69 and 70, the graphs of the two equations appear to be parallel. Yet, when the system is solved algebraically, you find that the system does have a solution. Find the solution and explain why it does not appear on the portion of the graph that is shown.

69. $\begin{cases} 100y - x = 200 \\ 99y - x = -198 \end{cases}$

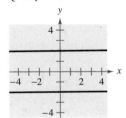

70. $\begin{cases} 21x - 20y = 0 \\ 13x - 12y = 120 \end{cases}$

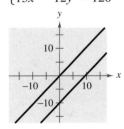

In Exercises 71 and 72, find the value of k such that the system of linear equations is inconsistent.

71. $\begin{cases} 4x - 8y = -3 \\ 2x + ky = 16 \end{cases}$ **72.** $\begin{cases} 15x + 3y = 6 \\ -10x + ky = 9 \end{cases}$

PROJECT: COLLEGE EXPENSES To work an extended application analyzing the average undergraduate tuition, room, and board charges at private degree-granting institutions in the United States from 1990 through 2007, visit this text's website at *academic.cengage.com*. (Data Source: U.S. Dept. of Education)

What you should learn

- Use back-substitution to solve linear systems in row-echelon form.
- Use Gaussian elimination to solve systems of linear equations.
- Solve nonsquare systems of linear equations.
- Use systems of linear equations in three or more variables to model and solve real-life problems.

Why you should learn it

Systems of linear equations in three or more variables can be used to model and solve real-life problems. For instance, in Exercise 83 on page 479, a system of equations can be used to determine the combination of scoring plays in Super Bowl XLIII.

6.3 MULTIVARIABLE LINEAR SYSTEMS

Row-Echelon Form and Back-Substitution

The method of elimination can be applied to a system of linear equations in more than two variables. In fact, this method easily adapts to computer use for solving linear systems with dozens of variables.

When elimination is used to solve a system of linear equations, the goal is to rewrite the system in a form to which back-substitution can be applied. To see how this works, consider the following two systems of linear equations.

System of Three Linear Equations in Three Variables: (See Example 3.)

$$\begin{cases} x - 2y + 3z = 9 \\ -x + 3y \quad\quad = -4 \\ 2x - 5y + 5z = 17 \end{cases}$$

Equivalent System in Row-Echelon Form: (See Example 1.)

$$\begin{cases} x - 2y + 3z = 9 \\ \quad\quad y + 3z = 5 \\ \quad\quad\quad\quad z = 2 \end{cases}$$

The second system is said to be in **row-echelon form,** which means that it has a "stair-step" pattern with leading coefficients of 1. After comparing the two systems, it should be clear that it is easier to solve the system in row-echelon form, using back-substitution.

Example 1   Using Back-Substitution in Row-Echelon Form

Solve the system of linear equations.

$$\begin{cases} x - 2y + 3z = 9 \\ \quad\quad y + 3z = 5 \\ \quad\quad\quad\quad z = 2 \end{cases}$$

Solution

From Equation 3, you know the value of z. To solve for y, substitute $z = 2$ into Equation 2 to obtain

$$y + 3(\ ) = 5$$
$$y = -1.$$

Then substitute $y = -1$ and $z = 2$ into Equation 1 to obtain

$$x - 2(\quad) + 3(\ ) = 9$$
$$x = 1.$$

The solution is $x = 1$, $y = -1$, and $z = 2$, which can be written as the **ordered triple** $(1, -1, 2)$. Check this in the original system of equations.

CHECK**Point**   Now try Exercise 11.

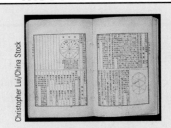

One of the most influential Chinese mathematics books was the *Chui-chang suan-shu* or *Nine Chapters on the Mathematical Art* (written in approximately 250 B.C.). Chapter Eight of the *Nine Chapters* contained solutions of systems of linear equations using positive and negative numbers. One such system was as follows.

$$\begin{cases} 3x + 2y + z = 39 \\ 2x + 3y + z = 34 \\ x + 2y + 3z = 26 \end{cases}$$

This system was solved using column operations on a matrix. Matrices (plural for matrix) will be discussed in the next chapter.

Gaussian Elimination

Two systems of equations are *equivalent* if they have the same solution set. To solve a system that is not in row-echelon form, first convert it to an *equivalent* system that is in row-echelon form by using the following operations.

> ## Operations That Produce Equivalent Systems
>
> Each of the following **row operations** on a system of linear equations produces an *equivalent* system of linear equations.
>
> **1.** Interchange two equations.
>
> **2.** Multiply one of the equations by a nonzero constant.
>
> **3.** Add a multiple of one of the equations to another equation to replace the latter equation.

To see how this is done, take another look at the method of elimination, as applied to a system of two linear equations.

Example 2   Using Gaussian Elimination to Solve a System

Solve the system of linear equations.

$$\begin{cases} 3x - 2y = -1 \\ x - y = 0 \end{cases}$$

Solution

There are two strategies that seem reasonable: eliminate the variable x or eliminate the variable y. The following steps show how to use the first strategy.

$$\begin{cases} x - y = 0 \\ 3x - 2y = -1 \end{cases}$$

$$\begin{cases} -3x + 3y = 0 \\ 3x - 2y = -1 \end{cases}$$

$$\begin{aligned} -3x + 3y &= 0 \\ \underline{3x - 2y} &= \underline{-1} \\ y &= -1 \end{aligned}$$

$$\begin{cases} x - y = 0 \\ y = -1 \end{cases}$$

Notice in the first step that interchanging rows is an easy way of obtaining a leading coefficient of 1. Now back-substitute $y = -1$ into Equation 2 and solve for x.

$$x - () = 0$$

$$x = -1$$

The solution is $x = -1$ and $y = -1$, which can be written as the ordered pair $(-1, -1)$.

CHECK*Point*   Now try Exercise 19.

Rewriting a system of linear equations in row-echelon form usually involves a chain of equivalent systems, each of which is obtained by using one of the three basic row operations listed on the previous page. This process is called **Gaussian elimination,** after the German mathematician Carl Friedrich Gauss (1777–1855).

| **Example 3**   **Using Gaussian Elimination to Solve a System**

Solve the system of linear equations.

$$\begin{cases} x - 2y + 3z = 9 \\ -x + 3y \quad\quad = -4 \\ 2x - 5y + 5z = 17 \end{cases}$$

Solution

Because the leading coefficient of the first equation is 1, you can begin by saving the x at the upper left and eliminating the other x-terms from the first column.

$$\begin{array}{r} x - 2y + 3z = 9 \\ \underline{-x + 3y \quad\quad = -4} \\ y + 3z = 5 \end{array}$$

$$\begin{cases} x - 2y + 3z = 9 \\ y + 3z = 5 \\ 2x - 5y + 5z = 17 \end{cases}$$

> Adding the first equation to the second equation produces a new second equation.

$$\begin{array}{r} -2x + 4y - 6z = -18 \\ \underline{2x - 5y + 5z = \quad\; 17} \\ -y - z = -1 \end{array}$$

$$\begin{cases} x - 2y + 3z = 9 \\ y + 3z = 5 \\ -y - z = -1 \end{cases}$$

> Adding -2 times the first equation to the third equation produces a new third equation.

Now that all but the first x have been eliminated from the first column, go to work on the second column. (You need to eliminate y from the third equation.)

$$\begin{cases} x - 2y + 3z = 9 \\ y + 3z = 5 \\ 2z = 4 \end{cases}$$

> Adding the second equation to the third equation produces a new third equation.

Finally, you need a coefficient of 1 for z in the third equation.

$$\begin{cases} x - 2y + 3z = 9 \\ y + 3z = 5 \\ z = 2 \end{cases}$$

> Multiplying the third equation by $\frac{1}{2}$ produces a new third equation.

This is the same system that was solved in Example 1, and, as in that example, you can conclude that the solution is

$$x = 1, \quad y = -1, \quad \text{and} \quad z = 2.$$

CHECK *Point* Now try Exercise 21.

⚠ WARNING / CAUTION

Arithmetic errors are often made when performing elementary row operations. You should note the operation performed in each step so that you can go back and check your work.

The next example involves an inconsistent system—one that has no solution. The key to recognizing an inconsistent system is that at some stage in the elimination process you obtain a false statement such as $0 = -2$.

Example 4 An Inconsistent System

Solve the system of linear equations.

$$\begin{cases} x - 3y + z = 1 \\ 2x - y - 2z = 2 \\ x + 2y - 3z = -1 \end{cases}$$

Solution

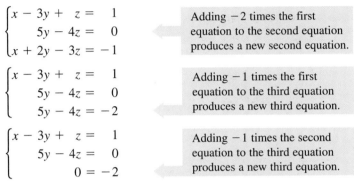

$$\begin{cases} x - 3y + z = 1 \\ 5y - 4z = 0 \\ x + 2y - 3z = -1 \end{cases}$$

Adding -2 times the first equation to the second equation produces a new second equation.

$$\begin{cases} x - 3y + z = 1 \\ 5y - 4z = 0 \\ 5y - 4z = -2 \end{cases}$$

Adding -1 times the first equation to the third equation produces a new third equation.

$$\begin{cases} x - 3y + z = 1 \\ 5y - 4z = 0 \\ 0 = -2 \end{cases}$$

Adding -1 times the second equation to the third equation produces a new third equation.

Because $0 = -2$ is a false statement, you can conclude that this system is inconsistent and has no solution. Moreover, because this system is equivalent to the original system, you can conclude that the original system also has no solution.

CHECK Point Now try Exercise 25.

As with a system of linear equations in two variables, the solution(s) of a system of linear equations in more than two variables must fall into one of three categories.

The Number of Solutions of a Linear System

For a system of linear equations, exactly one of the following is true.

1. There is exactly one solution.

2. There are infinitely many solutions.

3. There is no solution.

In Section 6.2, you learned that a system of two linear equations in two variables can be represented graphically as a pair of lines that are intersecting, coincident, or parallel. A system of three linear equations in three variables has a similar graphical representation—it can be represented as three planes in space that intersect in one point (exactly one solution) [see Figure 6.12], intersect in a line or a plane (infinitely many solutions) [see Figures 6.13 and 6.14], or have no points common to all three planes (no solution) [see Figures 6.15 and 6.16].

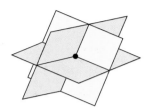

Solution: one point
FIGURE **6.12**

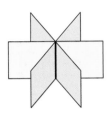

Solution: one line
FIGURE **6.13**

Solution: one plane
FIGURE **6.14**

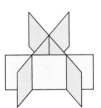

Solution: none
FIGURE **6.15**

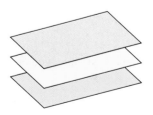

Solution: none
FIGURE **6.16**

Example 5   A System with Infinitely Many Solutions

Solve the system of linear equations.

$$\begin{cases} x + y - 3z = -1 \\ \quad\;\; y - z = \;\;\; 0 \\ -x + 2y \quad\;\;\; = \;\;\; 1 \end{cases}$$

Solution

$$\begin{cases} x + y - 3z = -1 \\ \quad\;\; y - z = \;\;\; 0 \\ \quad\; 3y - 3z = \;\;\; 0 \end{cases}$$

> Adding the first equation to the third equation produces a new third equation.

$$\begin{cases} x + y - 3z = -1 \\ \quad\;\; y - z = \;\;\; 0 \\ \qquad\qquad 0 = \;\;\; 0 \end{cases}$$

> Adding -3 times the second equation to the third equation produces a new third equation.

This result means that Equation 3 depends on Equations 1 and 2 in the sense that it gives no additional information about the variables. Because $0 = 0$ is a true statement, you can conclude that this system will have infinitely many solutions. However, it is incorrect to say simply that the solution is "infinite." You must also specify the correct form of the solution. So, the original system is equivalent to the system

$$\begin{cases} x + y - 3z = -1 \\ \quad\;\; y - z = \;\;\; 0 \end{cases}.$$

In the last equation, solve for y in terms of z to obtain $y = z$. Back-substituting $y = z$ in the first equation produces $x = 2z - 1$. Finally, letting $z = a$, where a is a real number, the solutions to the given system are all of the form $x = 2a - 1$, $y = a$, and $z = a$. So, every ordered triple of the form

$$(2a - 1, a, a)$$

is a solution of the system.

CHECK Point   Now try Exercise 29.

In Example 5, there are other ways to write the same infinite set of solutions. For instance, letting $x = b$, the solutions could have been written as

$$\left(b, \tfrac{1}{2}(b + 1), \tfrac{1}{2}(b + 1)\right).$$

To convince yourself that this description produces the same set of solutions, consider the following.

Substitution	Solution
$a = 0$	$(2(\;) - 1, \;, \;) = (-1, 0, 0)$
$b = -1$	$\left(\;, \tfrac{1}{2}(\;\; + 1), \tfrac{1}{2}(\;\; + 1)\right) = (-1, 0, 0)$
$a = 1$	$(2(\;) - 1, \;, \;) = (1, 1, 1)$
$b = 1$	$\left(\;, \tfrac{1}{2}(\;\; + 1), \tfrac{1}{2}(\;\; + 1)\right) = (1, 1, 1)$
$a = 2$	$(2(\;) - 1, \;, \;) = (3, 2, 2)$
$b = 3$	$\left(\;, \tfrac{1}{2}(\;\; + 1), \tfrac{1}{2}(\;\; + 1)\right) = (3, 2, 2)$

Study Tip

In Example 5, x and y are solved in terms of the third variable z. To write the correct form of the solution to the system that does not use any of the three variables of the system, let a represent any real number and let $z = a$. Then solve for x and y. The solution can then be written in terms of a, which is not one of the variables of the system.

Study Tip

When comparing descriptions of an infinite solution set, keep in mind that there is more than one way to describe the set.

Nonsquare Systems

So far, each system of linear equations you have looked at has been *square*, which means that the number of equations is equal to the number of variables. In a **nonsquare** system, the number of equations differs from the number of variables. A system of linear equations cannot have a unique solution unless there are at least as many equations as there are variables in the system.

Example 6 A System with Fewer Equations than Variables

Solve the system of linear equations.

$$\begin{cases} x - 2y + z = 2 \\ 2x - y - z = 1 \end{cases}$$

Solution

Begin by rewriting the system in row-echelon form.

$$\begin{cases} x - 2y + z = 2 \\ 3y - 3z = -3 \end{cases}$$

> Adding -2 times the first equation to the second equation produces a new second equation.

$$\begin{cases} x - 2y + z = 2 \\ y - z = -1 \end{cases}$$

> Multiplying the second equation by $\frac{1}{3}$ produces a new second equation.

Solve for y in terms of z, to obtain

$$y = z - 1.$$

By back-substituting $y = z - 1$ into Equation 1, you can solve for x, as follows.

$$x - 2y + z = 2$$
$$x - 2(\quad) + z = 2$$
$$x - 2z + 2 + z = 2$$
$$x = z$$

Finally, by letting $z = a$, where a is a real number, you have the solution

$$x = a, \quad y = a - 1, \quad \text{and} \quad z = a.$$

So, every ordered triple of the form

$$(a, a - 1, a)$$

is a solution of the system. Because there were originally three variables and only two equations, the system cannot have a unique solution.

CHECK*Point* Now try Exercise 33.

In Example 6, try choosing some values of a to obtain different solutions of the system, such as $(1, 0, 1)$, $(2, 1, 2)$, and $(3, 2, 3)$. Then check each of the solutions in the original system to verify that they are solutions of the original system.

Applications

Example 7   Vertical Motion

The height at time t of an object that is moving in a (vertical) line with constant acceleration a is given by the **position equation**

$$s = \tfrac{1}{2}at^2 + v_0 t + s_0.$$

The height s is measured in feet, the acceleration a is measured in feet per second squared, t is measured in seconds, v_0 is the initial velocity (at $t = 0$), and s_0 is the initial height. Find the values of a, v_0, and s_0 if $s = 52$ at $t = 1$, $s = 52$ at $t = 2$, and $s = 20$ at $t = 3$, and interpret the result. (See Figure 6.17.)

Solution

By substituting the three values of t and s into the position equation, you can obtain three linear equations in a, v_0, and s_0.

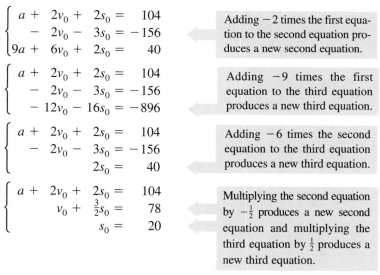

When $t = 1$: $\tfrac{1}{2}a(\ )^2 + v_0(\ ) + s_0 = $ ⟹ $a + 2v_0 + 2s_0 = 104$

When $t = 2$: $\tfrac{1}{2}a(\ )^2 + v_0(\ ) + s_0 = $ ⟹ $2a + 2v_0 + s_0 = 52$

When $t = 3$: $\tfrac{1}{2}a(\ )^2 + v_0(\ ) + s_0 = $ ⟹ $9a + 6v_0 + 2s_0 = 40$

This produces the following system of linear equations.

$$\begin{cases} a + 2v_0 + 2s_0 = 104 \\ 2a + 2v_0 + s_0 = 52 \\ 9a + 6v_0 + 2s_0 = 40 \end{cases}$$

Now solve the system using Gaussian elimination.

$$\begin{cases} a + 2v_0 + 2s_0 = 104 \\ \quad\ -2v_0 - 3s_0 = -156 \\ 9a + 6v_0 + 2s_0 = 40 \end{cases}$$ Adding -2 times the first equation to the second equation produces a new second equation.

$$\begin{cases} a + 2v_0 + 2s_0 = 104 \\ \quad\ -2v_0 - 3s_0 = -156 \\ \quad\ -12v_0 - 16s_0 = -896 \end{cases}$$ Adding -9 times the first equation to the third equation produces a new third equation.

$$\begin{cases} a + 2v_0 + 2s_0 = 104 \\ \quad\ -2v_0 - 3s_0 = -156 \\ \quad\ 2s_0 = 40 \end{cases}$$ Adding -6 times the second equation to the third equation produces a new third equation.

$$\begin{cases} a + 2v_0 + 2s_0 = 104 \\ \quad\ v_0 + \tfrac{3}{2}s_0 = 78 \\ \quad\ s_0 = 20 \end{cases}$$ Multiplying the second equation by $-\tfrac{1}{2}$ produces a new second equation and multiplying the third equation by $\tfrac{1}{2}$ produces a new third equation.

So, the solution of this system is $a = -32$, $v_0 = 48$, and $s_0 = 20$, which can be written as $(-32, 48, 20)$. This solution results in a position equation of $s = -16t^2 + 48t + 20$ and implies that the object was thrown upward at a velocity of 48 feet per second from a height of 20 feet.

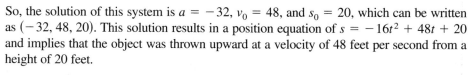   Now try Exercise 45.

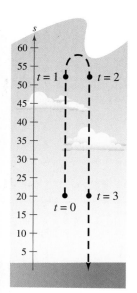

FIGURE **6.17**

Example 8 Data Analysis: Curve-Fitting

Find a quadratic equation

$$y = ax^2 + bx + c$$

whose graph passes through the points $(-1, 3)$, $(1, 1)$, and $(2, 6)$.

Solution

Because the graph of $y = ax^2 + bx + c$ passes through the points $(-1, 3)$, $(1, 1)$, and $(2, 6)$, you can write the following.

When $x = -1$, $y = 3$: $a(\quad)^2 + b(\quad) + c =$

When $x = 1$, $y = 1$: $a(\quad)^2 + b(\quad) + c =$

When $x = 2$, $y = 6$: $a(\quad)^2 + b(\quad) + c =$

This produces the following system of linear equations.

$$\begin{cases} a - b + c = 3 \\ a + b + c = 1 \\ 4a + 2b + c = 6 \end{cases}$$

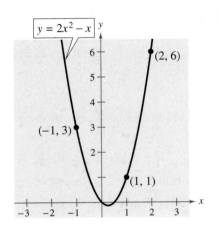

The solution of this system is $a = 2$, $b = -1$, and $c = 0$. So, the equation of the parabola is $y = 2x^2 - x$, as shown in Figure 6.18.

FIGURE **6.18**

CHECK *Point* Now try Exercise 49.

Example 9 Investment Analysis

An inheritance of $12,000 was invested among three funds: a money-market fund that paid 3% annually, municipal bonds that paid 4% annually, and mutual funds that paid 7% annually. The amount invested in mutual funds was $4000 more than the amount invested in municipal bonds. The total interest earned during the first year was $670. How much was invested in each type of fund?

Solution

Let x, y, and z represent the amounts invested in the money-market fund, municipal bonds, and mutual funds, respectively. From the given information, you can write the following equations.

$$\begin{cases} x + y + z = 12{,}000 \\ z = y + 4000 \\ 0.03x + 0.04y + 0.07z = 670 \end{cases}$$

Rewriting this system in standard form without decimals produces the following.

$$\begin{cases} x + y + z = 12{,}000 \\ -y + z = 4{,}000 \\ 3x + 4y + 7z = 67{,}000 \end{cases}$$

Using Gaussian elimination to solve this system yields $x = 2000$, $y = 3000$, and $z = 7000$. So, $2000 was invested in the money-market fund, $3000 was invested in municipal bonds, and $7000 was invested in mutual funds.

CHECK *Point* Now try Exercise 61.

6.3   EXERCISES

See www.CalcChat.com for worked-out solutions to odd-numbered exercises.

VOCABULARY: Fill in the blanks.

1. A system of equations that is in _____ form has a "stair-step" pattern with leading coefficients of 1.

2. A solution to a system of three linear equations in three unknowns can be written as an _____ _____, which has the form (x, y, z).

3. The process used to write a system of linear equations in row-echelon form is called _____ elimination.

4. Interchanging two equations of a system of linear equations is a _____ _____ that produces an equivalent system.

5. A system of equations is called _____ if the number of equations differs from the number of variables in the system.

6. The equation $s = \frac{1}{2}at^2 + v_0 t + s_0$ is called the _____ equation, and it models the height s of an object at time t that is moving in a vertical line with a constant acceleration a.

SKILLS AND APPLICATIONS

In Exercises 7–10, determine whether each ordered triple is a solution of the system of equations.

7. $\begin{cases} 6x - y + z = -1 \\ 4x - 3z = -19 \\ 2y + 5z = 25 \end{cases}$

   (a) $(2, 0, -2)$ (b) $(-3, 0, 5)$

   (c) $(0, -1, 4)$ (d) $(-1, 0, 5)$

8. $\begin{cases} 3x + 4y - z = 17 \\ 5x - y + 2z = -2 \\ 2x - 3y + 7z = -21 \end{cases}$

   (a) $(3, -1, 2)$ (b) $(1, 3, -2)$

   (c) $(4, 1, -3)$ (d) $(1, -2, 2)$

9. $\begin{cases} 4x + y - z = 0 \\ -8x - 6y + z = -\frac{7}{4} \\ 3x - y = -\frac{9}{4} \end{cases}$

   (a) $\left(\frac{1}{2}, -\frac{3}{4}, -\frac{7}{4}\right)$ (b) $\left(-\frac{3}{2}, \frac{5}{4}, -\frac{5}{4}\right)$

   (c) $\left(-\frac{1}{2}, \frac{3}{4}, -\frac{5}{4}\right)$ (d) $\left(-\frac{1}{2}, \frac{1}{6}, -\frac{3}{4}\right)$

10. $\begin{cases} -4x - y - 8z = -6 \\ y + z = 0 \\ 4x - 7y = 6 \end{cases}$

   (a) $(-2, -2, 2)$ (b) $\left(-\frac{33}{2}, -10, 10\right)$

   (c) $\left(\frac{1}{8}, -\frac{1}{2}, \frac{1}{2}\right)$ (d) $\left(-\frac{11}{2}, -4, 4\right)$

In Exercises 11–16, use back-substitution to solve the system of linear equations.

11. $\begin{cases} 2x - y + 5z = 24 \\ y + 2z = 6 \\ z = 8 \end{cases}$

12. $\begin{cases} 4x - 3y - 2z = 21 \\ 6y - 5z = -8 \\ z = -2 \end{cases}$

13. $\begin{cases} 2x + y - 3z = 10 \\ y + z = 12 \\ z = 2 \end{cases}$

14. $\begin{cases} x - y + 2z = 22 \\ 3y - 8z = -9 \\ z = -3 \end{cases}$

15. $\begin{cases} 4x - 2y + z = 8 \\ -y + z = 4 \\ z = 11 \end{cases}$

16. $\begin{cases} 5x - 8z = 22 \\ 3y - 5z = 10 \\ z = -4 \end{cases}$

In Exercises 17 and 18, perform the row operation and write the equivalent system.

17. Add Equation 1 to Equation 2.

$\begin{cases} x - 2y + 3z = 5 \\ -x + 3y - 5z = 4 \\ 2x - 3z = 0 \end{cases}$

What did this operation accomplish?

18. Add -2 times Equation 1 to Equation 3.

$\begin{cases} x - 2y + 3z = 5 \\ -x + 3y - 5z = 4 \\ 2x - 3z = 0 \end{cases}$

What did this operation accomplish?

In Exercises 19–44, solve the system of linear equations and check any solution algebraically.

19. $\begin{cases} x + y + z = 7 \\ 2x - y + z = 9 \\ 3x - z = 10 \end{cases}$

20. $\begin{cases} x + y + z = 5 \\ x - 2y + 4z = 13 \\ 3y + 4z = 13 \end{cases}$

21. $\begin{cases} 2x + 2z = 2 \\ 5x + 3y = 4 \\ 3y - 4z = 4 \end{cases}$

22. $\begin{cases} 2x + 4y + z = 1 \\ x - 2y - 3z = 2 \\ x + y - z = -1 \end{cases}$

23. $\begin{cases} 6y + 4z = -12 \\ 3x + 3y = 9 \\ 2x - 3z = 10 \end{cases}$

24. $\begin{cases} 2x + 4y - z = 7 \\ 2x - 4y + 2z = -6 \\ x + 4y + z = 0 \end{cases}$

25. $\begin{cases} 2x + y - z = 7 \\ x - 2y + 2z = -9 \\ 3x - y + z = 5 \end{cases}$

26. $\begin{cases} 5x - 3y + 2z = 3 \\ 2x + 4y - z = 7 \\ x - 11y + 4z = 3 \end{cases}$

27. $\begin{cases} 3x - 5y + 5z = 1 \\ 5x - 2y + 3z = 0 \\ 7x - y + 3z = 0 \end{cases}$ **28.** $\begin{cases} 2x + y + 3z = 1 \\ 2x + 6y + 8z = 3 \\ 6x + 8y + 18z = 5 \end{cases}$

29. $\begin{cases} x + 2y - 7z = -4 \\ 2x + y + z = 13 \\ 3x + 9y - 36z = -33 \end{cases}$

30. $\begin{cases} 2x + y - 3z = 4 \\ 4x + 2z = 10 \\ -2x + 3y - 13z = -8 \end{cases}$

31. $\begin{cases} 3x - 3y + 6z = 6 \\ x + 2y - z = 5 \\ 5x - 8y + 13z = 7 \end{cases}$ **32.** $\begin{cases} x + 2z = 5 \\ 3x - y - z = 1 \\ 6x - y + 5z = 16 \end{cases}$

33. $\begin{cases} x - 2y + 5z = 2 \\ 4x - z = 0 \end{cases}$ **34.** $\begin{cases} x - 3y + 2z = 18 \\ 5x - 13y + 12z = 80 \end{cases}$

35. $\begin{cases} 2x - 3y + z = -2 \\ -4x + 9y = 7 \end{cases}$

36. $\begin{cases} 2x + 3y + 3z = 7 \\ 4x + 18y + 15z = 44 \end{cases}$

37. $\begin{cases} x + 3w = 4 \\ 2y - z - w = 0 \\ 3y - 2w = 1 \\ 2x - y + 4z = 5 \end{cases}$

38. $\begin{cases} x + y + z + w = 6 \\ 2x + 3y - w = 0 \\ -3x + 4y + z + 2w = 4 \\ x + 2y - z + w = 0 \end{cases}$

39. $\begin{cases} x + 4z = 1 \\ x + y + 10z = 10 \\ 2x - y + 2z = -5 \end{cases}$ **40.** $\begin{cases} 2x - 2y - 6z = -4 \\ -3x + 2y + 6z = 1 \\ x - y - 5z = -3 \end{cases}$

41. $\begin{cases} 2x + 3y = 0 \\ 4x + 3y - z = 0 \\ 8x + 3y + 3z = 0 \end{cases}$ **42.** $\begin{cases} 4x + 3y + 17z = 0 \\ 5x + 4y + 22z = 0 \\ 4x + 2y + 19z = 0 \end{cases}$

43. $\begin{cases} 12x + 5y + z = 0 \\ 23x + 4y - z = 0 \end{cases}$ **44.** $\begin{cases} 2x - y - z = 0 \\ -2x + 6y + 4z = 2 \end{cases}$

VERTICAL MOTION In Exercises 45–48, an object moving vertically is at the given heights at the specified times. Find the position equation $s = \frac{1}{2}at^2 + v_0t + s_0$ for the object.

45. At $t = 1$ second, $s = 128$ feet

At $t = 2$ seconds, $s = 80$ feet

At $t = 3$ seconds, $s = 0$ feet

46. At $t = 1$ second, $s = 32$ feet

At $t = 2$ seconds, $s = 32$ feet

At $t = 3$ seconds, $s = 0$ feet

47. At $t = 1$ second, $s = 352$ feet

At $t = 2$ seconds, $s = 272$ feet

At $t = 3$ seconds, $s = 160$ feet

48. At $t = 1$ second, $s = 132$ feet

At $t = 2$ seconds, $s = 100$ feet

At $t = 3$ seconds, $s = 36$ feet

In Exercises 49–54, find the equation of the parabola

$$y = ax^2 + bx + c$$

that passes through the points. To verify your result, use a graphing utility to plot the points and graph the parabola.

49. $(0, 0), (2, -2), (4, 0)$ **50.** $(0, 3), (1, 4), (2, 3)$

51. $(2, 0), (3, -1), (4, 0)$ **52.** $(1, 3), (2, 2), (3, -3)$

53. $\left(\frac{1}{2}, 1\right), (1, 3), (2, 13)$

54. $(-2, -3), (-1, 0), \left(\frac{1}{2}, -3\right)$

In Exercises 55–58, find the equation of the circle

$$x^2 + y^2 + Dx + Ey + F = 0$$

that passes through the points. To verify your result, use a graphing utility to plot the points and graph the circle.

55. $(0, 0), (5, 5), (10, 0)$

56. $(0, 0), (0, 6), (3, 3)$

57. $(-3, -1), (2, 4), (-6, 8)$

58. $(0, 0), (0, -2), (3, 0)$

59. SPORTS In Super Bowl I, on January 15, 1967, the Green Bay Packers defeated the Kansas City Chiefs by a score of 35 to 10. The total points scored came from 13 different scoring plays, which were a combination of touchdowns, extra-point kicks, and field goals, worth 6, 1, and 3 points, respectively. The same number of touchdowns and extra-point kicks were scored. There were six times as many touchdowns as field goals. How many touchdowns, extra-point kicks, and field goals were scored during the game? (Source: SuperBowl.com)

60. SPORTS In the 2008 Women's NCAA Final Four Championship game, the University of Tennessee Lady Volunteers defeated the University of Stanford Cardinal by a score of 64 to 48. The Lady Volunteers won by scoring a combination of two-point baskets, three-point baskets, and one-point free throws. The number of two-point baskets was two more than the number of free throws. The number of free throws was two more than five times the number of three-point baskets. What combination of scoring accounted for the Lady Volunteers' 64 points? (Source: National Collegiate Athletic Association)

61. FINANCE A small corporation borrowed $775,000 to expand its clothing line. Some of the money was borrowed at 8%, some at 9%, and some at 10%. How much was borrowed at each rate if the annual interest owed was $67,500 and the amount borrowed at 8% was four times the amount borrowed at 10%?

62. FINANCE A small corporation borrowed $800,000 to expand its line of toys. Some of the money was borrowed at 8%, some at 9%, and some at 10%. How much was borrowed at each rate if the annual interest owed was $67,000 and the amount borrowed at 8% was five times the amount borrowed at 10%?

INVESTMENT PORTFOLIO In Exercises 63 and 64, consider an investor with a portfolio totaling $500,000 that is invested in certificates of deposit, municipal bonds, blue-chip stocks, and growth or speculative stocks. How much is invested in each type of investment?

63. The certificates of deposit pay 3% annually, and the municipal bonds pay 5% annually. Over a five-year period, the investor expects the blue-chip stocks to return 8% annually and the growth stocks to return 10% annually. The investor wants a combined annual return of 5% and also wants to have only one-fourth of the portfolio invested in stocks.

64. The certificates of deposit pay 2% annually, and the municipal bonds pay 4% annually. Over a five-year period, the investor expects the blue-chip stocks to return 10% annually and the growth stocks to return 14% annually. The investor wants a combined annual return of 6% and also wants to have only one-fourth of the portfolio invested in stocks.

65. AGRICULTURE A mixture of 5 pounds of fertilizer A, 13 pounds of fertilizer B, and 4 pounds of fertilizer C provides the optimal nutrients for a plant. Commercial brand X contains equal parts of fertilizer B and fertilizer C. Commercial brand Y contains one part of fertilizer A and two parts of fertilizer B. Commercial brand Z contains two parts of fertilizer A, five parts of fertilizer B, and two parts of fertilizer C. How much of each fertilizer brand is needed to obtain the desired mixture?

66. AGRICULTURE A mixture of 12 liters of chemical A, 16 liters of chemical B, and 26 liters of chemical C is required to kill a destructive crop insect. Commercial spray X contains 1, 2, and 2 parts, respectively, of these chemicals. Commercial spray Y contains only chemical C. Commercial spray Z contains only chemicals A and B in equal amounts. How much of each type of commercial spray is needed to get the desired mixture?

67. GEOMETRY The perimeter of a triangle is 110 feet. The longest side of the triangle is 21 feet longer than the shortest side. The sum of the lengths of the two shorter sides is 14 feet more than the length of the longest side. Find the lengths of the sides of the triangle.

68. GEOMETRY The perimeter of a triangle is 180 feet. The longest side of the triangle is 9 feet shorter than twice the shortest side. The sum of the lengths of the two shorter sides is 30 feet more than the length of the longest side. Find the lengths of the sides of the triangle.

In Exercises 69 and 70, find the values of x, y, and z in the figure.

69. **70.**

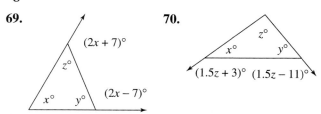

71. ADVERTISING A health insurance company advertises on television, on radio, and in the local newspaper. The marketing department has an advertising budget of $42,000 per month. A television ad costs $1000, a radio ad costs $200, and a newspaper ad costs $500. The department wants to run 60 ads per month, and have as many television ads as radio and newspaper ads combined. How many of each type of ad can the department run each month?

72. RADIO You work as a disc jockey at your college radio station. You are supposed to play 32 songs within two hours. You are to choose the songs from the latest rock, dance, and pop albums. You want to play twice as many rock songs as pop songs and four more pop songs than dance songs. How many of each type of song will you play?

73. ACID MIXTURE A chemist needs 10 liters of a 25% acid solution. The solution is to be mixed from three solutions whose concentrations are 10%, 20%, and 50%. How many liters of each solution will satisfy each condition?

(a) Use 2 liters of the 50% solution.

(b) Use as little as possible of the 50% solution.

(c) Use as much as possible of the 50% solution.

74. ACID MIXTURE A chemist needs 12 gallons of a 20% acid solution. The solution is to be mixed from three solutions whose concentrations are 10%, 15%, and 25%. How many gallons of each solution will satisfy each condition?

(a) Use 4 gallons of the 25% solution.

(b) Use as little as possible of the 25% solution.

(c) Use as much as possible of the 25% solution.

75. ELECTRICAL NETWORK Applying Kirchhoff's Laws to the electrical network in the figure, the currents I_1, I_2, and I_3 are the solution of the system

$$\begin{cases} I_1 - I_2 + I_3 = 0 \\ 3I_1 + 2I_2 \quad\quad = 7 \\ \quad\quad 2I_2 + 4I_3 = 8 \end{cases}$$

find the currents.

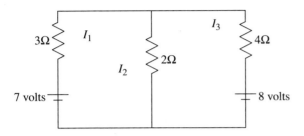

76. PULLEY SYSTEM A system of pulleys is loaded with 128-pound and 32-pound weights (see figure). The tensions t_1 and t_2 in the ropes and the acceleration a of the 32-pound weight are found by solving the system of equations

$$\begin{cases} t_1 - 2t_2 \quad\quad = 0 \\ t_1 \quad\quad - 2a = 128 \\ \quad t_2 + a = 32 \end{cases}$$

where t_1 and t_2 are measured in pounds and a is measured in feet per second squared.

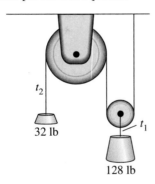

(a) Solve this system.

(b) The 32-pound weight in the pulley system is replaced by a 64-pound weight. The new pulley system will be modeled by the following system of equations.

$$\begin{cases} t_1 - 2t_2 \quad\quad = 0 \\ t_1 \quad\quad - 2a = 128 \\ \quad t_2 + a = 64 \end{cases}$$

Solve this system and use your answer for the acceleration to describe what (if anything) is happening in the pulley system.

FITTING A PARABOLA In Exercises 77–80, find the least squares regression parabola $y = ax^2 + bx + c$ for the points (x_1, y_1), (x_2, y_2), . . . , (x_n, y_n) by solving the following system of linear equations for a, b, and c. Then use the *regression* feature of a graphing utility to confirm the result. (If you are unfamiliar with summation notation, look at the discussion in Section 8.1 or in Appendix B at the website for this text at *academic.cengage.com*.)

$$nc + \left(\sum_{i=1}^{n} x_i\right)b + \left(\sum_{i=1}^{n} x_i^2\right)a = \sum_{i=1}^{n} y_i$$

$$\left(\sum_{i=1}^{n} x_i\right)c + \left(\sum_{i=1}^{n} x_i^2\right)b + \left(\sum_{i=1}^{n} x_i^3\right)a = \sum_{i=1}^{n} x_i y_i$$

$$\left(\sum_{i=1}^{n} x_i^2\right)c + \left(\sum_{i=1}^{n} x_i^3\right)b + \left(\sum_{i=1}^{n} x_i^4\right)a = \sum_{i=1}^{n} x_i^2 y_i$$

77.

(−2, 6), (−4, 5), (2, 6), (4, 2)

78.

(−1, 0), (2, 5), (−2, 0), (1, 2), (0, 1)

79.

(4, 12), (3, 6), (0, 0), (2, 2)

80.

(0, 10), (1, 9), (2, 6), (3, 0)

81. DATA ANALYSIS: WILDLIFE A wildlife management team studied the reproduction rates of deer in three tracts of a wildlife preserve. Each tract contained 5 acres. In each tract, the number of females x, and the percent of females y that had offspring the following year, were recorded. The results are shown in the table.

Number, x	Percent, y
100	75
120	68
140	55

(a) Use the technique demonstrated in Exercises 77–80 to set up a system of equations for the data and to find a least squares regression parabola that models the data.

(b) Use a graphing utility to graph the parabola and the data in the same viewing window.

(c) Use the model to create a table of estimated values of y. Compare the estimated values with the actual data.

(d) Use the model to estimate the percent of females that had offspring when there were 170 females.

(e) Use the model to estimate the number of females when 40% of the females had offspring.

82. DATA ANALYSIS: STOPPING DISTANCE   In testing a new automobile braking system, the speed x (in miles per hour) and the stopping distance y (in feet) were recorded in the table.

Speed, x	Stopping distance, y
30	55
40	105
50	188

(a) Use the technique demonstrated in Exercises 77–80 to set up a system of equations for the data and to find a least squares regression parabola that models the data.

(b) Graph the parabola and the data on the same set of axes.

(c) Use the model to estimate the stopping distance when the speed is 70 miles per hour.

83. SPORTS   In Super Bowl XLIII, on February 1, 2009, the Pittsburgh Steelers defeated the Arizona Cardinals by a score of 27 to 23. The total points scored came from 15 different scoring plays, which were a combination of touchdowns, extra-point kicks, field goals, and safeties, worth 6, 1, 3, and 2 points, respectively. There were three times as many touchdowns as field goals, and the number of extra-point kicks was equal to the number of touchdowns. How many touchdowns, extra-point kicks, field goals, and safeties were scored during the game?   (Source: National Football League)

84. SPORTS   In the 2008 Armed Forces Bowl, the University of Houston defeated the Air Force Academy by a score of 34 to 28. The total points scored came from 18 different scoring plays, which were a combination of touchdowns, extra-point kicks, field goals, and two-point conversions, worth 6, 1, 3, and 2 points, respectively. The number of touchdowns was one more than the number of extra-point kicks, and there were four times as many field goals as two-point conversions. How many touchdowns, extra-point kicks, field goals, and two-point conversions were scored during the game?   (Source: ESPN.com)

ƒ ADVANCED APPLICATIONS   In Exercises 85–88, find values of x, y, and λ that satisfy the system. These systems arise in certain optimization problems in calculus, and λ is called a Lagrange multiplier.

85. $\begin{cases} y + \lambda = 0 \\ x + \lambda = 0 \\ x + y - 10 = 0 \end{cases}$

86. $\begin{cases} 2x + \lambda = 0 \\ 2y + \lambda = 0 \\ x + y - 4 = 0 \end{cases}$

87. $\begin{cases} 2x - 2x\lambda = 0 \\ -2y + \lambda = 0 \\ y - x^2 = 0 \end{cases}$

88. $\begin{cases} 2 + 2y + 2\lambda = 0 \\ 2x + 1 + \lambda = 0 \\ 2x + y - 100 = 0 \end{cases}$

EXPLORATION

TRUE OR FALSE?   In Exercises 89 and 90, determine whether the statement is true or false. Justify your answer.

89. The system

$$\begin{cases} x + 3y - 6z = -16 \\ 2y - z = -1 \\ z = 3 \end{cases}$$

is in row-echelon form.

90. If a system of three linear equations is inconsistent, then its graph has no points common to all three equations.

91. THINK ABOUT IT   Are the following two systems of equations equivalent? Give reasons for your answer.

$\begin{cases} x + 3y - z = 6 \\ 2x - y + 2z = 1 \\ 3x + 2y - z = 2 \end{cases}$ $\begin{cases} x + 3y - z = 6 \\ -7y + 4z = 1 \\ -7y - 4z = -16 \end{cases}$

92. CAPSTONE   Find values of a, b, and c (if possible) such that the system of linear equations has (a) a unique solution, (b) no solution, and (c) an infinite number of solutions.

$$\begin{cases} x + y = 2 \\ y + z = 2 \\ x + z = 2 \\ ax + by + cz = 0 \end{cases}$$

In Exercises 93–96, find two systems of linear equations that have the ordered triple as a solution. (There are many correct answers.)

93. $(3, -4, 2)$

94. $(-5, -2, 1)$

95. $\left(-6, -\frac{1}{2}, -\frac{7}{4}\right)$

96. $\left(-\frac{3}{2}, 4, -7\right)$

PROJECT: EARNINGS PER SHARE   To work an extended application analyzing the earnings per share for Wal-Mart Stores, Inc. from 1992 through 2007, visit this text's website at *academic.cengage.com*.   (Data Source: Wal-Mart Stores, Inc.)

6.4 PARTIAL FRACTIONS

What you should learn

- Recognize partial fraction decompositions of rational expressions.
- Find partial fraction decompositions of rational expressions.

Why you should learn it

Partial fractions can help you analyze the behavior of a rational function. For instance, in Exercise 62 on page 487, you can analyze the exhaust temperatures of a diesel engine using partial fractions.

Introduction

In this section, you will learn to write a rational expression as the sum of two or more simpler rational expressions. For example, the rational expression

$$\frac{x + 7}{x^2 - x - 6}$$

can be written as the sum of two fractions with first-degree denominators. That is,

$$\frac{x + 7}{x^2 - x - 6} = \frac{2}{x - 3} + \frac{-1}{x + 2}.$$

Each fraction on the right side of the equation is a **partial fraction,** and together they make up the **partial fraction decomposition** of the left side.

Decomposition of $N(x)/D(x)$ into Partial Fractions

1. *Divide if improper:* If $N(x)/D(x)$ is an improper fraction [degree of $N(x) \geq$ degree of $D(x)$], divide the denominator into the numerator to obtain

$$\frac{N(x)}{D(x)} = (\text{polynomial}) + \frac{N_1(x)}{D(x)}$$

and apply Steps 2, 3, and 4 below to the proper rational expression $N_1(x)/D(x)$. Note that $N_1(x)$ is the remainder from the division of $N(x)$ by $D(x)$.

2. *Factor the denominator:* Completely factor the denominator into factors of the form

$$(px + q)^m \quad \text{and} \quad (ax^2 + bx + c)^n$$

where $(ax^2 + bx + c)$ is irreducible.

3. *Linear factors:* For *each* factor of the form $(px + q)^m$, the partial fraction decomposition must include the following sum of m fractions.

$$\frac{A_1}{(px + q)} + \frac{A_2}{(px + q)^2} + \cdots + \frac{A_m}{(px + q)^m}$$

4. *Quadratic factors:* For *each* factor of the form $(ax^2 + bx + c)^n$, the partial fraction decomposition must include the following sum of n fractions.

$$\frac{B_1x + C_1}{ax^2 + bx + c} + \frac{B_2x + C_2}{(ax^2 + bx + c)^2} + \cdots + \frac{B_nx + C_n}{(ax^2 + bx + c)^n}$$

Algebra Help

You can review how to find the degree of a polynomial (such as $x - 3$ and $x + 2$) in Section P.3.

Study Tip

Section P.5 shows you how to combine expressions such as

$$\frac{1}{x - 2} + \frac{-1}{x + 3} = \frac{5}{(x - 2)(x + 3)}.$$

The method of partial fraction decomposition shows you how to reverse this process and write

$$\frac{5}{(x - 2)(x + 3)} = \frac{1}{x - 2} + \frac{-1}{x + 3}.$$

Partial Fraction Decomposition

Algebraic techniques for determining the constants in the numerators of partial fractions are demonstrated in the examples that follow. Note that the techniques vary slightly, depending on the type of factors of the denominator: linear or quadratic, distinct or repeated.

Example 1   Distinct Linear Factors

Write the partial fraction decomposition of $\dfrac{x + 7}{x^2 - x - 6}$.

Solution

The expression is proper, so be sure to factor the denominator. Because $x^2 - x - 6 = (x - 3)(x + 2)$, you should include one partial fraction with a constant numerator for each linear factor of the denominator. Write the form of the decomposition as follows.

$$\frac{x + 7}{x^2 - x - 6} = \frac{A}{x - 3} + \frac{B}{x + 2}$$

Multiplying each side of this equation by the least common denominator, $(x - 3)(x + 2)$, leads to the **basic equation**

$$x + 7 = A(x + 2) + B(x - 3).$$

Because this equation is true for all x, you can substitute any *convenient* values of x that will help determine the constants A and B. Values of x that are especially convenient are ones that make the factors $(x + 2)$ and $(x - 3)$ equal to zero. For instance, let $x = -2$. Then

$$
\begin{aligned}
+ 7 &= A(\quad + 2) + B(\quad - 3) \\
5 &= A(0) + B(-5) \\
5 &= -5B \\
-1 &= B.
\end{aligned}
$$

To solve for A, let $x = 3$ and obtain

$$
\begin{aligned}
+ 7 &= A(\quad + 2) + B(\quad - 3) \\
10 &= A(5) + B(0) \\
10 &= 5A \\
2 &= A.
\end{aligned}
$$

So, the partial fraction decomposition is

$$\frac{x + 7}{x^2 - x - 6} = \frac{2}{x - 3} + \frac{-1}{x + 2}.$$

Check this result by combining the two partial fractions on the right side of the equation, or by using your graphing utility.

CHECK*Point*   Now try Exercise 23.

TECHNOLOGY

You can use a graphing utility to check the decomposition found in Example 1. To do this, graph

$$y_1 = \frac{x + 7}{x^2 - x - 6}$$

and

$$y_2 = \frac{2}{x - 3} + \frac{-1}{x + 2}$$

in the same viewing window. The graphs should be identical, as shown below.

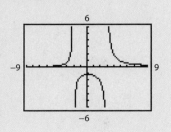

The next example shows how to find the partial fraction decomposition of a rational expression whose denominator has a *repeated* linear factor.

Example 2 Repeated Linear Factors

Write the partial fraction decomposition of $\dfrac{x^4 + 2x^3 + 6x^2 + 20x + 6}{x^3 + 2x^2 + x}$.

Solution

This rational expression is improper, so you should begin by dividing the numerator by the denominator to obtain

$$x + \frac{5x^2 + 20x + 6}{x^3 + 2x^2 + x}.$$

Because the denominator of the remainder factors as

$$x^3 + 2x^2 + x = x(x^2 + 2x + 1) = x(x + 1)^2$$

you should include one partial fraction with a constant numerator for each power of x and $(x + 1)$ and write the form of the decomposition as follows.

$$\frac{5x^2 + 20x + 6}{x(x + 1)^2} = \frac{A}{x} + \frac{B}{x + 1} + \frac{C}{(x + 1)^2}$$

Multiplying by the LCD, $x(x + 1)^2$, leads to the basic equation

$$5x^2 + 20x + 6 = A(x + 1)^2 + Bx(x + 1) + Cx.$$

Letting $x = -1$ eliminates the A- and B-terms and yields

$$5(\quad)^2 + 20(\quad) + 6 = A(\quad + 1)^2 + B(\quad)(\quad + 1) + C(\quad)$$
$$5 - 20 + 6 = 0 + 0 - C$$
$$C = 9.$$

Letting $x = 0$ eliminates the B- and C-terms and yields

$$5(\quad)^2 + 20(\quad) + 6 = A(\quad + 1)^2 + B(\quad)(\quad + 1) + C(\quad)$$
$$6 = A(1) + 0 + 0$$
$$6 = A.$$

At this point, you have exhausted the most convenient choices for x, so to find the value of B, use *any other value* for x along with the known values of A and C. So, using $x = 1$, $A = 6$, and $C = 9$,

$$5(\quad)^2 + 20(\quad) + 6 = (\quad + 1)^2 + B(\quad)(\quad + 1) + (\quad)$$
$$31 = 6(4) + 2B + 9$$
$$-2 = 2B$$
$$-1 = B.$$

So, the partial fraction decomposition is

$$\frac{x^4 + 2x^3 + 6x^2 + 20x + 6}{x^3 + 2x^2 + x} = x + \frac{6}{x} + \frac{-1}{x + 1} + \frac{9}{(x + 1)^2}.$$

CHECK *Point* Now try Exercise 49.

Algebra Help

You can review long division of polynomials in Section 3.3. You can review factoring of polynomials in Section P.4.

⚠ **WARNING / CAUTION**

To obtain the basic equation, be sure to multiply *each* fraction in the form of the decomposition by the LCD.

The procedure used to solve for the constants in Examples 1 and 2 works well when the factors of the denominator are linear. However, when the denominator contains irreducible quadratic factors, you should use a different procedure, which involves writing the right side of the basic equation in polynomial form and *equating the coefficients* of like terms. Then you can use a system of equations to solve for the coefficients.

Example 3   Distinct Linear and Quadratic Factors

Write the partial fraction decomposition of

$$\frac{3x^2 + 4x + 4}{x^3 + 4x}.$$

Solution

This expression is proper, so factor the denominator. Because the denominator factors as

$$x^3 + 4x = x(x^2 + 4)$$

you should include one partial fraction with a constant numerator and one partial fraction with a linear numerator and write the form of the decomposition as follows.

$$\frac{3x^2 + 4x + 4}{x^3 + 4x} = \frac{A}{x} + \frac{Bx + C}{x^2 + 4}$$

Multiplying by the LCD, $x(x^2 + 4)$, yields the basic equation

$$3x^2 + 4x + 4 = A(x^2 + 4) + (Bx + C)x.$$

Expanding this basic equation and collecting like terms produces

$$3x^2 + 4x + 4 = Ax^2 + 4A + Bx^2 + Cx$$

$$= (A + B)x^2 + Cx + 4A.$$

Finally, because two polynomials are equal if and only if the coefficients of like terms are equal, you can equate the coefficients of like terms on opposite sides of the equation.

$$3x^2 + 4x + 4 = (A + B)x^2 + Cx + 4A$$

You can now write the following system of linear equations.

$$\begin{cases} A + B & = 3 \\ \quad\quad C & = 4 \\ 4A & = 4 \end{cases}$$

From this system you can see that $A = 1$ and $C = 4$. Moreover, substituting $A = 1$ into Equation 1 yields

$$+ B = 3 \Longrightarrow B = 2.$$

So, the partial fraction decomposition is

$$\frac{3x^2 + 4x + 4}{x^3 + 4x} = \frac{1}{x} + \frac{2x + 4}{x^2 + 4}.$$

CHECK Point   Now try Exercise 33.

The next example shows how to find the partial fraction decomposition of a rational expression whose denominator has a *repeated* quadratic factor.

Example 4   Repeated Quadratic Factors

Write the partial fraction decomposition of $\dfrac{8x^3 + 13x}{(x^2 + 2)^2}$.

Solution

Include one partial fraction with a linear numerator for each power of $(x^2 + 2)$.

$$\frac{8x^3 + 13x}{(x^2 + 2)^2} = \frac{Ax + B}{x^2 + 2} + \frac{Cx + D}{(x^2 + 2)^2}$$

Multiplying by the LCD, $(x^2 + 2)^2$, yields the basic equation

$$\begin{aligned}
8x^3 + 13x &= (Ax + B)(x^2 + 2) + Cx + D \\
&= Ax^3 + 2Ax + Bx^2 + 2B + Cx + D \\
&= Ax^3 + Bx^2 + (2A + C)x + (2B + D).
\end{aligned}$$

Equating coefficients of like terms on opposite sides of the equation

$$8x^3 + 0x^2 + 13x + 0 = Ax^3 + Bx^2 + (2A + C)x + (2B + D)$$

produces the following system of linear equations.

$$\begin{cases}
A & & & = 8 \\
& B & & = 0 \\
2A + & & C & = 13 \\
& 2B + & D & = 0
\end{cases}$$

Finally, use the values $A = 8$ and $B = 0$ to obtain the following.

$$2(\ ) + C = 13$$
$$C = -3$$

$$2(\ ) + D = 0$$
$$D = 0$$

So, using $A = 8$, $B = 0$, $C = -3$, and $D = 0$, the partial fraction decomposition is

$$\frac{8x^3 + 13x}{(x^2 + 2)^2} = \frac{8x}{x^2 + 2} + \frac{-3x}{(x^2 + 2)^2}.$$

Check this result by combining the two partial fractions on the right side of the equation, or by using your graphing utility.

CHECK Point   Now try Exercise 55.

Guidelines for Solving the Basic Equation

Linear Factors

1. Substitute the *zeros* of the distinct linear factors into the basic equation.

2. For repeated linear factors, use the coefficients determined in Step 1 to rewrite the basic equation. Then substitute *other* convenient values of x and solve for the remaining coefficients.

Quadratic Factors

1. Expand the basic equation.

2. Collect terms according to powers of x.

3. Equate the coefficients of like terms to obtain equations involving A, B, C, and so on.

4. Use a system of linear equations to solve for A, B, C, ...

Keep in mind that for *improper* rational expressions such as

$$\frac{N(x)}{D(x)} = \frac{2x^3 + x^2 - 7x + 7}{x^2 + x - 2}$$

you must first divide before applying partial fraction decomposition.

CLASSROOM DISCUSSION

Error Analysis   You are tutoring a student in algebra. In trying to find a partial fraction decomposition, the student writes the following.

$$\frac{x^2 + 1}{x(x - 1)} = \frac{A}{x} + \frac{B}{x - 1}$$

$$\frac{x^2 + 1}{x(x - 1)} = \frac{A(x - 1)}{x(x - 1)} + \frac{Bx}{x(x - 1)}$$

$$x^2 + 1 = A(x - 1) + Bx$$

By substituting $x = 0$ and $x = 1$ into the basic equation, the student concludes that $A = -1$ and $B = 2$. However, in checking this solution, the student obtains the following.

$$\frac{-1}{x} + \frac{2}{x - 1} = \frac{(-1)(x - 1) + 2(x)}{x(x - 1)}$$

$$= \frac{x + 1}{x(x - 1)}$$

$$\neq \frac{x^2 + 1}{x(x - 1)}$$

What is wrong?

6.4 **EXERCISES**

See www.CalcChat.com for worked-out solutions to odd-numbered exercises.

VOCABULARY: Fill in the blanks.

1. The process of writing a rational expression as the sum or difference of two or more simpler rational expressions is called _____ _____ _____.

2. If the degree of the numerator of a rational expression is greater than or equal to the degree of the denominator, then the fraction is called _____.

3. Each fraction on the right side of the equation $\dfrac{x-1}{x^2-8x+15} = \dfrac{-1}{x-3} + \dfrac{2}{x-5}$ is a _____ _____.

4. The _____ _____ is obtained after multiplying each side of the partial fraction decomposition form by the least common denominator.

SKILLS AND APPLICATIONS

In Exercises 5–8, match the rational expression with the form of its decomposition. [The decompositions are labeled (a), (b), (c), and (d).]

(a) $\dfrac{A}{x} + \dfrac{B}{x+2} + \dfrac{C}{x-2}$ (b) $\dfrac{A}{x} + \dfrac{B}{x-4}$

(c) $\dfrac{A}{x} + \dfrac{B}{x^2} + \dfrac{C}{x-4}$ (d) $\dfrac{A}{x} + \dfrac{Bx+C}{x^2+4}$

5. $\dfrac{3x-1}{x(x-4)}$

6. $\dfrac{3x-1}{x^2(x-4)}$

7. $\dfrac{3x-1}{x(x^2+4)}$

8. $\dfrac{3x-1}{x(x^2-4)}$

In Exercises 9–18, write the form of the partial fraction decomposition of the rational expression. Do not solve for the constants.

9. $\dfrac{3}{x^2-2x}$

10. $\dfrac{x-2}{x^2+4x+3}$

11. $\dfrac{9}{x^3-7x^2}$

12. $\dfrac{x^2-3x+2}{4x^3+11x^2}$

13. $\dfrac{4x^2+3}{(x-5)^3}$

14. $\dfrac{6x+5}{(x+2)^4}$

15. $\dfrac{2x-3}{x^3+10x}$

16. $\dfrac{x-6}{2x^3+8x}$

17. $\dfrac{x-1}{x(x^2+1)^2}$

18. $\dfrac{x+4}{x^2(3x-1)^2}$

In Exercises 19–42, write the partial fraction decomposition of the rational expression. Check your result algebraically.

19. $\dfrac{1}{x^2+x}$

20. $\dfrac{3}{x^2-3x}$

21. $\dfrac{1}{2x^2+x}$

22. $\dfrac{5}{x^2+x-6}$

23. $\dfrac{3}{x^2+x-2}$

24. $\dfrac{x+1}{x^2-x-6}$

25. $\dfrac{1}{x^2-1}$

26. $\dfrac{1}{4x^2-9}$

27. $\dfrac{x^2+12x+12}{x^3-4x}$

28. $\dfrac{x+2}{x(x^2-9)}$

29. $\dfrac{3x}{(x-3)^2}$

30. $\dfrac{2x-3}{(x-1)^2}$

31. $\dfrac{4x^2+2x-1}{x^2(x+1)}$

32. $\dfrac{6x^2+1}{x^2(x-1)^2}$

33. $\dfrac{x^2+2x+3}{x^3+x}$

34. $\dfrac{2x}{x^3-1}$

35. $\dfrac{x}{x^3-x^2-2x+2}$

36. $\dfrac{x+6}{x^3-3x^2-4x+12}$

37. $\dfrac{2x^2+x+8}{(x^2+4)^2}$

38. $\dfrac{x^2}{x^4-2x^2-8}$

39. $\dfrac{x}{16x^4-1}$

40. $\dfrac{3}{x^4+x}$

41. $\dfrac{x^2+5}{(x+1)(x^2-2x+3)}$

42. $\dfrac{x^2-4x+7}{(x+1)(x^2-2x+3)}$

In Exercises 43–50, write the partial fraction decomposition of the improper rational expression.

43. $\dfrac{x^2-x}{x^2+x+1}$

44. $\dfrac{x^2-4x}{x^2+x+6}$

45. $\dfrac{2x^3-x^2+x+5}{x^2+3x+2}$

46. $\dfrac{x^3+2x^2-x+1}{x^2+3x-4}$

47. $\dfrac{x^4}{(x-1)^3}$

48. $\dfrac{16x^4}{(2x-1)^3}$

49. $\dfrac{x^4+2x^3+4x^2+8x+2}{x^3+2x^2+x}$

50. $\dfrac{2x^4+8x^3+7x^2-7x-12}{x^3+4x^2+4x}$

In Exercises 51–58, write the partial fraction decomposition of the rational expression. Use a graphing utility to check your result.

51. $\dfrac{5 - x}{2x^2 + x - 1}$

52. $\dfrac{3x^2 - 7x - 2}{x^3 - x}$

53. $\dfrac{4x^2 - 1}{2x(x + 1)^2}$

54. $\dfrac{3x + 1}{2x^3 + 3x^2}$

55. $\dfrac{x^2 + x + 2}{(x^2 + 2)^2}$

56. $\dfrac{x^3}{(x + 2)^2(x - 2)^2}$

57. $\dfrac{2x^3 - 4x^2 - 15x + 5}{x^2 - 2x - 8}$

58. $\dfrac{x^3 - x + 3}{x^2 + x - 2}$

GRAPHICAL ANALYSIS In Exercises 59 and 60, (a) write the partial fraction decomposition of the rational function, (b) identify the graph of the rational function and the graph of each term of its decomposition, and (c) state any relationship between the vertical asymptotes of the graph of the rational function and the vertical asymptotes of the graphs of the terms of the decomposition. To print an enlarged copy of the graph, go to the website *www.mathgraphs*.com.

59. $y = \dfrac{x - 12}{x(x - 4)}$

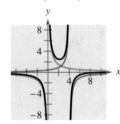

60. $y = \dfrac{2(4x - 3)}{x^2 - 9}$

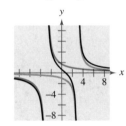

61. ENVIRONMENT The predicted cost C (in thousands of dollars) for a company to remove $p\%$ of a chemical from its waste water is given by the model

$$C = \frac{120p}{10,000 - p^2}, \quad 0 \le p < 100.$$

Write the partial fraction decomposition for the rational function. Verify your result by using the *table* feature of a graphing utility to create a table comparing the original function with the partial fractions.

62. THERMODYNAMICS The magnitude of the range R of exhaust temperatures (in degrees Fahrenheit) in an experimental diesel engine is approximated by the model

$$R = \frac{5000(4 - 3x)}{(11 - 7x)(7 - 4x)}, \quad 0 < x \le 1$$

where x is the relative load (in foot-pounds).

(a) Write the partial fraction decomposition of the equation.

(b) The decomposition in part (a) is the difference of two fractions. The absolute values of the terms give the expected maximum and minimum temperatures of the exhaust gases for different loads.

$$\text{Ymax} = |\text{1st term}| \qquad \text{Ymin} = |\text{2nd term}|$$

Write the equations for Ymax and Ymin.

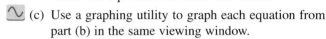 (c) Use a graphing utility to graph each equation from part (b) in the same viewing window.

(d) Determine the expected maximum and minimum temperatures for a relative load of 0.5.

EXPLORATION

TRUE OR FALSE? In Exercises 63–65, determine whether the statement is true or false. Justify your answer.

63. For the rational expression $\dfrac{x}{(x + 10)(x - 10)^2}$, the partial fraction decomposition is of the form $\dfrac{A}{x + 10} + \dfrac{B}{(x - 10)^2}$.

64. For the rational expression $\dfrac{2x + 3}{x^2(x + 2)^2}$, the partial fraction decomposition is of the form $\dfrac{Ax + B}{x^2} + \dfrac{Cx + D}{(x + 2)^2}$.

65. When writing the partial fraction decomposition of the expression $\dfrac{x^3 + x - 2}{x^2 - 5x - 14}$, the first step is to divide the numerator by the denominator.

66. CAPSTONE Explain the similarities and differences in finding the partial fraction decompositions of proper rational expressions whose denominators factor into (a) distinct linear factors, (b) distinct quadratic factors, (c) repeated factors, and (d) linear and quadratic factors.

In Exercises 67–70, write the partial fraction decomposition of the rational expression. Check your result algebraically. Then assign a value to the constant a to check the result graphically.

67. $\dfrac{1}{a^2 - x^2}$

68. $\dfrac{1}{x(x + a)}$

69. $\dfrac{1}{y(a - y)}$

70. $\dfrac{1}{(x + 1)(a - x)}$

71. WRITING Describe two ways of solving for the constants in a partial fraction decomposition.

6.5 SYSTEMS OF INEQUALITIES

What you should learn

- Sketch the graphs of inequalities in two variables.
- Solve systems of inequalities.
- Use systems of inequalities in two variables to model and solve real-life problems.

Why you should learn it

You can use systems of inequalities in two variables to model and solve real-life problems. For instance, in Exercise 83 on page 497, you will use a system of inequalities to analyze the retail sales of prescription drugs.

The Graph of an Inequality

The statements $3x - 2y < 6$ and $2x^2 + 3y^2 \geq 6$ are inequalities in two variables. An ordered pair (a, b) is a **solution of an inequality** in x and y if the inequality is true when a and b are substituted for x and y, respectively. The **graph of an inequality** is the collection of all solutions of the inequality. To sketch the graph of an inequality, begin by sketching the graph of the *corresponding equation*. The graph of the equation will normally separate the plane into two or more regions. In each such region, one of the following must be true.

1. *All* points in the region are solutions of the inequality.

2. *No* point in the region is a solution of the inequality.

So, you can determine whether the points in an entire region satisfy the inequality by simply testing *one* point in the region.

Sketching the Graph of an Inequality in Two Variables

1. Replace the inequality sign by an equal sign, and sketch the graph of the resulting equation. (Use a dashed line for $<$ or $>$ and a solid line for $\leq$ or $\geq$.)

2. Test one point in each of the regions formed by the graph in Step 1. If the point satisfies the inequality, shade the entire region to denote that every point in the region satisfies the inequality.

Example 1   Sketching the Graph of an Inequality

Sketch the graph of $y \geq x^2 - 1$.

Solution

Begin by graphing the corresponding equation $y = x^2 - 1$, which is a parabola, as shown in Figure 6.19. By testing a point *above* the parabola $(0, 0)$ and a point *below* the parabola $(0, -2)$, you can see that the points that satisfy the inequality are those lying above (or on) the parabola.

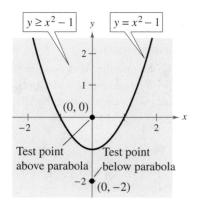

FIGURE 6.19

> ### WARNING / CAUTION
>
> Be careful when you are sketching the graph of an inequality in two variables. A dashed line means that all points on the line or curve *are not* solutions of the inequality. A solid line means that all points on the line or curve *are* solutions of the inequality.

CHECK*Point* Now try Exercise 7.

You can review the properties of inequalities in Section 1.7.

TECHNOLOGY

A graphing utility can be used to graph an inequality or a system of inequalities. For instance, to graph $y \geq x - 2$, enter $y = x - 2$ and use the *shade* feature of the graphing utility to shade the correct part of the graph. You should obtain the graph below. Consult the user's guide for your graphing utility for specific keystrokes.

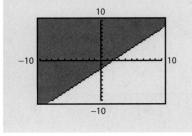

The inequality in Example 1 is a nonlinear inequality in two variables. Most of the following examples involve **linear inequalities** such as $ax + by < c$ (*a* and *b* are not both zero). The graph of a linear inequality is a half-plane lying on one side of the line $ax + by = c$.

Example 2   Sketching the Graph of a Linear Inequality

Sketch the graph of each linear inequality.

a. $x > -2$ **b.** $y \leq 3$

Solution

a. The graph of the corresponding equation $x = -2$ is a vertical line. The points that satisfy the inequality $x > -2$ are those lying to the right of this line, as shown in Figure 6.20.

b. The graph of the corresponding equation $y = 3$ is a horizontal line. The points that satisfy the inequality $y \leq 3$ are those lying below (or on) this line, as shown in Figure 6.21.

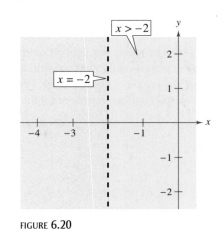

FIGURE **6.20**

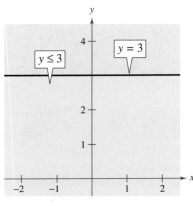

FIGURE **6.21**

CHECK**Point**   Now try Exercise 9.

Example 3   Sketching the Graph of a Linear Inequality

Sketch the graph of $x - y < 2$.

Solution

The graph of the corresponding equation $x - y = 2$ is a line, as shown in Figure 6.22. Because the origin $(0, 0)$ satisfies the inequality, the graph consists of the half-plane lying above the line. (Try checking a point below the line. Regardless of which point you choose, you will see that it does not satisfy the inequality.)

CHECK**Point**   Now try Exercise 15.

To graph a linear inequality, it can help to write the inequality in slope-intercept form. For instance, by writing $x - y < 2$ in the form

$$y > x - 2$$

you can see that the solution points lie *above* the line $x - y = 2$ (or $y = x - 2$), as shown in Figure 6.22.

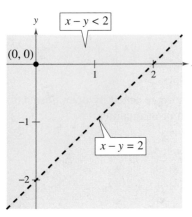

FIGURE **6.22**

Systems of Inequalities

Many practical problems in business, science, and engineering involve systems of linear inequalities. A **solution** of a system of inequalities in x and y is a point (x, y) that satisfies each inequality in the system.

To sketch the graph of a system of inequalities in two variables, first sketch the graph of each individual inequality (on the same coordinate system) and then find the region that is *common* to every graph in the system. This region represents the **solution set** of the system. For systems of *linear inequalities*, it is helpful to find the vertices of the solution region.

| Example 4   Solving a System of Inequalities

Sketch the graph (and label the vertices) of the solution set of the system.

$$\begin{cases} x - y < 2 \\ \quad x > -2 \\ \quad y \le 3 \end{cases}$$

Solution

The graphs of these inequalities are shown in Figures 6.22, 6.20, and 6.21, respectively, on page 489. The triangular region common to all three graphs can be found by super-imposing the graphs on the same coordinate system, as shown in Figure 6.23. To find the vertices of the region, solve the three systems of corresponding equations obtained by taking *pairs* of equations representing the boundaries of the individual regions.

> *Study Tip*
>
> Using different colored pencils to shade the solution of each inequality in a system will make identifying the solution of the system of inequalities easier.

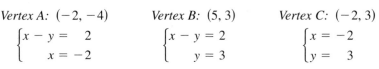

$$\textit{Vertex A: } (-2, -4) \qquad \textit{Vertex B: } (5, 3) \qquad \textit{Vertex C: } (-2, 3)$$
$$\begin{cases} x - y = 2 \\ \quad x = -2 \end{cases} \qquad \begin{cases} x - y = 2 \\ \quad y = 3 \end{cases} \qquad \begin{cases} x = -2 \\ y = \quad 3 \end{cases}$$

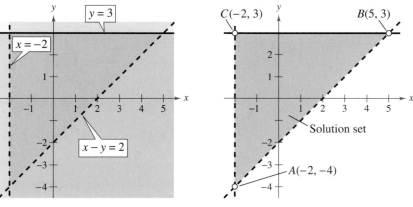

FIGURE **6.23**

Note in Figure 6.23 that the vertices of the region are represented by open dots. This means that the vertices *are not* solutions of the system of inequalities.

CHECK*Point*   Now try Exercise 41.

For the triangular region shown in Figure 6.23, each point of intersection of a pair of boundary lines corresponds to a vertex. With more complicated regions, two border lines can sometimes intersect at a point that is not a vertex of the region, as shown in Figure 6.24. To keep track of which points of intersection are actually vertices of the region, you should sketch the region and refer to your sketch as you find each point of intersection.

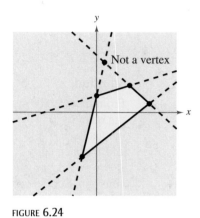

Not a vertex

FIGURE **6.24**

Example 5   Solving a System of Inequalities

Sketch the region containing all points that satisfy the system of inequalities.

$$\begin{cases} x^2 - y \le 1 \\ -x + y \le 1 \end{cases}$$

Solution

As shown in Figure 6.25, the points that satisfy the inequality

$$x^2 - y \le 1$$

are the points lying above (or on) the parabola given by

$$y = x^2 - 1.$$

The points satisfying the inequality

$$-x + y \le 1$$

are the points lying below (or on) the line given by

$$y = x + 1.$$

To find the points of intersection of the parabola and the line, solve the system of corresponding equations.

$$\begin{cases} x^2 - y = 1 \\ -x + y = 1 \end{cases}$$

Using the method of substitution, you can find the solutions to be $(-1, 0)$ and $(2, 3)$. So, the region containing all points that satisfy the system is indicated by the shaded region in Figure 6.25.

CHECK Point   Now try Exercise 43.

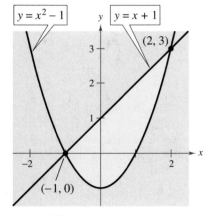

FIGURE **6.25**

When solving a system of inequalities, you should be aware that the system might have no solution *or* it might be represented by an unbounded region in the plane. These two possibilities are shown in Examples 6 and 7.

Example 6 A System with No Solution

Sketch the solution set of the system of inequalities.

$$\begin{cases} x + y > 3 \\ x + y < -1 \end{cases}$$

Solution

From the way the system is written, it is clear that the system has no solution, because the quantity $(x + y)$ cannot be both less than -1 and greater than 3. Graphically, the inequality $x + y > 3$ is represented by the half-plane lying above the line $x + y = 3$, and the inequality $x + y < -1$ is represented by the half-plane lying below the line $x + y = -1$, as shown in Figure 6.26. These two half-planes have no points in common. So, the system of inequalities has no solution.

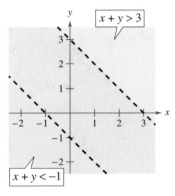

FIGURE **6.26**

CHECK**Point** Now try Exercise 45.

Example 7 An Unbounded Solution Set

Sketch the solution set of the system of inequalities.

$$\begin{cases} x + y < 3 \\ x + 2y > 3 \end{cases}$$

Solution

The graph of the inequality $x + y < 3$ is the half-plane that lies below the line $x + y = 3$, as shown in Figure 6.27. The graph of the inequality $x + 2y > 3$ is the half-plane that lies above the line $x + 2y = 3$. The intersection of these two half-planes is an *infinite wedge* that has a vertex at $(3, 0)$. So, the solution set of the system of inequalities is unbounded.

CHECK**Point** Now try Exercise 47.

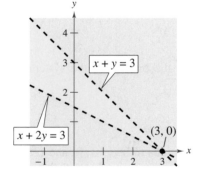

FIGURE **6.27**

Applications

Example 9 in Section 6.2 discussed the *equilibrium point* for a system of demand and supply equations. The next example discusses two related concepts that economists call **consumer surplus** and **producer surplus**. As shown in Figure 6.28, the consumer surplus is defined as the area of the region that lies *below* the demand curve, *above* the horizontal line passing through the equilibrium point, and to the right of the *p*-axis. Similarly, the producer surplus is defined as the area of the region that lies *above* the supply curve, *below* the horizontal line passing through the equilibrium point, and to the right of the *p*-axis. The consumer surplus is a measure of the amount that consumers would have been willing to pay *above what they actually paid*, whereas the producer surplus is a measure of the amount that producers would have been willing to receive *below what they actually received*.

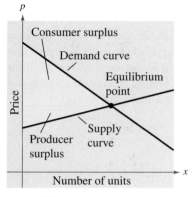

FIGURE **6.28**

Example 8   Consumer Surplus and Producer Surplus

The demand and supply equations for a new type of personal digital assistant are given by

$$\begin{cases} p = 150 - 0.00001x \\ p = 60 + 0.00002x \end{cases}$$

where *p* is the price (in dollars) and *x* represents the number of units. Find the consumer surplus and producer surplus for these two equations.

Solution

Begin by finding the equilibrium point (when supply and demand are equal) by solving the equation

$$60 + 0.00002x = 150 - 0.00001x.$$

In Example 9 in Section 6.2, you saw that the solution is $x = 3,000,000$ units, which corresponds to an equilibrium price of $p = \$120$. So, the consumer surplus and producer surplus are the areas of the following triangular regions.

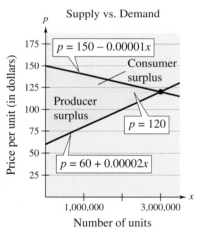

FIGURE **6.29**

Consumer Surplus	*Producer Surplus*
$\begin{cases} p \le 150 - 0.00001x \\ p \ge 120 \\ x \ge 0 \end{cases}$	$\begin{cases} p \ge 60 + 0.00002x \\ p \le 120 \\ x \ge 0 \end{cases}$

In Figure 6.29, you can see that the consumer and producer surpluses are defined as the areas of the shaded triangles.

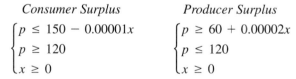

$$\begin{aligned} \text{Consumer surplus} &= \frac{1}{2}(\text{base})(\text{height}) \\ &= \frac{1}{2}()() = \$45,000,000 \end{aligned}$$

$$\begin{aligned} \text{Producer surplus} &= \frac{1}{2}(\text{base})(\text{height}) \\ &= \frac{1}{2}()() = \$90,000,000 \end{aligned}$$

CHECK*Point* ▸ Now try Exercise 71.

Example 9 Nutrition

The liquid portion of a diet is to provide at least 300 calories, 36 units of vitamin A, and 90 units of vitamin C. A cup of dietary drink X provides 60 calories, 12 units of vitamin A, and 10 units of vitamin C. A cup of dietary drink Y provides 60 calories, 6 units of vitamin A, and 30 units of vitamin C. Set up a system of linear inequalities that describes how many cups of each drink should be consumed each day to meet or exceed the minimum daily requirements for calories and vitamins.

Solution

Begin by letting x and y represent the following.

x = number of cups of dietary drink X

y = number of cups of dietary drink Y

To meet or exceed the minimum daily requirements, the following inequalities must be satisfied.

$$\begin{cases} 60x + 60y \geq 300 \\ 12x + 6y \geq 36 \\ 10x + 30y \geq 90 \\ \qquad\quad x \geq 0 \\ \qquad\quad y \geq 0 \end{cases}$$

The last two inequalities are included because x and y cannot be negative. The graph of this system of inequalities is shown in Figure 6.30. (More is said about this application in Example 6 in Section 6.6.)

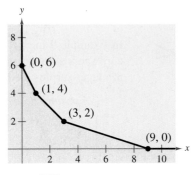

FIGURE 6.30

CHECK**Point** Now try Exercise 75.

CLASSROOM DISCUSSION

Creating a System of Inequalities Plot the points (0, 0), (4, 0), (3, 2), and (0, 2) in a coordinate plane. Draw the quadrilateral that has these four points as its vertices. Write a system of linear inequalities that has the quadrilateral as its solution. Explain how you found the system of inequalities.

6.5   EXERCISES

See www.CalcChat.com for worked-out solutions to odd-numbered exercises.

VOCABULARY: Fill in the blanks.

1. An ordered pair (a, b) is a _____ of an inequality in x and y if the inequality is true when a and b are substituted for x and y, respectively.

2. The _____ of an inequality is the collection of all solutions of the inequality.

3. The graph of a _____ inequality is a half-plane lying on one side of the line $ax + by = c$.

4. A _____ of a system of inequalities in x and y is a point (x, y) that satisfies each inequality in the system.

5. A _____ _____ of a system of inequalities in two variables is the region common to the graphs of every inequality in the system.

6. The area of the region that lies below the demand curve, above the horizontal line passing through the equilibrium point, to the right of the p-axis is called the _____ _____.

SKILLS AND APPLICATIONS

In Exercises 7–20, sketch the graph of the inequality.

7. $y < 5 - x^2$ **8.** $y^2 - x < 0$

9. $x \geq 6$ **10.** $x < -4$

11. $y > -7$ **12.** $10 \geq y$

13. $y < 2 - x$ **14.** $y > 4x - 3$

15. $2y - x \geq 4$ **16.** $5x + 3y \geq -15$

17. $(x + 1)^2 + (y - 2)^2 < 9$

18. $(x - 1)^2 + (y - 4)^2 > 9$

19. $y \leq \dfrac{1}{1 + x^2}$ **20.** $y > \dfrac{-15}{x^2 + x + 4}$

In Exercises 21–32, use a graphing utility to graph the inequality.

21. $y < \ln x$ **22.** $y \geq -2 - \ln(x + 3)$

23. $y < 4^{-x-5}$ **24.** $y \leq 2^{2x-0.5} - 7$

25. $y \geq \frac{5}{9}x - 2$ **26.** $y \leq 6 - \frac{3}{2}x$

27. $y < -3.8x + 1.1$ **28.** $y \geq -20.74 + 2.66x$

29. $x^2 + 5y - 10 \leq 0$ **30.** $2x^2 - y - 3 > 0$

31. $\frac{5}{2}y - 3x^2 - 6 \geq 0$ **32.** $-\frac{1}{10}x^2 - \frac{3}{8}y < -\frac{1}{4}$

In Exercises 33–36, write an inequality for the shaded region shown in the figure.

33.

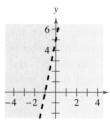

34.

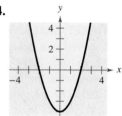

35.

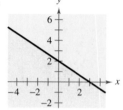

36.

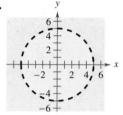

In Exercises 37–40, determine whether each ordered pair is a solution of the system of linear inequalities.

37. $\begin{cases} x \geq -4 \\ y > -3 \\ y \leq -8x - 3 \end{cases}$ (a) $(0, 0)$ (b) $(-1, -3)$
   (c) $(-4, 0)$ (d) $(-3, 11)$

38. $\begin{cases} -2x + 5y \geq 3 \\ y < 4 \\ -4x + 2y < 7 \end{cases}$ (a) $(0, 2)$ (b) $(-6, 4)$
   (c) $(-8, -2)$ (d) $(-3, 2)$

39. $\begin{cases} 3x + y > 1 \\ -y - \frac{1}{2}x^2 \leq -4 \\ -15x + 4y > 0 \end{cases}$ (a) $(0, 10)$ (b) $(0, -1)$
   (c) $(2, 9)$ (d) $(-1, 6)$

40. $\begin{cases} x^2 + y^2 \geq 36 \\ -3x + y \leq 10 \\ \frac{2}{3}x - y \geq 5 \end{cases}$ (a) $(-1, 7)$ (b) $(-5, 1)$
   (c) $(6, 0)$ (d) $(4, -8)$

In Exercises 41–54, sketch the graph and label the vertices of the solution set of the system of inequalities.

41. $\begin{cases} x + y \leq 1 \\ -x + y \leq 1 \\ y \geq 0 \end{cases}$ **42.** $\begin{cases} 3x + 4y < 12 \\ x > 0 \\ y > 0 \end{cases}$

43. $\begin{cases} x^2 + y \leq 7 \\ x \geq -2 \\ y \geq 0 \end{cases}$ **44.** $\begin{cases} 4x^2 + y \geq 2 \\ x \leq 1 \\ y \leq 1 \end{cases}$

45. $\begin{cases} 2x + y > 2 \\ 6x + 3y < 2 \end{cases}$

46. $\begin{cases} x - 7y > -36 \\ 5x + 2y > 5 \\ 6x - 5y > 6 \end{cases}$

47. $\begin{cases} -3x + 2y < 6 \\ x - 4y > -2 \\ 2x + y < 3 \end{cases}$

48. $\begin{cases} x - 2y < -6 \\ 5x - 3y > -9 \end{cases}$

49. $\begin{cases} x > y^2 \\ x < y + 2 \end{cases}$

50. $\begin{cases} x - y^2 > 0 \\ x - y > 2 \end{cases}$

51. $\begin{cases} x^2 + y^2 \le 36 \\ x^2 + y^2 \ge 9 \end{cases}$

52. $\begin{cases} x^2 + y^2 \le 25 \\ 4x - 3y \le 0 \end{cases}$

53. $\begin{cases} 3x + 4 \ge y^2 \\ x - y < 0 \end{cases}$

54. $\begin{cases} x < 2y - y^2 \\ 0 < x + y \end{cases}$

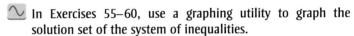 In Exercises 55–60, use a graphing utility to graph the solution set of the system of inequalities.

55. $\begin{cases} y \le \sqrt{3x} + 1 \\ y \ge x^2 + 1 \end{cases}$

56. $\begin{cases} y < -x^2 + 2x + 3 \\ y > x^2 - 4x + 3 \end{cases}$

57. $\begin{cases} y < x^3 - 2x + 1 \\ y > -2x \\ x \le 1 \end{cases}$

58. $\begin{cases} y \ge x^4 - 2x^2 + 1 \\ y \le 1 - x^2 \end{cases}$

59. $\begin{cases} x^2 y \ge 1 \\ 0 < x \le 4 \\ y \le 4 \end{cases}$

60. $\begin{cases} y \le e^{-x^2/2} \\ y \ge 0 \\ -2 \le x \le 2 \end{cases}$

In Exercises 61–70, derive a set of inequalities to describe the region.

61.
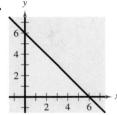

62.
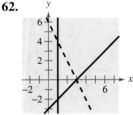

63.
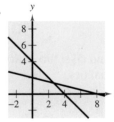

64.
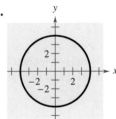

65.
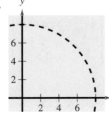

66.
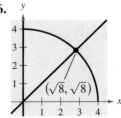

$(\sqrt{8}, \sqrt{8})$

67. Rectangle: vertices at $(4, 3)$, $(9, 3)$, $(9, 9)$, $(4, 9)$

68. Parallelogram: vertices at $(0, 0)$, $(4, 0)$, $(1, 4)$, $(5, 4)$

69. Triangle: vertices at $(0, 0)$, $(6, 0)$, $(1, 5)$

70. Triangle: vertices at $(-1, 0)$, $(1, 0)$, $(0, 1)$

SUPPLY AND DEMAND   In Exercises 71–74, (a) graph the systems representing the consumer surplus and producer surplus for the supply and demand equations and (b) find the consumer surplus and producer surplus.

	Demand	Supply
71.	$p = 50 - 0.5x$	$p = 0.125x$
72.	$p = 100 - 0.05x$	$p = 25 + 0.1x$
73.	$p = 140 - 0.00002x$	$p = 80 + 0.00001x$
74.	$p = 400 - 0.0002x$	$p = 225 + 0.0005x$

75. PRODUCTION   A furniture company can sell all the tables and chairs it produces. Each table requires 1 hour in the assembly center and $1\frac{1}{3}$ hours in the finishing center. Each chair requires $1\frac{1}{2}$ hours in the assembly center and $1\frac{1}{2}$ hours in the finishing center. The company's assembly center is available 12 hours per day, and its finishing center is available 15 hours per day. Find and graph a system of inequalities describing all possible production levels.

76. INVENTORY   A store sells two models of laptop computers. Because of the demand, the store stocks at least twice as many units of model A as of model B. The costs to the store for the two models are $800 and $1200, respectively. The management does not want more than $20,000 in computer inventory at any one time, and it wants at least four model A laptop computers and two model B laptop computers in inventory at all times. Find and graph a system of inequalities describing all possible inventory levels.

77. INVESTMENT ANALYSIS   A person plans to invest up to $20,000 in two different interest-bearing accounts. Each account is to contain at least $5000. Moreover, the amount in one account should be at least twice the amount in the other account. Find and graph a system of inequalities to describe the various amounts that can be deposited in each account.

78. TICKET SALES   For a concert event, there are $30 reserved seat tickets and $20 general admission tickets. There are 2000 reserved seats available, and fire regulations limit the number of paid ticket holders to 3000. The promoter must take in at least $75,000 in ticket sales. Find and graph a system of inequalities describing the different numbers of tickets that can be sold.

79. SHIPPING   A warehouse supervisor is told to ship at least 50 packages of gravel that weigh 55 pounds each and at least 40 bags of stone that weigh 70 pounds each. The maximum weight capacity of the truck to be used is 7500 pounds. Find and graph a system of inequalities describing the numbers of bags of stone and gravel that can be shipped.

80. TRUCK SCHEDULING   A small company that manufactures two models of exercise machines has an order for 15 units of the standard model and 16 units of the deluxe model. The company has trucks of two different sizes that can haul the products, as shown in the table.

Truck	Standard	Deluxe
Large	6	3
Medium	4	6

Find and graph a system of inequalities describing the numbers of trucks of each size that are needed to deliver the order.

81. NUTRITION   A dietitian is asked to design a special dietary supplement using two different foods. Each ounce of food X contains 20 units of calcium, 15 units of iron, and 10 units of vitamin B. Each ounce of food Y contains 10 units of calcium, 10 units of iron, and 20 units of vitamin B. The minimum daily requirements of the diet are 300 units of calcium, 150 units of iron, and 200 units of vitamin B.

(a) Write a system of inequalities describing the different amounts of food X and food Y that can be used.

(b) Sketch a graph of the region corresponding to the system in part (a).

(c) Find two solutions of the system and interpret their meanings in the context of the problem.

82. HEALTH   A person's maximum heart rate is $220 - x$, where x is the person's age in years for $20 \le x \le 70$. When a person exercises, it is recommended that the person strive for a heart rate that is at least 50% of the maximum and at most 75% of the maximum.   (Source: American Heart Association)

(a) Write a system of inequalities that describes the exercise target heart rate region.

(b) Sketch a graph of the region in part (a).

(c) Find two solutions to the system and interpret their meanings in the context of the problem.

83. DATA ANALYSIS: PRESCRIPTION DRUGS   The table shows the retail sales y (in billions of dollars) of prescription drugs in the United States from 2000 through 2007.   (Source: National Association of Chain Drug Stores)

Year	Retail sales, y
2000	145.6
2001	161.3
2002	182.7
2003	204.2
2004	220.1
2005	232.0
2006	250.6
2007	259.4

(a) Use the *regression* feature of a graphing utility to find a linear model for the data. Let t represent the year, with $t = 0$ corresponding to 2000.

(b) The total retail sales of prescription drugs in the United States during this eight-year period can be approximated by finding the area of the trapezoid bounded by the linear model you found in part (a) and the lines $y = 0, t = -0.5$, and $t = 7.5$. Use a graphing utility to graph this region.

(c) Use the formula for the area of a trapezoid to approximate the total retail sales of prescription drugs.

84. DATA ANALYSIS: MERCHANDISE   The table shows the retail sales y (in millions of dollars) for Aeropostale, Inc. from 2000 through 2007.   (Source: Aeropostale, Inc.)

Year	Retail sales, y
2000	213.4
2001	304.8
2002	550.9
2003	734.9
2004	964.2
2005	1204.3
2006	1413.2
2007	1590.9

(a) Use the *regression* feature of a graphing utility to find a linear model for the data. Let t represent the year, with $t = 0$ corresponding to 2000.

(b) The total retail sales for Aeropostale during this eight-year period can be approximated by finding the area of the trapezoid bounded by the linear model you found in part (a) and the lines $y = 0$, $t = -0.5$, and $t = 7.5$. Use a graphing utility to graph this region.

(c) Use the formula for the area of a trapezoid to approximate the total retail sales for Aeropostale.

85. PHYSICAL FITNESS FACILITY An indoor running track is to be constructed with a space for exercise equipment inside the track (see figure). The track must be at least 125 meters long, and the exercise space must have an area of at least 500 square meters.

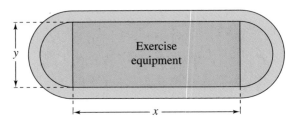

(a) Find a system of inequalities describing the requirements of the facility.

(b) Graph the system from part (a).

EXPLORATION

TRUE OR FALSE? In Exercises 86 and 87, determine whether the statement is true or false. Justify your answer.

86. The area of the figure defined by the system

$$\begin{cases} x \ge -3 \\ x \le 6 \\ y \le 5 \\ y \ge -6 \end{cases}$$

is 99 square units.

87. The graph below shows the solution of the system

$$\begin{cases} y \le 6 \\ -4x - 9y > 6. \\ 3x + y^2 \ge 2 \end{cases}$$

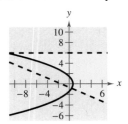

88. CAPSTONE

(a) Explain the difference between the graphs of the inequality $x \le -5$ on the real number line and on the rectangular coordinate system.

(b) After graphing the boundary of the inequality $x + y < 3$, explain how you decide on which side of the boundary the solution set of the inequality lies.

89. GRAPHICAL REASONING Two concentric circles have radii x and y, where $y > x$. The area between the circles must be at least 10 square units.

(a) Find a system of inequalities describing the constraints on the circles.

 (b) Use a graphing utility to graph the system of inequalities in part (a). Graph the line $y = x$ in the same viewing window.

(c) Identify the graph of the line in relation to the boundary of the inequality. Explain its meaning in the context of the problem.

90. The graph of the solution of the inequality $x + 2y < 6$ is shown in the figure. Describe how the solution set would change for each of the following.

(a) $x + 2y \le 6$

(b) $x + 2y > 6$

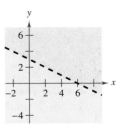

In Exercises 91–94, match the system of inequalities with the graph of its solution. [The graphs are labeled (a), (b), (c), and (d).]

(a)
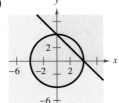

(b)
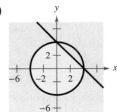

(c)
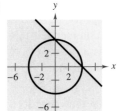

(d)
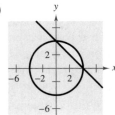

91. $\begin{cases} x^2 + y^2 \le 16 \\ x + y \ge 4 \end{cases}$

92. $\begin{cases} x^2 + y^2 \le 16 \\ x + y \le 4 \end{cases}$

93. $\begin{cases} x^2 + y^2 \ge 16 \\ x + y \ge 4 \end{cases}$

94. $\begin{cases} x^2 + y^2 \ge 16 \\ x + y \le 4 \end{cases}$

6.6   LINEAR PROGRAMMING

Linear Programming: A Graphical Approach

Many applications in business and economics involve a process called **optimization,** in which you are asked to find the minimum or maximum value of a quantity. In this section, you will study an optimization strategy called **linear programming.**

A two-dimensional linear programming problem consists of a linear **objective function** and a system of linear inequalities called **constraints.** The objective function gives the quantity that is to be maximized (or minimized), and the constraints determine the set of **feasible solutions.** For example, suppose you are asked to maximize the value of

$$z = ax + by$$

subject to a set of constraints that determines the shaded region in Figure 6.31.

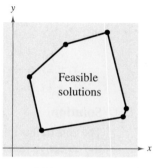

FIGURE 6.31

Because every point in the shaded region satisfies each constraint, it is not clear how you should find the point that yields a maximum value of z. Fortunately, it can be shown that if there is an optimal solution, it must occur at one of the vertices. This means that *you can find the maximum value of z by testing z at each of the vertices.*

> **Optimal Solution of a Linear Programming Problem**
>
> If a linear programming problem has a solution, it must occur at a vertex of the set of feasible solutions. If there is more than one solution, at least one of them must occur at such a vertex. In either case, the value of the objective function is unique.

Some guidelines for solving a linear programming problem in two variables are listed at the top of the next page.

Solving a Linear Programming Problem

1. Sketch the region corresponding to the system of constraints. (The points inside or on the boundary of the region are *feasible solutions*.)

2. Find the vertices of the region.

3. Test the objective function at each of the vertices and select the values of the variables that optimize the objective function. For a bounded region, both a minimum and a maximum value will exist. (For an unbounded region, *if* an optimal solution exists, it will occur at a vertex.)

Example 1 Solving a Linear Programming Problem

Find the maximum value of

$$z = 3x + 2y$$

subject to the following constraints.

$$x \geq 0$$
$$y \geq 0$$
$$x + 2y \leq 4$$
$$x - y \leq 1$$

Solution

The constraints form the region shown in Figure 6.32. At the four vertices of this region, the objective function has the following values.

At $(0, 0)$: $z = 3(\ ) + 2(\ ) = 0$
At $(0, 2)$: $z = 3(\ ) + 2(\ ) = 4$
At $(2, 1)$: $z = 3(\ ) + 2(\ ) = 8$
At $(1, 0)$: $z = 3(\ ) + 2(\ ) = 3$

So, the maximum value of z is 8, and this occurs when $x = 2$ and $y = 1$.

CHECK**Point** Now try Exercise 9.

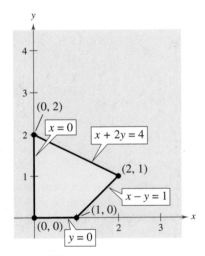

FIGURE 6.32

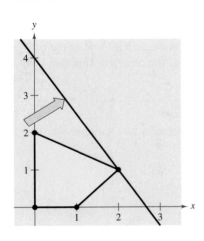

FIGURE 6.33

In Example 1, try testing some of the *interior* points in the region. You will see that the corresponding values of z are less than 8. Here are some examples.

At $(1, 1)$: $z = 3(\ ) + 2(\ ) = 5$ At $\left(\frac{1}{2}, \frac{3}{2}\right)$: $z = 3(\ ) + 2(\ ) = \frac{9}{2}$

To see why the maximum value of the objective function in Example 1 must occur at a vertex, consider writing the objective function in slope-intercept form

$$y = -\frac{3}{2}x + \frac{z}{2}$$

where $z/2$ is the y-intercept of the objective function. This equation represents a family of lines, each of slope $-\frac{3}{2}$. Of these infinitely many lines, you want the one that has the largest z-value while still intersecting the region determined by the constraints. In other words, of all the lines whose slope is $-\frac{3}{2}$, you want the one that has the largest y-intercept *and* intersects the given region, as shown in Figure 6.33. From the graph, you can see that such a line will pass through one (or more) of the vertices of the region.

The next example shows that the same basic procedure can be used to solve a problem in which the objective function is to be *minimized*.

Example 2   Minimizing an Objective Function

Find the minimum value of

$$z = 5x + 7y$$

where $x \geq 0$ and $y \geq 0$, subject to the following constraints.

$$2x + 3y \geq 6$$
$$3x - y \leq 15$$
$$-x + y \leq 4$$
$$2x + 5y \leq 27$$

Solution

FIGURE 6.34

The region bounded by the constraints is shown in Figure 6.34. By testing the objective function at each vertex, you obtain the following.

At $(0, 2)$: $z = 5(\ ) + 7(\ ) = 14$
At $(0, 4)$: $z = 5(\ ) + 7(\ ) = 28$
At $(1, 5)$: $z = 5(\ ) + 7(\ ) = 40$
At $(6, 3)$: $z = 5(\ ) + 7(\ ) = 51$
At $(5, 0)$: $z = 5(\ ) + 7(\ ) = 25$
At $(3, 0)$: $z = 5(\ ) + 7(\ ) = 15$

So, the minimum value of z is 14, and this occurs when $x = 0$ and $y = 2$.

CHECK**Point**   Now try Exercise 11.

Example 3   Maximizing an Objective Function

Find the maximum value of

$$z = 5x + 7y$$

where $x \geq 0$ and $y \geq 0$, subject to the following constraints.

$$2x + 3y \geq 6$$
$$3x - y \leq 15$$
$$-x + y \leq 4$$
$$2x + 5y \leq 27$$

Solution

This linear programming problem is identical to that given in Example 2 above, *except* that the objective function is maximized instead of minimized. Using the values of z at the vertices shown above, you can conclude that the maximum value of z is

$$z = 5(\ ) + 7(\ ) = 51$$

and occurs when $x = 6$ and $y = 3$.

CHECK**Point**   Now try Exercise 13.

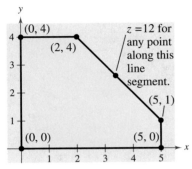

FIGURE **6.35**

It is possible for the maximum (or minimum) value in a linear programming problem to occur at *two* different vertices. For instance, at the vertices of the region shown in Figure 6.35, the objective function

$$z = 2x + 2y$$

has the following values.

At $(0, 0)$: $z = 2(\ ) + 2(\ ) = \ \ 0$
At $(0, 4)$: $z = 2(\ ) + 2(\ ) = \ \ 8$
At $(2, 4)$: $z = 2(\ ) + 2(\ ) = 12$
At $(5, 1)$: $z = 2(\ ) + 2(\ ) = 12$
At $(5, 0)$: $z = 2(\ ) + 2(\ ) = 10$

In this case, you can conclude that the objective function has a maximum value not only at the vertices $(2, 4)$ and $(5, 1)$; it also has a maximum value (of 12) at *any point on the line segment connecting these two vertices*. Note that the objective function in slope-intercept form $y = -x + \frac{1}{2}z$ has the same slope as the line through the vertices $(2, 4)$ and $(5, 1)$.

Some linear programming problems have no optimal solutions. This can occur if the region determined by the constraints is *unbounded*. Example 4 illustrates such a problem.

Algebra Help

The slope m of the nonvertical line through the points (x_1, y_1) and (x_2, y_2) is

$$m = \frac{y_2 - y_1}{x_2 - x_1}$$

where $x_1 \neq x_2$.

Example 4   An Unbounded Region

Find the maximum value of

$$z = 4x + 2y$$

where $x \geq 0$ and $y \geq 0$, subject to the following constraints.

$$x + 2y \geq 4$$
$$3x + \ y \geq 7$$
$$-x + 2y \leq 7$$

Solution

The region determined by the constraints is shown in Figure 6.36. For this unbounded region, there is no maximum value of z. To see this, note that the point $(x, 0)$ lies in the region for all values of $x \geq 4$. Substituting this point into the objective function, you get

$$z = 4(\ ) + 2(\ ) = 4x.$$

By choosing x to be large, you can obtain values of z that are as large as you want. So, there is no maximum value of z. However, there *is* a minimum value of z.

At $(1, 4)$: $z = 4(\ ) + 2(\ ) = 12$
At $(2, 1)$: $z = 4(\ ) + 2(\ ) = 10$
At $(4, 0)$: $z = 4(\ ) + 2(\ ) = 16$

So, the minimum value of z is 10, and this occurs when $x = 2$ and $y = 1$.

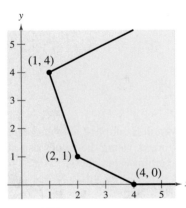

FIGURE **6.36**

CHECK*Point* Now try Exercise 15.

Applications

Example 5 shows how linear programming can be used to find the maximum profit in a business application.

Example 5   Optimal Profit

A candy manufacturer wants to maximize the combined profit for two types of boxed chocolates. A box of chocolate covered creams yields a profit of $1.50 per box, and a box of chocolate covered nuts yields a profit of $2.00 per box. Market tests and available resources have indicated the following constraints.

1. The combined production level should not exceed 1200 boxes per month.

2. The demand for a box of chocolate covered nuts is no more than half the demand for a box of chocolate covered creams.

3. The production level for chocolate covered creams should be less than or equal to 600 boxes plus three times the production level for chocolate covered nuts.

What is the maximum monthly profit? How many boxes of each type should be produced per month to yield the maximum profit?

Solution

Let x be the number of boxes of chocolate covered creams and let y be the number of boxes of chocolate covered nuts. So, the objective function (for the combined profit) is given by

$$P = 1.5x + 2y.$$

The three constraints translate into the following linear inequalities.

1. $x + y \leq 1200$ ⟹ $x + y \leq 1200$

2. $y \leq \frac{1}{2}x$ ⟹ $-x + 2y \leq 0$

3. $x \leq 600 + 3y$ ⟹ $x - 3y \leq 600$

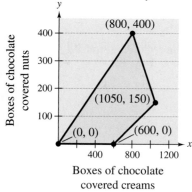

Maximum Monthly Profit

(800, 400)

(1050, 150)

(0, 0)   (600, 0)

Boxes of chocolate covered creams

FIGURE 6.37

Because neither x nor y can be negative, you also have the two additional constraints of $x \geq 0$ and $y \geq 0$. Figure 6.37 shows the region determined by the constraints. To find the maximum monthly profit, test the values of P at the vertices of the region.

At $(0, 0)$: $P = 1.5(\ )\ + 2(\ )\ =\ 0$
At $(800, 400)$: $P = 1.5(\ )\ + 2(\ )\ = 2000$
At $(1050, 150)$: $P = 1.5(\ )\ + 2(\ )\ = 1875$
At $(600, 0)$: $P = 1.5(\ )\ + 2(\ )\ =\ 900$

So, the maximum monthly profit is $2000, and it occurs when the monthly production consists of 800 boxes of chocolate covered creams and 400 boxes of chocolate covered nuts.

CHECK Point   Now try Exercise 35.

In Example 5, if the manufacturer improved the production of chocolate covered creams so that they yielded a profit of $2.50 per unit, the maximum monthly profit could then be found using the objective function $P = 2.5x + 2y$. By testing the values of P at the vertices of the region, you would find that the maximum monthly profit was $2925 and that it occurred when $x = 1050$ and $y = 150$.

Example 6 Optimal Cost

The liquid portion of a diet is to provide at least 300 calories, 36 units of vitamin A, and 90 units of vitamin C. A cup of dietary drink X costs $0.12 and provides 60 calories, 12 units of vitamin A, and 10 units of vitamin C. A cup of dietary drink Y costs $0.15 and provides 60 calories, 6 units of vitamin A, and 30 units of vitamin C. How many cups of each drink should be consumed each day to obtain an optimal cost and still meet the daily requirements?

Solution

As in Example 9 in Section 6.5, let x be the number of cups of dietary drink X and let y be the number of cups of dietary drink Y.

$$
\begin{aligned}
\text{For calories:} \quad & 60x + 60y \geq 300 \\
\text{For vitamin A:} \quad & 12x + 6y \geq 36 \\
\text{For vitamin C:} \quad & 10x + 30y \geq 90 \\
& x \geq 0 \\
& y \geq 0
\end{aligned}
$$

The cost C is given by $C = 0.12x + 0.15y$.

The graph of the region corresponding to the constraints is shown in Figure 6.38. Because you want to incur as little cost as possible, you want to determine the *minimum* cost. To determine the minimum cost, test C at each vertex of the region.

$$
\begin{aligned}
\text{At } (0, 6): \quad & C = 0.12() + 0.15() = 0.90 \\
\text{At } (1, 4): \quad & C = 0.12() + 0.15() = 0.72 \\
\text{At } (3, 2): \quad & C = 0.12() + 0.15() = 0.66 \\
\text{At } (9, 0): \quad & C = 0.12() + 0.15() = 1.08
\end{aligned}
$$

So, the minimum cost is $0.66 per day, and this occurs when 3 cups of drink X and 2 cups of drink Y are consumed each day.

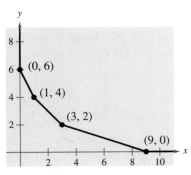

FIGURE **6.38**

CHECK**Point** Now try Exercise 37.

CLASSROOM DISCUSSION

Creating a Linear Programming Problem Sketch the region determined by the following constraints.

$$
\left.
\begin{aligned}
x + 2y &\leq 8 \\
x + y &\leq 5 \\
x &\geq 0 \\
y &\geq 0
\end{aligned}
\right\}
$$

Find, if possible, an objective function of the form $z = ax + by$ that has a maximum at each indicated vertex of the region.

a. $(0, 4)$ b. $(2, 3)$ c. $(5, 0)$ d. $(0, 0)$

Explain how you found each objective function.

6.6 EXERCISES

See www.CalcChat.com for worked-out solutions to odd-numbered exercises.

VOCABULARY: Fill in the blanks.

1. In the process called _____, you are asked to find the maximum or minimum value of a quantity.

2. One type of optimization strategy is called _____ _____.

3. The _____ function of a linear programming problem gives the quantity that is to be maximized or minimized.

4. The _____ of a linear programming problem determine the set of _____ _____.

5. The feasible solutions are _____ or _____ the boundary of the region corresponding to a system of constraints.

6. If a linear programming problem has a solution, it must occur at a _____ of the set of feasible solutions.

SKILLS AND APPLICATIONS

In Exercises 7–12, find the minimum and maximum values of the objective function and where they occur, subject to the indicated constraints. (For each exercise, the graph of the region determined by the constraints is provided.)

7. Objective function:

$z = 4x + 3y$

Constraints:

$x \geq 0$

$y \geq 0$

$x + y \leq 5$

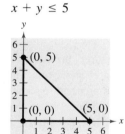

8. Objective function:

$z = 2x + 8y$

Constraints:

$x \geq 0$

$y \geq 0$

$2x + y \leq 4$

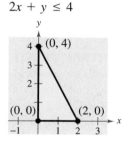

9. Objective function:

$z = 2x + 5y$

Constraints:

$x \geq 0$

$y \geq 0$

$x + 3y \leq 15$

$4x + y \leq 16$

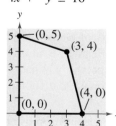

10. Objective function:

$z = 4x + 5y$

Constraints:

$x \geq 0$

$2x + 3y \geq 6$

$3x - y \leq 9$

$x + 4y \leq 16$

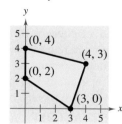

11. Objective function:

$z = 10x + 7y$

Constraints:

$0 \leq x \leq 60$

$0 \leq y \leq 45$

$5x + 6y \leq 420$

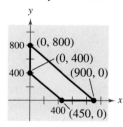

12. Objective function:

$z = 40x + 45y$

Constraints:

$x \geq 0$

$y \geq 0$

$8x + 9y \leq 7200$

$8x + 9y \geq 3600$

In Exercises 13–16, sketch the region determined by the constraints. Then find the minimum and maximum values of the objective function (if possible) and where they occur, subject to the indicated constraints.

13. Objective function:

$z = 3x + 2y$

Constraints:

$x \geq 0$

$y \geq 0$

$5x + 2y \leq 20$

$5x + y \geq 10$

14. Objective function:

$z = 5x + \frac{1}{2}y$

Constraints:

$x \geq 0$

$y \geq 0$

$\frac{1}{2}x + y \leq 8$

$x + \frac{1}{2}y \geq 4$

15. Objective function:

$z = 4x + 5y$

Constraints:

$x \geq 0$

$y \geq 0$

$x + y \geq 8$

$3x + 5y \geq 30$

16. Objective function:

$z = 5x + 4y$

Constraints:

$x \geq 0$

$y \geq 0$

$2x + 2y \geq 10$

$x + 2y \geq 6$

In Exercises 17–20, use a graphing utility to graph the region determined by the constraints. Then find the minimum and maximum values of the objective function and where they occur, subject to the constraints.

17. Objective function:

$z = 3x + y$

Constraints:

$x \geq 0$

$y \geq 0$

$x + 4y \leq 60$

$3x + 2y \geq 48$

18. Objective function:

$z = x$

Constraints:

$x \geq 0$

$y \geq 0$

$2x + 3y \leq 60$

$2x + y \leq 28$

$4x + y \leq 48$

19. Objective function:

$z = x + 4y$

Constraints:

(See Exercise 17.)

20. Objective function:

$z = y$

Constraints:

(See Exercise 18.)

In Exercises 21–24, find the minimum and maximum values of the objective function and where they occur, subject to the constraints $x \geq 0$, $y \geq 0$, $3x + y \leq 15$, and $4x + 3y \leq 30$.

21. $z = 2x + y$

22. $z = 5x + y$

23. $z = x + y$

24. $z = 3x + y$

In Exercises 25–28, find the minimum and maximum values of the objective function and where they occur, subject to the constraints $x \geq 0$, $y \geq 0$, $x + 4y \leq 20$, $x + y \leq 18$, and $2x + 2y \leq 21$.

25. $z = x + 5y$

26. $z = 2x + 4y$

27. $z = 4x + 5y$

28. $z = 4x + y$

In Exercises 29–34, the linear programming problem has an unusual characteristic. Sketch a graph of the solution region for the problem and describe the unusual characteristic. Find the minimum and maximum values of the objective function (if possible) and where they occur.

29. Objective function:

$z = 2.5x + y$

Constraints:

$x \geq 0$

$y \geq 0$

$3x + 5y \leq 15$

$5x + 2y \leq 10$

30. Objective function:

$z = x + y$

Constraints:

$x \geq 0$

$y \geq 0$

$-x + y \leq 1$

$-x + 2y \leq 4$

31. Objective function:

$z = -x + 2y$

Constraints:

$x \geq 0$

$y \geq 0$

$x \leq 10$

$x + y \leq 7$

32. Objective function:

$z = x + y$

Constraints:

$x \geq 0$

$y \geq 0$

$-x + y \leq 0$

$-3x + y \geq 3$

33. Objective function:

$z = 3x + 4y$

Constraints:

$x \geq 0$

$y \geq 0$

$x + y \leq 1$

$2x + y \leq 4$

34. Objective function:

$z = x + 2y$

Constraints:

$x \geq 0$

$y \geq 0$

$x + 2y \leq 4$

$2x + y \leq 4$

35. OPTIMAL PROFIT A merchant plans to sell two models of MP3 players at prices of $225 and $250. The $225 model yields a profit of $30 per unit and the $250 model yields a profit of $31 per unit. The merchant estimates that the total monthly demand will not exceed 275 units. The merchant does not want to invest more than $63,000 in inventory for these products. What is the optimal inventory level for each model? What is the optimal profit?

36. OPTIMAL PROFIT A manufacturer produces two models of elliptical cross-training exercise machines. The times for assembling, finishing, and packaging model X are 3 hours, 3 hours, and 0.8 hour, respectively. The times for model Y are 4 hours, 2.5 hours, and 0.4 hour. The total times available for assembling, finishing, and packaging are 6000 hours, 4200 hours, and 950 hours, respectively. The profits per unit are $300 for model X and $375 for model Y. What is the optimal production level for each model? What is the optimal profit?

37. OPTIMAL COST An animal shelter mixes two brands of dog food. Brand X costs $25 per bag and contains two units of nutritional element A, two units of element B, and two units of element C. Brand Y costs $20 per bag and contains one unit of nutritional element A, nine units of element B, and three units of element C. The minimum required amounts of nutrients A, B, and C are 12 units, 36 units, and 24 units, respectively. What is the optimal number of bags of each brand that should be mixed? What is the optimal cost?

38. OPTIMAL COST A humanitarian agency can use two models of vehicles for a refugee rescue mission. Each model A vehicle costs $1000 and each model B vehicle costs $1500. Mission strategies and objectives indicate the following constraints.

- A total of at least 20 vehicles must be used.

- A model A vehicle can hold 45 boxes of supplies. A model B vehicle can hold 30 boxes of supplies. The agency must deliver at least 690 boxes of supplies to the refugee camp.

- A model A vehicle can hold 20 refugees. A model B vehicle can hold 32 refugees. The agency must rescue at least 520 refugees.

What is the optimal number of vehicles of each model that should be used? What is the optimal cost?

39. OPTIMAL REVENUE An accounting firm has 780 hours of staff time and 272 hours of reviewing time available each week. The firm charges $1600 for an audit and $250 for a tax return. Each audit requires 60 hours of staff time and 16 hours of review time. Each tax return requires 10 hours of staff time and 4 hours of review time. What numbers of audits and tax returns will yield an optimal revenue? What is the optimal revenue?

40. OPTIMAL REVENUE The accounting firm in Exercise 39 lowers its charge for an audit to $1400. What numbers of audits and tax returns will yield an optimal revenue? What is the optimal revenue?

41. MEDIA SELECTION A company has budgeted a maximum of $1,000,000 for national advertising of an allergy medication. Each minute of television time costs $100,000 and each one-page newspaper ad costs $20,000. Each television ad is expected to be viewed by 20 million viewers, and each newspaper ad is expected to be seen by 5 million readers. The company's market research department recommends that at most 80% of the advertising budget be spent on television ads. What is the optimal amount that should be spent on each type of ad? What is the optimal total audience?

42. OPTIMAL COST According to AAA (Automobile Association of America), on March 27, 2009, the national average price per gallon of regular unleaded (87-octane) gasoline was $2.03, and the price of premium unleaded (93-octane) gasoline was $2.23.

(a) Write an objective function that models the cost of the blend of mid-grade unleaded gasoline (89-octane).

(b) Determine the constraints for the objective function in part (a).

(c) Sketch a graph of the region determined by the constraints from part (b).

(d) Determine the blend of regular and premium unleaded gasoline that results in an optimal cost of mid-grade unleaded gasoline.

(e) What is the optimal cost?

(f) Is the cost lower than the national average of $2.15 per gallon for mid-grade unleaded gasoline?

43. INVESTMENT PORTFOLIO An investor has up to $250,000 to invest in two types of investments. Type A pays 8% annually and type B pays 10% annually. To have a well-balanced portfolio, the investor imposes the following conditions. At least one-fourth of the total portfolio is to be allocated to type A investments and at least one-fourth of the portfolio is to be allocated to type B investments. What is the optimal amount that should be invested in each type of investment? What is the optimal return?

44. INVESTMENT PORTFOLIO An investor has up to $450,000 to invest in two types of investments. Type A pays 6% annually and type B pays 10% annually. To have a well-balanced portfolio, the investor imposes the following conditions. At least one-half of the total portfolio is to be allocated to type A investments and at least one-fourth of the portfolio is to be allocated to type B investments. What is the optimal amount that should be invested in each type of investment? What is the optimal return?

EXPLORATION

TRUE OR FALSE? In Exercises 45–47, determine whether the statement is true or false. Justify your answer.

45. If an objective function has a maximum value at the vertices $(4, 7)$ and $(8, 3)$, you can conclude that it also has a maximum value at the points $(4.5, 6.5)$ and $(7.8, 3.2)$.

46. If an objective function has a minimum value at the vertex $(20, 0)$, you can conclude that it also has a minimum value at the point $(0, 0)$.

47. When solving a linear programming problem, if the objective function has a maximum value at more than one vertex, you can assume that there are an infinite number of points that will produce the maximum value.

48. CAPSTONE Using the constraint region shown below, determine which of the following objective functions has (a) a maximum at vertex A, (b) a maximum at vertex B, (c) a maximum at vertex C, and (d) a minimum at vertex C.

(i) $z = 2x + y$

(ii) $z = 2x - y$

(iii) $z = -x + 2y$

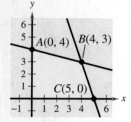

6 CHAPTER SUMMARY

What Did You Learn?	Explanation/Examples	Review Exercises
Section 6.1 Use the method of substitution to solve systems of linear equations in two variables *(p. 444)*.	**Method of Substitution** 1. *Solve* one of the equations for one variable in terms of the other. 2. *Substitute* the expression found in Step 1 into the other equation to obtain an equation in one variable. 3. *Solve* the equation obtained in Step 2. 4. *Back-substitute* the value obtained in Step 3 into the expression obtained in Step 1 to find the value of the other variable. 5. *Check* that the solution satisfies *each* of the original equations.	1–6
Use the method of substitution to solve systems of nonlinear equations in two variables *(p. 447)*.	The method of substitution (see steps above) can be used to solve systems in which one or both of the equations are nonlinear. (See Examples 3 and 4.)	7–10
Use a graphical approach to solve systems of equations in two variables *(p. 448)*.	One intersection point Two intersection points No intersection points	11–18
Use systems of equations to model and solve real-life problems *(p. 449)*.	A system of equations can be used to find the break-even point for a company. (See Example 6.)	19–24
Section 6.2 Use the method of elimination to solve systems of linear equations in two variables *(p. 455)*.	**Method of Elimination** 1. *Obtain coefficients* for x (or y) that differ only in sign. 2. *Add* the equations to eliminate one variable. 3. *Solve* the equation obtained in Step 2. 4. *Back-substitute* the value obtained in Step 3 into either of the original equations and solve for the other variable. 5. *Check* that the solution satisfies *each* of the original equations.	25–32
Interpret graphically the numbers of solutions of systems of linear equations in two variables *(p. 458)*.	Lines intersect at one point; exactly one solution Lines coincide; infinitely many solutions Lines are parallel; no solution	33–36
Use systems of linear equations in two variables to model and solve real-life problems *(p. 461)*.	A system of linear equations in two variables can be used to find the equilibrium point for a particular market. (See Example 9.)	37, 38

What Did You Learn?	Explanation/Examples	Review Exercises
Section 6.3 Use back-substitution to solve linear systems in row-echelon form (*p. 467*).	**Row-Echelon Form** $$\begin{cases} x - 2y + 3z = 9 \\ -x + 3y = -4 \\ 2x - 5y + 5z = 17 \end{cases} \rightarrow \begin{cases} x - 2y + 3z = 9 \\ y + 3z = 5 \\ z = 2 \end{cases}$$	39–42
Use Gaussian elimination to solve systems of linear equations (*p. 468*).	To produce an equivalent system of linear equations, use row operations by (1) interchanging two equations, (2) multiplying one equation by a nonzero constant, or (3) adding a multiple of one of the equations to another equation to replace the latter equation.	43–48
Solve nonsquare systems of linear equations (*p. 472*).	In a nonsquare system, the number of equations differs from the number of variables. A system of linear equations cannot have a unique solution unless there are at least as many equations as there are variables.	49, 50
Use systems of linear equations in three or more variables to model and solve real-life problems (*p. 473*).	A system of linear equations in three variables can be used to find the position equation of an object that is moving in a (vertical) line with constant acceleration. (See Example 7.)	51–60
Section 6.4 Recognize partial fraction decompositions of rational expressions (*p. 480*).	$$\frac{9}{x^3 - 6x^2} = \frac{9}{x^2(x-6)} = \frac{A}{x} + \frac{B}{x^2} + \frac{C}{x-6}$$	61–64
Find partial fraction decompositions of rational expressions (*p. 481*).	The techniques used for determining constants in the numerators of partial fractions vary slightly, depending on the type of factors of the denominator: linear or quadratic, distinct or repeated.	65–72
Section 6.5 Sketch the graphs of inequalities in two variables (*p. 488*).		73–78
Solve systems of inequalities (*p. 490*).	$$\begin{cases} x^2 + y \le 5 \\ x \ge -1 \\ y \ge 0 \end{cases}$$	79–86
Use systems of inequalities in two variables to model and solve real-life problems (*p. 493*).	A system of inequalities in two variables can be used to find the consumer surplus and producer surplus for given demand and supply functions. (See Example 8.)	87–92
Section 6.6 Solve linear programming problems (*p. 499*).	To solve a linear programming problem, (1) sketch the region corresponding to the system of constraints, (2) find the vertices of the region, and (3) test the objective function at each of the vertices and select the values of the variables that optimize the objective function.	93–98
Use linear programming to model and solve real-life problems (*p. 503*).	Linear programming can be used to find the maximum profit in business applications. (See Example 5.)	99–103

6 REVIEW EXERCISES

See www.CalcChat.com for worked-out solutions to odd-numbered exercises.

 6.1 In Exercises 1–10, solve the system by the method of substitution.

1. $\begin{cases} x + y = 2 \\ x - y = 0 \end{cases}$
2. $\begin{cases} 2x - 3y = 3 \\ x - y = 0 \end{cases}$

3. $\begin{cases} 4x - y - 1 = 0 \\ 8x + y - 17 = 0 \end{cases}$
4. $\begin{cases} 10x + 6y + 14 = 0 \\ x + 9y + 7 = 0 \end{cases}$

5. $\begin{cases} 0.5x + y = 0.75 \\ 1.25x - 4.5y = -2.5 \end{cases}$
6. $\begin{cases} -x + \frac{2}{5}y = \frac{3}{5} \\ -x + \frac{1}{5}y = -\frac{4}{5} \end{cases}$

7. $\begin{cases} x^2 - y^2 = 9 \\ x - y = 1 \end{cases}$
8. $\begin{cases} x^2 + y^2 = 169 \\ 3x + 2y = 39 \end{cases}$

9. $\begin{cases} y = 2x^2 \\ y = x^4 - 2x^2 \end{cases}$
10. $\begin{cases} x = y + 3 \\ x = y^2 + 1 \end{cases}$

In Exercises 11–14, solve the system graphically.

11. $\begin{cases} 2x - y = 10 \\ x + 5y = -6 \end{cases}$
12. $\begin{cases} 8x - 3y = -3 \\ 2x + 5y = 28 \end{cases}$

13. $\begin{cases} y = 2x^2 - 4x + 1 \\ y = x^2 - 4x + 3 \end{cases}$
14. $\begin{cases} y^2 - 2y + x = 0 \\ x + y = 0 \end{cases}$

 In Exercises 15–18, use a graphing utility to solve the system of equations. Find the solution accurate to two decimal places.

15. $\begin{cases} y = -2e^{-x} \\ 2e^x + y = 0 \end{cases}$
16. $\begin{cases} x^2 + y^2 = 100 \\ 2x - 3y = -12 \end{cases}$

17. $\begin{cases} y = 2 + \log x \\ y = \frac{3}{4}x + 5 \end{cases}$
18. $\begin{cases} y = \ln(x - 1) - 3 \\ y = 4 - \frac{1}{2}x \end{cases}$

19. **BREAK-EVEN ANALYSIS** You set up a scrapbook business and make an initial investment of $50,000. The unit cost of a scrapbook kit is $12 and the selling price is $25. How many kits must you sell to break even?

20. **CHOICE OF TWO JOBS** You are offered two sales jobs at a pharmaceutical company. One company offers an annual salary of $55,000 plus a year-end bonus of 1.5% of your total sales. The other company offers an annual salary of $52,000 plus a year-end bonus of 2% of your total sales. What amount of sales will make the second offer better? Explain.

21. **GEOMETRY** The perimeter of a rectangle is 480 meters and its length is 150% of its width. Find the dimensions of the rectangle.

22. **GEOMETRY** The perimeter of a rectangle is 68 feet and its width is $\frac{8}{9}$ times its length. Find the dimensions of the rectangle.

23. **GEOMETRY** The perimeter of a rectangle is 40 inches. The area of the rectangle is 96 square inches. Find the dimensions of the rectangle.

 24. **BODY MASS INDEX** Body Mass Index (BMI) is a measure of body fat based on height and weight. The 75th percentile BMI for females, ages 9 to 20, is growing more slowly than that for males of the same age range. Models that represent the 75th percentile BMI for males and females, ages 9 to 20, are given by

$B = 0.73a + 11$

$B = 0.61a + 12.8$

where B is the BMI (kg/m²) and a represents the age, with $a = 9$ corresponding to 9 years of age. Use a graphing utility to determine whether the BMI for males ever exceeds the BMI for females. (Source: National Center for Health Statistics)

6.2 In Exercises 25–32, solve the system by the method of elimination.

25. $\begin{cases} 2x - y = 2 \\ 6x + 8y = 39 \end{cases}$
26. $\begin{cases} 40x + 30y = 24 \\ 20x - 50y = -14 \end{cases}$

27. $\begin{cases} 0.2x + 0.3y = 0.14 \\ 0.4x + 0.5y = 0.20 \end{cases}$
28. $\begin{cases} 12x + 42y = -17 \\ 30x - 18y = 19 \end{cases}$

29. $\begin{cases} 3x - 2y = 0 \\ 3x + 2(y + 5) = 10 \end{cases}$
30. $\begin{cases} 7x + 12y = 63 \\ 2x + 3(y + 2) = 21 \end{cases}$

31. $\begin{cases} 1.25x - 2y = 3.5 \\ 5x - 8y = 14 \end{cases}$
32. $\begin{cases} 1.5x + 2.5y = 8.5 \\ 6x + 10y = 24 \end{cases}$

In Exercises 33–36, match the system of linear equations with its graph. Describe the number of solutions and state whether the system is consistent or inconsistent. [The graphs are labeled (a), (b), (c), and (d).]

(a)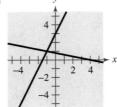

(b)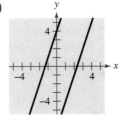

(c)

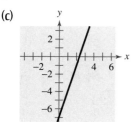

(d)

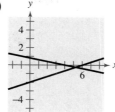

33. $\begin{cases} x + 5y = 4 \\ x - 3y = 6 \end{cases}$

34. $\begin{cases} -3x + y = -7 \\ 9x - 3y = 21 \end{cases}$

35. $\begin{cases} 3x - y = 7 \\ -6x + 2y = 8 \end{cases}$

36. $\begin{cases} 2x - y = -3 \\ x + 5y = 4 \end{cases}$

SUPPLY AND DEMAND In Exercises 37 and 38, find the equilibrium point of the demand and supply equations.

Demand	Supply
37. $p = 37 - 0.0002x$	$p = 22 + 0.00001x$
38. $p = 120 - 0.0001x$	$p = 45 + 0.0002x$

 6.3 In Exercises 39–42, use back-substitution to solve the system of linear equations.

39. $\begin{cases} x - 4y + 3z = 3 \\ -y + z = -1 \\ z = -5 \end{cases}$

40. $\begin{cases} x - 7y + 8z = 85 \\ y - 9z = -35 \\ z = 3 \end{cases}$

41. $\begin{cases} 4x - 3y - 2z = -65 \\ 8y - 7z = -14 \\ z = 10 \end{cases}$

42. $\begin{cases} 5x - 7z = 9 \\ 3y - 8z = -4 \\ z = -7 \end{cases}$

In Exercises 43–48, use Gaussian elimination to solve the system of equations.

43. $\begin{cases} x + 2y + 6z = 4 \\ -3x + 2y - z = -4 \\ 4x + 2z = 16 \end{cases}$

44. $\begin{cases} x + 3y - z = 13 \\ 2x - 5z = 23 \\ 4x - y - 2z = 14 \end{cases}$

45. $\begin{cases} x - 2y + z = -6 \\ 2x - 3y = -7 \\ -x + 3y - 3z = 11 \end{cases}$

46. $\begin{cases} 2x + 6z = -9 \\ 3x - 2y + 11z = -16 \\ 3x - y + 7z = -11 \end{cases}$

47. $\begin{cases} x + 4w = 1 \\ 3y + z - w = 4 \\ 2y - 3w = 2 \\ 4x - y + 2z = 5 \end{cases}$

48. $\begin{cases} x + y + z + w = 6 \\ 3x + 4y - w = 3 \\ -2x + 3y + z + 3w = 6 \\ x + 4y - z + 2w = 7 \end{cases}$

In Exercises 49 and 50, solve the nonsquare system of equations.

49. $\begin{cases} 5x - 12y + 7z = 16 \\ 3x - 7y + 4z = 9 \end{cases}$

50. $\begin{cases} 2x + 5y - 19z = 34 \\ 3x + 8y - 31z = 54 \end{cases}$

In Exercises 51 and 52, find the equation of the parabola $y = ax^2 + bx + c$ that passes through the points. To verify your result, use a graphing utility to plot the points and graph the parabola.

51.

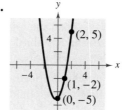

52.

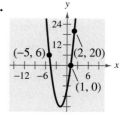

In Exercises 53 and 54, find the equation of the circle

$$x^2 + y^2 + Dx + Ey + F = 0$$

that passes through the points. To verify your result, use a graphing utility to plot the points and graph the circle.

53.

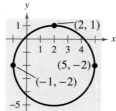

54.

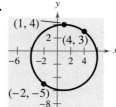

55. DATA ANALYSIS: ONLINE SHOPPING The table shows the projected online retail sales y (in billions of dollars) in the United States from 2010 through 2012. (Source: Forrester Research, Inc.)

Year	Online retail sales, y
2010	267.8
2011	301.0
2012	334.7

(a) Use the technique demonstrated in Exercises 77–80 in Section 6.3 to set up a system of equations for the data and to find a least squares regression parabola that models the data. Let x represent the year, with $x = 10$ corresponding to 2010.

(b) Use a graphing utility to graph the parabola and the data in the same viewing window. How well does the model fit the data?

(c) Use the model to estimate the online retail sales in the United States in 2015. Does your answer seem reasonable?

56. AGRICULTURE A mixture of 6 gallons of chemical A, 8 gallons of chemical B, and 13 gallons of chemical C is required to kill a destructive crop insect. Commercial spray X contains 1, 2, and 2 parts, respectively, of these chemicals. Commercial spray Y contains only chemical C. Commercial spray Z contains chemicals A, B, and C in equal amounts. How much of each type of commercial spray is needed to get the desired mixture?

57. INVESTMENT ANALYSIS An inheritance of $40,000 was divided among three investments yielding $3500 in interest per year. The interest rates for the three investments were 7%, 9%, and 11%. Find the amount placed in each investment if the second and third were $3000 and $5000 less than the first, respectively.

58. VERTICAL MOTION An object moving vertically is at the given heights at the specified times. Find the position equation

$$s = \frac{1}{2}at^2 + v_0 t + s_0$$

for the object.

(a) At $t = 1$ second, $s = 134$ feet
At $t = 2$ seconds, $s = 86$ feet
At $t = 3$ seconds, $s = 6$ feet

(b) At $t = 1$ second, $s = 184$ feet
At $t = 2$ seconds, $s = 116$ feet
At $t = 3$ seconds, $s = 16$ feet

59. SPORTS Pebble Beach Golf Links in Pebble Beach, California is an 18-hole course that consists of par-3 holes, par-4 holes, and par-5 holes. There are two more par-4 holes than twice the number of par-5 holes, and the number of par-3 holes is equal to the number of par-5 holes. Find the numbers of par-3, par-4, and par-5 holes on the course. (Source: Pebble Beach Resorts)

60. SPORTS St Andrews Golf Course in St Andrews, Scotland is one of the oldest golf courses in the world. It is an 18-hole course that consists of par-3 holes, par-4 holes, and par-5 holes. There are seven times as many par-4 holes as par-5 holes, and the sum of the numbers of par-3 and par-5 holes is four. Find the numbers of par-3, par-4, and par-5 holes on the course. (Source: St Andrews Links Trust)

6.4 In Exercises 61–64, write the form of the partial fraction decomposition for the rational expression. Do not solve for the constants.

61. $\dfrac{3}{x^2 + 20x}$

62. $\dfrac{x - 8}{x^2 - 3x - 28}$

63. $\dfrac{3x - 4}{x^3 - 5x^2}$

64. $\dfrac{x - 2}{x(x^2 + 2)^2}$

In Exercises 65–72, write the partial fraction decomposition of the rational expression.

65. $\dfrac{4 - x}{x^2 + 6x + 8}$

66. $\dfrac{-x}{x^2 + 3x + 2}$

67. $\dfrac{x^2}{x^2 + 2x - 15}$

68. $\dfrac{9}{x^2 - 9}$

69. $\dfrac{x^2 + 2x}{x^3 - x^2 + x - 1}$

70. $\dfrac{4x}{3(x - 1)^2}$

71. $\dfrac{3x^2 + 4x}{(x^2 + 1)^2}$

72. $\dfrac{4x^2}{(x - 1)(x^2 + 1)}$

6.5 In Exercises 73–78, sketch the graph of the inequality.

73. $y \le 5 - \frac{1}{2}x$

74. $3y - x \ge 7$

75. $y - 4x^2 > -1$

76. $y \le \dfrac{3}{x^2 + 2}$

77. $(x - 1)^2 + (y - 3)^2 < 16$

78. $x^2 + (y + 5)^2 > 1$

In Exercises 79–86, sketch the graph and label the vertices of the solution set of the system of inequalities.

79. $\begin{cases} x + 2y \le 160 \\ 3x + y \le 180 \\ x \ge 0 \\ y \ge 0 \end{cases}$

80. $\begin{cases} 2x + 3y \le 24 \\ 2x + y \le 16 \\ x \ge 0 \\ y \ge 0 \end{cases}$

81. $\begin{cases} 3x + 2y \ge 24 \\ x + 2y \ge 12 \\ 2 \le x \le 15 \\ y \le 15 \end{cases}$

82. $\begin{cases} 2x + y \ge 16 \\ x + 3y \ge 18 \\ 0 \le x \le 25 \\ 0 \le y \le 25 \end{cases}$

83. $\begin{cases} y < x + 1 \\ y > x^2 - 1 \end{cases}$

84. $\begin{cases} y \le 6 - 2x - x^2 \\ y \ge x + 6 \end{cases}$

85. $\begin{cases} 2x - 3y \ge 0 \\ 2x - y \le 8 \\ y \ge 0 \end{cases}$

86. $\begin{cases} x^2 + y^2 \le 9 \\ (x - 3)^2 + y^2 \le 9 \end{cases}$

87. INVENTORY COSTS A warehouse operator has 24,000 square feet of floor space in which to store two products. Each unit of product I requires 20 square feet of floor space and costs $12 per day to store. Each unit of product II requires 30 square feet of floor space and costs $8 per day to store. The total storage cost per day cannot exceed $12,400. Find and graph a system of inequalities describing all possible inventory levels.

88. NUTRITION A dietitian is asked to design a special dietary supplement using two different foods. Each ounce of food X contains 12 units of calcium, 10 units of iron, and 20 units of vitamin B. Each ounce of food Y contains 15 units of calcium, 20 units of iron, and 12 units of vitamin B. The minimum daily requirements of the diet are 300 units of calcium, 280 units of iron, and 300 units of vitamin B.

(a) Write a system of inequalities describing the different amounts of food X and food Y that can be used.

(b) Sketch a graph of the region in part (a).

(c) Find two solutions to the system and interpret their meanings in the context of the problem.

SUPPLY AND DEMAND In Exercises 89 and 90, (a) graph the systems representing the consumer surplus and producer surplus for the supply and demand equations and (b) find the consumer surplus and producer surplus.

	Demand	*Supply*
89.	$p = 160 - 0.0001x$	$p = 70 + 0.0002x$
90.	$p = 130 - 0.0002x$	$p = 30 + 0.0003x$

91. GEOMETRY Derive a set of inequalities to describe the region of a rectangle with vertices at $(3, 1)$, $(7, 1)$, $(7, 10)$, and $(3, 10)$.

92. DATA ANALYSIS: COMPUTER SALES The table shows the sales y (in billions of dollars) for Dell, Inc. from 2000 through 2007. (Source: Dell, Inc.)

Year	Sales, y
2000	31.9
2001	31.2
2002	35.4
2003	41.4
2004	49.2
2005	55.9
2006	57.4
2007	61.1

(a) Use the *regression* feature of a graphing utility to find a linear model for the data. Let t represent the year, with $t = 0$ corresponding to 2000.

(b) The total sales for Dell during this eight-year period can be approximated by finding the area of the trapezoid bounded by the linear model you found in part (a) and the lines $y = 0$, $t = -0.5$, and $t = 7.5$. Use a graphing utility to graph this region.

(c) Use the formula for the area of a trapezoid to approximate the total retail sales for Dell.

6.6 In Exercises 93–98, sketch the region determined by the constraints. Then find the minimum and maximum values of the objective function (if possible) and where they occur, subject to the indicated constraints.

93. Objective function:
$z = 3x + 4y$
Constraints:
$x \geq 0$
$y \geq 0$
$2x + 5y \leq 50$
$4x + y \leq 28$

94. Objective function:
$z = 10x + 7y$
Constraints:
$x \geq 0$
$y \geq 0$
$2x + y \geq 100$
$x + y \geq 75$

95. Objective function:
$z = 1.75x + 2.25y$
Constraints:
$x \geq 0$
$y \geq 0$
$2x + y \geq 25$
$3x + 2y \geq 45$

96. Objective function:
$z = 50x + 70y$
Constraints:
$x \geq 0$
$y \geq 0$
$x + 2y \leq 1500$
$5x + 2y \leq 3500$

97. Objective function:
$z = 5x + 11y$
Constraints:
$x \geq 0$
$y \geq 0$
$x + 3y \leq 12$
$3x + 2y \leq 15$

98. Objective function:
$z = -2x + y$
Constraints:
$x \geq 0$
$y \geq 0$
$x + y \geq 7$
$5x + 2y \geq 20$

99. OPTIMAL REVENUE A student is working part time as a hairdresser to pay college expenses. The student may work no more than 24 hours per week. Haircuts cost $25 and require an average of 20 minutes, and permanents cost $70 and require an average of 1 hour and 10 minutes. What combination of haircuts and/or permanents will yield an optimal revenue? What is the optimal revenue?

100. OPTIMAL PROFIT A shoe manufacturer produces a walking shoe and a running shoe yielding profits of $18 and $24, respectively. Each shoe must go through three processes, for which the required times per unit are shown in the table.

	Process I	Process II	Process III
Hours for walking shoe	4	1	1
Hours for running shoe	2	2	1
Hours available per day	24	9	8

What is the optimal production level for each type of shoe? What is the optimal profit?

101. OPTIMAL PROFIT A manufacturer produces two models of bicycles. The times (in hours) required for assembling, painting, and packaging each model are shown in the table.

Process	Hours, model A	Hours, model B
Assembling	2	2.5
Painting	4	1
Packaging	1	0.75

The total times available for assembling, painting, and packaging are 4000 hours, 4800 hours, and 1500 hours, respectively. The profits per unit are $45 for model A and $50 for model B. What is the optimal production level for each model? What is the optimal profit?

102. OPTIMAL COST A pet supply company mixes two brands of dry dog food. Brand X costs $15 per bag and contains eight units of nutritional element A, one unit of nutritional element B, and two units of nutritional element C. Brand Y costs $30 per bag and contains two units of nutritional element A, one unit of nutritional element B, and seven units of nutritional element C. Each bag of mixed dog food must contain at least 16 units, 5 units, and 20 units of nutritional elements A, B, and C, respectively. Find the numbers of bags of brands X and Y that should be mixed to produce a mixture meeting the minimum nutritional requirements and having an optimal cost. What is the optimal cost?

103. OPTIMAL COST Regular unleaded gasoline and premium unleaded gasoline have octane ratings of 87 and 93, respectively. For the week of March 23, 2009, regular unleaded gasoline in Houston, Texas averaged $1.85 per gallon. For the same week, premium unleaded gasoline averaged $2.10 per gallon. Determine the blend of regular and premium unleaded gasoline that results in an optimal cost of mid-grade unleaded (89-octane) gasoline. What is the optimal cost? (Source: Energy Information Administration)

EXPLORATION

TRUE OR FALSE? In Exercises 104–106, determine whether the statement is true or false. Justify your answer.

104. If a system of equations consists of a circle and a parabola, it is possible for the system to have three solutions.

105. The system

$$\begin{cases} y \leq 5 \\ y \geq -2 \\ y \geq \frac{7}{2}x - 9 \\ y \geq -\frac{7}{2}x + 26 \end{cases}$$

represents the region covered by an isosceles trapezoid.

106. It is possible for an objective function of a linear programming problem to have exactly 10 maximum value points.

In Exercises 107–110, find a system of linear equations having the ordered pair as a solution. (There are many correct answers.)

107. $(-8, 10)$ **108.** $(5, -4)$

109. $\left(\frac{4}{3}, 3\right)$ **110.** $\left(-2, \frac{11}{5}\right)$

In Exercises 111–114, find a system of linear equations having the ordered triple as a solution. (There are many correct answers.)

111. $(4, -1, 3)$

112. $(-3, 5, 6)$

113. $\left(5, \frac{3}{2}, 2\right)$

114. $\left(-\frac{1}{2}, -2, -\frac{3}{4}\right)$

115. WRITING Explain what is meant by an inconsistent system of linear equations.

116. How can you tell graphically that a system of linear equations in two variables has no solution? Give an example.

6 CHAPTER TEST

See www.CalcChat.com for worked-out solutions to odd-numbered exercises.

Take this test as you would take a test in class. When you are finished, check your work against the answers given in the back of the book.

In Exercises 1–3, solve the system by the method of substitution.

1. $\begin{cases} x + y = -9 \\ 5x - 8y = 20 \end{cases}$ **2.** $\begin{cases} y = x - 1 \\ y = (x - 1)^3 \end{cases}$ **3.** $\begin{cases} 2x - y^2 = 0 \\ x - y = 4 \end{cases}$

In Exercises 4–6, solve the system graphically.

4. $\begin{cases} 3x - 6y = 0 \\ 3x + 6y = 18 \end{cases}$ **5.** $\begin{cases} y = 9 - x^2 \\ y = x + 3 \end{cases}$ **6.** $\begin{cases} y - \ln x = 12 \\ 7x - 2y + 11 = -6 \end{cases}$

In Exercises 7–10, solve the linear system by the method of elimination.

7. $\begin{cases} 3x + 4y = -26 \\ 7x - 5y = 11 \end{cases}$ **8.** $\begin{cases} 1.4x - y = 17 \\ 0.8x + 6y = -10 \end{cases}$

9. $\begin{cases} x - 2y + 3z = 11 \\ 2x - z = 3 \\ 3y + z = -8 \end{cases}$ **10.** $\begin{cases} 3x + 2y + z = 17 \\ -x + y + z = 4 \\ x - y - z = 3 \end{cases}$

In Exercises 11–14, write the partial fraction decomposition of the rational expression.

11. $\dfrac{2x + 5}{x^2 - x - 2}$ **12.** $\dfrac{3x^2 - 2x + 4}{x^2(2 - x)}$ **13.** $\dfrac{x^2 + 5}{x^3 - x}$ **14.** $\dfrac{x^2 - 4}{x^3 + 2x}$

In Exercises 15–17, sketch the graph and label the vertices of the solution of the system of inequalities.

15. $\begin{cases} 2x + y \le 4 \\ 2x - y \ge 0 \\ x \ge 0 \end{cases}$ **16.** $\begin{cases} y < -x^2 + x + 4 \\ y > 4x \end{cases}$ **17.** $\begin{cases} x^2 + y^2 \le 36 \\ x \ge 2 \\ y \ge -4 \end{cases}$

18. Find the maximum and minimum values of the objective function $z = 20x + 12y$, and where they occur, subject to the following constraints.

$$\left.\begin{array}{r} x \ge 0 \\ y \ge 0 \\ x + 4y \le 32 \\ 3x + 2y \le 36 \end{array}\right\}$$

19. A total of \$50,000 is invested in two funds paying 4% and 5.5% simple interest. The yearly interest is \$2390. How much is invested at each rate?

20. Find the equation of the parabola $y = ax^2 + bx + c$ passing through the points $(0, 6)$, $(-2, 2)$, and $\left(3, \frac{9}{2}\right)$.

21. A manufacturer produces two types of television stands. The amounts (in hours) of time for assembling, staining, and packaging the two models are shown in the table at the left. The total amounts of time available for assembling, staining, and packaging are 4000, 8950, and 2650 hours, respectively. The profits per unit are \$30 (model I) and \$40 (model II). What is the optimal inventory level for each model? What is the optimal profit?

	Model I	Model II
Assembling	0.5	0.75
Staining	2.0	1.5
Packaging	0.5	0.5

TABLE FOR **21**

PROOFS IN MATHEMATICS

An **indirect proof** can be useful in proving statements of the form "p implies q." Recall that the conditional statement $p \to q$ is false only when p is true and q is false. To prove a conditional statement indirectly, assume that p is true and q is false. If this assumption leads to an impossibility, then you have proved that the conditional statement is true. An indirect proof is also called a **proof by contradiction.**

You can use an indirect proof to prove the following conditional statement,

"If a is a positive integer and a^2 is divisible by 2, then a is divisible by 2,"

as follows. First, assume that p, "a is a positive integer and a^2 is divisible by 2," is true and q, "a is divisible by 2," is false. This means that a is not divisible by 2. If so, a is odd and can be written as $a = 2n + 1$, where n is an integer.

$$a = 2n + 1$$
$$a^2 = 4n^2 + 4n + 1$$
$$a^2 = 2(2n^2 + 2n) + 1$$

So, by the definition of an odd integer, a^2 is odd. This contradicts the assumption, and you can conclude that a is divisible by 2.

Example   Using an Indirect Proof

Use an indirect proof to prove that $\sqrt{2}$ is an irrational number.

Solution

Begin by assuming that $\sqrt{2}$ is *not* an irrational number. Then $\sqrt{2}$ can be written as the quotient of two integers a and $b (b \neq 0)$ that have no common factors.

$$\sqrt{2} = \frac{a}{b}$$
$$2 = \frac{a^2}{b^2}$$
$$2b^2 = a^2$$

This implies that 2 is a factor of a^2. So, 2 is also a factor of a, and a can be written as $2c$, where c is an integer.

$$2b^2 = (\quad)^2$$
$$2b^2 = 4c^2$$
$$b^2 = 2c^2$$

This implies that 2 is a factor of b^2 and also a factor of b. So, 2 is a factor of both a and b. This contradicts the assumption that a and b have no common factors. So, you can conclude that $\sqrt{2}$ is an irrational number.

PROBLEM SOLVING

This collection of thought-provoking and challenging exercises further explores and expands upon concepts learned in this chapter.

1. A theorem from geometry states that if a triangle is inscribed in a circle such that one side of the triangle is a diameter of the circle, then the triangle is a right triangle. Show that this theorem is true for the circle

   $$x^2 + y^2 = 100$$

   and the triangle formed by the lines

   $y = 0$, $y = \frac{1}{2}x + 5$, and $y = -2x + 20$.

2. Find k_1 and k_2 such that the system of equations has an infinite number of solutions.

   $$\begin{cases} 3x - 5y = 8 \\ 2x + k_1 y = k_2 \end{cases}$$

3. Consider the following system of linear equations in x and y.

   $$\begin{cases} ax + by = e \\ cx + dy = f \end{cases}$$

   Under what conditions will the system have exactly one solution?

4. Graph the lines determined by each system of linear equations. Then use Gaussian elimination to solve each system. At each step of the elimination process, graph the corresponding lines. What do you observe?

   (a) $\begin{cases} x - 4y = -3 \\ 5x - 6y = 13 \end{cases}$

   (b) $\begin{cases} 2x - 3y = 7 \\ -4x + 6y = -14 \end{cases}$

5. A system of two equations in two unknowns is solved and has a finite number of solutions. Determine the maximum number of solutions of the system satisfying each condition.

   (a) Both equations are linear.

   (b) One equation is linear and the other is quadratic.

   (c) Both equations are quadratic.

6. In the 2008 presidential election, approximately 125.2 million voters divided their votes between Barack Obama and John McCain. Obama received approximately 8.5 million more votes than McCain. Write and solve a system of equations to find the total number of votes cast for each candidate. Let D represent the number of votes cast for Obama, and let R represent the number of votes cast for McCain. (Source: CNN.com)

7. The Vietnam Veterans Memorial (or "The Wall") in Washington, D.C. was designed by Maya Ying Lin when she was a student at Yale University. This monument has two vertical, triangular sections of black granite with a common side (see figure). The bottom of each section is level with the ground. The tops of the two sections can be approximately modeled by the equations

   $$-2x + 50y = 505 \quad \text{and} \quad 2x + 50y = 505$$

   when the x-axis is superimposed at the base of the wall. Each unit in the coordinate system represents 1 foot. How high is the memorial at the point where the two sections meet? How long is each section?

   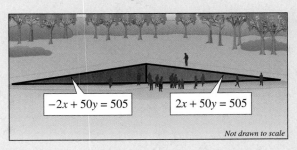

   Not drawn to scale

8. Weights of atoms and molecules are measured in atomic mass units (u). A molecule of C_2H_6 (ethane) is made up of two carbon atoms and six hydrogen atoms and weighs 30.070 u. A molecule of C_3H_8 (propane) is made up of three carbon atoms and eight hydrogen atoms and weighs 44.097 u. Find the weights of a carbon atom and a hydrogen atom.

9. Connecting a DVD player to a television set requires a cable with special connectors at both ends. You buy a six-foot cable for $15.50 and a three-foot cable for $10.25. Assuming that the cost of a cable is the sum of the cost of the two connectors and the cost of the cable itself, what is the cost of a four-foot cable? Explain your reasoning.

10. A hotel 35 miles from an airport runs a shuttle service to and from the airport. The 9:00 A.M. bus leaves for the airport traveling at 30 miles per hour. The 9:15 A.M. bus leaves for the airport traveling at 40 miles per hour. Write a system of linear equations that represents distance as a function of time for each bus. Graph and solve the system. How far from the airport will the 9:15 A.M. bus catch up to the 9:00 A.M. bus?

517

11. Solve each system of equations by letting $X = 1/x$, $Y = 1/y$, and $Z = 1/z$.

(a)
$$\begin{cases} \dfrac{12}{x} - \dfrac{12}{y} = 7 \\ \dfrac{3}{x} + \dfrac{4}{y} = 0 \end{cases}$$

(b)
$$\begin{cases} \dfrac{2}{x} + \dfrac{1}{y} - \dfrac{3}{z} = 4 \\ \dfrac{4}{x} \phantom{+\dfrac{1}{y}} + \dfrac{2}{z} = 10 \\ -\dfrac{2}{x} + \dfrac{3}{y} - \dfrac{13}{z} = -8 \end{cases}$$

12. What values should be given to a, b, and c so that the linear system shown has $(-1, 2, -3)$ as its only solution?

$$\begin{cases} x + 2y - 3z = a \\ -x - y + z = b \\ 2x + 3y - 2z = c \end{cases}$$

13. The following system has one solution: $x = 1$, $y = -1$, and $z = 2$.

$$\begin{cases} 4x - 2y + 5z = 16 \\ x + y = 0 \\ -x - 3y + 2z = 6 \end{cases}$$

Solve the system given by (a) Equation 1 and Equation 2, (b) Equation 1 and Equation 3, and (c) Equation 2 and Equation 3. (d) How many solutions does each of these systems have?

14. Solve the system of linear equations algebraically.

$$\begin{aligned} x_1 - x_2 + 2x_3 + 2x_4 + 6x_5 &= 6 \\ 3x_1 - 2x_2 + 4x_3 + 4x_4 + 12x_5 &= 14 \\ -x_2 - x_3 - x_4 - 3x_5 &= -3 \\ 2x_1 - 2x_2 + 4x_3 + 5x_4 + 15x_5 &= 10 \\ 2x_1 - 2x_2 + 4x_3 + 4x_4 + 13x_5 &= 13 \end{aligned}$$

15. Each day, an average adult moose can process about 32 kilograms of terrestrial vegetation (twigs and leaves) and aquatic vegetation. From this food, it needs to obtain about 1.9 grams of sodium and 11,000 calories of energy. Aquatic vegetation has about 0.15 gram of sodium per kilogram and about 193 calories of energy per kilogram, whereas terrestrial vegetation has minimal sodium and about four times as much energy as aquatic vegetation. Write and graph a system of inequalities that describes the amounts t and a of terrestrial and aquatic vegetation, respectively, for the daily diet of an average adult moose. (Source: Biology by Numbers)

16. For a healthy person who is 4 feet 10 inches tall, the recommended minimum weight is about 91 pounds and increases by about 3.65 pounds for each additional inch of height. The recommended maximum weight is about 119 pounds and increases by about 4.85 pounds for each additional inch of height. (Source: U.S. Department of Agriculture)

(a) Let x be the number of inches by which a person's height exceeds 4 feet 10 inches and let y be the person's weight in pounds. Write a system of inequalities that describes the possible values of x and y for a healthy person.

(b) Use a graphing utility to graph the system of inequalities from part (a).

(c) What is the recommended weight range for someone 6 feet tall?

17. The cholesterol in human blood is necessary, but too much cholesterol can lead to health problems. A blood cholesterol test gives three readings: LDL ("bad") cholesterol, HDL ("good") cholesterol, and total cholesterol (LDL + HDL). It is recommended that your LDL cholesterol level be less than 130 milligrams per deciliter, your HDL cholesterol level be at least 60 milligrams per deciliter, and your total cholesterol level be no more than 200 milligrams per deciliter. (Source: American Heart Association)

(a) Write a system of linear inequalities for the recommended cholesterol levels. Let x represent HDL cholesterol and let y represent LDL cholesterol.

(b) Graph the system of inequalities from part (a). Label any vertices of the solution region.

(c) Are the following cholesterol levels within recommendations? Explain your reasoning.

LDL: 120 milligrams per deciliter

HDL: 90 milligrams per deciliter

Total: 210 milligrams per deciliter

(d) Give an example of cholesterol levels in which the LDL cholesterol level is too high but the HDL and total cholesterol levels are acceptable.

(e) Another recommendation is that the ratio of total cholesterol to HDL cholesterol be less than 5. Find a point in your solution region from part (b) that meets this recommendation, and explain why it meets the recommendation.

Matrices and Determinants

7

In Mathematics

Matrices are used to model and solve a variety of problems. For instance, you can use matrices to solve systems of linear equations.

In Real Life

Matrices are used to model inventory levels, electrical networks, investment portfolios, and other real-life situations. For instance, you can use a matrix to model the number of people in the United States who participate in snowboarding. (See Exercise 114, page 533.)

Graham Heywood/istockphoto.com

IN CAREERS

There are many careers that use matrices. Several are listed below.

- Bank Teller
  Exercise 110, page 532

- Political Analyst
  Exercise 70, page 547

- Small Business Owner
  Exercises 67–70, pages 556 and 557

- Florist
  Exercise 72, page 557

What you should learn

- Write matrices and identify their orders.
- Perform elementary row operations on matrices.
- Use matrices and Gaussian elimination to solve systems of linear equations.
- Use matrices and Gauss-Jordan elimination to solve systems of linear equations.

Why you should learn it

You can use matrices to solve systems of linear equations in two or more variables. For instance, in Exercise 113 on page 532, you will use a matrix to find a model for the path of a ball thrown by a baseball player.

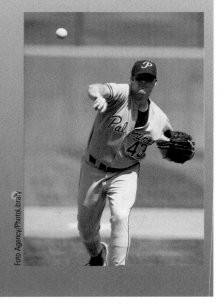

7.1 MATRICES AND SYSTEMS OF EQUATIONS

Matrices

In this section, you will study a streamlined technique for solving systems of linear equations. This technique involves the use of a rectangular array of real numbers called a **matrix.** The plural of matrix is *matrices.*

Definition of Matrix

If m and n are positive integers, an $m \times n$ (read "m by n") matrix is a rectangular array

$$\begin{bmatrix} a_{11} & a_{12} & a_{13} & \cdots & a_{1n} \\ a_{21} & a_{22} & a_{23} & \cdots & a_{2n} \\ a_{31} & a_{32} & a_{33} & \cdots & a_{3n} \\ \vdots & \vdots & \vdots & & \vdots \\ a_{m1} & a_{m2} & a_{m3} & \cdots & a_{mn} \end{bmatrix}$$

in which each **entry,** a_{ij}, of the matrix is a number. An $m \times n$ matrix has m rows and n columns. Matrices are usually denoted by capital letters.

The entry in the ith row and jth column is denoted by the *double subscript* notation a_{ij}. For instance, a_{23} refers to the entry in the second row, third column. A matrix having m rows and n columns is said to be of **order** $m \times n$. If $m = n$, the matrix is **square** of order $m \times m$ (or $n \times n$). For a square matrix, the entries $a_{11}, a_{22}, a_{33}, \ldots$ are the **main diagonal** entries.

Example 1 Order of Matrices

Determine the order of each matrix.

a. $[2]$
b. $\begin{bmatrix} 1 & -3 & 0 & \frac{1}{2} \end{bmatrix}$

c. $\begin{bmatrix} 0 & 0 \\ 0 & 0 \end{bmatrix}$
d. $\begin{bmatrix} 5 & 0 \\ 2 & -2 \\ -7 & 4 \end{bmatrix}$

Solution

a. This matrix has *one* row and *one* column. The order of the matrix is 1×1.

b. This matrix has *one* row and *four* columns. The order of the matrix is 1×4.

c. This matrix has *two* rows and *two* columns. The order of the matrix is 2×2.

d. This matrix has *three* rows and *two* columns. The order of the matrix is 3×2.

CHECK*Point* Now try Exercise 9.

A matrix that has only one row is called a **row matrix,** and a matrix that has only one column is called a **column matrix.**

A matrix derived from a system of linear equations (each written in standard form with the constant term on the right) is the **augmented matrix** of the system. Moreover, the matrix derived from the coefficients of the system (but not including the constant terms) is the **coefficient matrix** of the system.

System: $\begin{cases} x - 4y + 3z = 5 \\ -x + 3y - z = -3 \\ 2x - 4z = 6 \end{cases}$

Augmented Matrix: $\begin{bmatrix} 1 & -4 & 3 & \vdots & 5 \\ -1 & 3 & -1 & \vdots & -3 \\ 2 & 0 & -4 & \vdots & 6 \end{bmatrix}$

Coefficient Matrix: $\begin{bmatrix} 1 & -4 & 3 \\ -1 & 3 & -1 \\ 2 & 0 & -4 \end{bmatrix}$

Note the use of 0 for the missing coefficient of the *y*-variable in the third equation, and also note the fourth column of constant terms in the augmented matrix.

When forming either the coefficient matrix or the augmented matrix of a system, you should begin by vertically aligning the variables in the equations and using zeros for the coefficients of the missing variables.

Example 2   Writing an Augmented Matrix

Write the augmented matrix for the system of linear equations.

$$\begin{cases} x + 3y - w = 9 \\ -y + 4z + 2w = -2 \\ x - 5z - 6w = 0 \\ 2x + 4y - 3z = 4 \end{cases}$$

What is the order of the augmented matrix?

Solution

Begin by rewriting the linear system and aligning the variables.

$$\begin{cases} x + 3y - w = 9 \\ -y + 4z + 2w = -2 \\ x - 5z - 6w = 0 \\ 2x + 4y - 3z = 4 \end{cases}$$

Next, use the coefficients and constant terms as the matrix entries. Include zeros for the coefficients of the missing variables.

$$\begin{bmatrix} 1 & 3 & 0 & -1 & \vdots & 9 \\ 0 & -1 & 4 & 2 & \vdots & -2 \\ 1 & 0 & -5 & -6 & \vdots & 0 \\ 2 & 4 & -3 & 0 & \vdots & 4 \end{bmatrix}$$

The augmented matrix has four rows and five columns, so it is a 4×5 matrix. The notation R_n is used to designate each row in the matrix. For example, Row 1 is represented by R_1.

CHECK **Point** ▶ Now try Exercise 17.

Elementary Row Operations

In Section 6.3, you studied three operations that can be used on a system of linear equations to produce an equivalent system.

1. Interchange two equations.

2. Multiply an equation by a nonzero constant.

3. Add a multiple of an equation to another equation.

In matrix terminology, these three operations correspond to **elementary row operations.** An elementary row operation on an augmented matrix of a given system of linear equations produces a new augmented matrix corresponding to a new (but equivalent) system of linear equations. Two matrices are **row-equivalent** if one can be obtained from the other by a sequence of elementary row operations.

Elementary Row Operations

1. Interchange two rows.

2. Multiply a row by a nonzero constant.

3. Add a multiple of a row to another row.

Although elementary row operations are simple to perform, they involve a lot of arithmetic. Because it is easy to make a mistake, you should get in the habit of noting the elementary row operations performed in each step so that you can go back and check your work.

Example 3   Elementary Row Operations

a. Interchange the first and second rows of the original matrix.

Original Matrix

$$\begin{bmatrix} 0 & 1 & 3 & 4 \\ -1 & 2 & 0 & 3 \\ 2 & -3 & 4 & 1 \end{bmatrix}$$

New Row-Equivalent Matrix

$$\begin{bmatrix} -1 & 2 & 0 & 3 \\ 0 & 1 & 3 & 4 \\ 2 & -3 & 4 & 1 \end{bmatrix}$$

b. Multiply the first row of the original matrix by $\frac{1}{2}$.

Original Matrix

$$\begin{bmatrix} 2 & -4 & 6 & -2 \\ 1 & 3 & -3 & 0 \\ 5 & -2 & 1 & 2 \end{bmatrix}$$

New Row-Equivalent Matrix

$$\begin{bmatrix} 1 & -2 & 3 & -1 \\ 1 & 3 & -3 & 0 \\ 5 & -2 & 1 & 2 \end{bmatrix}$$

c. Add -2 times the first row of the original matrix to the third row.

Original Matrix

$$\begin{bmatrix} 1 & 2 & -4 & 3 \\ 0 & 3 & -2 & -1 \\ 2 & 1 & 5 & -2 \end{bmatrix}$$

New Row-Equivalent Matrix

$$\begin{bmatrix} 1 & 2 & -4 & 3 \\ 0 & 3 & -2 & -1 \\ 0 & -3 & 13 & -8 \end{bmatrix}$$

Note that the elementary row operation is written beside the row that is *changed.*

CHECK *Point*   Now try Exercise 37.

In Example 3 in Section 6.3, you used Gaussian elimination with back-substitution to solve a system of linear equations. The next example demonstrates the matrix version of Gaussian elimination. The two methods are essentially the same. The basic difference is that with matrices you do not need to keep writing the variables.

Example 4   Comparing Linear Systems and Matrix Operations

⚠ WARNING / CAUTION

Arithmetic errors are often made when elementary row operations are performed. Note the operation you perform in each step so that you can go back and check your work.

Linear System

$$\begin{cases} x - 2y + 3z = 9 \\ -x + 3y \quad\;\; = -4 \\ 2x - 5y + 5z = 17 \end{cases}$$

Add the first equation to the second equation.

$$\begin{cases} x - 2y + 3z = 9 \\ \quad\;\; y + 3z = 5 \\ 2x - 5y + 5z = 17 \end{cases}$$

Add -2 times the first equation to the third equation.

$$\begin{cases} x - 2y + 3z = 9 \\ \quad\;\; y + 3z = 5 \\ \quad -y - z = -1 \end{cases}$$

Add the second equation to the third equation.

$$\begin{cases} x - 2y + 3z = 9 \\ \quad\;\; y + 3z = 5 \\ \quad\quad\;\; 2z = 4 \end{cases}$$

Study Tip

Remember that you should check a solution by substituting the values of x, y, and z into each equation of the original system. For example, you can check the solution to Example 4 as follows.

Equation 1:
$- 2(\;\;) + 3(\;\;) = 9$ ✓

Equation 2:
$-\;\; + 3(\;\;) = -4$ ✓

Equation 3:
$2(\;) - 5(\;\;) + 5(\;) = 17$ ✓

Multiply the third equation by $\frac{1}{2}$.

$$\begin{cases} x - 2y + 3z = 9 \\ \quad\;\; y + 3z = 5 \\ \quad\quad\;\; z = 2 \end{cases}$$

Associated Augmented Matrix

$$\begin{bmatrix} 1 & -2 & 3 & \vdots & 9 \\ -1 & 3 & 0 & \vdots & -4 \\ 2 & -5 & 5 & \vdots & 17 \end{bmatrix}$$

Add the first row to the second row $(R_1 + R_2)$.

$$\begin{bmatrix} 1 & -2 & 3 & \vdots & 9 \\ 0 & 1 & 3 & \vdots & 5 \\ 2 & -5 & 5 & \vdots & 17 \end{bmatrix}$$

Add -2 times the first row to the third row $(-2R_1 + R_3)$.

$$\begin{bmatrix} 1 & -2 & 3 & \vdots & 9 \\ 0 & 1 & 3 & \vdots & 5 \\ 0 & -1 & -1 & \vdots & -1 \end{bmatrix}$$

Add the second row to the third row $(R_2 + R_3)$.

$$\begin{bmatrix} 1 & -2 & 3 & \vdots & 9 \\ 0 & 1 & 3 & \vdots & 5 \\ 0 & 0 & 2 & \vdots & 4 \end{bmatrix}$$

Multiply the third row by $\frac{1}{2}$ $\left(\frac{1}{2}R_3\right)$.

$$\begin{bmatrix} 1 & -2 & 3 & \vdots & 9 \\ 0 & 1 & 3 & \vdots & 5 \\ 0 & 0 & 1 & \vdots & 2 \end{bmatrix}$$

At this point, you can use back-substitution to find x and y.

$$y + 3(\;\;) = 5$$
$$y = -1$$
$$x - 2(\;\;) + 3(\;\;) = 9$$
$$x = 1$$

The solution is $x = 1$, $y = -1$, and $z = 2$.

CHECK *Point* Now try Exercise 39.

The last matrix in Example 4 is said to be in **row-echelon form.** The term *echelon* refers to the stair-step pattern formed by the nonzero elements of the matrix. To be in this form, a matrix must have the following properties.

Row-Echelon Form and Reduced Row-Echelon Form

A matrix in **row-echelon form** has the following properties.

1. Any rows consisting entirely of zeros occur at the bottom of the matrix.

2. For each row that does not consist entirely of zeros, the first nonzero entry is 1 (called a **leading 1**).

3. For two successive (nonzero) rows, the leading 1 in the higher row is farther to the left than the leading 1 in the lower row.

A matrix in *row-echelon form* is in **reduced row-echelon form** if every column that has a leading 1 has zeros in every position above and below its leading 1.

It is worth noting that the row-echelon form of a matrix is not unique. That is, two different sequences of elementary row operations may yield different row-echelon forms. However, the *reduced* row-echelon form of a given matrix is unique.

Example 5 Row-Echelon Form

Determine whether each matrix is in row-echelon form. If it is, determine whether the matrix is in reduced row-echelon form.

a. $\begin{bmatrix} 2 & -1 & 4 \\ & 0 & 3 \\ & & -2 \end{bmatrix}$

b. $\begin{bmatrix} 1 & 2 & -1 & 2 \\ 0 & 0 & 0 & 0 \\ 0 & 1 & 2 & -4 \end{bmatrix}$

c. $\begin{bmatrix} -5 & 2 & -1 & 3 \\ & & 3 & -2 \\ & & & 4 \\ & & & \end{bmatrix}$

d. $\begin{bmatrix} & 0 & 0 & -1 \\ & & 0 & 2 \\ & & & 3 \end{bmatrix}$

e. $\begin{bmatrix} 1 & 2 & -3 & 4 \\ 0 & 2 & 1 & -1 \\ 0 & 0 & 1 & -3 \end{bmatrix}$

f. $\begin{bmatrix} & 0 & 5 \\ & & 3 \\ & & \end{bmatrix}$

Solution

The matrices in (a), (c), (d), and (f) are in row-echelon form. The matrices in (d) and (f) are in *reduced* row-echelon form because every column that has a leading 1 has zeros in every position above and below its leading 1. The matrix in (b) is not in row-echelon form because a row of all zeros does not occur at the bottom of the matrix. The matrix in (e) is not in row-echelon form because the first nonzero entry in Row 2 is not a leading 1.

CHECK *Point* Now try Exercise 41.

Every matrix is row-equivalent to a matrix in row-echelon form. For instance, in Example 5, you can change the matrix in part (e) to row-echelon form by multiplying its second row by $\frac{1}{2}$.

Gaussian Elimination with Back-Substitution

Gaussian elimination with back-substitution works well for solving systems of linear equations by hand or with a computer. For this algorithm, the order in which the elementary row operations are performed is important. You should operate from left to right by columns, using elementary row operations to obtain zeros in all entries directly below the leading 1's.

| **Example 6** | **Gaussian Elimination with Back-Substitution** |

Solve the system
$$\begin{cases} y + z - 2w = -3 \\ x + 2y - z = 2 \\ 2x + 4y + z - 3w = -2 \\ x - 4y - 7z - w = -19 \end{cases}.$$

Solution

$$\begin{bmatrix} 0 & 1 & 1 & -2 & \vdots & -3 \\ 1 & 2 & -1 & 0 & \vdots & 2 \\ 2 & 4 & 1 & -3 & \vdots & -2 \\ 1 & -4 & -7 & -1 & \vdots & -19 \end{bmatrix}$$

$$\begin{bmatrix} & 2 & -1 & 0 & \vdots & 2 \\ 0 & 1 & 1 & -2 & \vdots & -3 \\ 2 & 4 & 1 & -3 & \vdots & -2 \\ 1 & -4 & -7 & -1 & \vdots & -19 \end{bmatrix}$$

$$\begin{bmatrix} & 2 & -1 & 0 & \vdots & 2 \\ & 1 & 1 & -2 & \vdots & -3 \\ & 0 & 3 & -3 & \vdots & -6 \\ & -6 & -6 & -1 & \vdots & -21 \end{bmatrix}$$

$$\begin{bmatrix} & 2 & -1 & 0 & \vdots & 2 \\ & & 1 & -2 & \vdots & -3 \\ & & 3 & -3 & \vdots & -6 \\ & & 0 & -13 & \vdots & -39 \end{bmatrix}$$

$$\begin{bmatrix} & 2 & -1 & 0 & \vdots & 2 \\ & & 1 & -2 & \vdots & -3 \\ & & & -1 & \vdots & -2 \\ & & & & \vdots & 3 \end{bmatrix}$$

The matrix is now in row-echelon form, and the corresponding system is

$$\begin{cases} x + 2y - z = 2 \\ y + z - 2w = -3 \\ z - w = -2 \\ w = 3 \end{cases}.$$

Using back-substitution, you can determine that the solution is $x = -1$, $y = 2$, $z = 1$, and $w = 3$.

CHECK *Point*   Now try Exercise 63.

The procedure for using Gaussian elimination with back-substitution is summarized below.

> ## Gaussian Elimination with Back-Substitution
>
> 1. Write the augmented matrix of the system of linear equations.
> 2. Use elementary row operations to rewrite the augmented matrix in row-echelon form.
> 3. Write the system of linear equations corresponding to the matrix in row-echelon form, and use back-substitution to find the solution.

When solving a system of linear equations, remember that it is possible for the system to have no solution. If, in the elimination process, you obtain a row of all zeros except for the last entry, it is unnecessary to continue the elimination process. You can simply conclude that the system has no solution, or is *inconsistent*.

Example 7   A System with No Solution

Solve the system $\begin{cases} x - y + 2z = 4 \\ x + z = 6 \\ 2x - 3y + 5z = 4 \\ 3x + 2y - z = 1 \end{cases}$.

Solution

$$\begin{bmatrix} 1 & -1 & 2 & \vdots & 4 \\ 1 & 0 & 1 & \vdots & 6 \\ 2 & -3 & 5 & \vdots & 4 \\ 3 & 2 & -1 & \vdots & 1 \end{bmatrix}$$

$$\begin{bmatrix} 1 & -1 & 2 & \vdots & 4 \\ 0 & 1 & -1 & \vdots & 2 \\ 0 & -1 & 1 & \vdots & -4 \\ 0 & 5 & -7 & \vdots & -11 \end{bmatrix}$$

$$\begin{bmatrix} 1 & -1 & 2 & \vdots & 4 \\ 0 & 1 & -1 & \vdots & 2 \\ 0 & 0 & 0 & \vdots & -2 \\ 0 & 5 & -7 & \vdots & -11 \end{bmatrix}$$

Note that the third row of this matrix consists entirely of zeros except for the last entry. This means that the original system of linear equations is inconsistent. You can see why this is true by converting back to a system of linear equations.

$$\begin{cases} x - y + 2z = 4 \\ y - z = 2 \\ 0 = -2 \\ 5y - 7z = -11 \end{cases}$$

Because the third equation is not possible, the system has no solution.

CHECK*Point*   Now try Exercise 81.

Gauss-Jordan Elimination

With Gaussian elimination, elementary row operations are applied to a matrix to obtain a (row-equivalent) row-echelon form of the matrix. A second method of elimination, called **Gauss-Jordan elimination,** after Carl Friedrich Gauss and Wilhelm Jordan (1842–1899), continues the reduction process until a *reduced* row-echelon form is obtained. This procedure is demonstrated in Example 8.

Example 8   Gauss-Jordan Elimination

Use Gauss-Jordan elimination to solve the system $\begin{cases} x - 2y + 3z = 9 \\ -x + 3y = -4. \\ 2x - 5y + 5z = 17 \end{cases}$

Solution

In Example 4, Gaussian elimination was used to obtain the row-echelon form of the linear system above.

$$\begin{bmatrix} 1 & -2 & 3 & \vdots & 9 \\ 0 & 1 & 3 & \vdots & 5 \\ 0 & 0 & 1 & \vdots & 2 \end{bmatrix}$$

Now, apply elementary row operations until you obtain zeros above each of the leading 1's, as follows.

$$\begin{bmatrix} 1 & & 9 & \vdots & 19 \\ 0 & 1 & 3 & \vdots & 5 \\ 0 & 0 & 1 & \vdots & 2 \end{bmatrix}$$

$$\begin{bmatrix} 1 & & & \vdots & 1 \\ 0 & 1 & & \vdots & -1 \\ 0 & 0 & 1 & \vdots & 2 \end{bmatrix}$$

The matrix is now in reduced row-echelon form. Converting back to a system of linear equations, you have

$$\begin{cases} x = 1 \\ y = -1. \\ z = 2 \end{cases}$$

Study Tip

The advantage of using Gauss-Jordan elimination to solve a system of linear equations is that the solution of the system is easily found without using back-substitution, as illustrated in Example 8.

Now you can simply read the solution, $x = 1$, $y = -1$, and $z = 2$, which can be written as the ordered triple $(1, -1, 2)$.

CHECK Point   Now try Exercise 71.

The elimination procedures described in this section sometimes result in fractional coefficients. For instance, in the elimination procedure for the system

$$\begin{cases} 2x - 5y + 5z = 17 \\ 3x - 2y + 3z = 11 \\ -3x + 3y = -6 \end{cases}$$

you may be inclined to multiply the first row by $\frac{1}{2}$ to produce a leading 1, which will result in working with fractional coefficients. You can sometimes avoid fractions by judiciously choosing the order in which you apply elementary row operations.

Recall from Chapter 6 that when there are fewer equations than variables in a system of equations, then the system has either no solution or infinitely many solutions.

Example 9 A System with an Infinite Number of Solutions

Solve the system.

$$\begin{cases} 2x + 4y - 2z = 0 \\ 3x + 5y \quad\quad = 1 \end{cases}$$

Solution

$$\begin{bmatrix} 2 & 4 & -2 & \vdots & 0 \\ 3 & 5 & 0 & \vdots & 1 \end{bmatrix}$$

$$\begin{bmatrix} 1 & 2 & -1 & \vdots & 0 \\ 3 & 5 & 0 & \vdots & 1 \end{bmatrix}$$

$$\begin{bmatrix} 1 & 2 & -1 & \vdots & 0 \\ 0 & -1 & 3 & \vdots & 1 \end{bmatrix}$$

$$\begin{bmatrix} 1 & 2 & -1 & \vdots & 0 \\ 0 & 1 & -3 & \vdots & -1 \end{bmatrix}$$

$$\begin{bmatrix} 1 & 0 & 5 & \vdots & 2 \\ 0 & 1 & -3 & \vdots & -1 \end{bmatrix}$$

The corresponding system of equations is

$$\begin{cases} x + 5z = 2 \\ y - 3z = -1 \end{cases}.$$

Solving for x and y in terms of z, you have

$$x = -5z + 2 \quad \text{and} \quad y = 3z - 1.$$

To write a solution of the system that does not use any of the three variables of the system, let a represent any real number and let

$$z = a.$$

Now substitute a for z in the equations for x and y.

$$x = -5z + 2 = -5 \quad + 2$$

$$y = 3z - 1 = 3 \quad - 1$$

So, the solution set can be written as an ordered triple with the form

$$(-5a + 2, 3a - 1, a)$$

where a is any real number. Remember that a solution set of this form represents an infinite number of solutions. Try substituting values for a to obtain a few solutions. Then check each solution in the original system of equations.

CHECK*Point* Now try Exercise 79.

Study Tip

In Example 9, x and y are solved for in terms of the third variable z. To write a solution of the system that does not use any of the three variables of the system, let a represent any real number and let $z = a$. Then solve for x and y. The solution can then be written in terms of a, which is not one of the variables of the system.

7.1   EXERCISES

See www.CalcChat.com for worked-out solutions to odd-numbered exercises.

VOCABULARY: Fill in the blanks.

1. A rectangular array of real numbers that can be used to solve a system of linear equations is called a _____.

2. A matrix is _____ if the number of rows equals the number of columns.

3. For a square matrix, the entries $a_{11}, a_{22}, a_{33}, \ldots, a_{nn}$ are the _____ _____ entries.

4. A matrix with only one row is called a _____ matrix, and a matrix with only one column is called a _____ matrix.

5. The matrix derived from a system of linear equations is called the _____ matrix of the system.

6. The matrix derived from the coefficients of a system of linear equations is called the _____ matrix of the system.

7. Two matrices are called _____ if one of the matrices can be obtained from the other by a sequence of elementary row operations.

8. A matrix in row-echelon form is in _____ _____ _____ if every column that has a leading 1 has zeros in every position above and below its leading 1.

SKILLS AND APPLICATIONS

In Exercises 9–14, determine the order of the matrix.

9. $\begin{bmatrix} 7 & 0 \end{bmatrix}$

10. $\begin{bmatrix} 5 & -3 & 8 & 7 \end{bmatrix}$

11. $\begin{bmatrix} 2 \\ 36 \\ 3 \end{bmatrix}$

12. $\begin{bmatrix} -3 & 7 & 15 & 0 \\ 0 & 0 & 3 & 3 \\ 1 & 1 & 6 & 7 \end{bmatrix}$

13. $\begin{bmatrix} 33 & 45 \\ -9 & 20 \end{bmatrix}$

14. $\begin{bmatrix} -7 & 6 & 4 \\ 0 & -5 & 1 \end{bmatrix}$

25. $\begin{bmatrix} 9 & 12 & 3 & 0 & \vdots & 0 \\ -2 & 18 & 5 & 2 & \vdots & 10 \\ 1 & 7 & -8 & 0 & \vdots & -4 \\ 3 & 0 & 2 & 0 & \vdots & -10 \end{bmatrix}$

26. $\begin{bmatrix} 6 & 2 & -1 & -5 & \vdots & -25 \\ -1 & 0 & 7 & 3 & \vdots & 7 \\ 4 & -1 & -10 & 6 & \vdots & 23 \\ 0 & 8 & 1 & -11 & \vdots & -21 \end{bmatrix}$

In Exercises 15–20, write the augmented matrix for the system of linear equations.

15. $\begin{cases} 4x - 3y = -5 \\ -x + 3y = 12 \end{cases}$

16. $\begin{cases} 7x + 4y = 22 \\ 5x - 9y = 15 \end{cases}$

17. $\begin{cases} x + 10y - 2z = 2 \\ 5x - 3y + 4z = 0 \\ 2x + y = 6 \end{cases}$

18. $\begin{cases} -x - 8y + 5z = 8 \\ -7x - 15z = -38 \\ 3x - y + 8z = 20 \end{cases}$

19. $\begin{cases} 7x - 5y + z = 13 \\ 19x - 8z = 10 \end{cases}$

20. $\begin{cases} 9x + 2y - 3z = 20 \\ -25y + 11z = -5 \end{cases}$

In Exercises 21–26, write the system of linear equations represented by the augmented matrix. (Use variables $x, y, z,$ and w, if applicable.)

21. $\begin{bmatrix} 1 & 2 & \vdots & 7 \\ 2 & -3 & \vdots & 4 \end{bmatrix}$

22. $\begin{bmatrix} 7 & -5 & \vdots & 0 \\ 8 & 3 & \vdots & -2 \end{bmatrix}$

23. $\begin{bmatrix} 2 & 0 & 5 & \vdots & -12 \\ 0 & 1 & -2 & \vdots & 7 \\ 6 & 3 & 0 & \vdots & 2 \end{bmatrix}$

24. $\begin{bmatrix} 4 & -5 & -1 & \vdots & 18 \\ -11 & 0 & 6 & \vdots & 25 \\ 3 & 8 & 0 & \vdots & -29 \end{bmatrix}$

In Exercises 27–34, fill in the blank(s) using elementary row operations to form a row-equivalent matrix.

27. $\begin{bmatrix} 1 & 4 & 3 \\ 2 & 10 & 5 \end{bmatrix}$
$\begin{bmatrix} 1 & 4 & 3 \\ 0 & & -1 \end{bmatrix}$

28. $\begin{bmatrix} 3 & 6 & 8 \\ 4 & -3 & 6 \end{bmatrix}$
$\begin{bmatrix} 1 & & \frac{8}{3} \\ 4 & -3 & 6 \end{bmatrix}$

29. $\begin{bmatrix} 1 & 1 & 1 \\ 5 & -2 & 4 \end{bmatrix}$
$\begin{bmatrix} 1 & 1 & 1 \\ 0 & & -1 \end{bmatrix}$

30. $\begin{bmatrix} -3 & 3 & 12 \\ 18 & -8 & 4 \end{bmatrix}$
$\begin{bmatrix} 1 & -1 & \\ 18 & -8 & 4 \end{bmatrix}$

31. $\begin{bmatrix} 1 & 5 & 4 & -1 \\ 0 & 1 & -2 & 2 \\ 0 & 0 & 1 & -7 \end{bmatrix}$
$\begin{bmatrix} 1 & 0 & & \\ 0 & 1 & -2 & 2 \\ 0 & 0 & 1 & -7 \end{bmatrix}$

32. $\begin{bmatrix} 1 & 0 & 6 & 1 \\ 0 & -1 & 0 & 7 \\ 0 & 0 & -1 & 3 \end{bmatrix}$
$\begin{bmatrix} 1 & 0 & 6 & 1 \\ 0 & 1 & 0 & \\ 0 & 0 & 1 & \end{bmatrix}$

33. $\begin{bmatrix} 1 & 1 & 4 & -1 \\ 3 & 8 & 10 & 3 \\ -2 & 1 & 12 & 6 \end{bmatrix}$ **34.** $\begin{bmatrix} 2 & 4 & 8 & 3 \\ 1 & -1 & -3 & 2 \\ 2 & 6 & 4 & 9 \end{bmatrix}$

$\begin{bmatrix} 1 & 1 & 4 & -1 \\ 0 & 5 & & \\ 0 & 3 & & \end{bmatrix}$ $\begin{bmatrix} 1 & & & \\ 1 & -1 & -3 & 2 \\ 2 & 6 & 4 & 9 \end{bmatrix}$

$\begin{bmatrix} 1 & 1 & 4 & -1 \\ 0 & 1 & -\frac{2}{5} & \frac{6}{5} \\ 0 & 3 & & \end{bmatrix}$ $\begin{bmatrix} 1 & 2 & 4 & \frac{3}{2} \\ 0 & & -7 & \frac{1}{2} \\ 0 & 2 & & \end{bmatrix}$

In Exercises 35–38, identify the elementary row operation(s) being performed to obtain the new row-equivalent matrix.

Original Matrix   *New Row-Equivalent Matrix*

35. $\begin{bmatrix} -2 & 5 & 1 \\ 3 & -1 & -8 \end{bmatrix}$ $\begin{bmatrix} 13 & 0 & -39 \\ 3 & -1 & -8 \end{bmatrix}$

Original Matrix   *New Row-Equivalent Matrix*

36. $\begin{bmatrix} 3 & -1 & -4 \\ -4 & 3 & 7 \end{bmatrix}$ $\begin{bmatrix} 3 & -1 & -4 \\ 5 & 0 & -5 \end{bmatrix}$

Original Matrix   *New Row-Equivalent Matrix*

37. $\begin{bmatrix} 0 & -1 & -5 & 5 \\ -1 & 3 & -7 & 6 \\ 4 & -5 & 1 & 3 \end{bmatrix}$ $\begin{bmatrix} -1 & 3 & -7 & 6 \\ 0 & -1 & -5 & 5 \\ 0 & 7 & -27 & 27 \end{bmatrix}$

Original Matrix   *New Row-Equivalent Matrix*

38. $\begin{bmatrix} -1 & -2 & 3 & -2 \\ 2 & -5 & 1 & -7 \\ 5 & 4 & -7 & 6 \end{bmatrix}$ $\begin{bmatrix} -1 & -2 & 3 & -2 \\ 0 & -9 & 7 & -11 \\ 0 & -6 & 8 & -4 \end{bmatrix}$

39. Perform the sequence of row operations on the matrix. What did the operations accomplish?

$\begin{bmatrix} 1 & 2 & 3 \\ 2 & -1 & -4 \\ 3 & 1 & -1 \end{bmatrix}$

  (a) Add -2 times R_1 to R_2.

  (b) Add -3 times R_1 to R_3.

  (c) Add -1 times R_2 to R_3.

  (d) Multiply R_2 by $-\frac{1}{5}$.

  (e) Add -2 times R_2 to R_1.

40. Perform the sequence of row operations on the matrix. What did the operations accomplish?

$\begin{bmatrix} 7 & 1 \\ 0 & 2 \\ -3 & 4 \\ 4 & 1 \end{bmatrix}$

  (a) Add R_3 to R_4.

  (b) Interchange R_1 and R_4.

  (c) Add 3 times R_1 to R_3.

  (d) Add -7 times R_1 to R_4.

  (e) Multiply R_2 by $\frac{1}{2}$.

  (f) Add the appropriate multiples of R_2 to $R_1, R_3,$ and R_4.

In Exercises 41–44, determine whether the matrix is in row-echelon form. If it is, determine if it is also in reduced row-echelon form.

41. $\begin{bmatrix} 1 & 0 & 0 & 0 \\ 0 & 1 & 1 & 5 \\ 0 & 0 & 0 & 0 \end{bmatrix}$ **42.** $\begin{bmatrix} 1 & 3 & 0 & 0 \\ 0 & 0 & 1 & 8 \\ 0 & 0 & 0 & 0 \end{bmatrix}$

43. $\begin{bmatrix} 1 & 0 & 0 & 1 \\ 0 & 1 & 0 & -1 \\ 0 & 0 & 0 & 2 \end{bmatrix}$ **44.** $\begin{bmatrix} 1 & 0 & 1 & 0 \\ 0 & 1 & 0 & 2 \\ 0 & 0 & 1 & 0 \end{bmatrix}$

In Exercises 45–48, write the matrix in row-echelon form. (Remember that the row-echelon form of a matrix is not unique.)

45. $\begin{bmatrix} 1 & 1 & 0 & 5 \\ -2 & -1 & 2 & -10 \\ 3 & 6 & 7 & 14 \end{bmatrix}$ **46.** $\begin{bmatrix} 1 & 2 & -1 & 3 \\ 3 & 7 & -5 & 14 \\ -2 & -1 & -3 & 8 \end{bmatrix}$

47. $\begin{bmatrix} 1 & -1 & -1 & 1 \\ 5 & -4 & 1 & 8 \\ -6 & 8 & 18 & 0 \end{bmatrix}$ **48.** $\begin{bmatrix} 1 & -3 & 0 & -7 \\ -3 & 10 & 1 & 23 \\ 4 & -10 & 2 & -24 \end{bmatrix}$

 In Exercises 49–54, use the matrix capabilities of a graphing utility to write the matrix in *reduced* row-echelon form.

49. $\begin{bmatrix} 3 & 3 & 3 \\ -1 & 0 & -4 \\ 2 & 4 & -2 \end{bmatrix}$ **50.** $\begin{bmatrix} 1 & 3 & 2 \\ 5 & 15 & 9 \\ 2 & 6 & 10 \end{bmatrix}$

51. $\begin{bmatrix} 1 & 2 & 3 & -5 \\ 1 & 2 & 4 & -9 \\ -2 & -4 & -4 & 3 \\ 4 & 8 & 11 & -14 \end{bmatrix}$

52. $\begin{bmatrix} -2 & 3 & -1 & -2 \\ 4 & -2 & 5 & 8 \\ 1 & 5 & -2 & 0 \\ 3 & 8 & -10 & -30 \end{bmatrix}$

53. $\begin{bmatrix} -3 & 5 & 1 & 12 \\ 1 & -1 & 1 & 4 \end{bmatrix}$ **54.** $\begin{bmatrix} 5 & 1 & 2 & 4 \\ -1 & 5 & 10 & -32 \end{bmatrix}$

In Exercises 55–58, write the system of linear equations represented by the augmented matrix. Then use back-substitution to solve. (Use variables $x, y,$ and z, if applicable.)

55. $\begin{bmatrix} 1 & -2 & \vdots & 4 \\ 0 & 1 & \vdots & -3 \end{bmatrix}$ **56.** $\begin{bmatrix} 1 & 5 & \vdots & 0 \\ 0 & 1 & \vdots & -1 \end{bmatrix}$

57. $\begin{bmatrix} 1 & -1 & 2 & \vdots & 4 \\ 0 & 1 & -1 & \vdots & 2 \\ 0 & 0 & 1 & \vdots & -2 \end{bmatrix}$ **58.** $\begin{bmatrix} 1 & 2 & -2 & \vdots & -1 \\ 0 & 1 & 1 & \vdots & 9 \\ 0 & 0 & 1 & \vdots & -3 \end{bmatrix}$

In Exercises 59–62, an augmented matrix that represents a system of linear equations (in variables x, y, and z, if applicable) has been reduced using Gauss-Jordan elimination. Write the solution represented by the augmented matrix.

59. $\begin{bmatrix} 1 & 0 & \vdots & 3 \\ 0 & 1 & \vdots & -4 \end{bmatrix}$ **60.** $\begin{bmatrix} 1 & 0 & \vdots & -6 \\ 0 & 1 & \vdots & 10 \end{bmatrix}$

61. $\begin{bmatrix} 1 & 0 & 0 & \vdots & -4 \\ 0 & 1 & 0 & \vdots & -10 \\ 0 & 0 & 1 & \vdots & 4 \end{bmatrix}$ **62.** $\begin{bmatrix} 1 & 0 & 0 & \vdots & 5 \\ 0 & 1 & 0 & \vdots & -3 \\ 0 & 0 & 1 & \vdots & 0 \end{bmatrix}$

In Exercises 63–84, use matrices to solve the system of equations (if possible). Use Gaussian elimination with back-substitution or Gauss-Jordan elimination.

63. $\begin{cases} x + 2y = 7 \\ 2x + y = 8 \end{cases}$ **64.** $\begin{cases} 2x + 6y = 16 \\ 2x + 3y = 7 \end{cases}$

65. $\begin{cases} 3x - 2y = -27 \\ x + 3y = 13 \end{cases}$ **66.** $\begin{cases} -x + y = 4 \\ 2x - 4y = -34 \end{cases}$

67. $\begin{cases} -2x + 6y = -22 \\ x + 2y = -9 \end{cases}$ **68.** $\begin{cases} 5x - 5y = -5 \\ -2x - 3y = 7 \end{cases}$

69. $\begin{cases} 8x - 4y = 7 \\ 5x + 2y = 1 \end{cases}$ **70.** $\begin{cases} x - 3y = 5 \\ -2x + 6y = -10 \end{cases}$

71. $\begin{cases} x - 3z = -2 \\ 3x + y - 2z = 5 \\ 2x + 2y + z = 4 \end{cases}$ **72.** $\begin{cases} 2x - y + 3z = 24 \\ 2y - z = 14 \\ 7x - 5y = 6 \end{cases}$

73. $\begin{cases} -x + y - z = -14 \\ 2x - y + z = 21 \\ 3x + 2y + z = 19 \end{cases}$ **74.** $\begin{cases} 2x + 2y - z = 2 \\ x - 3y + z = -28 \\ -x + y = 14 \end{cases}$

75. $\begin{cases} x + 2y - 3z = -28 \\ 4y + 2z = 0 \\ -x + y - z = -5 \end{cases}$ **76.** $\begin{cases} 3x - 2y + z = 15 \\ -x + y + 2z = -10 \\ x - y - 4z = 14 \end{cases}$

77. $\begin{cases} x + 2y = 0 \\ -x - y = 0 \end{cases}$ **78.** $\begin{cases} x + 2y = 0 \\ 2x + 4y = 0 \end{cases}$

79. $\begin{cases} x + 2y + z = 8 \\ 3x + 7y + 6z = 26 \end{cases}$ **80.** $\begin{cases} x + y + 4z = 5 \\ 2x + y - z = 9 \end{cases}$

81. $\begin{cases} -x + y = -22 \\ 3x + 4y = 4 \\ 4x - 8y = 32 \end{cases}$ **82.** $\begin{cases} x + 2y = 0 \\ x + y = 6 \\ 3x - 2y = 8 \end{cases}$

83. $\begin{cases} 3x + 2y - z + w = 0 \\ x - y + 4z + 2w = 25 \\ -2x + y + 2z - w = 2 \\ x + y + z + w = 6 \end{cases}$

84. $\begin{cases} x - 4y + 3z - 2w = 9 \\ 3x - 2y + z - 4w = -13 \\ -4x + 3y - 2z + w = -4 \\ -2x + y - 4z + 3w = -10 \end{cases}$

In Exercises 85–90, use the matrix capabilities of a graphing utility to reduce the augmented matrix corresponding to the system of equations, and solve the system.

85. $\begin{cases} 3x + 3y + 12z = 6 \\ x + y + 4z = 2 \\ 2x + 5y + 20z = 10 \\ -x + 2y + 8z = 4 \end{cases}$ **86.** $\begin{cases} 2x + 10y + 2z = 6 \\ x + 5y + 2z = 6 \\ x + 5y + z = 3 \\ -3x - 15y - 3z = -9 \end{cases}$

87. $\begin{cases} 2x + y - z + 2w = -6 \\ 3x + 4y + w = 1 \\ x + 5y + 2z + 6w = -3 \\ 5x + 2y - z - w = 3 \end{cases}$

88. $\begin{cases} x + 2y + 2z + 4w = 11 \\ 3x + 6y + 5z + 12w = 30 \\ x + 3y - 3z + 2w = -5 \\ 6x - y - z + w = -9 \end{cases}$

89. $\begin{cases} x + y + z + w = 0 \\ 2x + 3y + z - 2w = 0 \\ 3x + 5y + z = 0 \end{cases}$

90. $\begin{cases} x + 2y + z + 3w = 0 \\ x - y + w = 0 \\ y - z + 2w = 0 \end{cases}$

In Exercises 91–94, determine whether the two systems of linear equations yield the same solution. If so, find the solution using matrices.

91. (a) $\begin{cases} x - 2y + z = -6 \\ y - 5z = 16 \\ z = -3 \end{cases}$ (b) $\begin{cases} x + y - 2z = 6 \\ y + 3z = -8 \\ z = -3 \end{cases}$

92. (a) $\begin{cases} x - 3y + 4z = -11 \\ y - z = -4 \\ z = 2 \end{cases}$ (b) $\begin{cases} x + 4y = -11 \\ y + 3z = 4 \\ z = 2 \end{cases}$

93. (a) $\begin{cases} x - 4y + 5z = 27 \\ y - 7z = -54 \\ z = 8 \end{cases}$ (b) $\begin{cases} x - 6y + z = 15 \\ y + 5z = 42 \\ z = 8 \end{cases}$

94. (a) $\begin{cases} x + 3y - z = 19 \\ y + 6z = -18 \\ z = -4 \end{cases}$ (b) $\begin{cases} x - y + 3z = -15 \\ y - 2z = 14 \\ z = -4 \end{cases}$

In Exercises 95–98, use a system of equations to find the quadratic function $f(x) = ax^2 + bx + c$ that satisfies the equations. Solve the system using matrices.

95. $f(1) = 1, f(2) = -1, f(3) = -5$

96. $f(1) = 2, f(2) = 9, f(3) = 20$

97. $f(-2) = -15, f(-1) = 7, f(1) = -3$

98. $f(-2) = -3, f(1) = -3, f(2) = -11$

In Exercises 99–102, use a system of equations to find the cubic function $f(x) = ax^3 + bx^2 + cx + d$ that satisfies the equations. Solve the system using matrices.

99. $f(-1) = -5$
   $f(1) = -1$
   $f(2) = 1$
   $f(3) = 11$

100. $f(-1) = 4$
   $f(1) = 4$
   $f(2) = 16$
   $f(3) = 44$

101. $f(-2) = -7$
   $f(-1) = 2$
   $f(1) = -4$
   $f(2) = -7$

102. $f(-2) = -17$
   $f(-1) = -5$
   $f(1) = 1$
   $f(2) = 7$

103. Use the system

$$\begin{cases} x + 3y + z = 3 \\ x + 5y + 5z = 1 \\ 2x + 6y + 3z = 8 \end{cases}$$

to write two different matrices in row-echelon form that yield the same solution.

104. ELECTRICAL NETWORK The currents in an electrical network are given by the solution of the system

$$\begin{cases} I_1 - I_2 + I_3 = 0 \\ 3I_1 + 4I_2 = 18 \\ I_2 + 3I_3 = 6 \end{cases}$$

where I_1, I_2, and I_3 are measured in amperes. Solve the system of equations using matrices.

105. PARTIAL FRACTIONS Use a system of equations to write the partial fraction decomposition of the rational expression. Solve the system using matrices.

$$\frac{4x^2}{(x+1)^2(x-1)} = \frac{A}{x-1} + \frac{B}{x+1} + \frac{C}{(x+1)^2}$$

106. PARTIAL FRACTIONS Use a system of equations to write the partial fraction decomposition of the rational expression. Solve the system using matrices.

$$\frac{8x^2}{(x-1)^2(x+1)} = \frac{A}{x+1} + \frac{B}{x-1} + \frac{C}{(x-1)^2}$$

107. FINANCE A small shoe corporation borrowed $1,500,000 to expand its line of shoes. Some of the money was borrowed at 7%, some at 8%, and some at 10%. Use a system of equations to determine how much was borrowed at each rate if the annual interest was $130,500 and the amount borrowed at 10% was 4 times the amount borrowed at 7%. Solve the system using matrices.

108. FINANCE A small software corporation borrowed $500,000 to expand its software line. Some of the money was borrowed at 9%, some at 10%, and some at 12%. Use a system of equations to determine how much was borrowed at each rate if the annual interest was $52,000 and the amount borrowed at 10% was $2\frac{1}{2}$ times the amount borrowed at 9%. Solve the system using matrices.

109. TIPS A food server examines the amount of money earned in tips after working an 8-hour shift. The server has a total of $95 in denominations of $1, $5, $10, and $20 bills. The total number of paper bills is 26. The number of $5 bills is 4 times the number of $10 bills, and the number of $1 bills is 1 less than twice the number of $5 bills. Write a system of linear equations to represent the situation. Then use matrices to find the number of each denomination.

110. BANKING A bank teller is counting the total amount of money in a cash drawer at the end of a shift. There is a total of $2600 in denominations of $1, $5, $10, and $20 bills The total number of paper bills is 235. The number of $20 bills is twice the number of $1 bills, and the number of $5 bills is 10 more than the number of $1 bills. Write a system of linear equations to represent the situation. Then use matrices to find the number of each denomination.

In Exercises 111 and 112, use a system of equations to find the equation of the parabola $y = ax^2 + bx + c$ that passes through the points. Solve the system using matrices. Use a graphing utility to verify your results.

111.

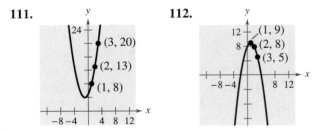

112.

113. MATHEMATICAL MODELING A video of the path of a ball thrown by a baseball player was analyzed with a grid covering the TV screen. The tape was paused three times, and the position of the ball was measured each time. The coordinates obtained are shown in the table. (x and y are measured in feet.)

Horizontal distance, x	Height, y
0	5.0
15	9.6
30	12.4

(a) Use a system of equations to find the equation of the parabola $y = ax^2 + bx + c$ that passes through the three points. Solve the system using matrices.

(b) Use a graphing utility to graph the parabola.

(c) Graphically approximate the maximum height of the ball and the point at which the ball struck the ground.

(d) Analytically find the maximum height of the ball and the point at which the ball struck the ground.

(e) Compare your results from parts (c) and (d).

114. DATA ANALYSIS: SNOWBOARDERS The table shows the numbers of people y (in millions) in the United States who participated in snowboarding in selected years from 2003 to 2007. (Source: National Sporting Goods Association)

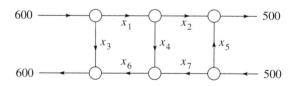

Year	Number, y
2003	6.3
2005	6.0
2007	5.1

(a) Use a system of equations to find the equation of the parabola $y = at^2 + bt + c$ that passes through the points. Let t represent the year, with $t = 3$ corresponding to 2003. Solve the system using matrices.

 (b) Use a graphing utility to graph the parabola.

(c) Use the equation in part (a) to estimate the number of people who participated in snowboarding in 2009. Does your answer seem reasonable? Explain.

(d) Do you believe that the equation can be used for years far beyond 2007? Explain.

NETWORK ANALYSIS In Exercises 115 and 116, answer the questions about the specified network. (In a network it is assumed that the total flow into each junction is equal to the total flow out of each junction.)

115. Water flowing through a network of pipes (in thousands of cubic meters per hour) is shown in the figure.

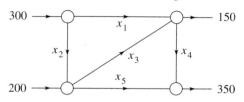

(a) Solve this system using matrices for the water flow represented by x_i, $i = 1, 2, \ldots, 7$.

(b) Find the network flow pattern when $x_6 = 0$ and $x_7 = 0$.

(c) Find the network flow pattern when $x_5 = 400$ and $x_6 = 500$.

116. The flow of traffic (in vehicles per hour) through a network of streets is shown in the figure.

(a) Solve this system using matrices for the traffic flow represented by x_i, $i = 1, 2, \ldots, 5$.

(b) Find the traffic flow when $x_2 = 200$ and $x_3 = 50$.

(c) Find the traffic flow when $x_2 = 150$ and $x_3 = 0$.

EXPLORATION

TRUE OR FALSE? In Exercises 117 and 118, determine whether the statement is true or false. Justify your answer.

117. $\begin{bmatrix} 5 & 0 & -2 & 7 \\ -1 & 3 & -6 & 0 \end{bmatrix}$ is a 4×2 matrix.

118. The method of Gaussian elimination reduces a matrix until a reduced row-echelon form is obtained.

119. THINK ABOUT IT The augmented matrix below represents system of linear equations (in variables x, y, and z) that has been reduced using Gauss-Jordan elimination. Write a system of equations with nonzero coefficients that is represented by the reduced matrix. (There are many correct answers.)

$$\begin{bmatrix} 1 & 0 & 3 & \vdots & -2 \\ 0 & 1 & 4 & \vdots & 1 \\ 0 & 0 & 0 & \vdots & 0 \end{bmatrix}$$

120. THINK ABOUT IT

(a) Describe the row-echelon form of an augmented matrix that corresponds to a system of linear equations that is inconsistent.

(b) Describe the row-echelon form of an augmented matrix that corresponds to a system of linear equations that has an infinite number of solutions.

121. Describe the three elementary row operations that can be performed on an augmented matrix.

122. CAPSTONE In your own words, describe the difference between a matrix in row-echelon form and a matrix in reduced row-echelon form. Include an example of each to support your explanation.

123. What is the relationship between the three elementary row operations performed on an augmented matrix and the operations that lead to equivalent systems of equations?

7.2 OPERATIONS WITH MATRICES

What you should learn

- Decide whether two matrices are equal.
- Add and subtract matrices and multiply matrices by scalars.
- Multiply two matrices.
- Use matrix operations to model and solve real-life problems.

Why you should learn it

Matrix operations can be used to model and solve real-life problems. For instance, in Exercise 76 on page 548, matrix operations are used to analyze annual health care costs.

Equality of Matrices

In Section 7.1, you used matrices to solve systems of linear equations. There is a rich mathematical theory of matrices, and its applications are numerous. This section and the next two introduce some fundamentals of matrix theory. It is standard mathematical convention to represent matrices in any of the following three ways.

Representation of Matrices

1. A matrix can be denoted by an uppercase letter such as A, B, or C.

2. A matrix can be denoted by a representative element enclosed in brackets, such as $[a_{ij}]$, $[b_{ij}]$, or $[c_{ij}]$.

3. A matrix can be denoted by a rectangular array of numbers such as

$$A = [a_{ij}] = \begin{bmatrix} a_{11} & a_{12} & a_{13} & \cdots & a_{1n} \\ a_{21} & a_{22} & a_{23} & \cdots & a_{2n} \\ a_{31} & a_{32} & a_{33} & \cdots & a_{3n} \\ \vdots & \vdots & \vdots & & \vdots \\ a_{m1} & a_{m2} & a_{m3} & \cdots & a_{mn} \end{bmatrix}.$$

Two matrices $A = [a_{ij}]$ and $B = [b_{ij}]$ are **equal** if they have the same order $(m \times n)$ and $a_{ij} = b_{ij}$ for $1 \le i \le m$ and $1 \le j \le n$. In other words, two matrices are equal if their corresponding entries are equal.

Example 1   Equality of Matrices

Solve for $a_{11}, a_{12}, a_{21},$ and a_{22} in the following matrix equation.

$$\begin{bmatrix} a_{11} & a_{12} \\ a_{21} & a_{22} \end{bmatrix} = \begin{bmatrix} 2 & -1 \\ -3 & 0 \end{bmatrix}$$

Solution

Because two matrices are equal only if their corresponding entries are equal, you can conclude that

$$a_{11} = 2, \quad a_{12} = -1, \quad a_{21} = -3, \quad \text{and} \quad a_{22} = 0.$$

CHECK Point Now try Exercise 7.

Be sure you see that for two matrices to be equal, they must have the same order *and* their corresponding entries must be equal. For instance,

$$\begin{bmatrix} 2 & -1 \\ \sqrt{4} & \frac{1}{2} \end{bmatrix} = \begin{bmatrix} 2 & -1 \\ 2 & 0.5 \end{bmatrix} \quad \text{but} \quad \begin{bmatrix} 2 & -1 \\ 3 & 4 \\ 0 & 0 \end{bmatrix} \neq \begin{bmatrix} 2 & -1 \\ 3 & 4 \end{bmatrix}.$$

Matrix Addition and Scalar Multiplication

In this section, three basic matrix operations will be covered. The first two are matrix addition and scalar multiplication. With matrix addition, you can add two matrices (of the same order) by adding their corresponding entries.

Definition of Matrix Addition

If $A = [a_{ij}]$ and $B = [b_{ij}]$ are matrices of order $m \times n$, their sum is the $m \times n$ matrix given by

$$A + B = [a_{ij} + b_{ij}].$$

The sum of two matrices of different orders is undefined.

Example 2   Addition of Matrices

a. $\begin{bmatrix} -1 & 2 \\ 0 & 1 \end{bmatrix} + \begin{bmatrix} 1 & 3 \\ -1 & 2 \end{bmatrix} = \begin{bmatrix} -1+1 & 2+3 \\ 0+(-1) & 1+2 \end{bmatrix}$

$$= \begin{bmatrix} 0 & 5 \\ -1 & 3 \end{bmatrix}$$

b. $\begin{bmatrix} 0 & 1 & -2 \\ 1 & 2 & 3 \end{bmatrix} + \begin{bmatrix} 0 & 0 & 0 \\ 0 & 0 & 0 \end{bmatrix} = \begin{bmatrix} 0 & 1 & -2 \\ 1 & 2 & 3 \end{bmatrix}$

c. $\begin{bmatrix} 1 \\ -3 \\ -2 \end{bmatrix} + \begin{bmatrix} -1 \\ 3 \\ 2 \end{bmatrix} = \begin{bmatrix} 0 \\ 0 \\ 0 \end{bmatrix}$

d. The sum of

$$A = \begin{bmatrix} 2 & 1 & 0 \\ 4 & 0 & -1 \\ 3 & -2 & 2 \end{bmatrix} \qquad \text{and}$$

$$B = \begin{bmatrix} 0 & 1 \\ -1 & 3 \\ 2 & 4 \end{bmatrix}$$

is undefined because A is of order 3×3 and B is of order 3×2.

CHECK Point   Now try Exercise 13(a).

In operations with matrices, numbers are usually referred to as **scalars.** In this text, scalars will always be real numbers. You can multiply a matrix A by a scalar c by multiplying each entry in A by c.

Definition of Scalar Multiplication

If $A = [a_{ij}]$ is an $m \times n$ matrix and c is a scalar, the **scalar multiple** of A by c is the $m \times n$ matrix given by

$$cA = [ca_{ij}].$$

The symbol $-A$ represents the negation of A, which is the scalar product $(-1)A$. Moreover, if A and B are of the same order, then $A - B$ represents the sum of A and $(-1)B$. That is,

$$A - B = A + (-1)B.$$

The order of operations for matrix expressions is similar to that for real numbers. In particular, you perform scalar multiplication before matrix addition and subtraction, as shown in Example 3(c).

Example 3   Scalar Multiplication and Matrix Subtraction

For the following matrices, find (a) $3A$, (b) $-B$, and (c) $3A - B$.

$$A = \begin{bmatrix} 2 & 2 & 4 \\ -3 & 0 & -1 \\ 2 & 1 & 2 \end{bmatrix} \quad \text{and} \quad B = \begin{bmatrix} 2 & 0 & 0 \\ 1 & -4 & 3 \\ -1 & 3 & 2 \end{bmatrix}$$

Solution

a. $3A = 3\begin{bmatrix} 2 & 2 & 4 \\ -3 & 0 & -1 \\ 2 & 1 & 2 \end{bmatrix}$

$$= \begin{bmatrix} (2) & (2) & (4) \\ (-3) & (0) & (-1) \\ (2) & (1) & (2) \end{bmatrix}$$

$$= \begin{bmatrix} 6 & 6 & 12 \\ -9 & 0 & -3 \\ 6 & 3 & 6 \end{bmatrix}$$

b. $-B = (-1)\begin{bmatrix} 2 & 0 & 0 \\ 1 & -4 & 3 \\ -1 & 3 & 2 \end{bmatrix}$

$$= \begin{bmatrix} -2 & 0 & 0 \\ -1 & 4 & -3 \\ 1 & -3 & -2 \end{bmatrix}$$

c. $3A - B = \begin{bmatrix} 6 & 6 & 12 \\ -9 & 0 & -3 \\ 6 & 3 & 6 \end{bmatrix} - \begin{bmatrix} 2 & 0 & 0 \\ 1 & -4 & 3 \\ -1 & 3 & 2 \end{bmatrix}$

$$= \begin{bmatrix} 4 & 6 & 12 \\ -10 & 4 & -6 \\ 7 & 0 & 4 \end{bmatrix}$$

CHECK Point   Now try Exercise 13(b), (c), and (d).

It is often convenient to rewrite the scalar multiple cA by factoring c out of every entry in the matrix. For instance, in the following example, the scalar $\frac{1}{2}$ has been factored out of the matrix.

$$\begin{bmatrix} \frac{1}{2} & -\frac{3}{2} \\ \frac{5}{2} & \frac{1}{2} \end{bmatrix} = \begin{bmatrix} \frac{1}{2}(1) & \frac{1}{2}(-3) \\ \frac{1}{2}(5) & \frac{1}{2}(1) \end{bmatrix} = \frac{1}{2}\begin{bmatrix} 1 & -3 \\ 5 & 1 \end{bmatrix}$$

The properties of matrix addition and scalar multiplication are similar to those of addition and multiplication of real numbers.

Properties of Matrix Addition and Scalar Multiplication

Let A, B, and C be $m \times n$ matrices and let c and d be scalars.

1. $A + B = B + A$

2. $A + (B + C) = (A + B) + C$

3. $(cd)A = c(dA)$

4. $1A = A$

5. $c(A + B) = cA + cB$

6. $(c + d)A = cA + dA$

Note that the Associative Property of Matrix Addition allows you to write expressions such as $A + B + C$ without ambiguity because the same sum occurs no matter how the matrices are grouped. This same reasoning applies to sums of four or more matrices.

Example 4   **Addition of More than Two Matrices**

By adding corresponding entries, you obtain the following sum of four matrices.

$$\begin{bmatrix} 1 \\ 2 \\ -3 \end{bmatrix} + \begin{bmatrix} -1 \\ -1 \\ 2 \end{bmatrix} + \begin{bmatrix} 0 \\ 1 \\ 4 \end{bmatrix} + \begin{bmatrix} 2 \\ -3 \\ -2 \end{bmatrix} = \begin{bmatrix} 2 \\ -1 \\ 1 \end{bmatrix}$$

CHECK*Point*   Now try Exercise 19.

Example 5   **Using the Distributive Property**

Perform the indicated matrix operations.

$$3\left(\begin{bmatrix} -2 & 0 \\ 4 & 1 \end{bmatrix} + \begin{bmatrix} 4 & -2 \\ 3 & 7 \end{bmatrix} \right)$$

Solution

$$3\left(\begin{bmatrix} -2 & 0 \\ 4 & 1 \end{bmatrix} + \begin{bmatrix} 4 & -2 \\ 3 & 7 \end{bmatrix} \right) = 3\begin{bmatrix} -2 & 0 \\ 4 & 1 \end{bmatrix} + 3\begin{bmatrix} 4 & -2 \\ 3 & 7 \end{bmatrix}$$

$$= \begin{bmatrix} -6 & 0 \\ 12 & 3 \end{bmatrix} + \begin{bmatrix} 12 & -6 \\ 9 & 21 \end{bmatrix}$$

$$= \begin{bmatrix} 6 & -6 \\ 21 & 24 \end{bmatrix}$$

CHECK*Point*   Now try Exercise 21.   ∎

In Example 5, you could add the two matrices first and then multiply the matrix by 3, as follows. Notice that you obtain the same result.

$$3\left(\begin{bmatrix} -2 & 0 \\ 4 & 1 \end{bmatrix} + \begin{bmatrix} 4 & -2 \\ 3 & 7 \end{bmatrix} \right) = 3\begin{bmatrix} 2 & -2 \\ 7 & 8 \end{bmatrix} = \begin{bmatrix} 6 & -6 \\ 21 & 24 \end{bmatrix}$$

One important property of addition of real numbers is that the number 0 is the additive identity. That is, $c + 0 = c$ for any real number c. For matrices, a similar property holds. That is, if A is an $m \times n$ matrix and O is the $m \times n$ **zero matrix** consisting entirely of zeros, then $A + O = A$.

In other words, O is the **additive identity** for the set of all $m \times n$ matrices. For example, the following matrices are the additive identities for the sets of all 2×3 and 2×2 matrices.

$$O = \begin{bmatrix} 0 & 0 & 0 \\ 0 & 0 & 0 \end{bmatrix} \quad \text{and} \quad O = \begin{bmatrix} 0 & 0 \\ 0 & 0 \end{bmatrix}$$

The algebra of real numbers and the algebra of matrices have many similarities. For example, compare the following solutions.

Real Numbers (Solve for x.)	$m \times n$ *Matrices* (Solve for X.)
$x + a = b$	$X + A = B$
$x + a + (-a) = b + (-a)$	$X + A + (-A) = B + (-A)$
$x + 0 = b - a$	$X + O = B - A$
$x = b - a$	$X = B - A$

The algebra of real numbers and the algebra of matrices also have important differences, which will be discussed later.

> **WARNING / CAUTION**
>
> Remember that matrices are denoted by capital letters. So, when you solve for X, you are solving for a *matrix* that makes the matrix equation true.

Example 6 Solving a Matrix Equation

Solve for X in the equation $3X + A = B$, where

$$A = \begin{bmatrix} 1 & -2 \\ 0 & 3 \end{bmatrix} \quad \text{and} \quad B = \begin{bmatrix} -3 & 4 \\ 2 & 1 \end{bmatrix}.$$

Solution

Begin by solving the matrix equation for X to obtain

$$3X = B - A$$

$$X = \frac{1}{3}(B - A).$$

Now, using the matrices A and B, you have

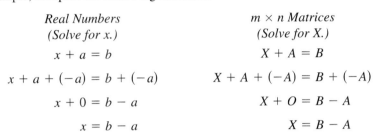

$$X = \frac{1}{3}\left(\qquad - \qquad \right)$$

$$= \frac{1}{3}\begin{bmatrix} -4 & 6 \\ 2 & -2 \end{bmatrix}$$

$$= \begin{bmatrix} -\frac{4}{3} & 2 \\ \frac{2}{3} & -\frac{2}{3} \end{bmatrix}.$$

CHECK*Point* Now try Exercise 31.

Matrix Multiplication

Another basic matrix operation is **matrix multiplication.** At first glance, the definition may seem unusual. You will see later, however, that this definition of the product of two matrices has many practical applications.

Definition of Matrix Multiplication

If $A = [a_{ij}]$ is an $m \times n$ matrix and $B = [b_{ij}]$ is an $n \times p$ matrix, the product AB is an $m \times p$ matrix

$$AB = [c_{ij}]$$

where $c_{ij} = a_{i1}b_{1j} + a_{i2}b_{2j} + a_{i3}b_{3j} + \cdots + a_{in}b_{nj}.$

The definition of matrix multiplication indicates a *row-by-column* multiplication, where the entry in the *i*th row and *j*th column of the product AB is obtained by multiplying the entries in the *i*th row of A by the corresponding entries in the *j*th column of B and then adding the results. So for the product of two matrices to be defined, the number of columns of the first matrix must equal the number of rows of the second matrix. The general pattern for matrix multiplication is as follows.

$$\begin{bmatrix} a_{11} & a_{12} & a_{13} & \cdots & a_{1n} \\ a_{21} & a_{22} & a_{23} & \cdots & a_{2n} \\ a_{31} & a_{32} & a_{33} & \cdots & a_{3n} \\ \vdots & \vdots & \vdots & & \vdots \\ \vdots & \vdots & \vdots & & \vdots \\ a_{m1} & a_{m2} & a_{m3} & \cdots & a_{mn} \end{bmatrix} \begin{bmatrix} b_{11} & b_{12} & \cdots & & \cdots & b_{1p} \\ b_{21} & b_{22} & \cdots & & \cdots & b_{2p} \\ b_{31} & b_{32} & \cdots & & \cdots & b_{3p} \\ \vdots & \vdots & & & & \vdots \\ b_{n1} & b_{n2} & \cdots & & \cdots & b_{np} \end{bmatrix} = \begin{bmatrix} c_{11} & c_{12} & \cdots & c_{1j} & \cdots & c_{1p} \\ c_{21} & c_{22} & \cdots & c_{2j} & \cdots & c_{2p} \\ \vdots & \vdots & & \vdots & & \vdots \\ c_{i1} & c_{i2} & \cdots & c_{ij} & \cdots & c_{ip} \\ \vdots & \vdots & & \vdots & & \vdots \\ c_{m1} & c_{m2} & \cdots & c_{mj} & \cdots & c_{mp} \end{bmatrix}$$

Example 7   Finding the Product of Two Matrices

Find the product AB using $A = \begin{bmatrix} -1 & 3 \\ 4 & -2 \\ 5 & 0 \end{bmatrix}$ and $B = \begin{bmatrix} -3 & 2 \\ -4 & 1 \end{bmatrix}.$

Solution

To find the entries of the product, multiply each row of A by each column of B.

$$AB = \begin{bmatrix} -1 & 3 \\ 4 & -2 \\ 5 & 0 \end{bmatrix} \begin{bmatrix} -3 \\ -4 \end{bmatrix}$$

$$= \begin{bmatrix} (-1)(-3) + (3)(-4) & (-1)(\ ) + (3)(\ ) \\ (4)(-3) + (-2)(-4) & (4)(\ ) + (-2)(\ ) \\ (5)(-3) + (0)(-4) & (5)(\ ) + (0)(\ ) \end{bmatrix}$$

$$= \begin{bmatrix} -9 & 1 \\ -4 & 6 \\ -15 & 10 \end{bmatrix}$$

CHECK *Point*   Now try Exercise 35.

Study Tip

In Example 7, the product AB is defined because the number of columns of A is equal to the number of rows of B. Also, note that the product AB has order 3×2.

Be sure you understand that for the product of two matrices to be defined, the number of *columns* of the first matrix must equal the number of *rows* of the second matrix. That is, the middle two indices must be the same. The outside two indices give the order of the product, as shown below.

$$A \quad \times \quad B \quad = \quad AB$$

Example 8 Finding the Product of Two Matrices

Find the product AB where

$$A = \begin{bmatrix} 1 & 0 & 3 \\ 2 & -1 & -2 \end{bmatrix} \quad \text{and} \quad B = \begin{bmatrix} -2 & 4 \\ 1 & 0 \\ -1 & 1 \end{bmatrix}.$$

Solution

Note that the order of A is 2×3 and the order of B is 3×2. So, the product AB has order 2×2.

$$AB = \begin{bmatrix} 1 & 0 & 3 \\ 2 & -1 & -2 \end{bmatrix} \begin{bmatrix} -2 \\ 1 \\ -1 \end{bmatrix}$$

$$= \begin{bmatrix} 1(-2) + & 0(1) + & 3(-1) & 1(\ ) + & 0(\ ) + & 3(\ ) \\ 2(-2) + & (-1)(1) + & (-2)(-1) & 2(\ ) + & (-1)(\ ) + & (-2)(\ ) \end{bmatrix}$$

$$= \begin{bmatrix} -5 & 7 \\ -3 & 6 \end{bmatrix}$$

CHECKPoint   Now try Exercise 33.

Example 9 Patterns in Matrix Multiplication

a. $\begin{bmatrix} 3 & 4 \\ -2 & 5 \end{bmatrix} \begin{bmatrix} 1 & 0 \\ 0 & 1 \end{bmatrix} = \begin{bmatrix} 3 & 4 \\ -2 & 5 \end{bmatrix}$

b. $\begin{bmatrix} 6 & 2 & 0 \\ 3 & -1 & 2 \\ 1 & 4 & 6 \end{bmatrix} \begin{bmatrix} 1 \\ 2 \\ -3 \end{bmatrix} = \begin{bmatrix} 10 \\ -5 \\ -9 \end{bmatrix}$

c. The product AB for the following matrices is not defined.

$$A = \begin{bmatrix} -2 & 1 \\ 1 & -3 \\ 1 & 4 \end{bmatrix} \quad \text{and} \quad B = \begin{bmatrix} -2 & 3 & 1 & 4 \\ 0 & 1 & -1 & 2 \\ 2 & -1 & 0 & 1 \end{bmatrix}$$

CHECKPoint   Now try Exercise 39.

| Example 10 | **Patterns in Matrix Multiplication**

a. $\begin{bmatrix} 1 & -2 & -3 \end{bmatrix} \begin{bmatrix} 2 \\ -1 \\ 1 \end{bmatrix} = \begin{bmatrix} 1 \end{bmatrix}$ **b.** $\begin{bmatrix} 2 \\ -1 \\ 1 \end{bmatrix} \begin{bmatrix} 1 & -2 & -3 \end{bmatrix} = \begin{bmatrix} 2 & -4 & -6 \\ -1 & 2 & 3 \\ 1 & -2 & -3 \end{bmatrix}$

CHECK *Point*   Now try Exercise 51.

In Example 10, note that the two products are different. Even if both AB and BA are defined, matrix multiplication is not, in general, commutative. That is, for most matrices, $AB \neq BA$. This is one way in which the algebra of real numbers and the algebra of matrices differ.

Properties of Matrix Multiplication

Let A, B, and C be matrices and let c be a scalar.

1. $A(BC) = (AB)C$

2. $A(B + C) = AB + AC$

3. $(A + B)C = AC + BC$

4. $c(AB) = (cA)B = A(cB)$

Definition of Identity Matrix

The $n \times n$ matrix that consists of 1's on its main diagonal and 0's elsewhere is called the **identity matrix of order $n \times n$** and is denoted by

$$I_n = \begin{bmatrix} 1 & 0 & 0 & \ldots & 0 \\ 0 & 1 & 0 & \ldots & 0 \\ 0 & 0 & 1 & \ldots & 0 \\ \vdots & \vdots & \vdots & & \vdots \\ 0 & 0 & 0 & \ldots & 1 \end{bmatrix}.$$

Note that an identity matrix must be *square*. When the order is understood to be $n \times n$, you can denote I_n simply by I.

If A is an $n \times n$ matrix, the identity matrix has the property that $AI_n = A$ and $I_n A = A$. For example,

$$\begin{bmatrix} 3 & -2 & 5 \\ 1 & 0 & 4 \\ -1 & 2 & -3 \end{bmatrix} \begin{bmatrix} 1 & 0 & 0 \\ 0 & 1 & 0 \\ 0 & 0 & 1 \end{bmatrix} = \begin{bmatrix} 3 & -2 & 5 \\ 1 & 0 & 4 \\ -1 & 2 & -3 \end{bmatrix}$$

and

$$\begin{bmatrix} 1 & 0 & 0 \\ 0 & 1 & 0 \\ 0 & 0 & 1 \end{bmatrix} \begin{bmatrix} 3 & -2 & 5 \\ 1 & 0 & 4 \\ -1 & 2 & -3 \end{bmatrix} = \begin{bmatrix} 3 & -2 & 5 \\ 1 & 0 & 4 \\ -1 & 2 & -3 \end{bmatrix}.$$

Applications

Matrix multiplication can be used to represent a system of linear equations. Note how the system

$$\begin{cases} a_{11}x_1 + a_{12}x_2 + a_{13}x_3 = b_1 \\ a_{21}x_1 + a_{22}x_2 + a_{23}x_3 = b_2 \\ a_{31}x_1 + a_{32}x_2 + a_{33}x_3 = b_3 \end{cases}$$

can be written as the matrix equation $AX = B$, where A is the *coefficient matrix* of the system, and X and B are column matrices.

$$\begin{bmatrix} a_{11} & a_{12} & a_{13} \\ a_{21} & a_{22} & a_{23} \\ a_{31} & a_{32} & a_{33} \end{bmatrix} \begin{bmatrix} x_1 \\ x_2 \\ x_3 \end{bmatrix} = \begin{bmatrix} b_1 \\ b_2 \\ b_3 \end{bmatrix}$$

$$\times \qquad = $$

Study Tip

The column matrix B is also called a *constant* matrix. Its entries are the constant terms in the system of equations.

Example 11   Solving a System of Linear Equations

Consider the following system of linear equations.

$$\begin{cases} x_1 - 2x_2 + x_3 = -4 \\ x_2 + 2x_3 = 4 \\ 2x_1 + 3x_2 - 2x_3 = 2 \end{cases}$$

a. Write this system as a matrix equation, $AX = B$.

b. Use Gauss-Jordan elimination on the augmented matrix $[A \vdots B]$ to solve for the matrix X.

Study Tip

The notation $[A \vdots B]$ represents the augmented matrix formed when matrix B is adjoined to matrix A. The notation $[I \vdots X]$ represents the reduced row-echelon form of the augmented matrix that yields the *solution* of the system.

Solution

a. In matrix form, $AX = B$, the system can be written as follows.

$$\begin{bmatrix} 1 & -2 & 1 \\ 0 & 1 & 2 \\ 2 & 3 & -2 \end{bmatrix} \begin{bmatrix} x_1 \\ x_2 \\ x_3 \end{bmatrix} = \begin{bmatrix} -4 \\ 4 \\ 2 \end{bmatrix}$$

b. The augmented matrix is formed by adjoining matrix B to matrix A.

$$[A \vdots B] = \begin{bmatrix} 1 & -2 & 1 & \vdots & -4 \\ 0 & 1 & 2 & \vdots & 4 \\ 2 & 3 & -2 & \vdots & 2 \end{bmatrix}$$

Using Gauss-Jordan elimination, you can rewrite this equation as

$$[I \vdots X] = \begin{bmatrix} 1 & 0 & 0 & \vdots & -1 \\ 0 & 1 & 0 & \vdots & 2 \\ 0 & 0 & 1 & \vdots & 1 \end{bmatrix}.$$

So, the solution of the system of linear equations is $x_1 = -1$, $x_2 = 2$, and $x_3 = 1$, and the solution of the matrix equation is

$$X = \begin{bmatrix} x_1 \\ x_2 \\ x_3 \end{bmatrix} = \begin{bmatrix} -1 \\ 2 \\ 1 \end{bmatrix}.$$

CHECK*Point*   Now try Exercise 61.

| **Example 12**   **Softball Team Expenses**

Two softball teams submit equipment lists to their sponsors.

	Women's Team	Men's Team
Bats	12	15
Balls	45	38
Gloves	15	17

Each bat costs $80, each ball costs $6, and each glove costs $60. Use matrices to find the total cost of equipment for each team.

Solution

The equipment lists E and the costs per item C can be written in matrix form as

$$E = \begin{bmatrix} 12 & 15 \\ 45 & 38 \\ 15 & 17 \end{bmatrix}$$

and

$$C = \begin{bmatrix} 80 & 6 & 60 \end{bmatrix}.$$

The total cost of equipment for each team is given by the product

$$CE = \begin{bmatrix} 80 & 6 & 60 \end{bmatrix} \begin{bmatrix} 12 & 15 \\ 45 & 38 \\ 15 & 17 \end{bmatrix}$$

$$= \begin{bmatrix} 80(12) + 6(45) + 60(15) & 80(15) + 6(38) + 60(17) \end{bmatrix}$$

$$= \begin{bmatrix} 2130 & 2448 \end{bmatrix}.$$

So, the total cost of equipment for the women's team is $2130 and the total cost of equipment for the men's team is $2448.

CHECK **Point**   Now try Exercise 69.

Study Tip

Notice in Example 12 that you cannot find the total cost using the product EC because EC is not defined. That is, the number of columns of E (2 columns) does not equal the number of rows of C (1 row).

CLASSROOM DISCUSSION

Problem Posing   Write a matrix multiplication application problem that uses the matrix

$$A = \begin{bmatrix} 20 & 42 & 33 \\ 17 & 30 & 50 \end{bmatrix}.$$

Exchange problems with another student in your class. Form the matrices that represent the problem, and solve the problem. Interpret your solution in the context of the problem. Check with the creator of the problem to see if you are correct. Discuss other ways to represent and/or approach the problem.

7.2   EXERCISES

See www.CalcChat.com for worked-out solutions to odd-numbered exercises.

VOCABULARY

In Exercises 1–4, fill in the blanks.

1. Two matrices are _____ if all of their corresponding entries are equal.
2. When performing matrix operations, real numbers are often referred to as _____.
3. A matrix consisting entirely of zeros is called a _____ matrix and is denoted by _____.
4. The $n \times n$ matrix consisting of 1's on its main diagonal and 0's elsewhere is called the _____ matrix of order $n \times n$.

In Exercises 5 and 6, match the matrix property with the correct form. A, B, and C are matrices of order $m \times n$, and c and d are scalars.

5. (a) $1A = A$
   (b) $A + (B + C) = (A + B) + C$
   (c) $(c + d)A = cA + dA$
   (d) $(cd)A = c(dA)$
   (e) $A + B = B + A$

   (i)   Distributive Property
   (ii) Commutative Property of Matrix Addition
   (iii) Scalar Identity Property
   (iv) Associative Property of Matrix Addition
   (v)   Associative Property of Scalar Multiplication

6. (a) $A + O = A$
   (b) $c(AB) = A(cB)$
   (c) $A(B + C) = AB + AC$
   (d) $A(BC) = (AB)C$

   (i)   Distributive Property
   (ii) Additive Identity of Matrix Addition
   (iii) Associative Property of Matrix Multiplication
   (iv) Associative Property of Scalar Multiplication

SKILLS AND APPLICATIONS

In Exercises 7–10, find x and y.

7. $\begin{bmatrix} x & -2 \\ 7 & y \end{bmatrix} = \begin{bmatrix} -4 & -2 \\ 7 & 22 \end{bmatrix}$ 8. $\begin{bmatrix} -5 & x \\ y & 8 \end{bmatrix} = \begin{bmatrix} -5 & 13 \\ 12 & 8 \end{bmatrix}$

9. $\begin{bmatrix} 16 & 4 & 5 & 4 \\ -3 & 13 & 15 & 6 \\ 0 & 2 & 4 & 0 \end{bmatrix} = \begin{bmatrix} 16 & 4 & 2x+1 & 4 \\ -3 & 13 & & 15 & 3x \\ 0 & 2 & 3y-5 & 0 \end{bmatrix}$

10. $\begin{bmatrix} x+2 & 8 & -3 \\ 1 & 2y & 2x \\ 7 & -2 & y+2 \end{bmatrix} = \begin{bmatrix} 2x+6 & 8 & -3 \\ 1 & 18 & -8 \\ 7 & -2 & 11 \end{bmatrix}$

In Exercises 11–18, if possible, find (a) $A + B$, (b) $A - B$, (c) $3A$, and (d) $3A - 2B$.

11. $A = \begin{bmatrix} 1 & -1 \\ 2 & -1 \end{bmatrix}$, $B = \begin{bmatrix} 2 & -1 \\ -1 & 8 \end{bmatrix}$

12. $A = \begin{bmatrix} 1 & 2 \\ 2 & 1 \end{bmatrix}$, $B = \begin{bmatrix} -3 & -2 \\ 4 & 2 \end{bmatrix}$

13. $A = \begin{bmatrix} 8 & -1 \\ 2 & 3 \\ -4 & 5 \end{bmatrix}$, $B = \begin{bmatrix} 1 & 6 \\ -1 & -5 \\ 1 & 10 \end{bmatrix}$

14. $A = \begin{bmatrix} 1 & -1 & 3 \\ 0 & 6 & 9 \end{bmatrix}$, $B = \begin{bmatrix} -2 & 0 & -5 \\ -3 & 4 & -7 \end{bmatrix}$

15. $A = \begin{bmatrix} 4 & 5 & -1 & 3 & 4 \\ 1 & 2 & -2 & -1 & 0 \end{bmatrix}$,
    $B = \begin{bmatrix} 1 & 0 & -1 & 1 & 0 \\ -6 & 8 & 2 & -3 & -7 \end{bmatrix}$

16. $A = \begin{bmatrix} -1 & 4 & 0 \\ 3 & -2 & 2 \\ 5 & 4 & -1 \\ 0 & 8 & -6 \\ -4 & -1 & 0 \end{bmatrix}$, $B = \begin{bmatrix} -3 & 5 & 1 \\ 2 & -4 & -7 \\ 10 & -9 & -1 \\ 3 & 2 & -4 \\ 0 & 1 & -2 \end{bmatrix}$

17. $A = \begin{bmatrix} 6 & 0 & 3 \\ -1 & -4 & 0 \end{bmatrix}$, $B = \begin{bmatrix} 8 & -1 \\ 4 & -3 \end{bmatrix}$

18. $A = \begin{bmatrix} 3 \\ 2 \\ -1 \end{bmatrix}$, $B = \begin{bmatrix} -4 & 6 & 2 \end{bmatrix}$

In Exercises 19–24, evaluate the expression.

19. $\begin{bmatrix} -5 & 0 \\ 3 & -6 \end{bmatrix} + \begin{bmatrix} 7 & 1 \\ -2 & -1 \end{bmatrix} + \begin{bmatrix} -10 & -8 \\ 14 & 6 \end{bmatrix}$

20. $\begin{bmatrix} 6 & 8 \\ -1 & 0 \end{bmatrix} + \begin{bmatrix} 0 & 5 \\ -3 & -1 \end{bmatrix} + \begin{bmatrix} -11 & -7 \\ 2 & -1 \end{bmatrix}$

21. $4\left(\begin{bmatrix} -4 & 0 & 1 \\ 0 & 2 & 3 \end{bmatrix} - \begin{bmatrix} 2 & 1 & -2 \\ 3 & -6 & 0 \end{bmatrix} \right)$

22. $\frac{1}{2}([5 \quad -2 \quad 4 \quad 0] + [14 \quad 6 \quad -18 \quad 9])$

23. $-3\left(\begin{bmatrix} 0 & -3 \\ 7 & 2 \end{bmatrix} + \begin{bmatrix} -6 & 3 \\ 8 & 1 \end{bmatrix}\right) - 2\begin{bmatrix} 4 & -4 \\ 7 & -9 \end{bmatrix}$

24. $-\begin{bmatrix} 4 & 11 \\ -2 & -1 \\ 9 & 3 \end{bmatrix} + \frac{1}{6}\left(\begin{bmatrix} -5 & -1 \\ 3 & 4 \\ 0 & 13 \end{bmatrix} + \begin{bmatrix} 7 & 5 \\ -9 & -1 \\ 6 & -1 \end{bmatrix}\right)$

In Exercises 25–28, use the matrix capabilities of a graphing utility to evaluate the expression. Round your results to three decimal places, if necessary.

25. $\frac{3}{7}\begin{bmatrix} 2 & 5 \\ -1 & -4 \end{bmatrix} + 6\begin{bmatrix} -3 & 0 \\ 2 & 2 \end{bmatrix}$

26. $55\left(\begin{bmatrix} 14 & -11 \\ -22 & 19 \end{bmatrix} + \begin{bmatrix} -22 & 20 \\ 13 & 6 \end{bmatrix}\right)$

27. $-\begin{bmatrix} 3.211 & 6.829 \\ -1.004 & 4.914 \\ 0.055 & -3.889 \end{bmatrix} - \begin{bmatrix} -1.630 & -3.090 \\ 5.256 & 8.335 \\ -9.768 & 4.251 \end{bmatrix}$

28. $-\begin{bmatrix} 10 & 15 \\ -20 & 10 \\ 12 & 4 \end{bmatrix} + \frac{1}{8}\left(\begin{bmatrix} -13 & 11 \\ 7 & 0 \\ 6 & 9 \end{bmatrix} + \begin{bmatrix} -3 & 13 \\ -3 & 8 \\ -14 & 15 \end{bmatrix}\right)$

In Exercises 29–32, solve for X in the equation, given

$$A = \begin{bmatrix} -2 & -1 \\ 1 & 0 \\ 3 & -4 \end{bmatrix} \quad \text{and} \quad B = \begin{bmatrix} 0 & 3 \\ 2 & 0 \\ -4 & -1 \end{bmatrix}.$$

29. $X = 3A - 2B$

30. $2X = 2A - B$

31. $2X + 3A = B$

32. $2A + 4B = -2X$

In Exercises 33–40, if possible, find AB and state the order of the result.

33. $A = \begin{bmatrix} 2 & 1 \\ -3 & 4 \\ 1 & 6 \end{bmatrix}, \quad B = \begin{bmatrix} 0 & -1 & 0 \\ 4 & 0 & 2 \\ 8 & -1 & 7 \end{bmatrix}$

34. $A = \begin{bmatrix} 0 & -1 & 2 \\ 6 & 0 & 3 \\ 7 & -1 & 8 \end{bmatrix}, \quad B = \begin{bmatrix} 2 & -1 \\ 4 & -5 \\ 1 & 6 \end{bmatrix}$

35. $A = \begin{bmatrix} -1 & 6 \\ -4 & 5 \\ 0 & 3 \end{bmatrix}, \quad B = \begin{bmatrix} 2 & 3 \\ 0 & 9 \end{bmatrix}$

36. $A = \begin{bmatrix} 1 & 0 & 0 \\ 0 & 4 & 0 \\ 0 & 0 & -2 \end{bmatrix}, \quad B = \begin{bmatrix} 3 & 0 & 0 \\ 0 & -1 & 0 \\ 0 & 0 & 5 \end{bmatrix}$

37. $A = \begin{bmatrix} 5 & 0 & 0 \\ 0 & -8 & 0 \\ 0 & 0 & 7 \end{bmatrix}, \quad B = \begin{bmatrix} \frac{1}{5} & 0 & 0 \\ 0 & -\frac{1}{8} & 0 \\ 0 & 0 & \frac{1}{2} \end{bmatrix}$

38. $A = \begin{bmatrix} 0 & 0 & 5 \\ 0 & 0 & -3 \\ 0 & 0 & 4 \end{bmatrix}, \quad B = \begin{bmatrix} 6 & -11 & 4 \\ 8 & 16 & 4 \\ 0 & 0 & 0 \end{bmatrix}$

39. $A = \begin{bmatrix} 10 \\ 12 \end{bmatrix}, \quad B = [6 \quad -2 \quad 1 \quad 6]$

40. $A = \begin{bmatrix} 1 & 0 & 3 & -2 \\ 6 & 13 & 8 & -17 \end{bmatrix}, \quad B = \begin{bmatrix} 1 & 6 \\ 4 & 2 \end{bmatrix}$

In Exercises 41–46, use the matrix capabilities of a graphing utility to find AB, if possible.

41. $A = \begin{bmatrix} 7 & 5 & -4 \\ -2 & 5 & 1 \\ 10 & -4 & -7 \end{bmatrix}, \quad B = \begin{bmatrix} 2 & -2 & 3 \\ 8 & 1 & 4 \\ -4 & 2 & -8 \end{bmatrix}$

42. $A = \begin{bmatrix} 11 & -12 & 4 \\ 14 & 10 & 12 \\ 6 & -2 & 9 \end{bmatrix}, \quad B = \begin{bmatrix} 12 & 10 \\ -5 & 12 \\ 15 & 16 \end{bmatrix}$

43. $A = \begin{bmatrix} -3 & 8 & -6 & 8 \\ -12 & 15 & 9 & 6 \\ 5 & -1 & 1 & 5 \end{bmatrix}, \quad B = \begin{bmatrix} 3 & 1 & 6 \\ 24 & 15 & 14 \\ 16 & 10 & 21 \\ 8 & -4 & 10 \end{bmatrix}$

44. $A = \begin{bmatrix} -2 & 4 & 8 \\ 21 & 5 & 6 \\ 13 & 2 & 6 \end{bmatrix}, \quad B = \begin{bmatrix} 2 & 0 \\ -7 & 15 \\ 32 & 14 \\ 0.5 & 1.6 \end{bmatrix}$

45. $A = \begin{bmatrix} 9 & 10 & -38 & 18 \\ 100 & -50 & 250 & 75 \end{bmatrix},$

$B = \begin{bmatrix} 52 & -85 & 27 & 45 \\ 40 & -35 & 60 & 82 \end{bmatrix}$

46. $A = \begin{bmatrix} 16 & -18 \\ -4 & 13 \\ -9 & 21 \end{bmatrix}, \quad B = \begin{bmatrix} -7 & 20 & -1 \\ 7 & 15 & 26 \end{bmatrix}$

In Exercises 47–52, if possible, find (a) AB, (b) BA, and (c) A^2. (*Note:* $A^2 = AA$.)

47. $A = \begin{bmatrix} 1 & 2 \\ 4 & 2 \end{bmatrix}, \quad B = \begin{bmatrix} 2 & -1 \\ -1 & 8 \end{bmatrix}$

48. $A = \begin{bmatrix} 6 & 3 \\ -2 & -4 \end{bmatrix}, \quad B = \begin{bmatrix} -2 & 0 \\ 2 & 4 \end{bmatrix}$

49. $A = \begin{bmatrix} 3 & -1 \\ 1 & 3 \end{bmatrix}, \quad B = \begin{bmatrix} 1 & -3 \\ 3 & 1 \end{bmatrix}$

50. $A = \begin{bmatrix} 1 & -1 \\ 1 & 1 \end{bmatrix}, \quad B = \begin{bmatrix} 1 & 3 \\ -3 & 1 \end{bmatrix}$

51. $A = \begin{bmatrix} 7 \\ 8 \\ -1 \end{bmatrix}, \quad B = [1 \quad 1 \quad 2]$

52. $A = [3 \quad 2 \quad 1], \quad B = \begin{bmatrix} 2 \\ 3 \\ 0 \end{bmatrix}$

In Exercises 53–56, evaluate the expression. Use the matrix capabilities of a graphing utility to verify your answer.

53. $\begin{bmatrix} 3 & 1 \\ 0 & -2 \end{bmatrix} \begin{bmatrix} 1 & 0 \\ -2 & 2 \end{bmatrix} \begin{bmatrix} 1 & 0 \\ 2 & 4 \end{bmatrix}$

54. $-3\left(\begin{bmatrix} 6 & 5 & -1 \\ 1 & -2 & 0 \end{bmatrix} \begin{bmatrix} 0 & 3 \\ -1 & -3 \\ 4 & 1 \end{bmatrix} \right)$

55. $\begin{bmatrix} 0 & 2 & -2 \\ 4 & 1 & 2 \end{bmatrix} \left(\begin{bmatrix} 4 & 0 \\ 0 & -1 \\ -1 & 2 \end{bmatrix} + \begin{bmatrix} -2 & 3 \\ -3 & 5 \\ 0 & -3 \end{bmatrix} \right)$

56. $\begin{bmatrix} 3 \\ -1 \\ 5 \\ 7 \end{bmatrix} \left(\begin{bmatrix} 5 & -6 \end{bmatrix} + \begin{bmatrix} 7 & -1 \end{bmatrix} + \begin{bmatrix} -8 & 9 \end{bmatrix} \right)$

In Exercises 57–64, (a) write the system of linear equations as a matrix equation, $AX = B$, and (b) use Gauss-Jordan elimination on the augmented matrix $[A \ \vdots \ B]$ to solve for the matrix X.

57. $\begin{cases} -x_1 + x_2 = 4 \\ -2x_1 + x_2 = 0 \end{cases}$

58. $\begin{cases} 2x_1 + 3x_2 = 5 \\ x_1 + 4x_2 = 10 \end{cases}$

59. $\begin{cases} -2x_1 - 3x_2 = -4 \\ 6x_1 + x_2 = -36 \end{cases}$

60. $\begin{cases} -4x_1 + 9x_2 = -13 \\ x_1 - 3x_2 = 12 \end{cases}$

61. $\begin{cases} x_1 - 2x_2 + 3x_3 = 9 \\ -x_1 + 3x_2 - x_3 = -6 \\ 2x_1 - 5x_2 + 5x_3 = 17 \end{cases}$

62. $\begin{cases} x_1 + x_2 - 3x_3 = -1 \\ -x_1 + 2x_2 = 1 \\ x_1 - x_2 + x_3 = 2 \end{cases}$

63. $\begin{cases} x_1 - 5x_2 + 2x_3 = -20 \\ -3x_1 + x_2 - x_3 = 8 \\ -2x_2 + 5x_3 = -16 \end{cases}$

64. $\begin{cases} x_1 - x_2 + 4x_3 = 17 \\ x_1 + 3x_2 = -11 \\ -6x_2 + 5x_3 = 40 \end{cases}$

65. MANUFACTURING A corporation has three factories, each of which manufactures acoustic guitars and electric guitars. The number of units of guitars produced at factory j in one day is represented by a_{ij} in the matrix

$$A = \begin{bmatrix} 70 & 50 & 25 \\ 35 & 100 & 70 \end{bmatrix}.$$

Find the production levels if production is increased by 20%.

66. MANUFACTURING A corporation has four factories, each of which manufactures sport utility vehicles and pickup trucks. The number of units of vehicle i produced at factory j in one day is represented by a_{ij} in the matrix

$$A = \begin{bmatrix} 100 & 90 & 70 & 30 \\ 40 & 20 & 60 & 60 \end{bmatrix}.$$

Find the production levels if production is increased by 10%.

67. AGRICULTURE A fruit grower raises two crops, apples and peaches. Each of these crops is sent to three different outlets for sale. These outlets are The Farmer's Market, The Fruit Stand, and The Fruit Farm. The numbers of bushels of apples sent to the three outlets are 125, 100, and 75, respectively. The numbers of bushels of peaches sent to the three outlets are 100, 175, and 125, respectively. The profit per bushel for apples is $3.50 and the profit per bushel for peaches is $6.00.

(a) Write a matrix A that represents the number of bushels of each crop i that are shipped to each outlet j. State what each entry a_{ij} of the matrix represents.

(b) Write a matrix B that represents the profit per bushel of each fruit. State what each entry b_{ij} of the matrix represents.

(c) Find the product BA and state what each entry of the matrix represents.

68. REVENUE An electronics manufacturer produces three models of LCD televisions, which are shipped to two warehouses. The numbers of units of model i that are shipped to warehouse j are represented by a_{ij} in the matrix

$$A = \begin{bmatrix} 5,000 & 4,000 \\ 6,000 & 10,000 \\ 8,000 & 5,000 \end{bmatrix}.$$

The prices per unit are represented by the matrix

$$B = [\$699.95 \quad \$899.95 \quad \$1099.95].$$

Compute BA and interpret the result.

69. INVENTORY A company sells five models of computers through three retail outlets. The inventories are represented by S.

$$S = \begin{bmatrix} 3 & 2 & 2 & 3 & 0 \\ 0 & 2 & 3 & 4 & 3 \\ 4 & 2 & 1 & 3 & 2 \end{bmatrix}$$

The wholesale and retail prices are represented by T.

$$T = \begin{bmatrix} \$840 & \$1100 \\ \$1200 & \$1350 \\ \$1450 & \$1650 \\ \$2650 & \$3000 \\ \$3050 & \$3200 \end{bmatrix}$$

Compute ST and interpret the result.

70. VOTING PREFERENCES   The matrix

$$P = \begin{bmatrix} 0.6 & 0.1 & 0.1 \\ 0.2 & 0.7 & 0.1 \\ 0.2 & 0.2 & 0.8 \end{bmatrix}$$

is called a *stochastic matrix*. Each entry $p_{ij}(i \neq j)$ represents the proportion of the voting population that changes from party i to party j, and p_{ii} represents the proportion that remains loyal to the party from one election to the next. Compute and interpret P^2.

71. VOTING PREFERENCES   Use a graphing utility to find P^3, P^4, P^5, P^6, P^7, and P^8 for the matrix given in Exercise 70. Can you detect a pattern as P is raised to higher powers?

72. LABOR/WAGE REQUIREMENTS   A company that manufactures boats has the following labor-hour and wage requirements.

Labor per boat

$$S = \begin{bmatrix} 1.0\,h & 0.5\,h & 0.2\,h \\ 1.6\,h & 1.0\,h & 0.2\,h \\ 2.5\,h & 2.0\,h & 1.4\,h \end{bmatrix}$$

Wages per hour

$$T = \begin{bmatrix} \$15 & \$13 \\ \$12 & \$11 \\ \$11 & \$10 \end{bmatrix}$$

Compute ST and interpret the result.

73. PROFIT   At a local dairy mart, the numbers of gallons of skim milk, 2% milk, and whole milk sold over the weekend are represented by A.

$$A = \begin{bmatrix} 40 & 64 & 52 \\ 60 & 82 & 76 \\ 76 & 96 & 84 \end{bmatrix}$$

The selling prices (in dollars per gallon) and the profits (in dollars per gallon) for the three types of milk sold by the dairy mart are represented by B.

$$B = \begin{bmatrix} \$3.45 & \$1.20 \\ \$3.65 & \$1.30 \\ \$3.85 & \$1.45 \end{bmatrix}$$

(a) Compute AB and interpret the result.

(b) Find the dairy mart's total profit from milk sales for the weekend.

74. PROFIT   At a convenience store, the numbers of gallons of 87-octane, 89-octane, and 93-octane gasoline sold over the weekend are represented by A.

$$A = \begin{bmatrix} 580 & 840 & 320 \\ 560 & 420 & 160 \\ 860 & 1020 & 540 \end{bmatrix}$$

The selling prices (in dollars per gallon) and the profits (in dollars per gallon) for the three grades of gasoline sold by the convenience store are represented by B.

$$B = \begin{bmatrix} \$2.00 & \$0.08 \\ \$2.10 & \$0.09 \\ \$2.20 & \$0.10 \end{bmatrix}$$

(a) Compute AB and interpret the result.

(b) Find the convenience store's profit from gasoline sales for the weekend.

75. EXERCISE   The numbers of calories burned by individuals of different body weights performing different types of aerobic exercises for a 20-minute time period are shown in matrix A.

$$A = \begin{bmatrix} 109 & 136 \\ 127 & 159 \\ 64 & 79 \end{bmatrix}$$

(a) A 120-pound person and a 150-pound person bicycled for 40 minutes, jogged for 10 minutes, and walked for 60 minutes. Organize the time they spent exercising in a matrix B.

(b) Compute BA and interpret the result.

76. HEALTH CARE The health care plans offered this year by a local manufacturing plant are as follows. For individuals, the comprehensive plan costs $694.32, the HMO standard plan costs $451.80, and the HMO Plus plan costs $489.48. For families, the comprehensive plan costs $1725.36, the HMO standard plan costs $1187.76, and the HMO Plus plan costs $1248.12. The plant expects the costs of the plans to change next year as follows. For individuals, the costs for the comprehensive, HMO standard, and HMO Plus plans will be $683.91, $463.10, and $499.27, respectively. For families, the costs for the comprehensive, HMO standard, and HMO Plus plans will be $1699.48, $1217.45, and $1273.08, respectively.

(a) Organize the information using two matrices A and B, where A represents the health care plan costs for this year and B represents the health care plan costs for next year. State what each entry of each matrix represents.

(b) Compute $A - B$ and interpret the result.

(c) The employees receive monthly paychecks from which the health care plan costs are deducted. Use the matrices from part (a) to write matrices that show how much will be deducted from each employees' paycheck this year and next year.

(d) Suppose instead that the costs of the health care plans increase by 4% next year. Write a matrix that shows the new monthly payments.

EXPLORATION

TRUE OR FALSE? In Exercises 77 and 78, determine whether the statement is true or false. Justify your answer.

77. Two matrices can be added only if they have the same order.

78. Matrix multiplication is commutative.

THINK ABOUT IT In Exercises 79–86, let matrices A, B, C, and D be of orders 2×3, 2×3, 3×2, and 2×2, respectively. Determine whether the matrices are of proper order to perform the operation(s). If so, give the order of the answer.

79. $A + 2C$

80. $B - 3C$

81. AB

82. BC

83. $BC - D$

84. $CB - D$

85. $D(A - 3B)$

86. $(BC - D)A$

87. Consider matrices A, B, and C below. Perform the indicated operations and compare the results.

$$A = \begin{bmatrix} 3 & -1 \\ 4 & 7 \end{bmatrix}, \ B = \begin{bmatrix} -2 & 0 \\ 8 & 1 \end{bmatrix}, \ C = \begin{bmatrix} 5 & 2 \\ 2 & -6 \end{bmatrix}$$

(a) Find $A + B$ and $B + A$.

(b) Find $A + B$, then add C to the resulting matrix. Find $B + C$, then add A to the resulting matrix.

(c) Find $2A$ and $2B$, then add the two resulting matrices. Find $A + B$, then multiply the resulting matrix by 2.

88. Use the following matrices to find AB, BA, $(AB)C$, and $A(BC)$. What do your results tell you about matrix multiplication, commutativity, and associativity?

$$A = \begin{bmatrix} 1 & 2 \\ 3 & 4 \end{bmatrix}, \ B = \begin{bmatrix} 0 & 1 \\ 2 & 3 \end{bmatrix}, \ C = \begin{bmatrix} 3 & 0 \\ 0 & 1 \end{bmatrix}$$

89. THINK ABOUT IT If a, b, and c are real numbers such that $c \neq 0$ and $ac = bc$, then $a = b$. However, if A, B, and C are nonzero matrices such that $AC = BC$, then A is *not necessarily* equal to B. Illustrate this using the following matrices.

$$A = \begin{bmatrix} 0 & 1 \\ 0 & 1 \end{bmatrix}, \ B = \begin{bmatrix} 1 & 0 \\ 1 & 0 \end{bmatrix}, \ C = \begin{bmatrix} 2 & 3 \\ 2 & 3 \end{bmatrix}$$

90. THINK ABOUT IT If a and b are real numbers such that $ab = 0$, then $a = 0$ or $b = 0$. However, if A and B are matrices such that $AB = O$, it is *not necessarily* true that $A = O$ or $B = O$. Illustrate this using the following matrices.

$$A = \begin{bmatrix} 3 & 3 \\ 4 & 4 \end{bmatrix}, \ B = \begin{bmatrix} 1 & -1 \\ -1 & 1 \end{bmatrix}$$

91. Let A and B be unequal diagonal matrices of the same order. (A **diagonal matrix** is a square matrix in which each entry not on the main diagonal is zero.) Determine the products AB for several pairs of such matrices. Make a conjecture about a quick rule for such products.

92. Let $i = \sqrt{-1}$ and let

$$A = \begin{bmatrix} i & 0 \\ 0 & i \end{bmatrix} \ \text{and} \ B = \begin{bmatrix} 0 & -i \\ i & 0 \end{bmatrix}.$$

(a) Find A^2, A^3, and A^4. Identify any similarities with i^2, i^3, and i^4.

(b) Find and identify B^2.

93. Find two matrices A and B such that $AB = BA$.

94. CAPSTONE Let matrices A and B be of orders 3×2 and 2×2, respectively. Answer the following questions and explain your reasoning.

(a) Is it possible that $A = B$?

(b) Is $A + B$ defined?

(c) Is AB defined? If so, is it possible that $AB = BA$?

7.3 THE INVERSE OF A SQUARE MATRIX

What you should learn

- Verify that two matrices are inverses of each other.
- Use Gauss-Jordan elimination to find the inverses of matrices.
- Use a formula to find the inverses of 2×2 matrices.
- Use inverse matrices to solve systems of linear equations.

Why you should learn it

You can use inverse matrices to model and solve real-life problems. For instance, in Exercise 73 on page 557, an inverse matrix is used to find a quadratic model for the enrollment projections for public universities in the United States.

The Inverse of a Matrix

This section further develops the algebra of matrices. To begin, consider the real number equation $ax = b$. To solve this equation for x, multiply each side of the equation by a^{-1} (provided that $a \neq 0$).

$$ax = b$$
$$(a^{-1}a)x = a^{-1}b$$
$$(1)x = a^{-1}b$$
$$x = a^{-1}b$$

The number a^{-1} is called the *multiplicative inverse of a* because $a^{-1}a = 1$. The definition of the multiplicative **inverse of a matrix** is similar.

Definition of the Inverse of a Square Matrix

Let A be an $n \times n$ matrix and let I_n be the $n \times n$ identity matrix. If there exists a matrix A^{-1} such that

$$AA^{-1} = I_n = A^{-1}A$$

then A^{-1} is called the **inverse** of A. The symbol A^{-1} is read "A inverse."

Example 1   The Inverse of a Matrix

Show that B is the inverse of A, where

$$A = \begin{bmatrix} -1 & 2 \\ -1 & 1 \end{bmatrix} \quad \text{and} \quad B = \begin{bmatrix} 1 & -2 \\ 1 & -1 \end{bmatrix}.$$

Solution

To show that B is the inverse of A, show that $AB = I = BA$, as follows.

$$AB = \begin{bmatrix} -1 & 2 \\ -1 & 1 \end{bmatrix}\begin{bmatrix} 1 & -2 \\ 1 & -1 \end{bmatrix} = \begin{bmatrix} -1+2 & 2-2 \\ -1+1 & 2-1 \end{bmatrix} = \begin{bmatrix} 1 & 0 \\ 0 & 1 \end{bmatrix}$$

$$BA = \begin{bmatrix} 1 & -2 \\ 1 & -1 \end{bmatrix}\begin{bmatrix} -1 & 2 \\ -1 & 1 \end{bmatrix} = \begin{bmatrix} -1+2 & 2-2 \\ -1+1 & 2-1 \end{bmatrix} = \begin{bmatrix} 1 & 0 \\ 0 & 1 \end{bmatrix}$$

As you can see, $AB = I = BA$. This is an example of a square matrix that has an inverse. Note that not all square matrices have inverses.

CHECK **Point**   Now try Exercise 5.

Recall that it is not always true that $AB = BA$, even if both products are defined. However, if A and B are both square matrices and $AB = I_n$, it can be shown that $BA = I_n$. So, in Example 1, you need only to check that $AB = I_2$.

Finding Inverse Matrices

If a matrix A has an inverse, A is called **invertible** (or **nonsingular**); otherwise, A is called **singular.** A nonsquare matrix cannot have an inverse. To see this, note that if A is of order $m \times n$ and B is of order $n \times m$ (where $m \neq n$), the products AB and BA are of different orders and so cannot be equal to each other. Not all square matrices have inverses (see the matrix at the bottom of page 552). If, however, a matrix does have an inverse, that inverse is unique. Example 2 shows how to use a system of equations to find the inverse of a matrix.

Example 2 Finding the Inverse of a Matrix

Find the inverse of

$$A = \begin{bmatrix} 1 & 4 \\ -1 & -3 \end{bmatrix}.$$

Solution

To find the inverse of A, try to solve the matrix equation $AX = I$ for X.

$$\begin{bmatrix} 1 & 4 \\ -1 & -3 \end{bmatrix}\begin{bmatrix} x_{11} & x_{12} \\ x_{21} & x_{22} \end{bmatrix} = \begin{bmatrix} 1 & 0 \\ 0 & 1 \end{bmatrix}$$

$$\begin{bmatrix} x_{11} + 4x_{21} & x_{12} + 4x_{22} \\ -x_{11} - 3x_{21} & -x_{12} - 3x_{22} \end{bmatrix} = \begin{bmatrix} 1 & 0 \\ 0 & 1 \end{bmatrix}$$

Equating corresponding entries, you obtain two systems of linear equations.

$$\begin{cases} x_{11} + 4x_{21} = 1 \\ -x_{11} - 3x_{21} = 0 \end{cases}$$

$$\begin{cases} x_{12} + 4x_{22} = 0 \\ -x_{12} - 3x_{22} = 1 \end{cases}$$

Solve the first system using elementary row operations to determine that $x_{11} = -3$ and $x_{21} = 1$. From the second system you can determine that $x_{12} = -4$ and $x_{22} = 1$. Therefore, the inverse of A is

$$X = A^{-1}$$

$$= \begin{bmatrix} -3 & -4 \\ 1 & 1 \end{bmatrix}.$$

You can use matrix multiplication to check this result.

Check

$$AA^{-1} = \begin{bmatrix} 1 & 4 \\ -1 & -3 \end{bmatrix}\begin{bmatrix} -3 & -4 \\ 1 & 1 \end{bmatrix} = \begin{bmatrix} 1 & 0 \\ 0 & 1 \end{bmatrix} \checkmark$$

$$A^{-1}A = \begin{bmatrix} -3 & -4 \\ 1 & 1 \end{bmatrix}\begin{bmatrix} 1 & 4 \\ -1 & -3 \end{bmatrix} = \begin{bmatrix} 1 & 0 \\ 0 & 1 \end{bmatrix} \checkmark$$

CHECK*Point* Now try Exercise 15.

In Example 2, note that the two systems of linear equations have the *same coefficient matrix A*. Rather than solve the two systems represented by

$$\begin{bmatrix} 1 & 4 & \vdots & 1 \\ -1 & -3 & \vdots & 0 \end{bmatrix}$$

and

$$\begin{bmatrix} 1 & 4 & \vdots & 0 \\ -1 & -3 & \vdots & 1 \end{bmatrix}$$

separately, you can solve them *simultaneously* by *adjoining* the identity matrix to the coefficient matrix to obtain

$$\begin{bmatrix} 1 & 4 & \vdots & 1 & 0 \\ -1 & -3 & \vdots & 0 & 1 \end{bmatrix}.$$

This "doubly augmented" matrix can be represented as $[A \vdots I]$. By applying Gauss-Jordan elimination to this matrix, you can solve *both* systems with a single elimination process.

$$\begin{bmatrix} 1 & 4 & \vdots & 1 & 0 \\ -1 & -3 & \vdots & 0 & 1 \end{bmatrix}$$

$$\begin{bmatrix} 1 & 4 & \vdots & 1 & 0 \\ 0 & 1 & \vdots & 1 & 1 \end{bmatrix}$$

$$\begin{bmatrix} 1 & 0 & \vdots & -3 & -4 \\ 0 & 1 & \vdots & 1 & 1 \end{bmatrix}$$

So, from the "doubly augmented" matrix $[A \vdots I]$, you obtain the matrix $[I \vdots A^{-1}]$.

$$\begin{bmatrix} 1 & 4 & \vdots & 1 & 0 \\ -1 & -3 & \vdots & 0 & 1 \end{bmatrix} \implies \begin{bmatrix} 1 & 0 & \vdots & -3 & -4 \\ 0 & 1 & \vdots & 1 & 1 \end{bmatrix}$$

This procedure (or algorithm) works for any square matrix that has an inverse.

TECHNOLOGY

Most graphing utilities can find the inverse of a square matrix. To do so, you may have to use the inverse key $\boxed{x^{-1}}$. Consult the user's guide for your graphing utility for specific keystrokes.

Finding an Inverse Matrix

Let A be a square matrix of order n.

1. Write the $n \times 2n$ matrix that consists of the given matrix A on the left and the $n \times n$ identity matrix I on the right to obtain $[A \vdots I]$.

2. If possible, row reduce A to I using elementary row operations on the *entire* matrix $[A \vdots I]$. The result will be the matrix $[I \vdots A^{-1}]$. If this is not possible, A is not invertible.

3. Check your work by multiplying to see that $AA^{-1} = I = A^{-1}A$.

Example 3 Finding the Inverse of a Matrix

Find the inverse of $A = \begin{bmatrix} 1 & -1 & 0 \\ 1 & 0 & -1 \\ 6 & -2 & -3 \end{bmatrix}$.

Solution

Begin by adjoining the identity matrix to A to form the matrix

$$[A \ \vdots \ I] = \begin{bmatrix} 1 & -1 & 0 & \vdots & 1 & 0 & 0 \\ 1 & 0 & -1 & \vdots & 0 & 1 & 0 \\ 6 & -2 & -3 & \vdots & 0 & 0 & 1 \end{bmatrix}.$$

Use elementary row operations to obtain the form $[I \ \vdots \ A^{-1}]$, as follows.

$$\begin{bmatrix} 1 & -1 & 0 & \vdots & 1 & 0 & 0 \\ 0 & 1 & -1 & \vdots & -1 & 1 & 0 \\ 0 & 4 & -3 & \vdots & -6 & 0 & 1 \end{bmatrix}$$

$$\begin{bmatrix} 1 & 0 & -1 & \vdots & 0 & 1 & 0 \\ 0 & 1 & -1 & \vdots & -1 & 1 & 0 \\ 0 & 0 & 1 & \vdots & -2 & -4 & 1 \end{bmatrix}$$

$$\begin{bmatrix} 1 & 0 & 0 & \vdots & -2 & -3 & 1 \\ 0 & 1 & 0 & \vdots & -3 & -3 & 1 \\ 0 & 0 & 1 & \vdots & -2 & -4 & 1 \end{bmatrix} = [I \ \vdots \ A^{-1}]$$

So, the matrix A is invertible and its inverse is

$$A^{-1} = \begin{bmatrix} -2 & -3 & 1 \\ -3 & -3 & 1 \\ -2 & -4 & 1 \end{bmatrix}.$$

Confirm this result by multiplying A and A^{-1} to obtain I, as follows.

Check

$$AA^{-1} = \begin{bmatrix} 1 & -1 & 0 \\ 1 & 0 & -1 \\ 6 & -2 & -3 \end{bmatrix} \begin{bmatrix} -2 & -3 & 1 \\ -3 & -3 & 1 \\ -2 & -4 & 1 \end{bmatrix} = \begin{bmatrix} 1 & 0 & 0 \\ 0 & 1 & 0 \\ 0 & 0 & 1 \end{bmatrix} = I$$

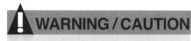 Now try Exercise 19.

> **⚠ WARNING / CAUTION**
>
> Be sure to check your solution because it is easy to make algebraic errors when using elementary row operations.

The process shown in Example 3 applies to any $n \times n$ matrix A. When using this algorithm, if the matrix A does not reduce to the identity matrix, then A does not have an inverse. For instance, the following matrix has no inverse.

$$A = \begin{bmatrix} 1 & 2 & 0 \\ 3 & -1 & 2 \\ -2 & 3 & -2 \end{bmatrix}$$

To confirm that matrix A above has no inverse, adjoin the identity matrix to A to form $[A \ \vdots \ I]$ and perform elementary row operations on the matrix. After doing so, you will see that it is impossible to obtain the identity matrix I on the left. Therefore, A is not invertible.

The Inverse of a 2 × 2 Matrix

Using Gauss-Jordan elimination to find the inverse of a matrix works well (even as a computer technique) for matrices of order 3×3 or greater. For 2×2 matrices, however, many people prefer to use a formula for the inverse rather than Gauss-Jordan elimination. This simple formula, which works *only* for 2×2 matrices, is explained as follows. If A is a 2×2 matrix given by

$$A = \begin{bmatrix} a & b \\ c & d \end{bmatrix}$$

then A is invertible if and only if $ad - bc \neq 0$. Moreover, if $ad - bc \neq 0$, the inverse is given by

$$A^{-1} = \frac{1}{ad - bc} \begin{bmatrix} d & -b \\ -c & a \end{bmatrix}.$$

The denominator $ad - bc$ is called the **determinant** of the 2×2 matrix A. You will study determinants in the next section.

Example 4   Finding the Inverse of a 2 × 2 Matrix

If possible, find the inverse of each matrix.

a. $A = \begin{bmatrix} 3 & -1 \\ -2 & 2 \end{bmatrix}$

b. $B = \begin{bmatrix} 3 & -1 \\ -6 & 2 \end{bmatrix}$

Solution

a. For the matrix A, apply the formula for the inverse of a 2×2 matrix to obtain

$$ad - bc = (\ )(\ ) - (\quad)(\quad)$$
$$= 4.$$

Because this quantity is not zero, the inverse is formed by interchanging the entries on the main diagonal, changing the signs of the other two entries, and multiplying by the scalar $\frac{1}{4}$, as follows.

$$A^{-1} = \frac{1}{4} \begin{bmatrix} \quad \\ \quad \end{bmatrix}$$

$$= \begin{bmatrix} \frac{1}{2} & \frac{1}{4} \\ \frac{1}{2} & \frac{3}{4} \end{bmatrix}$$

b. For the matrix B, you have

$$ad - bc = (\ )(\ ) - (\quad)(\quad)$$
$$= 0$$

which means that B is not invertible.

CHECK*Point*   Now try Exercise 35.

Systems of Linear Equations

You know that a system of linear equations can have exactly one solution, infinitely many solutions, or no solution. If the coefficient matrix A of a *square* system (a system that has the same number of equations as variables) is invertible, the system has a unique solution, which is defined as follows.

A System of Equations with a Unique Solution

If A is an invertible matrix, the system of linear equations represented by $AX = B$ has a unique solution given by

$$X = A^{-1}B.$$

TECHNOLOGY

To solve a system of equations with a graphing utility, enter the matrices A and B in the matrix editor. Then, using the inverse key, solve for X.

A ⓧ⁻¹ B ⟨ENTER⟩

The screen will display the solution, matrix X.

Example 5   Solving a System Using an Inverse Matrix

You are going to invest \$10,000 in AAA-rated bonds, AA-rated bonds, and B-rated bonds and want an annual return of \$730. The average yields are 6% on AAA bonds, 7.5% on AA bonds, and 9.5% on B bonds. You will invest twice as much in AAA bonds as in B bonds. Your investment can be represented as

$$\begin{cases} x + y + z = 10{,}000 \\ 0.06x + 0.075y + 0.095z = 730 \\ x - 2z = 0 \end{cases}$$

where x, y, and z represent the amounts invested in AAA, AA, and B bonds, respectively. Use an inverse matrix to solve the system.

Solution

Begin by writing the system in the matrix form $AX = B$.

$$\begin{bmatrix} 1 & 1 & 1 \\ 0.06 & 0.075 & 0.095 \\ 1 & 0 & -2 \end{bmatrix} \begin{bmatrix} x \\ y \\ z \end{bmatrix} = \begin{bmatrix} 10{,}000 \\ 730 \\ 0 \end{bmatrix}$$

Then, use Gauss-Jordan elimination to find A^{-1}.

$$A^{-1} = \begin{bmatrix} 15 & -200 & -2 \\ -21.5 & 300 & 3.5 \\ 7.5 & -100 & -1.5 \end{bmatrix}$$

Finally, multiply B by A^{-1} on the left to obtain the solution.

$$X = A^{-1}B$$

$$= \begin{bmatrix} 15 & -200 & -2 \\ -21.5 & 300 & 3.5 \\ 7.5 & -100 & -1.5 \end{bmatrix} \begin{bmatrix} 10{,}000 \\ 730 \\ 0 \end{bmatrix} = \begin{bmatrix} 4000 \\ 4000 \\ 2000 \end{bmatrix}$$

The solution of the system is $x = 4000$, $y = 4000$, and $z = 2000$. So, you will invest \$4000 in AAA bonds, \$4000 in AA bonds, and \$2000 in B bonds.

CHECK Point   Now try Exercise 63.

7.3   EXERCISES

See www.CalcChat.com for worked-out solutions to odd-numbered exercises.

VOCABULARY: Fill in the blanks.

1. In a _____ matrix, the number of rows equals the number of columns.

2. If there exists an $n \times n$ matrix A^{-1} such that $AA^{-1} = I_n = A^{-1}A$, then A^{-1} is called the _____ of A.

3. If a matrix A has an inverse, it is called invertible or _____; if it does not have an inverse, it is called _____.

4. If A is an invertible matrix, the system of linear equations represented by $AX = B$ has a unique solution given by $X = $ _____.

SKILLS AND APPLICATIONS

In Exercises 5–12, show that B is the inverse of A.

5. $A = \begin{bmatrix} 2 & 1 \\ 5 & 3 \end{bmatrix}, B = \begin{bmatrix} 3 & -1 \\ -5 & 2 \end{bmatrix}$

6. $A = \begin{bmatrix} 1 & -1 \\ -1 & 2 \end{bmatrix}, B = \begin{bmatrix} 2 & 1 \\ 1 & 1 \end{bmatrix}$

7. $A = \begin{bmatrix} 1 & 2 \\ 3 & 4 \end{bmatrix}, B = \begin{bmatrix} -2 & 1 \\ \frac{3}{2} & -\frac{1}{2} \end{bmatrix}$

8. $A = \begin{bmatrix} 1 & -1 \\ 2 & 3 \end{bmatrix}, B = \begin{bmatrix} \frac{3}{5} & \frac{1}{5} \\ -\frac{2}{5} & \frac{1}{5} \end{bmatrix}$

9. $A = \begin{bmatrix} 2 & -17 & 11 \\ -1 & 11 & -7 \\ 0 & 3 & -2 \end{bmatrix}, B = \begin{bmatrix} 1 & 1 & 2 \\ 2 & 4 & -3 \\ 3 & 6 & -5 \end{bmatrix}$

10. $A = \begin{bmatrix} -4 & 1 & 5 \\ -1 & 2 & 4 \\ 0 & -1 & -1 \end{bmatrix}, B = \begin{bmatrix} -\frac{1}{2} & 1 & \frac{3}{2} \\ \frac{1}{4} & -1 & -\frac{11}{4} \\ -\frac{1}{4} & 1 & \frac{7}{4} \end{bmatrix}$

11. $A = \begin{bmatrix} -2 & 2 & 3 \\ 1 & -1 & 0 \\ 0 & 1 & 4 \end{bmatrix}, B = \frac{1}{3}\begin{bmatrix} -4 & -5 & 3 \\ -4 & -8 & 3 \\ 1 & 2 & 0 \end{bmatrix}$

12. $A = \begin{bmatrix} -1 & 1 & 0 & -1 \\ 1 & -1 & 1 & 0 \\ -1 & 1 & 2 & 0 \\ 0 & -1 & 1 & 1 \end{bmatrix},$

$B = \frac{1}{3}\begin{bmatrix} -3 & 1 & 1 & -3 \\ -3 & -1 & 2 & -3 \\ 0 & 1 & 1 & 0 \\ -3 & -2 & 1 & 0 \end{bmatrix}$

In Exercises 13–24, find the inverse of the matrix (if it exists).

13. $\begin{bmatrix} 2 & 0 \\ 0 & 3 \end{bmatrix}$

14. $\begin{bmatrix} 1 & 2 \\ 3 & 7 \end{bmatrix}$

15. $\begin{bmatrix} 1 & -2 \\ 2 & -3 \end{bmatrix}$

16. $\begin{bmatrix} -7 & 33 \\ 4 & -19 \end{bmatrix}$

17. $\begin{bmatrix} 3 & 1 \\ 4 & 2 \end{bmatrix}$

18. $\begin{bmatrix} 4 & -1 \\ -3 & 1 \end{bmatrix}$

19. $\begin{bmatrix} 1 & 1 & 1 \\ 3 & 5 & 4 \\ 3 & 6 & 5 \end{bmatrix}$

20. $\begin{bmatrix} 1 & 2 & 2 \\ 3 & 7 & 9 \\ -1 & -4 & -7 \end{bmatrix}$

21. $\begin{bmatrix} -5 & 0 & 0 \\ 2 & 0 & 0 \\ -1 & 5 & 7 \end{bmatrix}$

22. $\begin{bmatrix} 1 & 0 & 0 \\ 3 & 0 & 0 \\ 2 & 5 & 5 \end{bmatrix}$

23. $\begin{bmatrix} -8 & 0 & 0 & 0 \\ 0 & 1 & 0 & 0 \\ 0 & 0 & 4 & 0 \\ 0 & 0 & 0 & -5 \end{bmatrix}$

24. $\begin{bmatrix} 1 & 3 & -2 & 0 \\ 0 & 2 & 4 & 6 \\ 0 & 0 & -2 & 1 \\ 0 & 0 & 0 & 5 \end{bmatrix}$

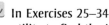 In Exercises 25–34, use the matrix capabilities of a graphing utility to find the inverse of the matrix (if it exists).

25. $\begin{bmatrix} 1 & 2 & -1 \\ 3 & 7 & -10 \\ -5 & -7 & -15 \end{bmatrix}$

26. $\begin{bmatrix} 10 & 5 & -7 \\ -5 & 1 & 4 \\ 3 & 2 & -2 \end{bmatrix}$

27. $\begin{bmatrix} 1 & 1 & 2 \\ 3 & 1 & 0 \\ -2 & 0 & 3 \end{bmatrix}$

28. $\begin{bmatrix} 3 & 2 & 2 \\ 2 & 2 & 2 \\ -4 & 4 & 3 \end{bmatrix}$

29. $\begin{bmatrix} -\frac{1}{2} & \frac{3}{4} & \frac{1}{4} \\ 1 & 0 & -\frac{3}{2} \\ 0 & -1 & \frac{1}{2} \end{bmatrix}$

30. $\begin{bmatrix} -\frac{5}{6} & \frac{1}{3} & \frac{11}{6} \\ 0 & \frac{2}{3} & 2 \\ 1 & -\frac{1}{2} & -\frac{5}{2} \end{bmatrix}$

31. $\begin{bmatrix} 0.1 & 0.2 & 0.3 \\ -0.3 & 0.2 & 0.2 \\ 0.5 & 0.4 & 0.4 \end{bmatrix}$

32. $\begin{bmatrix} 0.6 & 0 & -0.3 \\ 0.7 & -1 & 0.2 \\ 1 & 0 & -0.9 \end{bmatrix}$

33. $\begin{bmatrix} -1 & 0 & 1 & 0 \\ 0 & 2 & 0 & -1 \\ 2 & 0 & -1 & 0 \\ 0 & -1 & 0 & 1 \end{bmatrix}$

34. $\begin{bmatrix} 1 & -2 & -1 & -2 \\ 3 & -5 & -2 & -3 \\ 2 & -5 & -2 & -5 \\ -1 & 4 & 4 & 11 \end{bmatrix}$

In Exercises 35–40, use the formula on page 553 to find the inverse of the 2×2 matrix (if it exists).

35. $\begin{bmatrix} 2 & 3 \\ -1 & 5 \end{bmatrix}$

36. $\begin{bmatrix} 1 & -2 \\ -3 & 2 \end{bmatrix}$

37. $\begin{bmatrix} -4 & -6 \\ 2 & 3 \end{bmatrix}$

38. $\begin{bmatrix} -12 & 3 \\ 5 & -2 \end{bmatrix}$

39. $\begin{bmatrix} \frac{7}{2} & -\frac{3}{4} \\ \frac{1}{5} & \frac{4}{5} \end{bmatrix}$
40. $\begin{bmatrix} -\frac{1}{4} & \frac{9}{4} \\ \frac{5}{3} & \frac{8}{9} \end{bmatrix}$

In Exercises 41–44, use the inverse matrix found in Exercise 15 to solve the system of linear equations.

41. $\begin{cases} x - 2y = 5 \\ 2x - 3y = 10 \end{cases}$
42. $\begin{cases} x - 2y = 0 \\ 2x - 3y = 3 \end{cases}$

43. $\begin{cases} x - 2y = 4 \\ 2x - 3y = 2 \end{cases}$
44. $\begin{cases} x - 2y = 1 \\ 2x - 3y = -2 \end{cases}$

In Exercises 45 and 46, use the inverse matrix found in Exercise 19 to solve the system of linear equations.

45. $\begin{cases} x + y + z = 0 \\ 3x + 5y + 4z = 5 \\ 3x + 6y + 5z = 2 \end{cases}$
46. $\begin{cases} x + y + z = -1 \\ 3x + 5y + 4z = 2 \\ 3x + 6y + 5z = 0 \end{cases}$

In Exercises 47 and 48, use the inverse matrix found in Exercise 34 to solve the system of linear equations.

47. $\begin{cases} x_1 - 2x_2 - x_3 - 2x_4 = 0 \\ 3x_1 - 5x_2 - 2x_3 - 3x_4 = 1 \\ 2x_1 - 5x_2 - 2x_3 - 5x_4 = -1 \\ -x_1 + 4x_2 + 4x_3 + 11x_4 = 2 \end{cases}$

48. $\begin{cases} x_1 - 2x_2 - x_3 - 2x_4 = 1 \\ 3x_1 - 5x_2 - 2x_3 - 3x_4 = -2 \\ 2x_1 - 5x_2 - 2x_3 - 5x_4 = 0 \\ -x_1 + 4x_2 + 4x_3 + 11x_4 = -3 \end{cases}$

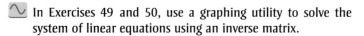 In Exercises 49 and 50, use a graphing utility to solve the system of linear equations using an inverse matrix.

49. $\begin{aligned} x_1 + 2x_2 - x_3 + 3x_4 - x_5 &= -3 \\ x_1 - 3x_2 + x_3 + 2x_4 - x_5 &= -3 \\ 2x_1 + x_2 + x_3 - 3x_4 + x_5 &= 6 \\ x_1 - x_2 + 2x_3 + x_4 - x_5 &= 2 \\ 2x_1 + x_2 - x_3 + 2x_4 + x_5 &= -3 \end{aligned}$

50. $\begin{aligned} x_1 + x_2 - x_3 + 3x_4 - x_5 &= 3 \\ 2x_1 + x_2 + x_3 + x_4 + x_5 &= 4 \\ x_1 + x_2 - x_3 + 2x_4 - x_5 &= 3 \\ 2x_1 + x_2 + 4x_3 + x_4 - x_5 &= -1 \\ 3x_1 + x_2 + x_3 - 2x_4 + x_5 &= 5 \end{aligned}$

In Exercises 51–58, use an inverse matrix to solve (if possible) the system of linear equations.

51. $\begin{cases} 3x + 4y = -2 \\ 5x + 3y = 4 \end{cases}$
52. $\begin{cases} 18x + 12y = 13 \\ 30x + 24y = 23 \end{cases}$

53. $\begin{cases} -0.4x + 0.8y = 1.6 \\ 2x - 4y = 5 \end{cases}$
54. $\begin{cases} 0.2x - 0.6y = 2.4 \\ -x + 1.4y = -8.8 \end{cases}$

55. $\begin{cases} -\frac{1}{4}x + \frac{3}{8}y = -2 \\ \frac{3}{2}x + \frac{3}{4}y = -12 \end{cases}$
56. $\begin{cases} \frac{5}{6}x - y = -20 \\ \frac{4}{3}x - \frac{7}{2}y = -51 \end{cases}$

57. $\begin{cases} 4x - y + z = -5 \\ 2x + 2y + 3z = 10 \\ 5x - 2y + 6z = 1 \end{cases}$
58. $\begin{cases} 4x - 2y + 3z = -2 \\ 2x + 2y + 5z = 16 \\ 8x - 5y - 2z = 4 \end{cases}$

In Exercises 59–62, use the matrix capabilities of a graphing utility to solve (if possible) the system of linear equations.

59. $\begin{cases} 5x - 3y + 2z = 2 \\ 2x + 2y - 3z = 3 \\ x - 7y + 8z = -4 \end{cases}$

60. $\begin{cases} 2x + 3y + 5z = 4 \\ 3x + 5y + 9z = 7 \\ 5x + 9y + 17z = 13 \end{cases}$

61. $\begin{cases} 3x - 2y + z = -29 \\ -4x + y - 3z = 37 \\ x - 5y + z = -24 \end{cases}$

62. $\begin{cases} -8x + 7y - 10z = -151 \\ 12x + 3y - 5z = 86 \\ 15x - 9y + 2z = 187 \end{cases}$

INVESTMENT PORTFOLIO In Exercises 63–66, consider a person who invests in AAA-rated bonds, A-rated bonds, and B-rated bonds. The average yields are 6.5% on AAA bonds, 7% on A bonds, and 9% on B bonds. The person invests twice as much in B bonds as in A bonds. Let x, y, and z represent the amounts invested in AAA, A, and B bonds, respectively.

$\begin{cases} x + y + z = \text{(total investment)} \\ 0.065x + 0.07y + 0.09z = \text{(annual return)} \\ 2y - z = 0 \end{cases}$

Use the inverse of the coefficient matrix of this system to find the amount invested in each type of bond.

	Total Investment	Annual Return
63.	$10,000	$705
64.	$10,000	$760
65.	$12,000	$835
66.	$500,000	$38,000

PRODUCTION In Exercises 67–70, a small home business creates muffins, bones, and cookies for dogs. In addition to other ingredients, each muffin requires 2 units of beef, 3 units of chicken, and 2 units of liver. Each bone requires 1 unit of beef, 1 unit of chicken, and 1 unit of liver. Each cookie requires 2 units of beef, 1 unit of chicken, and 1.5 units of liver. Find the numbers of muffins, bones, and cookies that the company can create with the given amounts of ingredients.

67. 700 units of beef
    500 units of chicken
    600 units of liver

68. 525 units of beef
    480 units of chicken
    500 units of liver

69. 800 units of beef
    750 units of chicken
    725 units of liver

70. 1000 units of beef
    950 units of chicken
    900 units of liver

71. COFFEE A coffee manufacturer sells a 10-pound package that contains three flavors of coffee for $26. French vanilla coffee costs $2 per pound, hazelnut flavored coffee costs $2.50 per pound, and Swiss chocolate flavored coffee costs $3 per pound. The package contains the same amount of hazelnut as Swiss chocolate. Let f represent the number of pounds of French vanilla, h represent the number of pounds of hazelnut, and s represent the number of pounds of Swiss chocolate.

(a) Write a system of linear equations that represents the situation.

(b) Write a matrix equation that corresponds to your system.

(c) Solve your system of linear equations using an inverse matrix. Find the number of pounds of each flavor of coffee in the 10-pound package.

72. FLOWERS A florist is creating 10 centerpieces for the tables at a wedding reception. Roses cost $2.50 each, lilies cost $4 each, and irises cost $2 each. The customer has a budget of $300 allocated for the centerpieces and wants each centerpiece to contain 12 flowers, with twice as many roses as the number of irises and lilies combined.

(a) Write a system of linear equations that represents the situation.

(b) Write a matrix equation that corresponds to your system.

(c) Solve your system of linear equations using an inverse matrix. Find the number of flowers of each type that the florist can use to create the 10 centerpieces.

73. ENROLLMENT The table shows the enrollment projections (in millions) for public universities in the United States for the years 2010 through 2012. (Source: U.S. National Center for Education Statistics, *Digest of Education Statistics*)

Year	Enrollment projections
2010	13.89
2011	14.04
2012	14.20

(a) The data can be modeled by the quadratic function $y = at^2 + bt + c$. Create a system of linear equations for the data. Let t represent the year, with $t = 10$ corresponding to 2010.

(b) Use the matrix capabilities of a graphing utility to find the inverse matrix to solve the system from part (a) and find the least squares regression parabola $y = at^2 + bt + c$.

(c) Use the graphing utility to graph the parabola with the data.

(d) Do you believe the model is a reasonable predictor of future enrollments? Explain.

EXPLORATION

74. CAPSTONE If A is a 2×2 matrix $A = \begin{bmatrix} a & b \\ c & d \end{bmatrix}$, then A is invertible if and only if $ad - bc \neq 0$. If $ad - bc \neq 0$, verify that the inverse is

$$A^{-1} = \frac{1}{ad - bc} \begin{bmatrix} d & -b \\ -c & a \end{bmatrix}.$$

TRUE OR FALSE? In Exercises 75 and 76, determine whether the statement is true or false. Justify your answer.

75. Multiplication of an invertible matrix and its inverse is commutative.

76. If you multiply two square matrices and obtain the identity matrix, you can assume that the matrices are inverses of one another.

77. WRITING Explain how to determine whether the inverse of a 2×2 matrix exists. If so, explain how to find the inverse.

78. WRITING Explain in your own words how to write a system of three linear equations in three variables as a matrix equation, $AX = B$, as well as how to solve the system using an inverse matrix.

79. Consider matrices of the form

$$A = \begin{bmatrix} a_{11} & 0 & 0 & 0 & \cdots & 0 \\ 0 & a_{22} & 0 & 0 & \cdots & 0 \\ 0 & 0 & a_{33} & 0 & \cdots & 0 \\ \vdots & \vdots & \vdots & \vdots & \cdots & \vdots \\ 0 & 0 & 0 & 0 & \cdots & a_{nn} \end{bmatrix}.$$

(a) Write a 2×2 matrix and a 3×3 matrix in the form of A. Find the inverse of each.

(b) Use the result of part (a) to make a conjecture about the inverses of matrices in the form of A.

PROJECT: VIEWING TELEVISION To work an extended application analyzing the average amounts of time spent viewing television in the United States, visit this text's website at *academic.cengage.com*. (Data Source: The Nielsen Company)

7.4 THE DETERMINANT OF A SQUARE MATRIX

What you should learn

- Find the determinants of 2 × 2 matrices.
- Find minors and cofactors of square matrices.
- Find the determinants of square matrices.

Why you should learn it

Determinants are often used in other branches of mathematics. For instance, Exercises 85–90 on page 565 show some types of determinants that are useful when changes in variables are made in calculus.

The Determinant of a 2 × 2 Matrix

Every *square* matrix can be associated with a real number called its **determinant.** Determinants have many uses, and several will be discussed in this and the next section. Historically, the use of determinants arose from special number patterns that occur when systems of linear equations are solved. For instance, the system

$$\begin{cases} a_1x + b_1y = c_1 \\ a_2x + b_2y = c_2 \end{cases}$$

has a solution

$$x = \frac{c_1b_2 - c_2b_1}{a_1b_2 - a_2b_1} \quad \text{and} \quad y = \frac{a_1c_2 - a_2c_1}{a_1b_2 - a_2b_1}$$

provided that $a_1b_2 - a_2b_1 \neq 0$. Note that the denominators of the two fractions are the same. This denominator is called the *determinant* of the coefficient matrix of the system.

Coefficient Matrix   *Determinant*

$$A = \begin{bmatrix} a_1 & b_1 \\ a_2 & b_2 \end{bmatrix} \qquad \det(A) = a_1b_2 - a_2b_1$$

The determinant of the matrix A can also be denoted by vertical bars on both sides of the matrix, as indicated in the following definition.

Definition of the Determinant of a 2 × 2 Matrix

The **determinant** of the matrix

$$A = \begin{bmatrix} a_1 & b_1 \\ a_2 & b_2 \end{bmatrix}$$

is given by

$$\det(A) = |A| = \begin{vmatrix} a_1 & b_1 \\ a_2 & b_2 \end{vmatrix} = a_1b_2 - a_2b_1.$$

In this text, $\det(A)$ and $|A|$ are used interchangeably to represent the determinant of A. Although vertical bars are also used to denote the absolute value of a real number, the context will show which use is intended.

A convenient method for remembering the formula for the determinant of a 2 × 2 matrix is shown in the following diagram.

$$\det(A) = \begin{vmatrix} a_1 & b_1 \\ a_2 & b_2 \end{vmatrix} = a_1b_2 - a_2b_1$$

Note that the determinant is the difference of the products of the two diagonals of the matrix.

| Example 1   The Determinant of a 2 × 2 Matrix

Find the determinant of each matrix.

a. $A = \begin{bmatrix} 2 & -3 \\ 1 & 2 \end{bmatrix}$

b. $B = \begin{bmatrix} 2 & 1 \\ 4 & 2 \end{bmatrix}$

c. $C = \begin{bmatrix} 0 & \frac{3}{2} \\ 2 & 4 \end{bmatrix}$

Solution

a. $\det(A) = \begin{vmatrix} 2 & -3 \\ 1 & 2 \end{vmatrix}$

$= 2(2) - 1(-3)$

$= 4 + 3 = 7$

b. $\det(B) = \begin{vmatrix} 2 & 1 \\ 4 & 2 \end{vmatrix}$

$= 2(2) - 4(1)$

$= 4 - 4 = 0$

c. $\det(C) = \begin{vmatrix} 0 & \frac{3}{2} \\ 2 & 4 \end{vmatrix}$

$= 0(4) - 2\left(\frac{3}{2}\right)$

$= 0 - 3 = -3$

CHECK*Point*   Now try Exercise 9.

Notice in Example 1 that the determinant of a matrix can be positive, zero, or negative.

The determinant of a matrix of order 1×1 is defined simply as the entry of the matrix. For instance, if $A = \begin{bmatrix} -2 \end{bmatrix}$, then $\det(A) = -2$.

TECHNOLOGY

Most graphing utilities can evaluate the determinant of a matrix. For instance, you can evaluate the determinant of

$$A = \begin{bmatrix} 2 & -3 \\ 1 & 2 \end{bmatrix}$$

by entering the matrix as [A] and then choosing the *determinant* feature. The result should be 7, as in Example 1(a). Try evaluating the determinants of other matrices. Consult the user's guide for your graphing utility for specific keystrokes.

Minors and Cofactors

To define the determinant of a square matrix of order 3×3 or higher, it is convenient to introduce the concepts of **minors** and **cofactors.**

Sign Pattern for Cofactors

$$\begin{bmatrix} + & - & + \\ - & + & - \\ + & - & + \end{bmatrix}$$

$$\begin{bmatrix} + & - & + & - \\ - & + & - & + \\ + & - & + & - \\ - & + & - & + \end{bmatrix}$$

$$\begin{bmatrix} + & - & + & - & + & \cdots \\ - & + & - & + & - & \cdots \\ + & - & + & - & + & \cdots \\ - & + & - & + & - & \cdots \\ + & - & + & - & + & \cdots \\ \vdots & \vdots & \vdots & \vdots & \vdots & \end{bmatrix}$$

> ### Minors and Cofactors of a Square Matrix
>
> If A is a square matrix, the **minor** M_{ij} of the entry a_{ij} is the determinant of the matrix obtained by deleting the ith row and jth column of A. The **cofactor** C_{ij} of the entry a_{ij} is
>
> $$C_{ij} = (-1)^{i+j} M_{ij}.$$

In the sign pattern for cofactors at the left, notice that *odd* positions (where $i + j$ is odd) have negative signs and *even* positions (where $i + j$ is even) have positive signs.

Example 2   Finding the Minors and Cofactors of a Matrix

Find all the minors and cofactors of

$$A = \begin{bmatrix} 0 & 2 & 1 \\ 3 & -1 & 2 \\ 4 & 0 & 1 \end{bmatrix}.$$

Solution

To find the minor M_{11}, delete the first row and first column of A and evaluate the determinant of the resulting matrix.

$$\begin{bmatrix} 0 & 2 & 1 \\ 3 & -1 & 2 \\ 4 & 0 & 1 \end{bmatrix}, \quad M_{11} = \begin{vmatrix} -1 & 2 \\ 0 & 1 \end{vmatrix} = -1(1) - 0(2) = -1$$

Similarly, to find M_{12}, delete the first row and second column.

$$\begin{bmatrix} 0 & 2 & 1 \\ 3 & -1 & 2 \\ 4 & 0 & 1 \end{bmatrix}, \quad M_{12} = \begin{vmatrix} 3 & 2 \\ 4 & 1 \end{vmatrix} = 3(1) - 4(2) = -5$$

Continuing this pattern, you obtain the minors.

$$M_{11} = -1 \qquad M_{12} = -5 \qquad M_{13} = 4$$
$$M_{21} = 2 \qquad M_{22} = -4 \qquad M_{23} = -8$$
$$M_{31} = 5 \qquad M_{32} = -3 \qquad M_{33} = -6$$

Now, to find the cofactors, combine these minors with the checkerboard pattern of signs for a 3×3 matrix shown at the upper left.

$$C_{11} = -1 \qquad C_{12} = 5 \qquad C_{13} = 4$$
$$C_{21} = -2 \qquad C_{22} = -4 \qquad C_{23} = 8$$
$$C_{31} = 5 \qquad C_{32} = 3 \qquad C_{33} = -6$$

CHECK*Point* Now try Exercise 29.

The Determinant of a Square Matrix

The definition below is called *inductive* because it uses determinants of matrices of order $n - 1$ to define determinants of matrices of order n.

Determinant of a Square Matrix

If A is a square matrix (of order 2×2 or greater), the determinant of A is the sum of the entries in any row (or column) of A multiplied by their respective cofactors. For instance, expanding along the first row yields

$$|A| = a_{11}C_{11} + a_{12}C_{12} + \cdots + a_{1n}C_{1n}.$$

Applying this definition to find a determinant is called **expanding by cofactors.**

Try checking that for a 2×2 matrix

$$A = \begin{bmatrix} a_1 & b_1 \\ a_2 & b_2 \end{bmatrix}$$

this definition of the determinant yields $|A| = a_1 b_2 - a_2 b_1$, as previously defined.

Example 3   The Determinant of a Matrix of Order 3 × 3

Find the determinant of

$$A = \begin{bmatrix} 0 & 2 & 1 \\ 3 & -1 & 2 \\ 4 & 0 & 1 \end{bmatrix}.$$

Solution

Note that this is the same matrix that was in Example 2. There you found the cofactors of the entries in the first row to be

$$C_{11} = -1, \quad C_{12} = 5, \quad \text{and} \quad C_{13} = 4.$$

So, by the definition of a determinant, you have

$$|A| = a_{11}C_{11} + a_{12}C_{12} + a_{13}C_{13}$$
$$= (\quad) + (\ ) + (\ )$$
$$= 14.$$

CHECK*Point*   Now try Exercise 39.

In Example 3, the determinant was found by expanding by the cofactors in the first row. You could have used any row or column. For instance, you could have expanded along the second row to obtain

$$|A| = a_{21}C_{21} + a_{22}C_{22} + a_{23}C_{23}$$
$$= (\quad) + (\quad)(\quad) + (\ )$$
$$= 14.$$

When expanding by cofactors, you do not need to find cofactors of zero entries, because zero times its cofactor is zero.

$$a_{ij}C_{ij} = (0)C_{ij} = 0$$

So, the row (or column) containing the most zeros is usually the best choice for expansion by cofactors. This is demonstrated in the next example.

Example 4   The Determinant of a Matrix of Order 4 × 4

Find the determinant of

$$A = \begin{bmatrix} 1 & -2 & 3 & 0 \\ -1 & 1 & 0 & 2 \\ 0 & 2 & 0 & 3 \\ 3 & 4 & 0 & 2 \end{bmatrix}.$$

Solution

After inspecting this matrix, you can see that three of the entries in the third column are zeros. So, you can eliminate some of the work in the expansion by using the third column.

$$|A| = 3(C_{13}) + 0(C_{23}) + 0(C_{33}) + 0(C_{43})$$

Because C_{23}, C_{33}, and C_{43} have zero coefficients, you need only find the cofactor C_{13}. To do this, delete the first row and third column of A and evaluate the determinant of the resulting matrix.

$$C_{13} = (-1)^{1+3} \begin{vmatrix} -1 & 1 & 2 \\ 0 & 2 & 3 \\ 3 & 4 & 2 \end{vmatrix}$$

$$= \begin{vmatrix} -1 & 1 & 2 \\ 0 & 2 & 3 \\ 3 & 4 & 2 \end{vmatrix}$$

Expanding by cofactors in the second row yields

$$C_{13} = 0(-1)^3 \begin{vmatrix} 1 & 2 \\ 4 & 2 \end{vmatrix} + 2(-1)^4 \begin{vmatrix} -1 & 2 \\ 3 & 2 \end{vmatrix} + 3(-1)^5 \begin{vmatrix} -1 & 1 \\ 3 & 4 \end{vmatrix}$$

$$= 0 + 2(1)(-8) + 3(-1)(-7)$$

$$= 5.$$

So, you obtain

$$|A| = 3C_{13}$$

$$= 3(\ )$$

$$= 15.$$

CHECK**Point** Now try Exercise 49.

Try using a graphing utility to confirm the result of Example 4.

7.4   EXERCISES

See www.CalcChat.com for worked-out solutions to odd-numbered exercises.

VOCABULARY: Fill in the blanks.

1. Both $\det(A)$ and $|A|$ represent the _____ of the matrix A.

2. The _____ M_{ij} of the entry a_{ij} is the determinant of the matrix obtained by deleting the ith row and jth column of the square matrix A.

3. The _____ C_{ij} of the entry a_{ij} of the square matrix A is given by $(-1)^{i+j}M_{ij}$.

4. The method of finding the determinant of a matrix of order 2×2 or greater is called _____ by _____.

SKILLS AND APPLICATIONS

In Exercises 5–20, find the determinant of the matrix.

5. $\begin{bmatrix} 4 \end{bmatrix}$

6. $\begin{bmatrix} -10 \end{bmatrix}$

7. $\begin{bmatrix} 8 & 4 \\ 2 & 3 \end{bmatrix}$

8. $\begin{bmatrix} -9 & 0 \\ 6 & 2 \end{bmatrix}$

9. $\begin{bmatrix} 6 & 2 \\ -5 & 3 \end{bmatrix}$

10. $\begin{bmatrix} 3 & -3 \\ 4 & -8 \end{bmatrix}$

11. $\begin{bmatrix} -7 & 0 \\ 3 & 0 \end{bmatrix}$

12. $\begin{bmatrix} 4 & -3 \\ 0 & 0 \end{bmatrix}$

13. $\begin{bmatrix} 2 & 6 \\ 0 & 3 \end{bmatrix}$

14. $\begin{bmatrix} 2 & -3 \\ -6 & 9 \end{bmatrix}$

15. $\begin{bmatrix} -3 & -2 \\ -6 & -1 \end{bmatrix}$

16. $\begin{bmatrix} 4 & 7 \\ -2 & 5 \end{bmatrix}$

17. $\begin{bmatrix} -7 & 6 \\ \frac{1}{2} & 3 \end{bmatrix}$

18. $\begin{bmatrix} 0 & 6 \\ -3 & 2 \end{bmatrix}$

19. $\begin{bmatrix} -\frac{1}{2} & \frac{1}{3} \\ -6 & \frac{1}{3} \end{bmatrix}$

20. $\begin{bmatrix} \frac{2}{3} & \frac{4}{3} \\ -1 & -\frac{1}{3} \end{bmatrix}$

In Exercises 21–24, use the matrix capabilities of a graphing utility to find the determinant of the matrix.

21. $\begin{bmatrix} 0.3 & 0.2 & 0.2 \\ 0.2 & 0.2 & 0.2 \\ -0.4 & 0.4 & 0.3 \end{bmatrix}$

22. $\begin{bmatrix} 0.1 & 0.2 & 0.3 \\ -0.3 & 0.2 & 0.2 \\ 0.5 & 0.4 & 0.4 \end{bmatrix}$

23. $\begin{bmatrix} 0.9 & 0.7 & 0 \\ -0.1 & 0.3 & 1.3 \\ -2.2 & 4.2 & 6.1 \end{bmatrix}$

24. $\begin{bmatrix} 0.1 & 0.1 & -4.3 \\ 7.5 & 6.2 & 0.7 \\ 0.3 & 0.6 & -1.2 \end{bmatrix}$

In Exercises 25–32, find all (a) minors and (b) cofactors of the matrix.

25. $\begin{bmatrix} 4 & 5 \\ 3 & -6 \end{bmatrix}$

26. $\begin{bmatrix} 0 & 10 \\ 3 & -4 \end{bmatrix}$

27. $\begin{bmatrix} 3 & 1 \\ -2 & -4 \end{bmatrix}$

28. $\begin{bmatrix} -6 & 5 \\ 7 & -2 \end{bmatrix}$

29. $\begin{bmatrix} 4 & 0 & 2 \\ -3 & 2 & 1 \\ 1 & -1 & 1 \end{bmatrix}$

30. $\begin{bmatrix} 1 & -1 & 0 \\ 3 & 2 & 5 \\ 4 & -6 & 4 \end{bmatrix}$

31. $\begin{bmatrix} -4 & 6 & 3 \\ 7 & -2 & 8 \\ 1 & 0 & -5 \end{bmatrix}$

32. $\begin{bmatrix} -2 & 9 & 4 \\ 7 & -6 & 0 \\ 6 & 7 & -6 \end{bmatrix}$

In Exercises 33–38, find the determinant of the matrix by the method of expansion by cofactors. Expand using the indicated row or column.

33. $\begin{bmatrix} -3 & 2 & 1 \\ 4 & 5 & 6 \\ 2 & -3 & 1 \end{bmatrix}$

   (a) Row 1
   (b) Column 2

34. $\begin{bmatrix} -3 & 4 & 2 \\ 6 & 3 & 1 \\ 4 & -7 & -8 \end{bmatrix}$

   (a) Row 2
   (b) Column 3

35. $\begin{bmatrix} 5 & 0 & -3 \\ 0 & 12 & 4 \\ 1 & 6 & 3 \end{bmatrix}$

   (a) Row 2
   (b) Column 2

36. $\begin{bmatrix} 10 & -5 & 5 \\ 30 & 0 & 10 \\ 0 & 10 & 1 \end{bmatrix}$

   (a) Row 3
   (b) Column 1

37. $\begin{bmatrix} 6 & 0 & -3 & 5 \\ 4 & 13 & 6 & -8 \\ -1 & 0 & 7 & 4 \\ 8 & 6 & 0 & 2 \end{bmatrix}$

   (a) Row 2
   (b) Column 2

38. $\begin{bmatrix} 10 & 8 & 3 & -7 \\ 4 & 0 & 5 & -6 \\ 0 & 3 & 2 & 7 \\ 1 & 0 & -3 & 2 \end{bmatrix}$

   (a) Row 3
   (b) Column 1

In Exercises 39–54, find the determinant of the matrix. Expand by cofactors on the row or column that appears to make the computations easiest.

39. $\begin{bmatrix} 2 & -1 & 0 \\ 4 & 2 & 1 \\ 4 & 2 & 1 \end{bmatrix}$

40. $\begin{bmatrix} -2 & 2 & 3 \\ 1 & -1 & 0 \\ 0 & 1 & 4 \end{bmatrix}$

41. $\begin{bmatrix} 6 & 3 & -7 \\ 0 & 0 & 0 \\ 4 & -6 & 3 \end{bmatrix}$

42. $\begin{bmatrix} 1 & 1 & 2 \\ 3 & 1 & 0 \\ -2 & 0 & 3 \end{bmatrix}$

43. $\begin{bmatrix} -1 & 8 & -3 \\ 0 & 3 & -6 \\ 0 & 0 & 3 \end{bmatrix}$

44. $\begin{bmatrix} 1 & 0 & 0 \\ -1 & -1 & 0 \\ 4 & 11 & 5 \end{bmatrix}$

45. $\begin{bmatrix} 1 & 4 & -2 \\ 3 & 2 & 0 \\ -1 & 4 & 3 \end{bmatrix}$ **46.** $\begin{bmatrix} 2 & -1 & 3 \\ 1 & 4 & 4 \\ 1 & 0 & 2 \end{bmatrix}$

47. $\begin{bmatrix} 2 & 4 & 6 \\ 0 & 3 & 1 \\ 0 & 0 & -5 \end{bmatrix}$ **48.** $\begin{bmatrix} -3 & 0 & 0 \\ 7 & 11 & 0 \\ 1 & 2 & 2 \end{bmatrix}$

49. $\begin{bmatrix} 2 & 6 & 6 & 2 \\ 2 & 7 & 3 & 6 \\ 1 & 5 & 0 & 1 \\ 3 & 7 & 0 & 7 \end{bmatrix}$ **50.** $\begin{bmatrix} 3 & 6 & -5 & 4 \\ -2 & 0 & 6 & 0 \\ 1 & 1 & 2 & 2 \\ 0 & 3 & -1 & -1 \end{bmatrix}$

51. $\begin{bmatrix} 5 & 3 & 0 & 6 \\ 4 & 6 & 4 & 12 \\ 0 & 2 & -3 & 4 \\ 0 & 1 & -2 & 2 \end{bmatrix}$ **52.** $\begin{bmatrix} 1 & 4 & 3 & 2 \\ -5 & 6 & 2 & 1 \\ 0 & 0 & 0 & 0 \\ 3 & -2 & 1 & 5 \end{bmatrix}$

53. $\begin{bmatrix} 3 & 2 & 4 & -1 & 5 \\ -2 & 0 & 1 & 3 & 2 \\ 1 & 0 & 0 & 4 & 0 \\ 6 & 0 & 2 & -1 & 0 \\ 3 & 0 & 5 & 1 & 0 \end{bmatrix}$

54. $\begin{bmatrix} 5 & 2 & 0 & 0 & -2 \\ 0 & 1 & 4 & 3 & 2 \\ 0 & 0 & 2 & 6 & 3 \\ 0 & 0 & 3 & 4 & 1 \\ 0 & 0 & 0 & 0 & 2 \end{bmatrix}$

In Exercises 55–62, use the matrix capabilities of a graphing utility to evaluate the determinant.

55. $\begin{vmatrix} 3 & 8 & -7 \\ 0 & -5 & 4 \\ 8 & 1 & 6 \end{vmatrix}$ **56.** $\begin{vmatrix} 5 & -8 & 0 \\ 9 & 7 & 4 \\ -8 & 7 & 1 \end{vmatrix}$

57. $\begin{vmatrix} 7 & 0 & -14 \\ -2 & 5 & 4 \\ -6 & 2 & 12 \end{vmatrix}$ **58.** $\begin{vmatrix} 3 & 0 & 0 \\ -2 & 5 & 0 \\ 12 & 5 & 7 \end{vmatrix}$

59. $\begin{vmatrix} 1 & -1 & 8 & 4 \\ 2 & 6 & 0 & -4 \\ 2 & 0 & 2 & 6 \\ 0 & 2 & 8 & 0 \end{vmatrix}$ **60.** $\begin{vmatrix} 0 & -3 & 8 & 2 \\ 8 & 1 & -1 & 6 \\ -4 & 6 & 0 & 9 \\ -7 & 0 & 0 & 14 \end{vmatrix}$

61. $\begin{vmatrix} 3 & -2 & 4 & 3 & 1 \\ -1 & 0 & 2 & 1 & 0 \\ 5 & -1 & 0 & 3 & 2 \\ 4 & 7 & -8 & 0 & 0 \\ 1 & 2 & 3 & 0 & 2 \end{vmatrix}$

62. $\begin{vmatrix} -2 & 0 & 0 & 0 & 0 \\ 0 & 3 & 0 & 0 & 0 \\ 0 & 0 & -1 & 0 & 0 \\ 0 & 0 & 0 & 2 & 0 \\ 0 & 0 & 0 & 0 & -4 \end{vmatrix}$

In Exercises 63–70, find (a) $|A|$, (b) $|B|$, (c) AB, and (d) $|AB|$.

63. $A = \begin{bmatrix} -1 & 0 \\ 0 & 3 \end{bmatrix}$, $B = \begin{bmatrix} 2 & 0 \\ 0 & -1 \end{bmatrix}$

64. $A = \begin{bmatrix} -2 & 1 \\ 4 & -2 \end{bmatrix}$, $B = \begin{bmatrix} 1 & 2 \\ 0 & -1 \end{bmatrix}$

65. $A = \begin{bmatrix} 4 & 0 \\ 3 & -2 \end{bmatrix}$, $B = \begin{bmatrix} -1 & 1 \\ -2 & 2 \end{bmatrix}$

66. $A = \begin{bmatrix} 5 & 4 \\ 3 & -1 \end{bmatrix}$, $B = \begin{bmatrix} 0 & 6 \\ 1 & -2 \end{bmatrix}$

67. $A = \begin{bmatrix} 0 & 1 & 2 \\ -3 & -2 & 1 \\ 0 & 4 & 1 \end{bmatrix}$, $B = \begin{bmatrix} 3 & -2 & 0 \\ 1 & -1 & 2 \\ 3 & 1 & 1 \end{bmatrix}$

68. $A = \begin{bmatrix} 3 & 2 & 0 \\ -1 & -3 & 4 \\ -2 & 0 & 1 \end{bmatrix}$, $B = \begin{bmatrix} -3 & 0 & 1 \\ 0 & 2 & -1 \\ -2 & -1 & 1 \end{bmatrix}$

69. $A = \begin{bmatrix} -1 & 2 & 1 \\ 1 & 0 & 1 \\ 0 & 1 & 0 \end{bmatrix}$, $B = \begin{bmatrix} -1 & 0 & 0 \\ 0 & 2 & 0 \\ 0 & 0 & 3 \end{bmatrix}$

70. $A = \begin{bmatrix} 2 & 0 & 1 \\ 1 & -1 & 2 \\ 3 & 1 & 0 \end{bmatrix}$, $B = \begin{bmatrix} 2 & -1 & 4 \\ 0 & 1 & 3 \\ 3 & -2 & 1 \end{bmatrix}$

In Exercises 71–76, evaluate the determinant(s) to verify the equation.

71. $\begin{vmatrix} w & x \\ y & z \end{vmatrix} = - \begin{vmatrix} y & z \\ w & x \end{vmatrix}$ **72.** $\begin{vmatrix} w & cx \\ y & cz \end{vmatrix} = c \begin{vmatrix} w & x \\ y & z \end{vmatrix}$

73. $\begin{vmatrix} w & x \\ y & z \end{vmatrix} = \begin{vmatrix} w & x + cw \\ y & z + cy \end{vmatrix}$

74. $\begin{vmatrix} w & x \\ cw & cx \end{vmatrix} = 0$

75. $\begin{vmatrix} 1 & x & x^2 \\ 1 & y & y^2 \\ 1 & z & z^2 \end{vmatrix} = (y - x)(z - x)(z - y)$

76. $\begin{vmatrix} a + b & a & a \\ a & a + b & a \\ a & a & a + b \end{vmatrix} = b^2(3a + b)$

In Exercises 77–84, solve for x.

77. $\begin{vmatrix} x & 2 \\ 1 & x \end{vmatrix} = 2$ **78.** $\begin{vmatrix} x & 4 \\ -1 & x \end{vmatrix} = 20$

79. $\begin{vmatrix} x & 1 \\ 2 & x - 2 \end{vmatrix} = -1$ **80.** $\begin{vmatrix} x + 1 & 2 \\ -1 & x \end{vmatrix} = 4$

81. $\begin{vmatrix} x - 1 & 2 \\ 3 & x - 2 \end{vmatrix} = 0$ **82.** $\begin{vmatrix} x - 2 & -1 \\ -3 & x \end{vmatrix} = 0$

83. $\begin{vmatrix} x + 3 & 2 \\ 1 & x + 2 \end{vmatrix} = 0$ **84.** $\begin{vmatrix} x + 4 & -2 \\ 7 & x - 5 \end{vmatrix} = 0$

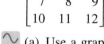 In Exercises 85–90, evaluate the determinant in which the entries are functions. Determinants of this type occur when changes of variables are made in calculus.

85. $\begin{vmatrix} 4u & -1 \\ -1 & 2v \end{vmatrix}$

86. $\begin{vmatrix} 3x^2 & -3y^2 \\ 1 & 1 \end{vmatrix}$

87. $\begin{vmatrix} e^{2x} & e^{3x} \\ 2e^{2x} & 3e^{3x} \end{vmatrix}$

88. $\begin{vmatrix} e^{-x} & xe^{-x} \\ -e^{-x} & (1-x)e^{-x} \end{vmatrix}$

89. $\begin{vmatrix} x & \ln x \\ 1 & 1/x \end{vmatrix}$

90. $\begin{vmatrix} x & x \ln x \\ 1 & 1 + \ln x \end{vmatrix}$

EXPLORATION

TRUE OR FALSE? In Exercises 91 and 92, determine whether the statement is true or false. Justify your answer.

91. If a square matrix has an entire row of zeros, the determinant will always be zero.

92. If two columns of a square matrix are the same, the determinant of the matrix will be zero.

93. Find square matrices A and B to demonstrate that $|A + B| \neq |A| + |B|$.

94. Consider square matrices in which the entries are consecutive integers. An example of such a matrix is

$$\begin{bmatrix} 4 & 5 & 6 \\ 7 & 8 & 9 \\ 10 & 11 & 12 \end{bmatrix}.$$

(a) Use a graphing utility to evaluate the determinants of four matrices of this type. Make a conjecture based on the results.

(b) Verify your conjecture.

95. **WRITING** Write a brief paragraph explaining the difference between a square matrix and its determinant.

96. **THINK ABOUT IT** If A is a matrix of order 3×3 such that $|A| = 5$, is it possible to find $|2A|$? Explain.

PROPERTIES OF DETERMINANTS In Exercises 97–99, a property of determinants is given (A and B are square matrices). State how the property has been applied to the given determinants and use a graphing utility to verify the results.

97. If B is obtained from A by interchanging two rows of A or interchanging two columns of A, then $|B| = -|A|$.

(a) $\begin{vmatrix} 1 & 3 & 4 \\ -7 & 2 & -5 \\ 6 & 1 & 2 \end{vmatrix} = -\begin{vmatrix} 1 & 4 & 3 \\ -7 & -5 & 2 \\ 6 & 2 & 1 \end{vmatrix}$

(b) $\begin{vmatrix} 1 & 3 & 4 \\ -2 & 2 & 0 \\ 1 & 6 & 2 \end{vmatrix} = -\begin{vmatrix} 1 & 6 & 2 \\ -2 & 2 & 0 \\ 1 & 3 & 4 \end{vmatrix}$

98. If B is obtained from A by adding a multiple of a row of A to another row of A or by adding a multiple of a column of A to another column of A, then $|B| = |A|$.

(a) $\begin{vmatrix} 1 & -3 \\ 5 & 2 \end{vmatrix} = \begin{vmatrix} 1 & -3 \\ 0 & 17 \end{vmatrix}$

(b) $\begin{vmatrix} 5 & 4 & 2 \\ 2 & -3 & 4 \\ 7 & 6 & 3 \end{vmatrix} = \begin{vmatrix} 1 & 10 & -6 \\ 2 & -3 & 4 \\ 7 & 6 & 3 \end{vmatrix}$

99. If B is obtained from A by multiplying a row by a nonzero constant c or by multiplying a column by a nonzero constant c, then $|B| = c|A|$.

(a) $\begin{vmatrix} 5 & 10 \\ 2 & -3 \end{vmatrix} = 5\begin{vmatrix} 1 & 2 \\ 2 & -3 \end{vmatrix}$

(b) $\begin{vmatrix} 1 & 8 & -3 \\ 3 & -12 & 6 \\ 7 & 4 & 9 \end{vmatrix} = 12\begin{vmatrix} 1 & 2 & -1 \\ 3 & -3 & 2 \\ 7 & 1 & 3 \end{vmatrix}$

100. **CAPSTONE** If A is an $n \times n$ matrix, explain how to find the following.

(a) The minor M_{ij} of the entry a_{ij}

(b) The cofactor C_{ij} of the entry a_{ij}

(c) The determinant of A

In Exercises 101–104, evaluate the determinant.

101. $\begin{vmatrix} 1 & 0 & 0 \\ 0 & 5 & 0 \\ 0 & 0 & 2 \end{vmatrix}$

102. $\begin{vmatrix} -2 & 0 & 0 & 0 \\ 0 & 2 & 0 & 0 \\ 0 & 0 & 1 & 0 \\ 0 & 0 & 0 & 3 \end{vmatrix}$

103. $\begin{vmatrix} -1 & 2 & -5 \\ 0 & 3 & 4 \\ 0 & 0 & 3 \end{vmatrix}$

104. $\begin{vmatrix} 1 & 0 & 0 \\ -4 & -1 & 0 \\ 5 & 1 & 5 \end{vmatrix}$

105. **CONJECTURE** A **triangular matrix** is a square matrix with all zero entries either below or above its main diagonal. A square matrix is **upper triangular** if it has all zero entries below its main diagonal and is **lower triangular** if it has all zero entries above its main diagonal. A matrix that is both upper and lower triangular is called diagonal. That is, a **diagonal matrix** is a square matrix in which all entries above and below the main diagonal are zero. In Exercises 101–104, you evaluated the determinants of triangular matrices. Make a conjecture based on your results.

106. Use the matrix capabilities of a graphing utility to find the determinant of A. What message appears on the screen? Why does the graphing utility display this message?

$$A = \begin{bmatrix} 1 & 2 \\ -1 & 0 \\ 3 & -2 \end{bmatrix}$$

7.5 APPLICATIONS OF MATRICES AND DETERMINANTS

What you should learn

- Use Cramer's Rule to solve systems of linear equations.
- Use determinants to find the areas of triangles.
- Use a determinant to test for collinear points and find an equation of a line passing through two points.
- Use matrices to encode and decode messages.

Why you should learn it

You can use Cramer's Rule to solve real-life problems. For instance, in Exercise 69 on page 577, Cramer's Rule is used to find a quadratic model for the per capita consumption of bottled water in the United States.

MAFORD/istockphoto.com

Cramer's Rule

So far, you have studied three methods for solving a system of linear equations: substitution, elimination with equations, and elimination with matrices. In this section, you will study one more method, **Cramer's Rule,** named after Gabriel Cramer (1704–1752). This rule uses determinants to write the solution of a system of linear equations. To see how Cramer's Rule works, take another look at the solution described at the beginning of Section 7.4. There, it was pointed out that the system

$$\begin{cases} a_1 x + b_1 y = c_1 \\ a_2 x + b_2 y = c_2 \end{cases}$$

has a solution

$$x = \frac{c_1 b_2 - c_2 b_1}{a_1 b_2 - a_2 b_1} \quad \text{and} \quad y = \frac{a_1 c_2 - a_2 c_1}{a_1 b_2 - a_2 b_1}$$

provided that $a_1 b_2 - a_2 b_1 \neq 0$. Each numerator and denominator in this solution can be expressed as a determinant, as follows.

$$x = \frac{c_1 b_2 - c_2 b_1}{a_1 b_2 - a_2 b_1} = \frac{\begin{vmatrix} c_1 & b_1 \\ c_2 & b_2 \end{vmatrix}}{\begin{vmatrix} a_1 & b_1 \\ a_2 & b_2 \end{vmatrix}} \qquad y = \frac{a_1 c_2 - a_2 c_1}{a_1 b_2 - a_2 b_1} = \frac{\begin{vmatrix} a_1 & c_1 \\ a_2 & c_2 \end{vmatrix}}{\begin{vmatrix} a_1 & b_1 \\ a_2 & b_2 \end{vmatrix}}$$

Relative to the original system, the denominator for x and y is simply the determinant of the *coefficient* matrix of the system. This determinant is denoted by D. The numerators for x and y are denoted by D_x and D_y, respectively. They are formed by using the column of constants as replacements for the coefficients of x and y, as follows.

Coefficient Matrix	D	D_x	D_y
$\begin{bmatrix} a_1 & b_1 \\ a_2 & b_2 \end{bmatrix}$	$\begin{vmatrix} a_1 & b_1 \\ a_2 & b_2 \end{vmatrix}$	$\begin{vmatrix} & b_1 \\ & b_2 \end{vmatrix}$	$\begin{vmatrix} a_1 & \\ a_2 & \end{vmatrix}$

For example, given the system

$$\begin{cases} 2x - 5y = 3 \\ -4x + 3y = 8 \end{cases}$$

the coefficient matrix, D, D_x, and D_y are as follows.

Coefficient Matrix	D	D_x	D_y
$\begin{bmatrix} 2 & -5 \\ -4 & 3 \end{bmatrix}$	$\begin{vmatrix} 2 & -5 \\ -4 & 3 \end{vmatrix}$	$\begin{vmatrix} 3 & -5 \\ 8 & 3 \end{vmatrix}$	$\begin{vmatrix} 2 & 3 \\ -4 & 8 \end{vmatrix}$

Cramer's Rule generalizes easily to systems of n equations in n variables. The value of each variable is given as the quotient of two determinants. The denominator is the determinant of the coefficient matrix, and the numerator is the determinant of the matrix formed by replacing the column corresponding to the variable (being solved for) with the column representing the constants. For instance, the solution for x_3 in the following system is shown.

$$\begin{cases} a_{11}x_1 + a_{12}x_2 + a_{13}x_3 = \\ a_{21}x_1 + a_{22}x_2 + a_{23}x_3 = \\ a_{31}x_1 + a_{32}x_2 + a_{33}x_3 = \end{cases} \qquad x_3 = \frac{|A_3|}{|A|} = \frac{\begin{vmatrix} a_{11} & a_{12} & \\ a_{21} & a_{22} & \\ a_{31} & a_{32} & \end{vmatrix}}{\begin{vmatrix} a_{11} & a_{12} & a_{13} \\ a_{21} & a_{22} & a_{23} \\ a_{31} & a_{32} & a_{33} \end{vmatrix}}$$

Cramer's Rule

If a system of n linear equations in n variables has a coefficient matrix A with a nonzero determinant $|A|$, the solution of the system is

$$x_1 = \frac{|A_1|}{|A|}, \quad x_2 = \frac{|A_2|}{|A|}, \quad \dots \quad , \quad x_n = \frac{|A_n|}{|A|}$$

where the ith column of A_i is the column of constants in the system of equations. If the determinant of the coefficient matrix is zero, the system has either no solution or infinitely many solutions.

Example 1   Using Cramer's Rule for a 2 × 2 System

Use Cramer's Rule to solve the system of linear equations.

$$\begin{cases} 4x - 2y = \\ 3x - 5y = \end{cases}$$

Solution

To begin, find the determinant of the coefficient matrix.

$$D = \begin{vmatrix} 4 & -2 \\ 3 & -5 \end{vmatrix} = -20 - (-6) = -14$$

Because this determinant is not zero, you can apply Cramer's Rule.

$$x = \frac{D_x}{D} = \frac{\begin{vmatrix} & -2 \\ & -5 \end{vmatrix}}{-14} = \frac{-50 - (-22)}{-14} = \frac{-28}{-14} = 2$$

$$y = \frac{D_y}{D} = \frac{\begin{vmatrix} 4 & \\ 3 & \end{vmatrix}}{-14} = \frac{44 - 30}{-14} = \frac{14}{-14} = -1$$

So, the solution is $x = 2$ and $y = -1$. Check this in the original system.

CHECK*Point*   Now try Exercise 7.

Example 2 Using Cramer's Rule for a 3 × 3 System

Use Cramer's Rule to solve the system of linear equations.

$$\begin{cases} -x + 2y - 3z = \\ 2x \quad\quad + z = \\ 3x - 4y + 4z = \end{cases}$$

Solution

To find the determinant of the coefficient matrix

$$\begin{bmatrix} -1 & 2 & -3 \\ 2 & 0 & 1 \\ 3 & -4 & 4 \end{bmatrix}$$

expand along the second row, as follows.

$$D = 2(-1)^3 \begin{vmatrix} 2 & -3 \\ -4 & 4 \end{vmatrix} + 0(-1)^4 \begin{vmatrix} -1 & -3 \\ 3 & 4 \end{vmatrix} + 1(-1)^5 \begin{vmatrix} -1 & 2 \\ 3 & -4 \end{vmatrix}$$

$$= -2(-4) + 0 - 1(-2)$$

$$= 10$$

Because this determinant is not zero, you can apply Cramer's Rule.

$$x = \frac{D_x}{D} = \frac{\begin{vmatrix} & 2 & -3 \\ & 0 & 1 \\ & -4 & 4 \end{vmatrix}}{10} = \frac{8}{10} = \frac{4}{5}$$

$$y = \frac{D_y}{D} = \frac{\begin{vmatrix} -1 & & -3 \\ 2 & & 1 \\ 3 & & 4 \end{vmatrix}}{10} = \frac{-15}{10} = -\frac{3}{2}$$

$$z = \frac{D_z}{D} = \frac{\begin{vmatrix} -1 & 2 & \\ 2 & 0 & \\ 3 & -4 & \end{vmatrix}}{10} = \frac{-16}{10} = -\frac{8}{5}$$

The solution is $\left(\frac{4}{5}, -\frac{3}{2}, -\frac{8}{5}\right)$. Check this in the original system as follows.

Check

$$-(\ ) + 2(\ ) - 3(\ ) \overset{?}{=} 1$$

$$-\frac{4}{5} - 3 + \frac{24}{5} = 1 \qquad\qquad ✓$$

$$2(\ ) \qquad + (\ ) \overset{?}{=} 0$$

$$\frac{8}{5} \qquad - \frac{8}{5} = 0 \qquad\qquad ✓$$

$$3(\ ) - 4(\ ) + 4(\ ) \overset{?}{=} 2$$

$$\frac{12}{5} + 6 - \frac{32}{5} = 2 \qquad\qquad ✓$$

CHECK *Point* Now try Exercise 13.

Remember that Cramer's Rule does not apply when the determinant of the coefficient matrix is zero. This would create division by zero, which is undefined.

Area of a Triangle

Another application of matrices and determinants is finding the area of a triangle whose vertices are given as points in a coordinate plane.

> ### Area of a Triangle
>
> The area of a triangle with vertices (x_1, y_1), (x_2, y_2), and (x_3, y_3) is
>
> $$\text{Area} = \pm \frac{1}{2} \begin{vmatrix} x_1 & y_1 & 1 \\ x_2 & y_2 & 1 \\ x_3 & y_3 & 1 \end{vmatrix}$$
>
> where the symbol $\pm$ indicates that the appropriate sign should be chosen to yield a positive area.

Example 3   Finding the Area of a Triangle

Find the area of a triangle whose vertices are $(1, 0)$, $(2, 2)$, and $(4, 3)$, as shown in Figure 7.1.

Solution

Let $(x_1, y_1) = (1, 0)$, $(x_2, y_2) = (2, 2)$, and $(x_3, y_3) = (4, 3)$. Then, to find the area of the triangle, evaluate the determinant.

$$\begin{vmatrix} x_1 & y_1 & 1 \\ x_2 & y_2 & 1 \\ x_3 & y_3 & 1 \end{vmatrix} = \begin{vmatrix} 1 & 0 & 1 \\ 2 & 2 & 1 \\ 4 & 3 & 1 \end{vmatrix}$$

$$= 1(-1)^2 \begin{vmatrix} 2 & 1 \\ 3 & 1 \end{vmatrix} + 0(-1)^3 \begin{vmatrix} 2 & 1 \\ 4 & 1 \end{vmatrix} + 1(-1)^4 \begin{vmatrix} 2 & 2 \\ 4 & 3 \end{vmatrix}$$

$$= 1(-1) + 0 + 1(-2)$$

$$= -3.$$

Using this value, you can conclude that the area of the triangle is

$$\text{Area} = -\frac{1}{2} \begin{vmatrix} 1 & 0 & 1 \\ 2 & 2 & 1 \\ 4 & 3 & 1 \end{vmatrix}$$

$$= -\frac{1}{2}(-3)$$

$$= \frac{3}{2} \text{ square units.}$$

CHECK *Point*   Now try Exercise 25.

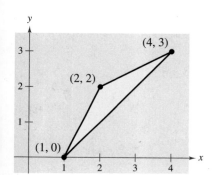

FIGURE 7.1

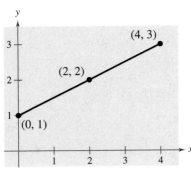

FIGURE 7.2

Lines in a Plane

What if the three points in Example 3 had been on the same line? What would have happened had the area formula been applied to three such points? The answer is that the determinant would have been zero. Consider, for instance, the three collinear points $(0, 1)$, $(2, 2)$, and $(4, 3)$, as shown in Figure 7.2. The area of the "triangle" that has these three points as vertices is

$$\frac{1}{2}\begin{vmatrix} 0 & 1 & 1 \\ 2 & 2 & 1 \\ 4 & 3 & 1 \end{vmatrix} = \frac{1}{2}\left[0(-1)^2\begin{vmatrix} 2 & 1 \\ 3 & 1 \end{vmatrix} + 1(-1)^3\begin{vmatrix} 2 & 1 \\ 4 & 1 \end{vmatrix} + 1(-1)^4\begin{vmatrix} 2 & 2 \\ 4 & 3 \end{vmatrix} \right]$$

$$= \frac{1}{2}[0 - 1(-2) + 1(-2)]$$

$$= 0.$$

The result is generalized as follows.

> ### Test for Collinear Points
>
> Three points (x_1, y_1), (x_2, y_2), and (x_3, y_3) are **collinear** (lie on the same line) if and only if
>
> $$\begin{vmatrix} x_1 & y_1 & 1 \\ x_2 & y_2 & 1 \\ x_3 & y_3 & 1 \end{vmatrix} = 0.$$

Example 4   Testing for Collinear Points

Determine whether the points $(-2, -2)$, $(1, 1)$, and $(7, 5)$ are collinear. (See Figure 7.3.)

Solution

Letting $(x_1, y_1) = (-2, -2)$, $(x_2, y_2) = (1, 1)$, and $(x_3, y_3) = (7, 5)$, you have

$$\begin{vmatrix} x_1 & y_1 & 1 \\ x_2 & y_2 & 1 \\ x_3 & y_3 & 1 \end{vmatrix} = \begin{vmatrix} -2 & -2 & 1 \\ 1 & 1 & 1 \\ 7 & 5 & 1 \end{vmatrix}$$

$$= -2(-1)^2\begin{vmatrix} 1 & 1 \\ 5 & 1 \end{vmatrix} + (-2)(-1)^3\begin{vmatrix} 1 & 1 \\ 7 & 1 \end{vmatrix} + 1(-1)^4\begin{vmatrix} 1 & 1 \\ 7 & 5 \end{vmatrix}$$

$$= -2(-4) + 2(-6) + 1(-2)$$

$$= -6.$$

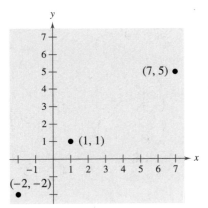

FIGURE 7.3

Because the value of this determinant is *not* zero, you can conclude that the three points do not lie on the same line. Moreover, the area of the triangle with vertices at these points is $\left(-\frac{1}{2}\right)(-6) = 3$ square units.

CHECK*Point* Now try Exercise 39.

The test for collinear points can be adapted to another use. That is, if you are given two points on a rectangular coordinate system, you can find an equation of the line passing through the two points, as follows.

Two-Point Form of the Equation of a Line

An equation of the line passing through the distinct points (x_1, y_1) and (x_2, y_2) is given by

$$\begin{vmatrix} x & y & 1 \\ x_1 & y_1 & 1 \\ x_2 & y_2 & 1 \end{vmatrix} = 0.$$

Example 5   Finding an Equation of a Line

Find an equation of the line passing through the two points $(2, 4)$ and $(-1, 3)$, as shown in Figure 7.4.

Solution

Let $(x_1, y_1) = (2, 4)$ and $(x_2, y_2) = (-1, 3)$. Applying the determinant formula for the equation of a line produces

$$\begin{vmatrix} x & y & 1 \\ 2 & 4 & 1 \\ -1 & 3 & 1 \end{vmatrix} = 0.$$

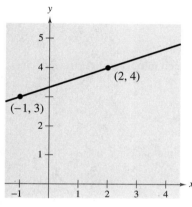

FIGURE **7.4**

To evaluate this determinant, you can expand by cofactors along the first row to obtain the following.

$$x(-1)^2 \begin{vmatrix} 4 & 1 \\ 3 & 1 \end{vmatrix} + y(-1)^3 \begin{vmatrix} 2 & 1 \\ -1 & 1 \end{vmatrix} + 1(-1)^4 \begin{vmatrix} 2 & 4 \\ -1 & 3 \end{vmatrix} = 0$$

$$x(1)(1) + y(-1)(3) + (1)(1)(10) = 0$$

$$x - 3y + 10 = 0$$

So, an equation of the line is

$$x - 3y + 10 = 0.$$

CHECK *Point*   Now try Exercise 47.

Note that this method of finding the equation of a line works for all lines, including horizontal and vertical lines. For instance, the equation of the vertical line through $(2, 0)$ and $(2, 2)$ is

$$\begin{vmatrix} x & y & 1 \\ 2 & 0 & 1 \\ 2 & 2 & 1 \end{vmatrix} = 0$$

$$4 - 2x = 0$$

$$x = 2.$$

Cryptography

A **cryptogram** is a message written according to a secret code. (The Greek word *kryptos* means "hidden.") Matrix multiplication can be used to encode and decode messages. To begin, you need to assign a number to each letter in the alphabet (with 0 assigned to a blank space), as follows.

0 = _	9 = I	18 = R
1 = A	10 = J	19 = S
2 = B	11 = K	20 = T
3 = C	12 = L	21 = U
4 = D	13 = M	22 = V
5 = E	14 = N	23 = W
6 = F	15 = O	24 = X
7 = G	16 = P	25 = Y
8 = H	17 = Q	26 = Z

Then the message is converted to numbers and partitioned into **uncoded row matrices,** each having n entries, as demonstrated in Example 6.

Example 6   Forming Uncoded Row Matrices

Write the uncoded row matrices of order 1×3 for the message

MEET ME MONDAY.

Solution

Partitioning the message (including blank spaces, but ignoring punctuation) into groups of three produces the following uncoded row matrices.

$$\begin{bmatrix} 13 & 5 & 5 \end{bmatrix} \begin{bmatrix} 20 & 0 & 13 \end{bmatrix} \begin{bmatrix} 5 & 0 & 13 \end{bmatrix} \begin{bmatrix} 15 & 14 & 4 \end{bmatrix} \begin{bmatrix} 1 & 25 & 0 \end{bmatrix}$$

Note that a blank space is used to fill out the last uncoded row matrix.

CHECKPoint   Now try Exercise 55(a).

To encode a message, use the techniques demonstrated in Section 7.3 to choose an $n \times n$ invertible matrix such as

$$A = \begin{bmatrix} 1 & -2 & 2 \\ -1 & 1 & 3 \\ 1 & -1 & -4 \end{bmatrix}$$

and multiply the uncoded row matrices by A (on the right) to obtain **coded row matrices.** Here is an example.

Uncoded Matrix Encoding Matrix A   Coded Matrix

$$\begin{bmatrix} 13 & 5 & 5 \end{bmatrix} \begin{bmatrix} 1 & -2 & 2 \\ -1 & 1 & 3 \\ 1 & -1 & -4 \end{bmatrix} = \begin{bmatrix} 13 & -26 & 21 \end{bmatrix}$$

Example 7   Encoding a Message

Use the following invertible matrix to encode the message MEET ME MONDAY.

$$A = \begin{bmatrix} 1 & -2 & 2 \\ -1 & 1 & 3 \\ 1 & -1 & -4 \end{bmatrix}$$

Solution

The coded row matrices are obtained by multiplying each of the uncoded row matrices found in Example 6 by the matrix A, as follows.

Uncoded Matrix *Encoding Matrix A*   *Coded Matrix*

$$\begin{bmatrix} 13 & 5 & 5 \end{bmatrix} \begin{bmatrix} 1 & -2 & 2 \\ -1 & 1 & 3 \\ 1 & -1 & -4 \end{bmatrix} = \begin{bmatrix} 13 & -26 & 21 \end{bmatrix}$$

$$\begin{bmatrix} 20 & 0 & 13 \end{bmatrix} \begin{bmatrix} 1 & -2 & 2 \\ -1 & 1 & 3 \\ 1 & -1 & -4 \end{bmatrix} = \begin{bmatrix} 33 & -53 & -12 \end{bmatrix}$$

$$\begin{bmatrix} 5 & 0 & 13 \end{bmatrix} \begin{bmatrix} 1 & -2 & 2 \\ -1 & 1 & 3 \\ 1 & -1 & -4 \end{bmatrix} = \begin{bmatrix} 18 & -23 & -42 \end{bmatrix}$$

$$\begin{bmatrix} 15 & 14 & 4 \end{bmatrix} \begin{bmatrix} 1 & -2 & 2 \\ -1 & 1 & 3 \\ 1 & -1 & -4 \end{bmatrix} = \begin{bmatrix} 5 & -20 & 56 \end{bmatrix}$$

$$\begin{bmatrix} 1 & 25 & 0 \end{bmatrix} \begin{bmatrix} 1 & -2 & 2 \\ -1 & 1 & 3 \\ 1 & -1 & -4 \end{bmatrix} = \begin{bmatrix} -24 & 23 & 77 \end{bmatrix}$$

So, the sequence of coded row matrices is

$$\begin{bmatrix} 13 & -26 & 21 \end{bmatrix} \begin{bmatrix} 33 & -53 & -12 \end{bmatrix} \begin{bmatrix} 18 & -23 & -42 \end{bmatrix} \begin{bmatrix} 5 & -20 & 56 \end{bmatrix} \begin{bmatrix} -24 & 23 & 77 \end{bmatrix}.$$

Finally, removing the matrix notation produces the following cryptogram.

$$13 \;-26 \; 21 \; 33 \;-53 \;-12 \; 18 \;-23 \;-42 \; 5 \;-20 \; 56 \;-24 \; 23 \; 77$$

CHECK*Point* Now try Exercise 55(b).

For those who do not know the encoding matrix A, decoding the cryptogram found in Example 7 is difficult. But for an authorized receiver who knows the encoding matrix A, decoding is simple. The receiver just needs to multiply the coded row matrices by A^{-1} (on the right) to retrieve the uncoded row matrices. Here is an example.

$$\begin{bmatrix} 13 & -26 & 21 \end{bmatrix} \begin{bmatrix} -1 & -10 & -8 \\ -1 & -6 & -5 \\ 0 & -1 & -1 \end{bmatrix} = \begin{bmatrix} 13 & 5 & 5 \end{bmatrix}$$

During World War II, Navajo soldiers created a code using their native language to send messages between battalions. Native words were assigned to represent characters in the English alphabet, and they created a number of expressions for important military terms, such as *iron-fish* to mean *submarine*. Without the Navajo Code Talkers, the Second World War might have had a very different outcome.

Example 8 Decoding a Message

Use the inverse of the matrix

$$A = \begin{bmatrix} 1 & -2 & 2 \\ -1 & 1 & 3 \\ 1 & -1 & -4 \end{bmatrix}$$

to decode the cryptogram

$$13 \; -26 \; 21 \; 33 \; -53 \; -12 \; 18 \; -23 \; -42 \; 5 \; -20 \; 56 \; -24 \; 23 \; 77.$$

Solution

First find A^{-1} by using the techniques demonstrated in Section 7.3. A^{-1} is the decoding matrix. Then partition the message into groups of three to form the coded row matrices. Finally, multiply each coded row matrix by A^{-1} (on the right).

Coded Matrix   Decoding Matrix A^{-1} Decoded Matrix

$$\begin{bmatrix} 13 & -26 & 21 \end{bmatrix} \begin{bmatrix} -1 & -10 & -8 \\ -1 & -6 & -5 \\ 0 & -1 & -1 \end{bmatrix} = \begin{bmatrix} 13 & 5 & 5 \end{bmatrix}$$

$$\begin{bmatrix} 33 & -53 & -12 \end{bmatrix} \begin{bmatrix} -1 & -10 & -8 \\ -1 & -6 & -5 \\ 0 & -1 & -1 \end{bmatrix} = \begin{bmatrix} 20 & 0 & 13 \end{bmatrix}$$

$$\begin{bmatrix} 18 & -23 & -42 \end{bmatrix} \begin{bmatrix} -1 & -10 & -8 \\ -1 & -6 & -5 \\ 0 & -1 & -1 \end{bmatrix} = \begin{bmatrix} 5 & 0 & 13 \end{bmatrix}$$

$$\begin{bmatrix} 5 & -20 & 56 \end{bmatrix} \begin{bmatrix} -1 & -10 & -8 \\ -1 & -6 & -5 \\ 0 & -1 & -1 \end{bmatrix} = \begin{bmatrix} 15 & 14 & 4 \end{bmatrix}$$

$$\begin{bmatrix} -24 & 23 & 77 \end{bmatrix} \begin{bmatrix} -1 & -10 & -8 \\ -1 & -6 & -5 \\ 0 & -1 & -1 \end{bmatrix} = \begin{bmatrix} 1 & 25 & 0 \end{bmatrix}$$

So, the message is as follows.

$$\begin{bmatrix} 13 & 5 & 5 \end{bmatrix} \begin{bmatrix} 20 & 0 & 13 \end{bmatrix} \begin{bmatrix} 5 & 0 & 13 \end{bmatrix} \begin{bmatrix} 15 & 14 & 4 \end{bmatrix} \begin{bmatrix} 1 & 25 & 0 \end{bmatrix}$$

CHECK*Point* Now try Exercise 63.

CLASSROOM DISCUSSION

Cryptography Use your school's library, the Internet, or some other reference source to research information about another type of cryptography. Write a short paragraph describing how mathematics is used to code and decode messages.

7.5   EXERCISES

See www.CalcChat.com for worked-out solutions to odd-numbered exercises.

VOCABULARY: Fill in the blanks.

1. The method of using determinants to solve a system of linear equations is called _____ _____.
2. Three points are _____ if the points lie on the same line.
3. The area A of a triangle with vertices (x_1, y_1), (x_2, y_2), and (x_3, y_3) is given by _____.
4. A message written according to a secret code is called a _____.
5. To encode a message, choose an invertible matrix A and multiply the _____ row matrices by A (on the right) to obtain _____ row matrices.
6. If a message is encoded using an invertible matrix A, then the message can be decoded by multiplying the coded row matrices by _____ (on the right).

SKILLS AND APPLICATIONS

In Exercises 7–16, use Cramer's Rule to solve (if possible) the system of equations.

7. $\begin{cases} -7x + 11y = -1 \\ 3x - 9y = 9 \end{cases}$

8. $\begin{cases} 4x - 3y = -10 \\ 6x + 9y = 12 \end{cases}$

9. $\begin{cases} 3x + 2y = -2 \\ 6x + 4y = 4 \end{cases}$

10. $\begin{cases} 6x - 5y = 17 \\ -13x + 3y = -76 \end{cases}$

11. $\begin{cases} -0.4x + 0.8y = 1.6 \\ 0.2x + 0.3y = 2.2 \end{cases}$

12. $\begin{cases} 2.4x - 1.3y = 14.63 \\ -4.6x + 0.5y = -11.51 \end{cases}$

13. $\begin{cases} 4x - y + z = -5 \\ 2x + 2y + 3z = 10 \\ 5x - 2y + 6z = 1 \end{cases}$

14. $\begin{cases} 4x - 2y + 3z = -2 \\ 2x + 2y + 5z = 16 \\ 8x - 5y - 2z = 4 \end{cases}$

15. $\begin{cases} x + 2y + 3z = -3 \\ -2x + y - z = 6 \\ 3x - 3y + 2z = -11 \end{cases}$

16. $\begin{cases} 5x - 4y + z = -14 \\ -x + 2y - 2z = 10 \\ 3x + y + z = 1 \end{cases}$

In Exercises 17–20, use a graphing utility and Cramer's Rule to solve (if possible) the system of equations.

17. $\begin{cases} 3x + 3y + 5z = 1 \\ 3x + 5y + 9z = 2 \\ 5x + 9y + 17z = 4 \end{cases}$

18. $\begin{cases} x + 2y - z = -7 \\ 2x - 2y - 2z = -8 \\ -x + 3y + 4z = 8 \end{cases}$

19. $\begin{cases} 2x - y + z = 5 \\ x - 2y - z = 1 \\ 3x + y + z = 4 \end{cases}$

20. $\begin{cases} 3x - y - 3z = 1 \\ 2x + y + 2z = -4 \\ x + y - z = 5 \end{cases}$

In Exercises 21–32, use a determinant and the given vertices of a triangle to find the area of the triangle.

21.
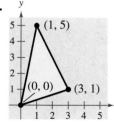

22.
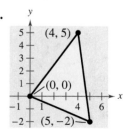

23.
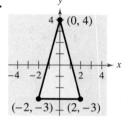

24.
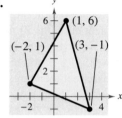

25.
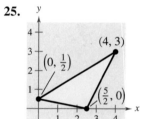

26.
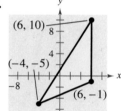

27. $(-2, 4), (2, 3), (-1, 5)$
28. $(0, -2), (-1, 4), (3, 5)$
29. $(-3, 5), (2, 6), (3, -5)$
30. $(-2, 4), (1, 5), (3, -2)$
31. $(-4, 2), \left(0, \frac{7}{2}\right), \left(3, -\frac{1}{2}\right)$
32. $\left(\frac{9}{2}, 0\right), (2, 6), \left(0, -\frac{3}{2}\right)$

In Exercises 33 and 34, find a value of y such that the triangle with the given vertices has an area of 4 square units.

33. $(-5, 1), (0, 2), (-2, y)$
34. $(-4, 2), (-3, 5), (-1, y)$

In Exercises 35 and 36, find a value of y such that the triangle with the given vertices has an area of 6 square units.

35. $(-2, -3), (1, -1), (-8, y)$
36. $(1, 0), (5, -3), (-3, y)$

37. **AREA OF A REGION** A large region of forest has been infested with gypsy moths. The region is roughly triangular, as shown in the figure on the next page. From the northernmost vertex A of the region, the distances to the other vertices are 25 miles south and 10 miles east (for vertex B), and 20 miles south and 28 miles east (for vertex C). Use a graphing utility to approximate the number of square miles in this region.

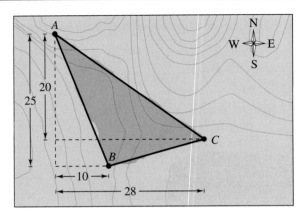

FIGURE FOR 37

38. AREA OF A REGION You own a triangular tract of land, as shown in the figure. To estimate the number of square feet in the tract, you start at one vertex, walk 65 feet east and 50 feet north to the second vertex, and then walk 85 feet west and 30 feet north to the third vertex. Use a graphing utility to determine how many square feet there are in the tract of land.

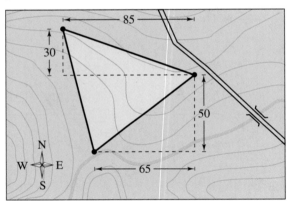

In Exercises 39–44, use a determinant to determine whether the points are collinear.

39. $(3, -1), (0, -3), (12, 5)$ **40.** $(3, -5), (6, 1), (4, 2)$
41. $\left(2, -\frac{1}{2}\right), (-4, 4), (6, -3)$ **42.** $\left(0, \frac{1}{2}\right), (2, -1), \left(-4, \frac{7}{2}\right)$
43. $(0, 2), (1, 2.4), (-1, 1.6)$ **44.** $(2, 3), (3, 3.5), (-1, 2)$

In Exercises 45 and 46, find y such that the points are collinear.

45. $(2, -5), (4, y), (5, -2)$ **46.** $(-6, 2), (-5, y), (-3, 5)$

In Exercises 47–52, use a determinant to find an equation of the line passing through the points.

47. $(0, 0), (5, 3)$ **48.** $(0, 0), (-2, 2)$
49. $(-4, 3), (2, 1)$ **50.** $(10, 7), (-2, -7)$
51. $\left(-\frac{1}{2}, 3\right), \left(\frac{5}{2}, 1\right)$ **52.** $\left(\frac{2}{3}, 4\right), (6, 12)$

In Exercises 53 and 54, (a) write the uncoded 1×2 row matrices for the message. (b) Then encode the message using the encoding matrix.

Message	Encoding Matrix
53. COME HOME SOON	$\begin{bmatrix} 1 & 2 \\ 3 & 5 \end{bmatrix}$
54. HELP IS ON THE WAY	$\begin{bmatrix} -2 & 3 \\ -1 & 1 \end{bmatrix}$

In Exercises 55 and 56, (a) write the uncoded 1×3 row matrices for the message. (b) Then encode the message using the encoding matrix.

Message	Encoding Matrix
55. CALL ME TOMORROW	$\begin{bmatrix} 1 & -1 & 0 \\ 1 & 0 & -1 \\ -6 & 2 & 3 \end{bmatrix}$
56. PLEASE SEND MONEY	$\begin{bmatrix} 4 & 2 & 1 \\ -3 & -3 & -1 \\ 3 & 2 & 1 \end{bmatrix}$

In Exercises 57–60, write a cryptogram for the message using the matrix A.

$$A = \begin{bmatrix} 1 & 2 & 2 \\ 3 & 7 & 9 \\ -1 & -4 & -7 \end{bmatrix}$$

57. LANDING SUCCESSFUL
58. ICEBERG DEAD AHEAD
59. HAPPY BIRTHDAY
60. OPERATION OVERLOAD

In Exercises 61–64, use A^{-1} to decode the cryptogram.

61. $A = \begin{bmatrix} 1 & 2 \\ 3 & 5 \end{bmatrix}$

11 21 64 112 25 50 29 53 23 46
40 75 55 92

62. $A = \begin{bmatrix} 2 & 3 \\ 3 & 4 \end{bmatrix}$

85 120 6 8 10 15 84 117 42 56 90
125 60 80 30 45 19 26

63. $A = \begin{bmatrix} 1 & -1 & 0 \\ 1 & 0 & -1 \\ -6 & 2 & 3 \end{bmatrix}$

9 -1 -9 38 -19 -19 28 -9 -19
-80 25 41 -64 21 31 9 -5 -4

64. $A = \begin{bmatrix} 3 & -4 & 2 \\ 0 & 2 & 1 \\ 4 & -5 & 3 \end{bmatrix}$

112 −140 83 19 −25 13 72 −76 61 95
−118 71 20 21 38 35 −23 36 42 −48 32

In Exercises 65 and 66, decode the cryptogram by using the inverse of the matrix A.

$A = \begin{bmatrix} 1 & 2 & 2 \\ 3 & 7 & 9 \\ -1 & -4 & -7 \end{bmatrix}$

65. 20 17 −15 −12 −56 −104 1 −25 −65
62 143 181

66. 13 −9 −59 61 112 106 −17 −73 −131 11
24 29 65 144 172

67. The following cryptogram was encoded with a 2×2 matrix.

8 21 −15 −10 −13 −13 5 10 5 25 5 19
−1 6 20 40 −18 −18 1 16

The last word of the message is _RON. What is the message?

68. The following cryptogram was encoded with a 2×2 matrix.

5 2 25 11 −2 −7 −15 −15 32 14 −8
−13 38 19 −19 −19 37 16

The last word of the message is _SUE. What is the message?

69. DATA ANALYSIS: BOTTLED WATER The table shows the per capita consumption of bottled water y (in gallons) in the United States from 2000 through 2007. (Source: Economic Research Service, U.S. Department of Agriculture)

Year	Consumption, y
2000	16.7
2001	18.2
2002	20.1
2003	21.6
2004	23.2
2005	25.5
2006	27.7
2007	29.1

(a) Use the technique demonstrated in Exercises 77–80 in Section 6.3 to create a system of linear equations for the data. Let t represent the year, with $t = 0$ corresponding to 2000.

(b) Use Cramer's Rule to solve the system from part (a) and find the least squares regression parabola $y = at^2 + bt + c$.

(c) Use a graphing utility to graph the parabola from part (b).

(d) Use the graph from part (c) to estimate when the per capita consumption of bottled water will exceed 35 gallons.

70. HAIR PRODUCTS A hair product company sells three types of hair products for $30, $20, and $10 per unit. In one year, the total revenue for the three products was $800,000, which corresponded to the sale of 40,000 units. The company sold half as many units of the $30 product as units of the $20 product. Use Cramer's Rule to solve a system of linear equations to find how many units of each product were sold.

EXPLORATION

TRUE OR FALSE? In Exercises 71–74, determine whether the statement is true or false. Justify your answer.

71. In Cramer's Rule, the numerator is the determinant of the coefficient matrix.

72. You cannot use Cramer's Rule when solving a system of linear equations if the determinant of the coefficient matrix is zero.

73. In a system of linear equations, if the determinant of the coefficient matrix is zero, the system has no solution.

74. The points $(-5, -13)$, $(0, 2)$, and $(3, 11)$ are collinear.

75. WRITING Use your school's library, the Internet, or some other reference source to research a few current real-life uses of cryptography. Write a short summary of these uses. Include a description of how messages are encoded and decoded in each case.

76. CAPSTONE

(a) State Cramer's Rule for solving a system of linear equations.

(b) At this point in the text, you have learned several methods for solving systems of linear equations. Briefly describe which method(s) you find easiest to use and which method(s) you find most difficult to use.

77. Use determinants to find the area of a triangle with vertices $(3, -1)$, $(7, -1)$, and $(7, 5)$. Confirm your answer by plotting the points in a coordinate plane and using the formula

Area $= \frac{1}{2}$(base)(height).

7 CHAPTER SUMMARY

What Did You Learn?	Explanation/Examples	Review Exercises
Write matrices and identify their orders *(p. 520)*.	$\begin{bmatrix} -1 & 1 \\ 4 & 7 \end{bmatrix}$ $\begin{bmatrix} -2 & 3 & 0 \end{bmatrix}$ $\begin{bmatrix} 4 & -3 \\ 5 & 0 \\ -2 & 1 \end{bmatrix}$ $\begin{bmatrix} 8 \\ -8 \end{bmatrix}$	1–8
Perform elementary row operations on matrices *(p. 522)*.	**Elementary Row Operations** **1.** Interchange two rows. **2.** Multiply a row by a nonzero constant. **3.** Add a multiple of a row to another row.	9, 10
Use matrices and Gaussian elimination to solve systems of linear equations *(p. 525)*.	**Gaussian Elimination with Back-Substitution** **1.** Write the augmented matrix of the system of linear equations. **2.** Use elementary row operations to rewrite the augmented matrix in row-echelon form. **3.** Write the system of linear equations corresponding to the matrix in row-echelon form, and use back-substitution to find the solution.	11–28
Use matrices and Gauss-Jordan elimination to solve systems of linear equations *(p. 527)*.	Gauss-Jordan elimination continues the reduction process on a matrix in row-echelon form until a *reduced* row-echelon form is obtained. (See Example 8.)	29–36
Decide whether two matrices are equal *(p. 534)*.	Two matrices are equal if their corresponding entries are equal.	37–40
Add and subtract matrices and multiply matrices by scalars *(p. 535)*.	**Definition of Matrix Addition** If $A = [a_{ij}]$ and $B = [b_{ij}]$ are matrices of order $m \times n$, their sum is the $m \times n$ matrix given by $A + B = [a_{ij} + b_{ij}]$. **Definition of Scalar Multiplication** If $A = [a_{ij}]$ is an $m \times n$ matrix and c is a scalar, the scalar multiple of A by c is the $m \times n$ matrix given by $cA = [c a_{ij}]$.	41–54
Multiply two matrices *(p. 539)*.	**Matrix Multiplication** If $A = [a_{ij}]$ is an $m \times n$ matrix and $B = [b_{ij}]$ is an $n \times p$ matrix, the product AB is an $m \times p$ matrix $$AB = [c_{ij}]$$ where $c_{ij} = a_{i1}b_{1j} + a_{i2}b_{2j} + a_{i3}b_{3j} + \cdots + a_{in}b_{nj}$.	55–68
Use matrix operations to model and solve real-life problems *(p. 542)*.	Matrix operations can be used to find the total cost of equipment for two softball teams. (See Example 12.)	69–72
Verify that two matrices are inverses of each other *(p. 549)*.	**Inverse of a Square Matrix** Let A be an $n \times n$ matrix and let I_n be the $n \times n$ identity matrix. If there exists a matrix A^{-1} such that $$AA^{-1} = I_n = A^{-1}A$$ then A^{-1} is the inverse of A.	73–76

Section 7.1 · *Section 7.2* · *Section 7.3*

What Did You Learn?	Explanation/Examples	Review Exercises
Section 7.3		
Use Gauss-Jordan elimination to find the inverses of matrices (p. 550).	**Finding an Inverse Matrix** Let A be a square matrix of order n. **1.** Write the $n \times 2n$ matrix that consists of the given matrix A on the left and the $n \times n$ identity matrix I on the right to obtain $[A \vdots I]$. **2.** If possible, row reduce A to I using elementary row operations on the *entire* matrix $[A \vdots I]$. The result will be the matrix $[I \vdots A^{-1}]$. If this is not possible, A is not invertible. **3.** Check your work to see that $AA^{-1} = I = A^{-1}A$.	77–84
Use a formula to find the inverses of 2×2 matrices (p. 553).	If $A = \begin{bmatrix} a & b \\ c & d \end{bmatrix}$ and $ad - bc \neq 0$, then $$A^{-1} = \frac{1}{ad - bc}\begin{bmatrix} d & -b \\ -c & a \end{bmatrix}.$$	85–92
Use inverse matrices to solve systems of linear equations (p. 554).	If A is an invertible matrix, the system of linear equations represented by $AX = B$ has a unique solution given by $X = A^{-1}B$.	93–110
Section 7.4		
Find the determinants of 2×2 matrices (p. 558).	The determinant of the matrix $A = \begin{bmatrix} a_1 & b_1 \\ a_2 & b_2 \end{bmatrix}$ is given by $$\det(A) = |A| = \begin{vmatrix} a_1 & b_1 \\ a_2 & b_2 \end{vmatrix} = a_1b_2 - a_2b_1.$$	111–114
Find minors and cofactors of square matrices (p. 560).	If A is a square matrix, the minor M_{ij} of the entry a_{ij} is the determinant of the matrix obtained by deleting the ith row and jth column of A. The cofactor C_{ij} of the entry a_{ij} is $C_{ij} = (-1)^{i+j}M_{ij}$.	115–118
Find the determinants of square matrices (p. 561).	If A is a square matrix (of order 2×2 or greater), the determinant of A is the sum of the entries in any row (or column) of A multiplied by their respective cofactors.	119–128
Section 7.5		
Use Cramer's Rule to solve systems of linear equations (p. 566).	Cramer's Rule uses determinants to write the solution of a system of linear equations.	129–132
Use determinants to find the areas of triangles (p. 569).	The area of a triangle with vertices (x_1, y_1), (x_2, y_2), and (x_3, y_3) is $$\text{Area} = \pm\frac{1}{2}\begin{vmatrix} x_1 & y_1 & 1 \\ x_2 & y_2 & 1 \\ x_3 & y_3 & 1 \end{vmatrix}$$ where the symbol $\pm$ indicates that the appropriate sign should be chosen to yield a positive area.	133–136
Use a determinant to test for collinear points and find an equation of a line passing through two points (p. 570).	Three points (x_1, y_1), (x_2, y_2), and (x_3, y_3) are collinear (lie on the same line) if and only if $$\begin{vmatrix} x_1 & y_1 & 1 \\ x_2 & y_2 & 1 \\ x_3 & y_3 & 1 \end{vmatrix} = 0.$$	137–142
Use matrices to encode and decode messages (p. 572).	The inverse of a matrix can be used to decode a cryptogram. (See Example 8.)	143–146

7 REVIEW EXERCISES

See www.CalcChat.com for worked-out solutions to odd-numbered exercises.

7.1 In Exercises 1–4, determine the order of the matrix.

1. $\begin{bmatrix} -4 \\ 0 \\ 5 \end{bmatrix}$

2. $\begin{bmatrix} 3 & -1 & 0 & 6 \\ -2 & 7 & 1 & 4 \end{bmatrix}$

3. $[3]$

4. $\begin{bmatrix} 6 & 2 & -5 & 8 & 0 \end{bmatrix}$

In Exercises 5 and 6, write the augmented matrix for the system of linear equations.

5. $\begin{cases} 3x - 10y = 15 \\ 5x + 4y = 22 \end{cases}$

6. $\begin{cases} 8x - 7y + 4z = 12 \\ 3x - 5y + 2z = 20 \\ 5x + 3y - 3z = 26 \end{cases}$

In Exercises 7 and 8, write the system of linear equations represented by the augmented matrix. (Use variables x, y, z, and w, if applicable.)

7. $\begin{bmatrix} 5 & 1 & 7 & \vdots & -9 \\ 4 & 2 & 0 & \vdots & 10 \\ 9 & 4 & 2 & \vdots & 3 \end{bmatrix}$

8. $\begin{bmatrix} 13 & 16 & 7 & 3 & \vdots & 2 \\ 1 & 21 & 8 & 5 & \vdots & 12 \\ 4 & 10 & -4 & 3 & \vdots & -1 \end{bmatrix}$

In Exercises 9 and 10, write the matrix in row-echelon form. (Remember that the row-echelon form of a matrix is not unique.)

9. $\begin{bmatrix} 0 & 1 & 1 \\ 1 & 2 & 3 \\ 2 & 2 & 2 \end{bmatrix}$

10. $\begin{bmatrix} 4 & 8 & 16 \\ 3 & -1 & 2 \\ -2 & 10 & 12 \end{bmatrix}$

In Exercises 11–14, write the system of linear equations represented by the augmented matrix. Then use back-substitution to solve the system. (Use variables x, y, and z.)

11. $\begin{bmatrix} 1 & 2 & 3 & \vdots & 9 \\ 0 & 1 & -2 & \vdots & 2 \\ 0 & 0 & 1 & \vdots & 0 \end{bmatrix}$

12. $\begin{bmatrix} 1 & 3 & -9 & \vdots & 4 \\ 0 & 1 & -1 & \vdots & 10 \\ 0 & 0 & 1 & \vdots & -2 \end{bmatrix}$

13. $\begin{bmatrix} 1 & -5 & 4 & \vdots & 1 \\ 0 & 1 & 2 & \vdots & 3 \\ 0 & 0 & 1 & \vdots & 4 \end{bmatrix}$

14. $\begin{bmatrix} 1 & -8 & 0 & \vdots & -2 \\ 0 & 1 & -1 & \vdots & -7 \\ 0 & 0 & 1 & \vdots & 1 \end{bmatrix}$

In Exercises 15–28, use matrices and Gaussian elimination with back-substitution to solve the system of equations (if possible).

15. $\begin{cases} 5x + 4y = 2 \\ -x + y = -22 \end{cases}$

16. $\begin{cases} 2x - 5y = 2 \\ 3x - 7y = 1 \end{cases}$

17. $\begin{cases} 0.3x - 0.1y = -0.13 \\ 0.2x - 0.3y = -0.25 \end{cases}$

18. $\begin{cases} 0.2x - 0.1y = 0.07 \\ 0.4x - 0.5y = -0.01 \end{cases}$

19. $\begin{cases} -x + 2y = 3 \\ 2x - 4y = 6 \end{cases}$

20. $\begin{cases} -x + 2y = 3 \\ 2x - 4y = -6 \end{cases}$

21. $\begin{cases} x - 2y + z = 7 \\ 2x + y - 2z = -4 \\ -x + 3y + 2z = -3 \end{cases}$

22. $\begin{cases} x - 2y + z = 4 \\ 2x + y - 2z = -24 \\ -x + 3y + 2z = 20 \end{cases}$

23. $\begin{cases} 2x + y + 2z = 4 \\ 2x + 2y = 5 \\ 2x - y + 6z = 2 \end{cases}$

24. $\begin{cases} x + 2y + 6z = 1 \\ 2x + 5y + 15z = 4 \\ 3x + y + 3z = -6 \end{cases}$

25. $\begin{cases} 2x + 3y + z = 10 \\ 2x - 3y - 3z = 22 \\ 4x - 2y + 3z = -2 \end{cases}$

26. $\begin{cases} 2x + 3y + 3z = 3 \\ 6x + 6y + 12z = 13 \\ 12x + 9y - z = 2 \end{cases}$

27. $\begin{cases} 2x + y + z = 6 \\ -2y + 3z - w = 9 \\ 3x + 3y - 2z - 2w = -11 \\ x + z + 3w = 14 \end{cases}$

28. $\begin{cases} x + 2y + w = 3 \\ -3y + 3z = 0 \\ 4x + 4y + z + 2w = 0 \\ 2x + z = 3 \end{cases}$

In Exercises 29–34, use matrices and Gauss-Jordan elimination to solve the system of equations.

29. $\begin{cases} x + 2y - z = 3 \\ x - y - z = -3 \\ 2x + y + 3z = 10 \end{cases}$

30. $\begin{cases} x - 3y + z = 2 \\ 3x - y - z = -6 \\ -x + y - 3z = -2 \end{cases}$

31. $\begin{cases} -x + y + 2z = 1 \\ 2x + 3y + z = -2 \\ 5x + 4y + 2z = 4 \end{cases}$

32. $\begin{cases} 4x + 4y + 4z = 5 \\ 4x - 2y - 8z = 1 \\ 5x + 3y + 8z = 6 \end{cases}$

33. $\begin{cases} 2x - y + 9z = -8 \\ -x - 3y + 4z = -15 \\ 5x + 2y - z = 17 \end{cases}$

34. $\begin{cases} -3x + y + 7z = -20 \\ 5x - 2y - z = 34 \\ -x + y + 4z = -8 \end{cases}$

In Exercises 35 and 36, use the matrix capabilities of a graphing utility to reduce the augmented matrix corresponding to the system of equations, and solve the system.

35. $\begin{cases} 3x - y + 5z - 2w = -44 \\ x + 6y + 4z - w = 1 \\ 5x - y + z + 3w = -15 \\ 4y - z - 8w = 58 \end{cases}$

36. $\begin{cases} 4x + 12y + 2z = 20 \\ x + 6y + 4z = 12 \\ x + 6y + z = 8 \\ -2x - 10y - 2z = -10 \end{cases}$

7.2 In Exercises 37–40, find x and y.

37. $\begin{bmatrix} -1 & x \\ y & 9 \end{bmatrix} = \begin{bmatrix} -1 & 12 \\ -7 & 9 \end{bmatrix}$

38. $\begin{bmatrix} -1 & 0 \\ x & 5 \\ -4 & y \end{bmatrix} = \begin{bmatrix} -1 & 0 \\ 8 & 5 \\ -4 & 0 \end{bmatrix}$

39. $\begin{bmatrix} x+3 & -4 & 4y \\ 0 & -3 & 2 \\ -2 & y+5 & 6x \end{bmatrix} = \begin{bmatrix} 5x-1 & -4 & 44 \\ 0 & -3 & 2 \\ -2 & 16 & 6 \end{bmatrix}$

40. $\begin{bmatrix} -9 & 4 & 2 & -5 \\ 0 & -3 & 7 & -4 \\ 6 & -1 & 1 & 0 \end{bmatrix} = \begin{bmatrix} -9 & 4 & x-10 & -5 \\ 0 & -3 & 7 & 2y \\ \frac{1}{2}x & -1 & 1 & 0 \end{bmatrix}$

In Exercises 41–44, if possible, find (a) $A + B$, (b) $A - B$, (c) $4A$, and (d) $A + 3B$.

41. $A = \begin{bmatrix} 2 & -2 \\ 3 & 5 \end{bmatrix}$, $B = \begin{bmatrix} -3 & 10 \\ 12 & 8 \end{bmatrix}$

42. $A = \begin{bmatrix} 4 & 3 \\ -6 & 1 \\ 10 & 1 \end{bmatrix}$, $B = \begin{bmatrix} 3 & 11 \\ 15 & 25 \\ 20 & 29 \end{bmatrix}$

43. $A = \begin{bmatrix} 5 & 4 \\ -7 & 2 \\ 11 & 2 \end{bmatrix}$, $B = \begin{bmatrix} 0 & 3 \\ 4 & 12 \\ 20 & 40 \end{bmatrix}$

44. $A = \begin{bmatrix} 6 & -5 & 7 \end{bmatrix}$, $B = \begin{bmatrix} -1 \\ 4 \\ 8 \end{bmatrix}$

In Exercises 45–48, perform the matrix operations. If it is not possible, explain why.

45. $\begin{bmatrix} 7 & 3 \\ -1 & 5 \end{bmatrix} + \begin{bmatrix} 10 & -20 \\ 14 & -3 \end{bmatrix}$

46. $\begin{bmatrix} -11 & 16 & 19 \\ -7 & -2 & 1 \end{bmatrix} - \begin{bmatrix} 6 & 0 \\ 8 & -4 \\ -2 & 10 \end{bmatrix}$

47. $-2 \begin{bmatrix} 1 & 2 \\ 5 & -4 \\ 6 & 0 \end{bmatrix} + 8 \begin{bmatrix} 7 & 1 \\ 1 & 2 \\ 1 & 4 \end{bmatrix}$

48. $-\begin{bmatrix} 8 & -1 & 8 \\ -2 & 4 & 12 \\ 0 & -6 & 0 \end{bmatrix} - 5 \begin{bmatrix} -2 & 0 & -4 \\ 3 & -1 & 1 \\ 6 & 12 & -8 \end{bmatrix}$

In Exercises 49 and 50, use the matrix capabilities of a graphing utility to evaluate the expression.

49. $3 \begin{bmatrix} 8 & -2 & 5 \\ 1 & 3 & -1 \end{bmatrix} + 6 \begin{bmatrix} 4 & -2 & -3 \\ 2 & 7 & 6 \end{bmatrix}$

50. $-5 \begin{bmatrix} 2 & 0 \\ 7 & -2 \\ 8 & 2 \end{bmatrix} + 4 \begin{bmatrix} 4 & -2 \\ 6 & 11 \\ -1 & 3 \end{bmatrix}$

In Exercises 51–54, solve for X in the equation, given

$$A = \begin{bmatrix} -4 & 0 \\ 1 & -5 \\ -3 & 2 \end{bmatrix} \quad \text{and} \quad B = \begin{bmatrix} 1 & 2 \\ -2 & 1 \\ 4 & 4 \end{bmatrix}.$$

51. $X = 2A - 3B$ **52.** $6X = 4A + 3B$

53. $3X + 2A = B$ **54.** $2A - 5B = 3X$

In Exercises 55–58, find AB, if possible.

55. $A = \begin{bmatrix} 2 & -2 \\ 3 & 5 \end{bmatrix}$, $B = \begin{bmatrix} -3 & 10 \\ 12 & 8 \end{bmatrix}$

56. $A = \begin{bmatrix} 5 & 4 \\ -7 & 2 \\ 11 & 2 \end{bmatrix}$, $B = \begin{bmatrix} 4 & 12 \\ 20 & 40 \\ 15 & 30 \end{bmatrix}$

57. $A = \begin{bmatrix} 5 & 4 \\ -7 & 2 \\ 11 & 2 \end{bmatrix}$, $B = \begin{bmatrix} 4 & 12 \\ 20 & 40 \end{bmatrix}$

58. $A = \begin{bmatrix} 6 & -5 & 7 \end{bmatrix}$, $B = \begin{bmatrix} -1 \\ 4 \\ 8 \end{bmatrix}$

In Exercises 59–66, perform the matrix operations, if possible. If it is not possible, explain why.

59. $\begin{bmatrix} 1 & 2 \\ 5 & -4 \\ 6 & 0 \end{bmatrix} \begin{bmatrix} 6 & -2 & 8 \\ 4 & 0 & 0 \end{bmatrix}$

60. $\begin{bmatrix} 1 & 5 & 6 \\ 2 & -4 & 0 \end{bmatrix} \begin{bmatrix} 6 & -2 & 8 \\ 4 & 0 & 0 \end{bmatrix}$

61. $\begin{bmatrix} 1 & 5 & 6 \\ 2 & -4 & 0 \end{bmatrix} \begin{bmatrix} 6 & 4 \\ -2 & 0 \\ 8 & 0 \end{bmatrix}$

62. $\begin{bmatrix} 1 & 3 & 2 \\ 0 & 2 & -4 \\ 0 & 0 & 3 \end{bmatrix} \begin{bmatrix} 4 & -3 & 2 \\ 0 & 3 & -1 \\ 0 & 0 & 2 \end{bmatrix}$

63. $\begin{bmatrix} 1 & 2 & -1 \\ 0 & 4 & -2 \\ 1 & 1 & 3 \end{bmatrix} \begin{bmatrix} 1 & -1 & 2 \end{bmatrix}$

64. $\begin{bmatrix} 4 & -2 & 6 \end{bmatrix} \begin{bmatrix} -2 & 1 \\ 0 & -3 \\ 2 & 0 \end{bmatrix}$

65. $\begin{bmatrix} 2 & 1 \\ 6 & 0 \end{bmatrix} \left(\begin{bmatrix} 4 & 2 \\ -3 & 1 \end{bmatrix} + \begin{bmatrix} -2 & 4 \\ 0 & 4 \end{bmatrix} \right)$

66. $-3 \begin{bmatrix} 1 & -1 \\ 4 & 2 \end{bmatrix} \left(\begin{bmatrix} 0 & 3 \\ 1 & 2 \end{bmatrix} \begin{bmatrix} 1 & 0 \\ 5 & -3 \end{bmatrix} \right)$

In Exercises 67 and 68, use the matrix capabilities of a graphing utility to find the product.

67. $\begin{bmatrix} 4 & 1 \\ 11 & -7 \\ 12 & 3 \end{bmatrix} \begin{bmatrix} 3 & -5 & 6 \\ 2 & -2 & -2 \end{bmatrix}$

68. $\begin{bmatrix} -2 & 3 & 10 \\ 4 & -2 & 2 \end{bmatrix} \begin{bmatrix} 1 & 1 \\ -5 & 2 \\ 3 & 2 \end{bmatrix}$

69. MANUFACTURING A tire corporation has three factories, each of which manufactures two models of tires. The number of units of model i produced at factory j in one day is represented by a_{ij} in the matrix

$$A = \begin{bmatrix} 80 & 120 & 140 \\ 40 & 100 & 80 \end{bmatrix}.$$

Find the production levels if production is decreased by 5%.

70. MANUFACTURING A power tool company has four manufacturing plants, each of which produces three types of cordless power tools. The number of units of cordless power tool i produced at plant j in one day is represented by a_{ij} in the matrix

$$A = \begin{bmatrix} 80 & 70 & 90 & 40 \\ 50 & 30 & 80 & 20 \\ 90 & 60 & 100 & 50 \end{bmatrix}.$$

Find the production levels if production is increased by 20%.

71. MANUFACTURING An electronics manufacturing company produces three different models of headphones that are shipped to two warehouses. The number of units of model i that are shipped to warehouse j is represented by a_{ij} in the matrix

$$A = \begin{bmatrix} 8200 & 7400 \\ 6500 & 9800 \\ 5400 & 4800 \end{bmatrix}.$$

The price per unit is represented by the matrix

$$B = \begin{bmatrix} \$79.99 & \$109.95 & \$189.99 \end{bmatrix}.$$

Compute BA and interpret the result.

72. CELL PHONE CHARGES The pay-as-you-go charges (in dollars per minute) of two cellular telephone companies for calls inside the coverage area, regional roaming calls, and calls outside the coverage area are represented by C.

$$C = \begin{bmatrix} 0.07 & 0.095 \\ 0.10 & 0.08 \\ 0.28 & 0.25 \end{bmatrix}$$

Each month, you plan to use 120 minutes on calls inside the coverage area, 80 minutes on regional roaming calls, and 20 minutes on calls outside the coverage area.

(a) Write a matrix T that represents the times spent on the phone for each type of call.

(b) Compute TC and interpret the result.

7.3 In Exercises 73–76, show that B is the inverse of A.

73. $A = \begin{bmatrix} -4 & -1 \\ 7 & 2 \end{bmatrix}$, $B = \begin{bmatrix} -2 & -1 \\ 7 & 4 \end{bmatrix}$

74. $A = \begin{bmatrix} 5 & -1 \\ 11 & -2 \end{bmatrix}$, $B = \begin{bmatrix} -2 & 1 \\ -11 & 5 \end{bmatrix}$

75. $A = \begin{bmatrix} 1 & 1 & 0 \\ 1 & 0 & 1 \\ 6 & 2 & 3 \end{bmatrix}$, $B = \begin{bmatrix} -2 & -3 & 1 \\ 3 & 3 & -1 \\ 2 & 4 & -1 \end{bmatrix}$

76. $A = \begin{bmatrix} 1 & -1 & 0 \\ -1 & 0 & -1 \\ 8 & -4 & 2 \end{bmatrix}$, $B = \begin{bmatrix} -2 & 1 & \frac{1}{2} \\ -3 & 1 & \frac{1}{2} \\ 2 & -2 & -\frac{1}{2} \end{bmatrix}$

In Exercises 77–80, find the inverse of the matrix (if it exists).

77. $\begin{bmatrix} -6 & 5 \\ -5 & 4 \end{bmatrix}$

78. $\begin{bmatrix} -3 & -5 \\ 2 & 3 \end{bmatrix}$

79. $\begin{bmatrix} 2 & 0 & 3 \\ -1 & 1 & 1 \\ 2 & -2 & 1 \end{bmatrix}$

80. $\begin{bmatrix} 0 & -2 & 1 \\ -5 & -2 & -3 \\ 7 & 3 & 4 \end{bmatrix}$

In Exercises 81–84, use the matrix capabilities of a graphing utility to find the inverse of the matrix (if it exists).

81. $\begin{bmatrix} -1 & -2 & -2 \\ 3 & 7 & 9 \\ 1 & 4 & 7 \end{bmatrix}$

82. $\begin{bmatrix} 1 & 4 & 6 \\ 2 & -3 & 1 \\ -1 & 18 & 16 \end{bmatrix}$

83. $\begin{bmatrix} 1 & 3 & 1 & 6 \\ 4 & 4 & 2 & 6 \\ 3 & 4 & 1 & 2 \\ -1 & 2 & -1 & -2 \end{bmatrix}$

84. $\begin{bmatrix} 8 & 0 & 2 & 8 \\ 4 & -2 & 0 & -2 \\ 1 & 2 & 1 & 4 \\ -1 & 4 & 1 & 1 \end{bmatrix}$

In Exercises 85–92, use the formula below to find the inverse of the matrix, if it exists.

$$A^{-1} = \frac{1}{ad - bc}\begin{bmatrix} d & -b \\ -c & a \end{bmatrix}$$

85. $\begin{bmatrix} -7 & 2 \\ -8 & 2 \end{bmatrix}$

86. $\begin{bmatrix} 10 & 4 \\ 7 & 3 \end{bmatrix}$

87. $\begin{bmatrix} -12 & 6 \\ 10 & -5 \end{bmatrix}$

88. $\begin{bmatrix} -18 & -15 \\ -6 & -5 \end{bmatrix}$

89. $\begin{bmatrix} -\frac{1}{2} & 20 \\ \frac{3}{10} & -6 \end{bmatrix}$

90. $\begin{bmatrix} -\frac{3}{4} & \frac{5}{2} \\ -\frac{4}{5} & -\frac{8}{3} \end{bmatrix}$

91. $\begin{bmatrix} 0.5 & 0.1 \\ -0.2 & -0.4 \end{bmatrix}$

92. $\begin{bmatrix} 1.6 & -3.2 \\ 1.2 & -2.4 \end{bmatrix}$

In Exercises 93–104, use an inverse matrix to solve (if possible) the system of linear equations.

93. $\begin{cases} -x + 4y = 8 \\ 2x - 7y = -5 \end{cases}$

94. $\begin{cases} 5x - y = 13 \\ -9x + 2y = -24 \end{cases}$

95. $\begin{cases} -3x + 10y = 8 \\ 5x - 17y = -13 \end{cases}$

96. $\begin{cases} 4x - 2y = -10 \\ -19x + 9y = 47 \end{cases}$

97. $\begin{cases} \frac{1}{2}x + \frac{1}{3}y = 2 \\ -3x + 2y = 0 \end{cases}$

98. $\begin{cases} -\frac{5}{6}x + \frac{3}{8}y = -2 \\ 4x - 3y = 0 \end{cases}$

99. $\begin{cases} 0.3x + 0.7y = 10.2 \\ 0.4x + 0.6y = 7.6 \end{cases}$

100. $\begin{cases} 3.5x - 4.5y = 8 \\ 2.5x - 7.5y = 25 \end{cases}$

101. $\begin{cases} 3x + 2y - z = 6 \\ x - y + 2z = -1 \\ 5x + y + z = 7 \end{cases}$

102. $\begin{cases} -x + 4y - 2z = 12 \\ 2x - 9y + 5z = -25 \\ -x + 5y - 4z = 10 \end{cases}$

103. $\begin{cases} -2x + y + 2z = -13 \\ -x - 4y + z = -11 \\ -y - z = 0 \end{cases}$

104. $\begin{cases} 3x - y + 5z = -14 \\ -x + y + 6z = 8 \\ -8x + 4y - z = 44 \end{cases}$

In Exercises 105–110, use the matrix capabilities of a graphing utility to solve (if possible) the system of linear equations.

105. $\begin{cases} x + 2y = -1 \\ 3x + 4y = -5 \end{cases}$

106. $\begin{cases} x + 3y = 23 \\ -6x + 2y = -18 \end{cases}$

107. $\begin{cases} \frac{6}{5}x - \frac{4}{7}y = \frac{6}{5} \\ -\frac{12}{5}x + \frac{12}{7}y = -\frac{17}{5} \end{cases}$

108. $\begin{cases} 5x + 10y = 7 \\ 2x + y = -98 \end{cases}$

109. $\begin{cases} -3x - 3y - 4z = 2 \\ y + z = -1 \\ 4x + 3y + 4z = -1 \end{cases}$

110. $\begin{cases} x - 3y - 2z = 8 \\ -2x + 7y + 3z = -19 \\ x - y - 3z = 3 \end{cases}$

7.4 In Exercises 111–114, find the determinant of the matrix.

111. $\begin{bmatrix} 8 & 5 \\ 2 & -4 \end{bmatrix}$

112. $\begin{bmatrix} -9 & 11 \\ 7 & -4 \end{bmatrix}$

113. $\begin{bmatrix} 50 & -30 \\ 10 & 5 \end{bmatrix}$

114. $\begin{bmatrix} 14 & -24 \\ 12 & -15 \end{bmatrix}$

In Exercises 115–118, find all (a) minors and (b) cofactors of the matrix.

115. $\begin{bmatrix} 2 & -1 \\ 7 & 4 \end{bmatrix}$

116. $\begin{bmatrix} 3 & 6 \\ 5 & -4 \end{bmatrix}$

117. $\begin{bmatrix} 3 & 2 & -1 \\ -2 & 5 & 0 \\ 1 & 8 & 6 \end{bmatrix}$

118. $\begin{bmatrix} 8 & 3 & 4 \\ 6 & 5 & -9 \\ -4 & 1 & 2 \end{bmatrix}$

In Exercises 119–128, find the determinant of the matrix. Expand by cofactors on the row or column that appears to make the computations easiest.

119. $\begin{bmatrix} -2 & 0 & 0 \\ 2 & -1 & 0 \\ -1 & 1 & -3 \end{bmatrix}$

120. $\begin{bmatrix} 0 & 1 & -2 \\ 0 & 1 & 2 \\ -1 & -1 & 3 \end{bmatrix}$

121. $\begin{bmatrix} 4 & 1 & -1 \\ 2 & 3 & 2 \\ 1 & -1 & 0 \end{bmatrix}$

122. $\begin{bmatrix} -1 & -2 & 1 \\ 2 & 3 & 0 \\ -5 & -1 & 3 \end{bmatrix}$

123. $\begin{bmatrix} -2 & 4 & 1 \\ -6 & 0 & 2 \\ 5 & 3 & 4 \end{bmatrix}$

124. $\begin{bmatrix} 1 & 1 & 4 \\ -4 & 1 & 2 \\ 0 & 1 & -1 \end{bmatrix}$

125. $\begin{bmatrix} 1 & 2 & -1 & 0 \\ 1 & 2 & -4 & 1 \\ 2 & -4 & -3 & 1 \\ 2 & 0 & 0 & 0 \end{bmatrix}$

126. $\begin{bmatrix} 1 & -2 & 1 & 2 \\ 4 & 1 & 4 & 1 \\ 2 & 3 & 3 & 0 \\ 0 & -2 & -4 & 2 \end{bmatrix}$

127. $\begin{bmatrix} 3 & 0 & -4 & 0 \\ 0 & 8 & 1 & 2 \\ 6 & 1 & 8 & 2 \\ 0 & 3 & -4 & 1 \end{bmatrix}$

128. $\begin{bmatrix} -5 & 6 & 0 & 0 \\ 0 & 1 & -1 & 2 \\ -3 & 4 & -5 & 1 \\ 1 & 6 & 0 & 3 \end{bmatrix}$

7.5 In Exercises 129–132, use Cramer's Rule to solve (if possible) the system of equations.

129. $\begin{cases} 5x - 2y = 6 \\ -11x + 3y = -23 \end{cases}$

130. $\begin{cases} 3x + 8y = -7 \\ 9x - 5y = 37 \end{cases}$

131. $\begin{cases} -2x + 3y - 5z = -11 \\ 4x - y + z = -3 \\ -x - 4y + 6z = 15 \end{cases}$

132. $\begin{cases} 5x - 2y + z = 15 \\ 3x - 3y - z = -7 \\ 2x - y - 7z = -3 \end{cases}$

In Exercises 133–136, use a determinant and the given vertices of a triangle to find the area of the triangle.

133.

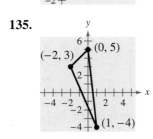

134.

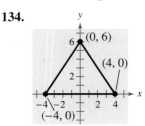

135.

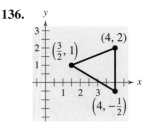

136.

In Exercises 137 and 138, use a determinant to determine whether the points are collinear.

137. $(-1, 7), (3, -9), (-3, 15)$

138. $(0, -5), (-2, -6), (8, -1)$

In Exercises 139–142, use a determinant to find an equation of the line passing through the points.

139. $(-4, 0), (4, 4)$
140. $(2, 5), (6, -1)$
141. $\left(-\frac{5}{2}, 3\right), \left(\frac{7}{2}, 1\right)$
142. $(-0.8, 0.2), (0.7, 3.2)$

In Exercises 143 and 144, (a) write the uncoded 1×3 row matrices for the message, and (b) encode the message using the encoding matrix.

Message	Encoding Matrix

143. LOOK OUT BELOW $\begin{bmatrix} 2 & -2 & 0 \\ 3 & 0 & -3 \\ -6 & 2 & 3 \end{bmatrix}$

144. HEAD DUE WEST $\begin{bmatrix} 1 & 2 & 2 \\ 3 & 7 & 9 \\ -1 & -4 & -7 \end{bmatrix}$

In Exercises 145 and 146, decode the cryptogram by using the inverse of the matrix

$A = \begin{bmatrix} -5 & 4 & -3 \\ 10 & -7 & 6 \\ 8 & -6 & 5 \end{bmatrix}$.

145. -5 11 -2 370 -265 225 -57 48 -33 32
-15 20 245 -171 147

146. 145 -105 92 264 -188 160 23 -16 15
129 -84 78 -9 8 -5 159 -118 100 219
-152 133 370 -265 225 -105 84 -63

EXPLORATION

TRUE OR FALSE? In Exercises 147 and 148, determine whether the statement is true or false. Justify your answer.

147. It is possible to find the determinant of a 4×5 matrix.

148. $\begin{vmatrix} a_{11} & a_{12} & a_{13} \\ a_{21} & a_{22} & a_{23} \\ a_{31} + c_1 & a_{32} + c_2 & a_{33} + c_3 \end{vmatrix} =$

$\begin{vmatrix} a_{11} & a_{12} & a_{13} \\ a_{21} & a_{22} & a_{23} \\ a_{31} & a_{32} & a_{33} \end{vmatrix} + \begin{vmatrix} a_{11} & a_{12} & a_{13} \\ a_{21} & a_{22} & a_{23} \\ c_1 & c_2 & c_3 \end{vmatrix}$

149. Use the matrix capabilities of a graphing utility to find the inverse of the matrix

$A = \begin{bmatrix} 1 & -3 \\ -2 & 6 \end{bmatrix}$.

What message appears on the screen? Why does the graphing utility display this message?

150. Under what conditions does a matrix have an inverse?

151. **WRITING** What is meant by the cofactor of an entry of a matrix? How are cofactors used to find the determinant of the matrix?

152. Three people were asked to solve a system of equations using an augmented matrix. Each person reduced the matrix to row-echelon form. The reduced matrices were

$\begin{bmatrix} 1 & 2 & \vdots & 3 \\ 0 & 1 & \vdots & 1 \end{bmatrix}, \begin{bmatrix} 1 & 0 & \vdots & 1 \\ 0 & 1 & \vdots & 1 \end{bmatrix}$,

and

$\begin{bmatrix} 1 & 2 & \vdots & 3 \\ 0 & 0 & \vdots & 0 \end{bmatrix}$.

Can all three be right? Explain.

153. **THINK ABOUT IT** Describe the row-echelon form of an augmented matrix that corresponds to a system of linear equations that has a unique solution.

154. Solve the equation for λ.

$\begin{vmatrix} 2 - \lambda & 5 \\ 3 & -8 - \lambda \end{vmatrix} = 0$

7 CHAPTER TEST   See www.CalcChat.com for worked-out solutions to odd-numbered exercises.

Take this test as you would take a test in class. When you are finished, check your work against the answers given in the back of the book.

In Exercises 1 and 2, write the matrix in reduced row-echelon form.

1. $\begin{bmatrix} 1 & -1 & 5 \\ 6 & 2 & 3 \\ 5 & 3 & -3 \end{bmatrix}$

2. $\begin{bmatrix} 1 & 0 & -1 & 2 \\ -1 & 1 & 1 & -3 \\ 1 & 1 & -1 & 1 \\ 3 & 2 & -3 & 4 \end{bmatrix}$

3. Write the augmented matrix corresponding to the system of equations and solve the system.

$$\begin{cases} 4x + 3y - 2z = 14 \\ -x - y + 2z = -5 \\ 3x + y - 4z = 8 \end{cases}$$

4. Find (a) $A - B$, (b) $3A$, (c) $3A - 2B$, and (d) AB (if possible).

$$A = \begin{bmatrix} 6 & 5 \\ -5 & -5 \end{bmatrix}, \quad B = \begin{bmatrix} 5 & 0 \\ -5 & -1 \end{bmatrix}$$

In Exercises 5 and 6, find the inverse of the matrix (if it exists).

5. $\begin{bmatrix} -4 & 3 \\ 5 & -2 \end{bmatrix}$

6. $\begin{bmatrix} -2 & 4 & -6 \\ 2 & 1 & 0 \\ 4 & -2 & 5 \end{bmatrix}$

7. Use the result of Exercise 5 to solve the system.

$$\begin{cases} -4x + 3y = 6 \\ 5x - 2y = 24 \end{cases}$$

In Exercises 8–10, evaluate the determinant of the matrix.

8. $\begin{bmatrix} -6 & 4 \\ 10 & 12 \end{bmatrix}$

9. $\begin{bmatrix} \frac{5}{2} & \frac{13}{4} \\ -8 & \frac{6}{5} \end{bmatrix}$

10. $\begin{bmatrix} 6 & -7 & 2 \\ 3 & -2 & 0 \\ 1 & 5 & 1 \end{bmatrix}$

In Exercises 11 and 12, use Cramer's Rule to solve (if possible) the system of equations.

11. $\begin{cases} 7x + 6y = 9 \\ -2x - 11y = -49 \end{cases}$

12. $\begin{cases} 6x - y + 2z = -4 \\ -2x + 3y - z = 10 \\ 4x - 4y + z = -18 \end{cases}$

13. Use a determinant to find the area of the triangle in the figure.

14. Find the uncoded 1×3 row matrices for the message KNOCK ON WOOD. Then encode the message using the matrix A below.

$$A = \begin{bmatrix} 1 & -1 & 0 \\ 1 & 0 & -1 \\ 6 & -2 & -3 \end{bmatrix}$$

15. One hundred liters of a 50% solution is obtained by mixing a 60% solution with a 20% solution. How many liters of each solution must be used to obtain the desired mixture?

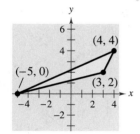

FIGURE FOR 13

PROOFS IN MATHEMATICS

Proofs without words are pictures or diagrams that give a visual understanding of why a theorem or statement is true. They can also provide a starting point for writing a formal proof. The following proof shows that a 2×2 determinant is the area of a parallelogram.

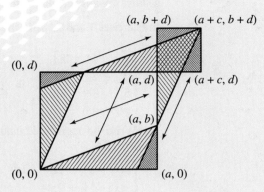

$$\begin{vmatrix} a & b \\ c & d \end{vmatrix} = ad - bc = \| \square \| - \| \square \| = \| \square \|$$

The following is a color-coded version of the proof along with a brief explanation of why this proof works.

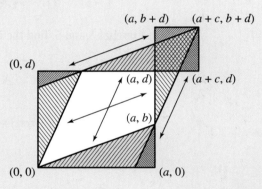

$$\begin{vmatrix} a & b \\ c & d \end{vmatrix} = ad - bc = \| \square \| - \| \square \| = \| \square \|$$

Area of $\square$ = Area of orange $\triangle$ + Area of yellow $\triangle$ + Area of blue $\triangle$ + Area of pink $\triangle$ + Area of white quadrilateral

Area of $\square$ = Area of orange $\triangle$ + Area of pink $\triangle$ + Area of green quadrilateral

Area of $\square$ = Area of white quadrilateral + Area of blue $\triangle$ + Area of yellow $\triangle$ − Area of green quadrilateral
= Area of $\square$ − Area of $\square$

From "Proof Without Words" by Solomon W. Golomb, *Mathematics Magazine*, March 1985, Vol. 58, No. 2, pg. 107. Reprinted with permission.

PROBLEM SOLVING

This collection of thought-provoking and challenging exercises further explores and expands upon concepts learned in this chapter.

1. The columns of matrix T show the coordinates of the vertices of a triangle. Matrix A is a transformation matrix.

$$A = \begin{bmatrix} 0 & -1 \\ 1 & 0 \end{bmatrix} \quad T = \begin{bmatrix} 1 & 2 & 3 \\ 1 & 4 & 2 \end{bmatrix}$$

(a) Find AT and AAT. Then sketch the original triangle and the two transformed triangles. What transformation does A represent?

(b) Given the triangle determined by AAT, describe the transformation process that produces the triangle determined by AT and then the triangle determined by T.

2. The matrices show the number of people (in thousands) who lived in each region of the United States in 2000 and the number of people (in thousands) projected to live in each region in 2015. The regional populations are separated into three age categories. (Source: U.S. Census Bureau)

2000

0–17	18–64	65 +
13,048	33,174	7,372
16,648	39,486	8,259
25,567	62,232	12,438
4,935	11,208	2,030
12,097	28,037	4,892

2015

0–17	18–64	65 +
12,441	35,288	8,837
16,363	42,249	9,957
29,372	73,495	17,574
6,016	14,231	3,338
12,826	33,294	7,085

(a) The total population in 2000 was approximately 281,422,000 and the projected total population in 2015 is 322,366,000. Rewrite the matrices to give the information as percents of the total population.

(b) Write a matrix that gives the projected change in the percent of the population in each region and age group from 2000 to 2015.

(c) Based on the result of part (b), which region(s) and age group(s) are projected to show relative growth from 2000 to 2015?

3. Determine whether the matrix is idempotent. A square matrix is **idempotent** if $A^2 = A$.

(a) $\begin{bmatrix} 1 & 0 \\ 0 & 0 \end{bmatrix}$ (b) $\begin{bmatrix} 0 & 1 \\ 1 & 0 \end{bmatrix}$

(c) $\begin{bmatrix} 2 & 3 \\ -1 & -2 \end{bmatrix}$ (d) $\begin{bmatrix} 2 & 3 \\ 1 & 2 \end{bmatrix}$

4. Let $A = \begin{bmatrix} 1 & 2 \\ -2 & 1 \end{bmatrix}$.

(a) Show that $A^2 - 2A + 5I = O$, where I is the identity matrix of order 2.

(b) Show that $A^{-1} = \frac{1}{5}(2I - A)$.

(c) Show in general that for any square matrix satisfying

$$A^2 - 2A + 5I = O$$

the inverse of A is given by

$$A^{-1} = \frac{1}{5}(2I - A).$$

5. Two competing companies offer satellite television to a city with 100,000 households. Gold Satellite System has 25,000 subscribers and Galaxy Satellite Network has 30,000 subscribers. (The other 45,000 households do not subscribe.) The percent changes in satellite subscriptions each year are shown in the matrix below.

$$\begin{bmatrix} 0.70 & 0.15 & 0.15 \\ 0.20 & 0.80 & 0.15 \\ 0.10 & 0.05 & 0.70 \end{bmatrix}$$

(a) Find the number of subscribers each company will have in 1 year using matrix multiplication. Explain how you obtained your answer.

(b) Find the number of subscribers each company will have in 2 years using matrix multiplication. Explain how you obtained your answer.

(c) Find the number of subscribers each company will have in 3 years using matrix multiplication. Explain how you obtained your answer.

(d) What is happening to the number of subscribers to each company? What is happening to the number of nonsubscribers?

6. Find x such that the matrix is equal to its own inverse.

$$A = \begin{bmatrix} 3 & x \\ -2 & -3 \end{bmatrix}$$

7. Find x such that the matrix is singular.

$$A = \begin{bmatrix} 4 & x \\ -2 & -3 \end{bmatrix}$$

8. Find an example of a singular 2×2 matrix satisfying $A^2 = A$.

587

9. Verify the following equation.

$$\begin{vmatrix} 1 & 1 & 1 \\ a & b & c \\ a^2 & b^2 & c^2 \end{vmatrix} = (a - b)(b - c)(c - a)$$

10. Verify the following equation.

$$\begin{vmatrix} 1 & 1 & 1 \\ a & b & c \\ a^3 & b^3 & c^3 \end{vmatrix} = (a - b)(b - c)(c - a)(a + b + c)$$

11. Verify the following equation.

$$\begin{vmatrix} x & 0 & c \\ -1 & x & b \\ 0 & -1 & a \end{vmatrix} = ax^2 + bx + c$$

12. Use the equation given in Exercise 11 as a model to find a determinant that is equal to $ax^3 + bx^2 + cx + d$.

13. The atomic masses of three compounds are shown in the table. Use a linear system and Cramer's Rule to find the atomic masses of sulfur (S), nitrogen (N), and fluorine (F).

Compound	Formula	Atomic Mass
Tetrasulfur tetranitride	S_4N_4	184
Sulfur hexafluoride	SF_6	146
Dinitrogen tetrafluoride	N_2F_4	104

14. A walkway lighting package includes a transformer, a certain length of wire, and a certain number of lights on the wire. The price of each lighting package depends on the length of wire and the number of lights on the wire. Use the following information to find the cost of a transformer, the cost per foot of wire, and the cost of a light. Assume that the cost of each item is the same in each lighting package.

- A package that contains a transformer, 25 feet of wire, and 5 lights costs $20.
- A package that contains a transformer, 50 feet of wire, and 15 lights costs $35.
- A package that contains a transformer, 100 feet of wire, and 20 lights costs $50.

15. The **transpose** of a matrix, denoted A^T, is formed by writing its columns as rows. Find the transpose of each matrix and verify that $(AB)^T = B^T A^T$.

$$A = \begin{bmatrix} -1 & 1 & -2 \\ 2 & 0 & 1 \end{bmatrix}, \quad B = \begin{bmatrix} -3 & 0 \\ 1 & 2 \\ 1 & -1 \end{bmatrix}$$

16. Use the inverse of matrix A to decode the cryptogram.

$$A = \begin{bmatrix} 1 & -2 & 2 \\ 1 & 1 & -3 \\ 1 & -1 & 4 \end{bmatrix}$$

```
23   13  -34   31  -34   63   25  -17   61
24   14  -37   41  -17   -8   20  -29   40
38  -56  116   13  -11    1   22   -3   -6
41  -53   85   28  -32   16
```

17. A code breaker intercepted the encoded message below.

```
45  -35   38  -30   18  -18   35  -30   81  -60
42  -28   75  -55    2   -2   22  -21   15
-10
```

Let

$$A^{-1} = \begin{bmatrix} w & x \\ y & z \end{bmatrix}.$$

(a) You know that $\begin{bmatrix} 45 & -35 \end{bmatrix} A^{-1} = \begin{bmatrix} 10 & 15 \end{bmatrix}$ and that $\begin{bmatrix} 38 & -30 \end{bmatrix} A^{-1} = \begin{bmatrix} 8 & 14 \end{bmatrix}$, where A^{-1} is the inverse of the encoding matrix A. Write and solve two systems of equations to find w, x, y, and z.

(b) Decode the message.

 18. Let

$$A = \begin{bmatrix} 6 & 4 & 1 \\ 0 & 2 & 3 \\ 1 & 1 & 2 \end{bmatrix}.$$

Use a graphing utility to find A^{-1}. Compare $|A^{-1}|$ with $|A|$. Make a conjecture about the determinant of the inverse of a matrix.

19. Let A be an $n \times n$ matrix each of whose rows adds up to zero. Find $|A|$.

20. Consider matrices of the form

$$A = \begin{bmatrix} 0 & a_{12} & a_{13} & a_{14} & \cdots & a_{1n} \\ 0 & 0 & a_{23} & a_{24} & \cdots & a_{2n} \\ 0 & 0 & 0 & a_{34} & \cdots & a_{3n} \\ \vdots & \vdots & \vdots & \vdots & \cdots & \vdots \\ 0 & 0 & 0 & 0 & \cdots & a_{(n-1)n} \\ 0 & 0 & 0 & 0 & \cdots & 0 \end{bmatrix}.$$

(a) Write a 2×2 matrix and a 3×3 matrix in the form of A.

(b) Use a graphing utility to raise each of the matrices to higher powers. Describe the result.

(c) Use the result of part (b) to make a conjecture about powers of A if A is a 4×4 matrix. Use a graphing utility to test your conjecture.

(d) Use the results of parts (b) and (c) to make a conjecture about powers of A if A is an $n \times n$ matrix.

APPENDIX A   ERRORS AND THE ALGEBRA OF CALCULUS

What you should learn

- Avoid common algebraic errors.
- Recognize and use algebraic techniques that are common in calculus.

Why you should learn it

An efficient command of algebra is critical in mastering this course and in the study of calculus.

Algebraic Errors to Avoid

This section contains five lists of common algebraic errors: errors involving parentheses, errors involving fractions, errors involving exponents, errors involving radicals, and errors involving dividing out. Many of these errors are made because they seem to be the *easiest* things to do. For instance, the operations of subtraction and division are often believed to be commutative and associative. The following examples illustrate the fact that subtraction and division are neither commutative nor associative.

Not commutative	*Not associative*
$4 - 3 \neq 3 - 4$	$8 - (6 - 2) \neq (8 - 6) - 2$
$15 \div 5 \neq 5 \div 15$	$20 \div (4 \div 2) \neq (20 \div 4) \div 2$

Errors Involving Parentheses

Potential Error	Correct Form	Comment
$a - (x - b) = a - x - b$	$a - (x - b) = a - x + b$	
$(a + b)^2 = a^2 + b^2$	$(a + b)^2 = a^2 + 2ab + b^2$	
$\left(\dfrac{1}{2}a\right)\left(\dfrac{1}{2}b\right) = \dfrac{1}{2}(ab)$	$\left(\dfrac{1}{2}a\right)\left(\dfrac{1}{2}b\right) = \dfrac{1}{4}(ab) = \dfrac{ab}{4}$	
$(3x + 6)^2 = 3(x + 2)^2$	$(3x + 6)^2 = [3(x + 2)]^2$ $= 3^2(x + 2)^2$	

Errors Involving Fractions

Potential Error	Correct Form	Comment
$\dfrac{a}{x + b} = \dfrac{a}{x} + \dfrac{a}{b}$	Leave as $\dfrac{a}{x + b}$.	
$\dfrac{\left(\dfrac{x}{a}\right)}{b} = \dfrac{bx}{a}$	$\dfrac{\left(\dfrac{x}{a}\right)}{b} = \left(\dfrac{x}{a}\right)\left(\dfrac{1}{b}\right) = \dfrac{x}{ab}$	
$\dfrac{1}{a} + \dfrac{1}{b} = \dfrac{1}{a + b}$	$\dfrac{1}{a} + \dfrac{1}{b} = \dfrac{b + a}{ab}$	
$\dfrac{1}{3x} = \dfrac{1}{3}x$	$\dfrac{1}{3x} = \dfrac{1}{3} \cdot \dfrac{1}{x}$	
$(1/3)x = \dfrac{1}{3x}$	$(1/3)x = \dfrac{1}{3} \cdot x = \dfrac{x}{3}$	
$(1/x) + 2 = \dfrac{1}{x + 2}$	$(1/x) + 2 = \dfrac{1}{x} + 2 = \dfrac{1 + 2x}{x}$	

Errors Involving Exponents

Potential Error	Correct Form	Comment
$(x^2)^3 = x^5$	$(x^2)^3 = x^{2 \cdot 3} = x^6$	
$x^2 \cdot x^3 = x^6$	$x^2 \cdot x^3 = x^{2+3} = x^5$	
$2x^3 = (2x)^3$	$2x^3 = 2(x^3)$	
$\dfrac{1}{x^2 - x^3} = x^{-2} - x^{-3}$	Leave as $\dfrac{1}{x^2 - x^3}$.	

Errors Involving Radicals

Potential Error	Correct Form	Comment
$\sqrt{5x} = 5\sqrt{x}$	$\sqrt{5x} = \sqrt{5}\sqrt{x}$	
$\sqrt{x^2 + a^2} = x + a$	Leave as $\sqrt{x^2 + a^2}$.	
$\sqrt{-x + a} = -\sqrt{x - a}$	Leave as $\sqrt{-x + a}$.	

Errors Involving Dividing Out

Potential Error	Correct Form	Comment
$\dfrac{a + bx}{a} = 1 + bx$	$\dfrac{a + bx}{a} = \dfrac{a}{a} + \dfrac{bx}{a} = 1 + \dfrac{b}{a}x$	
$\dfrac{a + ax}{a} = a + x$	$\dfrac{a + ax}{a} = \dfrac{a(1 + x)}{a} = 1 + x$	
$1 + \dfrac{x}{2x} = 1 + \dfrac{1}{x}$	$1 + \dfrac{x}{2x} = 1 + \dfrac{1}{2} = \dfrac{3}{2}$	

A good way to avoid errors is to *work slowly*, *write neatly*, and *talk to yourself*. Each time you write a step, ask yourself why the step is algebraically legitimate. You can justify the step below because *dividing the numerator and denominator by the same nonzero number produces an equivalent fraction.*

$$\frac{2x}{6} = \frac{2 \cdot x}{2 \cdot 3} = \frac{x}{3}$$

Example 1   Using the Property for Adding Fractions

Describe and correct the error. $\dfrac{1}{2x} + \dfrac{1}{3x} = \dfrac{1}{5x}$

Solution

When adding fractions, use the property for adding fractions: $\dfrac{1}{a} + \dfrac{1}{b} = \dfrac{b + a}{ab}$.

$$\frac{1}{2x} + \frac{1}{3x} = \frac{3x + 2x}{6x^2} = \frac{5x}{6x^2} = \frac{5}{6x}$$

CHECK*Point*   Now try Exercise 19.

Some Algebra of Calculus

In calculus it is often necessary to take a simplified algebraic expression and rewrite it. See the following lists, taken from a standard calculus text.

Unusual Factoring

Expression	Useful Calculus Form	Comment
$\dfrac{5x^4}{8}$	$\dfrac{5}{8}x^4$	
$\dfrac{x^2 + 3x}{-6}$	$-\dfrac{1}{6}(x^2 + 3x)$	
$2x^2 - x - 3$	$2\left(x^2 - \dfrac{x}{2} - \dfrac{3}{2}\right)$	
$\dfrac{x}{2}(x + 1)^{-1/2} + (x + 1)^{1/2}$	$\dfrac{(x + 1)^{-1/2}}{2}[x + 2(x + 1)]$	

Writing with Negative Exponents

Expression	Useful Calculus Form	Comment
$\dfrac{9}{5x^3}$	$\dfrac{9}{5}x^{-3}$	
$\dfrac{7}{\sqrt{2x - 3}}$	$7(2x - 3)^{-1/2}$	

Writing a Fraction as a Sum

Expression	Useful Calculus Form	Comment
$\dfrac{x + 2x^2 + 1}{\sqrt{x}}$	$x^{1/2} + 2x^{3/2} + x^{-1/2}$	
$\dfrac{1 + x}{x^2 + 1}$	$\dfrac{1}{x^2 + 1} + \dfrac{x}{x^2 + 1}$	
$\dfrac{2x}{x^2 + 2x + 1}$	$\dfrac{2x + 2 - 2}{x^2 + 2x + 1}$	
	$= \dfrac{2x + 2}{x^2 + 2x + 1} - \dfrac{2}{(x + 1)^2}$	
$\dfrac{x^2 - 2}{x + 1}$	$x - 1 - \dfrac{1}{x + 1}$	
$\dfrac{x + 7}{x^2 - x - 6}$	$\dfrac{2}{x - 3} - \dfrac{1}{x + 2}$	

Inserting Factors and Terms

Expression	Useful Calculus Form	Comment
$(2x - 1)^3$	$\dfrac{1}{2}(2x - 1)^3(2)$	
$7x^2(4x^3 - 5)^{1/2}$	$\dfrac{7}{12}(4x^3 - 5)^{1/2}(12x^2)$	
$\dfrac{4x^2}{9} - 4y^2 = 1$	$\dfrac{x^2}{9/4} - \dfrac{y^2}{1/4} = 1$	
$\dfrac{x}{x + 1}$	$\dfrac{x + 1 - 1}{x + 1} = 1 - \dfrac{1}{x + 1}$	

The next five examples demonstrate many of the steps in the preceding lists.

Example 2   Factors Involving Negative Exponents

Factor $x(x + 1)^{-1/2} + (x + 1)^{1/2}$.

Solution

When multiplying factors with like bases, you add exponents. When factoring, you are undoing multiplication, and so you *subtract* exponents.

$$x(x + 1)^{-1/2} + (x + 1)^{1/2} = (x + 1)^{-1/2}[x(x + 1)^0 + (x + 1)^1]$$
$$= (x + 1)^{-1/2}[x + (x + 1)]$$
$$= (x + 1)^{-1/2}(2x + 1)$$

CHECK Point   Now try Exercise 29.

Another way to simplify the expression in Example 2 is to multiply the expression by a fractional form of 1 and then use the Distributive Property.

$$x(x + 1)^{-1/2} + (x + 1)^{1/2} = [x(x + 1)^{-1/2} + (x + 1)^{1/2}] \cdot$$

$$= \frac{x(x + 1)^0 + (x + 1)^1}{(x + 1)^{1/2}} = \frac{2x + 1}{\sqrt{x + 1}}$$

Example 3   Inserting Factors in an Expression

Insert the required factor: $\dfrac{x + 2}{(x^2 + 4x - 3)^2} = (\quad)\dfrac{1}{(x^2 + 4x - 3)^2}(2x + 4)$.

Solution

The expression on the right side of the equation is twice the expression on the left side. To make both sides equal, insert a factor of $\frac{1}{2}$.

$$\frac{x + 2}{(x^2 + 4x - 3)^2} = \left(\frac{1}{2}\right)\frac{1}{(x^2 + 4x - 3)^2}(2x + 4)$$

CHECK Point   Now try Exercise 31.

Example 4   Rewriting Fractions

Explain why the two expressions are equivalent.

$$\frac{4x^2}{9} - 4y^2 = \frac{x^2}{\frac{9}{4}} - \frac{y^2}{\frac{1}{4}}$$

Solution

To write the expression on the left side of the equation in the form given on the right side, multiply the numerators and denominators of both terms by $\frac{1}{4}$.

$$\frac{4x^2}{9} - 4y^2 = \frac{4x^2}{9} \qquad - 4y^2 \qquad = \frac{x^2}{\frac{9}{4}} - \frac{y^2}{\frac{1}{4}}$$

CHECK Point   Now try Exercise 35.

Example 5   Rewriting with Negative Exponents

Rewrite each expression using negative exponents.

a. $\dfrac{-4x}{(1 - 2x^2)^2}$ **b.** $\dfrac{2}{5x^3} - \dfrac{1}{\sqrt{x}} + \dfrac{3}{5(4x)^2}$

Solution

a. $\dfrac{-4x}{(1 - 2x^2)^2} = -4x(1 - 2x^2)^{-2}$

b. Begin by writing the second term in exponential form.

$$\frac{2}{5x^3} - \frac{1}{\sqrt{x}} + \frac{3}{5(4x)^2} = \frac{2}{5x^3} - \frac{1}{x^{1/2}} + \frac{3}{5(4x)^2}$$

$$= \frac{2}{5}x^{-3} - x^{-1/2} + \frac{3}{5}(4x)^{-2}$$

CHECK Point   Now try Exercise 47.

Example 6   Writing a Fraction as a Sum of Terms

Rewrite each fraction as the sum of three terms.

a. $\dfrac{x^2 - 4x + 8}{2x}$ **b.** $\dfrac{x + 2x^2 + 1}{\sqrt{x}}$

Solution

a. $\dfrac{x^2 - 4x + 8}{2x} = \dfrac{x^2}{2x} - \dfrac{4x}{2x} + \dfrac{8}{2x}$

$\qquad\qquad\qquad = \dfrac{x}{2} - 2 + \dfrac{4}{x}$

b. $\dfrac{x + 2x^2 + 1}{\sqrt{x}} = \dfrac{x}{x^{1/2}} + \dfrac{2x^2}{x^{1/2}} + \dfrac{1}{x^{1/2}}$

$\qquad\qquad\qquad = x^{1/2} + 2x^{3/2} + x^{-1/2}$

CHECK Point   Now try Exercise 51.

A EXERCISES

See www.CalcChat.com for worked-out solutions to odd-numbered exercises.

VOCABULARY: Fill in the blanks.

1. To write the expression $\dfrac{3}{x^5}$ with negative exponents, move x^5 to the _____ and change the sign of the exponent.

2. When dividing fractions, multiply by the _____.

SKILLS AND APPLICATIONS

In Exercises 3–22, describe and correct the error.

3. $2x - (3y + 4) = 2x - 3y + 4$

4. $5z + 3(x - 2) = 5z + 3x - 2$

5. $\dfrac{4}{16x - (2x + 1)} = \dfrac{4}{14x + 1}$

6. $\dfrac{1 - x}{(5 - x)(-x)} = \dfrac{x - 1}{x(x - 5)}$

7. $(5z)(6z) = 30z$

8. $x(yz) = (xy)(xz)$

9. $a\left(\dfrac{x}{y}\right) = \dfrac{ax}{ay}$

10. $(4x)^2 = 4x^2$

11. $\sqrt{x + 9} = \sqrt{x} + 3$

12. $\sqrt{25 - x^2} = 5 - x$

13. $\dfrac{2x^2 + 1}{5x} = \dfrac{2x + 1}{5}$

14. $\dfrac{6x + y}{6x - y} = \dfrac{x + y}{x - y}$

15. $\dfrac{1}{a^{-1} + b^{-1}} = \left(\dfrac{1}{a + b}\right)^{-1}$

16. $\dfrac{1}{x + y^{-1}} = \dfrac{y}{x + 1}$

17. $(x^2 + 5x)^{1/2} = x(x + 5)^{1/2}$

18. $x(2x - 1)^2 = (2x^2 - x)^2$

19. $\dfrac{3}{x} + \dfrac{4}{y} = \dfrac{7}{x + y}$

20. $\dfrac{1}{2y} = (1/2)y$

21. $\dfrac{x}{2y} + \dfrac{y}{3} = \dfrac{x + y}{2y + 3}$

22. $5 + (1/y) = \dfrac{1}{5 + y}$

In Exercises 23–44, insert the required factor in the parentheses.

23. $\dfrac{5x + 3}{4} = \dfrac{1}{4}()$

24. $\dfrac{7x^2}{10} = \dfrac{7}{10}()$

25. $\frac{2}{3}x^2 + \frac{1}{3}x + 5 = \frac{1}{3}()$

26. $\frac{3}{4}x + \frac{1}{2} = \frac{1}{4}()$

27. $x^2(x^3 - 1)^4 = ()(x^3 - 1)^4(3x^2)$

28. $x(1 - 2x^2)^3 = ()(1 - 2x^2)^3(-4x)$

29. $2(y - 5)^{1/2} + y(y - 5)^{-1/2} = (y - 5)^{-1/2}()$

30. $3t(6t + 1)^{-1/2} + (6t + 1)^{1/2} = (6t + 1)^{-1/2}()$

31. $\dfrac{4x + 6}{(x^2 + 3x + 7)^3} = ()\dfrac{1}{(x^2 + 3x + 7)^3}(2x + 3)$

32. $\dfrac{x + 1}{(x^2 + 2x - 3)^2} = ()\dfrac{1}{(x^2 + 2x - 3)^2}(2x + 2)$

33. $\dfrac{3}{x} + \dfrac{5}{2x^2} - \dfrac{3}{2}x = ()(6x + 5 - 3x^3)$

34. $\dfrac{(x - 1)^2}{169} + (y + 5)^2 = \dfrac{(x - 1)^3}{169()} + (y + 5)^2$

35. $\dfrac{25x^2}{36} + \dfrac{4y^2}{9} = \dfrac{x^2}{()} + \dfrac{y^2}{()}$

36. $\dfrac{5x^2}{9} - \dfrac{16y^2}{49} = \dfrac{x^2}{()} - \dfrac{y^2}{()}$

37. $\dfrac{x^2}{3/10} - \dfrac{y^2}{4/5} = \dfrac{10x^2}{()} - \dfrac{5y^2}{()}$

38. $\dfrac{x^2}{5/8} + \dfrac{y^2}{6/11} = \dfrac{8x^2}{()} + \dfrac{11y^2}{()}$

39. $x^{1/3} - 5x^{4/3} = x^{1/3}()$

40. $3(2x + 1)x^{1/2} + 4x^{3/2} = x^{1/2}()$

41. $(1 - 3x)^{4/3} - 4x(1 - 3x)^{1/3} = (1 - 3x)^{1/3}()$

42. $\dfrac{1}{2\sqrt{x}} + 5x^{3/2} - 10x^{5/2} = \dfrac{1}{2\sqrt{x}}()$

43. $\dfrac{1}{10}(2x + 1)^{5/2} - \dfrac{1}{6}(2x + 1)^{3/2} = \dfrac{(2x + 1)^{3/2}}{15}()$

44. $\dfrac{3}{7}(t + 1)^{7/3} - \dfrac{3}{4}(t + 1)^{4/3} = \dfrac{3(t + 1)^{4/3}}{28}()$

In Exercises 45–50, write the expression using negative exponents.

45. $\dfrac{7}{(x + 3)^5}$

46. $\dfrac{2 - x}{(x + 1)^{3/2}}$

47. $\dfrac{2x^5}{(3x + 5)^4}$

48. $\dfrac{x + 1}{x(6 - x)^{1/2}}$

49. $\dfrac{4}{3x} + \dfrac{4}{x^4} - \dfrac{7x}{\sqrt[3]{2x}}$

50. $\dfrac{x}{x - 2} + \dfrac{1}{x^2} + \dfrac{8}{3(9x)^3}$

In Exercises 51–56, write the fraction as a sum of two or more terms.

51. $\dfrac{x^2 + 6x + 12}{3x}$

52. $\dfrac{x^3 - 5x^2 + 4}{x^2}$

53. $\dfrac{4x^3 - 7x^2 + 1}{x^{1/3}}$

54. $\dfrac{2x^5 - 3x^3 + 5x - 1}{x^{3/2}}$

55. $\dfrac{3 - 5x^2 - x^4}{\sqrt{x}}$

56. $\dfrac{x^3 - 5x^4}{3x^2}$

In Exercises 57–68, simplify the expression.

57. $\dfrac{-2(x^2 - 3)^{-3}(2x)(x + 1)^3 - 3(x + 1)^2(x^2 - 3)^{-2}}{[(x + 1)^3]^2}$

58. $\dfrac{x^5(-3)(x^2 + 1)^{-4}(2x) - (x^2 + 1)^{-3}(5)x^4}{(x^5)^2}$

59. $\dfrac{(6x + 1)^3(27x^2 + 2) - (9x^3 + 2x)(3)(6x + 1)^2(6)}{[(6x + 1)^3]^2}$

60. $\dfrac{(4x^2 + 9)^{1/2}(2) - (2x + 3)(\frac{1}{2})(4x^2 + 9)^{-1/2}(8x)}{[(4x^2 + 9)^{1/2}]^2}$

61. $\dfrac{(x + 2)^{3/4}(x + 3)^{-2/3} - (x + 3)^{1/3}(x + 2)^{-1/4}}{[(x + 2)^{3/4}]^2}$

62. $(2x - 1)^{1/2} - (x + 2)(2x - 1)^{-1/2}$

63. $\dfrac{2(3x - 1)^{1/3} - (2x + 1)(\frac{1}{3})(3x - 1)^{-2/3}(3)}{(3x - 1)^{2/3}}$

64. $\dfrac{(x + 1)(\frac{1}{2})(2x - 3x^2)^{-1/2}(2 - 6x) - (2x - 3x^2)^{1/2}}{(x + 1)^2}$

65. $\dfrac{1}{(x^2 + 4)^{1/2}} \cdot \dfrac{1}{2}(x^2 + 4)^{-1/2}(2x)$

66. $\dfrac{1}{x^2 - 6}(2x) + \dfrac{1}{2x + 5}(2)$

67. $(x^2 + 5)^{1/2}(\frac{3}{2})(3x - 2)^{1/2}(3)$
$+ (3x - 2)^{3/2}(\frac{1}{2})(x^2 + 5)^{-1/2}(2x)$

68. $(3x + 2)^{-1/2}(3)(x - 6)^{1/2}(1)$
$+ (x - 6)^3(-\frac{1}{2})(3x + 2)^{-3/2}(3)$

69. **ATHLETICS**   An athlete has set up a course for training as part of her regimen in preparation for an upcoming triathlon. She is dropped off by a boat 2 miles from the nearest point on shore. The finish line is 4 miles down the coast and 2 miles inland (see figure). She can swim 2 miles per hour and run 6 miles per hour. The time t (in hours) required for her to reach the finish line can be approximated by the model

$$t = \dfrac{\sqrt{x^2 + 4}}{2} + \dfrac{\sqrt{(4 - x)^2 + 4}}{6}$$

where x is the distance down the coast (in miles) to the point at which she swims and then leaves the water to start her run.

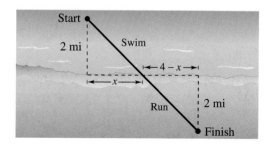

(a) Find the times required for the triathlete to finish when she swims to the points $x = 0.5$, $x = 1.0, \ldots, x = 3.5$, and $x = 4.0$ miles down the coast.

(b) Use your results from part (a) to determine the distance down the coast that will yield the minimum amount of time required for the triathlete to reach the finish line.

(c) The expression below was obtained using calculus. It can be used to find the minimum amount of time required for the triathlete to reach the finish line. Simplify the expression.

$$\tfrac{1}{2}x(x^2 + 4)^{-1/2} + \tfrac{1}{6}(x - 4)(x^2 - 8x + 20)^{-1/2}$$

70. (a) Verify that $y_1 = y_2$ analytically.

$$y_1 = x^2\left(\dfrac{1}{3}\right)(x^2 + 1)^{-2/3}(2x) + (x^2 + 1)^{1/3}(2x)$$

$$y_2 = \dfrac{2x(4x^2 + 3)}{3(x^2 + 1)^{2/3}}$$

(b) Complete the table and demonstrate the equality in part (a) numerically.

x	-2	-1	$-\frac{1}{2}$	0	1	2	$\frac{5}{2}$
y_1							
y_2							

EXPLORATION

71. **WRITING**   Write a paragraph explaining to a classmate why $\dfrac{1}{(x - 2)^{1/2} + x^4} \neq (x - 2)^{-1/2} + x^{-4}$.

72. **CAPSTONE**   You are taking a course in calculus, and for one of the homework problems you obtain the following answer.

$$\dfrac{1}{10}(2x - 1)^{5/2} + \dfrac{1}{6}(2x - 1)^{3/2}$$

The answer in the back of the book is $\frac{1}{15}(2x - 1)^{3/2}(3x + 1)$. Show how the second answer can be obtained from the first. Then use the same technique to simplify each of the following expressions.

(a) $\dfrac{2}{3}x(2x - 3)^{3/2} - \dfrac{2}{15}(2x - 3)^{5/2}$

(b) $\dfrac{2}{3}x(4 + x)^{3/2} - \dfrac{2}{15}(4 + x)^{5/2}$

ANSWERS TO ODD-NUMBERED EXERCISES AND TESTS

Chapter P

Section P.1   *(page 12)*

1. rational   3. origin   5. composite
7. variables; constants   9. coefficient
11. (a) 5, 1, 2   (b) 0, 5, 1, 2   (c) $-9, 5, 0, 1, -4, 2, -11$
    (d) $-\frac{7}{2}, \frac{2}{3}, -9, 5, 0, 1, -4, 2, -11$ (e) $\sqrt{2}$
13. (a) 1   (b) 1   (c) $-13, 1, -6$
    (d) $2.01, -13, 1, -6, 0.666\ldots$ (e) $0.010110111\ldots$
15. (a) $\frac{6}{3}, 8$ (b) $\frac{6}{3}, 8$ (c) $\frac{6}{3}, -1, 8, -22$
    (d) $-\frac{1}{3}, \frac{6}{3}, -7.5, -1, 8, -22$ (e) $-\pi, \frac{1}{2}\sqrt{2}$

17. (a) 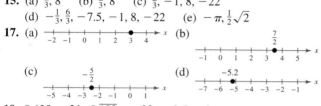   (b)

(c)   (d)

19. 0.625   21. $0.\overline{123}$ 23. $-2.5 < 2$
25.
$-4 > -8$

27.   29.
$\frac{3}{2} < 7$ $\frac{5}{6} > \frac{2}{3}$

31. (a) $x \le 5$ denotes the set of all real numbers less than or equal to 5.
    (b)   (c) Unbounded

33. (a) $x < 0$ denotes the set of all real numbers less than 0.
    (b)   (c) Unbounded

35. (a) $[4, \infty)$ denotes the set of all real numbers greater than or equal to 4.
    (b)   (c) Unbounded

37. (a) $-2 < x < 2$ denotes the set of all real numbers greater than -2 and less than 2.
    (b)   (c) Bounded

39. (a) $-1 \le x < 0$ denotes the set of all real numbers greater than or equal to -1 and less than 0.
    (b)   (c) Bounded

41. (a) $[-2, 5)$ denotes the set of all real numbers greater than or equal to -2 and less than 5.
    (b)   (c) Bounded

Inequality	*Interval*
43. $y \ge 0$	$[0, \infty)$
45. $-2 < x \le 4$	$(-2, 4]$
47. $10 \le t \le 22$	$[10, 22]$
49. $W > 65$	$(65, \infty)$

51. 10   53. 5   55. -1 57. -1 59. -1
61. $|-3| > -|-3|$ 63. $-5 = -|5|$
65. $-|-2| = -|2|$ 67. 51   69. $\frac{5}{2}$ 71. $\frac{128}{75}$
73. $|x - 5| \le 3$ 75. $|y| \ge 6$
77. $|57 - 236| = 179$ mi
79. $|\$113{,}356 - \$112{,}700| = \$656 > \$500$
    $0.05(\$112{,}700) = \$5635$
    Because the actual expense differs from the budget by more than $500, there is failure to meet the "budget variance test."
81. $|\$37{,}335 - \$37{,}640| = \$305 < \$500$
    $0.05(\$37{,}640) = \$1882$
    Because the difference between the actual expense and the budget is less than $500 and less than 5% of the budgeted amount, there is compliance with the "budget variance test."
83. $1453.2 billion; $107.4 billion
85. $2025.5 billion; $236.3 billion
87. $1880.3 billion; $412.7 billion
89. $7x$ and 4 are the terms; 7 is the coefficient.
91. $\sqrt{3}x^2$, $-8x$, and -11 are the terms; $\sqrt{3}$ and -8 are the coefficients.
93. $4x^3$, $x/2$, and -5 are the terms; 4 and $\frac{1}{2}$ are the coefficients.
95. (a) -10 (b) -6 97. (a) 14   (b) 2
99. (a) Division by 0 is undefined.   (b) 0
101. Commutative Property of Addition
103. Multiplicative Inverse Property
105. Distributive Property
107. Multiplicative Identity Property
109. Associative Property of Addition
111. Distributive Property   113. $\frac{1}{2}$ 115. $\frac{3}{8}$ 117. 48
119. $\dfrac{5x}{12}$ 121. (a) Negative   (b) Negative
123. (a)

n	1	0.5	0.01	0.0001	0.000001
$5/n$	5	10	500	50,000	5,000,000

(b) The value of $5/n$ approaches infinity as n approaches 0.
125. True. Because $b < 0$, $a - b$ subtracts a negative number from (or adds a positive number to) a positive number. The sum of two positive numbers is positive.
127. False. If $a < b$, then $\dfrac{1}{a} > \dfrac{1}{b}$, where $a \ne 0$ and $b \ne 0$.
129. (a) No. If one variable is negative and the other is positive, the expressions are unequal.
    (b) No. $|u + v| \le |u| + |v|$
    The expressions are equal when u and v have the same sign. If u and v differ in sign, $|u + v|$ is less than $|u| + |v|$.
131. The only even prime number is 2, because its only factors are itself and 1.
133. Yes. $|a| = -a$ if $a < 0$.

Section P.2 *(page 25)*

1. exponent; base 3. square root 5. index; radicand
7. like radicals 9. rationalizing 11. (a) 27 (b) 81
13. (a) 1 (b) -9 15. (a) 243 (b) $-\frac{3}{4}$
17. (a) $\frac{5}{6}$ (b) 4 19. -1600 21. 2.125 23. -24
25. 6 27. -54 29. -5 31. (a) $-125z^3$ (b) $5x^6$

33. (a) $24y^2$ (b) $3x^2$ 35. (a) $\frac{7}{x}$ (b) $\frac{4}{3}(x+y)^2$

37. (a) $\frac{x^2}{y^2}$ (b) $\frac{b^5}{a^5}$ 39. (a) 1 (b) $\frac{1}{4x^4}$

41. (a) $-2x^3$ (b) $\frac{10}{x}$ 43. (a) 3^{3n} (b) $\frac{b^5}{a^5}$

45. 1.02504×10^4 47. -1.25×10^{-4}
49. 5.73×10^7 mi^2 51. 8.99×10^{-5} g/cm^3
53. 125,000 55. -0.002718 57. 15,000,000°C
59. 0.00009 m 61. (a) 6.8×10^5 (b) 6.0×10^4
63. (a) 954.448 (b) 3.077×10^{10} 65. (a) 3 (b) $\frac{3}{2}$
67. (a) $\frac{1}{8}$ (b) $\frac{27}{8}$ 69. (a) -4 (b) 2
71. (a) 7.550 (b) -7.225 73. (a) -0.011 (b) 0.005
75. (a) 67,082.039 (b) 39.791 77. (a) 2 (b) $2\sqrt[5]{3}x$

79. (a) $2\sqrt{5}$ (b) $4\sqrt[3]{2}$ 81. (a) $6x\sqrt{2x}$ (b) $\frac{18\sqrt{z}}{z^2}$

83. (a) $2x\sqrt[3]{2x^2}$ (b) $\frac{5|x|\sqrt{3}}{y^2}$ 85. (a) $34\sqrt{2}$ (b) $22\sqrt{2}$

87. (a) $2\sqrt{x}$ (b) $4\sqrt{y}$ 89. (a) $13\sqrt{x+1}$ (b) $18\sqrt{5x}$
91. $\sqrt{5} + \sqrt{3} > \sqrt{5+3}$ 93. $5 > \sqrt{3^2 + 2^2}$

95. $\frac{\sqrt{3}}{3}$ 97. $\frac{\sqrt{14}+2}{2}$ 99. $\frac{2}{\sqrt{2}}$ 101. $\frac{2}{3(\sqrt{5}-\sqrt{3})}$

103. $2.5^{1/2}$ 105. $\sqrt[4]{81}$ 107. $(-216)^{1/3}$

109. $81^{3/4}$ 111. $\frac{2}{|x|}$ 113. $\frac{1}{x^3}$, $x > 0$

115. (a) $\sqrt{3}$ (b) $\sqrt[3]{(x+1)^2}$ 117. (a) $2\sqrt[4]{2}$ (b) $\sqrt[8]{2x}$

119. $\frac{\pi}{2} \approx 1.57$ sec

121. (a)

h	0	1	2	3	4	5	6
t	0	2.93	5.48	7.67	9.53	11.08	12.32

h	7	8	9	10	11	12
t	13.29	14.00	14.50	14.80	14.93	14.96

(b) $t \to 8.64\sqrt{3} \approx 14.96$
123. True. When dividing variables, you subtract exponents.

125. $a^0 = 1$, $a \neq 0$, using the property $\frac{a^m}{a^n} = a^{m-n}$:

$\frac{a^m}{a^m} = a^{m-m} = a^0 = 1$.

127. No. A number is in scientific notation when there is only one nonzero digit to the left of the decimal point.
129. No. Rationalizing the denominator produces a number equivalent to the original fraction; squaring does not.

Section P.3 *(page 33)*

1. n; a_n; a_0 3. monomial; binomial; trinomial
5. First terms; Outer terms; Inner terms; Last terms 6. c
7. a 8. b 9. d 10. e 11. b 12. a 13. f
14. c 15. $-2x^3 + 4x^2 - 3x + 20$ 17. $-15x^4 + 1$
19. (a) $-\frac{1}{2}x^5 + 14x$ (b) Degree: 5; Leading coefficient: $-\frac{1}{2}$
(c) Binomial
21. (a) $-3x^4 + x^2 - 4$
(b) Degree: 4; Leading coefficient: -3 (c) Trinomial
23. (a) $-x^6 + 3$ (b) Degree: 6; Leading coefficient: -1
(c) Binomial
25. (a) 3 (b) Degree: 0; Leading coefficient: 3
(c) Monomial
27. (a) $-4x^5 + 6x^4 + 1$
(b) Degree: 5; Leading coefficient: -4 (c) Trinomial
29. (a) $4x^3y$ (b) Degree: 4; Leading coefficient: 4
(c) Monomial
31. Polynomial: $-3x^3 + 2x + 8$
33. Not a polynomial because it includes a term with a negative exponent
35. Polynomial: $-y^4 + y^3 + y^2$ 37. $-2x - 10$
39. $5t^3 - 5t + 1$ 41. $8.3x^3 + 29.7x^2 + 11$ 43. $12z + 8$
45. $3x^3 - 6x^2 + 3x$ 47. $-15z^2 + 5z$ 49. $-4x^4 + 4x$
51. $-4.5t^3 - 15t$ 53. $-0.2x^2 - 34x$
55. $4x^3 - 2x^2 + 4$ 57. $5x^2 - 4x + 11$
59. $2x^2 + 17x + 21$ 61. $x^4 + 2x^2 + 9$
63. $x^2 + 7x + 12$ 65. $6x^2 - 7x - 5$
67. $x^2 - 100$ 69. $x^2 - 4y^2$ 71. $4x^2 + 12x + 9$
73. $x^3 + 3x^2 + 3x + 1$ 75. $8x^3 - 12x^2y + 6xy^2 - y^3$
77. $16x^6 - 24x^3 + 9$ 79. $x^4 + x^2 + 1$
81. $-3x^4 - x^3 - 12x^2 - 19x - 5$
83. $m^2 - n^2 - 6m + 9$
85. $x^2 + 2xy + y^2 - 6x - 6y + 9$ 87. $4r^4 - 25$
89. $\frac{1}{16}x^2 - \frac{5}{2}x + 25$ 91. $\frac{1}{25}x^2 - 9$
93. $5.76x^2 + 14.4x + 9$ 95. $2.25x^2 - 16$ 97. $2x^2 + 2x$
99. $u^4 - 16$ 101. $x - y$ 103. $x^2 - 2\sqrt{5}x + 5$
105. (a) $P = 22x - 25,000$ (b) \$85,000
107. (a) $500r^2 + 1000r + 500$
(b)

r	$2\frac{1}{2}\%$	3%	4%
$500(1 + r)^2$	\$525.31	\$530.45	\$540.80

r	$4\frac{1}{2}\%$	5%
$500(1 + r)^2$	\$546.01	\$551.25

109. (a) $V = 4x^3 - 88x^2 + 468x$
(b)

x (cm)	1	2	3
V (cm^3)	384	616	720

111. (a) $3x^2 + 8x$ (b) $30x^2$
(c) $x^2 + \frac{7}{2}x$ (d) $\frac{3}{2}x^2 + 14x + 30$
113. $44x + 308$

115. (a) Estimates will vary.

(b) The difference in safe load decreases in magnitude as the span increases.

117. $(x + 1)(x + 4) = x(x + 4) + 1(x + 4)$

Distributive Property

119. False. $(4x^2 + 1)(3x + 1) = 12x^3 + 4x^2 + 3x + 1$

121. $m + n$

123. The student omitted the middle term when squaring the binomial. $(x - 3)^2 = x^2 - 6x + 9 \neq x^2 + 9$

125. No. $(x^2 + 1) + (-x^2 + 3) = 4$, which is not a second-degree polynomial. (Examples will vary.)

127. $(3 + 4)^2 = 49 \neq 25 = 3^2 + 4^2$.

If either x or y is zero, then $(x + y)^2 = x^2 + y^2$.

Section P.4 　*(page 42)*

1. factoring　　**3.** factoring by grouping　　**5.** 40

7. $6x^2y$　　**9.** $4(x + 4)$　　**11.** $2x(x^2 - 3)$

13. $(x - 5)(3x + 8)$　　**15.** $(x + 3)(x - 1)$

17. $\frac{1}{2}(x + 8)$　　**19.** $\frac{1}{2}x(x^2 + 4x - 10)$

21. $\frac{2}{3}(x - 6)(x - 3)$　　**23.** $(x + 9)(x - 9)$

25. $3(4y - 3)(4y + 3)$　　**27.** $\left(4x + \frac{1}{3}\right)\left(4x - \frac{1}{3}\right)$

29. $(x + 1)(x - 3)$　　**31.** $(3u + 2v)(3u - 2v)$

33. $(x - 2)^2$　　**35.** $(2t + 1)^2$　　**37.** $(5y - 1)^2$

39. $(3u + 4v)^2$　　**41.** $\left(x - \frac{2}{3}\right)^2$

43. $\frac{1}{9}(6x - 1)^2$　　**45.** $(x - 2)(x^2 + 2x + 4)$

47. $(y + 4)(y^2 - 4y + 16)$　　**49.** $\frac{1}{27}(3x - 2)(9x^2 + 6x + 4)$

51. $(2t - 1)(4t^2 + 2t + 1)$　　**53.** $(u + 3v)(u^2 - 3uv + 9v^2)$

55. $(x - y + 2)(x^2 + xy + 4x + y^2 + 2y + 4)$

57. $(x + 2)(x - 1)$　　**59.** $(s - 3)(s - 2)$

61. $-(y + 5)(y - 4)$　　**63.** $(x - 20)(x - 10)$

65. $(3x - 2)(x - 1)$　　**67.** $(5x + 1)(x + 5)$

69. $-(3z - 2)(3z + 1)$　　**71.** $(x - 1)(x^2 + 2)$

73. $(2x - 1)(x^2 - 3)$　　**75.** $(3 + x)(2 - x^3)$

77. $(3x^2 - 1)(2x + 1)$　　**79.** $(x + 2)(3x + 4)$

81. $(2x - 1)(3x + 2)$　　**83.** $(3x - 1)(5x - 2)$

85. $6(x + 3)(x - 3)$　　**87.** $x^2(x - 1)$

89. $x(x - 4)(x + 4)$　　**91.** $(x - 1)^2$　　**93.** $(1 - 2x)^2$

95. $-2x(x + 1)(x - 2)$　　**97.** $\frac{1}{81}(x + 36)(x - 18)$

99. $(3x + 1)(x^2 + 5)$　　**101.** $x(x - 4)(x^2 + 1)$

103. $(x - 2)(x + 2)(2x + 1)$　　**105.** $\frac{1}{4}(x^2 + 3)(x + 12)$

107. $(t + 6)(t - 8)$　　**109.** $(x + 2)(x + 4)(x - 2)(x - 4)$

111. $5(x + 2)(x^2 - 2x + 4)$　　**113.** $(3 - 4x)(23 - 60x)$

115. $5(1 - x)^2(3x + 2)(4x + 3)$

117. $(x - 2)^2(x + 1)^3(7x - 5)$

119. $3(x^2 + 1)^4(x^4 - x^2 + 1)^4(3x + 2)^2(33x^6 + 20x^5 + 3)$

121. b　　**122.** c　　**123.** a　　**124.** d

125.

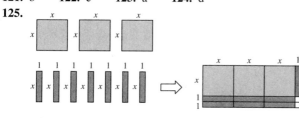

127.

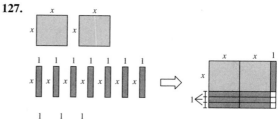

129. $4\pi(r + 1)$　　**131.** $4(6 - x)(6 + x)$

133. $4x^3(2x + 1)^3(2x^2 + 2x + 1)$

135. $(2x - 5)^3(5x - 4)^2(70x - 107)$　　**137.** $-\dfrac{8}{(5x - 1)^2}$

139. $-14, 14, -2, 2$　　**141.** $-51, 51, -15, 15, -27, 27$

143. Two possible answers: $2, -12$

145. Two possible answers: $-2, -4$

147. (a) $\pi h(R - r)(R + r)$　　(b) $V = 2\pi\left[\left(\dfrac{R + r}{2}\right)(R - r)\right]h$

(c) $13.38h$ bags

(d)

h	$\frac{1}{2}$	1	$\frac{3}{2}$	2	$\frac{5}{2}$	3
Number of bags	6.69	13.38	20.07	26.76	33.45	40.14

h	$\frac{7}{2}$	4	$\frac{9}{2}$	5	$\frac{11}{2}$	6
Number of bags	46.83	53.52	60.21	66.90	73.59	80.28

149. True. $a^2 - b^2 = (a + b)(a - b)$

151. A 3 was not factored out of the second binomial.

153. $(x^n + y^n)(x^n - y^n)$

155. Answers will vary. Sample answer: $x^2 - 3$

157. $(u + v)(u - v)(u^2 + uv + v^2)(u^2 - uv + v^2)$

$(x - 1)(x + 1)(x^2 + x + 1)(x^2 - x + 1)$

$(x - 2)(x + 2)(x^2 - 2x + 4)(x^2 + 2x + 4)$

Section P.5 　*(page 51)*

1. domain　　**3.** complex　　**5.** equivalent

7. All real numbers x　　**9.** All nonnegative real numbers x

11. All real numbers x such that $x \neq 3$

13. All real numbers x such that $x \neq 1$

15. All real numbers x such that $x \neq 3$

17. All real numbers x such that $x \geq -7$

19. All real numbers x such that $x \geq \frac{5}{2}$

21. All real numbers x such that $x > 3$　　**23.** $3x, \ x \neq 0$

25. $\dfrac{3x}{2}, \ x \neq 0$　　**27.** $\dfrac{3y}{y + 1}, \ x \neq 0$　　**29.** $\dfrac{-4y}{5}, \ y \neq \dfrac{1}{2}$

31. $-\dfrac{1}{2}, \ x \neq 5$　　**33.** $y - 4, \ y \neq -4$

35. $\dfrac{x(x + 3)}{x - 2}, \ x \neq -2$　　**37.** $\dfrac{y - 4}{y + 6}, \ y \neq 3$

39. $\dfrac{-(x^2 + 1)}{(x + 2)}, \ x \neq 2$　　**41.** $z - 2$

43. When simplifying fractions, you can only divide out common factors, not terms.

45.

x	0	1	2	3	4	5	6
$\dfrac{x^2 - 2x - 3}{x - 3}$	1	2	3	Undef.	5	6	7
$x + 1$	1	2	3	4	5	6	7

The expressions are equivalent except at $x = 3$.

47. $\dfrac{\pi}{4}$, $r \neq 0$ **49.** $\dfrac{1}{5(x - 2)}$, $x \neq 1$ **51.** $\dfrac{r + 1}{r}$, $r \neq 1$

53. $\dfrac{t - 3}{(t + 3)(t - 2)}$, $t \neq -2$

55. $\dfrac{(x + 6)(x + 1)}{x^2}$, $x \neq 6, -1$ **57.** $\dfrac{6x + 13}{x + 3}$

59. $\dfrac{x + 5}{x - 1}$ **61.** $-\dfrac{2}{x - 2}$ **63.** $\dfrac{-2x^2 + 3x + 8}{(2x + 1)(x + 2)}$

65. $-\dfrac{x^2 + 3}{(x + 1)(x - 2)(x - 3)}$ **67.** $\dfrac{2 - x}{x^2 + 1}$, $x \neq 0$

69. The error is incorrect subtraction in the numerator.

71. $\dfrac{1}{2}$, $x \neq 2$ **73.** $x(x + 1)$, $x \neq -1, 0$

75. $\dfrac{2x - 1}{2x}$, $x > 0$ **77.** $\dfrac{x^7 - 2}{x^2}$ **79.** $\dfrac{-1}{(x^2 + 1)^5}$

81. $\dfrac{2x^3 - 2x^2 - 5}{(x - 1)^{1/2}}$ **83.** $\dfrac{3x - 1}{3}$, $x \neq 0$

85. $\dfrac{-1}{x(x + h)}$, $h \neq 0$ **87.** $\dfrac{-1}{(x - 4)(x + h - 4)}$, $h \neq 0$

89. $\dfrac{1}{\sqrt{x + 2} + \sqrt{x}}$ **91.** $\dfrac{1}{\sqrt{t + 3} + \sqrt{3}}$, $t \neq 0$

93. $\dfrac{1}{\sqrt{x + h + 1} + \sqrt{x + 1}}$, $h \neq 0$

95. $\dfrac{x}{2(2x + 1)}$, $x \neq 0$

97. (a) $\dfrac{1}{50}$ min   (b) $\dfrac{x}{50}$ min   (c) $\dfrac{120}{50} = 2.4$ min

99. (a) 6.39%   (b) $\dfrac{288(MN - P)}{N(MN + 12P)}$; 6.39%

101. (a)

t	0	2	4	6	8	10	12
T	75	55.9	48.3	45	43.3	42.3	41.7

t	14	16	18	20	22
T	41.3	41.1	40.9	40.7	40.6

(b) The model is approaching a T-value of 40.

103. False. In order for the simplified expression to be equivalent to the original expression, the domain of the simplified expression needs to be restricted. If n is even, $x \neq -1, 1$. If n is odd, $x \neq 1$.

105. Completely factor each polynomial in the numerator and in the denominator. Then conclude that there are no common factors.

Section P.6   *(page 61)*

1. (a) v   (b) vi   (c) i   (d) iv   (e) iii   (f) ii

3. Distance Formula

5. A: $(2, 6)$, B: $(-6, -2)$, C: $(4, -4)$, D: $(-3, 2)$

7.

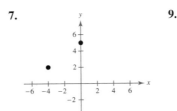

9.

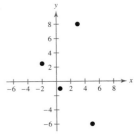

11. $(-3, 4)$ **13.** $(-5, -5)$ **15.** Quadrant IV

17. Quadrant II   **19.** Quadrant III or IV

21. Quadrant III   **23.** Quadrant I or III

25.

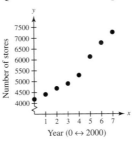

27. 8   **29.** 5   **31.** 13   **33.** $\sqrt{61}$ **35.** $\dfrac{\sqrt{277}}{6}$

37. 8.47   **39.** (a) 4, 3, 5   (b) $4^2 + 3^2 = 5^2$

41. (a) 10, 3, $\sqrt{109}$ (b) $10^2 + 3^2 = \left(\sqrt{109}\right)^2$

43. $\left(\sqrt{5}\right)^2 + \left(\sqrt{45}\right)^2 = \left(\sqrt{50}\right)^2$

45. Distances between the points: $\sqrt{29}$, $\sqrt{58}$, $\sqrt{29}$

47. (a)

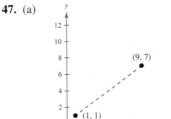

(b) 10   (c) $(5, 4)$

49. (a)

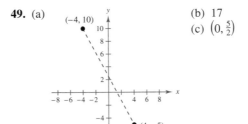

(b) 17   (c) $\left(0, \dfrac{5}{2}\right)$

51. (a)

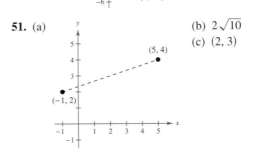

(b) $2\sqrt{10}$ (c) $(2, 3)$

53. (a)

(b) $\dfrac{\sqrt{82}}{3}$

(c) $\left(-1, \dfrac{7}{6}\right)$

55. (a)

(b) $\sqrt{110.97}$

(c) $(1.25, 3.6)$

57. $30\sqrt{41} \approx 192$ km   **59.** \$4415 million

61. $(0, 1), (4, 2), (1, 4)$ **63.** $(-3, 6), (2, 10), (2, 4), (-3, 4)$

65. \$3.87/gal; 2007   **67.** (a) About 9.6%   (b) About 28.6%

69. The number of performers elected each year seems to be nearly steady except for the first few years. Five performers will be elected in 2010.

71. \$24,331 million

73. (a)

(b) 2008

(c) Answers will vary. Sample answer: Technology now enables us to transport information in many ways other than by mail. The Internet is one example.

75. $(2x_m - x_1, 2y_m - y_1)$

77. $\left(\dfrac{3x_1 + x_2}{4}, \dfrac{3y_1 + y_2}{4}\right), \left(\dfrac{x_1 + x_2}{2}, \dfrac{y_1 + y_2}{2}\right),$
$\left(\dfrac{x_1 + 3x_2}{4}, \dfrac{y_1 + 3y_2}{4}\right)$

79.

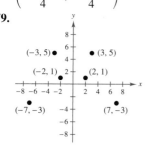

(a) The point is reflected through the y-axis.
(b) The point is reflected through the x-axis.
(c) The point is reflected through the origin.

81. False. The Midpoint Formula would be used 15 times.
83. No. It depends on the magnitudes of the quantities measured.
85. Use the Midpoint Formula to prove that the diagonals of the parallelogram bisect each other.
$$\left(\dfrac{b + a}{2}, \dfrac{c + 0}{2}\right) = \left(\dfrac{a + b}{2}, \dfrac{c}{2}\right)$$
$$\left(\dfrac{a + b + 0}{2}, \dfrac{c + 0}{2}\right) = \left(\dfrac{a + b}{2}, \dfrac{c}{2}\right)$$

Review Exercises *(page 68)*

1. (a) 11   (b) 11   (c) 11, -14
   (d) 11, $-14, -\dfrac{8}{9}, \dfrac{5}{2}, 0.4$ (e) $\sqrt{6}$

3. (a) $0.8\overline{3}$ (b) 0.875

$\dfrac{5}{6} < \dfrac{7}{8}$

5. The set consists of all real numbers less than or equal to 7.

7. 122   **9.** $|x - 7| \geq 4$ **11.** $|y + 30| < 5$

13. (a) -7 (b) -19 **15.** (a) -1 (b) -3

17. Associative Property of Addition

19. Additive Identity Property

21. Commutative Property of Addition   **23.** -11

25. $\dfrac{1}{12}$ **27.** -144 **29.** $\dfrac{47x}{60}$

31. (a) $192x^{11}$ (b) $\dfrac{y^5}{2}, y \neq 0$ **33.** (a) $-8z^3$ (b) $\dfrac{1}{y^2}$

35. (a) a^2b^2 (b) $3a^3b^2$ **37.** (a) $\dfrac{1}{625a^4}$ (b) $64x^2$

39. 5.015×10^8 **41.** 484,000,000   **43.** (a) 9   (b) 343

45. (a) 216   (b) 32   **47.** (a) $2\sqrt{2}$ (b) $26\sqrt{2}$

49. Radicals cannot be combined by addition or subtraction unless the index and the radicand are the same.

51. $\dfrac{\sqrt{3}}{4}$ **53.** $2 + \sqrt{3}$ **55.** $\dfrac{3}{\sqrt{7} - 1}$

57. 64   **59.** $6x^{9/10}$

61. $-11x^2 + 3$; Degree: 2; Leading coefficient: -11

63. $-12x^2 - 4$; Degree: 2; Leading coefficient: -12

65. $-3x^2 - 7x + 1$ **67.** $2x^3 - 10x^2 + 12x$

69. $2x^3 - x^2 + 3x - 9$ **71.** $15x^2 - 27x - 6$

73. $4x^2 - 12x + 9$ **75.** 41   **77.** $2500r^2 + 5000r + 2500$

79. $x^2 + 28x + 192$ **81.** $x(x + 1)(x - 1)$

83. $(5x + 7)(5x - 7)$ **85.** $(x - 4)(x^2 + 4x + 16)$

87. $(x + 10)(2x + 1)$ **89.** $(x - 1)(x^2 + 2)$

91. All real numbers except $x = -6$

93. $\dfrac{x - 8}{15}, x \neq -8$ **95.** $\dfrac{1}{x^2}, x \neq \pm 2$

97. $\dfrac{3x}{(x - 1)(x^2 + x + 1)}$ **99.** $\dfrac{3ax^2}{(a^2 - x)(a - x)}$

101. $\dfrac{-1}{2x(x + h)}, h \neq 0$

CHAPTER P

103.

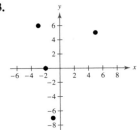

105. Quadrant IV

107. (a)

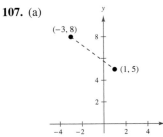

(b) 5
(c) $\left(-1, \frac{13}{2}\right)$

109. (a)

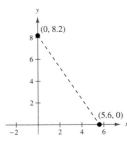

(b) $\sqrt{98.6}$
(c) $(2.8, 4.1)$

111. $(0, 0), (2, 0), (0, -5), (2, -5)$ **113.** $6.275 million
115. False. There is also a cross-product term when a binomial sum is squared.
$(x + a)^2 = x^2 + 2ax + a^2$
117. No. When $x = b/a$, the expression is undefined.

Chapter Test   *(page 71)*

1. $-\frac{10}{3} > -|-4|$ **2.** 9.15
3. Additive Identity Property
4. (a) -18 (b) $\frac{4}{9}$ (c) $-\frac{27}{125}$ (d) $\frac{8}{729}$
5. (a) 25   (b) $\frac{3\sqrt{6}}{2}$ (c) 1.8×10^5 (d) 2.7×10^{13}
6. (a) $12z^8$ (b) $(u - 2)^{-7}$ (c) $\frac{3x^2}{y^2}$
7. (a) $15z\sqrt{2z}$ (b) $4x^{14/15}$ (c) $\frac{2\sqrt[3]{2v}}{v^2}$
8. $-2x^5 - x^4 + 3x^3 + 3$; Degree: 5; Leading coefficient: -2
9. $2x^2 - 3x - 5$ **10.** $x^2 - 5$ **11.** $5, x \neq 4$
12. $\frac{x - 1}{2x}$, $x \neq \pm 1$
13. (a) $x^2(2x + 1)(x - 2)$ (b) $(x - 2)(x + 2)^2$
14. (a) $4\sqrt[3]{4}$ (b) $-4\left(1 + \sqrt{2}\right)$
15. All real numbers x except $x = 1$
16. $\frac{4}{y + 4}$, $y \neq 2$ **17.** $545

18.

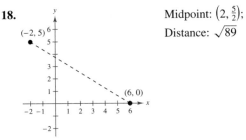

Midpoint: $\left(2, \frac{5}{2}\right)$;
Distance: $\sqrt{89}$

19. $\frac{5}{6}\sqrt{3}\,x^2$

Problem Solving   *(page 73)*

1. (a) Men's: 1,150,347 mm³;
            696,910 mm³
        Women's: 696,910 mm³;
            448,921 mm³
    (b) Men's: 1.04×10^{-5} kg/mm³;
            6.31×10^{-6} kg/mm³
        Women's: 8.91×10^{-6} kg/mm³;
            5.74×10^{-6} kg/mm³
    (c) No. Iron has a greater density than cork.
3. 1.62 oz   **5.** Answers will vary.
7. $r \approx 0.28$ **9.** 9.57 ft²
11. $y_1(0) = 0$, $y_2(0) = 2$
    $y_2 = \dfrac{x(2 - 3x^2)}{\sqrt{1 - x^2}}$
13. (a) $(2, -1), (3, 0)$ (b) $\left(-\frac{4}{3}, -2\right), \left(-\frac{2}{3}, -1\right)$

Chapter 1

Section 1.1   *(page 84)*

1. solution or solution point   **3.** intercepts
5. circle; (h, k); r **7.** (a) Yes   (b) Yes
9. (a) Yes   (b) No   **11.** (a) Yes   (b) No
13. (a) No   (b) Yes
15.

x	-1	0	1	2	$\frac{5}{2}$
y	7	5	3	1	0
(x, y)	$(-1, 7)$	$(0, 5)$	$(1, 3)$	$(2, 1)$	$\left(\frac{5}{2}, 0\right)$

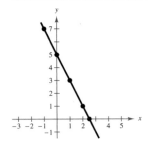

17.

x	-1	0	1	2	3
y	4	0	-2	-2	0
(x, y)	$(-1, 4)$	$(0, 0)$	$(1, -2)$	$(2, -2)$	$(3, 0)$

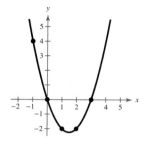

19. x-intercept: $(3, 0)$
y-intercept: $(0, 9)$

21. x-intercept: $(-2, 0)$
y-intercept: $(0, 2)$

23. x-intercept: $(1, 0)$
y-intercept: $(0, 2)$

25. y-axis symmetry   **27.** Origin symmetry

29. Origin symmetry   **31.** x-axis symmetry

33.

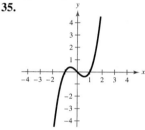

35.

37. x-intercept: $\left(\frac{1}{3}, 0\right)$
y-intercept: $(0, 1)$
No symmetry

39. x-intercepts: $(0, 0), (2, 0)$
y-intercept: $(0, 0)$
No symmetry

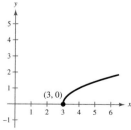

41. x-intercept: $\left(\sqrt[3]{-3}, 0\right)$
y-intercept: $(0, 3)$
No symmetry

43. x-intercept: $(3, 0)$
y-intercept: None
No symmetry

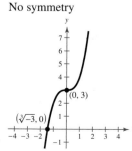

45. x-intercept: $(6, 0)$
y-intercept: $(0, 6)$
No symmetry

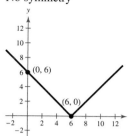

47. x-intercept: $(-1, 0)$
y-intercepts: $(0, \pm 1)$
x-axis symmetry

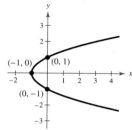

49.

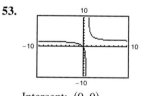

Intercepts: $(10, 0), (0, 5)$

51.

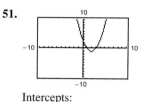

Intercepts:
$(3, 0), (1, 0), (0, 3)$

53.

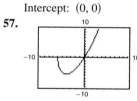

Intercept: $(0, 0)$

55.

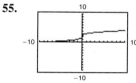

Intercepts: $(-8, 0), (0, 2)$

57.

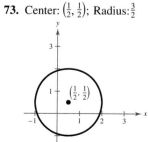

Intercepts: $(0, 0), (-6, 0)$

59.

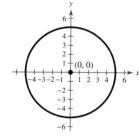

Intercepts: $(-3, 0), (0, 3)$

61. $x^2 + y^2 = 16$ **63.** $(x - 2)^2 + (y + 1)^2 = 16$

65. $(x + 1)^2 + (y - 2)^2 = 5$

67. $(x - 3)^2 + (y - 4)^2 = 25$

69. Center: $(0, 0)$; Radius: 5   **71.** Center: $(1, -3)$; Radius: 3

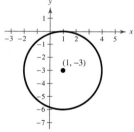

73. Center: $\left(\frac{1}{2}, \frac{1}{2}\right)$; Radius: $\frac{3}{2}$

75.

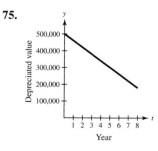

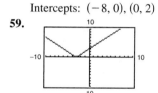

CHAPTER 1

77. (a)

(b) Answers will vary.

(c)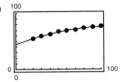

(d) $x = 86\frac{2}{3}$, $y = 86\frac{2}{3}$

(e) A regulation NFL playing field is 120 yards long and $53\frac{1}{3}$ yards wide. The actual area is 6400 square yards.

79. (a)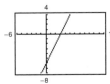

The model fits the data very well.

(b) 75.66 yr

(c) 1993

(d) The projection given by the model, 77.2 years, is less.

(e) Answers will vary.

81. (a) $a = 1, b = 0$ (b) $a = 0, b = 1$

Section 1.2 *(page 92)*

1. equation   **3.** identities; conditional   **5.** extraneous

7. (a) No   (b) No   (c) Yes   (d) No

9. (a) Yes   (b) Yes   (c) No   (d) No

11. (a) Yes   (b) No   (c) No   (d) No

13. (a) Yes   (b) No   (c) No   (d) No

15. (a) No   (b) No   (c) No   (d) No

17. (a) Yes   (b) No   (c) Yes   (d) No

19. Identity   **21.** Conditional equation   **23.** Identity

25. Identity   **27.** Conditional equation

29. Conditional equation

31. Original equation

Subtract 32 from each side.

Simplify.

Divide each side by 4.

Simplify.

33. 4   **35.** -9 **37.** 12   **39.** 1   **41.** No solution

43. -4 **45.** $-\frac{6}{5}$ **47.** 9   **49.** $-\frac{96}{23}$ **51.** $\frac{20}{81}$

53. No solution. The x-terms sum to zero.   **55.** 10   **57.** 4

59. 3   **61.** 0   **63.** No solution. The variable is divided out.

65. No solution. The solution is extraneous.   **67.** 5

69. No solution. The solution is extraneous.   **71.** 0

73. All real numbers

75.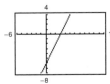

$x = 3$

77.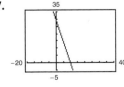

$x = 10$

79.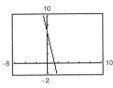

$x = \frac{7}{5}$

81. x-intercept: $\left(\frac{12}{5}, 0\right)$

y-intercept: $(0, 12)$

83. x-intercept: $\left(-\frac{1}{2}, 0\right)$ **85.** x-intercept: $(5, 0)$

y-intercept: $(0, -3)$ y-intercept: $\left(0, \frac{10}{3}\right)$

87. x-intercept: $(-20, 0)$ **89.** x-intercept: $(1.6, 0)$

y-intercept: $\left(0, \frac{8}{3}\right)$ y-intercept: $(0, -0.3)$

91. Substituting $x = 2$ in the equation yields a zero in the denominator, so $x = 2$ is an extraneous solution.

93. 138.889   **95.** 19.993

97. $\dfrac{1}{3 - a}$, $a \neq 3$ **99.** $\dfrac{5}{4 + a}$, $a \neq -4$

101. $\dfrac{18}{36 + a}$, $a \neq -36$ **103.** $\dfrac{-17}{10 - 2a}$, $a \neq 5$

105. $h = 10$ ft

107. (a) 61.2 in.

(b) Yes. The estimated height of a male with a 19-inch femur is 69.4 inches.

(c)

Height, x	Female femur length	Male femur length
60	15.48	14.79
70	19.80	19.28
80	24.12	23.77
90	28.44	28.26
100	32.76	32.75
110	37.08	37.24

100 in.

(d) $x \approx 100.59$; There would not be a problem because it is not likely for either a male or a female to be 100 inches (8 feet 4 inches) tall.

109. (a)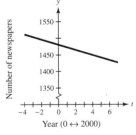

y-intercept: $(0, 1480.7)$

(b) y-intercept: $(0, 1480.7)$ (c) 2014; Answers will vary.

111. 23,437.5 mi

113. False. $x(3 - x) = 10$

$3x - x^2 = 10$

The equation cannot be written in the form $ax + b = 0$.

115. False. The equation is an identity.

117. False. $x = 4$ is a solution.

119. (a)

x	-1	0	1	2	3	4
$3.2x - 5.8$	-9	-5.8	-2.6	0.6	3.8	7

(b) $1 < x < 2$. The expression changes from negative to positive in this interval.

(c)

x	1.5	1.6	1.7	1.8	1.9	2
$3.2x - 5.8$	-1	-0.68	-0.36	-0.04	0.28	0.6

(d) $1.8 < x < 1.9$. To improve accuracy, evaluate the expression in this interval and determine where the sign changes.

121. (a) 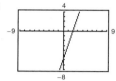 (b) $(2, 0)$

(c) The x-intercept is the solution of the equation $3x - 6 = 0$.

Section 1.3   (page 103)

1. mathematical modeling   **3.** $A = \pi r^2$ **5.** $V = s^3$

7. $A = P\left(1 + \dfrac{r}{n}\right)^{12t}$ **9.** A number increased by 4

11. A number divided by 5

13. A number decreased by 4 is divided by 5.

15. Negative 3 is multiplied by a number increased by 2.

17. 4 is multiplied by a number decreased by 1 and the product is divided by that number.

19. $n + (n + 1) = 2n + 1$

21. $(2n - 1)(2n + 1) = 4n^2 - 1$ **23.** $55t$ **25.** $0.20x$

27. $6x$ **29.** $2500 + 40x$ **31.** $0.30L$ **33.** $N = p(672)$

35. $4x + 8x = 12x$ **37.** 262, 263   **39.** 37, 185

41. $-5, -4$ **43.** 13.5   **45.** 27%   **47.** 2400

49. First salesperson: \$516.89; Second salesperson: \$608.11

51. \$47,267.19   **53.** 85.4% increase   **55.** 36.8% increase

57. (a)

(b) $l = 1.5w$; $p = 5w$
(c) 7.5 m × 5 m

59. 97   **61.** 5 h   **63.** About 46.3 mi/h

65. About 8.33 min   **67.** 1044 ft

69. (a)   (b) 42 ft

71. \$8000   **73.** Red maple: \$25,000; Dogwood: \$15,000

75. (a) Solution 1: 25 gal; Solution 2: 75 gal
(b) Solution 1: 4 L; Solution 2: 1 L
(c) Solution 1: 5 qt; Solution 2: 5 qt
(d) Solution 1: 18.75 gal; Solution 2: 6.25 gal

77. About 0.48 gal   **79.** 7529 units

81. $\dfrac{2A}{b}$ **83.** $\dfrac{S}{1 + R}$ **85.** $\dfrac{3V}{4\pi a^2}$

87. $\dfrac{2(h - v_0 t)}{t^2}$ **89.** $\dfrac{CC_2}{C_2 - C}$ **91.** $\dfrac{L - a + d}{d}$

93. $x = 6$ feet from the 50-pound child

95. $\sqrt[3]{\dfrac{4.47}{\pi}} \approx 1.12$ in.   **97.** 18°C   **99.** 122°F

101. False. The expression should be $\dfrac{z^3 - 8}{z^2 - 9}$.

103. (a) Negative; answers will vary.
(b) Positive; answers will vary.

105. Answers will vary. Sample answer: $x + 7 = 4$.

Section 1.4   (page 117)

1. quadratic equation

3. factoring; square roots; completing; square; Quadratic Formula

5. position equation   **7.** $2x^2 + 5x - 3 = 0$

9. $x^2 - 6x + 6 = 0$ **11.** $3x^2 - 60x - 10 = 0$

13. $0, -\dfrac{1}{2}$ **15.** $4, -2$ **17.** -5 **19.** $3, -\dfrac{1}{2}$

21. $2, -6$ **23.** $-\dfrac{20}{3}, -4$ **25.** ± 7 **27.** $\pm\sqrt{11}$

29. $\pm 3\sqrt{3}$ **31.** 8, 16   **33.** $-2 \pm \sqrt{14}$

35. $\dfrac{1 \pm 3\sqrt{2}}{2}$ **37.** 2   **39.** $4, -8$ **41.** $-3 \pm \sqrt{7}$

43. $1 \pm \dfrac{\sqrt{6}}{3}$ **45.** $1 \pm 2\sqrt{2}$ **47.** $\dfrac{-5 \pm \sqrt{89}}{4}$

49. $\dfrac{1}{(x + 1)^2 + 4}$ **51.** $\dfrac{4}{(x + 2)^2 - 7}$ **53.** $\dfrac{\frac{1}{4}}{(x + \frac{1}{2})^2 + 2}$

55. $\dfrac{1}{\sqrt{9 - (x - 3)^2}}$

57. (a)   (b) and (c) $x = -1, -5$
(d) The answers are the same.

59. (a)   (b) and (c) $x = 3, 1$
(d) The answers are the same.

61. (a)   (b) and (c) $x = -\dfrac{1}{2}, \dfrac{3}{2}$
(d) The answers are the same.

63. (a) 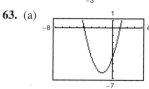   (b) and (c) $x = 1, -4$
(d) The answers are the same.

65. No real solution **67.** Two real solutions

69. No real solution **71.** Two real solutions

73. $\frac{1}{2}, -1$ **75.** $\frac{1}{4}, -\frac{3}{4}$ **77.** $1 \pm \sqrt{3}$ **79.** $-6 \pm 2\sqrt{5}$

81. $-4 \pm 2\sqrt{5}$ **83.** $\frac{2}{3} \pm \frac{\sqrt{7}}{3}$ **85.** $-\frac{5}{3}$

87. $-\frac{1}{2} \pm \sqrt{2}$ **89.** $\frac{2}{7}$ **91.** $2 \pm \frac{\sqrt{6}}{2}$ **93.** $6 \pm \sqrt{11}$

95. $-\frac{3}{8} \pm \frac{\sqrt{265}}{8}$ **97.** $0.976, -0.643$

99. $1.355, -14.071$ **101.** $1.687, -0.488$

103. $-0.290, -2.200$ **105.** $1 \pm \sqrt{2}$ **107.** $6, -12$

109. $\frac{1}{2} \pm \sqrt{3}$ **111.** $-\frac{1}{2}$

113. (a) $w(w + 14) = 1632$ (b) $w = 34$ ft, $l = 48$ ft

115. 6 in. $\times$ 6 in. $\times$ 3 in. **117.** 19.098 ft; 9.5 trips

119. (a) About 39.5 sec (b) About 5.5 mi

121. (a) $s = -16t^2 + 146\frac{2}{3}t + 6.25$

(b) $s(3) = 302.25$ ft; $s(4) = 336.92$ ft; $s(5) = 339.58$ ft;

During the interval $3 \le t \le 5$, the baseball's speed decreased due to gravity.

(c) Assuming the ball is not caught and drops to the ground, the baseball is in the air about 9.209 seconds.

123. (a)

t	1	2	3	4	5	6	7	8
P	5.68	5.83	6.00	6.19	6.40	6.63	6.89	7.16

The average admission price reached or surpassed $6.50 in 2005.

(b) Answers will vary. (c) $9.23. Answers will vary.

125. (a) $x^2 + 15^2 = l^2$ (b) $30\sqrt{6} \approx 73.5$ ft

127. $\frac{9\sqrt{2}}{2} \approx 6.36$ cm **129.** 50,000 units

131. 258 units **133.** 653 units

135. (a)

t	0	1	2	3	4
D	5.600	5.842	6.148	6.518	6.952

t	5	6	7	8
D	7.450	8.012	8.638	9.328

The public debt reached or surpassed $7 trillion in 2004.

(b) Answers will vary.

(c) $14,812 trillion; Answers will vary.

137. (a) 15.508 ft $\times$ 16.508 ft (b) 63,897.6 lb

(c) 1532.017 min or 25.5 h

139. False. $b^2 - 4ac < 0$, so the quadratic equation has no real solution.

141. Yes. The student should have subtracted $15x$ from both sides to make the right side of the equation equal to zero. Factoring out an x shows that there are two solutions, $x = 0$ and $x = 6$.

143. (a) and (b) $x = -5, -\frac{10}{3}$

(c) The method used in part (a) requires fewer algebraic steps.

145. Answers will vary. Sample answer: $x^2 + 2x - 15 = 0$

147. Answers will vary. Sample answer: $x^2 - 22x + 112 = 0$

149. Answers will vary. Sample answer: $x^2 - 2x - 1 = 0$

151. (a) positive; 4 (b) zero; 0 (c) negative; -4

zero solutions; two solutions

153. (a) and (b) Proofs

Section 1.5 *(page 127)*

1. (a) iii (b) i (c) ii **3.** principal square

5. $a = -12, b = 7$ **7.** $a = 6, b = 5$ **9.** $8 + 5i$

11. $2 - 3\sqrt{3}i$ **13.** $4\sqrt{5}i$ **15.** 14 **17.** $-1 - 10i$

19. $0.3i$ **21.** $10 - 3i$ **23.** 1 **25.** $3 - 3\sqrt{2}i$

27. $-14 + 20i$ **29.** $\frac{1}{6} + \frac{7}{6}i$ **31.** $5 + i$

33. $108 + 12i$ **35.** 24 **37.** $-13 + 84i$ **39.** -10

41. $9 - 2i, 85$ **43.** $-1 + \sqrt{5}i, 6$ **45.** $-2\sqrt{5}i, 20$

47. $\sqrt{6}, 6$ **49.** $-3i$ **51.** $\frac{8}{41} + \frac{10}{41}i$ **53.** $\frac{12}{13} + \frac{5}{13}i$

55. $-4 - 9i$ **57.** $-\frac{120}{1681} - \frac{27}{1681}i$ **59.** $-\frac{1}{2} - \frac{5}{2}i$

61. $\frac{62}{949} + \frac{297}{949}i$ **63.** $-2\sqrt{3}$ **65.** -15

67. $\left(21 + 5\sqrt{2}\right) + \left(7\sqrt{5} - 3\sqrt{10}\right)i$ **69.** $1 \pm i$

71. $-2 \pm \frac{1}{2}i$ **73.** $-\frac{5}{2}, -\frac{3}{2}$ **75.** $2 \pm \sqrt{2}i$

77. $\frac{5}{7} \pm \frac{5\sqrt{15}}{7}$ **79.** $-1 + 6i$ **81.** $-14i$

83. $-432\sqrt{2}i$ **85.** i **87.** 81

89. (a) $z_1 = 9 + 16i, z_2 = 20 - 10i$

(b) $z = \frac{11{,}240}{877} + \frac{4630}{877}i$

91. (a) 16 (b) 16 (c) 16 (d) 16

93. False. If the complex number is real, the number equals its conjugate.

95. False.

$i^{44} - i^{150} - i^{74} - i^{109} + i^{61} = 1 - 1 + 1 - i + i = 1$

97. $i, -1, -i, 1, i, -1, -i, 1$; The pattern repeats the first four results. Divide the exponent by 4.

If the remainder is 1, the result is i.

If the remainder is 2, the result is -1.

If the remainder is 3, the result is $-i$.

If the remainder is 0, the result is 1.

99. $\sqrt{-6}\sqrt{-6} = \sqrt{6}i\sqrt{6}i = 6i^2 = -6$

101. Proof

Section 1.6 *(page 136)*

1. polynomial **3.** quadratic type **5.** $0, \pm\frac{\sqrt{21}}{3}$

7. $\pm 3, \pm 3i$ **9.** $-8, 4 \pm 4\sqrt{3}i$ **11.** $-3, 0$

13. $3, 1, -1$ **15.** $\pm 1, \frac{1}{2} \pm \frac{\sqrt{3}}{2}i$ **17.** $\pm\sqrt{3}, \pm 1$

19. $\pm\frac{1}{2}, \pm 4$ **21.** $1, -2, 1 \pm \sqrt{3}i, -\frac{1}{2} \pm \frac{\sqrt{3}i}{2}$

23. $-\frac{1}{5}, -\frac{1}{3}$ **25.** $-\frac{2}{3}, -4$ **27.** $\frac{1}{4}$ **29.** $1, -\frac{125}{8}$

31. (a)

(b) $(0, 0), (3, 0), (-1, 0)$

(c) $x = 0, 3, -1$

(d) The x-intercepts and the solutions are the same.

33. (a)

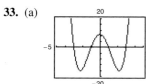

(b) $(\pm 3, 0), (\pm 1, 0)$
(c) $x = \pm 3, \pm 1$
(d) The x-intercepts and the solutions are the same.

35. 48   **37.** 26   **39.** -16 **41.** 2, -5 **43.** 0
45. 9   **47.** $\frac{101}{4}$ **49.** 14   **51.** 9   **53.** $-3 \pm 16\sqrt{2}$
55. $\pm\sqrt{14}$ **57.** 1

59. (a)

(b) $(5, 0), (6, 0)$
(c) $x = 5, 6$
(d) The x-intercepts and the solutions are the same.

61. (a)

(b) $(0, 0), (4, 0)$
(c) $x = 0, 4$
(d) The x-intercepts and the solutions are the same.

63. $2, -\frac{3}{2}$ **65.** $\frac{-3 \pm \sqrt{21}}{6}$ **67.** 5, -6

69. $\frac{1 \pm \sqrt{31}}{3}$ **71.** 8, -3 **73.** $2\sqrt{6}, -6$

75. $3, \frac{-1 - \sqrt{17}}{2}$

77. (a)

(b) $(-1, 0)$ (c) $x = -1$
(d) The x-intercept and the solution are the same.

79. (a)

(b) $(1, 0), (-3, 0)$
(c) $x = 1, -3$
(d) The x-intercepts and the solutions are the same.

81. ± 1.038 **83.** $-1.143, 0.968$ **85.** 16.756
87. $-2.280, -0.320$ **89.** $x^2 - 3x - 28 = 0$
91. $21x^2 + 31x - 42 = 0$ **93.** $x^3 - 4x^2 - 3x + 12 = 0$
95. $x^2 + 1 = 0$ **97.** $x^4 - 1 = 0$ **99.** 34 students
101. 191.5 mi/h   **103.** 4%
105. (a) 2003   (b) During 2011; Answers will vary.
107. (a)

x	5	10	15	20
T	162.56	192.31	212.68	228.20

x	25	30	35	40
T	240.62	250.83	259.38	266.60

(b) About 15 lb/in.²
(c) $x \approx 14.81$
(d) Answers will vary.

109. 500 units   **111.** 90 ft
113. (a)

The height $h \approx 11.4$ when $S = 350$.
(b)

h	8	9	10	11	12	13
S	284.3	302.6	321.9	341.8	362.5	383.6

The height h is between 11 and 12 inches when $S = 350$.
(c) $h = 11.4$ when $S = 350$.
(d) Solving graphically or numerically yields an approximate solution. An exact solution is obtained algebraically.
115. $\frac{21 + \sqrt{585}}{2} \approx 22.6$ h

$\frac{27 + \sqrt{585}}{2} \approx 25.6$ h

117. $g = \frac{\mu s v^2}{R}$ **119.** False. See Example 7 on page 133.

121. True. There is no value that satisfies this equation.
123. 6, -4 **125.** ± 15 **127.** $a = 9, b = 9$
129. $a = 4, b = 24$

Section 1.7   *(page 146)*

1. solution set   **3.** negative   **5.** double
7. (a) $0 \le x < 9$ (b) Bounded
9. (a) $-1 \le x \le 5$ (b) Bounded
11. (a) $x > 11$ (b) Unbounded
13. (a) $x < -2$ (b) Unbounded
15. b   **16.** h   **17.** e   **18.** d
19. f   **20.** a   **21.** g   **22.** c
23. (a) Yes   (b) No   (c) Yes   (d) No
25. (a) Yes   (b) No   (c) No   (d) Yes
27. (a) Yes   (b) Yes   (c) Yes   (d) No
29. $x < 3$ **31.** $x < \frac{3}{2}$

33. $x \ge 12$ **35.** $x > 2$

37. $x \ge \frac{2}{7}$ **39.** $x < 5$

41. $x \ge 4$ **43.** $x \ge 2$

45. $x \ge -4$ **47.** $-1 < x < 3$

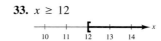

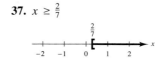
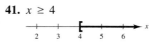
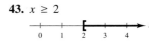
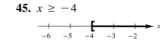
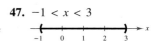

CHAPTER 1

49. $-2 < x \le 5$

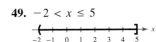

51. $-\frac{9}{2} < x < \frac{15}{2}$

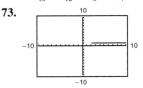

53. $-\frac{3}{4} < x < -\frac{1}{4}$

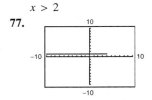

55. $10.5 \le x \le 13.5$

57. $-5 < x < 5$

59. $x < -2, \ x > 2$

61. No solution

63. $14 \le x \le 26$

65. $x \le -\frac{3}{2}, \ x \ge 3$

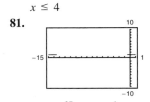

67. $x \le -5, \ x \ge 11$

69. $4 < x < 5$

71. $x \le -\frac{29}{2}, \ x \ge -\frac{11}{2}$

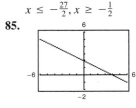

73.

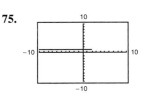

$x > 2$

75.

$x \le 2$

77.

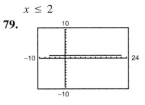

$x \le 4$

79.

$-6 \le x \le 22$

81.

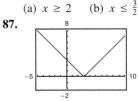

$x \le -\frac{27}{2}, \ x \ge -\frac{1}{2}$

83.

(a) $x \ge 2$ (b) $x \le \frac{3}{2}$

85.

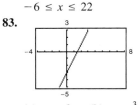

(a) $-2 \le x \le 4$
(b) $x \le 4$

87.

(a) $1 \le x \le 5$
(b) $x \le -1, \ x \ge 7$

89. $[5, \infty)$ **91.** $[-3, \infty)$ **93.** $\left(-\infty, \frac{7}{2}\right]$

95. All real numbers within eight units of 10

97. $|x| \le 3$ **99.** $|x - 7| \ge 3$ **101.** $|x - 12| < 10$

103. $|x + 3| > 4$ **105.** $4.10 \le E \le 4.25$ **107.** $p \le 0.45$

109. $100 \le r \le 170$ **111.** $9.00 + 0.75x > 13.50; \ x > 6$

113. $r > 3.125\%$ **115.** $x \ge 36$ **117.** $160 \le x \le 280$

119. (a)

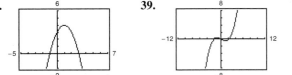

(b) $x \ge 129$

121. (a) $1.47 \le t \le 10.18$ (Between 1991 and 2000)
     (b) $t > 21.19$ (2011)

123. $106.864 \text{ in.}^2 \le \text{area} \le 109.464 \text{ in.}^2$

125. You might be undercharged or overcharged by \$0.21.

127. $13.7 < t < 17.5$

129. $20 \le h \le 80$ **131.** False. c has to be greater than zero.

133. b   **135.** Sample answer: $x > 5$

Section 1.8   *(page 157)*

1. positive; negative   **3.** zeros; undefined values

5. (a) No   (b) Yes   (c) Yes   (d) No

7. (a) Yes   (b) No   (c) No   (d) Yes

9. $-\frac{2}{3}, 1$ **11.** $4, 5$

13. $(-3, 3)$ **15.** $[-7, 3]$

17. $(-\infty, -5] \cup [1, \infty)$ **19.** $(-3, 2)$

21. $(-3, 1)$ **23.** $\left(-\infty, -\frac{4}{3}\right) \cup (5, \infty)$

25. $(-\infty, -3) \cup (6, \infty)$ **27.** $(-1, 1) \cup (3, \infty)$

29. $x = \frac{1}{2}$

31. $(-\infty, 0) \cup \left(0, \frac{3}{2}\right)$ **33.** $[-2, 0] \cup [2, \infty)$ **35.** $[-2, \infty)$

37.

(a) $x \le -1, \ x \ge 3$
(b) $0 \le x \le 2$

39.

(a) $-2 \le x \le 0,$
     $2 \le x < \infty$
(b) $x \le 4$

41. $(-\infty, 0) \cup \left(\frac{1}{4}, \infty\right)$

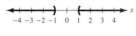

43. $\left(-\infty, \frac{5}{3}\right] \cup (5, \infty)$

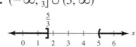

45. $(-\infty, -1) \cup (4, \infty)$

47. $(-5, 3) \cup (11, \infty)$

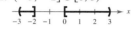

49. $\left(-\frac{3}{4}, 3\right) \cup [6, \infty)$

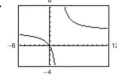

51. $(-3, -2) \cup [0, 3)$

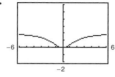

53. $(-\infty, -1) \cup (1, \infty)$

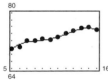

55.

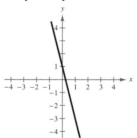

57.

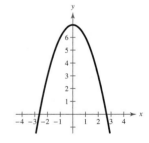

(a) $0 \le x < 2$

(b) $2 < x \le 4$

(a) $|x| \ge 2$

(b) $-\infty < x < \infty$

59. $[-2, 2]$ **61.** $(-\infty, 4] \cup [5, \infty)$ **63.** $(-5, 0] \cup (7, \infty)$

65. $(-3.51, 3.51)$ **67.** $(-0.13, 25.13)$ **69.** $(2.26, 2.39)$

71. (a) $t = 10$ sec (b) 4 sec $< t < 6$ sec

73. 13.8 m $\le L \le 36.2$ m

75. $40{,}000 \le x \le 50{,}000$; $\$50.00 \le p \le \$55.00$

77. (a) and (c)

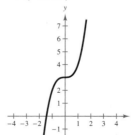

The model fits the data well.

(b) $N = -0.00412t^4 + 0.1705t^3 - 2.538t^2 + 16.55t + 31.5$

(d) 2003 to 2006

(e) No; The model decreases sharply after 2006.

79. $R_1 \ge 2$ ohms

81. True. The test intervals are $(-\infty, -3)$, $(-3, 1)$, $(1, 4)$, and $(4, \infty)$.

83. (a) $(-\infty, -4] \cup [4, \infty)$

(b) If $a > 0$ and $c > 0$, $b \le -2\sqrt{ac}$ or $b \ge 2\sqrt{ac}$.

85. (a) $\left(-\infty, -2\sqrt{30}\right] \cup \left[2\sqrt{30}, \infty\right)$

(b) If $a > 0$ and $c > 0$, $b \le -2\sqrt{ac}$ or $b \ge 2\sqrt{ac}$.

87.

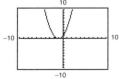

For part (b), the y-values that are less than or equal to 0 occur only at $x = -1$.

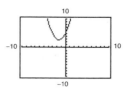

For part (c), there are no y-values that are less than 0.

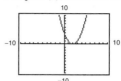

For part (d), the y-values that are greater than 0 occur for all values of x except 2.

Review Exercises *(page 162)*

1.

x	-2	-1	0	1	2
y	9	5	1	-3	-7

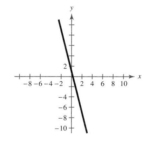

3. x-intercepts: $(1, 0)$, $(5, 0)$

y-intercept: $(0, 5)$

5. No symmetry

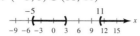

7. y-axis symmetry

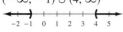

9. No symmetry

11. y-axis symmetry

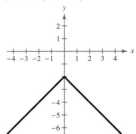

13. Center: $(0, 0)$;
Radius: 3

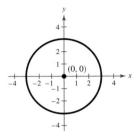

15. Center: $(-2, 0)$;
Radius: 4

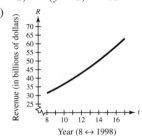

17. $(x - 2)^2 + (y + 3)^2 = 13$

19. (a)

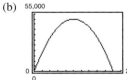

(b) 2003

21. Identity **23.** Identity **25.** 5 **27.** 9 **29.** -30

31. x-intercept: $\left(\frac{1}{3}, 0\right)$ **33.** x-intercept: $(4, 0)$
y-intercept: $(0, -1)$ y-intercept: $(0, -8)$

35. x-intercept: $\left(\frac{4}{3}, 0\right)$
y-intercept: $\left(0, \frac{2}{3}\right)$

37. $h = 10$ in. **39.** September: \$325,000; October: \$364,000

41. Nine **43.** $\frac{20}{7}$ L ≈ 2.857 L **45.** $h = \dfrac{3V}{\pi r^2}$

47. $-\frac{5}{2}, 3$ **49.** $\pm\sqrt{2}$ **51.** $-8, -18$

53. $-6 \pm \sqrt{11}$ **55.** $-\dfrac{5}{4} \pm \dfrac{\sqrt{241}}{4}$

57. (a) $x = 0, 20$
(b)
55,000

(c) $x = 10$

0 ⌞____⌟ 22
0

59. $4 + 3i$ **61.** $-1 + 3i$ **63.** $3 + 7i$ **65.** $12 + 30i$

67. $-5 - 6i$ **69.** $\dfrac{21}{13} - \dfrac{1}{13}i$ **71.** $\pm\dfrac{\sqrt{3}}{3}i$ **73.** $1 \pm 3i$

75. $0, \frac{12}{5}$ **77.** $\pm\sqrt{2}, \pm\sqrt{3}$ **79.** No solution

81. $-124, 126$ **83.** $\pm\sqrt{10}$ **85.** $-5, 15$

87. $1, 3$ **89.** 143,203 units

91. $-7 < x \leq 2$; Bounded **93.** $x \leq -10$; Unbounded

95. $x < -18$ **97.** $\left[\frac{32}{15}, \infty\right)$ **99.** $(-\infty, -1) \cup (7, \infty)$

101. 353.44 cm^2 $\leq$ area $\leq$ 392.04 cm^2 **103.** $(-3, 9)$

105. $\left(-\frac{4}{3}, \frac{1}{2}\right)$ **107.** $[-5, -1) \cup (1, \infty)$ **109.** 4.9%

111. False. $\sqrt{-18}\sqrt{-2} = \left(3\sqrt{2}\,i\right)\left(\sqrt{2}\,i\right) = 6i^2 = -6$
and $\sqrt{(-18)(-2)} = \sqrt{36} = 6$

113. Some solutions to certain types of equations may be extraneous solutions, which do not satisfy the original equations. So, checking is crucial.

Chapter Test *(page 165)*

1. No symmetry

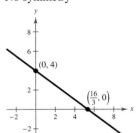

2. y-axis symmetry

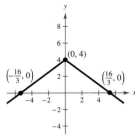

3. No symmetry

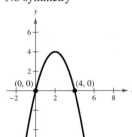

4. Origin symmetry

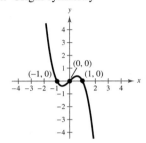

5. No symmetry

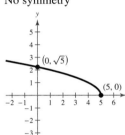

6. x-axis symmetry

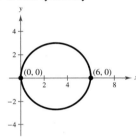

7. $\frac{128}{11}$ **8.** $-3, 5$ **9.** No solution

10. $\pm\sqrt{2}, \pm\sqrt{3}\,i$ **11.** 4 **12.** $-2, \frac{8}{3}$

13. $-\frac{11}{2} \leq x < 3$ **14.** $x < -6$ or $0 < x < 4$

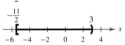

15. $x < -4$ or $x > \frac{3}{2}$ **16.** $x \leq -5$ or $x \geq -\frac{5}{3}$

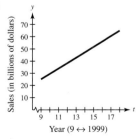

17. (a) $-3 + 5i$ (b) 26 **18.** $2 - i$

19. (a)

(b) \$86.83 billion
(c) Answers will vary.

20. $r \approx 4.774$ in. **21.** $93\frac{3}{4}$ km/h **22.** $a = 80, b = 20$

Problem Solving   *(page 167)*

1.

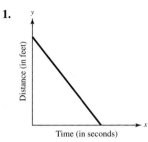

3. (a) Answers will vary.
   Sample answer:
   $A = \pi ab$
   $b = 20 - a$, since $a + b = 20$
   $A = \pi a(20 - a)$

(b)

a	4	7	10	13	16
A	64π	91π	100π	91π	64π

(c) $\dfrac{10\pi + 10\sqrt{\pi(\pi - 3)}}{\pi} \approx 12.12$ or

   $\dfrac{10\pi - 10\sqrt{\pi(\pi - 3)}}{\pi} \approx 7.88$

(d)

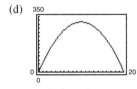

(e) $(0, 0)$, $(20, 0)$
   They represent the minimum and maximum values of a.

(f) 100π; $a = 10$; $b = 10$

5. (a) About 60.6 sec   (b) About 146.2 sec
   (c) The speed at which water drains decreases as the amount
       of water in the bathtub decreases.

7. (a) Answers will vary. Sample answers: 5, 12, 13; 8, 15, 17.
   (b) Yes; yes; yes
   (c) The product of the three numbers in a Pythagorean Triple
       is divisible by 60.

9. $x_1 + x_2 = -\dfrac{b}{a}$; $x_1 \cdot x_2 = \dfrac{c}{a}$

11. (a) $\dfrac{1}{2} - \dfrac{1}{2}i$ (b) $\dfrac{3}{10} + \dfrac{1}{10}i$ (c) $-\dfrac{1}{34} - \dfrac{2}{17}i$

13. (a) Yes   (b) No   (c) Yes

15. $(-\infty, -2) \cup (-1, 2) \cup (2, \infty)$

Chapter 2

Section 2.1   *(page 179)*

1. linear   **3.** parallel   **5.** rate or rate of change
7. general   **9.** (a) L_2 (b) L_3 (c) L_1

11.

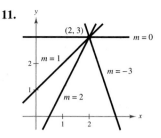

13. $\dfrac{3}{2}$ **15.** -4

17. $m = 5$
   y-intercept: $(0, 3)$

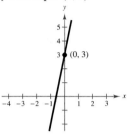

19. $m = -\dfrac{1}{2}$
   y-intercept: $(0, 4)$

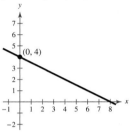

21. m is undefined.
   There is no y-intercept.

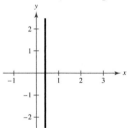

23. $m = -\dfrac{7}{6}$
   y-intercept: $(0, 5)$

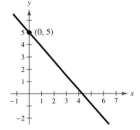

25. $m = 0$
   y-intercept: $(0, 3)$

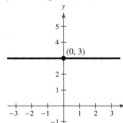

27. m is undefined.
   There is no y-intercept.

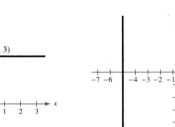

29.

$m = -\dfrac{3}{2}$

31.

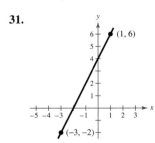

$m = 2$

33.
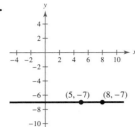
$m = 0$

35.
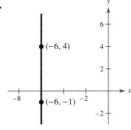
m is undefined.

37.
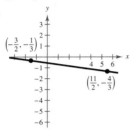
$m = -\frac{1}{7}$

39.
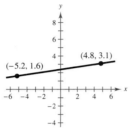
$m = 0.15$

41. $(0, 1), (3, 1), (-1, 1)$ **43.** $(6, -5), (7, -4), (8, -3)$
45. $(-8, 0), (-8, 2), (-8, 3)$ **47.** $(-4, 6), (-3, 8), (-2, 10)$
49. $(9, -1), (11, 0), (13, 1)$

51. $y = 3x - 2$
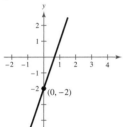

53. $y = -2x$
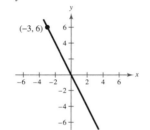

55. $y = -\frac{1}{3}x + \frac{4}{3}$
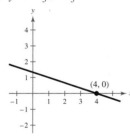

57. $y = -\frac{1}{2}x - 2$
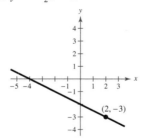

59. $x = 6$
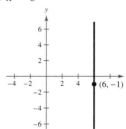

61. $y = \frac{5}{2}$
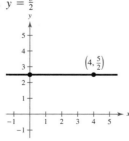

63. $y = 5x + 27.3$
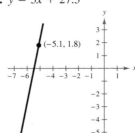

65. $y = -\frac{3}{5}x + 2$
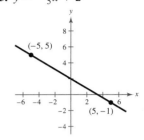

67. $x = -8$
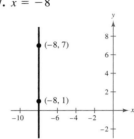

69. $y = -\frac{1}{2}x + \frac{3}{2}$
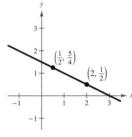

71. $y = -\frac{6}{5}x - \frac{18}{25}$
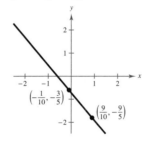

73. $y = 0.4x + 0.2$
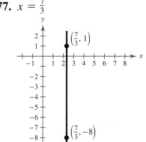

75. $y = -1$
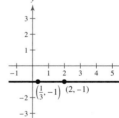

77. $x = \frac{7}{3}$
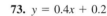

79. Parallel **81.** Neither **83.** Perpendicular
85. Parallel **87.** (a) $y = 2x - 3$ (b) $y = -\frac{1}{2}x + 2$
89. (a) $y = -\frac{3}{4}x + \frac{3}{8}$ (b) $y = \frac{4}{3}x + \frac{127}{72}$
91. (a) $y = 0$ (b) $x = -1$
93. (a) $x = 3$ (b) $y = -2$
95. (a) $y = x + 4.3$ (b) $y = -x + 9.3$
97. $3x + 2y - 6 = 0$ **99.** $12x + 3y + 2 = 0$
101. $x + y - 3 = 0$
103. Line (b) is perpendicular to line (c).
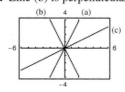

105. Line (a) is parallel to line (b).
Line (c) is perpendicular to line (a) and line (b).

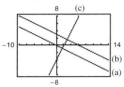

107. $3x - 2y - 1 = 0$ **109.** $80x + 12y + 139 = 0$

111. (a) Sales increasing 135 units/yr   (b) No change in sales
(c) Sales decreasing 40 units/yr

113. (a) The average salary increased the greatest from 2006 to 2008 and increased the least from 2002 to 2004.
(b) $m = 2350.75$
(c) The average salary increased \$2350.75 per year over the 12 years between 1996 and 2008.

115. 12 ft   **117.** $V(t) = 3790 - 125t$

119. V-intercept: initial cost; Slope: annual depreciation

121. $V = -175t + 875$ **123.** $S = 0.8L$

125. $W = 0.07S + 2500$

127. $y = 0.03125t + 0.92875$; $y(22) \approx \$1.62$; $y(24) \approx \$1.68$

129. (a) $y(t) = 442.625t + 40{,}571$
(b) $y(10) = 44{,}997$; $y(15) = 47{,}210$
(c) $m = 442.625$; Each year, enrollment increases by about 443 students.

131. (a) $C = 18t + 42{,}000$ (b) $R = 30t$
(c) $P = 12t - 42{,}000$ (d) $t = 3500$ h

133. (a)

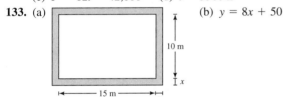

(b) $y = 8x + 50$

(c)

(d) $m = 8$, 8 m

135. (a) and (b)

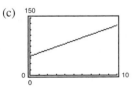

(c) Answers will vary. Sample answer: $y = 2.39x + 44.9$
(d) Answers will vary. Sample answer: The y-intercept indicates that in 2000 there were 44.9 thousand doctors of osteopathic medicine. The slope means that the number of doctors increases by 2.39 thousand each year.
(e) The model is accurate.
(f) Answers will vary. Sample answer: 73.6 thousand

137. False. The slope with the greatest magnitude corresponds to the steepest line.

139. Find the distance between each two points and use the Pythagorean Theorem.

141. No. The slope cannot be determined without knowing the scale on the y-axis. The slopes could be the same.

143. The line $y = 4x$ rises most quickly, and the line $y = -4x$ falls most quickly. The greater the magnitude of the slope (the absolute value of the slope), the faster the line rises or falls.

145. No. The slopes of two perpendicular lines have opposite signs (assume that neither line is vertical or horizontal).

Section 2.2   *(page 194)*

1. domain; range; function   **3.** independent; dependent

5. implied domain   **7.** Yes   **9.** No

11. Yes, each input value has exactly one output value.

13. No, the input values 7 and 10 each have two different output values.

15. (a) Function
(b) Not a function, because the element 1 in A corresponds to two elements, -2 and 1, in B.
(c) Function
(d) Not a function, because not every element in A is matched with an element in B.

17. Each is a function. For each year there corresponds one and only one circulation.

19. Not a function   **21.** Function   **23.** Function

25. Not a function   **27.** Not a function   **29.** Function

31. Function   **33.** Not a function   **35.** Function

37. (a) -1 (b) -9 (c) $2x - 5$

39. (a) 36π (b) $\frac{9}{2}\pi$ (c) $\frac{32}{3}\pi r^3$

41. (a) 15   (b) $4t^2 - 19t + 27$ (c) $4t^2 - 3t - 10$

43. (a) 1   (b) 2.5   (c) $3 - 2|x|$

45. (a) $-\dfrac{1}{9}$ (b) Undefined   (c) $\dfrac{1}{y^2 + 6y}$

47. (a) 1   (b) -1 (c) $\dfrac{|x - 1|}{x - 1}$

49. (a) -1 (b) 2   (c) 6   **51.** (a) -7 (b) 4   (c) 9

53.

x	-2	-1	0	1	2
$f(x)$	1	-2	-3	-2	1

55.

t	-5	-4	-3	-2	-1
$h(t)$	1	$\frac{1}{2}$	0	$\frac{1}{2}$	1

57.

x	-2	-1	0	1	2
$f(x)$	5	$\frac{9}{2}$	4	1	0

59. 5   **61.** $\frac{4}{3}$ **63.** ± 3 **65.** $0, \pm 1$ **67.** $-1, 2$

69. $0, \pm 2$ **71.** All real numbers x

73. All real numbers t except $t = 0$

75. All real numbers y such that $y \geq 10$

77. All real numbers x except $x = 0, -2$

CHAPTER 2

79. All real numbers s such that $s \geq 1$ except $s = 4$

81. All real numbers x such that $x > 0$

83. $\{(-2, 4), (-1, 1), (0, 0), (1, 1), (2, 4)\}$

85. $\{(-2, 4), (-1, 3), (0, 2), (1, 3), (2, 4)\}$ **87.** $A = \dfrac{P^2}{16}$

89. (a) The maximum volume is 1024 cubic centimeters.

(b)

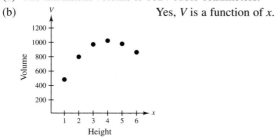

Yes, V is a function of x.

(c) $V = x(24 - 2x)^2, \quad 0 < x < 12$

91. $A = \dfrac{x^2}{2(x - 2)}, \quad x > 2$

93. Yes, the ball will be at a height of 6 feet.

95.
1998: \$136,164	2003: \$180,419
1999: \$140,971	2004: \$195,900
2000: \$147,800	2005: \$216,900
2001: \$156,651	2006: \$224,000
2002: \$167,524	2007: \$217,200

97. (a) $C = 12.30x + 98,000$ (b) $R = 17.98x$

(c) $P = 5.68x - 98,000$

99. (a) $R = \dfrac{240n - n^2}{20}, \quad n \geq 80$

(b)

n	90	100	110	120	130	140	150
$R(n)$	\$675	\$700	\$715	\$720	\$715	\$700	\$675

The revenue is maximum when 120 people take the trip.

101. (a)

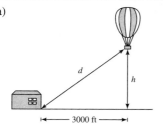

(b) $h = \sqrt{d^2 - 3000^2}, \quad d \geq 3000$

103. $3 + h, \quad h \neq 0$ **105.** $3x^2 + 3xh + h^2 + 3, \quad h \neq 0$

107. $-\dfrac{x + 3}{9x^2}, \quad x \neq 3$ **109.** $\dfrac{\sqrt{5x} - 5}{x - 5}$

111. $g(x) = cx^2; \quad c = -2$ **113.** $r(x) = \dfrac{c}{x}; \quad c = 32$

115. False. A function is a special type of relation.

117. False. The range is $[-1, \infty)$.

119. Domain of $f(x)$: all real numbers $x \geq 1$

Domain of $g(x)$: all real numbers $x > 1$

Notice that the domain of $f(x)$ includes $x = 1$ and the domain of $g(x)$ does not because you cannot divide by 0.

121. No; x is the independent variable, f is the name of the function.

123. (a) Yes. The amount you pay in sales tax will increase as the price of the item purchased increases.

(b) No. The length of time that you study will not necessarily determine how well you do on an exam.

Section 2.3 *(page 207)*

1. ordered pairs **3.** zeros **5.** maximum **7.** odd

9. Domain: $(-\infty, -1] \cup [1, \infty)$

Range: $[0, \infty)$

11. Domain: $[-4, 4]$

Range: $[0, 4]$

13. Domain: $(-\infty, \infty)$; Range: $[-4, \infty)$

(a) 0 (b) -1 (c) 0 (d) -2

15. Domain: $(-\infty, \infty)$; Range: $(-2, \infty)$

(a) 0 (b) 1 (c) 2 (d) 3

17. Function **19.** Not a function **21.** Function

23. $-\frac{5}{2}, 6$ **25.** 0 **27.** $0, \pm\sqrt{2}$ **29.** $\pm\frac{1}{2}, 6$ **31.** $\frac{1}{2}$

33.

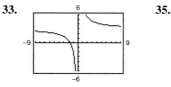

$-\frac{5}{3}$

35.

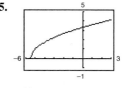

$-\frac{11}{2}$

37.

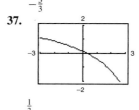

$\frac{1}{3}$

39. Increasing on $(-\infty, \infty)$

41. Increasing on $(-\infty, 0)$ and $(2, \infty)$

Decreasing on $(0, 2)$

43. Increasing on $(1, \infty)$; Decreasing on $(-\infty, -1)$

Constant on $(-1, 1)$

45. Increasing on $(-\infty, 0)$ and $(2, \infty)$; Constant on $(0, 2)$

47.

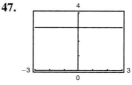

Constant on $(-\infty, \infty)$

49.

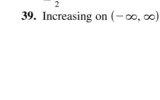

Decreasing on $(-\infty, 0)$

Increasing on $(0, \infty)$

51.

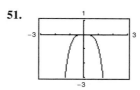

Increasing on $(-\infty, 0)$

Decreasing on $(0, \infty)$

53.

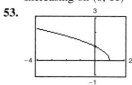

Decreasing on $(-\infty, 1)$

55.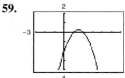

Increasing on $(0, \infty)$

57.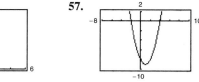

Relative minimum: $(1, -9)$

59.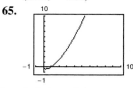

Relative maximum: $(1.5, 0.25)$

61.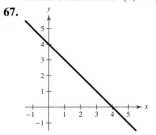

Relative maximum: $(-1.79, 8.21)$
Relative minimum: $(1.12, -4.06)$

63.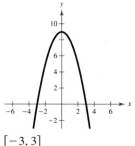

Relative maximum: $(-2, 20)$
Relative minimum: $(1, -7)$

65.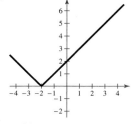

Relative minimum: $(0.33, -0.38)$

67.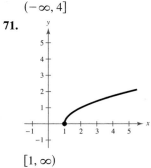

$(-\infty, 4]$

69.

$[-3, 3]$

71.

$[1, \infty)$

73.

$f(x) < 0$ for all x

75. The average rate of change from $x_1 = 0$ to $x_2 = 3$ is -2.
77. The average rate of change from $x_1 = 1$ to $x_2 = 5$ is 18.
79. The average rate of change from $x_1 = 1$ to $x_2 = 3$ is 0.
81. The average rate of change from $x_1 = 3$ to $x_2 = 11$ is $-\frac{1}{4}$.
83. Even; y-axis symmetry   **85.** Odd; origin symmetry
87. Neither; no symmetry   **89.** Neither; no symmetry

91.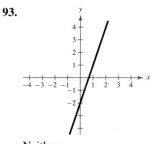

Even

93.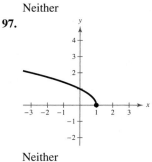

Neither

95.

Even

97.

Neither

99.

Neither
101. $h = -x^2 + 4x - 3$ **103.** $h = 2x - x^2$
105. $L = \frac{1}{2}y^2$ **107.** $L = 4 - y^2$
109. (a) 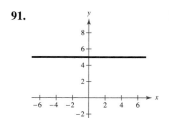   (b) 30 W

111. (a) Ten thousands   (b) Ten millions   (c) Percents
113. (a)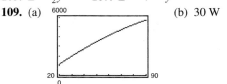

(b) The average rate of change from 1970 to 2005 is 0.705.
The enrollment rate of children in preschool has slowly been increasing each year.
115. (a) $s = -16t^2 + 64t + 6$
(b) 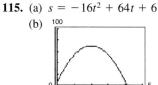   (c) Average rate of change $= 16$

CHAPTER 2

(d) The slope of the secant line is positive.

(e) Secant line: $16t + 6$

(f)

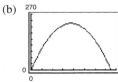

117. (a) $s = -16t^2 + 120t$

(b)

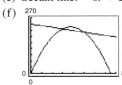

(c) Average rate of change $= -8$

(d) The slope of the secant line is negative.

(e) Secant line: $-8t + 240$

(f)

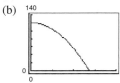

119. (a) $s = -16t^2 + 120$

(b)

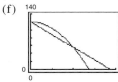

(c) Average rate of change $= -32$

(d) The slope of the secant line is negative.

(e) Secant line: $-32t + 120$

(f)

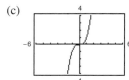

121. False. The function $f(x) = \sqrt{x^2 + 1}$ has a domain of all real numbers.

123. (a) Even. The graph is a reflection in the x-axis.

(b) Even. The graph is a reflection in the y-axis.

(c) Even. The graph is a vertical translation of f.

(d) Neither. The graph is a horizontal translation of f.

125. (a) $\left(\frac{3}{2}, 4\right)$ (b) $\left(\frac{3}{2}, -4\right)$

127. (a) $(-4, 9)$ (b) $(-4, -9)$

129. (a) $(-x, -y)$ (b) $(-x, y)$

131. (a)

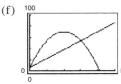

(b)

(c)

(d)

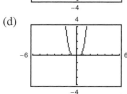

(e)

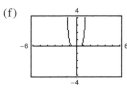

(f)

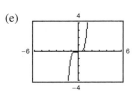

All the graphs pass through the origin. The graphs of the odd powers of x are symmetric with respect to the origin, and the graphs of the even powers are symmetric with respect to the y-axis. As the powers increase, the graphs become flatter in the interval $-1 < x < 1$.

133. 60 ft/sec; As the time traveled increases, the distance increases rapidly, causing the average speed to increase with each time increment. From $t = 0$ to $t = 4$, the average speed is less than from $t = 4$ to $t = 9$. Therefore, the overall average from $t = 0$ to $t = 9$ falls below the average found in part (b).

135. Answers will vary.

Section 2.4   *(page 217)*

1. g   **2.** i   **3.** h   **4.** a   **5.** b   **6.** e   **7.** f

8. c   **9.** d

11. (a) $f(x) = -2x + 6$

(b)

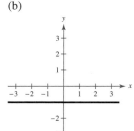

13. (a) $f(x) = -3x + 11$

(b)

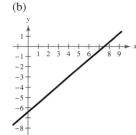

15. (a) $f(x) = -1$

(b)

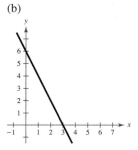

17. (a) $f(x) = \frac{6}{7}x - \frac{45}{7}$

(b)

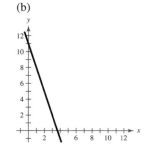

19.

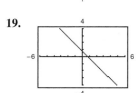

21.

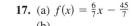

23.

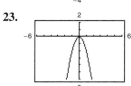

25.

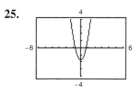

27. 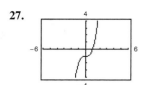　**29.**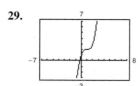

31. 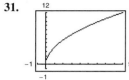　**33.**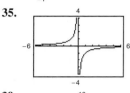

35. 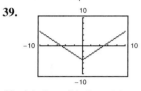　**37.**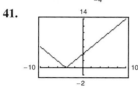

39. 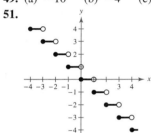　**41.**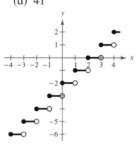

43. (a) 2　(b) 2　(c) −4　(d) 3
45. (a) 1　(b) 3　(c) 7　(d) −19
47. (a) 6　(b) −11　(c) 6　(d) −22
49. (a) −10　(b) −4　(c) −1　(d) 41

51. 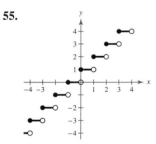　**53.**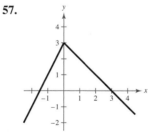

55. 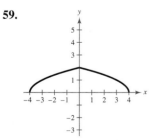　**57.**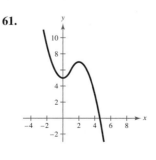

59. 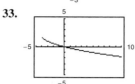　**61.**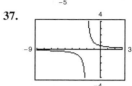

63.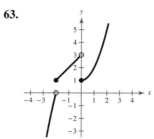

65. (a)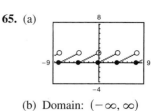

(b) Domain: $(-\infty, \infty)$
Range: $[0, 2)$
(c) Sawtooth pattern

67. (a)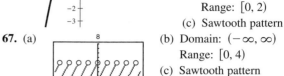

(b) Domain: $(-\infty, \infty)$
Range: $[0, 4)$
(c) Sawtooth pattern

69. (a)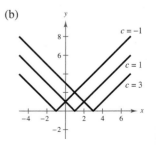

(b) $57.15

Cost of overnight delivery (in dollars)
Weight (in pounds)

71. (a) $W(30) = 420$; $W(40) = 560$;
$W(45) = 665$; $W(50) = 770$

(b) $W(h) = \begin{cases} 14h, & 0 < h \le 45 \\ 21(h-45) + 630, & h > 45 \end{cases}$

73. (a)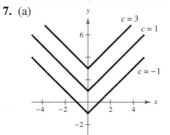

$f(x) = \begin{cases} 0.505x^2 - 1.47x + 6.3, & 1 \le x \le 6 \\ -1.97x + 26.3, & 6 < x \le 12 \end{cases}$

Answers will vary. Sample answer: The domain is determined by inspection of a graph of the data with the two models.

(b) $f(5) = 11.575$, $f(11) = 4.63$; These values represent the revenue for the months of May and November, respectively.

(c) These values are quite close to the actual data values.

75. False. A linear equation could be a horizontal or vertical line.

Section 2.5　(page 224)

1. rigid　**3.** nonrigid
5. vertical stretch; vertical shrink
7. (a) 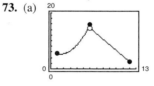　(b)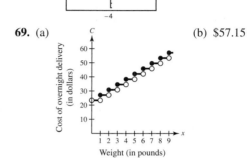

(c)

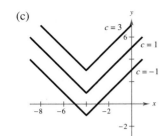

(g)

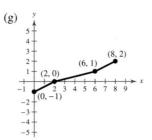

9. (a)

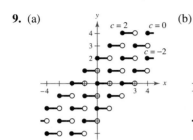

(b)

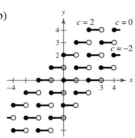

13. (a)

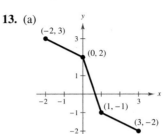

(b)

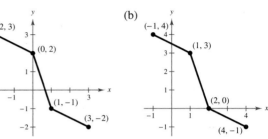

(c)

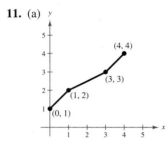

(c)

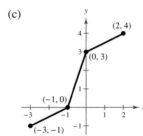

(d)

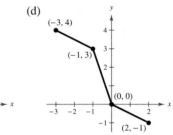

11. (a)

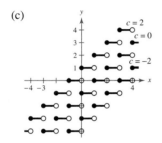

(b)

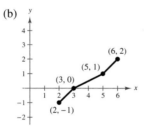

(e)

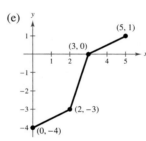

(f)

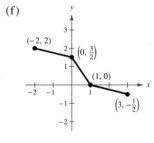

(c)

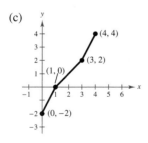

(d)

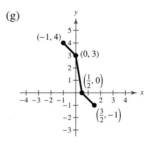

(g)

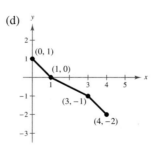

(e)

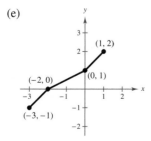

(f)

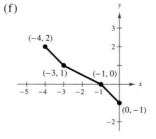

15. (a) $y = x^2 - 1$ (b) $y = 1 - (x + 1)^2$
(c) $y = -(x - 2)^2 + 6$ (d) $y = (x - 5)^2 - 3$
17. (a) $y = |x| + 5$ (b) $y = -|x + 3|$
(c) $y = |x - 2| - 4$ (d) $y = -|x - 6| - 1$
19. Horizontal shift of $y = x^3$; $y = (x - 2)^3$
21. Reflection in the x-axis of $y = x^2$; $y = -x^2$
23. Reflection in the x-axis and vertical shift of $y = \sqrt{x}$; $y = 1 - \sqrt{x}$

25. (a) $f(x) = x^2$
   (b) Reflection in the x-axis and vertical shift 12 units upward
   (c)

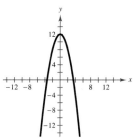

   (d) $g(x) = 12 - f(x)$
27. (a) $f(x) = x^3$
   (b) Vertical shift seven units upward
   (c)

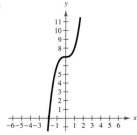

   (d) $g(x) = f(x) + 7$
29. (a) $f(x) = x^2$
   (b) Vertical shrink of two-thirds and vertical shift four units upward
   (c)

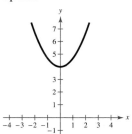

   (d) $g(x) = \frac{2}{3}f(x) + 4$
31. (a) $f(x) = x^2$
   (b) Reflection in the x-axis, horizontal shift five units to the left, and vertical shift two units upward
   (c)

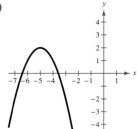

   (d) $g(x) = 2 - f(x + 5)$
33. (a) $f(x) = x^2$
   (b) Vertical stretch of two, horizontal shift four units to the right, and vertical shift three units upward

(c)

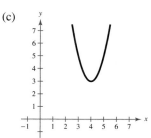

   (d) $g(x) = 3 + 2f(x - 4)$

35. (a) $f(x) = \sqrt{x}$
   (b) Horizontal shrink of one-third
   (c)

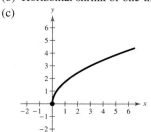

   (d) $g(x) = f(3x)$

37. (a) $f(x) = x^3$
   (b) Vertical shift two units upward and horizontal shift one unit to the right
   (c)

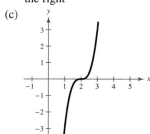

   (d) $g(x) = f(x - 1) + 2$

39. (a) $f(x) = x^3$
   (b) Vertical stretch of three and horizontal shift two units to the right
   (c)

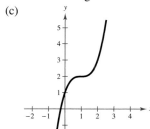

   (d) $g(x) = 3f(x - 2)$

41. (a) $f(x) = |x|$
   (b) Reflection in the x-axis and vertical shift two units downward
   (c)

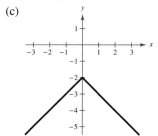

   (d) $g(x) = -f(x) - 2$

CHAPTER 2

43. (a) $f(x) = |x|$

(b) Reflection in the x-axis, horizontal shift four units to the left, and vertical shift eight units upward

(c) 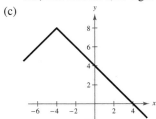   (d) $g(x) = -f(x + 4) + 8$

45. (a) $f(x) = |x|$

(b) Reflection in the x-axis, vertical stretch of two, horizontal shift one unit to the right, and vertical shift four units downward

(c) 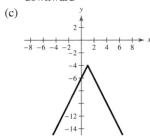   (d) $g(x) = -2f(x - 1) - 4$

47. (a) $f(x) = [\![x]\!]$

(b) Reflection in the x-axis and vertical shift three units upward

(c) 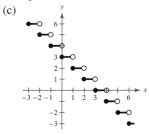   (d) $g(x) = 3 - f(x)$

49. (a) $f(x) = \sqrt{x}$

(b) Horizontal shift nine units to the right

(c) 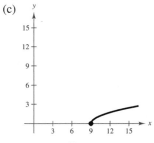   (d) $g(x) = f(x - 9)$

51. (a) $f(x) = \sqrt{x}$

(b) Reflection in the y-axis, horizontal shift seven units to the right, and vertical shift two units downward

(c) 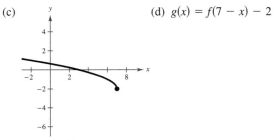   (d) $g(x) = f(7 - x) - 2$

53. (a) $f(x) = \sqrt{x}$

(b) Horizontal stretch and vertical shift four units downward

(c) 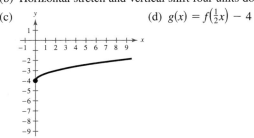   (d) $g(x) = f\left(\tfrac{1}{2}x\right) - 4$

55. $g(x) = (x - 3)^2 - 7$ **57.** $g(x) = (x - 13)^3$

59. $g(x) = -|x| + 12$ **61.** $g(x) = -\sqrt{-x + 6}$

63. (a) $y = -3x^2$ (b) $y = 4x^2 + 3$

65. (a) $y = -\tfrac{1}{2}|x|$ (b) $y = 3|x| - 3$

67. Vertical stretch of $y = x^3$; $y = 2x^3$

69. Reflection in the x-axis and vertical shrink of $y = x^2$; $y = -\tfrac{1}{2}x^2$

71. Reflection in the y-axis and vertical shrink of $y = \sqrt{x}$; $y = \tfrac{1}{2}\sqrt{-x}$

73. $y = -(x - 2)^3 + 2$ **75.** $y = -\sqrt{x} - 3$

77. (a)   (b)

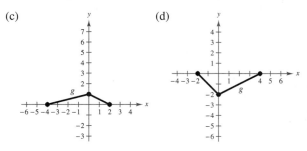

(e) 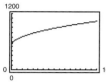   (f)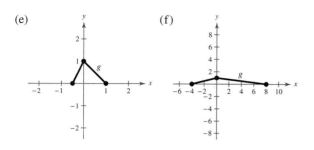

79. (a) Vertical stretch of 128.0 and a vertical shift of 527 units upward

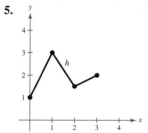

(b) 32; Each year, the total number of miles driven by vans, pickups, and SUVs increases by an average of 32 billion miles.

(c) $f(t) = 527 + 128\sqrt{t + 10}$; The graph is shifted 10 units to the left.

(d) 1127 billion miles; Answers will vary. Sample answer: Yes, because the number of miles driven has been steadily increasing.

81. False. The graph of $y = f(-x)$ is a reflection of the graph of $f(x)$ in the y-axis.

83. True. $|-x| = |x|$

85. (a) $g(t) = \frac{3}{4}f(t)$ (b) $g(t) = f(t) + 10{,}000$
(c) $g(t) = f(t - 2)$

87. $(-2, 0), (-1, 1), (0, 2)$

89. No. $g(x) = -x^4 - 2$. Yes. $h(x) = -(x - 3)^4$.

Section 2.6   *(page 234)*

1. addition; subtraction; multiplication; division   **3.** $g(x)$

5. 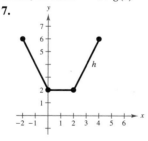   **7.**

9. (a) $2x$ (b) 4   (c) $x^2 - 4$
(d) $\frac{x + 2}{x - 2}$; all real numbers x except $x = 2$

11. (a) $x^2 + 4x - 5$ (b) $x^2 - 4x + 5$ (c) $4x^3 - 5x^2$
(d) $\frac{x^2}{4x - 5}$; all real numbers x except $x = \frac{5}{4}$

13. (a) $x^2 + 6 + \sqrt{1 - x}$ (b) $x^2 + 6 - \sqrt{1 - x}$
(c) $(x^2 + 6)\sqrt{1 - x}$
(d) $\frac{(x^2 + 6)\sqrt{1 - x}}{1 - x}$; all real numbers x such that $x < 1$

15. (a) $\frac{x + 1}{x^2}$ (b) $\frac{x - 1}{x^2}$ (c) $\frac{1}{x^3}$
(d) x; all real numbers x except $x = 0$

17. 3   **19.** 5   **21.** $9t^2 - 3t + 5$ **23.** 74

25. 26   **27.** $\frac{3}{5}$

29. 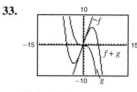   **31.**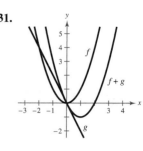

33. 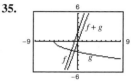   **35.**

$f(x), g(x)$ $f(x), f(x)$

37. (a) $(x - 1)^2$ (b) $x^2 - 1$ (c) $x - 2$

39. (a) x (b) x (c) $x^9 + 3x^6 + 3x^3 + 2$

41. (a) $\sqrt{x^2 + 4}$ (b) $x + 4$
Domains of f and $g \circ f$: all real numbers x such that $x \geq -4$
Domains of g and $f \circ g$: all real numbers x

43. (a) $x + 1$ (b) $\sqrt{x^2 + 1}$
Domains of f and $g \circ f$: all real numbers x
Domains of g and $f \circ g$: all real numbers x such that $x \geq 0$

45. (a) $|x + 6|$ (b) $|x| + 6$
Domains of $f, g, f \circ g$, and $g \circ f$: all real numbers x

47. (a) $\frac{1}{x + 3}$ (b) $\frac{1}{x} + 3$
Domains of f and $g \circ f$: all real numbers x except $x = 0$
Domain of g: all real numbers x
Domain of $f \circ g$: all real numbers x except $x = -3$

49. (a) 3   (b) 0   **51.** (a) 0   (b) 4

53. $f(x) = x^2, g(x) = 2x + 1$

55. $f(x) = \sqrt[3]{x}, \ g(x) = x^2 - 4$

57. $f(x) = \frac{1}{x}, \ g(x) = x + 2$ **59.** $f(x) = \frac{x + 3}{4 + x}, \ g(x) = -x^2$

61. (a) $T = \frac{3}{4}x + \frac{1}{15}x^2$

(b)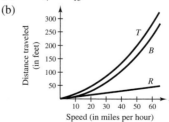

(c) The braking function $B(x)$. As x increases, $B(x)$ increases at a faster rate than $R(x)$.

63. (a) $c(t) = \dfrac{b(t) - d(t)}{p(t)} \times 100$

(b) $c(5)$ is the percent change in the population due to births and deaths in the year 2005.

65. (a) $(N + M)(t) = 0.227t^3 - 4.11t^2 + 14.6t + 544$, which represents the total number of Navy and Marines personnel combined.

$(N + M)(0) = 544$

$(N + M)(6) \approx 533$

$(N + M)(12) \approx 520$

(b) $(N - M)(t) = 0.157t^3 - 3.65t^2 + 11.2t + 200$, which represents the difference between the number of Navy personnel and the number of Marines personnel.

$(N - M)(0) = 200$

$(N - M)(6) \approx 170$

$(N - M)(12) \approx 80$

67. $(B - D)(t) = -0.197t^3 + 10.17t^2 - 128.0t + 2043$, which represents the change in the United States population.

69. (a) For each time t there corresponds one and only one temperature T.

(b) $60°$, $72°$

(c) All the temperature changes occur 1 hour later.

(d) The temperature is decreased by 1 degree.

(e) $T(t) = \begin{cases} 60, & 0 \le t \le 6 \\ 12t - 12, & 6 < t < 7 \\ 72, & 7 \le t \le 20 \\ -12t + 312, & 20 < t < 21 \\ 60, & 21 \le t \le 24 \end{cases}$

71. $(A \circ r)(t) = 0.36\pi t^2$; $(A \circ r)(t)$ represents the area of the circle at time t.

73. (a) $N(T(t)) = 30(3t^2 + 2t + 20)$; This represents the number of bacteria in the food as a function of time.

(b) About 653 bacteria   (c) 2.846 h

75. $g(f(x))$ represents 3 percent of an amount over \$500,000.

77. False. $(f \circ g)(x) = 6x + 1$ and $(g \circ f)(x) = 6x + 6$

79. (a) $O(M(Y)) = 2(6 + \frac{1}{2}Y) = 12 + Y$

(b) Middle child is 8 years old; youngest child is 4 years old.

81. Proof

83. (a) Proof

(b) $\frac{1}{2}[f(x) + f(-x)] + \frac{1}{2}[f(x) - f(-x)]$

$\quad = \frac{1}{2}[f(x) + f(-x) + f(x) - f(-x)]$

$\quad = \frac{1}{2}[2f(x)]$

$\quad = f(x)$

(c) $f(x) = (x^2 + 1) + (-2x)$

$k(x) = \dfrac{-1}{(x + 1)(x - 1)} + \dfrac{x}{(x + 1)(x - 1)}$

Section 2.7   *(page 244)*

1. inverse   **3.** range; domain   **5.** one-to-one

7. $f^{-1}(x) = \frac{1}{6}x$ **9.** $f^{-1}(x) = x - 9$

11. $f^{-1}(x) = \dfrac{x - 1}{3}$ **13.** $f^{-1}(x) = x^3$

15. c   **16.** b   **17.** a   **18.** d

19. $f(g(x)) = f\left(-\dfrac{2x + 6}{7}\right) = -\dfrac{7}{2}\left(-\dfrac{2x + 6}{7}\right) - 3 = x$

$g(f(x)) = g\left(-\dfrac{7}{2}x - 3\right) = -\dfrac{2\left(-\frac{7}{2}x - 3\right) + 6}{7} = x$

21. $f(g(x)) = f(\sqrt[3]{x - 5}) = (\sqrt[3]{x - 5})^3 + 5 = x$

$g(f(x)) = g(x^3 + 5) = \sqrt[3]{(x^3 + 5) - 5} = x$

23. (a) $f(g(x)) = f\left(\dfrac{x}{2}\right) = 2\left(\dfrac{x}{2}\right) = x$

$g(f(x)) = g(2x) = \dfrac{(2x)}{2} = x$

(b)
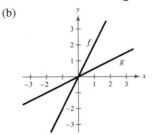

25. (a) $f(g(x)) = f\left(\dfrac{x - 1}{7}\right) = 7\left(\dfrac{x - 1}{7}\right) + 1 = x$

$g(f(x)) = g(7x + 1) = \dfrac{(7x + 1) - 1}{7} = x$

(b)
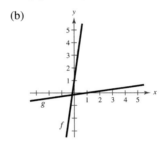

27. (a) $f(g(x)) = f(\sqrt[3]{8x}) = \dfrac{(\sqrt[3]{8x})^3}{8} = x$

$g(f(x)) = g\left(\dfrac{x^3}{8}\right) = \sqrt[3]{8\left(\dfrac{x^3}{8}\right)} = x$

(b)
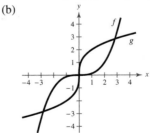

29. (a) $f(g(x)) = f(x^2 + 4)$, $x \ge 0$

$\quad = \sqrt{(x^2 + 4) - 4} = x$

$g(f(x)) = g(\sqrt{x - 4})$

$\quad = (\sqrt{x - 4})^2 + 4 = x$

(b)
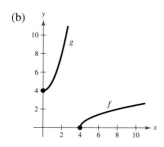

31. (a) $f(g(x)) = f(\sqrt{9-x}), \ x \le 9$
$$= 9 - (\sqrt{9-x})^2 = x$$
$g(f(x)) = g(9-x^2), \ x \ge 0$
$$= \sqrt{9-(9-x^2)} = x$$

(b)
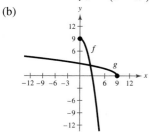

33. (a) $f(g(x)) = f\left(-\dfrac{5x+1}{x-1}\right) = \dfrac{-\left(\dfrac{5x+1}{x-1}\right)-1}{-\left(\dfrac{5x+1}{x-1}\right)+5}$
$$= \dfrac{-5x-1-x+1}{-5x-1+5x-5} = x$$
$g(f(x)) = g\left(\dfrac{x-1}{x+5}\right) = \dfrac{-5\left(\dfrac{x-1}{x+5}\right)-1}{\dfrac{x-1}{x+5}-1}$
$$= \dfrac{-5x+5-x-5}{x-1-x-5} = x$$

(b)
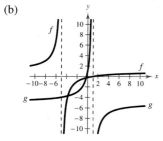

35. No
37.

x	-2	0	2	4	6	8
$f^{-1}(x)$	-2	-1	0	1	2	3

39. Yes   **41.** No
43.

The function has an inverse.

45.
The function does not have inverse.

47.
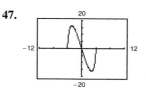

The function does not have an inverse.

49. (a) $f^{-1}(x) = \dfrac{x+3}{2}$

(b)
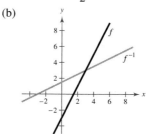

(c) The graph of f^{-1} is the reflection of the graph of f in the line $y = x$.
(d) The domains and ranges of f and f^{-1} are all real numbers.

51. (a) $f^{-1}(x) = \sqrt[5]{x+2}$

(b)
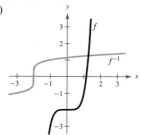

(c) The graph of f^{-1} is the reflection of the graph of f in the line $y = x$.
(d) The domains and ranges of f and f^{-1} are all real numbers.

53. (a) $f^{-1}(x) = \sqrt{4-x^2}, \ 0 \le x \le 2$

(b)
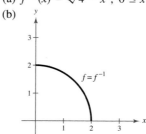

(c) The graph of f^{-1} is the same as the graph of f.
(d) The domains and ranges of f and f^{-1} are all real numbers x such that $0 \le x \le 2$.

55. (a) $f^{-1}(x) = \dfrac{4}{x}$ (b)
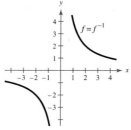

(c) The graph of f^{-1} is the same as the graph of f.

(d) The domains and ranges of f and f^{-1} are all real numbers x except $x = 0$.

57. (a) $f^{-1}(x) = \dfrac{2x + 1}{x - 1}$

(b)

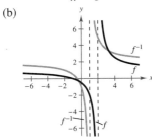

(c) The graph of f^{-1} is the reflection of the graph of f in the line $y = x$.

(d) The domain of f and the range of f^{-1} are all real numbers x except $x = 2$. The domain of f^{-1} and the range of f are all real numbers x except $x = 1$.

59. (a) $f^{-1}(x) = x^3 + 1$

(b)

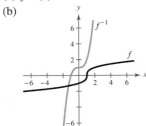

(c) The graph of f^{-1} is the reflection of the graph of f in the line $y = x$.

(d) The domains and ranges of f and f^{-1} are all real numbers.

61. (a) $f^{-1}(x) = \dfrac{5x - 4}{6 - 4x}$

(b)

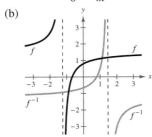

(c) The graph of f^{-1} is the reflection of the graph of f in the line $y = x$.

(d) The domain of f and the range of f^{-1} are all real numbers x except $x = -\frac{5}{4}$. The domain of f^{-1} and the range of f are all real numbers x except $x = \frac{3}{2}$.

63. No inverse   **65.** $g^{-1}(x) = 8x$ **67.** No inverse

69. $f^{-1}(x) = \sqrt{x} - 3e$ **71.** No inverse   **73.** No inverse

75. $f^{-1}(x) = \dfrac{x^2 - 3}{2}, \quad x \geq 0$

77. $f^{-1}(x) = \sqrt{x} + 2$

The domain of f and the range of f^{-1} are all real numbers x such that $x \geq 2$. The domain of f^{-1} and the range of f are all real numbers x such that $x \geq 0$.

79. $f^{-1}(x) = x - 2$

The domain of f and the range of f^{-1} are all real numbers x such that $x \geq -2$. The domain of f^{-1} and the range of f are all real numbers x such that $x \geq 0$.

81. $f^{-1}(x) = \sqrt{x} - 6$

The domain of f and the range of f^{-1} are all real numbers x such that $x \geq -6$. The domain of f^{-1} and the range of f are all real numbers x such that $x \geq 0$.

83. $f^{-1}(x) = \dfrac{\sqrt{-2(x - 5)}}{2}$

The domain of f and the range of f^{-1} are all real numbers x such that $x \geq 0$. The domain of f^{-1} and the range of f are all real numbers x such that $x \leq 5$.

85. $f^{-1}(x) = x + 3$

The domain of f and the range of f^{-1} are all real numbers x such that $x \geq 4$. The domain of f^{-1} and the range of f are all real numbers x such that $x \geq 1$.

87. 32   **89.** 600   **91.** $2\sqrt[3]{x + 3}$

93. $\dfrac{x + 1}{2}$ **95.** $\dfrac{x + 1}{2}$

97. (a) Yes; each European shoe size corresponds to exactly one U.S. shoe size.

(b) 45   (c) 10   (d) 41   (e) 13

99. (a) Yes

(b) S^{-1} represents the time in years for a given sales level.

(c) $S^{-1}(8430) = 6$

(d) No, because then the sales for 2007 and 2009 would be the same, so the function would no longer be one-to-one.

101. (a) $y = \dfrac{x - 10}{0.75}$

x = hourly wage; y = number of units produced

(b) 19 units

103. False. $f(x) = x^2$ has no inverse.   **105.** Proof

107.

x	1	3	4	6
y	1	2	6	7

x	1	2	6	7
$f^{-1}(x)$	1	3	4	6

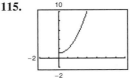

109. This situation could be represented by a one-to-one function if the runner does not stop to rest. The inverse function would represent the time in hours for a given number of miles completed.

111. This function could not be represented by a one-to-one function because it oscillates.

113. $k = \frac{1}{4}$

115.

There is an inverse function $f^{-1}(x) = \sqrt{x - 1}$ because the domain of f is equal to the range of f^{-1} and the range of f is equal to the domain of f^{-1}.

Review Exercises   *(page 250)*

1. slope: -2
   y-intercept: -7

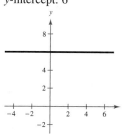

3. slope: 0
   y-intercept: 6

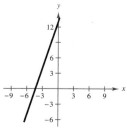

5. slope: $-\frac{5}{2}$
   y-intercept: -1

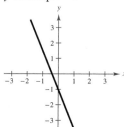

7. slope: 3
   y-intercept: 13

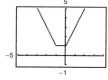

9.

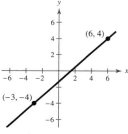

$m = \frac{8}{9}$

11.

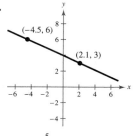

$m = -\frac{5}{11}$

13. $y = \frac{2}{3}x - 2$

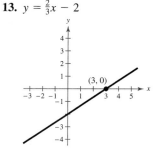

15. $y = -\frac{1}{2}x + 2$

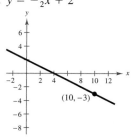

17. $x = 0$ **19.** $y = \frac{2}{7}x + \frac{2}{7}$

21. (a) $y = \frac{5}{4}x - \frac{23}{4}$ (b) $y = -\frac{4}{5}x + \frac{2}{5}$

23. $V = -850t + 21,000,\quad 10 \le t \le 15$

25. No   **27.** Yes

29. (a) 5   (b) 17   (c) $t^4 + 1$ (d) $t^2 + 2t + 2$

31. (a) -3 (b) -1 (c) 2   (d) 6

33. All real numbers x such
   that $-5 \le x \le 5$

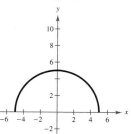

35. All real numbers x except
   $x = 3, -2$

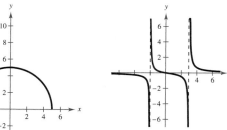

37. (a) 16 ft/sec   (b) 1.5 sec   (c) -16 ft/sec

39. $4x + 2h + 3,\quad h \ne 0$

41. Function   **43.** Not a function   **45.** $-3, 7$

47. $-\frac{3}{8}$ **49.** 0, 1

51.

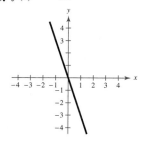

Increasing on $(0, \infty)$
Decreasing on $(-\infty, -1)$
Constant on $(-1, 0)$

53.

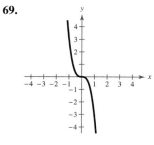

55.

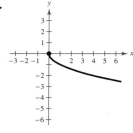
(0.1250, 0.000488)

57. 4   **59.** $\dfrac{1 - \sqrt{2}}{2}$

61. Neither even nor odd   **63.** Odd

65. $f(x) = -3x$ **67.**

69.   **71.**

73.

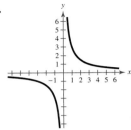

75.

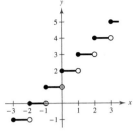

(c)

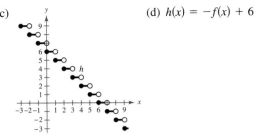

(d) $h(x) = -f(x) + 6$

77.

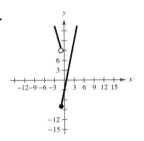

79. $y = x^3$

89. (a) $f(x) = |x|$
(b) Reflections in the x-axis and the y-axis, horizontal shift four units to the right, and vertical shift six units upward
(c) **(d)** $h(x) = -f(-x + 4) + 6$

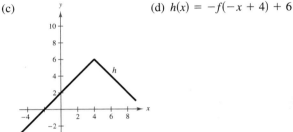

81. (a) $f(x) = x^2$ (b) Vertical shift nine units downward
(c) **(d)** $h(x) = f(x) - 9$

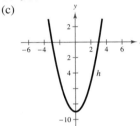

91. (a) $f(x) = [\![x]\!]$
(b) Horizontal shift nine units to the right and vertical stretch
(c) **(d)** $h(x) = 5f(x - 9)$

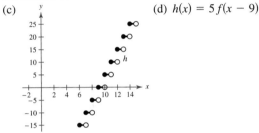

83. (a) $f(x) = \sqrt{x}$
(b) Reflection in the x-axis and vertical shift four units upward
(c) **(d)** $h(x) = -f(x) + 4$

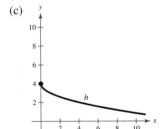

93. (a) $f(x) = \sqrt{x}$
(b) Reflection in the x-axis, vertical stretch, and horizontal shift four units to the right
(c) **(d)** $h(x) = -2f(x - 4)$

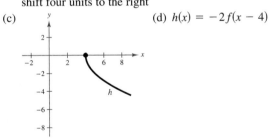

85. (a) $f(x) = x^2$
(b) Reflection in the x-axis, horizontal shift two units to the left, and vertical shift three units upward
(c) **(d)** $h(x) = -f(x + 2) + 3$

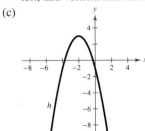

95. (a) $x^2 + 2x + 2$ (b) $x^2 - 2x + 4$
(c) $2x^3 - x^2 + 6x - 3$
(d) $\dfrac{x^2 + 3}{2x - 1}$; all real numbers x except $x = \dfrac{1}{2}$

97. (a) $x - \dfrac{8}{3}$ (b) $x - 8$
Domains of $f, g, f \circ g$, and $g \circ f$: all real numbers x

99. (a) $x + 3$ (b) $\sqrt[3]{x^3 + 3}$
Domains of $f, g, f \circ g$, and $g \circ f$: all real numbers x

101. $f(x) = x^3, g(x) = 1 - 2x$

103. (a) $(r + c)(t) = 178.8t + 856$; This represents the average annual expenditures for both residental and cellular phone services.

87. (a) $f(x) = [\![x]\!]$
(b) Reflection in the x-axis and vertical shift six units upward

(b)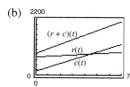

(c) $(r + c)(13) = \$3180.40$

105. $f^{-1}(x) = \dfrac{x - 8}{3}$

$f(f^{-1}(x)) = 3\left(\dfrac{x - 8}{3}\right) + 8 = x$

$f^{-1}(f(x)) = \dfrac{3x + 8 - 8}{3} = x$

107. $f^{-1}(x) = \sqrt[3]{x + 1}$

$f(f^{-1}(x)) = (\sqrt[3]{x + 1})^3 - 1 = x$

$f^{-1}(f(x)) = \sqrt[3]{(x^3 - 1) + 1} = x$

109. The function has an inverse.

111. 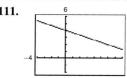   **113.**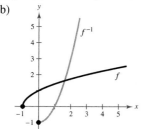

The function has an inverse.   The function has an inverse.

115. (a) $f^{-1}(x) = 2x + 6$

(b)

(c) The graphs are reflections of each other in the line $y = x$.

(d) Both f and f^{-1} have domains and ranges that are all real numbers.

117. (a) $f^{-1}(x) = x^2 - 1,\ x \geq 0$

(b)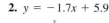

(c) The graphs are reflections of each other in the line $y = x$.

(d) The graph of f has a domain of all real numbers x such that $x \geq -1$ and a range of $[0, \infty)$. The graph of f^{-1} has a domain of all real numbers x such that $x \geq 0$ and a range of $[-1, \infty)$.

119. $x > 4;\ f^{-1}(x) = \sqrt{\dfrac{x}{2} + 4},\ x \neq 0$

121. False. The graph is reflected in the x-axis, shifted 9 units to the left, and then shifted 13 units downward.

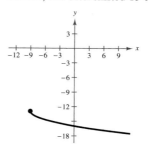

123. A function from a set A to a set B is a relation that assigns to each element x in the set A exactly one element y in the set B.

125. Answers will vary. Sample answer: $f(x) = x^3$

The domain and range of the function are the set of all real numbers.

The function is odd.

The graph is increasing on the interval $(-\infty, \infty)$.

The graph is symmetric with respect to the origin.

The graph has an intercept at $(0, 0)$.

$f(x) = x^3 + 1$

The domain and range of the function are the set of all real numbers.

The function is neither odd nor even.

The graph is increasing on the interval $(-\infty, \infty)$.

The graph is not symmetric.

The graph has intercepts at $(0, 1)$ and $(-1, 0)$.

Chapter Test   *(page 253)*

1. $y = -2x + 3$ **2.** $y = -1.7x + 5.9$

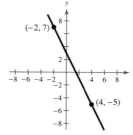   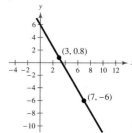

3. (a) $y = -\dfrac{5}{2}x + 4$ (b) $y = \dfrac{2}{5}x + 4$

4. (a) -9 (b) 1 (c) $|x - 4| - 15$

5. (a) $-\dfrac{1}{8}$ (b) $-\dfrac{1}{28}$ (c) $\dfrac{\sqrt{x}}{x^2 - 18x}$

6. All real numbers x

7. All real numbers x such that $x \leq 3$

8. (a)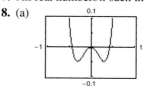

(b) Increasing on $(-0.31, 0),\ (0.31, \infty)$

Decreasing on $(-\infty, -0.31),\ (0, 0.31)$

(c) Even

9. (a)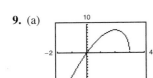

(b) Increasing on $(-\infty, 2)$
Decreasing on $(2, 3)$
(c) Neither even nor odd

10. (a)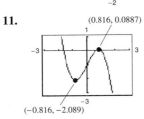

(b) Increasing on $(-5, \infty)$
Decreasing on $(-\infty, -5)$
(c) Neither even nor odd

11.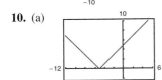

(0.816, 0.0887)
(−0.816, −2.089)

12. Average rate of change $= -3$

13.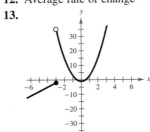

14. (a) $f(x) = [\![x]\!]$ (b) Vertical stretch of $y = [\![x]\!]$
(c)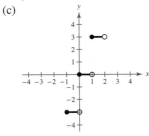

15. (a) $f(x) = \sqrt{x}$
(b) Reflection in the x-axis, horizontal shift, and vertical shift of $y = \sqrt{x}$
(c)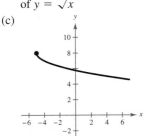

16. (a) $f(x) = x^3$
(b) Vertical stretch, reflections in the x-axis, vertical shift, and horizontal shift of $y = x^3$

(c)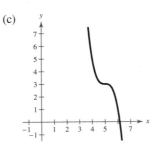

17. (a) $2x^2 - 4x - 2$ (b) $4x^2 + 4x - 12$
(c) $-3x^4 - 12x^3 + 22x^2 + 28x - 35$
(d) $\dfrac{3x^2 - 7}{-x^2 - 4x + 5}$, $x \neq 1, -5$
(e) $3x^4 + 24x^3 + 18x^2 - 120x + 68$
(f) $-9x^4 + 30x^2 - 16$

18. (a) $\dfrac{1 + 2x^{3/2}}{x}$, $x > 0$ (b) $\dfrac{1 - 2x^{3/2}}{x}$, $x > 0$
(c) $\dfrac{2\sqrt{x}}{x}$, $x > 0$ (d) $\dfrac{1}{2x^{3/2}}$, $x > 0$
(e) $\dfrac{\sqrt{x}}{2x}$, $x > 0$ (f) $\dfrac{2\sqrt{x}}{x}$, $x > 0$

19. $f^{-1}(x) = \sqrt[3]{x - 8}$ **20.** No inverse
21. $f^{-1}(x) = \left(\tfrac{1}{3}x\right)^{2/3}$, $x \geq 0$ **22.** \$153

Cumulative Test for Chapters P–2   *(page 254)*

1. $\dfrac{4x^3}{15y^5}$, $x \neq 0$ **2.** $3xy^2\sqrt{2x}$ **3.** $5x - 6$

4. $x^3 - x^2 - 5x + 6$ **5.** $\dfrac{s - 1}{(s + 1)(s + 3)}$

6. $(x + 3)(7 - x)$ **7.** $x(x + 1)(1 - 6x)$
8. $2(3x + 2)(9x^2 - 6x + 4)$
9. $4x^2 + 12x$ **10.** $\tfrac{3}{2}x^2 + 8x + \tfrac{5}{2}$

11. **12.**

13. **14.** $x = -\tfrac{13}{3}$

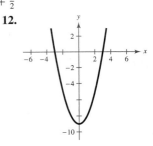

15. $x = \tfrac{27}{5}$ **16.** $x = \tfrac{23}{6}$ **17.** $1, 3$ **18.** $2 \pm \sqrt{10}$
19. ± 6 **20.** $\dfrac{-5 \pm \sqrt{97}}{6}$ **21.** $-\dfrac{3}{2} \pm \dfrac{\sqrt{69}}{6}$

22. ± 8 **23.** $0, -12, \pm 2i$ **24.** $0, 3$ **25.** ± 8

26. 6 **27.** $13, -5$ **28.** No solution

29. (a) Not a solution   (b) Not a solution   (c) Solution
(d) Solution

30. (a) Not a solution   (b) Solution   (c) Not a solution
(d) Not a solution

31. $-7 \le x \le 5$ **32.** $x > -\frac{1}{3}, x < -\frac{4}{3}$

33. $x \le -\frac{7}{5}, x \ge -1$

34. $x < \frac{1 - \sqrt{17}}{2}, x > \frac{1 + \sqrt{17}}{2}$

35. $y = 2x + 2$

36. For some values of x there correspond two values of y.

37. (a) $\frac{3}{2}$ (b) Division by 0 is undefined.   (c) $\frac{s + 2}{s}$

38. Neither   **39.** Neither   **40.** Even

41. (a) Vertical shrink by $\frac{1}{2}$
(b) Vertical shift two units upward
(c) Horizontal shift two units to the left

42. (a) $4x - 3$ (b) $-2x - 5$ (c) $3x^2 - 11x - 4$
(d) $\frac{x - 4}{3x + 1}$; Domain: all real numbers x except $x = -\frac{1}{3}$

43. (a) $\sqrt{x - 1} + x^2 + 1$ (b) $\sqrt{x - 1} - x^2 - 1$
(c) $x^2\sqrt{x - 1} + \sqrt{x - 1}$
(d) $\frac{\sqrt{x - 1}}{x^2 + 1}$; Domain: all real numbers x such that $x \ge 1$

44. (a) $2x + 12$ (b) $\sqrt{2x^2 + 6}$
Domain of $f \circ g$: all real numbers x such that $x \ge -6$
Domain of $g \circ f$: all real numbers x

45. (a) $|x| - 2$ (b) $|x - 2|$
Domains of $f \circ g$ and $g \circ f$: all real numbers x

46. $h(x)^{-1} = \frac{1}{3}(x + 4)$ **47.** $n = 9$

48. (a) $R(n) = -0.05n^2 + 13n, \quad n \ge 60$
(b)   130 passengers

49. 4; Answers will vary.

Problem Solving   *(page 257)*

1. (a) $W_1 = 2000 + 0.07S$ (b) $W_2 = 2300 + 0.05S$

(c) 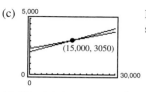   Both jobs pay the same monthly salary if sales equal \$15,000.
(15,000, 3050)

(d) No. Job 1 would pay \$3400 and job 2 would pay \$3300.

3. (a) The function will be even.
(b) The function will be odd.
(c) The function will be neither even nor odd.

5. $f(x) = a_{2n}x^{2n} + a_{2n-2}x^{2n-2} + \cdots + a_2x^2 + a_0$
$f(-x) = a_{2n}(-x)^{2n} + a_{2n-2}(-x)^{2n-2}$
$\qquad + \cdots + a_2(-x)^2 + a_0$
$\qquad = f(x)$

7. (a) $81\frac{2}{3}$ h   (b) $25\frac{5}{7}$ mi/h
(c) $y = \frac{-180}{7}x + 3400$
Domain: $0 \le x \le \frac{1190}{9}$
Range: $0 \le y \le 3400$
(d)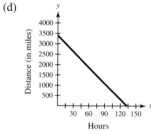

9. (a) $(f \circ g)(x) = 4x + 24$ (b) $(f \circ g)^{-1}(x) = \frac{1}{4}x - 6$
(c) $f^{-1}(x) = \frac{1}{4}x; g^{-1}(x) = x - 6$
(d) $(g^{-1} \circ f^{-1})(x) = \frac{1}{4}x - 6$; They are the same.
(e) $(f \circ g)(x) = 8x^3 + 1; (f \circ g)^{-1}(x) = \frac{1}{2}\sqrt[3]{x - 1}$;
$f^{-1}(x) = \sqrt[3]{x - 1}; g^{-1}(x) = \frac{1}{2}x$;
$(g^{-1} \circ f^{-1})(x) = \frac{1}{2}\sqrt[3]{x - 1}$
(f) Answers will vary.
(g) $(f \circ g)^{-1}(x) = (g^{-1} \circ f^{-1})(x)$

11. (a)   (b)
(c)   (d)

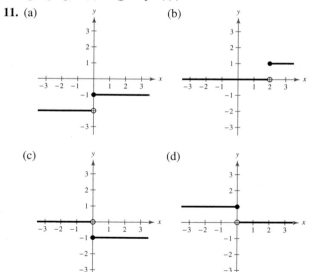

(e)

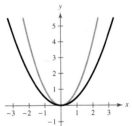

(f)

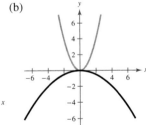

13. Proof

15. (a)

x	-4	-2	0	4
$f(f^{-1}(x))$	-4	-2	0	4

(b)

x	-3	-2	0	1
$(f + f^{-1})(x)$	5	1	-3	-5

(c)

x	-3	-2	0	1
$(f \cdot f^{-1})(x)$	4	0	2	6

(d)

x	-4	-3	0	4
$|f^{-1}(x)|$	2	1	1	3

Chapter 3

Section 3.1 *(page 266)*

1. polynomial   **3.** quadratic; parabola
5. positive; minimum   **7.** e   **8.** c   **9.** b   **10.** a
11. f   **12.** d
13. (a)

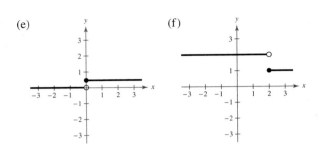

Vertical shrink

(b)

Vertical shrink and
reflection in the x-axis

(c)

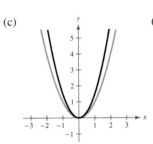

Vertical stretch

(d)

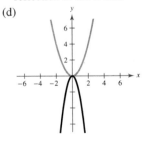

Vertical stretch and
reflection in the x-axis

15. (a)

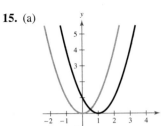

Horizontal shift one unit
to the right

(b)

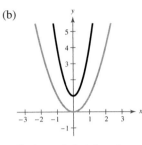

Horizontal shrink and
vertical shift one unit upward

(c)

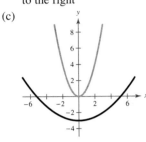

Horizontal stretch and
vertical shift three units
downward

(d)

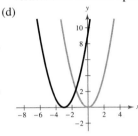

Horizontal shift three units
to the left

17.

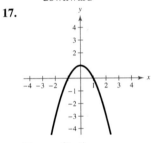

Vertex: $(0, 1)$
Axis of symmetry: y-axis
x-intercepts: $(-1, 0)$ $(1, 0)$

19.

Vertex: $(0, 7)$
Axis of symmetry: y-axis
No x-intercept

21.

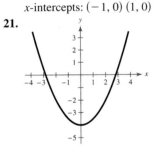

Vertex: $(0, -4)$
Axis of symmetry: y-axis
x-intercepts: $(\pm 2\sqrt{2}, 0)$

23.

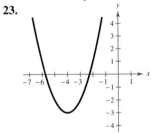

Vertex: $(-4, -3)$
Axis of symmetry: $x = -4$
x-intercepts: $(-4 \pm \sqrt{3}, 0)$

25.

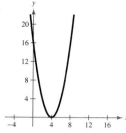

Vertex: $(4, 0)$
Axis of symmetry: $x = 4$
x-intercept: $(4, 0)$

27.

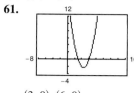

Vertex: $\left(\frac{1}{2}, 1\right)$
Axis of symmetry: $x = \frac{1}{2}$
No x-intercept

29.

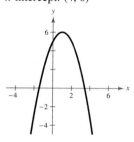

Vertex: $(1, 6)$
Axis of symmetry: $x = 1$
x-intercepts: $\left(1 \pm \sqrt{6}, 0\right)$

31.

Vertex: $\left(\frac{1}{2}, 20\right)$
Axis of symmetry: $x = \frac{1}{2}$
No x-intercept

33.

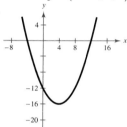

Vertex: $(4, -16)$
Axis of symmetry: $x = 4$
x-intercepts: $(-4, 0), (12, 0)$

35.

Vertex: $(-1, 4)$
Axis of symmetry: $x = -1$
x-intercepts: $(1, 0), (-3, 0)$

37.

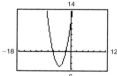

Vertex: $(-4, -5)$
Axis of symmetry: $x = -4$
x-intercepts: $\left(-4 \pm \sqrt{5}, 0\right)$

39.

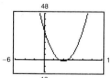

Vertex: $(4, -1)$
Axis of symmetry: $x = 4$
x-intercepts: $\left(4 \pm \frac{1}{2}\sqrt{2}, 0\right)$

41.

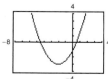

Vertex: $(-2, -3)$
Axis of symmetry: $x = -2$
x-intercepts: $\left(-2 \pm \sqrt{6}, 0\right)$

43. $y = -(x + 1)^2 + 4$ **45.** $y = -2(x + 2)^2 + 2$
47. $f(x) = (x + 2)^2 + 5$ **49.** $f(x) = 4(x - 1)^2 - 2$

51. $f(x) = \frac{3}{4}(x - 5)^2 + 12$ **53.** $f(x) = -\frac{24}{49}\left(x + \frac{1}{4}\right)^2 + \frac{3}{2}$
55. $f(x) = -\frac{16}{3}\left(x + \frac{5}{2}\right)^2$ **57.** $(5, 0), (-1, 0)$
59.

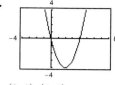

$(0, 0), (4, 0)$

61.

$(3, 0), (6, 0)$

63.

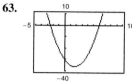

$\left(-\frac{5}{2}, 0\right), (6, 0)$

65. $f(x) = x^2 - 2x - 3$
$g(x) = -x^2 + 2x + 3$

67. $f(x) = x^2 - 10x$ **69.** $f(x) = 2x^2 + 7x + 3$
$g(x) = -x^2 + 10x$ $g(x) = -2x^2 - 7x - 3$
71. $55, 55$ **73.** $12, 6$ **75.** 16 ft **77.** 20 fixtures
79. (a) \$14,000,000; \$14,375,000; \$13,500,000
(b) \$24; \$14,400,000
Answers will vary.
81. (a) $A = \dfrac{8x(50 - x)}{3}$

(b)

x	5	10	15	20	25	30
a	600	$1066\frac{2}{3}$	1400	1600	$1666\frac{2}{3}$	1600

$x = 25$ ft, $y = 33\frac{1}{3}$ ft

(c)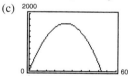

$x = 25$ ft, $y = 33\frac{1}{3}$ ft

(d) $A = -\frac{8}{3}(x - 25)^2 + \frac{5000}{3}$ (e) They are identical.
83. (a) $R = -100x^2 + 3500x$, $15 \le x \le 20$
(b) \$17.50; \$30,625
85. (a)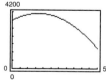

(b) 4075 cigarettes; Yes, the warning had an effect because the maximum consumption occurred in 1966.
(c) 7366 cigarettes per year; 20 cigarettes per day
87. True. The equation has no real solutions, so the graph has no x-intercepts.
89. True. The graph of a quadratic function with a negative leading coefficient will be a downward-opening parabola.
91. $b = \pm 20$ **93.** $b = \pm 8$
95. $f(x) = a\left(x + \dfrac{b}{2a}\right)^2 + \dfrac{4ac - b^2}{4a}$

CHAPTER 3

97. (a)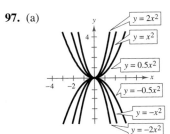

As $|a|$ increases, the parabola becomes narrower. For $a > 0$, the parabola opens upward. For $a < 0$, the parabola opens downward.

(b)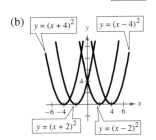

For $h < 0$, the vertex will be on the negative x-axis. For $h > 0$, the vertex will be on the positive x-axis. As $|h|$ increases, the parabola moves away from the origin.

(c)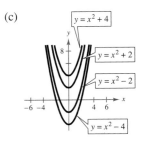

As $|k|$ increases, the vertex moves upward (for $k > 0$) or downward (for $k < 0$), away from the origin.

99. Yes. A graph of a quadratic equation whose vertex is on the x-axis has only one x-intercept.

Section 3.2   (page 279)

1. continuous   **3.** x^n

5. (a) solution; (b) $(x - a)$; (c) x-intercept   **7.** standard

9. c   **10.** g   **11.** h   **12.** f

13. a   **14.** e   **15.** d   **16.** b

17. (a)

(b)

(c)

(d)

19. (a) 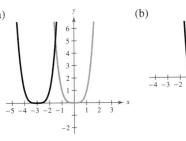   (b)

(c) 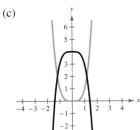   (d)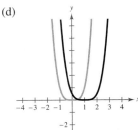

(e) 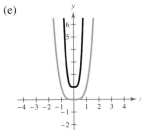   (f)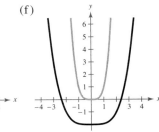

21. Falls to the left, rises to the right

23. Falls to the left, falls to the right

25. Rises to the left, falls to the right

27. Rises to the left, falls to the right

29. Falls to the left, falls to the right

31. 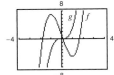   **33.**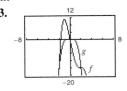

35. (a) ± 6

(b) Odd multiplicity; number of turning points: 1

(c)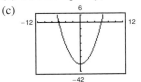

37. (a) 3

(b) Even multiplicity; number of turning points: 1

(c)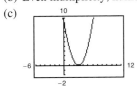

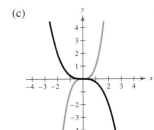

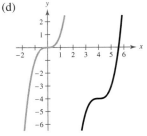

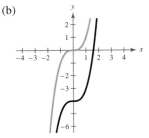

39. (a) $-2, 1$

(b) Odd multiplicity; number of turning points: 1

(c)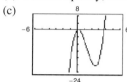

41. (a) $0, 2 \pm \sqrt{3}$

(b) Odd multiplicity; number of turning points: 2

(c)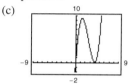

43. (a) $0, 4$

(b) 0, odd multiplicity; 4, even multiplicity; number of turning points: 2

(c)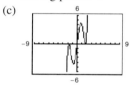

45. (a) $0, \pm\sqrt{3}$

(b) 0, odd multiplicity; $\pm\sqrt{3}$, even multiplicity; number of turning points: 4

(c)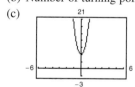

47. (a) No real zeros

(b) Number of turning points: 1

(c)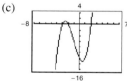

49. (a) $\pm 2, -3$

(b) Odd multiplicity; number of turning points: 2

(c)

51. (a)

(b) x-intercepts: $(0, 0), \left(\frac{5}{2}, 0\right)$ (c) $x = 0, \frac{5}{2}$

(d) The answers in part (c) match the x-intercepts.

53. (a)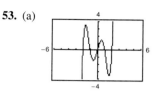

(b) x-intercepts: $(0, 0), (\pm 1, 0), (\pm 2, 0)$

(c) $x = 0, 1, -1, 2, -2$

(d) The answers in part (c) match the x-intercepts.

55. $f(x) = x^2 - 8x$ **57.** $f(x) = x^2 + 4x - 12$

59. $f(x) = x^3 + 9x^2 + 20x$

61. $f(x) = x^4 - 4x^3 - 9x^2 + 36x$ **63.** $f(x) = x^2 - 2x - 2$

65. $f(x) = x^2 + 6x + 9$ **67.** $f(x) = x^3 + 4x^2 - 5x$

69. $f(x) = x^3 - 3x$

71. $f(x) = x^4 + x^3 - 15x^2 + 23x - 10$

73. $f(x) = x^5 + 16x^4 + 96x^3 + 256x^2 + 256x$

75. (a) Falls to the left, rises to the right

(b) $0, 5, -5$ (c) Answers will vary.

(d)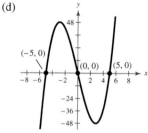

77. (a) Rises to the left, rises to the right

(b) No zeros   (c) Answers will vary.

(d)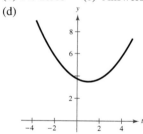

79. (a) Falls to the left, rises to the right

(b) $0, 2$ (c) Answers will vary.

(d)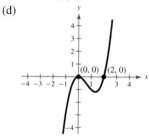

81. (a) Falls to the left, rises to the right

(b) $0, 2, 3$ (c) Answers will vary.

(d)

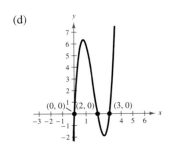

83. (a) Rises to the left, falls to the right
(b) −5, 0 (c) Answers will vary.
(d)

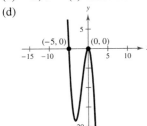

85. (a) Falls to the left, rises to the right
(b) 0, 4 (c) Answers will vary.
(d)

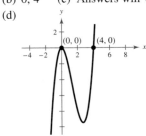

87. (a) Falls to the left, falls to the right
(b) ±2 (c) Answers will vary.
(d)

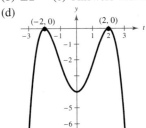

89.

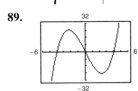

Zeros: 0, ±4,
odd multiplicity

91.

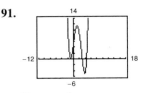

Zeros: −1,
even multiplicity;
3, $\frac{9}{2}$, odd multiplicity

93. $[-1, 0], [1, 2], [2, 3]$; about $-0.879, 1.347, 2.532$
95. $[-2, -1], [0, 1]$; about $-1.585, 0.779$
97. (a) $V(x) = x(36 - 2x)^2$ (b) Domain: $0 < x < 18$
(c)

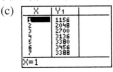

6 in. × 24 in. × 24 in.

(d)

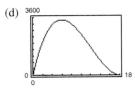

$x = 6$; The results are the same.
99. (a) $A = -2x^2 + 12x$ (b) $V = -384x^2 + 2304x$
(c) 0 in. $< x <$ 6 in.
(d)

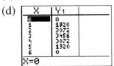

When $x = 3$, the volume is maximum at $V = 3456$;
dimensions of gutter are 3 in. × 6 in. × 3 in.
(e)

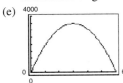

The maximum value is the same.
(f) No. Answers will vary.
101. (a)

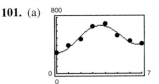

(b) The model fits the data well.
(c) Relative minima: $(0.21, 300.54), (6.62, 410.74)$
Relative maximum: $(3.62, 681.72)$
(d) Increasing: $(0.21, 3.62), (6.62, 7)$
Decreasing: $(0, 0.21), (3.62, 6.62)$
(e) Answers will vary.
103. (a) (b) $t \approx 15$

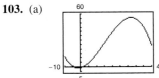

(c) Vertex: $(15.22, 2.54)$
(d) The results are approximately equal.
105. False. A fifth-degree polynomial can have at most four turning points.
107. True. The degree of the function is odd and its leading coefficient is negative, so the graph rises to the left and falls to the right.
109.

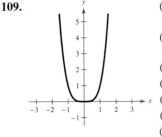

(a) Vertical shift of two units; Even
(b) Horizontal shift of two units; Neither
(c) Reflection in the y-axis; Even
(d) Reflection in the x-axis; Even
(e) Horizontal stretch; Even
(f) Vertical shrink; Even
(g) $g(x) = x^3, x \geq 0$; Neither
(h) $g(x) = x^{16}$; Even

111. (a)

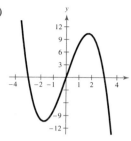

Zeros: 3
Relative minimum: 1
Relative maximum: 1
The number of zeros is the same as the degree, and the number of extrema is one less than the degree.

(b)

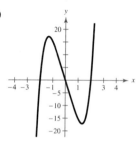

Zeros: 4
Relative minima: 2
Relative maximum: 1
The number of zeros is the same as the degree, and the number of extrema is one less than the degree.

(c)

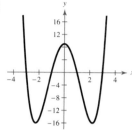

Zeros: 3
Relative minimum: 1
Relative maximum: 1
The number of zeros and the number of extrema are both less than the degree.

Section 3.3   *(page 290)*

1. $f(x)$: dividend; $d(x)$: divisor; $q(x)$: quotient; $r(x)$: remainder
3. improper   **5.** Factor   **7.** Answers will vary.
9. (a) and (b)   (c) Answers will vary.

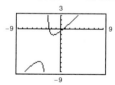

11. $2x + 4$, $x \neq -3$ **13.** $x^2 - 3x + 1$, $x \neq -\frac{5}{4}$
15. $x^3 + 3x^2 - 1$, $x \neq -2$ **17.** $x^2 + 3x + 9$, $x \neq 3$
19. $7 - \dfrac{11}{x + 2}$ **21.** $x - \dfrac{x + 9}{x^2 + 1}$ **23.** $2x - 8 + \dfrac{x - 1}{x^2 + 1}$
25. $x + 3 + \dfrac{6x^2 - 8x + 3}{(x - 1)^3}$ **27.** $3x^2 - 2x + 5$, $x \neq 5$
29. $6x^2 + 25x + 74 + \dfrac{248}{x - 3}$ **31.** $4x^2 - 9$, $x \neq -2$
33. $-x^2 + 10x - 25$, $x \neq -10$
35. $5x^2 + 14x + 56 + \dfrac{232}{x - 4}$
37. $10x^3 + 10x^2 + 60x + 360 + \dfrac{1360}{x - 6}$
39. $x^2 - 8x + 64$, $x \neq -8$
41. $-3x^3 - 6x^2 - 12x - 24 - \dfrac{48}{x - 2}$

43. $-x^3 - 6x^2 - 36x - 36 - \dfrac{216}{x - 6}$
45. $4x^2 + 14x - 30$, $x \neq -\frac{1}{2}$
47. $f(x) = (x - 4)(x^2 + 3x - 2) + 3$, $f(4) = 3$
49. $f(x) = \left(x + \frac{2}{3}\right)(15x^3 - 6x + 4) + \frac{34}{3}$, $f\left(-\frac{2}{3}\right) = \frac{34}{3}$
51. $f(x) = \left(x - \sqrt{2}\right)\left[x^2 + \left(3 + \sqrt{2}\right)x + 3\sqrt{2}\right] - 8$, $f\left(\sqrt{2}\right) = -8$
53. $f(x) = \left(x - 1 + \sqrt{3}\right)\left[-4x^2 + \left(2 + 4\sqrt{3}\right)x + \left(2 + 2\sqrt{3}\right)\right]$, $f\left(1 - \sqrt{3}\right) = 0$
55. (a) -2 (b) 1   (c) $-\frac{1}{4}$ (d) 5
57. (a) -35 (b) -22 (c) -10 (d) -211
59. $(x - 2)(x + 3)(x - 1)$; Solutions: 2, -3, 1
61. $(2x - 1)(x - 5)(x - 2)$; Solutions: $\frac{1}{2}$, 5, 2
63. $\left(x + \sqrt{3}\right)\left(x - \sqrt{3}\right)(x + 2)$; Solutions: $-\sqrt{3}$, $\sqrt{3}$, -2
65. $(x - 1)\left(x - 1 - \sqrt{3}\right)\left(x - 1 + \sqrt{3}\right)$; Solutions: 1, $1 + \sqrt{3}$, $1 - \sqrt{3}$
67. (a) Answers will vary.   (b) $2x - 1$
    (c) $f(x) = (2x - 1)(x + 2)(x - 1)$
    (d) $\frac{1}{2}$, -2, 1   (e)

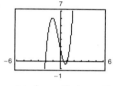

69. (a) Answers will vary.   (b) $(x - 1)$, $(x - 2)$
    (c) $f(x) = (x - 1)(x - 2)(x - 5)(x + 4)$
    (d) 1, 2, 5, -4 (e)

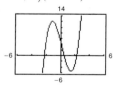

71. (a) Answers will vary.   (b) $x + 7$
    (c) $f(x) = (x + 7)(2x + 1)(3x - 2)$
    (d) -7, $-\frac{1}{2}$, $\frac{2}{3}$ (e)

73. (a) Answers will vary.   (b) $x - \sqrt{5}$
    (c) $f(x) = \left(x - \sqrt{5}\right)\left(x + \sqrt{5}\right)(2x - 1)$
    (d) $\pm\sqrt{5}$, $\frac{1}{2}$ (e)

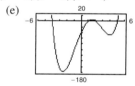

75. (a) Zeros are 2 and about ± 2.236.
    (b) $x = 2$ (c) $f(x) = (x - 2)\left(x - \sqrt{5}\right)\left(x + \sqrt{5}\right)$
77. (a) Zeros are -2, about 0.268, and about 3.732.
    (b) $t = -2$
    (c) $h(t) = (t + 2)\left[t - \left(2 + \sqrt{3}\right)\right]\left[t - \left(2 - \sqrt{3}\right)\right]$
79. (a) Zeros are 0, 3, 4, and about ± 1.414.
    (b) $x = 0$ (c) $h(x) = x(x - 4)(x - 3)\left(x + \sqrt{2}\right)\left(x - \sqrt{2}\right)$

CHAPTER 3

81. $2x^2 - x - 1$, $x \neq \frac{3}{2}$ **83.** $x^2 + 3x$, $x \neq -2, -1$

85. (a) and (b)

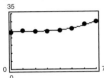

$A = 0.0349t^3 - 0.168t^2 + 0.42t + 23.4$

(c)

t	0	1	2	3
$A(t)$	23.4	23.7	23.8	24.1

t	4	5	6	7
$A(t)$	24.6	25.7	27.4	30.1

(d) \$45.7 billion; No, because the model will approach infinity quickly.

87. False. $-\frac{4}{7}$ is a zero of f.

89. True. The degree of the numerator is greater than the degree of the denominator.

91. $x^{2n} + 6x^n + 9$, $x^n \neq -3$ **93.** The remainder is 0.

95. $c = -210$ **97.** $k = 7$

99. (a) $x + 1$, $x \neq 1$ (b) $x^2 + x + 1$, $x \neq 1$
(c) $x^3 + x^2 + x + 1$, $x \neq 1$

In general, $\dfrac{x^n - 1}{x - 1} = x^{n-1} + x^{n-2} + \cdots + x + 1$, $x \neq 1$

Section 3.4   *(page 303)*

1. Fundamental Theorem of Algebra   **3.** Rational Zero

5. linear; quadratic; quadratic   **7.** Descartes's Rule of Signs

9. $0, 6$ **11.** $2, -4$ **13.** $-6, \pm i$ **15.** $\pm 1, \pm 2$

17. $\pm 1, \pm 3, \pm 5, \pm 9, \pm 15, \pm 45, \pm\frac{1}{2}, \pm\frac{3}{2}, \pm\frac{5}{2}, \pm\frac{9}{2}, \pm\frac{15}{2}, \pm\frac{45}{2}$

19. $1, 2, 3$ **21.** $1, -1, 4$ **23.** $-6, -1$

25. $\frac{1}{2}, -1$ **27.** $-2, 3, \pm\frac{2}{3}$ **29.** $-2, 1$ **31.** $-4, \frac{1}{2}, 1, 1$

33. (a) $\pm 1, \pm 2, \pm 4$
(b)

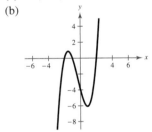

(c) $-2, -1, 2$

35. (a) $\pm 1, \pm 3, \pm\frac{1}{2}, \pm\frac{3}{2}, \pm\frac{1}{4}, \pm\frac{3}{4}$
(b)

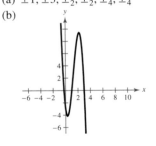

(c) $-\frac{1}{4}, 1, 3$

37. (a) $\pm 1, \pm 2, \pm 4, \pm 8, \pm\frac{1}{2}$
(b)

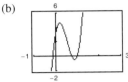

(c) $-\frac{1}{2}, 1, 2, 4$

39. (a) $\pm 1, \pm 3, \pm\frac{1}{2}, \pm\frac{3}{2}, \pm\frac{1}{4}, \pm\frac{3}{4}, \pm\frac{1}{8}, \pm\frac{3}{8}, \pm\frac{1}{16}, \pm\frac{3}{16}, \pm\frac{1}{32}, \pm\frac{3}{32}$
(b)

(c) $1, \frac{3}{4}, -\frac{1}{8}$

41. (a) ± 1, about ± 1.414 (b) $\pm 1, \pm\sqrt{2}$
(c) $f(x) = (x + 1)(x - 1)(x + \sqrt{2})(x - \sqrt{2})$

43. (a) $0, 3, 4$, about ± 1.414 (b) $0, 3, 4, \pm\sqrt{2}$
(c) $h(x) = x(x - 3)(x - 4)(x + \sqrt{2})(x - \sqrt{2})$

45. $x^3 - x^2 + 25x - 25$ **47.** $x^3 - 12x^2 + 46x - 52$

49. $3x^4 - 17x^3 + 25x^2 + 23x - 22$

51. (a) $(x^2 + 9)(x^2 - 3)$ (b) $(x^2 + 9)(x + \sqrt{3})(x - \sqrt{3})$
(c) $(x + 3i)(x - 3i)(x + \sqrt{3})(x - \sqrt{3})$

53. (a) $(x^2 - 2x - 2)(x^2 - 2x + 3)$
(b) $(x - 1 + \sqrt{3})(x - 1 - \sqrt{3})(x^2 - 2x + 3)$
(c) $(x - 1 + \sqrt{3})(x - 1 - \sqrt{3})(x - 1 + \sqrt{2}i)$
$(x - 1 - \sqrt{2}i)$

55. $\pm 2i, 1$ **57.** $\pm 5i, -\frac{1}{2}, 1$ **59.** $-3 \pm i, \frac{1}{4}$

61. $2, -3 \pm \sqrt{2}i, 1$ **63.** $\pm 6i$; $(x + 6i)(x - 6i)$

65. $1 \pm 4i$; $(x - 1 - 4i)(x - 1 + 4i)$

67. $\pm 2, \pm 2i$; $(x - 2)(x + 2)(x - 2i)(x + 2i)$

69. $1 \pm i$; $(z - 1 + i)(z - 1 - i)$

71. $-1, 2 \pm i$; $(x + 1)(x - 2 + i)(x - 2 - i)$

73. $-2, 1 \pm \sqrt{2}i$; $(x + 2)(x - 1 + \sqrt{2}i)(x - 1 - \sqrt{2}i)$

75. $-\frac{1}{5}, 1 \pm \sqrt{5}i$; $(5x + 1)(x - 1 + \sqrt{5}i)(x - 1 - \sqrt{5}i)$

77. $2, \pm 2i$; $(x - 2)^2(x + 2i)(x - 2i)$

79. $\pm i, \pm 3i$; $(x + i)(x - i)(x + 3i)(x - 3i)$

81. $-10, -7 \pm 5i$ **83.** $-\frac{3}{4}, 1 \pm \frac{1}{2}i$ **85.** $-2, -\frac{1}{2}, \pm i$

87. One positive zero   **89.** One negative zero

91. One positive zero, one negative zero

93. One or three positive zeros

95–97. Answers will vary.   **99.** $1, -\frac{1}{2}$ **101.** $-\frac{3}{4}$

103. $\pm 2, \pm\frac{3}{2}$ **105.** $\pm 1, \frac{1}{4}$ **107.** d   **108.** a

109. b   **110.** c

111. (a)

(b) $V(x) = x(9 - 2x)(15 - 2x)$
Domain: $0 < x < \frac{9}{2}$

(c)

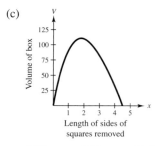

1.82 cm × 5.36 cm × 11.36 cm

(d) $\frac{1}{2}, \frac{7}{2}, 8$; 8 is not in the domain of V.

113. $x \approx 38.4$, or \$384,000

115. (a) $V(x) = x^3 + 9x^2 + 26x + 24 = 120$

(b) 4 ft × 5 ft × 6 ft

117. $x \approx 40$, or 4000 units

119. No. Setting $p = 9,000,000$ and solving the resulting equation yields imaginary roots.

121. False. The most complex zeros it can have is two, and the Linear Factorization Theorem guarantees that there are three linear factors, so one zero must be real.

123. r_1, r_2, r_3 **125.** $5 + r_1, 5 + r_2, 5 + r_3$

127. The zeros cannot be determined.

129. Answers will vary. There are infinitely many possible functions for f. Sample equation and graph:

$f(x) = -2x^3 + 3x^2 + 11x - 6$

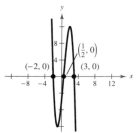

131. Answers will vary. Sample graph:

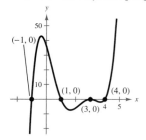

133. $f(x) = x^4 + 5x^2 + 4$ **135.** $f(x) = x^3 - 3x^2 + 4x - 2$

137. (a) $-2, 1, 4$

(b) The graph touches the x-axis at $x = 1$.

(c) The least possible degree of the function is 4, because there are at least four real zeros (1 is repeated) and a function can have at most the number of real zeros equal to the degree of the function. The degree cannot be odd by the definition of multiplicity.

(d) Positive. From the information in the table, it can be concluded that the graph will eventually rise to the left and rise to the right.

(e) $f(x) = x^4 - 4x^3 - 3x^2 + 14x - 8$

(f)

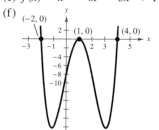

139. (a) Not correct because f has $(0, 0)$ as an intercept.

(b) Not correct because the function must be at least a fourth-degree polynomial.

(c) Correct function

(d) Not correct because k has $(-1, 0)$ as an intercept.

Section 3.5   *(page 314)*

1. variation; regression   **3.** least squares regression

5. directly proportional   **7.** directly proportional

9. combined

11.

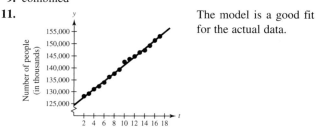

The model is a good fit for the actual data.

13.

$y = \frac{1}{4}x + 3$ $y = -\frac{1}{2}x + 3$

15.

17. (a) and (b)

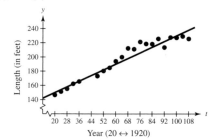

$y \approx t + 130$

(c) $y = 1.01t + 130.82$ (d) The models are similar.

(e) Part (b): 242 ft; Part (c): 243.94 ft

(f) Answers will vary.

19. (a)

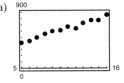

(b) $S = 38.3t + 224$

(c)

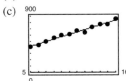

The model is a good fit.

(d) 2007: \$875.1 million; 2009: \$951.7 million

(e) Each year the annual gross ticket sales for Broadway shows in New York City increase by \$38.3 million.

21. Inversely

23.

x	2	4	6	8	10
$y = kx^2$	4	16	36	64	100

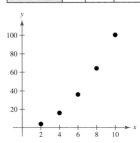

25.

x	2	4	6	8	10
$y = kx^2$	2	8	18	32	50

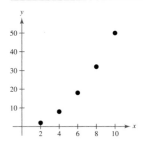

27.

x	2	4	6	8	10
$y = k/x^2$	$\frac{1}{2}$	$\frac{1}{8}$	$\frac{1}{18}$	$\frac{1}{32}$	$\frac{1}{50}$

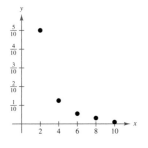

29.

x	2	4	6	8	10
$y = k/x^2$	$\frac{5}{2}$	$\frac{5}{8}$	$\frac{5}{18}$	$\frac{5}{32}$	$\frac{1}{10}$

31. $y = \dfrac{5}{x}$ **33.** $y = -\dfrac{7}{10}x$ **35.** $y = \dfrac{12}{5}x$

37. $y = 205x$ **39.** $I = 0.035P$

41. Model: $y = \frac{33}{13}x$; 25.4 cm, 50.8 cm

43. $y = 0.0368x$; \$8280   **45.** (a) 0.05 m   (b) $176\frac{2}{3}$ N

47. 39.47 lb   **49.** $A = kr^2$ **51.** $y = \dfrac{k}{x^2}$

53. $F = \dfrac{kg}{r^2}$ **55.** $P = \dfrac{k}{V}$ **57.** $F = \dfrac{km_1m_2}{r^2}$

59. The area of a triangle is jointly proportional to its base and height.

61. The area of an equilateral triangle varies directly as the square of one of its sides.

63. The volume of a sphere varies directly as the cube of its radius.

65. Average speed is directly proportional to the distance and inversely proportional to the time.

67. $A = \pi r^2$ **69.** $y = \dfrac{28}{x}$ **71.** $F = 14rs^3$

73. $z = \dfrac{2x^2}{3y}$ **75.** About 0.61 mi/h   **77.** 506 ft

79. 1470 J   **81.** The velocity is increased by one-third.

83. (a)

(b) Yes. $k_1 = 4200$, $k_2 = 3800$, $k_3 = 4200$, $k_4 = 4800$, $k_5 = 4500$

(c) $C = \dfrac{4300}{d}$

(d)

(e) About 1433 m

85. (a)

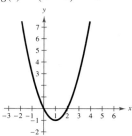

(b) $0.2857 \ \mu W/cm^2$

87. False. E is jointly proportional to the mass of an object and the square of its velocity.

89. (a) Good approximation   (b) Poor approximation
(c) Poor approximation   (d) Good approximation

91. As one variable increases, the other variable will also increase.

93. (a) y will change by a factor of one-fourth.
(b) y will change by a factor of four.

Review Exercises   *(page 322)*

1. (a)

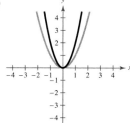

Vertical stretch

(b)

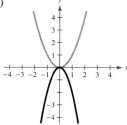

Vertical stretch and reflection in the x-axis

(c)

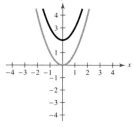

Vertical translation

(d)

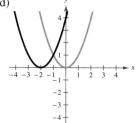

Horizontal translation

3. $g(x) = (x - 1)^2 - 1$

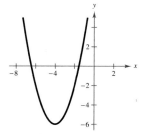

Vertex: $(1, -1)$
Axis of symmetry: $x = 1$
x-intercepts: $(0, 0), (2, 0)$

5. $f(x) = (x + 4)^2 - 6$

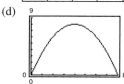

Vertex: $(-4, -6)$
Axis of symmetry: $x = -4$
x-intercepts: $\left(-4 \pm \sqrt{6}, 0\right)$

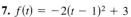

7. $f(t) = -2(t - 1)^2 + 3$

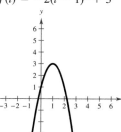

Vertex: $(1, 3)$
Axis of symmetry: $t = 1$
t-intercepts: $\left(1 \pm \dfrac{\sqrt{6}}{2}, 0\right)$

9. $h(x) = 4\left(x + \frac{1}{2}\right)^2 + 12$

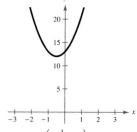

Vertex: $\left(-\frac{1}{2}, 12\right)$
Axis of symmetry: $x = -\frac{1}{2}$
No x-intercept

11. $h(x) = \left(x + \frac{5}{2}\right)^2 - \frac{41}{4}$

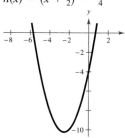

Vertex: $\left(-\frac{5}{2}, -\frac{41}{4}\right)$
Axis of symmetry: $x = -\frac{5}{2}$
x-intercepts: $\left(\dfrac{\pm\sqrt{41} - 5}{2}, 0\right)$

13. $f(x) = \frac{1}{3}\left(x + \frac{5}{2}\right)^2 - \frac{41}{12}$

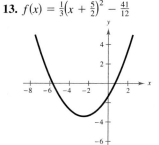

Vertex: $\left(-\frac{5}{2}, -\frac{41}{12}\right)$
Axis of symmetry: $x = -\frac{5}{2}$
x-intercepts: $\left(\dfrac{\pm\sqrt{41} - 5}{2}, 0\right)$

15. $f(x) = -\frac{1}{2}(x - 4)^2 + 1$ **17.** $f(x) = (x - 1)^2 - 4$

19. $y = -\frac{11}{36}\left(x + \frac{3}{2}\right)^2$

21. (a) $A = x\left(\dfrac{8 - x}{2}\right)$ (b) $0 < x < 8$

(c)

x	1	2	3	4	5	6
A	$\frac{7}{2}$	6	$\frac{15}{2}$	8	$\frac{15}{2}$	6

(d)

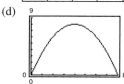

$x = 4, y = 2$

(e) $A = -\frac{1}{2}(x - 4)^2 + 8; \ x = 4, \ y = 2$

23. (a) $12,000; $13,750; $15,000

(b) Maximum revenue at $40; $16,000; Any price greater or less than $40 per unit will not yield as much revenue.

25. 1091 units

CHAPTER 3

27. 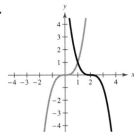 **29.**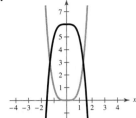

31.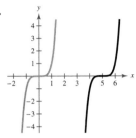

33. Falls to the left, falls to the right
35. Rises to the left, rises to the right
37. $-8, \frac{4}{3}$, odd multiplicity; turning points: 1
39. $0, \pm\sqrt{3}$, odd multiplicity; turning points: 2
41. $\frac{2}{3}$, odd multiplicity; 0, even multiplicity; turning points: 2
43. (a) Rises to the left, falls to the right (b) -1
  (c) Answers will vary.
  (d)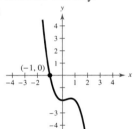

45. (a) Rises to the left, rises to the right (b) $-3, 0, 1$
  (c) Answers will vary.
  (d)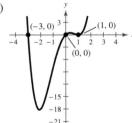

47. (a) $[-1, 0]$ (b) About -0.900
49. (a) $[-1, 0], [1, 2]$ (b) About -0.200, about 1.772
51. $6x + 3 + \dfrac{17}{5x - 3}$ **53.** $5x + 4, \; x \neq \dfrac{5}{2} \pm \dfrac{\sqrt{29}}{2}$
55. $x^2 - 3x + 2 - \dfrac{1}{x^2 + 2}$
57. $6x^3 + 8x^2 - 11x - 4 - \dfrac{8}{x - 2}$
59. $2x^2 - 9x - 6, \; x \neq 8$
61. (a) Yes (b) Yes (c) Yes (d) No

63. (a) -421 (b) -9
65. (a) Answers will vary.
  (b) $(x + 7), (x + 1)$
  (c) $f(x) = (x + 7)(x + 1)(x - 4)$
  (d) $-7, -1, 4$
  (e)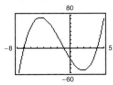

67. (a) Answers will vary. (b) $(x + 1), (x - 4)$
  (c) $f(x) = (x + 1)(x - 4)(x + 2)(x - 3)$
  (d) $-2, -1, 3, 4$
  (e)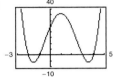

69. $0, 3$ **71.** $2, 9$ **73.** $-4, 6, \pm 2i$
75. $\pm 1, \pm 3, \pm 5, \pm 15, \pm\frac{1}{2}, \pm\frac{3}{2}, \pm\frac{5}{2}, \pm\frac{15}{2}, \pm\frac{1}{4}, \pm\frac{3}{4}, \pm\frac{5}{4}, \pm\frac{15}{4}$
77. $-6, -2, 5$ **79.** $1, 8$ **81.** $-4, 3$
83. $f(x) = 3x^4 - 14x^3 + 17x^2 - 42x + 24$
85. $4, \pm i$ **87.** $-3, \frac{1}{2}, 2 \pm i$
89. $0, 1, -5; f(x) = x(x - 1)(x + 5)$
91. $-4, 2 \pm 3i; g(x) = (x + 4)^2(x - 2 - 3i)(x - 2 + 3i)$
93. (a) 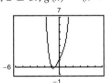 **95.** (a)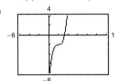
  (b) Two zeros (b) One zero
  (c) $-1, -0.54$ (c) 3.26
97. Two or no positive zeros, one negative zero
99. Answers will vary.
101. (a) 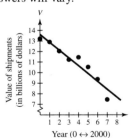 (b) The model fits the data well.

103. Model: $y = \frac{8}{5}x$; 3.2 km, 16 km **105.** A factor of 4
107. About 2 h, 26 min
109. False. A fourth-degree polynomial can have at most four zeros, and complex zeros occur in conjugate pairs.
111. Find the vertex of the quadratic function and write the function in standard form. If the leading coefficient is positive, the vertex is a minimum. If the leading coefficient is negative, the vertex is a maximum.

Chapter Test   *(page 326)*

1. (a) Reflection in the x-axis followed by a vertical translation
   (b) Horizontal translation
2. Vertex: $(-2, -1)$; Intercepts: $(0, 3), (-3, 0), (-1, 0)$
3. $y = (x - 3)^2 - 6$
4. (a) 50 ft
   (b) 5. Yes, changing the constant term results in a vertical translation of the graph and therefore changes the maximum height.
5. Rises to the left, falls to the right

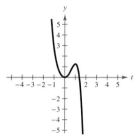

6. $3x + \dfrac{x - 1}{x^2 + 1}$ 7. $2x^3 + 4x^2 + 3x + 6 + \dfrac{9}{x - 2}$
8. $(2x - 5)(x - \sqrt{3})(x + \sqrt{3})$;
   Zeros: $\pm\sqrt{3}, \frac{5}{2}$
9. $-2, \frac{3}{2}$ 10. $\pm 1, -\frac{2}{3}$
11. $f(x) = x^4 - 7x^3 + 17x^2 - 15x$
12. $f(x) = x^4 - 6x^3 + 16x^2 - 24x + 16$
13. $1, -5, -\frac{2}{3}$ 14. $-2, 4, -1 \pm \sqrt{2}i$
15. $v = 6\sqrt{s}$ 16. $A = \dfrac{25}{6}xy$ 17. $b = \dfrac{48}{a}$
18. $S = 385t + 115$; This model is a fairly good fit.

Problem Solving   *(page 329)*

1. (a) (i) $6, -2$ (ii) $0, -5$ (iii) $-5, 2$
   (iv) 2 (v) $1 \pm \sqrt{7}$ (vi) $\dfrac{-3 \pm \sqrt{7}i}{2}$
   (b)
   (i) 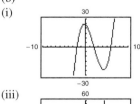   (ii)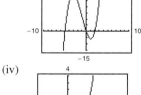
   (iii)   (iv)
   (v)   (vi)

   Graph (iii) touches the x-axis at $(2, 0)$, and all the other graphs pass through the x-axis at $(2, 0)$.
   (c) (i) $(6, 0), (-2, 0)$ (iv) No other x-intercepts
   (ii) $(0, 0), (-5, 0)$ (v) $(-1.6, 0), (3.6, 0)$
   (iii) $(-5, 0)$ (vi) No other x-intercepts

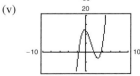

(d) When the function has two real zeros, the results are the same. When the function has one real zero, the graph touches the x-axis at the zero. When there are no real zeros, there is no x-intercept.
3. Answers will vary.   5. 2 in. $\times$ 2 in. $\times$ 5 in.
7. False. The statement would be true if $f(-1) = 2$.
9. (a) $m_1 = 5$; less than   (b) $m_2 = 3$; greater than
   (c) $m_3 = 4.1$; less than   (d) $m_h = h + 4$
   (e) $m_h = 3, 5, 4.1$; The values are the same.
   (f) $m_{\tan} = 4$ because $h = 0$.

11. (a) $A(x) = \dfrac{x^2}{16} + \dfrac{1}{\pi}\left(\dfrac{100 - x}{2}\right)^2$ (b) $0 \le x \le 100$
    (c) Maximum area at $x = 0$; Minimum area at $x = 56$
    (d) Answers will vary.

Chapter 4

Section 4.1   *(page 337)*

1. rational functions   3. horizontal asymptote
5. (a) Domain: all real numbers x except $x = 1$
   (b)

x	$f(x)$
0.5	-2
0.9	-10
0.99	-100
0.999	-1000

x	$f(x)$
1.5	2
1.1	10
1.01	100
1.001	1000

x	$f(x)$
-1.5	-0.4
-1.1	-0.48
-1.01	-0.498
-1.001	-0.4998

x	$f(x)$
-0.5	-0.67
-0.9	-0.53
-0.99	-0.503
-0.999	-0.5003

(c) $f(x) \to -\infty$ as $x \to 1^-$, $f(x) \to \infty$ as $x \to 1^+$
7. (a) Domain: all real numbers x except $x = \pm 1$
   (b)

x	$f(x)$
0.5	-1
0.9	-12.79
0.99	-147.8
0.999	-1498

x	$f(x)$
1.5	5.4
1.1	17.29
1.01	152.3
1.001	1502

x	$f(x)$
-1.5	5.4
-1.1	17.29
-1.01	152.3
-1.001	1502

x	$f(x)$
-0.5	-1
-0.9	-12.79
-0.99	-147.8
-0.999	-1498

(c) $f(x) \to \infty$ as $x \to -1^-$ and as $x \to 1^+$, $f(x) \to -\infty$ as $x \to -1^+$ and as $x \to 1^-$

CHAPTER 4

9. Domain: all real numbers x except $x = 0$
Vertical asymptote: $x = 0$
Horizontal asymptote: $y = 0$

11. Domain: all real numbers x except $x = 5$
Vertical asymptote: $x = 5$
Horizontal asymptote: $y = -1$

13. Domain: all real numbers x except $x = \pm 1$
Vertical asymptotes: $x = \pm 1$

15. Domain: all real numbers x
Horizontal asymptote: $y = 3$

17. d   **18.** a   **19.** c   **20.** b

21. 1   **23.** None   **25.** 6   **27.** 2

29. The domain is all real numbers x except $x = \pm 4$. There is a vertical asymptote at $x = -4$, and a horizontal asymptote at $y = 0$.

31. The domain is all real numbers x except $x = -1, 3$. There is a vertical asymptote at $x = 3$, and a horizontal asymptote at $y = 1$.

33. The domain is all real numbers x except $x = -1, \frac{1}{2}$. There is a vertical asymptote at $x = \frac{1}{2}$, and a horizontal asymptote at $y = \frac{1}{2}$.

35. The domain is all real numbers x except $x = \frac{2}{3}, 2$. There is a vertical asymptote at $x = 2$, and a horizontal asymptote at $y = 2$.

37. (a) Domain of f: all real numbers x except $x = -2$
Domain of g: all real numbers x
(b) $x - 2$; Vertical asymptote: none
(c)

x	-4	-3	-2.5	-2	-1.5	-1	0
$f(x)$	-6	-5	-4.5	Undef.	-3.5	-3	-2
$g(x)$	-6	-5	-4.5	-4	-3.5	-3	-2

(d) The functions differ only at $x = -2$, where f is undefined.

39. (a) Domain of f: all real numbers x except $x = 0, \frac{1}{2}$
Domain of g: all real numbers x except $x = 0$
(b) $\dfrac{1}{x}$; Vertical asymptote: $x = 0$
(c)

x	-1	-0.5	0	0.5	2	3	4
$f(x)$	-1	-2	Undef.	Undef.	$\frac{1}{2}$	$\frac{1}{3}$	$\frac{1}{4}$
$g(x)$	-1	-2	Undef.	2	$\frac{1}{2}$	$\frac{1}{3}$	$\frac{1}{4}$

(d) The functions differ only at $x = 0.5$, where f is undefined.

41. (a)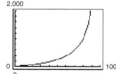
(b) $28.33 million; $170 million; $765 million
(c) No. The function is undefined at $p = 100$.

43. (a)

M	200	400	600	800	1000
t	0.472	0.596	0.710	0.817	0.916

M	1200	1400	1600	1800	2000
t	1.009	1.096	1.178	1.255	1.328

(b) The model is a good fit for the experimental times.
(c) $M \approx 1306$ g

45. (a)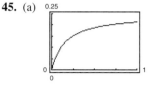
(b) About 0.247 mg

47. False. Polynomials do not have vertical asymptotes.

49. (a) 4   (b) Less than   (c) Greater than

51. (a) 2   (b) Greater than   (c) Less than

53. Sample answer: $f(x) = \dfrac{2x^2}{x^2 + 1}$

55. Answers will vary.
Sample answers: $f(x) = \dfrac{1}{x^2 + 15}; f(x) = \dfrac{1}{x - 15}$

Section 4.2   (page 345)

1. slant asymptote

3. 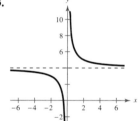   **5.**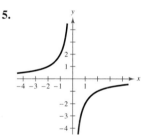

7. 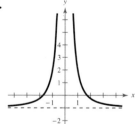   **9.**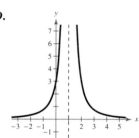

11. 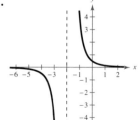   **13.**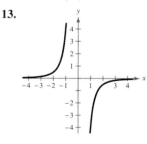

15. (a) The domain is all real numbers x except $x = -2$.

(b) y-intercept: $\left(0, \frac{1}{2}\right)$

(c) Vertical asymptote: $x = -2$

Horizontal asymptote: $y = 0$

(d)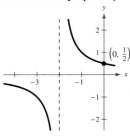

17. (a) The domain is all real numbers x except $x = -4$.

(b) y-intercept: $\left(0, -\frac{1}{4}\right)$

(c) Vertical asymptote: $x = -4$

Horizontal asymptote: $y = 0$

(d)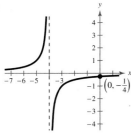

19. (a) The domain is all real numbers x except $x = -2$.

(b) x-intercept: $\left(-\frac{7}{2}, 0\right)$

y-intercept: $\left(0, \frac{7}{2}\right)$

(c) Vertical asymptote: $x = -2$

Horizontal asymptote: $y = 2$

(d)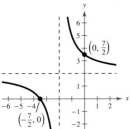

21. (a) The domain is all real numbers x except $x = -2$.

(b) x-intercept: $\left(-\frac{5}{2}, 0\right)$

y-intercept: $\left(0, \frac{5}{2}\right)$

(c) Vertical asymptote: $x = -2$

Horizontal asymptote: $y = 2$

(d)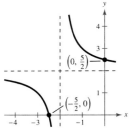

23. (a) The domain is all real numbers x.

(b) Intercept: $(0, 0)$

(c) Horizontal asymptote: $y = 1$

(d)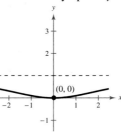

25. (a) The domain is all real numbers x except $x = \pm 3$.

(b) Intercept: $(0, 0)$

(c) Vertical asymptotes: $x = \pm 3$

Horizontal asymptote: $y = 1$

(d)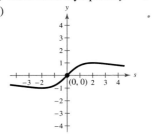

27. (a) The domain is all real numbers s.

(b) Intercept: $(0, 0)$

(c) Horizontal asymptote: $y = 0$

(d)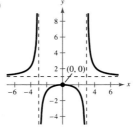

29. (a) The domain is all real numbers x except $x = 0, 4$.

(b) x-intercept: $(-1, 0)$

(c) Vertical asymptotes: $x = 0$, $x = 4$

Horizontal asymptote: $y = 0$

(d)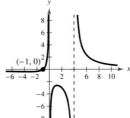

31. (a) The domain is all real numbers x except $x = 4, -1$.

(b) Intercept: $(0, 0)$

(c) Vertical asymptotes: $x = -1$, $x = 4$

Horizontal asymptote: $y = 0$

CHAPTER 4

(d)

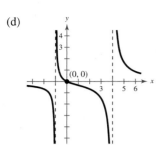

33. (a) The domain is all real numbers x except $x = \pm 2$.
(b) x-intercepts: $(1, 0)$ and $(4, 0)$
   y-intercept: $(0, -1)$
(c) Vertical asymptotes: $x = \pm 2$
   Horizontal asymptote: $y = 1$
(d)

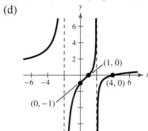

35. (a) The domain is all real numbers x except $x = -2, 7$.
(b) Intercept: $(0, 0)$
(c) Vertical asymptotes: $x = -2$, $x = 7$
   Horizontal asymptote: $y = 0$
(d)

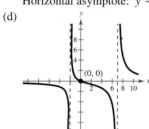

37. (a) The domain is all real numbers x except $x = \pm 1, 2$.
(b) x-intercepts: $(3, 0)$, $\left(-\frac{1}{2}, 0\right)$
   y-intercept: $\left(0, -\frac{3}{2}\right)$
(c) Vertical asymptotes: $x = 2$, $x = \pm 1$
   Horizontal asymptote: $y = 0$
(d)

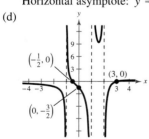

39. (a) The domain is all real numbers x except $x = 2, -3$.
(b) Intercept: $(0, 0)$
(c) Vertical asymptote: $x = 2$
   Horizontal asymptote: $y = 1$

(d)

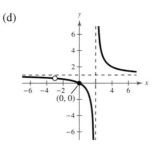

41. (a) The domain is all real numbers x except $x = -\frac{3}{2}, 2$.
(b) x-intercept: $\left(\frac{1}{2}, 0\right)$
   y-intercept: $\left(0, -\frac{1}{3}\right)$
(c) Vertical asymptote: $x = -\frac{3}{2}$
   Horizontal asymptote: $y = 1$
(d)

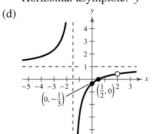

43. (a) The domain is all real numbers t except $t = 1$.
(b) t-intercept: $(-1, 0)$
   y-intercept: $(0, 1)$
(c) No asymptotes
(d)

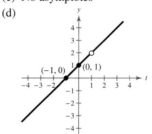

45. (a) Domain of f: all real numbers x except $x = -1$
   Domain of g: all real numbers x
(b) $x - 1$; Vertical asymptote: none
(c)

x	-3	-2	-1.5	-1	-0.5	0	1
$f(x)$	-4	-3	-2.5	Undef.	-1.5	-1	0
$g(x)$	-4	-3	-2.5	-2	-1.5	-1	0

(d)

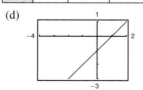

(e) Because there are only a finite number of pixels, the graphing utility may not attempt to evaluate the function where it does not exist.

47. (a) Domain of f: all real numbers x except $x = 0, 2$
Domain of g: all real numbers x except $x = 0$

(b) $\dfrac{1}{x}$; Vertical asymptote: $x = 0$

(c)

x	-0.5	0	0.5	1	1.5	2	3
$f(x)$	-2	Undef.	2	1	$\frac{2}{3}$	Undef.	$\frac{1}{3}$
$g(x)$	-2	Undef.	2	1	$\frac{2}{3}$	$\frac{1}{2}$	$\frac{1}{3}$

(d)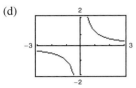

(e) Because there are only a finite number of pixels, the graphing utility may not attempt to evaluate the function where it does not exist.

49. (a) Domain: all real numbers x except $x = 0$

(b) x-intercepts: $(\pm 3, 0)$

(c) Vertical asymptote: $x = 0$
Slant asymptote: $y = x$

(d)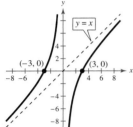

51. (a) Domain: all real numbers x except $x = 0$

(b) No intercepts

(c) Vertical asymptote: $x = 0$
Slant asymptote: $y = 2x$

(d)

53. (a) Domain: all real numbers x except $x = 0$

(b) No intercepts

(c) Vertical asymptote: $x = 0$
Slant asymptote: $y = x$

(d)

55. (a) Domain: all real numbers t except $t = -5$

(b) y-intercept: $\left(0, -\frac{1}{5}\right)$

(c) Vertical asymptote: $t = -5$
Slant asymptote: $y = -t + 5$

(d)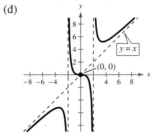

57. (a) Domain: all real numbers x except $x = \pm 2$

(b) Intercept: $(0, 0)$

(c) Vertical asymptotes: $x = \pm 2$
Slant asymptote: $y = x$

(d)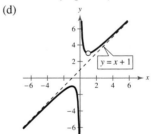

59. (a) Domain: all real numbers x except $x = 0, 1$

(b) No intercepts

(c) Vertical asymptote: $x = 0$
Slant asymptote: $y = x + 1$

(d)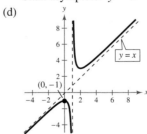

61. (a) Domain: all real numbers x except $x = 1$

(b) y-intercept: $(0, -1)$

(c) Vertical asymptote: $x = 1$
Slant asymptote: $y = x$

(d)

63. (a) Domain: all real numbers x except $x = -1, -2$

(b) y-intercept: $\left(0, \frac{1}{2}\right)$
x-intercepts: $\left(\frac{1}{2}, 0\right), (1, 0)$

CHAPTER 4

(c) Vertical asymptote: $x = -2$
Slant asymptote: $y = 2x - 7$

(d)

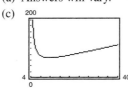

65.

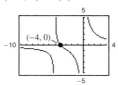

Domain: all real numbers x except $x = -3$
Vertical asymptote: $x = -3$
Slant asymptote: $y = x + 2$
$y = x + 2$

67.

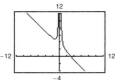

Domain: all real numbers x except $x = 0$
Vertical asymptote: $x = 0$
Slant asymptote: $y = -x + 3$
$y = -x + 3$

69. (a) $(-1, 0)$ (b) -1 **71.** (a) $(1, 0), (-1, 0)$ (b) ± 1

73. (a) $(-4, 0)$

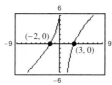

(b) -4

75. (a) $(3, 0), (-2, 0)$

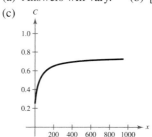

(b) $3, -2$

77. (a) Answers will vary. (b) $[0, 950]$

(c)

(d) Increases more slowly; 0.75

79. (a) Answers will vary.   (b) $(4, \infty)$

(c)

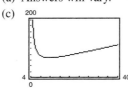

11.75 in. $\times$ 5.87 in.

81.

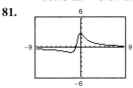

Minimum: $(-2, -1)$
Maximum: $(0, 3)$

83.

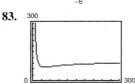

$x \approx 40.45$, or 4045 components

85. (a) Answers will vary.
(b) Vertical asymptote: $x = 25$
Horizontal asymptote: $y = 25$

(c)

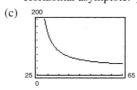

(d)

x	30	35	40	45	50	55	60
y	150	87.5	66.7	56.3	50	45.8	42.9

(e) Sample answer: No. You might expect the average speed for the round trip to be the average of the average speeds for the two parts of the trip.
(f) No. At 20 miles per hour you would use more time in one direction than is required for the round trip at an average speed of 50 miles per hour.

87. False. There are two distinct branches of the graph.

89. False. The degree of the numerator is 2 more than the degree of the denominator.

91.

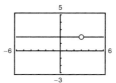

The fraction is not reduced.

93. Sample answer: If the degree of the numerator is exactly one more than the degree of the denominator, the graph of the function has a slant asymptote. To find the equation of a slant asymptote, use long division to expand the function.

Section 4.3 *(page 357)*

1. conic or conic section   **3.** parabola; directrix; focus
5. axis   **7.** major axis; center

9. hyberbola; foci   **11.** Not shown
12. c   **13.** e   **14.** a   **15.** Not shown   **16.** h
17. f   **18.** b   **19.** d   **20.** g
21. Vertex: $(0, 0)$
  Focus: $\left(0, \frac{1}{2}\right)$
23. Vertex: $(0, 0)$
  Focus: $\left(-\frac{3}{2}, 0\right)$

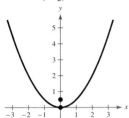

25. Vertex: $(0, 0)$
  Focus: $(0, -3)$

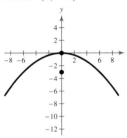

27. $y^2 = -8x$ **29.** $x^2 = 2y$ **31.** $x^2 = -4y$
33. $y^2 = 4x$ **35.** $y^2 = 9x$ **37.** $x^2 = 16y$
39. $x^2 = \frac{3}{2}y$; Focus: $\left(0, \frac{3}{8}\right)$ **41.** $y^2 = \frac{9}{5}x$; Focus: $\left(\frac{9}{20}, 0\right)$
43. $y^2 = 6x$

45. (a)

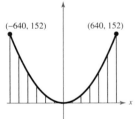

  (b) $y = \dfrac{19x^2}{51,200}$

(c)

Distance, x	0	200	400	500	600
Height, y	0	$14\frac{27}{32}$	$59\frac{3}{8}$	$92\frac{99}{128}$	$133\frac{19}{32}$

47. Center: $(0, 0)$
  Vertices: $(\pm 5, 0)$
49. Center: $(0, 0)$
  Vertices: $\left(\pm\frac{5}{3}, 0\right)$

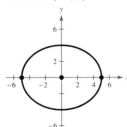
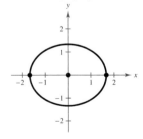

51. Center: $(0, 0)$
  Vertices: $(\pm 6, 0)$
53. Center: $(0, 0)$
  Vertices: $(0, \pm 1)$

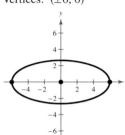

55. Center: $(0, 0)$
  Vertices: $(0, \pm 9)$

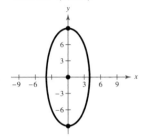

57. $\dfrac{x^2}{1} + \dfrac{y^2}{4} = 1$ **59.** $\dfrac{x^2}{4} + \dfrac{y^2}{9/4} = 1$ **61.** $\dfrac{x^2}{25} + \dfrac{y^2}{21} = 1$
63. $\dfrac{x^2}{49} + \dfrac{y^2}{24} = 1$ **65.** $\dfrac{21x^2}{400} + \dfrac{y^2}{25} = 1$
67. $(\pm\sqrt{5}, 0)$; Length of string: 6 ft
69. (a)

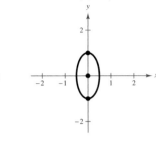

  (b) $y = \frac{3}{4}\sqrt{400 - x^2}$

(c) Yes, with clearance of 0.52 foot.

71.

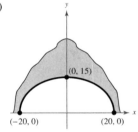

73.

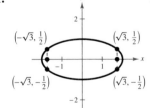

75. Center: $(0, 0)$
  Vertices: $(\pm 1, 0)$
77. Center: $(0, 0)$
  Vertices: $(0, \pm 1)$

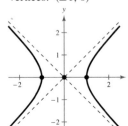
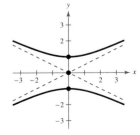

CHAPTER 4

79. Center: $(0, 0)$
Vertices: $(0, \pm 7)$

81. Center: $(0, 0)$
Vertices: $\left(0, \pm \frac{1}{2}\right)$

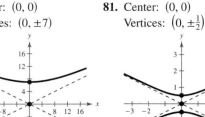

83. Center: $(0, 0)$
Vertices: $(0, \pm 6)$

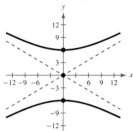

85. $\dfrac{y^2}{4} - \dfrac{x^2}{32} = 1$　**87.** $\dfrac{x^2}{1} - \dfrac{y^2}{9} = 1$

89. $\dfrac{17y^2}{1024} - \dfrac{17x^2}{64} = 1$　**91.** $\dfrac{y^2}{9} - \dfrac{x^2}{9/4} = 1$

93. (a) $\dfrac{x^2}{1} - \dfrac{y^2}{169/3} = 1$　(b) 2.403 ft　**95.** 10 mi

97. False. The equation represents a hyperbola:
$$\dfrac{x^2}{144} - \dfrac{y^2}{144} = 1.$$

99. False. If the graph crossed the directrix, there would exist points nearer the directrix than the focus.

101. (a) $A = \pi a(20 - a)$　(b) $\dfrac{x^2}{196} + \dfrac{y^2}{36} = 1$

(c)

a	8	9	10	11	12	13
A	301.6	311.0	314.2	311.0	301.6	285.9

$a = 10$, circle

(d)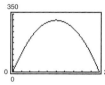

$a = 10$, circle

103. An ellipse is a circle if $a = b$.

105.

 Two intersecting lines

107–111. Answers will vary.

Section 4.4 *(page 367)*

1. b　**2.** d　**3.** e　**4.** c　**5.** a　**6.** f

7. Center: $(-2, 1)$, horizontal shift two units to the left and vertical shift one unit upward

9. Center: $(1, -3)$, horizontal shift one unit to the right and vertical shift three units downward

11. Center: $(-4, -2)$, horizontal shift four units to the left and vertical shift two units downward

13. Center: $(0, 0)$
Radius: 7

15. Center: $(4, 5)$
Radius: 6

17. Center: $(1, 0)$
Radius: $\sqrt{10}$

19. $(x - 1)^2 + (y + 3)^2 = 1$
Center: $(1, -3)$
Radius: 1

21. $(x - 4)^2 + y^2 = 16$
Center: $(4, 0)$
Radius: 4

23. $\left(x + \frac{3}{2}\right)^2 + (y - 3)^2 = 1$
Center: $\left(-\frac{3}{2}, 3\right)$
Radius: 1

25. Vertex: $(1, -2)$
Focus: $(1, -4)$
Directrix: $y = 0$

27. Vertex: $\left(5, -\frac{1}{2}\right)$
Focus: $\left(\frac{11}{2}, -\frac{1}{2}\right)$
Directrix: $x = \frac{9}{2}$

29. Vertex: $(1, 1)$
Focus: $(1, 2)$
Directrix: $y = 0$

31. Vertex: $(-2, -3)$
Focus: $(-4, -3)$
Directrix: $x = 0$

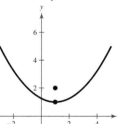

33. $(y - 2)^2 = -8(x - 3)$　**35.** $x^2 = 8(y - 4)$

37. $(y - 4)^2 = 16x$　**39.** 34,295 ft

41. (a) $S = -0.355t^2 + 4.33t + 0.7$

(b) $(6.099, 13.903)$; the maximum sales occurred in 2006

(c)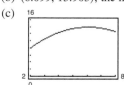

(d) 2006　(e) 2006

(f) Results are the same.

43. Center: $(1, 5)$
Foci: $(1, 9), (1, 1)$
Vertices: $(1, 10), (1, 0)$

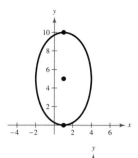

45. Center: $(-2, -4)$
Foci: $\left(\dfrac{-4 \pm \sqrt{3}}{2}, -4 \right)$
Vertices:
$(-3, -4), (-1, -4)$

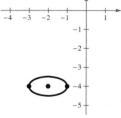

47. Center: $(2, 1)$
Foci: $\left(\dfrac{14}{5}, 1 \right), \left(\dfrac{6}{5}, 1 \right)$
Vertices: $(1, 1), (3, 1)$

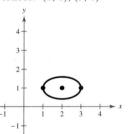

49. Center: $(-2, 3)$
Foci: $\left(-2, 3 \pm \sqrt{5} \right)$
Vertices: $(-2, 6), (-2, 0)$

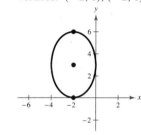

51. $\dfrac{(x - 3)^2}{1} + \dfrac{y^2}{9} = 1$ **53.** $\dfrac{(x - 2)^2}{16} + \dfrac{y^2}{12} = 1$

55. $\dfrac{x^2}{16} + \dfrac{(y - 4)^2}{12} = 1$ **57.** $\dfrac{(x - 2)^2}{4} + \dfrac{(y - 2)^2}{1} = 1$

59. $\dfrac{x^2}{25} + \dfrac{y^2}{16} = 1$ **61.** 2,756,170,000 mi; 4,583,830,000 mi

63. Center: $(2, -1)$
Foci: $(7, -1), (-3, -1)$
Vertices: $(6, -1), (-2, -1)$

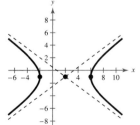

65. Center: $(2, -6)$
Foci: $\left(2, -6 \pm \sqrt{2} \right)$
Vertices: $(2, -5), (2, -7)$

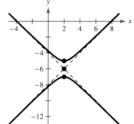

67. Center: $(-1, -3)$
Foci: $\left(-1 \pm \dfrac{5\sqrt{2}}{3}, -3 \right)$
Vertices: $\left(-1 \pm \sqrt{5}, -3 \right)$

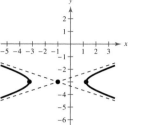

69. Center: $(2, -3)$
Foci: $\left(2 \pm \sqrt{10}, -3 \right)$
Vertices: $(3, -3), (1, -3)$

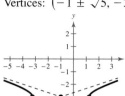

71. $\dfrac{(y - 1)^2}{1} - \dfrac{x^2}{3} = 1$ **73.** $\dfrac{(x - 4)^2}{4} - \dfrac{y^2}{12} = 1$

75. $\dfrac{y^2}{9} - \dfrac{4(x - 2)^2}{9} = 1$ **77.** $\dfrac{(x - 3)^2}{9} - \dfrac{(y - 2)^2}{4} = 1$

79. $(x - 3)^2 + (y + 2)^2 = 4$ **81.** $\dfrac{(y + 2)^2}{4} - \dfrac{x^2}{4} = 1$

Circle

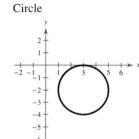

Hyperbola

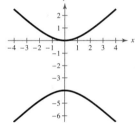

83. $\left(y + \dfrac{1}{4} \right)^2 = -8 \left(x - \dfrac{1}{16} \right)$ **85.** $\dfrac{(x + 2)^2}{16} + \dfrac{(y + 4)^2}{9} = 1$

Parabola

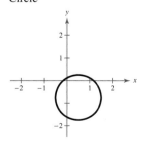

Ellipse

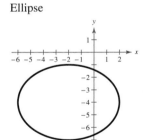

87. $\left(x - \dfrac{1}{2} \right)^2 + \left(y + \dfrac{3}{4} \right)^2 = 1$

Circle

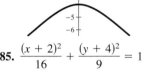

89. True. The conic is an ellipse.

CHAPTER 4

91. (a) Answers will vary.

(b)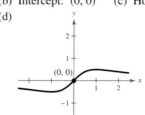

As e approaches 0, the ellipse becomes a circle.

Review Exercises *(page 372)*

1. Domain: all real numbers x except $x = -10$

3. Domain: all real numbers x except $x = 6, 4$

5. Vertical asymptote: $x = -3$
Horizontal asymptote: $y = 0$

7. Vertical asymptotes: $x = \pm 2$
Horizontal asymptote: $y = 1$

9. Vertical asymptote: $x = 6$
Horizontal asymptote: $y = 0$

11. $0.50 is the horizontal asymptote of the function.

13. (a) Domain: all real numbers x except $x = 0$

(b) No intercepts

(c) Vertical asymptote: $x = 0$
Horizontal asymptote: $y = 0$

(d)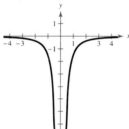

15. (a) Domain: all real numbers x except $x = 1$

(b) x-intercept: $(-2, 0)$
y-intercept: $(0, 2)$

(c) Vertical asymptote: $x = 1$
Horizontal asymptote: $y = -1$

(d)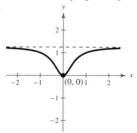

17. (a) Domain: all real numbers x (b) Intercept: $(0, 0)$

(c) Horizontal asymptote: $y = \frac{5}{4}$

(d)

19. (a) Domain: all real numbers x

(b) Intercept: $(0, 0)$ (c) Horizontal asymptote: $y = 0$

(d)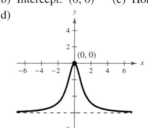

21. (a) Domain: all real numbers x

(b) Intercept: $(0, 0)$ (c) Horizontal asymptote: $y = -6$

(d)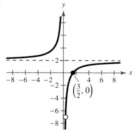

23. (a) Domain: all real numbers x except $x = 0, \frac{1}{3}$

(b) x-intercept: $\left(\frac{3}{2}, 0\right)$

(c) Vertical asymptote: $x = 0$
Horizontal asymptote: $y = 2$

(d)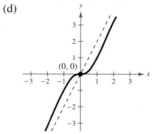

25. (a) Domain: all real numbers x

(b) Intercept: $(0, 0)$ (c) Slant asymptote: $y = 2x$

(d)

27. (a) Domain: all real numbers x except $x = -2$

(b) x-intercepts: $(2, 0), (-5, 0)$
y-intercept: $(0, -5)$

(c) Vertical asymptote: $x = -2$
Slant asymptote: $y = x + 1$

(d)

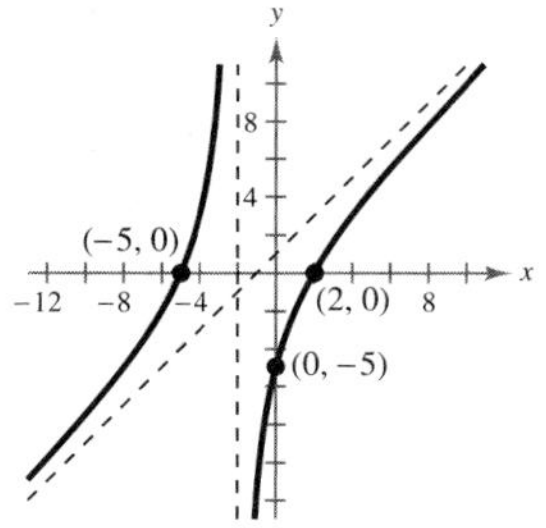

29. (a) Domain: all real numbers x except $x = \frac{4}{3}, -1$

(b) x-intercepts: $\left(\frac{2}{3}, 0\right), (1, 0)$

y-intercept: $\left(0, -\frac{1}{2}\right)$

(c) Vertical asymptote: $x = \frac{4}{3}$

Slant asymptote: $y = x - \frac{1}{3}$

(d)

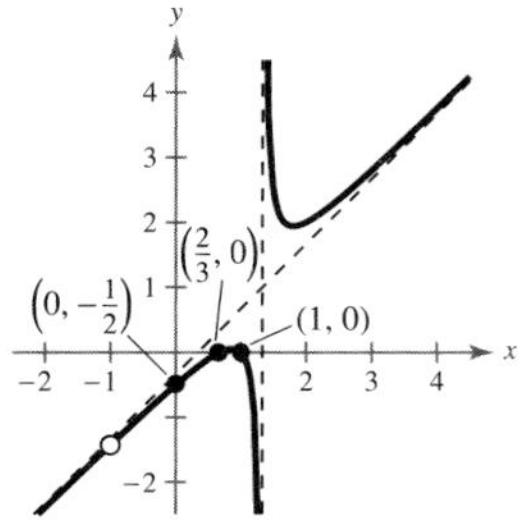

31. (a)

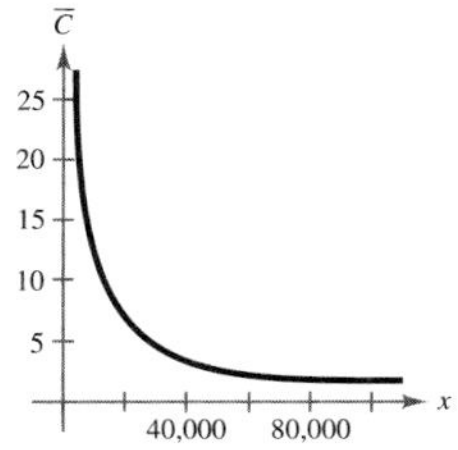

(b) \$100.90, \$10.90, \$1.90

(c) \$0.90 is the horizontal asymptote of the function.

33.

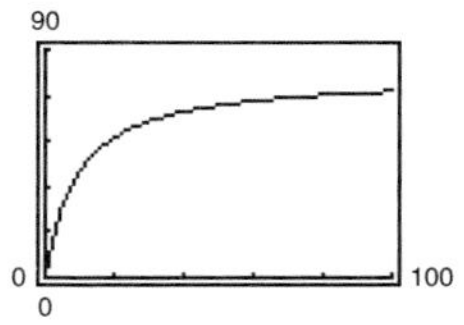

80.3 milligrams per square decimeter per hour

35. Parabola **37.** Hyperbola **39.** Parabola

41. Hyperbola **43.** $y^2 = 12x$ **45.** $y^2 = -24x$

47. $x^2 = 12y$ **49.** (0, 50)

51. $\frac{x^2}{81} + \frac{y^2}{9} = 1$ **53.** $\frac{2x^2}{9} + \frac{y^2}{36} = 1$

55. $\frac{x^2}{221} + \frac{y^2}{25} = 1$

57. The foci should be placed 3 feet on either side of the center and have the same height as the pillars.

59. $\frac{y^2}{1} - \frac{x^2}{24} = 1$ **61.** $\frac{x^2}{1} - \frac{y^2}{4} = 1$

63. $(x + 8)^2 = 28(y - 8)$ **65.** $(x - 4)^2 = -8(y - 2)$

67. $\frac{(x - 6)^2}{36} + \frac{(y - 3)^2}{9} = 1$ **69.** $\frac{(x - 2)^2}{25} + \frac{y^2}{21} = 1$

71. $\frac{x^2}{36} - \frac{(y - 7)^2}{9} = 1$ **73.** $\frac{(x + 2)^2}{64} - \frac{(y - 3)^2}{36} = 1$

75. $\frac{5(x - 4)^2}{16} - \frac{5y^2}{64} = 1$

77. $(x - 3)^2 = -2y$

Parabola

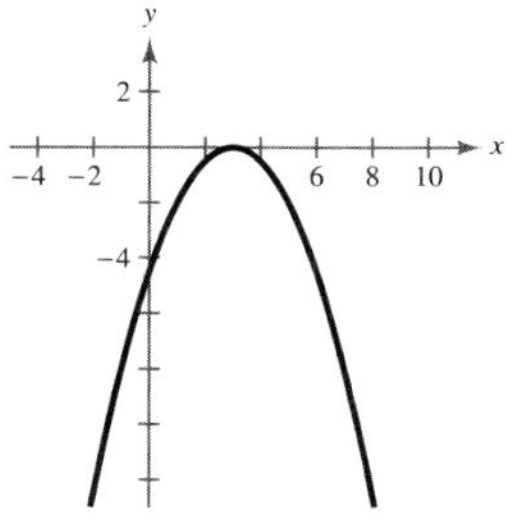

Shifted three units to the right from the origin

79. (1, 2)

Degenerate conic (a point)

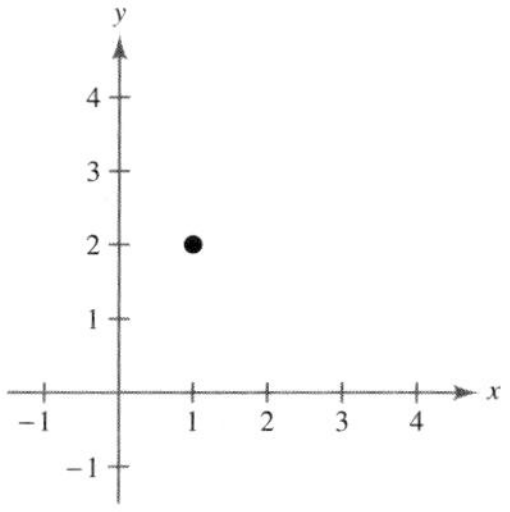

Shifted one unit to the right and two units upward from the origin.

81. $\frac{(x + 5)^2}{9} + \frac{(y - 1)^2}{1} = 1$

Ellipse

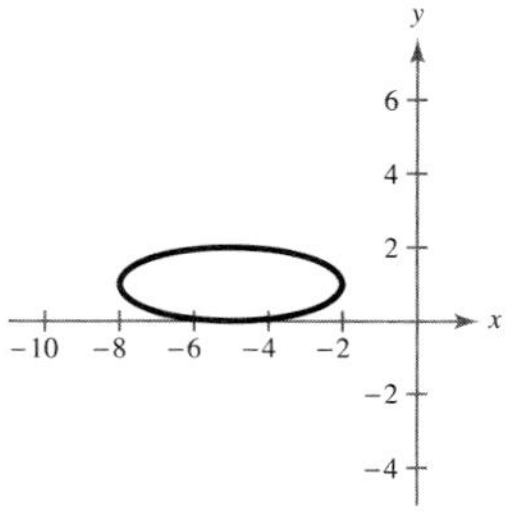

Shifted five units to the left and one unit upward from the origin.

83. $\frac{(x - 4)^2}{1} - \frac{(y - 4)^2}{9} = 1$

Hyperbola

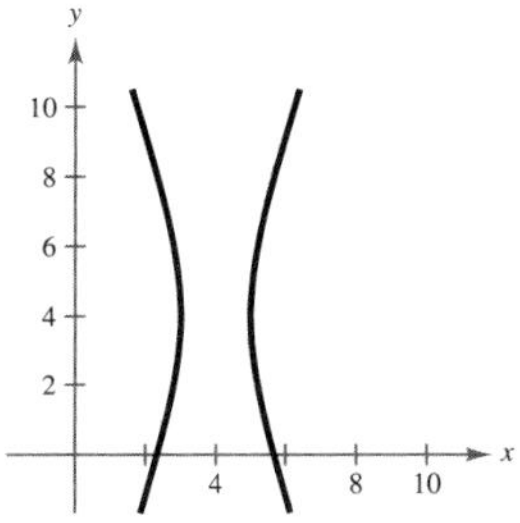

Shifted four units to the right and four units upward from the origin.

85. $8\sqrt{6}$ m

87. (a) Circle

(b) $(x - 100)^2 + y^2 = 62{,}500$

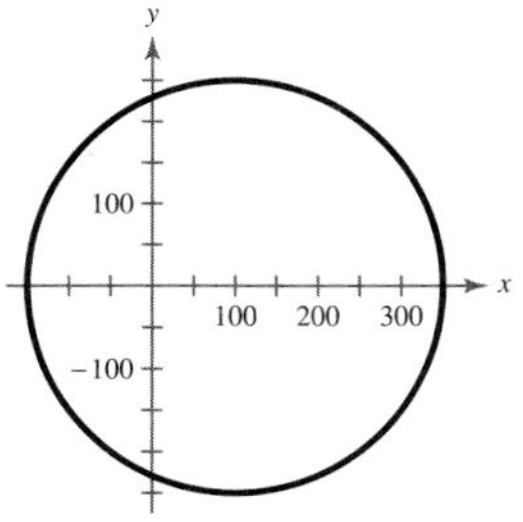

(c) Approximately 180.28 m

89. True. See Exercise 79.

Chapter Test *(page 375)*

1. Domain: all real numbers x except $x = -1$

Vertical asymptote: $x = -1$

Horizontal asymptote: $y = 3$

2. Domain: all real numbers x

Vertical asymptote: none

Horizontal asymptote: $y = -1$

3. Domain: all real numbers x except $x = 3$

No asymptotes

4. x-intercepts: $(-2, 0)$, $(2, 0)$
Vertical asymptote: $x = 0$
Horizontal asymptote: $y = -1$

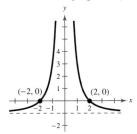

5. y-intercept: $(0, -2)$
Vertical asymptote: $x = 1$
Slant asymptote: $y = x + 1$

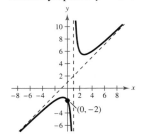

6. x-intercept: $(-1, 0)$
y-intercept: $\left(0, -\frac{1}{12}\right)$
Vertical asymptotes: $x = 3$, $x = -4$
Horizontal asymptote: $y = 0$

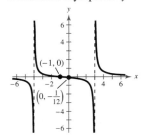

7. x-intercept: $\left(-\frac{3}{2}, 0\right)$
y-intercept: $\left(0, \frac{3}{4}\right)$
Vertical asymptote: $x = -4$
Horizontal asymptote: $y = 2$

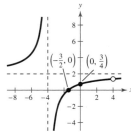

8. y-intercept: $(0, 1)$
Horizontal asymptote: $y = \frac{2}{5}$

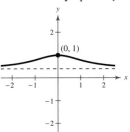

9. x-intercept: $\left(-\frac{1}{2}, 0\right)$
y-intercept: $(0, -2)$
Vertical asymptote: $x = -1$
Slant asymptote: $y = 2x - 5$

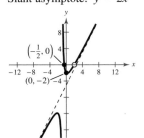

10. 6.24 in. $\times$ 12.49 in.

11. (a) Answers will vary.

(b) $A = \dfrac{x^2}{2(x - 2)}$, $x > 2$

(c)

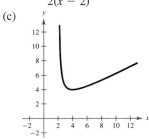

$A = 4$

12.

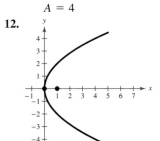

Vertex: $(0, 0)$
Focus: $(1, 0)$

13.

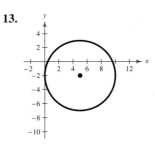

Center: $(5, -2)$

14.

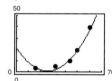

Vertex: $(5, -3)$
Focus: $\left(5, -\frac{5}{2}\right)$

15.

Vertices: $(\pm 1, 0)$
Foci: $\left(\pm\sqrt{5}, 0\right)$

16.

Vertices: $(0, \pm 2)$
Foci: $\left(0, \pm\sqrt{5}\right)$

17.

Center: $(1, -6)$
Vertices: $(4, -6), (-2, -6)$
Foci: $\left(1 \pm \sqrt{6}, -6\right)$

18. $\dfrac{(x-4)^2}{16} + \dfrac{(y-2)^2}{4} = 1$

19. $\dfrac{y^2}{9} - \dfrac{x^2}{4} = 1$

20. 34 m

21. Smallest distance: About 363,292 km
Greatest distance: About 405,508 km

Problem Solving   *(page 377)*

1. (a) iii   (b) ii   (c) iv   (d) i

3. (a) $y_1 \approx 0.031x^2 - 1.59x + 21.0$

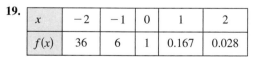

(b) $y_2 \approx \dfrac{1}{-0.007x + 0.44}$

(c) The models are a good fit for the original data.

(d) $y_1(25) = 0.625$; $y_2(25) = 3.774$
The rational model is the better fit for the original data.

(e) The reciprocal model should not be used to predict the near point for a person who is 70 years old because a negative value is obtained. The quadratic model is a better fit.

5. Answers will vary.   **7.** $y^2 = 6x$; About 2.04 in.

9. (a) $\dfrac{x_1}{2p}$

(b) $x^2 = 4y$
Tangent lines:
$y = x - 1$
$y = -x - 1$

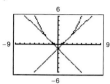

$x^2 = 8y$
Tangent lines:
$y = x - 2$
$y = -x - 2$

$x^2 = 12y$
Tangent lines:
$y = x - 3$
$y = -x - 3$

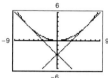

$x^2 = 16y$
Tangent lines:
$y = x - 4$
$y = -x - 4$

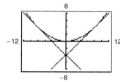

11. Proof

Chapter 5

Section 5.1   *(page 388)*

1. algebraic   **3.** One-to-One   **5.** $A = P\left(1 + \dfrac{r}{n}\right)^{nt}$

7. 0.863   **9.** 0.006   **11.** 1767.767

13. d   **14.** c   **15.** a   **16.** b

17.

x	-2	-1	0	1	2
$f(x)$	4	2	1	0.5	0.25

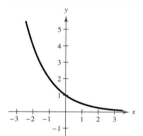

19.

x	-2	-1	0	1	2
$f(x)$	36	6	1	0.167	0.028

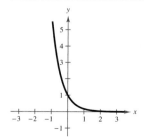

21.

x	−2	−1	0	1	2
f(x)	0.125	0.25	0.5	1	2

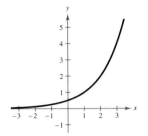

23. Shift the graph of *f* one unit upward.
25. Reflect the graph of *f* in the *x*-axis and shift three units upward.
27. Reflect the graph of *f* in the origin.

29. 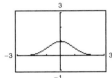   **31.**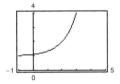

33. 0.472   **35.** 3.857×10^{-22} **37.** 7166.647

39.

x	−2	−1	0	1	2
f(x)	0.135	0.368	1	2.718	7.389

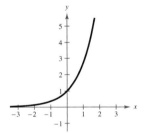

41.

x	−8	−7	−6	−5	−4
f(x)	0.055	0.149	0.406	1.104	3

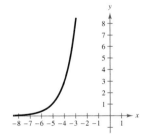

43.

x	−2	−1	0	1	2
f(x)	4.037	4.100	4.271	4.736	6

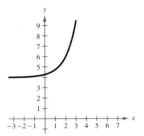

45. 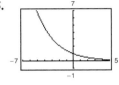   **47.**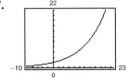

49.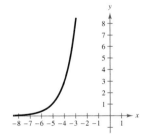

51. x = 2   **53.** x = −5   **55.** x = ⅓   **57.** x = 3, −1

59.

n	1	2	4	12
A	$1828.49	$1830.29	$1831.19	$1831.80

n	365	Continuous
A	$1832.09	$1832.10

61.

n	1	2	4	12
A	$5477.81	$5520.10	$5541.79	$5556.46

n	365	Continuous
A	$5563.61	$5563.85

63.

t	10	20	30
A	$17,901.90	$26,706.49	$39,841.40

t	40	50
A	$59,436.39	$88,668.67

65.

t	10	20	30
A	$22,986.49	$44,031.56	$84,344.25

t	40	50
A	$161,564.86	$309,484.08

67. $104,710.29   **69.** $35.45

71. (a)

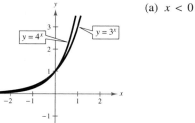

(b)

t	15	16	17	18	19	20
P (in millions)	40.19	40.59	40.99	41.39	41.80	42.21

t	21	22	23	24	25	26
P (in millions)	42.62	43.04	43.47	43.90	44.33	44.77

t	27	28	29	30
P (in millions)	45.21	45.65	46.10	46.56

(c) 2038

73. (a) 16 g   (b) 1.85 g

(c)

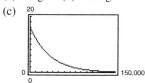

75. (a) $V(t) = 30,500\left(\frac{7}{8}\right)^t$ (b) $17,878.54

77. True. As $x \to -\infty$, $f(x) \to -2$ but never reaches -2.

79. $f(x) = h(x)$ **81.** $f(x) = g(x) = h(x)$

83.   (a) $x < 0$ (b) $x > 0$

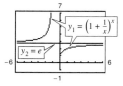

85.

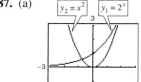

As the x-value increases, y_1 approaches the value of e.

87. (a)   (b)

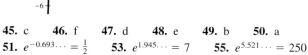

In both viewing windows, the constant raised to a variable power increases more rapidly than the variable raised to a constant power.

89. (a) $A = $5466.09   (b) $A = $5466.35
(c) $A = $5466.36   (d) $A = $5466.38
No. Answers will vary.

Section 5.2   *(page 398)*

1. logarithmic   **3.** natural; e **5.** $x = y$ **7.** $4^2 = 16$

9. $9^{-2} = \frac{1}{81}$ **11.** $32^{2/5} = 4$ **13.** $64^{1/2} = 8$

15. $\log_5 125 = 3$ **17.** $\log_{81} 3 = \frac{1}{4}$ **19.** $\log_6 \frac{1}{36} = -2$

21. $\log_{24} 1 = 0$ **23.** 6   **25.** 0   **27.** 2

29. -0.058 **31.** 1.097   **33.** 7   **35.** 1

37.

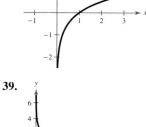

Domain: $(0, \infty)$
x-intercept: $(1, 0)$
Vertical asymptote: $x = 0$

39.

Domain: $(0, \infty)$
x-intercept: $(9, 0)$
Vertical asymptote: $x = 0$

41.

Domain: $(-2, \infty)$
x-intercept: $(-1, 0)$
Vertical asymptote: $x = -2$

43.

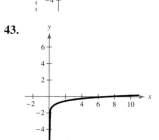

Domain: $(0, \infty)$
x-intercept: $(7, 0)$
Vertical asymptote: $x = 0$

45. c   **46.** f   **47.** d   **48.** e   **49.** b   **50.** a

51. $e^{-0.693\ldots} = \frac{1}{2}$ **53.** $e^{1.945\ldots} = 7$ **55.** $e^{5.521\ldots} = 250$

57. $e^0 = 1$ **59.** $\ln 54.598\ldots = 4$

61. $\ln 1.6487\ldots = \frac{1}{2}$ **63.** $\ln 0.406\ldots = -0.9$

65. $\ln 4 = x$ **67.** 2.913   **69.** -23.966 **71.** 5   **73.** $-\frac{5}{6}$

CHAPTER 5

75.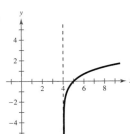

Domain: $(4, \infty)$
x-intercept: $(5, 0)$
Vertical asymptote: $x = 4$

77.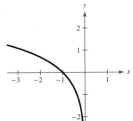

Domain: $(-\infty, 0)$
x-intercept: $(-1, 0)$
Vertical asymptote: $x = 0$

79. 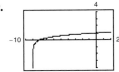 **81.**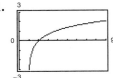

83.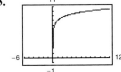

85. $x = 5$ **87.** $x = 7$ **89.** $x = 8$ **91.** $x = -5, 5$
93. (a) 30 yr; 10 yr (b) $323,179; $199,109
(c) $173,179; $49,109
(d) $x = 750$; The monthly payment must be greater than $750.
95. (a)

t	1	2	3	4	5	6
C	10.36	9.94	9.37	8.70	7.96	7.15

(b)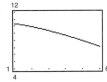

(c) No, the model begins to decrease rapidly, eventually producing negative values.
97. (a)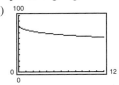

(b) 80 (c) 68.1 (d) 62.3
99. False. Reflecting $g(x)$ about the line $y = x$ will determine the graph of $f(x)$.

101. 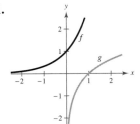 **103.**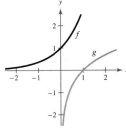

The functions f and g are inverses.

The functions f and g are inverses.

105.

x	-2	-1	0	1	2
$f(x) = 10^x$	$\frac{1}{100}$	$\frac{1}{10}$	1	10	100

x	$\frac{1}{100}$	$\frac{1}{10}$	1	10	100
$f(x) = \log x$	-2	-1	0	1	2

The domain of $f(x) = 10^x$ is equal to the range of $f(x) = \log x$ and vice versa. $f(x) = 10^x$ and $f(x) = \log x$ are inverses of each other.

107. (a)

x	1	5	10	10^2
$f(x)$	0	0.322	0.230	0.046

x	10^4	10^6
$f(x)$	0.00092	0.0000138

(b) 0
(c)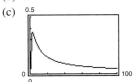

109. Answers will vary.
111. (a)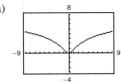

(b) Increasing: $(0, \infty)$
Decreasing: $(-\infty, 0)$
(c) Relative minimum: $(0, 0)$

Section 5.3 *(page 405)*

1. change-of-base **3.** $\dfrac{1}{\log_b a}$ **4.** c
5. a **6.** b **7.** (a) $\dfrac{\log 16}{\log 5}$ (b) $\dfrac{\ln 16}{\ln 5}$
9. (a) $\dfrac{\log x}{\log \frac{1}{5}}$ (b) $\dfrac{\ln x}{\ln \frac{1}{5}}$ **11.** (a) $\dfrac{\log \frac{3}{10}}{\log x}$ (b) $\dfrac{\ln \frac{3}{10}}{\ln x}$

13. (a) $\dfrac{\log x}{\log 2.6}$ (b) $\dfrac{\ln x}{\ln 2.6}$ **15.** 1.771   **17.** -2.000

19. -1.048 **21.** 2.633   **23.** $\frac{3}{2}$ **25.** $-3 - \log_5 2$

27. $6 + \ln 5$ **29.** 2   **31.** $\frac{3}{4}$ **33.** 4

35. -2 is not in the domain of $\log_2 x$.

37. 4.5   **39.** $-\frac{1}{2}$ **41.** 7   **43.** 2   **45.** $\ln 4 + \ln x$

47. $4 \log_8 x$ **49.** $1 - \log_5 x$ **51.** $\frac{1}{2} \ln z$

53. $\ln x + \ln y + 2 \ln z$ **55.** $\ln z + 2 \ln(z - 1)$

57. $\frac{1}{2} \log_2(a - 1) - 2 \log_2 3$ **59.** $\frac{1}{3} \ln x - \frac{1}{3} \ln y$

61. $2 \ln x + \frac{1}{2} \ln y - \frac{1}{2} \ln z$

63. $2 \log_5 x - 2 \log_5 y - 3 \log_5 z$ **65.** $\frac{3}{4} \ln x + \frac{1}{4}\ln(x^2 + 3)$

67. $\ln 2x$ **69.** $\log_4 \dfrac{z}{y}$ **71.** $\log_2 x^2 y^4$ **73.** $\log_3 \sqrt[4]{5x}$

75. $\log \dfrac{x}{(x + 1)^2}$ **77.** $\log \dfrac{xz^3}{y^2}$ **79.** $\ln \dfrac{x}{(x + 1)(x - 1)}$

81. $\ln \sqrt[3]{\dfrac{x(x + 3)^2}{x^2 - 1}}$ **83.** $\log_8 \dfrac{\sqrt[3]{y(y + 4)^2}}{y - 1}$

85. $\log_2 \frac{32}{4} = \log_2 32 - \log_2 4$; Property 2

87. $\beta = 10(\log I + 12)$; 60 dB   **89.** 70 dB

91. $\ln y = \frac{1}{4} \ln x$ **93.** $\ln y = -\frac{1}{4} \ln x + \ln \frac{5}{2}$

95. $y = 256.24 - 20.8 \ln x$

97. (a) and (b)   (c)

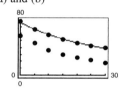   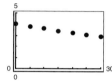

$T = 21 + e^{-0.037t + 3.997}$

The results are similar.

(d)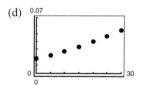

$T = 21 + \dfrac{1}{0.001t + 0.016}$

(e) Answers will vary.

99. Proof

101. False; $\ln 1 = 0$ **103.** False; $\ln(x - 2) \neq \ln x - \ln 2$

105. False; $u = v^2$

107. $f(x) = \dfrac{\log x}{\log 2} = \dfrac{\ln x}{\ln 2}$ **109.** $f(x) = \dfrac{\log x}{\log \frac{1}{2}} = \dfrac{\ln x}{\ln \frac{1}{2}}$

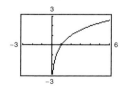   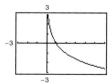

111. $f(x) = \dfrac{\log x}{\log 11.8} = \dfrac{\ln x}{\ln 11.8}$ **113.** $f(x) = h(x)$; Property 2

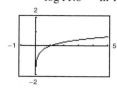   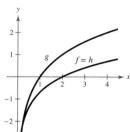

115.

$\ln 1 = 0$	$\ln 9 \approx 2.1972$
$\ln 2 \approx 0.6931$	$\ln 10 \approx 2.3025$
$\ln 3 \approx 1.0986$	$\ln 12 \approx 2.4848$
$\ln 4 \approx 1.3862$	$\ln 15 \approx 2.7080$
$\ln 5 \approx 1.6094$	$\ln 16 \approx 2.7724$
$\ln 6 \approx 1.7917$	$\ln 18 \approx 2.8903$
$\ln 8 \approx 2.0793$	$\ln 20 \approx 2.9956$

Section 5.4   (*page 415*)

1. solve

3. (a) One-to-One   (b) logarithmic; logarithmic

 (c) exponential; exponential

5. (a) Yes   (b) No

7. (a) No   (b) Yes   (c) Yes, approximate

9. (a) Yes, approximate   (b) No   (c) Yes

11. (a) No   (b) Yes   (c) Yes, approximate

13. 2   **15.** -5 **17.** 2   **19.** $\ln 2 \approx 0.693$

21. $e^{-1} \approx 0.368$ **23.** 64   **25.** (3, 8)   **27.** (9, 2)

29. $2, -1$ **31.** About 1.618, about -0.618

33. $\dfrac{\ln 5}{\ln 3} \approx 1.465$ **35.** $\ln 5 \approx 1.609$ **37.** $\ln 28 \approx 3.332$

39. $\dfrac{\ln 80}{2 \ln 3} \approx 1.994$ **41.** 2   **43.** 4

45. $3 - \dfrac{\ln 565}{\ln 2} \approx -6.142$ **47.** $\dfrac{1}{3} \log\left(\dfrac{3}{2}\right) \approx 0.059$

49. $1 + \dfrac{\ln 7}{\ln 5} \approx 2.209$ **51.** $\dfrac{\ln 12}{3} \approx 0.828$

53. $-\ln \frac{3}{5} \approx 0.511$ **55.** 0

57. $\dfrac{\ln \frac{8}{3}}{3 \ln 2} + \dfrac{1}{3} \approx 0.805$ **59.** $\ln 5 \approx 1.609$

61. $\ln 4 \approx 1.386$ **63.** $2 \ln 75 \approx 8.635$

65. $\dfrac{1}{2} \ln 1498 \approx 3.656$ **67.** $\dfrac{\ln 4}{365 \ln\left(1 + \frac{0.065}{365}\right)} \approx 21.330$

69. $\dfrac{\ln 2}{12 \ln\left(1 + \frac{0.10}{12}\right)} \approx 6.960$

71. 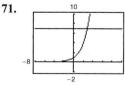   **73.**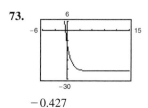

2.807 -0.427

75.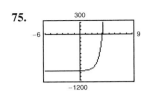

3.847

77.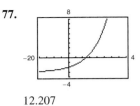

12.207

79.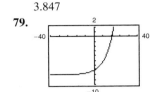

16.636

81. $e^{-3} \approx 0.050$ **83.** $e^7 \approx 1096.633$ **85.** $\dfrac{e^{2.4}}{2} \approx 5.512$

87. 1,000,000   **89.** $\dfrac{e^{10/3}}{5} \approx 5.606$ **91.** $e^2 - 2 \approx 5.389$

93. $e^{-2/3} \approx 0.513$ **95.** $\dfrac{e^{19/2}}{3} \approx 4453.242$

97. $2(3^{11/6}) \approx 14.988$ **99.** No solution

101. $1 + \sqrt{1 + e} \approx 2.928$ **103.** No solution   **105.** 7

107. $\dfrac{-1 + \sqrt{17}}{2} \approx 1.562$ **109.** 2

111. $\dfrac{725 + 125\sqrt{33}}{8} \approx 180.384$

113.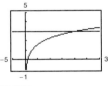

20.086

115.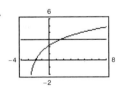

1.482

117. (a) 13.86 yr   (b) 21.97 yr
119. (a) 27.73 yr   (b) 43.94 yr
121. $-1, 0$ **123.** 1   **125.** $e^{-1/2} \approx 0.607$
127. $e^{-1} \approx 0.368$ **129.** (a) 210 coins   (b) 588 coins
131. (a)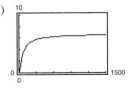

(b) $V = 6.7$; The yield will approach 6.7 million cubic feet per acre.
(c) 29.3 yr
133. 2003
135. (a) $y = 100$ and $y = 0$; The range falls between 0% and 100%.
(b) Males: 69.71 in.   Females: 64.51 in.

137. (a)

x	0.2	0.4	0.6	0.8	1.0
y	162.6	78.5	52.5	40.5	33.9

(b)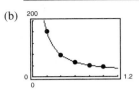

The model appears to fit the data well.
(c) 1.2 m
(d) No. According to the model, when the number of g's is less than 23, x is between 2.276 meters and 4.404 meters, which isn't realistic in most vehicles.
139. $\log_b uv = \log_b u + \log_b v$
True by Property 1 in Section 5.3.
141. $\log_b(u - v) = \log_b u - \log_b v$
False
$1.95 \approx \log(100 - 10) \neq \log 100 - \log 10 = 1$
143. Yes. See Exercise 103.
145. Yes. Time to double: $t = \dfrac{\ln 2}{r}$;

Time to quadruple: $t = \dfrac{\ln 4}{r} = 2\left(\dfrac{\ln 2}{r}\right)$
147. (a) 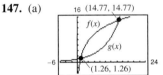   (b) $a = e^{1/e}$
(c) $1 < a < e^{1/e}$

Section 5.5   *(page 426)*

1. $y = ae^{bx}$; $y = ae^{-bx}$ **3.** normally distributed

5. $y = \dfrac{a}{1 + be^{-rx}}$ **7.** c   **8.** e   **9.** b

10. a   **11.** d   **12.** f

13. (a) $P = \dfrac{A}{e^{rt}}$ (b) $t = \dfrac{\ln\left(\dfrac{A}{P}\right)}{r}$

	Initial Investment	Annual % Rate	Time to Double	Amount After 10 years
15.	$1000	3.5%	19.8 yr	$1419.07
17.	$750	8.9438%	7.75 yr	$1834.37
19.	$500	11.0%	6.3 yr	$1505.00
21.	$6376.28	4.5%	15.4 yr	$10,000.00

23. $303,580.52
25. (a) 7.27 yr   (b) 6.96 yr   (c) 6.93 yr   (d) 6.93 yr
27.

r	2%	4%	6%	8%	10%	12%
t	54.93	27.47	18.31	13.73	10.99	9.16

29.

r	2%	4%	6%	8%	10%	12%
t	55.48	28.01	18.85	14.27	11.53	9.69

31.

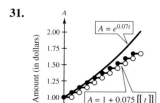

Continuous compounding

Half-life (years)	Initial Quantity	Amount After 1000 Years
33. 1599	10 g	6.48 g
35. 24,100	2.1 g	2.04 g
37. 5715	2.26 g	2 g

39. $y = e^{0.7675x}$ **41.** $y = 5e^{-0.4024x}$

43. (a)

Year	1970	1980	1990	2000	2007
Population	73.7	103.74	143.56	196.35	243.24

(b) 2014

(c) No; The population will not continue to grow at such a quick rate.

45. $k = 0.2988$; About 5,309,734 hits

47. (a) $k = 0.02603$; The population is increasing because $k > 0$.

(b) 449,910; 512,447   (c) 2014

49. About 800 bacteria

51. (a) About 12,180 yr old   (b) About 4797 yr old

53. (a) $V = -5400t + 23,300$ (b) $V = 23,300e^{-0.311t}$

(c)

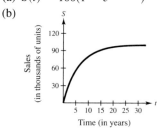

The exponential model depreciates faster.

(d)

t	1 yr	3 yr
$V = -5400t + 23,300$	17,900	7100
$V = 23,300e^{-0.311t}$	17,072	9166

(e) Answers will vary.

55. (a) $S(t) = 100(1 - e^{-0.1625t})$

(b)

(c) 55,625

57. (a)

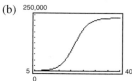

(b) 100

59. (a) 715; 90,880; 199,043

(b)

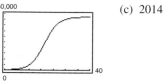

(c) 2014

(d) $235,000 = \dfrac{237,101}{1 + 1950e^{-0.355t}}$

$t \approx 34.63$

61. (a) 203 animals   (b) 13 mo

(c)

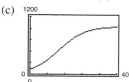

Horizontal asymptotes: $p = 0, p = 1000$. The population size will approach 1000 as time increases.

63. (a) $10^{8.5} \approx 316,227,766$ (b) $10^{5.4} \approx 251,189$

(c) $10^{6.1} \approx 1,258,925$

65. (a) 20 dB   (b) 70 dB   (c) 40 dB   (d) 120 dB

67. 95%   **69.** 4.64   **71.** 1.58×10^{-6} moles/L

73. $10^{5.1}$ **75.** 3:00 A.M.

77. (a)

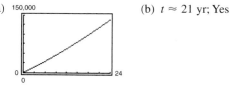

(b) $t \approx 21$ yr; Yes

79. False. The domain can be the set of real numbers for a logistic growth function.

81. False. The graph of $f(x)$ is the graph of $g(x)$ shifted upward five units.

83. Answers will vary.

Review Exercises   *(page 434)*

1. 0.164   **3.** 0.337   **5.** 1456.529

7. Shift the graph of f two units downward.

9. Reflect f in the y-axis and shift two units to the right.

11. Reflect f in the x-axis and shift one unit upward.

13. Reflect f in the x-axis and shift two units to the left.

CHAPTER 5

15.

x	-1	0	1	2	3
$f(x)$	8	5	4.25	4.063	4.016

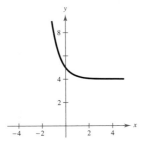

17.

x	-1	0	1	2	3
$f(x)$	4.008	4.04	4.2	5	9

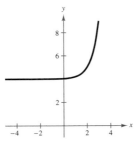

19.

x	-2	-1	0	1	2
$f(x)$	3.25	3.5	4	5	7

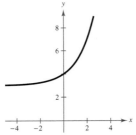

21. $x = 1$ **23.** $x = 4$ **25.** 2980.958 **27.** 0.183

29.

x	-2	-1	0	1	2
$h(x)$	2.72	1.65	1	0.61	0.37

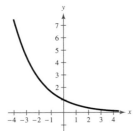

31.

x	-3	-2	-1	0	1
$f(x)$	0.37	1	2.72	7.39	20.09

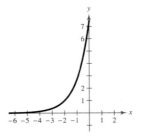

33.

n	1	2	4	12
A	\$6719.58	\$6734.28	\$6741.74	\$6746.77

n	365	Continuous
A	\$6749.21	\$6749.29

35. (a) 0.154 (b) 0.487 (c) 0.811 **37.** $\log_3 27 = 3$
39. $\ln 2.2255 \ldots = 0.8$ **41.** 3 **43.** -2
45. $x = 7$ **47.** $x = -5$
49. Domain: $(0, \infty)$ **51.** Domain: $(-5, \infty)$
x-intercept: $(1, 0)$ x-intercept: $(9995, 0)$
Vertical asymptote: $x = 0$ Vertical asymptote: $x = -5$

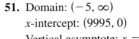
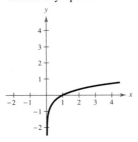

53. (a) 3.118 (b) -0.020
55. Domain: $(0, \infty)$ **57.** Domain: $(-\infty, 0), (0, \infty)$
x-intercept: $(e^{-3}, 0)$ x-intercepts: $(\pm 1, 0)$
Vertical asymptote: $x = 0$ Vertical asymptote: $x = 0$

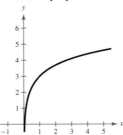
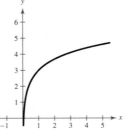

59. 53.4 in. **61.** 2.585 **63.** -2.322
65. $\log 2 + 2 \log 3 \approx 1.255$ **67.** $2 \ln 2 + \ln 5 \approx 2.996$
69. $1 + 2 \log_5 x$ **71.** $2 - \frac{1}{2} \log_3 x$
73. $2 \ln x + 2 \ln y + \ln z$ **75.** $\log_2 5x$

77. $\ln \dfrac{x}{\sqrt[4]{y}}$ **79.** $\log_3 \dfrac{\sqrt{x}}{(y+8)^2}$

81. (a) $0 \le h < 18{,}000$

(b)

Vertical asymptote: $h = 18{,}000$

(c) The plane is climbing at a slower rate, so the time required increases.

(d) 5.46 min

83. 3   **85.** $\ln 3 \approx 1.099$ **87.** $e^4 \approx 54.598$

89. $x = 1, 3$ **91.** $\dfrac{\ln 32}{\ln 2} = 5$

93.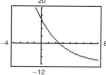

2.447

95. $\frac{1}{3}e^{8.2} \approx 1213.650$ **97.** $3e^2 \approx 22.167$

99. $e^8 \approx 2980.958$ **101.** No solution   **103.** 0.900

105. 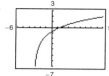   **107.**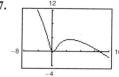

1.482   0, 0.416, 13.627

109. 31.4 yr   **111.** e   **112.** b   **113.** f

114. d   **115.** a   **116.** c   **117.** $y = 2e^{0.1014x}$

119. (a)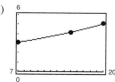

The model fits the data well.

(b) 2022; Answers will vary.

121. (a) 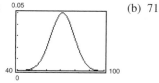   (b) 71

123. (a) 10^{-6} W/m²   (b) $10\sqrt{10}$ W/m²

(c) 1.259×10^{-12} W/m²

125. True by the inverse properties

Chapter Test   (page 437)

1. 2.366   **2.** 687.291   **3.** 0.497   **4.** 22.198

5.

x	-1	$-\frac{1}{2}$	0	$\frac{1}{2}$	1
$f(x)$	10	3.162	1	0.316	0.1

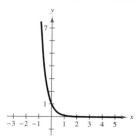

6.

x	-1	0	1	2	3
$f(x)$	-0.005	-0.028	-0.167	-1	-6

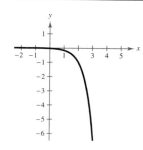

7.

x	-1	$-\frac{1}{2}$	0	$\frac{1}{2}$	1
$f(x)$	0.865	0.632	0	-1.718	-6.389

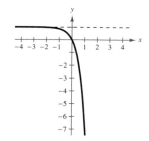

8. (a) -0.89 (b) 9.2

9.

x	$\frac{1}{2}$	1	$\frac{3}{2}$	2	4
$f(x)$	-5.699	-6	-6.176	-6.301	-6.602

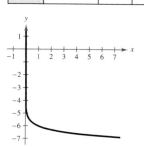

Vertical asymptote: $x = 0$

10.

x	5	7	9	11	13
$f(x)$	0	1.099	1.609	1.946	2.197

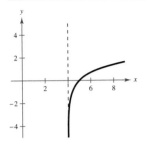

Vertical asymptote: $x = 4$

11.

x	-5	-3	-1	0	1
$f(x)$	1	2.099	2.609	2.792	2.946

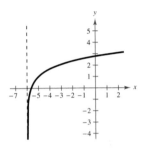

Vertical asymptote: $x = -6$

12. 1.945   **13.** -0.167 **14.** -11.047
15. $\log_2 3 + 4\log_2 |a|$ **16.** $\ln 5 + \frac{1}{2}\ln x - \ln 6$
17. $3\log(x-1) - 2\log y - \log z$ **18.** $\log_3 13y$

19. $\ln \dfrac{x^4}{y^4}$ **20.** $\ln\left(\dfrac{x^3 y^2}{x+3}\right)$ **21.** $x = -2$

22. $x = \dfrac{\ln 44}{-5} \approx -0.757$ **23.** $\dfrac{\ln 197}{4} \approx 1.321$

24. $e^{1/2} \approx 1.649$ **25.** $e^{-11/4} \approx 0.0639$ **26.** 20
27. $y = 2745e^{0.1570t}$ **28.** 55%
29. (a)

x	$\frac{1}{4}$	1	2	4	5	6
H	58.720	75.332	86.828	103.43	110.59	117.38

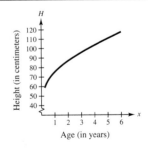

(b) 103 cm; 103.43 cm

Cumulative Test for Chapters 3–5   *(page 438)*

1. $y = -\frac{3}{4}(x+8)^2 + 5$
2.

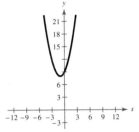

3.

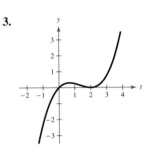

4.

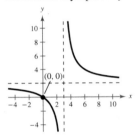

5. $-2, \pm 2i$ **6.** $-7, 0, 3$

7. $3x - 2 - \dfrac{3x-2}{2x^2+1}$ **8.** $3x^3 + 6x^2 + 14x + 23 + \dfrac{49}{x-2}$

9. 1.20   **10.** $x^4 + 3x^3 - 11x^2 + 9x + 70$
11. Domain: all real numbers x except $x = 3$
    Vertical asymptote: $x = 3$
    Horizontal asymptote: $y = 2$

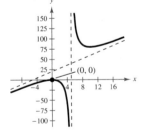

12. Domain: all real numbers x except $x = 5$
    Vertical asymptote: $x = 5$
    Slant asymptote: $y = 4x + 20$

13. Intercept: $(0, 0)$
Vertical asymptotes: $x = 1, -3$
Horizontal asymptote: $y = 0$

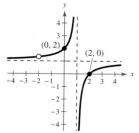

14. y-intercept: $(0, 2)$
x-intercept: $(2, 0)$
Vertical asymptote: $x = 1$
Horizontal asymptote: $y = 1$

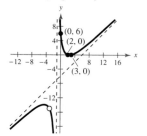

15. y-intercept: $(0, 6)$
x-intercepts: $(2, 0), (3, 0)$
Vertical asymptote: $x = -1$
Slant asymptote: $y = x - 6$

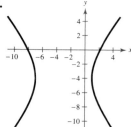

16.

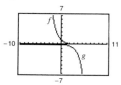

17.

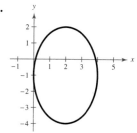

18. $(x - 3)^2 = \dfrac{3}{2}(y + 2)$ **19.** $\dfrac{(y - 2)^2}{\frac{4}{5}} - \dfrac{x^2}{\frac{16}{5}} = 1$

20. Reflect f in the x-axis and y-axis, and shift three units to the right.

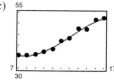

21. Reflect f in the x-axis, and shift four units upward.

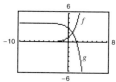

22. 1.991   **23.** -0.067 **24.** 1.717   **25.** 0.390
26. 0.906   **27.** -1.733 **28.** -4.087
29. $\ln(x + 4) + \ln(x - 4) - 4 \ln x, \ x > 4$

30. $\ln \dfrac{x^2}{\sqrt{x + 5}}, \ x > 0$ **31.** $\dfrac{\ln 12}{2} \approx 1.242$

32. $\dfrac{\ln 9}{\ln 4} + 5 \approx 6.585$ **33.** $\ln 6 \approx 1.792$ or $\ln 7 \approx 1.946$

34. $\frac{64}{5} = 12.8$ **35.** $\frac{1}{2}e^8 \approx 1490.479$
36. $e^6 - 2 \approx 401.429$
37.

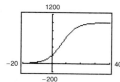

Horizontal asymptotes: $y = 0, \ y = 1000$
38. 2000
39. (a) and (c)

The model is a good fit for the data.
(b) $S = -0.0297t^3 + 1.175t^2 - 12.96t + 79.0$
(d) 25.3; Answers will vary.
40. $16,302.05   **41.** 6.3 h   **42.** 2015
43. (a) 300   (b) 570   (c) About 9 yr

Problem Solving   (page 441)

1.

$y = 0.5^x$ and $y = 1.2^x$
$0 < a \le e^{1/e}$
3. As $x \to \infty$, the graph of e^x increases at a greater rate than the graph of x^n.
5. Answers will vary.

CHAPTER 5

7. (a)  (b)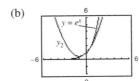

(c)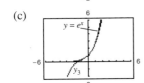

9.

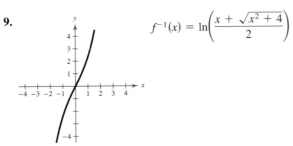

$$f^{-1}(x) = \ln\left(\frac{x + \sqrt{x^2 + 4}}{2}\right)$$

11. c   **13.** $t = \dfrac{\ln c_1 - \ln c_2}{\left(\dfrac{1}{k_2} - \dfrac{1}{k_1}\right)\ln\dfrac{1}{2}}$

15. (a) $y_1 = 252{,}606(1.0310)^t$

(b) $y_2 = 400.88t^2 - 1464.6t + 291{,}782$

(c)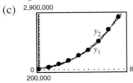

(d) The exponential model is a better fit. No, because the model is rapidly approaching infinity.

17. $1, e^2$

19. $y_4 = (x - 1) - \frac{1}{2}(x - 1)^2 + \frac{1}{3}(x - 1)^3 - \frac{1}{4}(x - 1)^4$

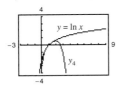

The pattern implies that
$\ln x = (x - 1) - \frac{1}{2}(x - 1)^2 + \frac{1}{3}(x - 1)^3 - \cdots$.

21.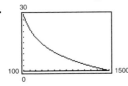

17.7 ft³/min

23. (a) 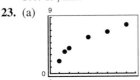   **25.** (a)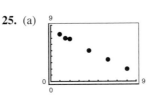

(b)–(e) Answers will vary.   (b)–(e) Answers will vary.

Chapter 6

Section 6.1   *(page 451)*

1. system; equations   **3.** solving   **5.** point; intersection

7. (a) No   (b) No   (c) No   (d) Yes

9. (a) No   (b) Yes   (c) No   (d) No

11. $(2, 2)$ **13.** $(2, 6), (-1, 3)$

15. $(-3, -4), (5, 0)$ **17.** $(0, 0), (2, -4)$

19. $(0, 1), (1, -1), (3, 1)$ **21.** $(6, 4)$ **23.** $\left(\frac{1}{2}, 3\right)$

25. $(1, 1)$ **27.** $\left(\frac{20}{3}, \frac{40}{3}\right)$ **29.** No solution

31. $(-2, 4), (0, 0)$ **33.** No solution   **35.** $(6, 2)$

37. $\left(-\frac{3}{2}, \frac{1}{2}\right)$ **39.** $(2, 2), (4, 0)$ **41.** $(1, 4), (4, 7)$

43. $\left(4, -\frac{1}{2}\right)$ **45.** No solution   **47.** $(4, 3), (-4, 3)$

49. 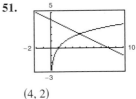   **51.**

$(0, 1)$ $(4, 2)$

53.

$(0, -13), (\pm 12, 5)$

55. $(1, 2)$ **57.** No solution   **59.** $(0.287, 1.751)$

61. $(-1, 0), (0, 1), (1, 0)$ **63.** $\left(\frac{1}{2}, 2\right), \left(-4, -\frac{1}{4}\right)$

65. 192 units   **67.** (a) 1013 units   (b) 5061 units

69. (a) 8 weeks

(b)

	1	2	3	4
$360 - 24x$	336	312	288	264
$24 + 18x$	42	60	78	96

	5	6	7	8
$360 - 24x$	240	216	192	168
$24 + 18x$	114	132	150	168

71. More than $16,666.67

73. (a) $\begin{cases} x + y = 25{,}000 \\ 0.06x + 0.085y = 2{,}000 \end{cases}$

(b) 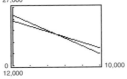   (c) $5000

Decreases; Interest is fixed.

75. (a) Solar: $0.0598t^3 - 1.719t^2 + 14.66t + 32.2$
Wind: $3.237t^2 - 51.97t + 247.9$

(b)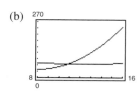

(c) Point of intersection: (10.9, 65.26); Consumption of solar and wind energy are equal at this point in time in the year 2000.

(d) Answers will vary

(e) Answers will vary.

77. (a) $T_1 = 26.560t^2 + 85.54t + 2468.5$
$T_2 = 794.14t + 14,124.6$

(b) 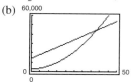 (c) 2038 (d) 2038

79. 60 cm × 80 cm **81.** 44 ft × 198 ft

83. 10 km × 12 km

85. False. To solve a system of equations by substitution, you can solve for either variable in one of the two equations and then back-substitute.

87.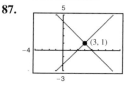

(3, 1); The point of intersection is equal to the solution found in Example 1.

89. For a linear system, the result will be a contradictory equation such as $0 = N$, where N is a nonzero real number. For a nonlinear system, there may be an equation with imaginary solutions.

91. (a) $y = 2x$ (b) $y = 0$ (c) $y = x - 2$

Section 6.2 *(page 463)*

1. elimination **3.** consistent; inconsistent

5. (2, 1) **7.** (1, −1)

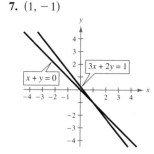

9. No solution **11.** $\left(a, \frac{3}{2}a - \frac{5}{2}\right)$

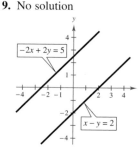

13. (4, 1) **15.** $\left(\frac{3}{2}, -\frac{1}{2}\right)$ **17.** (4, −1) **19.** $\left(\frac{12}{7}, \frac{18}{7}\right)$

21. No solution **23.** Infinitely many solutions: $\left(a, -\frac{1}{2} + \frac{5}{6}a\right)$

25. (101, 96) **27.** $\left(-\frac{6}{35}, \frac{43}{35}\right)$ **29.** (5, −2)

31. b; one solution; consistent **32.** c; one solution; consistent

33. a; infinitely many solutions; consistent

34. d; no solutions; inconsistent

35. (4, 1)

37. (2, −1) **39.** (6, −3) **41.** $\left(\frac{49}{4}, \frac{33}{4}\right)$

43. 550 mi/h, 50 mi/h **45.** (240, 404)

47. (2,000,000, 100)

49. Cheeseburger: 300 calories; French fries: 230 calories

51. (a) $\begin{cases} x + y = 30 \\ 0.25x + 0.5y = 12 \end{cases}$

(b)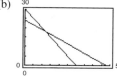

Decreases

(c) 25% solution: 12 L; 50% solution: 18 L

53. $18,000

55. (a)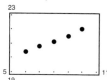

Pharmacy A: $P = 0.52t + 16.0$

Pharmacy B: $P = 0.39t + 18.0$

(b) Yes, in the year 2015

57. $y = 0.97x + 2.1$ **59.** $y = -2x + 8$

61. (a) $y = 14x + 19$ (b) 41.4 bushels/acre

63. False. Two lines that coincide have infinitely many points of intersection.

65. No. Two lines will intersect only once or will coincide, and if they coincide the system will have infinitely many solutions.

67. The method of elimination is much easier.

69. (39,600, 398). It is necessary to change the scale on the axes to see the point of intersection.

71. $k = -4$

CHAPTER 6

Section 6.3 *(page 475)*

1. row-echelon   3. Gaussian   5. nonsquare
7. (a) No   (b) No   (c) No   (d) Yes
9. (a) No   (b) No   (c) Yes   (d) No
11. $(-13, -10, 8)$ 13. $(3, 10, 2)$ 15. $\left(\frac{11}{4}, 7, 11\right)$
17. $\begin{cases} x - 2y + 3z = 5 \\ \quad\quad y - 2z = 9 \\ 2x \quad\quad - 3z = 0 \end{cases}$

First step in putting the system in row-echelon form.
19. $(4, 1, 2)$ 21. $(-4, 8, 5)$ 23. $(5, -2, 0)$
25. No solution   27. $\left(-\frac{1}{2}, 1, \frac{3}{2}\right)$
29. $(-3a + 10, 5a - 7, a)$ 31. $(-a + 3, a + 1, a)$
33. $(2a, 21a - 1, 8a)$ 35. $\left(-\frac{3}{2}a + \frac{1}{2}, -\frac{2}{3}a + 1, a\right)$
37. $(1, 1, 1, 1)$ 39. No solution   41. $(0, 0, 0)$
43. $(9a, -35a, 67a)$ 45. $s = -16t^2 + 144$
47. $s = -16t^2 - 32t + 400$
49. $y = \frac{1}{2}x^2 - 2x$ 51. $y = x^2 - 6x + 8$

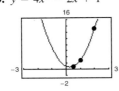   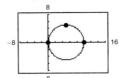

53. $y = 4x^2 - 2x + 1$ 55. $x^2 + y^2 - 10x = 0$

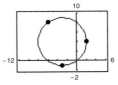

57. $x^2 + y^2 + 6x - 8y = 0$

59. 6 touchdowns, 6 extra-point kicks, 1 field goal
61. $300,000 at 8%
    $400,000 at 9%
    $75,000 at 10%
63. $187,500 + s$ in certificates of deposit
    $187,500 - s$ in municipal bonds
    $125,000 - s$ in blue-chip stocks
    s in growth stocks
65. Brand X = 4 lb
    Brand Y = 9 lb
    Brand Z = 9 lb
67. 48 ft, 35 ft, 27 ft   69. $x = 60°, y = 67°, z = 53°$
71. Television = 30 ads
    Radio = 10 ads
    Newspaper = 20 ads
73. (a) 1 L of 10%, 7 L of 20%, 2 L of 50%
    (b) 0 L of 10%, $8\frac{1}{3}$ L of 20%, $1\frac{2}{3}$ L of 50%
    (c) $6\frac{1}{4}$ L of 10%, 0 L of 20%, $3\frac{3}{4}$ L of 50%

75. $I_1 = 1, I_2 = 2, I_3 = 1$ 77. $y = -\frac{5}{24}x^2 - \frac{3}{10}x + \frac{41}{6}$
79. $y = x^2 - x$
81. (a) $y = -0.0075x^2 + 1.3x + 20$
    (b) 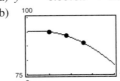   (c)

x	100	120	140
y	75	68	55

The values are the same.

(d) 24.25%   (e) 156 females
83. 6 touchdowns, 6 extra-point kicks, 6 field goals, 1 safety
85. $x = 5, y = 5, \lambda = -5$
87. $x = \pm\frac{\sqrt{2}}{2}, y = \frac{1}{2}, \lambda = 1$ or $x = 0, y = 0, \lambda = 0$
89. False. Equation 2 does not have a leading coefficient of 1.
91. No. Answers will vary.
93. Sample answers:
$\begin{cases} 2x + y - z = 0 \\ \quad\quad y + 2z = 0 \\ -x + 2y + z = -9 \end{cases}$ $\begin{cases} x + y + z = 1 \\ 2x \quad\quad - z = 4 \\ \quad\quad 4y + 8z = 0 \end{cases}$
95. Sample answers:
$\begin{cases} x + 2y + 4z = -14 \\ x - 12y \quad\quad = 0 \\ x \quad\quad - 8z = 8 \end{cases}$ $\begin{cases} 4x - 2y - 8z = -9 \\ -x \quad\quad + 4z = -1 \\ \quad -7y + 2z = 0 \end{cases}$

Section 6.4 *(page 486)*

1. partial fraction decomposition   3. partial fraction
5. b   6. c   7. d   8. a
9. $\dfrac{A}{x} + \dfrac{B}{x - 2}$ 11. $\dfrac{A}{x} + \dfrac{B}{x^2} + \dfrac{C}{x - 7}$
13. $\dfrac{A}{x - 5} + \dfrac{B}{(x - 5)^2} + \dfrac{C}{(x - 5)^3}$
15. $\dfrac{A}{x} + \dfrac{Bx + C}{x^2 + 10}$ 17. $\dfrac{A}{x} + \dfrac{Bx + C}{x^2 + 1} + \dfrac{Dx + E}{(x^2 + 1)^2}$
19. $\dfrac{1}{x} - \dfrac{1}{x + 1}$ 21. $\dfrac{1}{x} - \dfrac{2}{2x + 1}$ 23. $\dfrac{1}{x - 1} - \dfrac{1}{x + 2}$
25. $\dfrac{1}{2}\left(\dfrac{1}{x - 1} - \dfrac{1}{x + 1}\right)$ 27. $-\dfrac{3}{x} - \dfrac{1}{x + 2} + \dfrac{5}{x - 2}$
29. $\dfrac{3}{x - 3} + \dfrac{9}{(x - 3)^2}$ 31. $\dfrac{3}{x} - \dfrac{1}{x^2} + \dfrac{1}{x + 1}$
33. $\dfrac{3}{x} - \dfrac{2x - 2}{x^2 + 1}$ 35. $-\dfrac{1}{x - 1} + \dfrac{x + 2}{x^2 - 2}$
37. $\dfrac{2}{x^2 + 4} + \dfrac{x}{(x^2 + 4)^2}$
39. $\dfrac{1}{8}\left(\dfrac{1}{2x + 1} + \dfrac{1}{2x - 1} - \dfrac{4x}{4x^2 + 1}\right)$
41. $\dfrac{1}{x + 1} + \dfrac{2}{x^2 - 2x + 3}$ 43. $1 - \dfrac{2x + 1}{x^2 + x + 1}$
45. $2x - 7 + \dfrac{17}{x + 2} + \dfrac{1}{x + 1}$
47. $x + 3 + \dfrac{6}{x - 1} + \dfrac{4}{(x - 1)^2} + \dfrac{1}{(x - 1)^3}$
49. $x + \dfrac{2}{x} + \dfrac{1}{x + 1} + \dfrac{3}{(x + 1)^2}$

51. $\dfrac{3}{2x-1} - \dfrac{2}{x+1}$ **53.** $\dfrac{1}{2}\left[-\dfrac{1}{x} + \dfrac{5}{x+1} - \dfrac{3}{(x+1)^2} \right]$

55. $\dfrac{1}{x^2+2} + \dfrac{x}{(x^2+2)^2}$ **57.** $2x + \dfrac{1}{2}\left(\dfrac{3}{x-4} - \dfrac{1}{x+2} \right)$

59. (a) $\dfrac{3}{x} - \dfrac{2}{x-4}$

(b) $y = \dfrac{x-12}{x(x-4)}$ $y = \dfrac{3}{x},\quad y = -\dfrac{2}{x-4}$

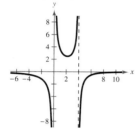   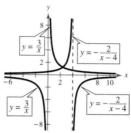

(c) The vertical asymptotes are the same.

61. $\dfrac{60}{100-p} - \dfrac{60}{100+p}$

63. False. The partial fraction decomposition is

$\dfrac{A}{x+10} + \dfrac{B}{x-10} + \dfrac{C}{(x-10)^2}.$

65. True. The expression is an improper rational expression.

67. $\dfrac{1}{2a}\left(\dfrac{1}{a+x} + \dfrac{1}{a-x} \right)$ **69.** $\dfrac{1}{a}\left(\dfrac{1}{y} + \dfrac{1}{a-y} \right)$

71. Answers will vary. Sample answer: You can substitute any convenient values of x that will help determine the constants. You can also find the basic equation, expand it, then equate coefficients of like terms.

Section 6.5   *(page 495)*

1. solution   **3.** linear   **5.** solution set

7.

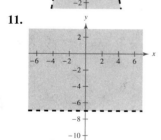

9.

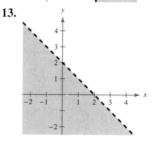

11.

13.

15.

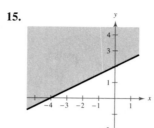

17.

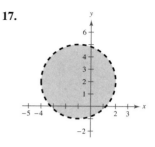

19.

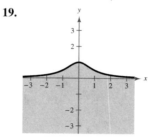

21.

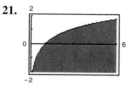

23.

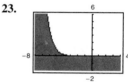

25.

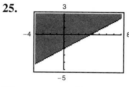

27.

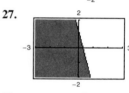

29.

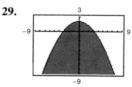

31.

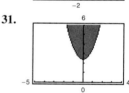

33. $y < 5x + 5$ **35.** $y \ge -\dfrac{2}{3}x + 2$

37. (a) No   (b) No   (c) Yes   (d) Yes

39. (a) Yes   (b) No   (c) Yes   (d) Yes

41.

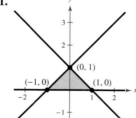

43.

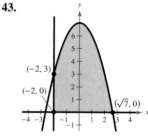

45.

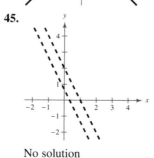

No solution

47.

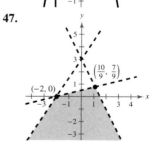

49.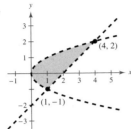

51.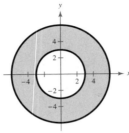

53.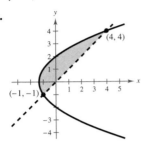

55.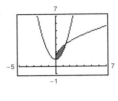

57.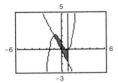

59.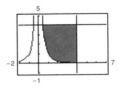

61. $\begin{cases} x \ge 0 \\ y \ge 0 \\ y \le 6 - x \end{cases}$

63. $\begin{cases} y \ge 4 - x \\ y \ge 2 - \frac{1}{4}x \\ x \ge 0, \ y \ge 0 \end{cases}$

65. $\begin{cases} x \ge 0 \\ y \ge 0 \\ x^2 + y^2 < 64 \end{cases}$

67. $\begin{cases} x \ge 4 \\ x \le 9 \\ y \ge 3 \\ y \le 9 \end{cases}$

69. $\begin{cases} y \ge 0 \\ y \le 5x \\ y \le -x + 6 \end{cases}$

71. (a)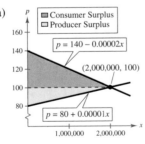

(b) Consumer surplus: $1600
Producer surplus: $400

73. (a)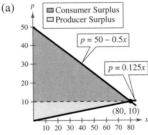

(b) Consumer surplus: $40,000,000
Producer surplus: $20,000,000

75. $\begin{cases} x + \frac{3}{2}y \le 12 \\ \frac{4}{3}x + \frac{3}{2}y \le 15 \\ x \ge 0 \\ y \ge 0 \end{cases}$

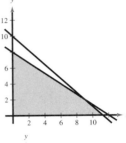

77. $\begin{cases} x + y \le 20{,}000 \\ y \ge 2x \\ x \ge 5000 \\ y \ge 5000 \end{cases}$

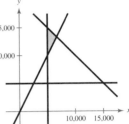

79. $\begin{cases} 55x + 70y \le 7500 \\ x \ge 50 \\ y \ge 40 \end{cases}$

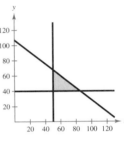

81. (a) $\begin{cases} 20x + 10y \ge 300 \\ 15x + 10y \ge 150 \\ 10x + 20y \ge 200 \\ x \ge 0 \\ y \ge 0 \end{cases}$

(b)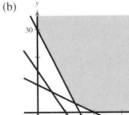

(c) Answers will vary.

83. (a) $y = 16.75t + 148.4$

(b)

(c) $1656.2 billion

85. (a) $\begin{cases} xy \ge 500 \\ 2x + \pi y \ge 125 \\ x \ge 0 \\ y \ge 0 \end{cases}$

(b)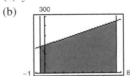

87. False. The graph shows the solution of the system
$$\begin{cases} y < 6 \\ -4x - 9y < 6. \\ 3x + y^2 \geq 2 \end{cases}$$

89. (a) $\begin{cases} \pi y^2 - \pi x^2 \geq 10 \\ \quad\quad y > x \\ \quad\quad x > 0 \end{cases}$ (b)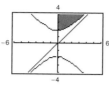

(c) The line is an asymptote to the boundary. The larger the circles, the closer the radii can be while still satisfying the constraint.

91. d **92.** b **93.** c **94.** a

Section 6.6 *(page 505)*

1. optimization **3.** objective **5.** inside; on
7. Minimum at $(0, 0)$: 0 **9.** Minimum at $(0, 0)$: 0
   Maximum at $(5, 0)$: 20   Maximum at $(3, 4)$: 26
11. Minimum at $(0, 0)$: 0
   Maximum at $(60, 20)$: 740

13. 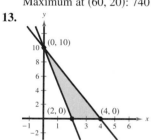   **15.**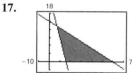

Minimum at $(2, 0)$: 6   Minimum at $(5, 3)$: 35
Maximum at $(0, 10)$: 20   No maximum

17. 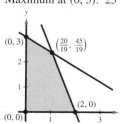   **19.**

Minimum at $(7.2, 13.2)$: 34.8   Minimum at $(16, 0)$: 16
Maximum at $(60, 0)$: 180   Maximum at any point on
                            the line segment connecting
                            $(7.2, 13.2)$ and $(60, 0)$: 60

21. Minimum at $(0, 0)$: 0 **23.** Minimum at $(0, 0)$: 0
   Maximum at $(3, 6)$: 12   Maximum at $(0, 10)$: 10
25. Minimum at $(0, 0)$: 0 **27.** Minimum at $(0, 0)$: 0
   Maximum at $(0, 5)$: 25   Maximum at $\left(\frac{22}{3}, \frac{19}{6}\right)$: $\frac{271}{6}$

29.

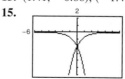

The maximum, 5, occurs at any point on the line segment connecting $(2, 0)$ and $\left(\frac{20}{19}, \frac{45}{19}\right)$. Minimum at $(0, 0)$: 0

31.

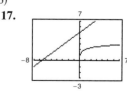

The constraint $x \leq 10$ is extraneous. Minimum at $(7, 0)$: -7; maximum at $(0, 7)$: 14

33.

The constraint $2x + y \leq 4$ is extraneous. Minimum at $(0, 0)$: 0; maximum at $(0, 1)$: 4

35. 230 units of the $225 model
   45 units of the $250 model
   Optimal profit: $8295
37. 3 bags of brand X **39.** 13 audits
   6 bags of brand Y   0 tax returns
   Optimal cost: $195   Optimal revenue: $20,800
41. $0 on TV ads
   $1,000,000 on newspaper ads
   Optimal audience: 250 million people
43. $62,500 to type A
   $187,500 to type B
   Optimal return: $23,750
45. True. The objective function has a maximum value at any point on the line segment connecting the two vertices.
47. True. If an objective function has a maximum value at more than one vertex, then any point on the line segment connecting the points will produce the maximum value.

Review Exercises *(page 510)*

1. $(1, 1)$ **3.** $\left(\frac{3}{2}, 5\right)$ **5.** $(0.25, 0.625)$ **7.** $(5, 4)$
9. $(0, 0), (2, 8), (-2, 8)$ **11.** $(4, -2)$
13. $(1.41, -0.66), (-1.41, 10.66)$
15.   **17.**

$(0, -2)$ No solution
19. 3847 units **21.** 96 m × 144 m **23.** 8 in. × 12 in.
25. $\left(\frac{5}{2}, 3\right)$ **27.** $(-0.5, 0.8)$ **29.** $(0, 0)$ **31.** $\left(\frac{8}{5}a + \frac{14}{5}, a\right)$
33. d, one solution, consistent
34. c, infinitely many solutions, consistent
35. b, no solution, inconsistent **36.** a, one solution, consistent
37. $\left(\frac{500,000}{7}, \frac{159}{7}\right)$ **39.** $(2, -4, -5)$ **41.** $(-6, 7, 10)$

CHAPTER 6

43. $\left(\frac{24}{5}, \frac{22}{5}, -\frac{8}{5}\right)$ **45.** $(3a + 4, 2a + 5, a)$ **47.** $(1, 1, 1, 0)$

49. $(a - 4, a - 3, a)$ **51.** $y = 2x^2 + x - 5$

53. $x^2 + y^2 - 4x + 4y - 1 = 0$

55. (a) $y = 0.25x^2 + 27.95x - 36.7$

(b)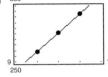

(c) $438.8 billion; yes

The model is a good fit.

57. $16,000 at 7%   **59.** 4 par-3 holes
$13,000 at 9%   10 par-4 holes
$11,000 at 11%   4 par-5 holes

61. $\dfrac{A}{x} + \dfrac{B}{x + 20}$ **63.** $\dfrac{A}{x} + \dfrac{B}{x^2} + \dfrac{C}{x - 5}$

65. $\dfrac{3}{x + 2} - \dfrac{4}{x + 4}$ **67.** $1 - \dfrac{25}{8(x + 5)} + \dfrac{9}{8(x - 3)}$

69. $\dfrac{1}{2}\left(\dfrac{3}{x - 1} - \dfrac{x - 3}{x^2 + 1}\right)$ **71.** $\dfrac{3}{x^2 + 1} + \dfrac{4x - 3}{(x^2 + 1)^2}$

73.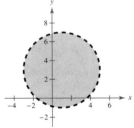

75.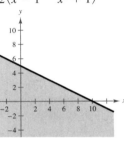

77.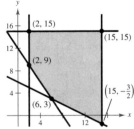

79.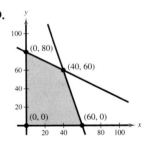

81.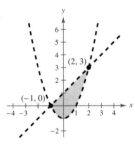

83.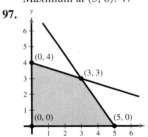

85.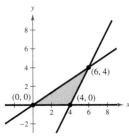

87. $\begin{cases} 20x + 30y \le 24{,}000 \\ 12x + 8y \le 12{,}400 \\ \quad x \ge 0 \\ \quad\quad y \ge 0 \end{cases}$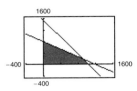

89. (a)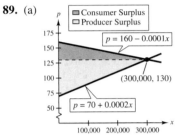

91. $\begin{cases} x \ge 3 \\ x \le 7 \\ y \ge 1 \\ y \le 10 \end{cases}$

(b) Consumer surplus: $4,500,000
Producer surplus: $9,000,000

93.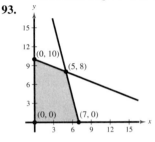

Minimum at $(0, 0)$: 0
Maximum at $(5, 8)$: 47

95.

Minimum at $(15, 0)$: 26.25
No maximum

97.

Minimum at $(0, 0)$: 0
Maximum at $(3, 3)$: 48

99. 72 haircuts, 0 permanents; Optimal revenue: $1800

101. 750 units of model A   **103.** $\frac{2}{3}$ regular unleaded
1000 units of model B $\frac{1}{3}$ premium unleaded
Optimal profit: $83,750   Optimal cost: $1.93

105. False. To represent a region covered by an isosceles trapezoid, the last two inequality signs should be $\le$.

107. $\begin{cases} 4x + y = -22 \\ \frac{1}{2}x + y = 6 \end{cases}$ **109.** $\begin{cases} 3x + y = 7 \\ -6x + 3y = 1 \end{cases}$

111. $\begin{cases} x + y + z = 6 \\ x + y - z = 0 \\ x - y - z = 2 \end{cases}$ **113.** $\begin{cases} 2x + 2y - 3z = 7 \\ x - 2y + z = 4 \\ -x + 4y - z = -1 \end{cases}$

115. An inconsistent system of linear equations has no solution.

Chapter Test   *(page 515)*

1. $(-4, -5)$ **2.** $(0, -1), (1, 0), (2, 1)$
3. $(8, 4), (2, -2)$

4.
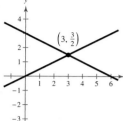

5.
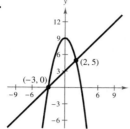

$\left(3, \frac{3}{2}\right)$

$(-3, 0), (2, 5)$

6.
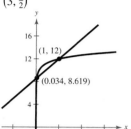

7. $(-2, -5)$ **8.** $(10, -3)$

$(1, 12), (0.034, 8.619)$

9. $(2, -3, 1)$ **10.** No solution

11. $-\dfrac{1}{x + 1} + \dfrac{3}{x - 2}$ **12.** $\dfrac{2}{x^2} + \dfrac{3}{2 - x}$

13. $-\dfrac{5}{x} + \dfrac{3}{x + 1} + \dfrac{3}{x - 1}$ **14.** $-\dfrac{2}{x} + \dfrac{3x}{x^2 + 2}$

15.
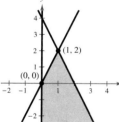

16.
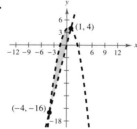

17.
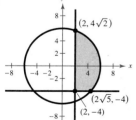

18. Maximum at $(12, 0)$: 240; Minimum at $(0, 0)$: 0
19. \$24,000 in 4% fund   **20.** $y = -\frac{1}{2}x^2 + x + 6$
$26,000 in 5.5% fund
21. 0 units of model I
5300 units of model II
Optimal profit: \$212,000

Problem Solving   *(page 517)*

1.
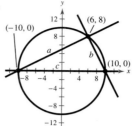

$a = 8\sqrt{5}, b = 4\sqrt{5}, c = 20$
$\left(8\sqrt{5}\right)^2 + \left(4\sqrt{5}\right)^2 = 20^2$
Therefore, the triangle is a right triangle.
3. $ad \neq bc$ **5.** (a) One   (b) Two   (c) Four
7. 10.1 ft; About 252.7 ft   **9.** \$12.00
11. (a) $(3, -4)$ (b) $\left(\dfrac{2}{-a + 5}, \dfrac{1}{4a - 1}, \dfrac{1}{a}\right)$
13. (a) $\left(\dfrac{-5a + 16}{6}, \dfrac{5a - 16}{6}, a\right)$
   (b) $\left(\dfrac{-11a + 36}{14}, \dfrac{13a - 40}{14}, a\right)$
   (c) $(-a + 3, a - 3, a)$ (d) Infinitely many

15. $\begin{cases} a + \quad t \leq 32 \\ 0.15a \quad \geq 1.9 \\ 193a + 772t \geq 11,000 \end{cases}$

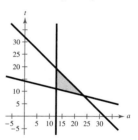

17. (a) $\begin{cases} x + y \leq 200 \\ x \quad\quad \geq 60 \\ 0 < y \leq 130 \end{cases}$ (b)

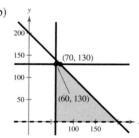

   (c) No, because the total cholesterol is greater than 200 milligrams per deciliter.
   (d) LDL: 135 mg/dL, HDL: 65 mg/dL,
      LDL + HDL: 200 mg/dL
   (e) $(75, 105)$; $\frac{180}{75} = 2.4 < 5$; Answers will vary.

Chapter 7

Section 7.1   *(page 529)*

1. matrix   **3.** main diagonal   **5.** augmented
7. row-equivalent   **9.** 1×2 **11.** 3×1 **13.** 2×2

15. $\begin{bmatrix} 4 & -3 & \vdots & -5 \\ -1 & 3 & \vdots & 12 \end{bmatrix}$ **17.** $\begin{bmatrix} 1 & 10 & -2 & \vdots & 2 \\ 5 & -3 & 4 & \vdots & 0 \\ 2 & 1 & 0 & \vdots & 6 \end{bmatrix}$

19. $\begin{bmatrix} 7 & -5 & 1 & \vdots & 13 \\ 19 & 0 & -8 & \vdots & 10 \end{bmatrix}$ **21.** $\begin{cases} x + 2y = 7 \\ 2x - 3y = 4 \end{cases}$

23. $\begin{cases} 2x & + 5z = -12 \\ y - 2z = 7 \\ 6x + 3y & = 2 \end{cases}$

25. $\begin{cases} 9x + 12y + 3z & = 0 \\ -2x + 18y + 5z + 2w = 10 \\ x + 7y - 8z & = -4 \\ 3x & + 2z = -10 \end{cases}$

27. $\begin{bmatrix} 1 & 4 & 3 \\ 0 & 2 & -1 \end{bmatrix}$ **29.** $\begin{bmatrix} 1 & 1 & 1 \\ 0 & -7 & -1 \end{bmatrix}$

31. $\begin{bmatrix} 1 & 0 & 14 & -11 \\ 0 & 1 & -2 & 2 \\ 0 & 0 & 1 & -7 \end{bmatrix}$

33. $\begin{bmatrix} 1 & 1 & 4 & -1 \\ 0 & 5 & -2 & 6 \\ 0 & 3 & 20 & 4 \end{bmatrix} \begin{bmatrix} 1 & 1 & 4 & -1 \\ 0 & 1 & -\frac{2}{5} & \frac{6}{5} \\ 0 & 3 & 20 & 4 \end{bmatrix}$

35. Add 5 times Row 2 to Row 1.

37. Interchange Row 1 and Row 2.

Add 4 times new Row 1 to Row 3.

39. (a) $\begin{bmatrix} 1 & 2 & 3 \\ 0 & -5 & -10 \\ 3 & 1 & -1 \end{bmatrix}$ (b) $\begin{bmatrix} 1 & 2 & 3 \\ 0 & -5 & -10 \\ 0 & -5 & -10 \end{bmatrix}$

(c) $\begin{bmatrix} 1 & 2 & 3 \\ 0 & -5 & -10 \\ 0 & 0 & 0 \end{bmatrix}$ (d) $\begin{bmatrix} 1 & 2 & 3 \\ 0 & 1 & 2 \\ 0 & 0 & 0 \end{bmatrix}$

(e) $\begin{bmatrix} 1 & 0 & -1 \\ 0 & 1 & 2 \\ 0 & 0 & 0 \end{bmatrix}$

The matrix is in reduced row-echelon form.

41. Reduced row-echelon form **43.** Not in row-echelon form

45. $\begin{bmatrix} 1 & 1 & 0 & 5 \\ 0 & 1 & 2 & 0 \\ 0 & 0 & 1 & -1 \end{bmatrix}$ **47.** $\begin{bmatrix} 1 & -1 & -1 & 1 \\ 0 & 1 & 6 & 3 \\ 0 & 0 & 0 & 0 \end{bmatrix}$

49. $\begin{bmatrix} 1 & 0 & 0 \\ 0 & 1 & 0 \\ 0 & 0 & 1 \end{bmatrix}$ **51.** $\begin{bmatrix} 1 & 2 & 0 & 0 \\ 0 & 0 & 1 & 0 \\ 0 & 0 & 0 & 1 \\ 0 & 0 & 0 & 0 \end{bmatrix}$

53. $\begin{bmatrix} 1 & 0 & 3 & 16 \\ 0 & 1 & 2 & 12 \end{bmatrix}$ **55.** $\begin{cases} x - 2y = 4 \\ y = -3 \end{cases}$

$(-2, -3)$

57. $\begin{cases} x - y + 2z = 4 \\ y - z = 2 \\ z = -2 \end{cases}$

$(8, 0, -2)$

59. $(3, -4)$ **61.** $(-4, -10, 4)$ **63.** $(3, 2)$

65. $(-5, 6)$ **67.** $(-1, -4)$ **69.** $\left(\frac{1}{2}, -\frac{3}{4}\right)$

71. $(4, -3, 2)$ **73.** $(7, -3, 4)$ **75.** $(-4, -3, 6)$

77. $(0, 0)$ **79.** $(5a + 4, -3a + 2, a)$ **81.** Inconsistent

83. $(3, -2, 5, 0)$ **85.** $(0, 2 - 4a, a)$ **87.** $(1, 0, 4, -2)$

89. $(-2a, a, a, 0)$ **91.** Yes; $(-1, 1, -3)$ **93.** No

95. $f(x) = -x^2 + x + 1$ **97.** $f(x) = -9x^2 - 5x + 11$

99. $f(x) = x^3 - 2x^2 + x - 1$

101. $f(x) = x^3 - 2x^2 - 4x + 1$

103. $\begin{bmatrix} 1 & 3 & \frac{3}{2} & \vdots & 4 \\ 0 & 1 & \frac{7}{4} & \vdots & -\frac{3}{2} \\ 0 & 0 & 1 & \vdots & 2 \end{bmatrix}, \begin{bmatrix} 1 & 3 & 1 & \vdots & 3 \\ 0 & 1 & 2 & \vdots & -1 \\ 0 & 0 & 1 & \vdots & 2 \end{bmatrix}$

105. $\dfrac{4x^2}{(x + 1)^2(x - 1)} = \dfrac{1}{x - 1} + \dfrac{3}{x + 1} - \dfrac{2}{(x + 1)^2}$

107. \$150,000 at 7%

\$750,000 at 8%

\$600,000 at 10%

109. $\begin{cases} x + 5y + 10z + 20w = 95 \\ x + y + z + w = 26 \\ y - 4z = 0 \\ x - 2y = -1 \end{cases}$

\$1 bills: 15

\$5 bills: 8

\$10 bills: 2

\$20 bills: 1

111. $y = x^2 + 2x + 5$

113. (a) $y = -0.004x^2 + 0.367x + 5$

(b)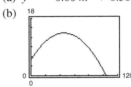

(c) 13 ft, 104 ft (d) 13.418 ft, 103.793 ft

(e) The results are similar.

115. (a) $x_1 = s, x_2 = t, x_3 = 600 - s, x_4 = s - t,$

$x_5 = 500 - t, x_6 = s, x_7 = t$

(b) $x_1 = 0, x_2 = 0, x_3 = 600, x_4 = 0, x_5 = 500,$

$x_6 = 0, x_7 = 0$

(c) $x_1 = 500, x_2 = 100, x_3 = 100, x_4 = 400,$

$x_5 = 400, x_6 = 500, x_7 = 100$

117. False. It is a 2×4 matrix.

119. Answers will vary. For example:

$\begin{cases} x + y + 7z = -1 \\ x + 2y + 11z = 0 \\ 2x + y + 10z = -3 \end{cases}$

121. Interchange two rows.

Multiply a row by a nonzero constant.

Add a multiple of a row to another row.

123. They are the same.

Section 7.2 *(page 544)*

1. equal **3.** zero; O

5. (a) iii (b) iv (c) i (d) v (e) ii

7. $x = -4, y = 22$ **9.** $x = 2, y = 3$

11. (a) $\begin{bmatrix} 3 & -2 \\ 1 & 7 \end{bmatrix}$ (b) $\begin{bmatrix} -1 & 0 \\ 3 & -9 \end{bmatrix}$ (c) $\begin{bmatrix} 3 & -3 \\ 6 & -3 \end{bmatrix}$

(d) $\begin{bmatrix} -1 & -1 \\ 8 & -19 \end{bmatrix}$

13. (a) $\begin{bmatrix} 9 & 5 \\ 1 & -2 \\ -3 & 15 \end{bmatrix}$ (b) $\begin{bmatrix} 7 & -7 \\ 3 & 8 \\ -5 & -5 \end{bmatrix}$ (c) $\begin{bmatrix} 24 & -3 \\ 6 & 9 \\ -12 & 15 \end{bmatrix}$

(d) $\begin{bmatrix} 22 & -15 \\ 8 & 19 \\ -14 & -5 \end{bmatrix}$

15. (a) $\begin{bmatrix} 5 & 5 & -2 & 4 & 4 \\ -5 & 10 & 0 & -4 & -7 \end{bmatrix}$

(b) $\begin{bmatrix} 3 & 5 & 0 & 2 & 4 \\ 7 & -6 & -4 & 2 & 7 \end{bmatrix}$

(c) $\begin{bmatrix} 12 & 15 & -3 & 9 & 12 \\ 3 & 6 & -6 & -3 & 0 \end{bmatrix}$

(d) $\begin{bmatrix} 10 & 15 & -1 & 7 & 12 \\ 15 & -10 & -10 & 3 & 14 \end{bmatrix}$

17. (a), (b), and (d) not possible   (c) $\begin{bmatrix} 18 & 0 & 9 \\ -3 & -12 & 0 \end{bmatrix}$

19. $\begin{bmatrix} -8 & -7 \\ 15 & -1 \end{bmatrix}$ 21. $\begin{bmatrix} -24 & -4 & 12 \\ -12 & 32 & 12 \end{bmatrix}$ 23. $\begin{bmatrix} 10 & 8 \\ -59 & 9 \end{bmatrix}$

25. $\begin{bmatrix} -17.143 & 2.143 \\ 11.571 & 10.286 \end{bmatrix}$ 27. $\begin{bmatrix} -1.581 & -3.739 \\ -4.252 & -13.249 \\ 9.713 & -0.362 \end{bmatrix}$

29. $\begin{bmatrix} -6 & -9 \\ -1 & 0 \\ 17 & -10 \end{bmatrix}$ 31. $\begin{bmatrix} 3 & 3 \\ -\frac{1}{2} & 0 \\ -\frac{13}{2} & \frac{11}{2} \end{bmatrix}$ 33. Not possible

35. $\begin{bmatrix} -2 & 51 \\ -8 & 33 \\ 0 & 27 \end{bmatrix}$ Order: 3×2

37. $\begin{bmatrix} 1 & 0 & 0 \\ 0 & 1 & 0 \\ 0 & 0 & \frac{7}{2} \end{bmatrix}$ Order: 3×3

39. $\begin{bmatrix} 60 & -20 & 10 & 60 \\ 72 & -24 & 12 & 72 \end{bmatrix}$ Order: 2×4

41. $\begin{bmatrix} 70 & -17 & 73 \\ 32 & 11 & 6 \\ 16 & -38 & 70 \end{bmatrix}$ 43. $\begin{bmatrix} 151 & 25 & 48 \\ 516 & 279 & 387 \\ 47 & -20 & 87 \end{bmatrix}$

45. Not possible

47. (a) $\begin{bmatrix} 0 & 15 \\ 6 & 12 \end{bmatrix}$ (b) $\begin{bmatrix} -2 & 2 \\ 31 & 14 \end{bmatrix}$ (c) $\begin{bmatrix} 9 & 6 \\ 12 & 12 \end{bmatrix}$

49. (a) $\begin{bmatrix} 0 & -10 \\ 10 & 0 \end{bmatrix}$ (b) $\begin{bmatrix} 0 & -10 \\ 10 & 0 \end{bmatrix}$ (c) $\begin{bmatrix} 8 & -6 \\ 6 & 8 \end{bmatrix}$

51. (a) $\begin{bmatrix} 7 & 7 & 14 \\ 8 & 8 & 16 \\ -1 & -1 & -2 \end{bmatrix}$ (b) $[13]$ (c) Not possible

53. $\begin{bmatrix} 5 & 8 \\ -4 & -16 \end{bmatrix}$ 55. $\begin{bmatrix} -4 & 10 \\ 3 & 14 \end{bmatrix}$

57. (a) $\begin{bmatrix} -1 & 1 \\ -2 & 1 \end{bmatrix}\begin{bmatrix} x_1 \\ x_2 \end{bmatrix} = \begin{bmatrix} 4 \\ 0 \end{bmatrix}$ (b) $\begin{bmatrix} 4 \\ 8 \end{bmatrix}$

59. (a) $\begin{bmatrix} -2 & -3 \\ 6 & 1 \end{bmatrix}\begin{bmatrix} x_1 \\ x_2 \end{bmatrix} = \begin{bmatrix} -4 \\ -36 \end{bmatrix}$ (b) $\begin{bmatrix} -7 \\ 6 \end{bmatrix}$

61. (a) $\begin{bmatrix} 1 & -2 & 3 \\ -1 & 3 & -1 \\ 2 & -5 & 5 \end{bmatrix}\begin{bmatrix} x_1 \\ x_2 \\ x_3 \end{bmatrix} = \begin{bmatrix} 9 \\ -6 \\ 17 \end{bmatrix}$ (b) $\begin{bmatrix} 1 \\ -1 \\ 2 \end{bmatrix}$

63. (a) $\begin{bmatrix} 1 & -5 & 2 \\ -3 & 1 & -1 \\ 0 & -2 & 5 \end{bmatrix}\begin{bmatrix} x_1 \\ x_2 \\ x_3 \end{bmatrix} = \begin{bmatrix} -20 \\ 8 \\ -16 \end{bmatrix}$ (b) $\begin{bmatrix} -1 \\ 3 \\ -2 \end{bmatrix}$

65. $\begin{bmatrix} 84 & 60 & 30 \\ 42 & 120 & 84 \end{bmatrix}$

67. (a) $A = \begin{bmatrix} 125 & 100 & 75 \\ 100 & 175 & 125 \end{bmatrix}$

The entries represent the numbers of bushels of each crop that are shipped to each outlet.

(b) $B = [\$3.50 \quad \$6.00]$

The entries represent the profits per bushel of each crop.

(c) $BA = [\$1037.50 \quad \$1400 \quad \$1012.50]$

The entries represent the profits from both crops at each of the three outlets.

69. $\begin{bmatrix} \$15,770 & \$18,300 \\ \$26,500 & \$29,250 \\ \$21,260 & \$24,150 \end{bmatrix}$

The entries represent the wholesale and retail values of the inventories at the three outlets.

71. $P^3 = \begin{bmatrix} 0.300 & 0.175 & 0.175 \\ 0.308 & 0.433 & 0.217 \\ 0.392 & 0.392 & 0.608 \end{bmatrix}$

$P^4 = \begin{bmatrix} 0.250 & 0.188 & 0.188 \\ 0.315 & 0.377 & 0.248 \\ 0.435 & 0.435 & 0.565 \end{bmatrix}$

$P^5 = \begin{bmatrix} 0.225 & 0.194 & 0.194 \\ 0.314 & 0.345 & 0.267 \\ 0.461 & 0.461 & 0.539 \end{bmatrix}$

$P^6 = \begin{bmatrix} 0.213 & 0.197 & 0.197 \\ 0.311 & 0.326 & 0.280 \\ 0.477 & 0.477 & 0.523 \end{bmatrix}$

$P^7 = \begin{bmatrix} 0.206 & 0.198 & 0.198 \\ 0.308 & 0.316 & 0.288 \\ 0.486 & 0.486 & 0.514 \end{bmatrix}$

$P^8 = \begin{bmatrix} 0.203 & 0.199 & 0.199 \\ 0.305 & 0.309 & 0.292 \\ 0.492 & 0.492 & 0.508 \end{bmatrix}$

Approaches the matrix $\begin{bmatrix} 0.2 & 0.2 & 0.2 \\ 0.3 & 0.3 & 0.3 \\ 0.5 & 0.5 & 0.5 \end{bmatrix}$

73. (a) $\begin{array}{c} \text{Sales \$} \quad \text{Profit} \\ \begin{bmatrix} 571.8 & 206.6 \\ 798.9 & 288.8 \\ 936 & 337.8 \end{bmatrix} \end{array}$

The entries represent the total sales and profits for milk on Friday, Saturday, and Sunday.

(b) $\$833.20$

75. (a) $[2 \quad 0.5 \quad 3]$

(b) 120 lb   150 lb
$[473.5 \quad 588.5]$
The entries represent the total calories burned.

CHAPTER 7

77. True. The sum of two matrices of different orders is undefined.

79. Not possible   **81.** Not possible

83. 2×2 **85.** 2×3

87. (a) $A + B = \begin{bmatrix} 1 & -1 \\ 12 & 8 \end{bmatrix}$, $B + A = \begin{bmatrix} 1 & -1 \\ 12 & 8 \end{bmatrix}$

(b) $(A + B) + C = \begin{bmatrix} 6 & 1 \\ 14 & 2 \end{bmatrix}$, $(B + C) + A = \begin{bmatrix} 6 & 1 \\ 14 & 2 \end{bmatrix}$

(c) $2A + 2B = \begin{bmatrix} 2 & -2 \\ 24 & 16 \end{bmatrix}$, $2(A + B) = \begin{bmatrix} 2 & -2 \\ 24 & 16 \end{bmatrix}$

89. $AC = BC = \begin{bmatrix} 2 & 3 \\ 2 & 3 \end{bmatrix}$

91. AB is a diagonal matrix whose entries are the products of the corresponding entries of A and B.

93. Answers will vary.

Section 7.3   *(page 555)*

1. square   **3.** nonsingular; singular

5–11. $AB = I$ and $BA = I$

13. $\begin{bmatrix} \frac{1}{2} & 0 \\ 0 & \frac{1}{3} \end{bmatrix}$ **15.** $\begin{bmatrix} -3 & 2 \\ -2 & 1 \end{bmatrix}$ **17.** $\begin{bmatrix} 1 & -\frac{1}{2} \\ -2 & \frac{3}{2} \end{bmatrix}$

19. $\begin{bmatrix} 1 & 1 & -1 \\ -3 & 2 & -1 \\ 3 & -3 & 2 \end{bmatrix}$ **21.** Does not exist

23. $\begin{bmatrix} -\frac{1}{8} & 0 & 0 & 0 \\ 0 & 1 & 0 & 0 \\ 0 & 0 & \frac{1}{4} & 0 \\ 0 & 0 & 0 & -\frac{1}{5} \end{bmatrix}$ **25.** $\begin{bmatrix} -175 & 37 & -13 \\ 95 & -20 & 7 \\ 14 & -3 & 1 \end{bmatrix}$

27. $\begin{bmatrix} -1.5 & 1.5 & 1 \\ 4.5 & -3.5 & -3 \\ -1 & 1 & 1 \end{bmatrix}$ **29.** $\begin{bmatrix} -12 & -5 & -9 \\ -4 & -2 & -4 \\ -8 & -4 & -6 \end{bmatrix}$

31. $\begin{bmatrix} 0 & -1.\overline{81} & 0.\overline{90} \\ -10 & 5 & 5 \\ 10 & -2.\overline{72} & -3.\overline{63} \end{bmatrix}$ **33.** $\begin{bmatrix} 1 & 0 & 1 & 0 \\ 0 & 1 & 0 & 1 \\ 2 & 0 & 1 & 0 \\ 0 & 1 & 0 & 2 \end{bmatrix}$

35. $\begin{bmatrix} \frac{5}{13} & -\frac{3}{13} \\ \frac{1}{13} & \frac{2}{13} \end{bmatrix}$ **37.** Does not exist   **39.** $\begin{bmatrix} \frac{16}{59} & \frac{15}{59} \\ -\frac{4}{59} & \frac{70}{59} \end{bmatrix}$

41. $(5, 0)$ **43.** $(-8, -6)$ **45.** $(3, 8, -11)$

47. $(2, 1, 0, 0)$ **49.** $(0, 1, 2, -1, 0)$ **51.** $(2, -2)$

53. No solution   **55.** $(-4, -8)$ **57.** $(-1, 3, 2)$

59. $\left(\frac{5}{16}a + \frac{13}{16}, \frac{19}{16}a + \frac{11}{16}, a\right)$ **61.** $(-7, 3, -2)$

63. \$7000 in AAA-rated bonds
  \$1000 in A-rated bonds
  \$2000 in B-rated bonds

65. \$9000 in AAA-rated bonds
  \$1000 in A-rated bonds
  \$2000 in B-rated bonds

67. 0 muffins, 300 bones, 200 cookies

69. 100 muffins, 300 bones, 150 cookies

71. (a) $\begin{cases} 2f + 2.5h + 3s = 26 \\ f + h + s = 10 \\ h - s = 0 \end{cases}$

(b) $\begin{bmatrix} 2 & 2.5 & 3 \\ 1 & 1 & 1 \\ 0 & 1 & -1 \end{bmatrix} \begin{bmatrix} f \\ h \\ s \end{bmatrix} = \begin{bmatrix} 26 \\ 10 \\ 0 \end{bmatrix}$

(c) 2 pounds of French vanilla, 4 pounds of hazelnut, 4 pounds of Swiss chocolate

73. (a) $\begin{cases} 100a + 10b + c = 13.89 \\ 121a + 11b + c = 14.04 \\ 144a + 12b + c = 14.20 \end{cases}$

(b) $y = 0.005t^2 + 0.045t + 12.94$

(c)

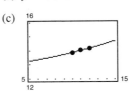

(d) For the immediate future it is, but not for long-term predictions.

75. True. If B is the inverse of A, then $AB = I = BA$.

77. Answers will vary.

79. (a) Answers will vary.

(b) $A^{-1} = \begin{bmatrix} \frac{1}{a_{11}} & 0 & 0 & \cdots & 0 \\ 0 & \frac{1}{a_{22}} & 0 & \cdots & 0 \\ 0 & 0 & \frac{1}{a_{33}} & \cdots & 0 \\ \vdots & \vdots & \vdots & \cdots & \vdots \\ 0 & 0 & 0 & \cdots & \frac{1}{a_{nn}} \end{bmatrix}$

Section 7.4   *(page 563)*

1. determinant   **3.** cofactor   **5.** 4   **7.** 16

9. 28   **11.** 0   **13.** 6   **15.** -9 **17.** -24

19. $\frac{11}{6}$ **21.** -0.002 **23.** -4.842

25. (a) $M_{11} = -6, M_{12} = 3, M_{21} = 5, M_{22} = 4$

(b) $C_{11} = -6, C_{12} = -3, C_{21} = -5, C_{22} = 4$

27. (a) $M_{11} = -4, M_{12} = -2, M_{21} = 1, M_{22} = 3$

(b) $C_{11} = -4, C_{12} = 2, C_{21} = -1, C_{22} = 3$

29. (a) $M_{11} = 3, M_{12} = -4, M_{13} = 1, M_{21} = 2, M_{22} = 2,$
$M_{23} = -4, M_{31} = -4, M_{32} = 10, M_{33} = 8$

(b) $C_{11} = 3, C_{12} = 4, C_{13} = 1, C_{21} = -2, C_{22} = 2,$
$C_{23} = 4, C_{31} = -4, C_{32} = -10, C_{33} = 8$

31. (a) $M_{11} = 10, M_{12} = -43, M_{13} = 2, M_{21} = -30,$
$M_{22} = 17, M_{23} = -6, M_{31} = 54, M_{32} = -53,$
$M_{33} = -34$

(b) $C_{11} = 10, C_{12} = 43, C_{13} = 2, C_{21} = 30, C_{22} = 17,$
$C_{23} = 6, C_{31} = 54, C_{32} = 53, C_{33} = -34$

33. (a) -75 (b) -75 **35.** (a) 96   (b) 96

37. (a) 170   (b) 170   **39.** 0   **41.** 0   **43.** -9

45. -58 **47.** -30 **49.** -168 **51.** 0   **53.** 412

55. -126 **57.** 0   **59.** -336 **61.** 410

63. (a) -3 (b) -2 (c) $\begin{bmatrix} -2 & 0 \\ 0 & -3 \end{bmatrix}$ (d) 6

65. (a) -8 (b) 0   (c) $\begin{bmatrix} -4 & 4 \\ 1 & -1 \end{bmatrix}$ (d) 0

67. (a) -21 (b) -19 (c) $\begin{bmatrix} 7 & 1 & 4 \\ -8 & 9 & -3 \\ 7 & -3 & 9 \end{bmatrix}$ (d) 399

69. (a) 2 (b) -6 (c) $\begin{bmatrix} 1 & 4 & 3 \\ -1 & 0 & 3 \\ 0 & 2 & 0 \end{bmatrix}$ (d) -12

71–75. Answers will vary. **77.** $x = \pm 2$

79. $x = 1 \pm \sqrt{2}$ **81.** $-1, 4$ **83.** $-1, -4$

85. $8uv - 1$ **87.** e^{5x} **89.** $1 - \ln x$

91. True. If an entire row is zero, then each cofactor in the expansion is multiplied by zero.

93. Answers will vary.

95. A square matrix is a square array of numbers. The determinant of a square matrix is a real number.

97. (a) Columns 2 and 3 of A were interchanged.
$|A| = -115 = -|B|$
(b) Rows 1 and 3 of A were interchanged.
$|A| = -40 = -|B|$

99. (a) Multiply Row 1 by 5.
(b) Multiply Column 2 by 4 and Column 3 by 3.

101. 10 **103.** -9

105. The determinant of a triangular matrix is the product of the terms in the diagonal.

Section 7.5 (page 575)

1. Cramer's Rule **3.** $A = \pm \dfrac{1}{2} \begin{vmatrix} x_1 & y_1 & 1 \\ x_2 & y_2 & 1 \\ x_3 & y_3 & 1 \end{vmatrix}$

5. uncoded; coded **7.** $(-3, -2)$

9. Not possible **11.** $\left(\frac{32}{7}, \frac{30}{7}\right)$ **13.** $(-1, 3, 2)$

15. $(-2, 1, -1)$ **17.** $\left(0, -\frac{1}{2}, \frac{1}{2}\right)$ **19.** $(1, -1, 2)$

21. 7 **23.** 14 **25.** $\frac{33}{8}$ **27.** $\frac{5}{2}$ **29.** 28

31. $\frac{41}{4}$ **33.** $y = \frac{16}{5}$ or $y = 0$ **35.** $y = -3$ or $y = -11$

37. 250 mi^2 **39.** Collinear **41.** Not collinear

43. Collinear **45.** $y = -3$ **47.** $3x - 5y = 0$

49. $x + 3y - 5 = 0$ **51.** $2x + 3y - 8 = 0$

53. (a) Uncoded: $[3 \quad 15], [13 \quad 5], [0 \quad 8], [15 \quad 13], [5 \quad 0],$
$[19 \quad 15], [15 \quad 14]$
(b) Encoded: 48 81 28 51 24 40 54 95 5
10 64 113 57 100

55. (a) Uncoded: $[3 \quad 1 \quad 12], [12 \quad 0 \quad 13], [5 \quad 0 \quad 20],$
$[15 \quad 13 \quad 15], [18 \quad 18 \quad 15], [23 \quad 0 \quad 0]$
(b) Encoded: -68 21 35 -66 14 39 -115
35 60 -62 15 32 -54 12 27
23 -23 0

57. 1 -25 -65 17 15 -9 -12 -62 -119
27 51 48 43 67 48 57 111 117

59. -5 -41 -87 91 207 257 11 -5 -41 40 80
84 76 177 227

61. HAPPY NEW YEAR **63.** CLASS IS CANCELED

65. SEND PLANES **67.** MEET ME TONIGHT RON

69. (a) $\begin{cases} 8c + 28b + 140a = 182.1 \\ 28c + 140b + 784a = 713.4 \\ 140c + 784b + 4676a = 3724.8 \end{cases}$

(b) $y = 0.034t^2 + 1.57t + 16.66$

(c)

(d) 2009

71. False. The denominator is the determinant of the coefficient matrix.

73. False. If the determinant of the coefficient matrix is zero, the system has either no solution or infinitely many solutions.

75. Answers will vary. **77.** 12

Review Exercises (page 580)

1. 3×1 **3.** 1×1 **5.** $\begin{bmatrix} 3 & -10 & \vdots & 15 \\ 5 & 4 & \vdots & 22 \end{bmatrix}$

7. $\begin{cases} 5x + y + 7z = -9 \\ 4x + 2y = 10 \\ 9x + 4y + 2z = 3 \end{cases}$ **9.** $\begin{bmatrix} 1 & 2 & 3 \\ 0 & 1 & 1 \\ 0 & 0 & 1 \end{bmatrix}$

11. $\begin{cases} x + 2y + 3z = 9 \\ y - 2z = 2 \\ z = 0 \end{cases}$ **13.** $\begin{cases} x - 5y + 4z = 1 \\ y + 2z = 3 \\ z = 4 \end{cases}$

$(5, 2, 0)$ $(-40, -5, 4)$

15. $(10, -12)$ **17.** $\left(-\frac{1}{5}, \frac{7}{10}\right)$ **19.** Inconsistent

21. $(1, -2, 2)$ **23.** $\left(-2a + \frac{3}{2}, 2a + 1, a\right)$ **25.** $(5, 2, -6)$

27. $(1, 0, 4, 3)$ **29.** $(1, 2, 2)$ **31.** $(2, -3, 3)$

33. $(2, 3, -1)$ **35.** $(2, 6, -10, -3)$ **37.** $x = 12, y = -7$

39. $x = 1, y = 11$

41. (a) $\begin{bmatrix} -1 & 8 \\ 15 & 13 \end{bmatrix}$ (b) $\begin{bmatrix} 5 & -12 \\ -9 & -3 \end{bmatrix}$

(c) $\begin{bmatrix} 8 & -8 \\ 12 & 20 \end{bmatrix}$ (d) $\begin{bmatrix} -7 & 28 \\ 39 & 29 \end{bmatrix}$

43. (a) $\begin{bmatrix} 5 & 7 \\ -3 & 14 \\ 31 & 42 \end{bmatrix}$ (b) $\begin{bmatrix} 5 & 1 \\ -11 & -10 \\ -9 & -38 \end{bmatrix}$

(c) $\begin{bmatrix} 20 & 16 \\ -28 & 8 \\ 44 & 8 \end{bmatrix}$ (d) $\begin{bmatrix} 5 & 13 \\ 5 & 38 \\ 71 & 122 \end{bmatrix}$

45. $\begin{bmatrix} 17 & -17 \\ 13 & 2 \end{bmatrix}$ **47.** $\begin{bmatrix} 54 & 4 \\ -2 & 24 \\ -4 & 32 \end{bmatrix}$ **49.** $\begin{bmatrix} 48 & -18 & -3 \\ 15 & 51 & 33 \end{bmatrix}$

51. $\begin{bmatrix} -11 & -6 \\ 8 & -13 \\ -18 & -8 \end{bmatrix}$ **53.** $\begin{bmatrix} 3 & \frac{2}{3} \\ -\frac{4}{3} & \frac{11}{3} \\ \frac{10}{3} & 0 \end{bmatrix}$ **55.** $\begin{bmatrix} -30 & 4 \\ 51 & 70 \end{bmatrix}$

57. $\begin{bmatrix} 100 & 220 \\ 12 & -4 \\ 84 & 212 \end{bmatrix}$ **59.** $\begin{bmatrix} 14 & -2 & 8 \\ 14 & -10 & 40 \\ 36 & -12 & 48 \end{bmatrix}$ **61.** $\begin{bmatrix} 44 & 4 \\ 20 & 8 \end{bmatrix}$

63. Not possible. The number of columns of the first matrix does not equal the number of rows of the second matrix.

65. $\begin{bmatrix} 1 & 17 \\ 12 & 36 \end{bmatrix}$ **67.** $\begin{bmatrix} 14 & -22 & 22 \\ 19 & -41 & 80 \\ 42 & -66 & 66 \end{bmatrix}$

69. $\begin{bmatrix} 76 & 114 & 133 \\ 38 & 95 & 76 \end{bmatrix}$

71. $[\$2,396,539 \quad \$2,581,388]$

The merchandise shipped to warehouse 1 is worth \$2,396,539 and the merchandise shipped to warehouse 2 is worth \$2,581,388.

73–75. $AB = I$ and $BA = I$

77. $\begin{bmatrix} 4 & -5 \\ 5 & -6 \end{bmatrix}$ **79.** $\begin{bmatrix} \frac{1}{2} & -1 & -\frac{1}{2} \\ \frac{1}{2} & -\frac{2}{3} & -\frac{5}{6} \\ 0 & \frac{2}{3} & \frac{1}{3} \end{bmatrix}$ **81.** $\begin{bmatrix} 13 & 6 & -4 \\ -12 & -5 & 3 \\ 5 & 2 & -1 \end{bmatrix}$

83. $\begin{bmatrix} -3 & 6 & -5.5 & 3.5 \\ 1 & -2 & 2 & -1 \\ 7 & -15 & 14.5 & -9.5 \\ -1 & 2.5 & -2.5 & 1.5 \end{bmatrix}$ **85.** $\begin{bmatrix} 1 & -1 \\ 4 & -\frac{7}{2} \end{bmatrix}$

87. Does not exist **89.** $\begin{bmatrix} 2 & \frac{20}{3} \\ \frac{1}{10} & \frac{1}{6} \end{bmatrix}$ **91.** $\begin{bmatrix} \frac{20}{9} & \frac{5}{9} \\ -\frac{10}{9} & -\frac{25}{9} \end{bmatrix}$

93. $(36, 11)$ **95.** $(-6, -1)$ **97.** $(2, 3)$

99. $(-8, 18)$ **101.** $(2, -1, -2)$ **103.** $(6, 1, -1)$

105. $(-3, 1)$ **107.** $\left(\frac{1}{6}, -\frac{7}{4}\right)$ **109.** $(1, 1, -2)$

111. -42 **113.** 550

115. (a) $M_{11} = 4, M_{12} = 7, M_{21} = -1, M_{22} = 2$
(b) $C_{11} = 4, C_{12} = -7, C_{21} = 1, C_{22} = 2$

117. (a) $M_{11} = 30, M_{12} = -12, M_{13} = -21,$
$M_{21} = 20, M_{22} = 19, M_{23} = 22, M_{31} = 5,$
$M_{32} = -2, M_{33} = 19$
(b) $C_{11} = 30, C_{12} = 12, C_{13} = -21,$
$C_{21} = -20, C_{22} = 19, C_{23} = -22,$
$C_{31} = 5, C_{32} = 2, C_{33} = 19$

119. -6 **121.** 15 **123.** 130 **125.** -8 **127.** 279

129. $(4, 7)$ **131.** $(-1, 4, 5)$ **133.** 16 **135.** 10

137. Collinear **139.** $x - 2y + 4 = 0$

141. $2x + 6y - 13 = 0$

143. (a) Uncoded: $[12 \quad 15 \quad 15], [11 \quad 0 \quad 15], [21 \quad 20 \quad 0],$
$[2 \quad 5 \quad 12], [15 \quad 23 \quad 0]$
(b) Encoded: $-21 \quad 6 \quad 0 \quad -68 \quad 8 \quad 45 \quad 102 \quad -42$
$-60 \quad -53 \quad 20 \quad 21 \quad 99 \quad -30 \quad -69$

145. SEE YOU FRIDAY

147. False. The matrix must be square.

149. An error message appears because $1(6) - (-2)(-3) = 0$.

151. If A is a square matrix, the cofactor C_{ij} of the entry a_{ij} is $(-1)^{i+j}M_{ij}$, where M_{ij} is the determinant obtained by deleting the ith row and jth column of A. The determinant of A is the sum of the entries of any row or column of A multiplied by their respective cofactors.

153. The part of the matrix corresponding to the coefficients of the system reduces to a matrix in which the number of rows with nonzero entries is the same as the number of variables.

Chapter Test (page 585)

1. $\begin{bmatrix} 1 & 0 & 0 \\ 0 & 1 & 0 \\ 0 & 0 & 1 \end{bmatrix}$ **2.** $\begin{bmatrix} 1 & 0 & -1 & 2 \\ 0 & 1 & 0 & -1 \\ 0 & 0 & 0 & 0 \\ 0 & 0 & 0 & 0 \end{bmatrix}$

3. $\begin{bmatrix} 4 & 3 & -2 & \vdots & 14 \\ -1 & -1 & 2 & \vdots & -5 \\ 3 & 1 & -4 & \vdots & 8 \end{bmatrix}, \left(1, 3, -\frac{1}{2}\right)$

4. (a) $\begin{bmatrix} 1 & 5 \\ 0 & -4 \end{bmatrix}$ (b) $\begin{bmatrix} 18 & 15 \\ -15 & -15 \end{bmatrix}$
(c) $\begin{bmatrix} 8 & 15 \\ -5 & -13 \end{bmatrix}$ (d) $\begin{bmatrix} 5 & -5 \\ 0 & 5 \end{bmatrix}$

5. $\begin{bmatrix} \frac{2}{7} & \frac{3}{7} \\ \frac{5}{7} & \frac{4}{7} \end{bmatrix}$ **6.** $\begin{bmatrix} -\frac{5}{2} & 4 & -3 \\ 5 & -7 & 6 \\ 4 & -6 & 5 \end{bmatrix}$

7. $(12, 18)$ **8.** -112 **9.** 29 **10.** 43

11. $(-3, 5)$ **12.** $(-2, 4, 6)$ **13.** 7

14. Uncoded: $[11 \quad 14 \quad 15], [3 \quad 11 \quad 0], [15 \quad 14 \quad 0], [23 \quad 15 \quad 15],$
$[4 \quad 0 \quad 0]$
Encoded: $115 \quad -41 \quad -59 \quad 14 \quad -3 \quad -11 \quad 29 \quad -15$
$-14 \quad 128 \quad -53 \quad -60 \quad 4 \quad -4 \quad 0$

15. 75 L of 60% solution, 25 L of 20% solution

Problem Solving (page 587)

1. (a) $AT = \begin{bmatrix} -1 & -4 & -2 \\ 1 & 2 & 3 \end{bmatrix}$
$AAT = \begin{bmatrix} -1 & -2 & -3 \\ -1 & -4 & -2 \end{bmatrix}$

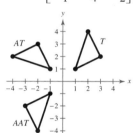

A represents a counterclockwise rotation.
(b) AAT is rotated clockwise $90°$ to obtain AT. AT is then rotated clockwise $90°$ to obtain T.

3. (a) Yes (b) No (c) No (d) No

5. (a) Gold Satellite System: 28,750 subscribers
Galaxy Satellite Network: 35,750 subscribers
Nonsubscribers: 35,500
Answers will vary.
(b) Gold Satellite System: 30,813 subscribers
Galaxy Satellite Network: 39,675 subscribers
Nonsubscribers: 29,513
Answers will vary.
(c) Gold Satellite System: 31,947 subscribers
Galaxy Satellite Network: 42,329 subscribers
Nonsubscribers: 25,724
Answers will vary.
(d) Satellite companies are increasing the number of subscribers, while the nonsubscribers are decreasing.

7. $x = 6$ **9–11.** Answers will vary.

13. Sulfur: 32 atomic mass units; Nitrogen: 14 atomic mass units; Fluorine: 19 atomic mass units

INDEX

FORMULAS FROM GEOMETRY

Triangle:

$h = a \sin \theta$

$\text{Area} = \dfrac{1}{2}bh$

$c^2 = a^2 + b^2 - 2ab \cos \theta$ (Law of Cosines)

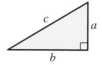

Right Triangle:

Pythagorean Theorem
$c^2 = a^2 + b^2$

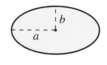

Equilateral Triangle:

$h = \dfrac{\sqrt{3}s}{2}$

$\text{Area} = \dfrac{\sqrt{3}s^2}{4}$

Parallelogram:

$\text{Area} = bh$

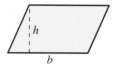

Trapezoid:

$\text{Area} = \dfrac{h}{2}(a + b)$

Circle:

$\text{Area} = \pi r^2$

$\text{Circumference} = 2\pi r$

Sector of Circle:

$\text{Area} = \dfrac{\theta r^2}{2}$

$s = r\theta$

θ in radians

Circular Ring:

$\text{Area} = \pi(R^2 - r^2)$

$\quad\quad\quad = 2\pi pw$

$p = \text{average radius,}$

$w = \text{width of ring}$

Sector of Circular Ring:

$\text{Area} = \theta pw$

$p = \text{average radius,}$

$w = \text{width of ring,}$

θ in radians

Ellipse:

$\text{Area} = \pi ab$

$\text{Circumference} \approx 2\pi \sqrt{\dfrac{a^2 + b^2}{2}}$

Cone:

$\text{Volume} = \dfrac{Ah}{3}$

$A = \text{area of base}$

Right Circular Cone:

$\text{Volume} = \dfrac{\pi r^2 h}{3}$

$\text{Lateral Surface Area} = \pi r \sqrt{r^2 + h^2}$

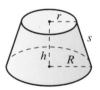

Frustum of Right Circular Cone:

$\text{Volume} = \dfrac{\pi(r^2 + rR + R^2)h}{3}$

$\text{Lateral Surface Area} = \pi s(R + r)$

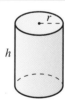

Right Circular Cylinder:

$\text{Volume} = \pi r^2 h$

$\text{Lateral Surface Area} = 2\pi rh$

Sphere:

$\text{Volume} = \dfrac{4}{3}\pi r^3$

$\text{Surface Area} = 4\pi r^2$

Wedge:

$A = B \sec \theta$

$A = \text{area of upper face,}$

$B = \text{area of base}$

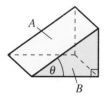

ALGEBRA

Factors and Zeros of Polynomials:

Given the polynomial $p(x) = a_n x^n + a_{n-1} x^{n-1} + \cdots + a_1 x + a_0$. If $p(b) = 0$, then b is a *zero* of the polynomial and a *solution* of the equation $p(x) = 0$. Furthermore, $(x - b)$ is a *factor* of the polynomial.

Fundamental Theorem of Algebra:

An nth degree polynomial has n (not necessarily distinct) zeros.

Quadratic Formula:

If $p(x) = ax^2 + bx + c$, $a \neq 0$ and $b^2 - 4ac \geq 0$, then the real zeros of p are $x = \left(-b \pm \sqrt{b^2 - 4ac}\right)/2a$.

Example

If $p(x) = x^2 + 3x - 1$, then $p(x) = 0$ if
$$x = \frac{-3 \pm \sqrt{13}}{2}.$$

Special Factors:

$x^2 - a^2 = (x - a)(x + a)$

$x^3 - a^3 = (x - a)(x^2 + ax + a^2)$

$x^3 + a^3 = (x + a)(x^2 - ax + a^2)$

$x^4 - a^4 = (x - a)(x + a)(x^2 + a^2)$

$x^4 + a^4 = \left(x^2 + \sqrt{2}\,ax + a^2\right)\left(x^2 - \sqrt{2}\,ax + a^2\right)$

$x^n - a^n = (x - a)(x^{n-1} + ax^{n-2} + \cdots + a^{n-1})$, for n odd

$x^n + a^n = (x + a)(x^{n-1} - ax^{n-2} + \cdots + a^{n-1})$, for n odd

$x^{2n} - a^{2n} = (x^n - a^n)(x^n + a^n)$

Examples

$x^2 - 9 = (x - 3)(x + 3)$

$x^3 - 8 = (x - 2)(x^2 + 2x + 4)$

$x^3 + 4 = \left(x + \sqrt[3]{4}\right)\left(x^2 - \sqrt[3]{4}x + \sqrt[3]{16}\right)$

$x^4 - 4 = \left(x - \sqrt{2}\right)\left(x + \sqrt{2}\right)(x^2 + 2)$

$x^4 + 4 = (x^2 + 2x + 2)(x^2 - 2x + 2)$

$x^5 - 1 = (x - 1)(x^4 + x^3 + x^2 + x + 1)$

$x^7 + 1 = (x + 1)(x^6 - x^5 + x^4 - x^3 + x^2 - x + 1)$

$x^6 - 1 = (x^3 - 1)(x^3 + 1)$

Binomial Theorem:

$(x + a)^2 = x^2 + 2ax + a^2$

$(x - a)^2 = x^2 - 2ax + a^2$

$(x + a)^3 = x^3 + 3ax^2 + 3a^2 x + a^3$

$(x - a)^3 = x^3 - 3ax^2 + 3a^2 x - a^3$

$(x + a)^4 = x^4 + 4ax^3 + 6a^2 x^2 + 4a^3 + a^4$

$(x - a)^4 = x^4 - 4ax^3 + 6a^2 x^2 - 4a^3 x + a^4$

$(x + a)^n = x^n + nax^{n-1} + \dfrac{n(n-1)}{2!}a^2 x^{n-2} + \cdots + na^{n-1}x + a^n$

$(x - a)^n = x^n - nax^{n-1} + \dfrac{n(n-1)}{2!}a^2 x^{n-2} - \cdots \pm na^{n-1}x \mp a^n$

Examples

$(x + 3)^2 = x^2 + 6x + 9$

$(x^2 - 5)^2 = x^4 - 10x^2 + 25$

$(x + 2)^3 = x^3 + 6x^2 + 12x + 8$

$(x - 1)^3 = x^3 - 3x^2 + 3x - 1$

$\left(x + \sqrt{2}\right)^4 = x^4 + 4\sqrt{2}x^3 + 12x^2 + 8\sqrt{2}x + 4$

$(x - 4)^4 = x^4 - 16x^3 + 96x^2 - 256x + 256$

$(x + 1)^5 = x^5 + 5x^4 + 10x^3 + 10x^2 + 5x + 1$

$(x - 1)^6 = x^6 - 6x^5 + 15x^4 - 20x^3 + 15x^2 - 6x + 1$

Rational Zero Test:

If $p(x) = a_n x^n + a_{n-1} x^{n-1} + \cdots + a_1 x + a_0$ has integer coefficients, then every *rational* zero of $p(x) = 0$ is of the form $x = r/s$, where r is a factor of a_0 and s is a factor of a_n.

Example

If $p(x) = 2x^4 - 7x^3 + 5x^2 - 7x + 3$, then the only possible *rational* zeros are $x = \pm 1, \pm\frac{1}{2}, \pm 3$, and $\pm\frac{3}{2}$. By testing, you find the two rational zeros to be $\frac{1}{2}$ and 3.

Factoring by Grouping:

$acx^3 + adx^2 + bcx + bd = ax^2(cx + d) + b(cx + d)$
$$= (ax^2 + b)(cx + d)$$

Example

$3x^3 - 2x^2 - 6x + 4 = x^2(3x - 2) - 2(3x - 2)$
$$= (x^2 - 2)(3x - 2)$$